Porsche Boxster (986)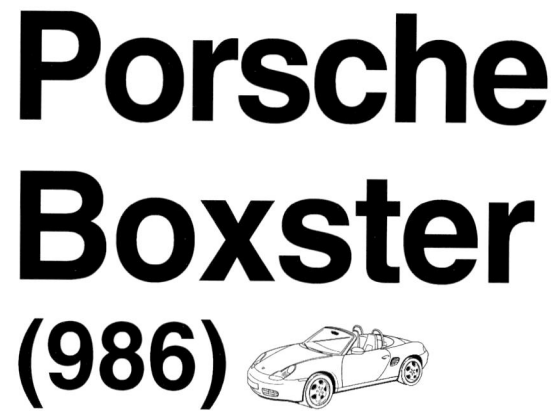

Service Manual
Boxster, Boxster S
1997, 1998, 1999, 2000, 2001, 2002, 2003, 2004

BentleyPublishers®
.com

 | Automotive Reference™

Bentley Publishers, a division of Robert Bentley, Inc.
1734 Massachusetts Avenue
Cambridge, MA 02138 USA
800-423-4595 / 617-547-4170

Information that makes the difference®

BentleyPublishers®
.com

Technical contact information

We welcome your feedback. Please submit corrections and additions to our Porsche technical discussion forum at:
http://www.BentleyPublishers.com

Updates and Corrections

We will evaluate submissions and post appropriate editorial changes online as updates or tech discussion. Appropriate updates and corrections will be added to the book in future printings. Check for updates and corrections for this book before beginning work on your vehicle. See the following web address for additional information:
http://www.BentleyPublishers.com/updates/

WARNING—Important Safety Notice

Do not use this manual unless you are familiar with basic automotive repair procedures and safe workshop practices. This manual illustrates the workshop procedures required for most service work. It is not a substitute for full and up-to-date information from the vehicle manufacturer or for proper training as an automotive technician. Note that it is not possible for us to anticipate all of the ways or conditions under which vehicles may be serviced or to provide cautions as to all of the possible hazards that may result.

The vehicle manufacturer will continue to issue service information updates and parts retrofits after the editorial closing of this manual. Some of these updates and retrofits will apply to procedures and specifications in this manual. We regret that we cannot supply updates to purchasers of this manual.

We have endeavored to ensure the accuracy of the information in this manual. Please note, however, that considering the vast quantity and the complexity of the service information involved, we cannot warrant the accuracy or completeness of the information contained in this manual.

FOR THESE REASONS, NEITHER THE PUBLISHER NOR THE AUTHOR MAKES ANY WARRANTIES, EXPRESS OR IMPLIED, THAT THE INFORMATION IN THIS BOOK IS FREE OF ERRORS OR OMISSIONS, AND WE EXPRESSLY DISCLAIM THE IMPLIED WARRANTIES OF MERCHANTABILITY AND OF FITNESS FOR A PARTICULAR PURPOSE, EVEN IF THE PUBLISHER OR AUTHOR HAVE BEEN ADVISED OF A PARTICULAR PURPOSE, AND EVEN IF A PARTICULAR PURPOSE IS INDICATED IN THE MANUAL. THE PUBLISHER AND AUTHOR ALSO DISCLAIM ALL LIABILITY FOR DIRECT, INDIRECT, INCIDENTAL OR CONSEQUENTIAL DAMAGES THAT RESULT FROM ANY USE OF THE EXAMPLES, INSTRUCTIONS OR OTHER INFORMATION IN THIS BOOK. IN NO EVENT SHALL OUR LIABILITY WHETHER IN TORT, CONTRACT OR OTHERWISE EXCEED THE COST OF THIS MANUAL.

Your common sense and good judgment are crucial to safe and successful service work. Read procedures through before starting them. Think about whether the condition of your car, your level of mechanical skill, or your level of reading comprehension might result in or contribute in some way to an occurrence which might cause you injury, damage your car, or result in an unsafe repair. If you have doubts for these or other reasons about your ability to perform safe repair work on your car, have the work done at an authorized Porsche dealer or other qualified shop.

Part numbers listed in this manual are for identification purposes only, not for ordering. Always check with your authorized Porsche dealer to verify part numbers and availability before beginning service work that may require new parts.

Before attempting any work on your Porsche, read **00 Warnings and Cautions** and any **WARNING** or **CAUTION** that accompanies a procedure in the service manual. Review the **WARNINGS** and **CAUTIONS** each time you prepare to work on your Porsche.

Special tools required to perform certain service operations are identified in the manual and are recommended for use. Use of tools other than those recommended in this manual may be detrimental to the car's safe operation as well as the safety of the person servicing the car.

This manual is prepared, published and distributed by Bentley Publishers, 1734 Massachusetts Avenue, Cambridge, Massachusetts 02138 USA. All information contained in this manual is based on the information available to the publisher at the time of editorial closing. Porsche has not reviewed and does not vouch for the accuracy or completeness of the technical specifications and work procedures described and given in this manual.

© 2010 Robert Bentley, Inc., Bentley Publishers is a trademark of Robert Bentley, Inc.

B® is a registered trademark of Bentley Publishers and Robert Bentley, Inc.

All rights reserved. The right is reserved to make changes at any time without notice. No part of this publication may be reproduced, stored in a retrieval system, or transmitted in any form or by any means, electronic, mechanical, photocopying, recording, or otherwise, without the prior written consent of the publisher. This includes text, figures, and tables. All rights reserved under Berne and Pan-American Copyright conventions. This manual is simultaneously published in Canada.

ISBN 978-0-8376-1645-2   Editorial closing 09 / 2005.   Job code: PB04-08

Library of Congress Cataloging-in-Publication Data for 2005 edition

Porsche Boxster (986) : service manual : Boxster, Boxster S 1997, 1998, 1999, 2000, 2001, 2002, 2003, 2004.

   p. cm.

Includes index.

ISBN-13: 978-0-8376-1333-8 (pbk. : alk. paper)

ISBN-10: 0-8376-1333-7

1. Boxster automobile--Maintenance and repair--Handbooks, manuals, etc. I. Robert Bentley, inc.

TL215.B586P67 2005

629.28'722--dc22

                   2005026341

Cover design and use of blue band on spine and back cover are trade dress and are trademarks of Bentley Publishers™. All rights reserved. The paper used in this publication is acid free and meets the requirements of the National Standard for Information Sciences–Permanence of Paper for Printed Library Materials. ∞

Manufactured in the United States of America.

202505R001

WARNING–Important Safety Notice ii
Foreword iv
Index .... rear of book

0 General Information, Maintenance

00	Warnings and Cautions	02	Boxster Familiarization
01	Vehicle Identification and VIN Decoder	03	Service and Maintenance

1 Engine

10	Engine Removal and Installation	17	Lubrication System
13	Engine Pulleys and Seals	19	Cooling System
15	Cylinder Head		

2 Engine Management, Exhaust, and Engine Electrical

20	Fuel Storage and Supply	27	Battery, Starter, Alternator
24	Fuel Injection	28	Ignition System
26	Exhaust System		

3 Clutch and Transmission

30	Clutch	34	Manual Transmission
32	Torque Converter	37	Automatic Transmission

4 Suspension, Brakes, and Steering

40	Front Suspension	46	Brakes–Mechanical
42	Rear Suspension	47	Brakes–Hydraulic
44	Wheels, Tires and Alignment	48	Steering
45	Antilock Brakes (ABS)		

5 Body–Assembly

50	Fenders, Drains	57	Doors and Locks
55	Trunk Lids		

6 Body–Components and Accessories

61	Convertible Top	66	Exterior Equipment
63	Bumpers	69	Seat Belts, Airbags
64	Door Windows		

7 Body–Interior Trim

70	Interior Trim		
72	Seats		

8 Heating and Air-conditioning

80	Heating		
87	Air-conditioning		

9 Electrical System

90	Instruments, Horns	94	Exterior Lights
91	Radio and Communication	96	Interior Lights, Anti-theft
92	Wipers and Washers	97	Fuses, Relays, Component Locations

Electrical Wiring Diagrams

EWD	Electrical Wiring Diagrams

Foreword

For the Porsche owner with basic mechanical skills and for independent auto service professionals, this book includes many of the specifications and procedures that were available in an authorized Porsche dealer service department as this manual went to press. The Porsche owner with no intention of working on his or her car will find that owning and referring to this manual makes it possible to be better informed and to discuss repairs with a professional automotive technician more knowledgeably.

If you are a Porsche owner intending to do maintenance and repair work, make sure you have screwdrivers, a set of metric wrenches and sockets, and metric Allen and Torx wrenches, since these basic hand tools are needed for most of the work described in this manual. Many procedures also require a torque wrench to ensure that fasteners are tightened properly and in accordance with specifications. In some cases, the text refers to special tools that are recommended or required to accomplish adjustments or repairs. These tools are often identified by their Porsche special tool number and illustrated.

> **Disclaimer**
>
> We have endeavored to ensure the accuracy of the information in this manual. When the vast array of data presented in the manual is taken into account, however, no claim to infallibility can be made. We therefore cannot be responsible for the result of any errors that may have crept into the text. Please also read **WARNING–Important Safety Notice** on the copyright page at the beginning of this book.
>
> Prior to starting a repair procedure, read the procedure, **Warnings and Cautions** and the warnings and cautions that accompany the procedure. Reading a procedure before beginning work helps you determine in advance the need for specific skills, identify hazards, prepare for appropriate capture and handling of hazardous materials, and the need for particular tools and replacement parts such as gaskets.

Bentley Publishers encourages comments from the readers of this manual with regard to errors, and suggestions for improvement of our product. These communications have been and will be carefully considered in the preparation of this and other manuals. If you identify inconsistencies in the manual, you may have found an error. Please contact the publisher and we will endeavor to post applicable corrections on our website. Review corrections (errata) that we have posted before beginning work. Please see the following web address:

`http://www.BentleyPublishers.com/updates/`

Porsche offers extensive warranties, especially on components of the fuel delivery and emission control systems. Therefore, before deciding to repair a Porsche that may be covered wholly or in part by any warranties issued by Porsche Car North America, Inc., consult your authorized Porsche dealer. You may find that the dealer can make the repair either free or at minimum cost. Regardless of its age, or whether it is under warranty, your Porsche is both an easy car to service and an easy car to get serviced. So if at any time a repair is needed that you feel is too difficult to do yourself, a trained Porsche technician is ready to do the job for you.

Bentley Publishers

00 Warnings and Cautions

Please read these warnings and cautions before proceeding with maintenance and repair work.

WARNINGS—
*See also **CAUTIONS** on next page.*

- Read the important safety notice on the copyright page at the beginning of the book.

- Some repairs may be beyond your capability. If you lack the skills, tools and equipment, or a suitable workplace for any procedure described in this manual, we suggest you leave such repairs to an authorized Porsche dealer service department or other qualified shop.

- Thoroughly read each procedure and the **WARNINGS** and **CAUTIONS** that accompany the procedure. Also review posted corrections at www.BentleyPublishers.com/updates/ before beginning work.

- If any procedure, tightening torque, wear limit, specification or data presented in this manual does not appear to be appropriate for a specific application, contact the publisher or the vehicle manufacturer for clarification before using the information in question.

- Porsche is constantly improving its cars. Sometimes these changes, both in parts and specifications, are made applicable to earlier models. Therefore, before starting any major jobs or repairs to components on which passenger safety may depend, consult your authorized Porsche dealer about technical bulletins that may have been issued.

- Do not reuse any fasteners that are worn or deformed in normal use. Many fasteners are designed to be used only once and become unreliable and may fail when used a second time. This includes, but is not limited to, nuts, bolts, washers, self-locking nuts or bolts, circlips and cotter pins. Replace these fasteners with new parts.

- Do not work under a lifted car unless it is solidly supported on stands designed for the purpose. Do not support a car on cinder blocks, hollow tiles or other props that may crumble under continuous load. Do not work under a car that is supported solely by a jack. Do not work under the car while the engine is running.

- If you are going to work under a car on the ground, make sure that the ground is level. Block the wheels to keep the car from rolling. Disconnect the battery negative (–) terminal (ground strap) to prevent others from starting the car while you are under it.

- Do not run the engine unless the work area is well ventilated. Carbon monoxide kills.

- Remove finger rings, bracelets and other jewelry so that they cannot cause electrical shorts, get caught in running machinery or be crushed by heavy parts.

- Tie long hair behind your head. Do not wear a necktie, a scarf, loose clothing, or a necklace when you work near machine tools or running engines. If your hair, clothing, or jewelry were to get caught in the machinery, severe injury could result.

- Do not attempt to work on your car if you do not feel well. You increase the danger of injury to yourself and others if you are tired, upset or have taken medication or any other substance that may keep you from being fully alert.

- Illuminate your work area adequately but safely. Use a portable safety light for working inside or under the car. Make sure the bulb is enclosed by a wire cage. The hot filament of an accidentally broken bulb can ignite spilled fuel, vapors or oil.

- Catch draining fuel, oil or brake fluid in suitable containers. Do not use food or beverage containers that might mislead someone into drinking from them. Store flammable fluids away from fire hazards. Wipe up spills at once but do not store the oily rags, which can ignite and burn spontaneously.

- Observe good workshop practices. Wear goggles when you operate machine tools or work with battery acid. Wear gloves or other protective clothing whenever the job requires working with harmful substances.

- Greases, lubricants and other automotive chemicals contain toxic substances, many of which are absorbed directly through the skin. Read the manufacturer's instructions and warnings carefully. Use hand and eye protection. Avoid direct skin contact.

- Disconnect the battery negative (–) terminal (ground strap) whenever you work on the fuel system or the electrical system. Do not smoke or work near heaters or other fire hazards. Keep an approved fire extinguisher handy.

- Friction components (such as brake pads or shoes or clutch discs) contain asbestos fibers or other friction materials. Do not create dust by grinding, sanding, or by cleaning with compressed air. Avoid breathing dust. Breathing any friction material dust can lead to serious diseases and may result in death.

- Batteries give off explosive hydrogen gas during charging. Keep sparks, lighted matches and open flame away from the top of the battery. If hydrogen gas escaping from the cap vents is ignited, it will ignite gas trapped in the cells and cause the battery to explode.

- Battery acid (electrolyte) can cause severe burns. Flush contact area with water and seek medical attention.

- Do not quick-charge the battery (for boost starting) for longer than one minute. Wait at least one minute before boosting the battery a second time.

- Connect and disconnect a battery charger only with the battery charger switched OFF.

- Connect and disconnect battery cables, jumper cables or a battery charger only with the ignition switched OFF. Do not disconnect the battery while the engine is running.

- Do not allow the battery charging voltage to exceed 16.5 volts. If the battery begins producing gas or boiling violently, reduce the charging rate. Boosting a sulfated battery at a high rate can cause an explosion.

continued on next page

Warnings and Cautions

Please read these warnings and cautions before proceeding with maintenance and repair work.

WARNINGS— (continued)

- The air-conditioning system is filled with a hazardous refrigerant. Make sure the system is serviced only by a trained technician using approved refrigerant recovery / recycling equipment, trained in related safety precautions, and familiar with regulations governing the discharging and disposal of automotive chemical refrigerants.

- Do not expose any part of the air-conditioning system to high temperatures such as open flame. Excessive heat increases system pressure and may cause the system to burst.

- Some aerosol tire inflators are highly flammable. Be extremely cautious when repairing a tire that may have been inflated using an aerosol tire inflator. Keep sparks, open flame or other sources of ignition away from the tire repair area. Inflate and deflate the tire at least four times before breaking the bead from the rim. Completely remove the tire from the rim before attempting any repair.

- Cars covered by this manual are equipped with a supplemental restraint system (SRS), that automatically deploys airbags and pyrotechnic seat belt reels in case of a frontal or side impact. These are explosive devices. Handled improperly or without adequate safeguards, they can be accidently activated and cause serious injury.

- The ignition system produces high voltages that can be fatal. Avoid contact with exposed terminals and use extreme care when working on a car with the engine running or the ignition switched ON.

- Place jack stands only at locations specified by the manufacturer. The vehicle lifting jack supplied with the vehicle is intended for tire changes only. Use a heavy duty floor jack to lift the vehicle before installing jack stands. See **03 Service and Maintenance**.

- Aerosol cleaners and solvents may contain hazardous or deadly vapors and are highly flammable. Use only in a well ventilated area. Do not use on hot surfaces (engines, brakes, etc.).

- Due to risk of personal injury, be sure the engine is cold before beginning work on the cooling system.

- Use extreme care when draining and disposing of engine coolant. Coolant is poisonous and lethal to humans and pets. Pets are attracted to coolant because of its sweet smell and taste. Seek medical attention immediately if coolant is ingested.

CAUTIONS—
*See also **WARNINGS** on previous page.*

- If you lack the skills, tools and equipment, or a suitable workshop for any procedure described in this manual, we suggest you leave such repairs to an authorized Porsche dealer or other qualified shop.

- Porsche is constantly improving its cars and sometimes these changes, both in parts and specifications, are made applicable to earlier models. Therefore, part numbers listed in this manual are for reference only. Always check with your authorized Porsche dealer parts department for the latest information.

- Before starting a job, make certain that you have the necessary tools and parts on hand. Read the instructions thoroughly and do not attempt shortcuts. Use tools appropriate to the work and use replacement parts meeting Porsche specifications. Makeshift tools, parts and procedures do not make good repairs

- Do not use pneumatic and electric tools to tighten fasteners, especially on light alloy parts. Use a torque wrench to tighten fasteners to the tightening torque specification listed.

- Be mindful of the environment. Before you drain the crankcase, find out the proper way to dispose of the oil. Do not pour oil onto the ground, down a drain, or into a stream, pond or lake. Dispose of in accordance with Federal, State and Local laws.

- Do not expose electronic control modules to temperatures from a paint-drying booth or a heat lamp in excess of 203° F (95° C). Do not expose control modules to temperatures in excess of 185° F (85° C) for more than two hours.

- Before doing any electrical welding on cars equipped with ABS, disconnect the battery negative (–) terminal (ground strap) and the ABS control module connector.

- Make sure ignition is OFF before disconnecting the battery.

- On cars with alarm siren with tilt sensor (option code M536), turn ignition ON prior to disconnecting the battery. Make sure no electrical consumers are on prior to disconnecting the battery. This prevents the alarm siren from sounding when the battery is reconnected.

- Label battery cables before disconnecting. On some models, battery cables are not color coded.

- Disconnecting the battery may erase fault code(s) stored in control module memory. Use Porsche diagnostic equipment or equivalent to check for fault codes prior to disconnecting the battery cables. If the Check Engine light (malfunction indicator light or MIL) is illuminated or any other system faults are detected (indicated by an illuminated warning light), see an authorized Porsche dealer.

- If a normal or rapid charger is used to charge battery, disconnect the battery and remove from the vehicle in order to avoid damaging paint and upholstery.

- Do not quick-charge the battery (for boost starting) for longer than one minute. Wait at least one minute before boosting the battery a second time.

- Slow-charge a sealed or "maintenance free" battery at an amperage rate that is approximately 10% of the battery's ampere-hour (Ah) rating.

01 Vehicle Identification and VIN Decoder

Vehicle Identification Number (VIN), decoding

Some of the information in this manual applies only to cars of a particular model year or range of years. For example, 1999 refers to the 1999 model year but does not necessarily match the calendar year in which the car was manufactured or sold. To be sure of the model year of a particular car, check the Vehicle Identification Number (VIN) on the car.

The VIN is a unique sequence of 17 characters assigned by Porsche to identify each individual car. When decoded, the VIN tells the country and year of manufacture; make, model and serial number; assembly plant and some equipment specifications.

The Porsche VIN is on a plate mounted on top of the dashboard, on the driver's side where the number can be seen through the windshield. The 10th character is the model year code. The letters I, O, Q and U are not used for model year designation. Examples: X for 1999, Y for 2000, 1 for 2001, 2 for 2002, etc. The table below explains some of the codes in the VIN for 1997 through 2004 Porsche Boxsters covered by this manual.

Sample VIN: WP0CA2982WU622902
position: 1 2 3 4 5 6 7 8 9 10 11 12-17

VIN position	Description		Decoding information
1	Manufacturing Country	W	Germany
2	Manufacturer	P	Dr. Ing. h.c. F. Porsche AG
3	Vehicle Type	O	Passenger vehicle
4	Body code	C	Cabriolet
5	Engine code	A	2.5 liter 6-cylinder (Boxster)
		A	2.7 liter 6-cylinder (Boxster)
		B	3.2 liter 6-cylinder (Boxster S)
6	Restraint system	2	Manual belts with dual SRS airbags
7-8	Model type	98	First 2 digits of Porsche model (Boxster = 986)
9	Check digit		0 - 9 or X, calculated by NHTSA
10	Model year	V	1997
		W	1998
		X	1999
		Y	2000
		1	2001
		2	2002
		3	2003
		4	2004
11	Assembly plant	S	Stuttgart, Germany
		U	Uusikaupunki, Finland (pronounced oo-see-COW-poon-key)
12-17	Serial number		Sequential production number for specific vehicle

02 Boxster Familiarization

General 02-1	Steering 02-15
Engine 02-2	Brakes 02-15
Engine component overview	**Body** 02-15
(2.5 liter engine, pulley end) 02-3	Body, general layout 02-16
Engine component overview	Body dimensions 02-16
(2.5 liter engine, flywheel end) 02-3	Roll bar 02-17
Crankcase and crankshaft 02-4	Convertible top 02-17
Engine lubrication 02-4	Hardtop 02-17
Connecting rods, pistons, piston rings 02-4	Airbags 02-18
Cylinder heads, camshafts, ...02-4	Headlights 02-18
Camshaft timing chain components	Instruments 02-18
(1997- 2002 models) 02-5	Climate control 02-18
VarioCam systems 02-5	**Model Year News** 02-19
Engine cooling 02-6	1998 model year 02-19
Cooling system component overview	1999 model year 02-20
(2.5 liter engine) 02-6	2000 model year 02-21
Fuel and Ignition Systems 02-7	2001 model year 02-25
Fuel supply system overview (2.5 liter engine) .02-7	2002 model year 02-26
Engine management 02-8	2003 model year 02-28
DME system overview (2.5 liter engine) 02-9	2004 model year 02-32
Fuel injection 02-9	**Option Codes** 02-33
Secondary air injection 02-11	**TABLES**
Exhaust system 02-11	a. Engine applications and specifications 02-3
Ignition system 02-12	b. Engine management systems 02-8
Clutch and Transmission 02-12	c. Manual transmission applications 02-12
Manual transmission 02-12	d. Automatic transmission applications 02-13
Tiptronic (automatic) transmission 02-13	e. Overview of differences between Boxster and
Suspension, Brakes and Steering 02-13	Boxster S interiors (2000 model year) 02-24
Front suspension components 02-14	f. Option codes 02-33
Rear suspension components 02-14	

General

This section contains introductory and technical information for the 1997 through 2004 Porsche Boxster and Boxster S.

In Porsche factory literature, the 1997 - 2004 Boxster and 2000 - 2004 Boxster S are referred to as Type 986.

02-2 Boxster Familiarization

Engine

◁ During its eight year production run in the US, various engines, transmissions and technical systems were introduced. These changes are detailed in this section, including a year-by-year summary of changes.

◁ Similar to the Boxster concept car that debuted in 1993 at the North American International Automobile Show, the 1997-2004 Boxster reflected many of the design elements found on the mid-engine Porsche road and race cars of decades past, including the 550 Spyder and RS60.

The Boxster was released in Europe in late 1996, and for the first time in the U.S. at the L.A. Auto Show in January 1997. The original car was delivered to the U.S. market with a 2.5 liter 6-cylinder engine. This water cooled, 4-valves-per-cylinder boxer engine produced a respectable 201 hp.

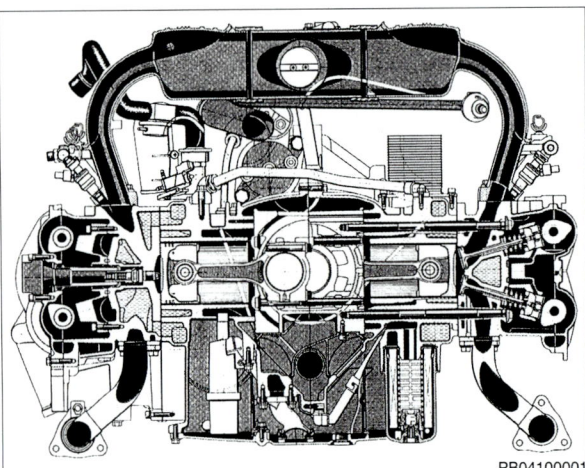

◁ The name Boxster comes from a combination of the words "boxer" and "roadster". The engine in the Boxster is a flat-six horizontally-opposed boxer engine.

A five-speed manual gearbox was fitted as standard. The optional Tiptronic S transmission allowed drivers the choice of either fully automatic gear changes or manual gear selection.

ENGINE

The M 96.20 2.5 liter water-cooled boxer engine was offered in the US from 1997 through 1999. In model year 2000, the 2.5 liter motor was replaced with a larger, more powerful 2.7 liter version featuring a twin-resonance induction system (similar to that in the 996) that helps boost horsepower and torque. Also in model year 2000, the 3.2 liter Boxster S model was introduced to the US market.

◁ Highlights of the M 96.20 engine:

- Aluminum alloy vertically split crankcase with LOKASIL® cylinder sleeves and iron-coated pistons
- Aluminum main bearing shells with cast iron nodulations
- Semi-dry sump lubrication
- Four overhead camshafts with variable intake camshaft timing (VarioCam)
- Four valves per cylinder
- Three piece cylinder heads
- Hydraulic valve lifters

Boxster Familiarization 02-3
Engine

Table a lists the engines used in 1997 - 2004 986 Boxster models. Additional engine technical information can be found later in **Model Year News** in this repair group

Table a. Engine applications and specifications

Model, year	Engine code	Displacement cc (cu. in.)	Compression ratio	Torque Nm (lb-ft) @ rpm	Horsepower Kw (hp) @ rpm	Bosch DME
Boxster 1997 - 1999	M 96.20	2480 (151.3)	11 : 1	245 (181) @ 4500	150 (204) @ 6000	M 5.2.2
Boxster 2000 - 2002	M 96.22	2687 (163.9)	11 : 1	260 (192) @ 4750	162 (217) @ 6400	ME 7.2
Boxster 2003 - 2004	M 96.23	2687 (163.9)	11 : 1	260 (192) @ 4700	168 (225) @ 6300	ME 7.8
Boxster S 2000 - 2002	M 96.21	3179 (194)	11 : 1	305 (225) @ 4500	185 (248) @ 6250	ME 7.2
Boxster S 2003 - 2004	M 96.24	3179 (194)	11 : 1	310 (229) @ 4600	191 (258) @ 6200	ME 7.8

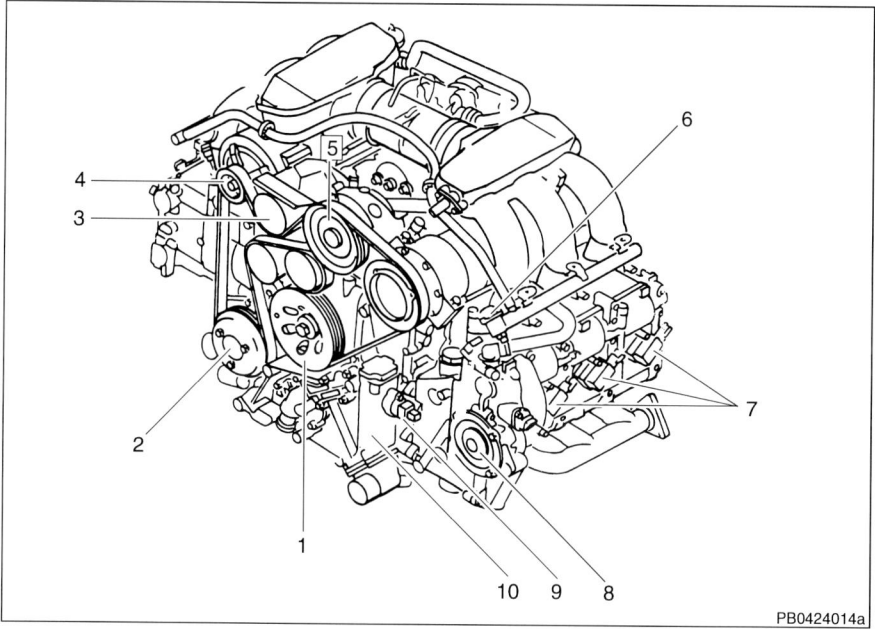

Engine component overview (2.5 liter engine, pulley end)

1. Crankshaft pulley
2. Coolant pump pulley
3. Belt tensioner
4. Alternator
5. Power steering pump
6. Fuel pressure regulator
7. Ignition coils 4 -6
8. Engine oil return pump (cyl. 4 - 6)
9. Engine coolant temperature (ECT) sensor
10. Coolant manifold

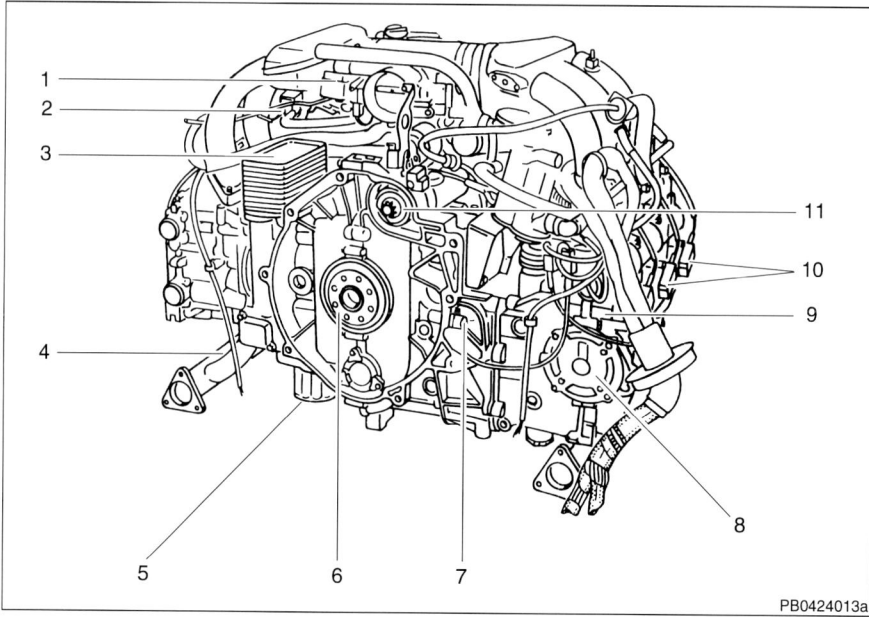

Engine component overview (2.5 liter engine, flywheel end)

1. Throttle position sensor
2. Fuel tank evaporative vent valve
3. Engine oil cooler (Oil-to-water heat exchanger)
4. Exhaust manifold
5. Oil filter housing
6. Crankshaft rear main seal
7. Crankshaft sensor (engine speed / reference sensor)
8. Engine oil return pump (cyl. 1 - 3)
9. Intake camshaft adjuster (VarioCam)
10. Ignition coils 1- 3
11. Starter

Engine

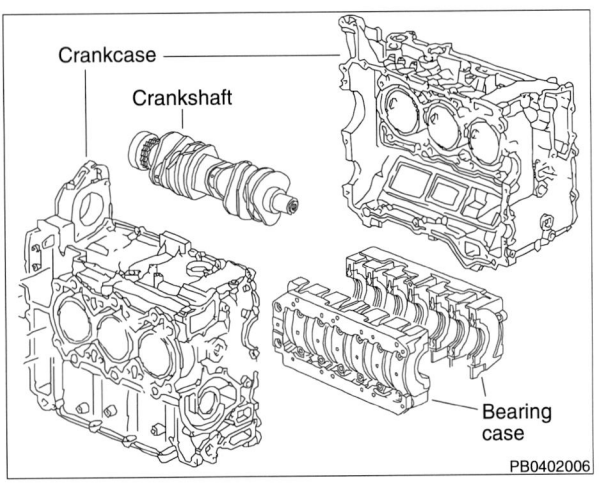

Crankcase and crankshaft

◁ The crankcase is made up of 4 pieces: the crankcase halves and the aluminum two-piece main bearing case. The main bearings include cast iron construction to control bearing clearance.

The drop-forged crankshaft rides in seven main bearings and features 12 counterweights. Main bearing 4 is the thrust bearing.

The cylinders are manufactured using a special manufacturing process termed LOKASIL®. A porous silicon sleeve (25% silicon / 75% air) is cast into the crankcase. During the casting only the silicon remains locally, hence the name LOKASIL®.

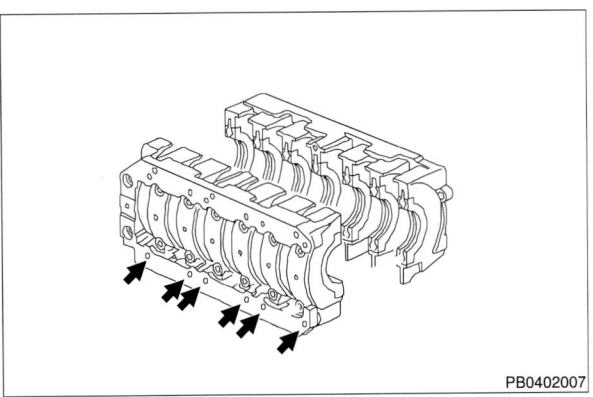

Engine lubrication

◁ The crankshaft bearing case includes spray nozzles **(arrows)** to cool the pistons. To ensure sufficient oil pressure at low engine speeds, the oil spray nozzles do not open until the main gallery oil pressure exceeds 1.8 bar (26 psi).

The main oil pump is driven off the intermediate shaft and draws oil out of the oil sump via an oil suction tube. Oil return pumps are used in each cylinder head to return the oil to the oil sump via individual oil separators. See **17 Lubrication System** for additional information

Connecting rods, pistons, piston rings

The forged steel connecting rods use replaceable split-shell bearings at the crankshaft end and solid bushings at the piston pin end. Each rod is laser-scored at the big end and then cracked after forging, creating perfectly matched halves. Centering of the cap on the rod is done through the structure of the break.

The pistons are cast aluminum and coated with iron to provide a wear resistant sliding surface. The piston pin is offset by 0.8 mm toward the intake side.

The pistons rings are of the three-ring type with two upper compression rings and a lower three-piece oil scraper ring.

Cylinder heads, camshafts

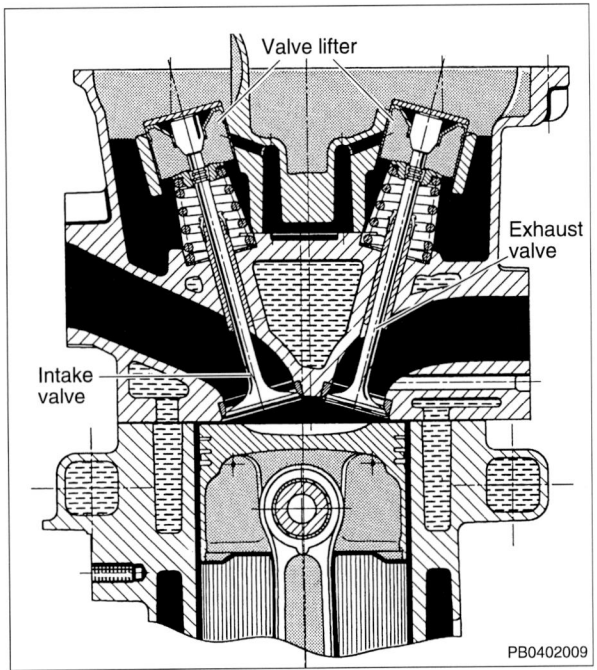

◁ Cylinder heads utilize four-valve-per-cylinder technology, hydraulic valve lifters, and adjustable intake camshaft timing (VarioCam).

Boxster Familiarization 02-5
Engine

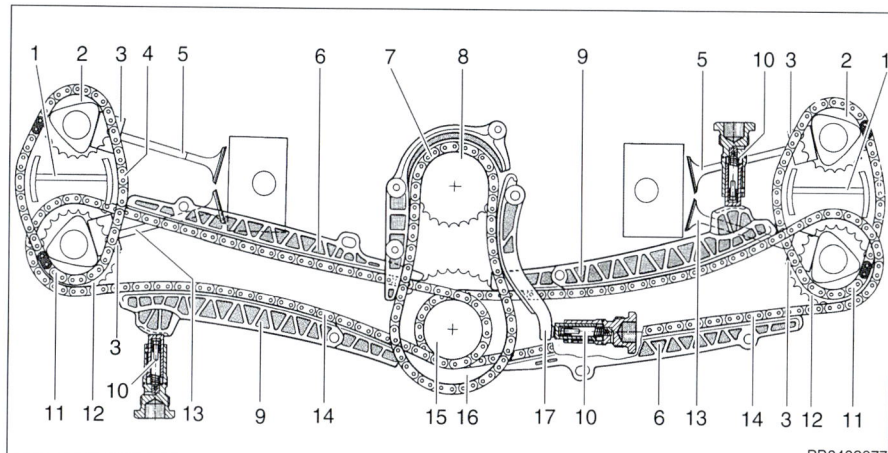

Camshaft timing chain components (1997- 2002 models)

1. VarioCam actuator
2. Intake camshaft sprocket
3. Hydraulic valve lifter
4. Single-row chain
5. Intake valve
6. Chain guide
7. Double-row chain
8. Crankshaft sprocket
9. Tensioning chain guide
10. Chain tensioner
11. Exhaust camshaft drive sprocket
12. Exhaust camshaft input sprocket
13. Exhaust valve
14. Double-row chain
15. Intermediate shaft output sprocket
16. Intermediate shaft input sprocket
17. Chain tensioner guide

VarioCam systems

Valve overlap occurs at the end of the engine's exhaust stroke and the beginning of the intake stroke when all valves are open. Valve overlap directly affects torque and emissions. Greater valve overlap enhances cylinder filling and boosts torque. Reduced valve overlap (late valve timing) increases power and also reduces exhaust hydrocarbon emissions at idle and low engine speeds. In previous engine designs, Porsche engineers had to compromise on camshaft timing, sacrificing some low-end torque for some high-end horsepower, or vice versa.

The Boxster engine design incorporates adjustable camshaft timing. This system, called VarioCam, boosts low-end and mid-range torque by varying valve overlap at different engine speeds. It is controlled by the Bosch DME engine management system.

◁ 1997 - 2002 models: Electric solenoid VarioCam adjustment units are mounted in the left and right cylinder heads. This design allows advanced camshaft timing to be switched ON or OFF.

02-6 Boxster Familiarization

Engine

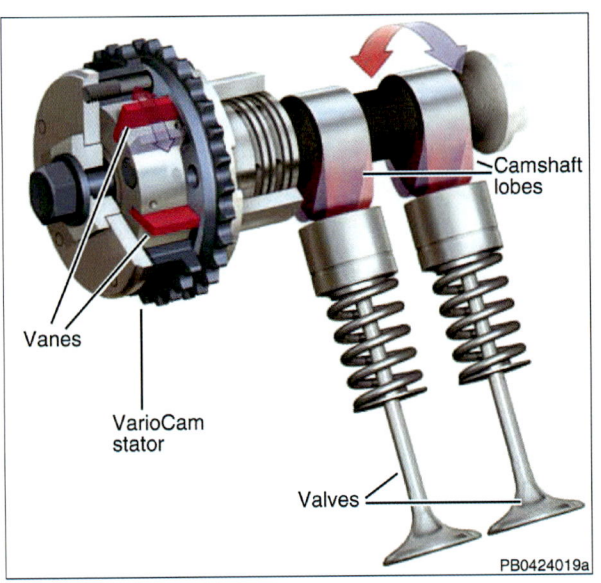

◁ 2003- 2004 models: Redesigned VarioCam system uses an oil filled stator with five internal vanes on the front of each intake camshaft. This type of camshaft timing adjuster allows for infinitely-varied adjustment within a range of 40° of crankshaft rotation.

Engine cooling

The Boxster cooling system employs a cross-flow design with two front mounted radiators. Boxster S models are fitted with an additional center mounted radiator. See **19 Cooling System** for more information.

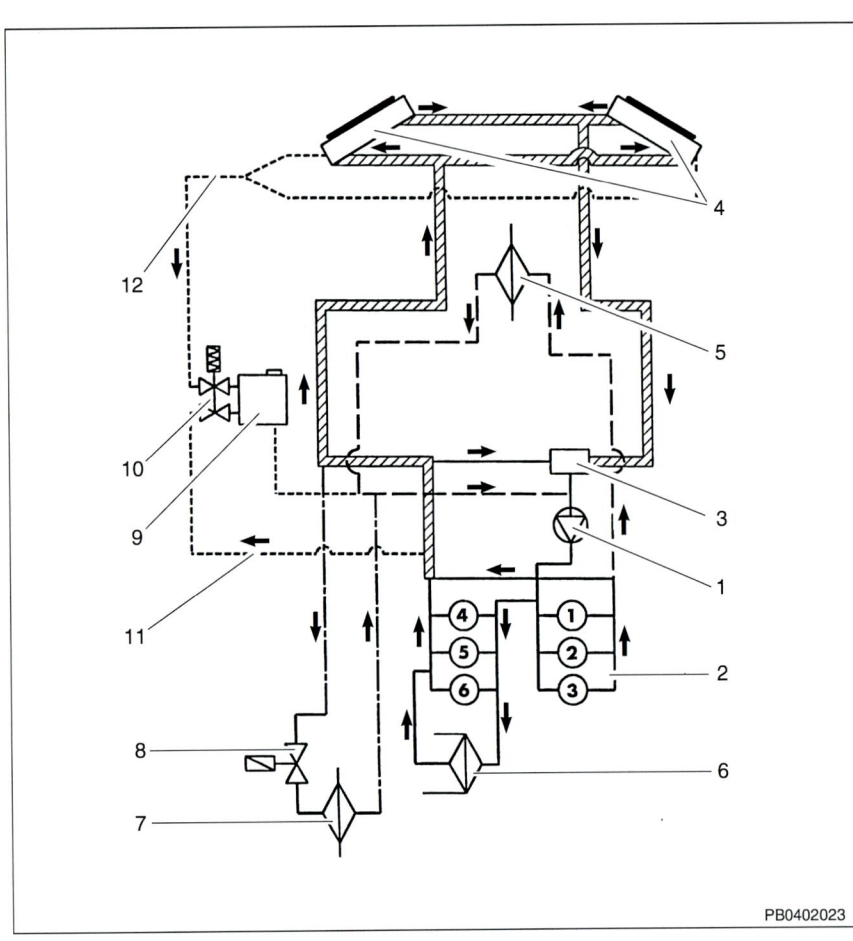

Cooling system component overview (2.5 liter engine)

1. Coolant pump
2. Cylinder block
3. Thermostat
4. Radiators
5. Heater core
6. Engine oil cooler
7. ATF cooler (Tiptronic only)
8. ATF cooler electric shut-off valve
9. Coolant expansion tank
10. Shut-off valve
11. Vent (engine)
12. Vent (radiator)

Boxster Familiarization 02-7

Fuel and Ignition Systems

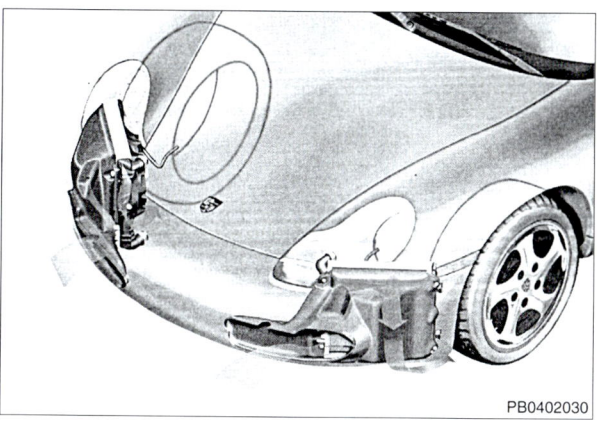

◁ The radiators are cooled using electric fans. Air is drawn in through ducts in the front bumper cover. The air flow from the fans also serves to provide cooling air to the brake calipers.

◁ The engine compartment cooling fan intake duct is a grilled opening in the right rear fender. The opening in the left rear fender is used for engine intake air.

FUEL AND IGNITION SYSTEMS

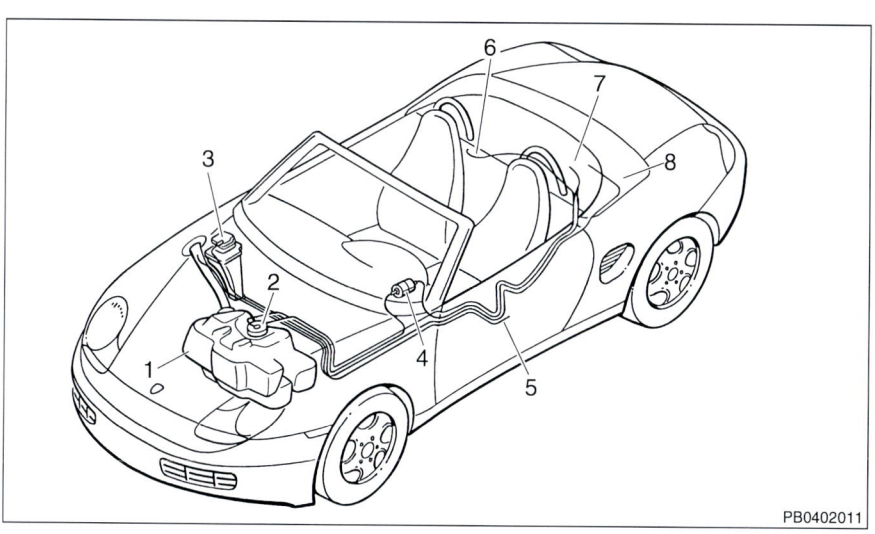

Fuel supply system overview (2.5 liter engine)

1. Fuel tank
2. Fuel pump and fuel level sender
3. Carbon canister
4. Fuel filter
5. Scavenging air line
6. Fuel supply line
7. Fuel return line
8. To carbon canister

Fuel is supplied to the fuel injection system via an in-tank fuel pump. Fuel pressure is regulated by a fuel pressure regulator.

1997 - 2001 models: The fuel pressure regulator is mounted on the fuel rail on the engine.

2002 - 2004 models: The fuel pressure regulator is integrated with the in-tank fuel pump together with the fuel filter.

See **20 Fuel Storage and Supply** for additional information.

02-8 Boxster Familiarization

Fuel and Ignition Systems

Engine management

There are three versions of engine management systems used on the 1997 through 2004 Boxster models, depending on engine application and model year. These systems, which control fuel injection and ignition functions, are referred to as DME (for digital motor electronics) or Motronic. See **Table b**

Table b. Engine management systems	
Year, model	DME system
1997 - 1999 Boxster 2.5 liter	Bosch M 5.2.2
2000 - 2002 Boxster 2.7 liter / Boxster S	Bosch ME 7.2
2003 - 2004 Boxster 2.7 liter / Boxster S	Bosch ME 7.8

The engine management systems used on the cars covered by this manual are compliant with second generation on-board diagnostics (OBD II). OBD II standards require sophisticated diagnostic software capable of recognizing and electronically storing hundreds of diagnostic troubles codes (DTCs) in the engine control module (ECM). DTCs can only be accessed using special test equipment (scan tool). The Porsche dealer is equipped with the specialized OBD II scan tool to quickly and efficiently locate engine management problems. Alternately, a generic scan tool may be used to access OBD II fault information. See **24 Fuel Injection** for OBD II information and fault codes.

Each DME system has the same basic components and operating principles. The most notable difference is that the 2000 and later cars are equipped with E-Gas, a motor driven throttle system without a throttle cable.

◁ The engine control module (ECM) uses signals from the following sensors as the primary inputs to electronically control fuel delivery and ignition timing:

- Oxygen sensors
- Mass air flow sensor
- Knock sensors
- Crankshaft sensor
- Camshaft sensors
- Air and coolant temperature sensors
- Throttle position sensor

Boxster Familiarization 02-9
Fuel and Ignition Systems

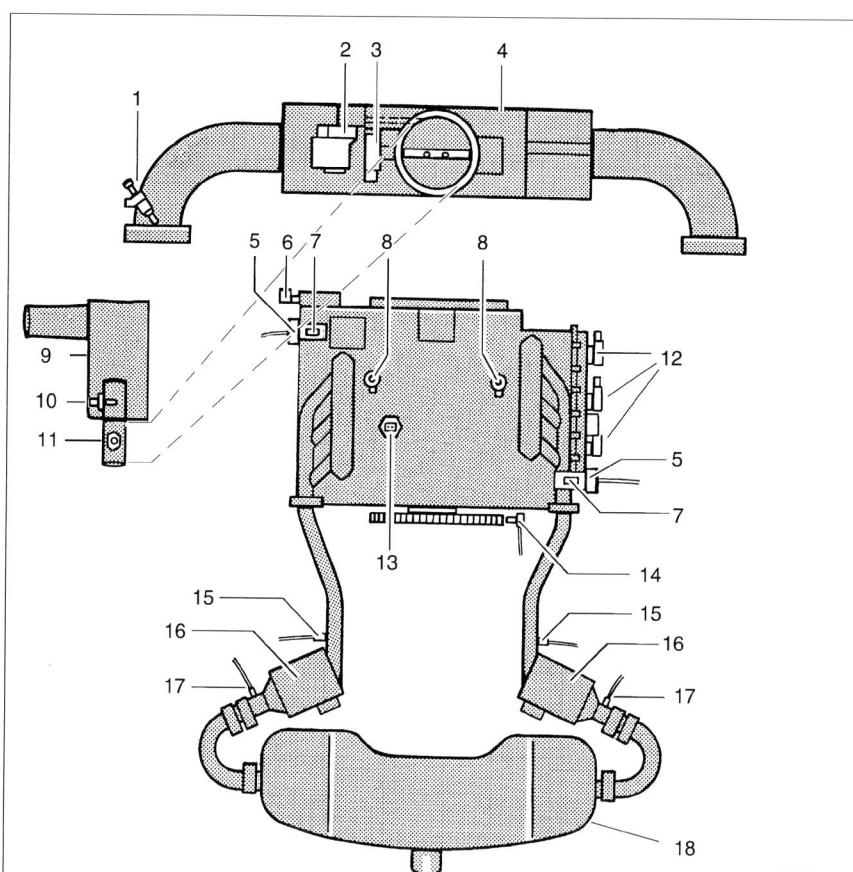

DME system overview (2.5 liter engine)

1. Fuel injector
2. Idle air control valve (IACV)
3. Throttle position sensor (TPS)
4. Intake manifold
5. VarioCam actuator
6. Engine coolant temperature (ECT) sensor
7. Camshaft sensor (CMP)
8. Knock sensor
9. Air filter housing
10. Intake air temperature (IAT) sensor
11. Mass air flow (MAF) sensor
12. Ignition coils
13. Oil temperature sensor
14. Crankshaft sensor
15. Oxygen sensor, precatalytic converter
16. Catalytic converter
17. Oxygen sensor, post-catalytic converter
18. Muffler

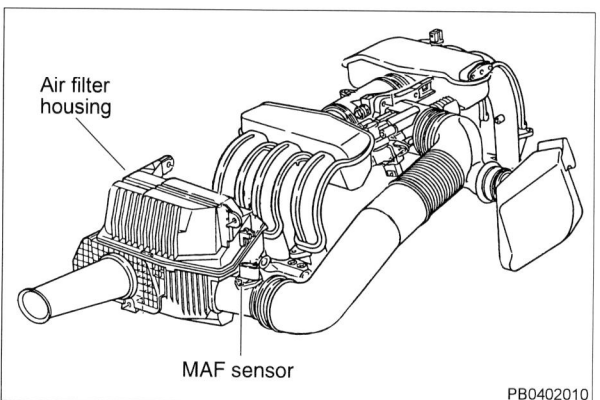

Fuel injection

The fuel injection system is completely electronic in operation. Air entering the engine passes through a pleated paper filter element in the air filter housing. Intake air mass is measured by the mass air flow (MAF) sensor. A reference current is used to heat a thin film in the sensor when the engine is running. The current used to heat the film is electronically converted into a voltage measurement corresponding to the mass of the intake air.

Additional sensors supply information about engine operating conditions.

The engine control module (ECM), using signals from the various sensors, calculates the amount of fuel needed for the correct air-fuel ratio and actuates the fuel injectors accordingly. The ECM meters fuel by changing the opening time (pulse width) of the fuel injectors.

To ensure that injector pulse width is the only factor that determines fuel metering, fuel pressure is maintained by a fuel pressure regulator. The fuel injectors are mounted to a common fuel supply called the fuel rail.

02-10 Boxster Familiarization

Fuel and Ignition Systems

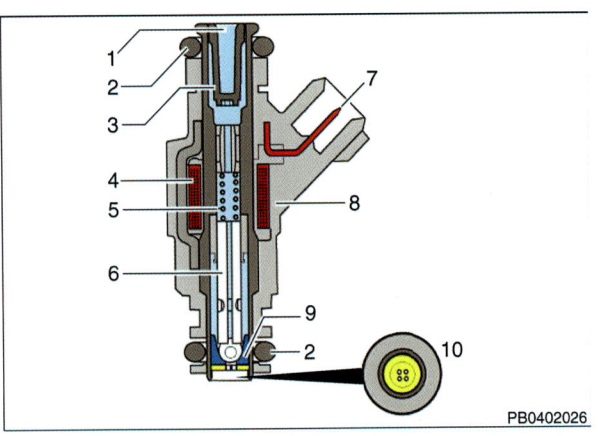

◂ Shown is the 4-hole spray disc fuel injector used on 2002 and later engines.

1. Fuel line connection
2. Sealing O-rings
3. Filter screen
4. Electromagnetic coil
5. Spring
6. Valve needle with magneto armature and sealing ball
7. Electrical connection
8. Valve housing
9. Valve seat
10. Spray disc (4 hole)

The ECM monitors engine speed to determine the duration of injector openings. Other signals to the ECM help determine injector pulse time for different operating conditions. A temperature sensor signals engine temperature for mixture adaptation. A throttle position sensor signals throttle position. The exhaust oxygen sensors signal information about combustion efficiency for control of the air-fuel mixture.

1997 - 1999 models: Idle speed control valve controls idle speed by bypassing varying amounts of air around closed throttle valve.

2000 - 2004 models: Idle speed is controlled via throttle control module.

Knock sensors monitor and control ignition knock through the ECM. A knock sensor functions as a microphone to convert mechanical vibration (knock) into electrical signals. The ECM is programmed to react to frequencies that are characteristic of engine knock and adapt the ignition timing point accordingly. See **28 Ignition System** for further details.

Idle speed, idle mixture (%CO), and ignition timing are not adjustable. The adaptive engine management system is designed to automatically compensate for changes in engine operating conditions.

◂ If the DME system adaptive limits are exceeded, the Check Engine light (malfunction indicator light or MIL) illuminates, indicating an emissions-related fault. For MIL diagnostics, see **24 Fuel Injection**.

The exhaust system uses four heated oxygen sensors, two per catalytic converter. Oxygen sensor signals are used by the DME system for mixture control and fuel system and catalytic converter efficiency monitoring required by OBD II standards.

NOTE—
- *2000 and later cars are equipped with additional precatalytic converters. See **26 Exhaust System** for additional information.*

Boxster Familiarization 02-11
Fuel and Ignition Systems

Secondary air injection

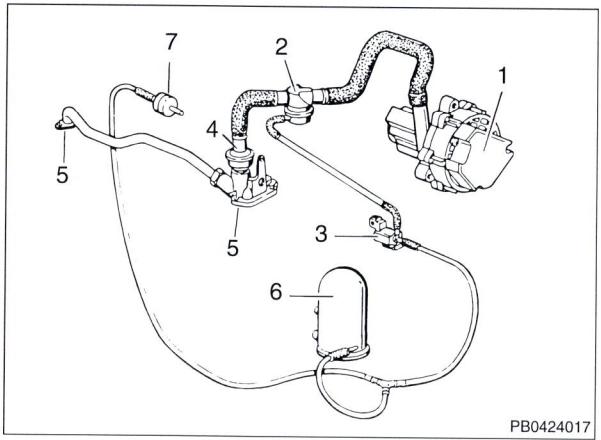

The Boxster engine is equipped with a secondary air injection system which uses an electric pump to force fresh air into the exhaust system upstream of the catalytic converter during engine warm-up. By providing extra oxygen to the unburned fuel in the exhaust, hydrocarbons oxidize and carbon monoxide combines with oxygen to form carbon dioxide and water. The air injection pump stops within a specified timed interval.

1. Secondary air injection pump
2. Air change-over valve
3. Electric vacuum switching valve
4. Non-return valve
5. To cylinder heads
6. Vacuum reservoir
7. To vacuum source

Exhaust system

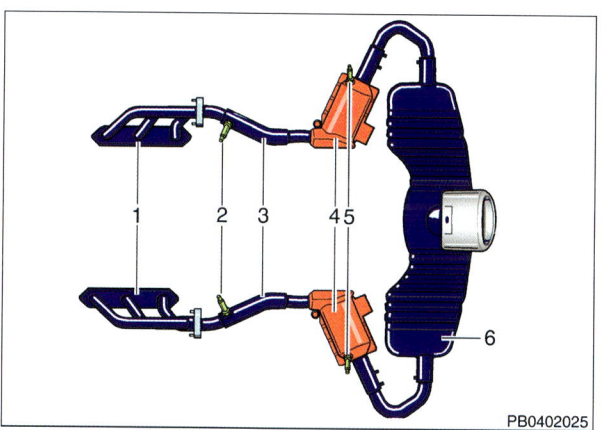

1997-1999 (2.5 liter) engine was equipped with 2 catalytic converters and 4 oxygen sensors:

1. Exhaust manifold
2. Oxygen sensor, precatalytic converter
3. Intermediate pipe
4. Catalytic converter
5. Oxygen sensor, post-catalytic converter
6. Muffler

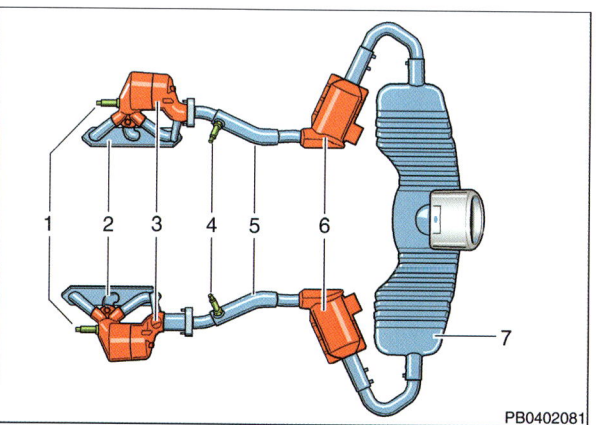

2000 - 2004 (2.7 liter and 3.2 liter) engines were equipped 2 warm-up catalytic converters (precatalyst), 2 main catalytic converters and 4 oxygen sensors.

1. Oxygen sensor, precatalytic converter
2. Exhaust manifold
3. Warm-up catalytic converter (precatalyst)
4. Oxygen sensor, post-catalytic converter
5. Intermediate pipe
6. Main catalytic converter
7. Muffler (Boxster S, 3.2 liter engine, equipped with dual tailpipe tip)

02-12 Boxster Familiarization

Clutch and Transmission

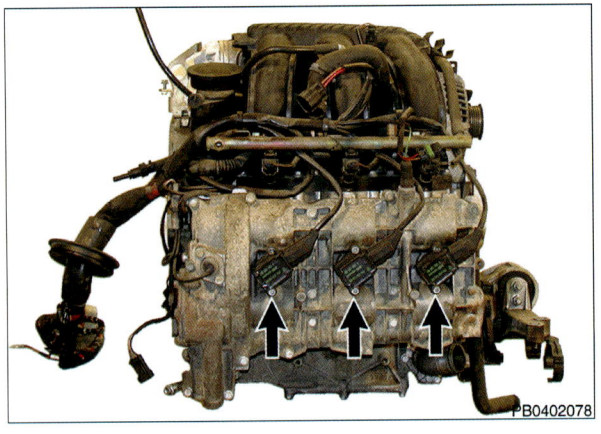

Ignition system

◁ The DME ignition system is a direct-ignition design with individual coils **(arrows)** at each cylinder. Each coil is individually controlled and monitored by the engine control module (ECM).

CLUTCH AND TRANSMISSION

Boxster models are equipped with a longitudinal, mid-engine drivetrain. The transmission is bolted directly to the rear of the engine, behind the passenger compartment. The clutch uses a single dry disc and dual-mass flywheel. See also:

- 30 Clutch
- 34 Manual Transmission
- 37 Automatic Transmission

Manual transmission

In order to suit multiple engine sizes and outputs, three different manual transmissions were offered in the 986 Boxster and Boxster S

Table c. Manual transmission applications and gear ratios			
Transmission type	5-speed G86.00 (Boxster 2.5 liter)	5-speed G86.01 (Boxster 2.7 liter)	6-speed G86.20 (Boxster S 3.2 liter)
Transmission code	CWA, DVY	EFD	n/a
1st gear	3.50	3.50	3.82
2nd gear	2.12	2.12	2.20
3rd gear	1.43	1.43	1.52
4th gear	1.03	1.09	1.22
5th gear	0.79	0.84	1.02
6th gear	-	-	0.89
Final drive ratio	3.89	3.56	3.44

◁ Manual transmissions can be identified by two sets of code stampings, one located at the top of the bellhousing, the second located on the underside of the transmission.

Code stampings convey valuable information about the transmission. The first three letters indicate the transmission type. The second five number code is the date that the transmission was built: Day (05), month (11), year (7). The code given in the example on the left indicates that transmission was built on November 5, 1997.

Boxster Familiarization 02-13
Suspension, Brakes and Steering

Tiptronic (automatic) transmission

◀ The 5-speed automatic transmission used in the Porsche Boxster is a model 5HP-19 transmission manufactured by ZF Getriebe, GmbH

This transmission differs from regular automatic transmissions in that it adapts to driving style for optimum performance and allows the driver to select and shift through the forward gears like a manual transmission, without the need for clutch actuation.

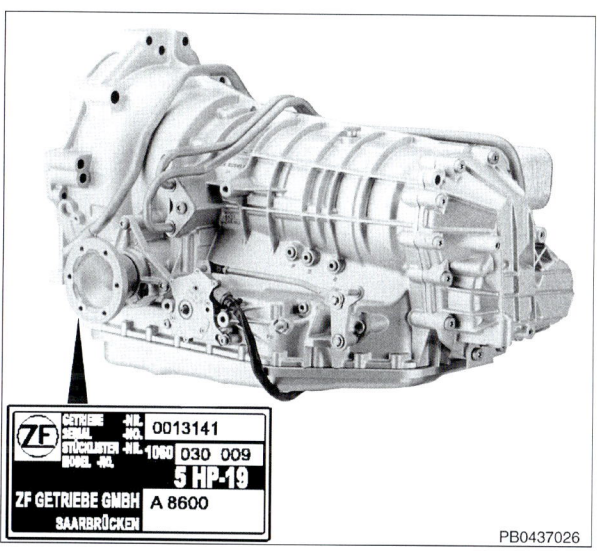

◀ Transmission identification plates are located on the top of transmission near the transmission fluid cooler, and on underside of torque converter bellhousing. Porsche identifies transmissions using five digit codes. See **Table d**

Table d. Automatic transmission applications	
Year, model	Transmission code
1997-1999 Boxster	A86.00
2000 - 2004 Boxster	A86.05
2000 - 2004 Boxster S	A86.20

SUSPENSION, BRAKES AND STEERING

◀ The mid-engine design and MacPherson strut front and rear suspension employ many lightweight and high performance components, including aluminum control arms, crossmember, and wheel bearing carriers.

02-14 Product Familiarization

Suspension, Brakes and Steering

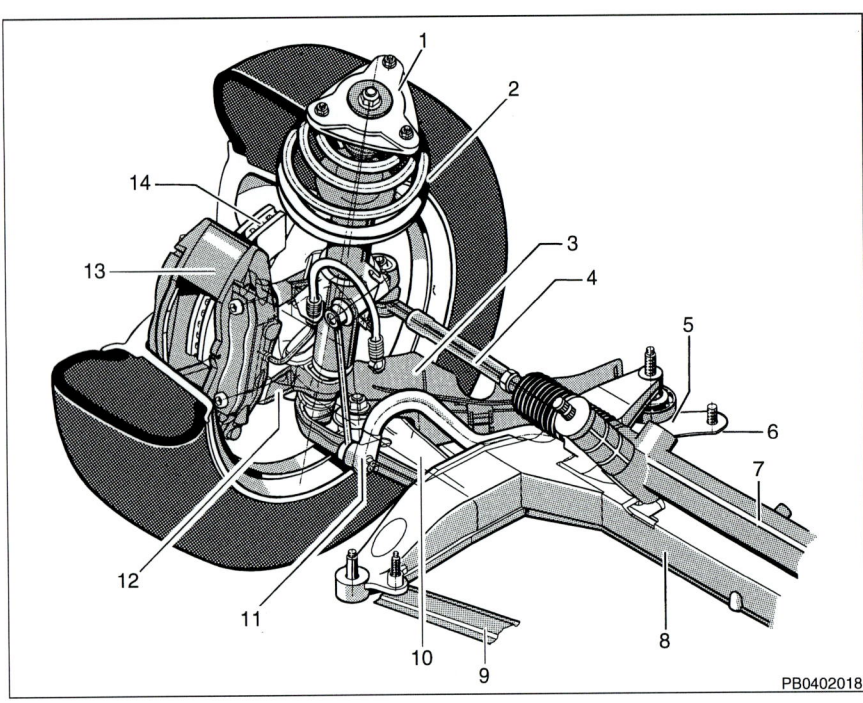

Front suspension components

1. Upper strut mount
2. Coil spring
3. Air duct
4. Tie rod
5. Diagonal control arm
6. Corner plate
7. Steering rack
8. Aluminum crossmember
9. Brace
10. Control arm
11. Sway bar
12. Wheel bearing carrier
13. Brake caliper
14. Brake rotor

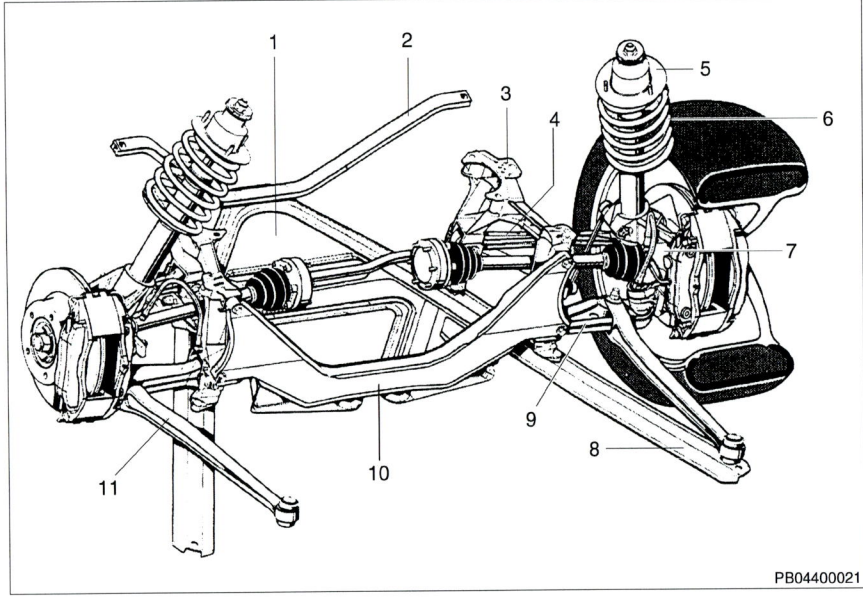

Rear suspension components

1. Chassis reinforcement plate
2. Reinforcement plate support
3. Rear suspension arm mount
4. Track arm
5. Strut bearing
6. Coil spring
7. Wheel bearing carrier
8. Diagonal brace
9. Control arm
10. Rear axle support
11. Trailing arm

◁ The rear suspension is fitted with a large bolt-on aluminum chassis reinforcement plate which is a structural component of the rear suspension.

CAUTION—
- *Do not alter the chassis reinforcement plate in any way.*
- *Do not drive the car with the chassis reinforcement plate removed.*

Boxster Familiarization 02-15

Body

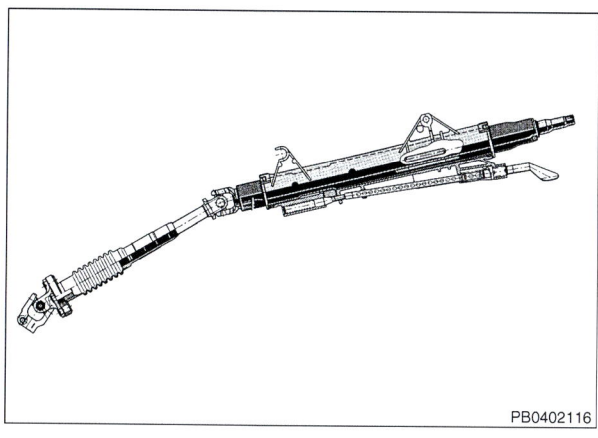

Steering

The power-assisted steering rack-and pinion has a 1: 16.9 ratio. The rack is hollow for reduced weight. The engine mounted power steering pump is belt driven off the crankshaft. The system uses Pentosin CH 11s hydraulic fluid.

◄ The steering column can be telescopically adjusted in and out by 40 mm (approx. 1.6 in).

Brakes

The brake system is a dual circuit design with a 10 inch vacuum brake booster. A brake proportioning valve (switching pressure of 25 bar / reduction factor of 0.46) is installed in the rear axle circuit at the master cylinder. Front and rear vented brake rotors and rotor air ducting is used to reduce braking temperatures at the wheels. Boxster S brake rotors are cross-drilled.

Boxster models are fitted with either standard ABS or an optional traction control system.

Body

◄ The 986 Boxster roadster body shape was optimized in the wind tunnel with a 0.31 coefficient of drag (Cd). A retractable rear spoiler automatically extends at speeds above 75 mph (120 kph), reducing axle lift by as much 30%. The spoiler retracts when speed falls below 50 mph (80 kph).

The undercarriage was also analyzed in the wind tunnel. The floor pan and other components were designed flat and smooth with multiple protective under panels.

NOTE—
- *Aerodynamic lift is an important consideration when developing an automotive chassis. When excess under-car lift is present, the car becomes lighter on its axles. This adversely affects vehicle balance, handling, and passenger safety. Through deliberate body design, aerodynamic downforce can counteract the aerodynamic lift. Automotive design engineers constantly seek the ideal balance of lift, down-force and drag, while maintaining a visually attractive design.*

The Boxster's mid-engine design allows for an optimum passenger compartment and two trunk, each with a volume of 130 liters (4.6 cu. ft.). The engine has a low design height, which allows for more convertible top storage space.

◄ The body shell is a unitized sheet steel design with welded box sections and bolt-on front fenders. Hot-dipped and electrolytically galvanized panels are used for optimum corrosion protection.

The body shell has an extremely rigid passenger compartment for passive safety. Appropriate deformation zones are integrated into the body design both front and rear.

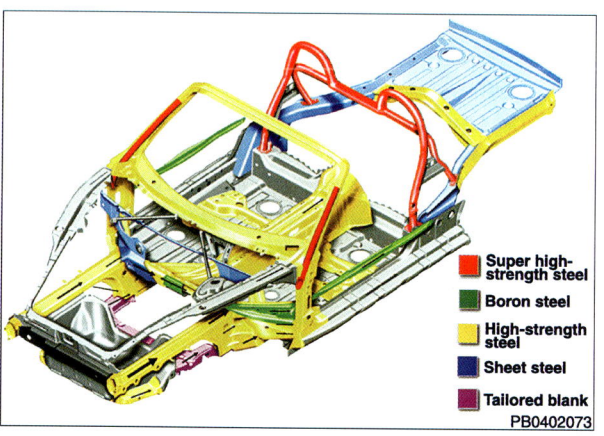

02-16 Boxster Familiarization

Body

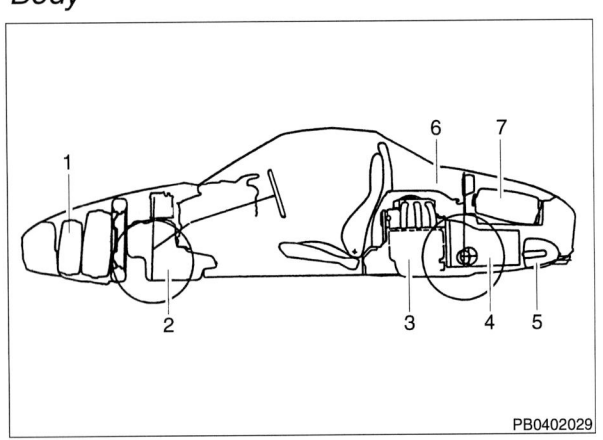

Body, general layout

1. Front trunk
2. Fuel tank
3. Engine
4. Transmission
5. Muffler
6. Convertible top storage area
7. Rear trunk

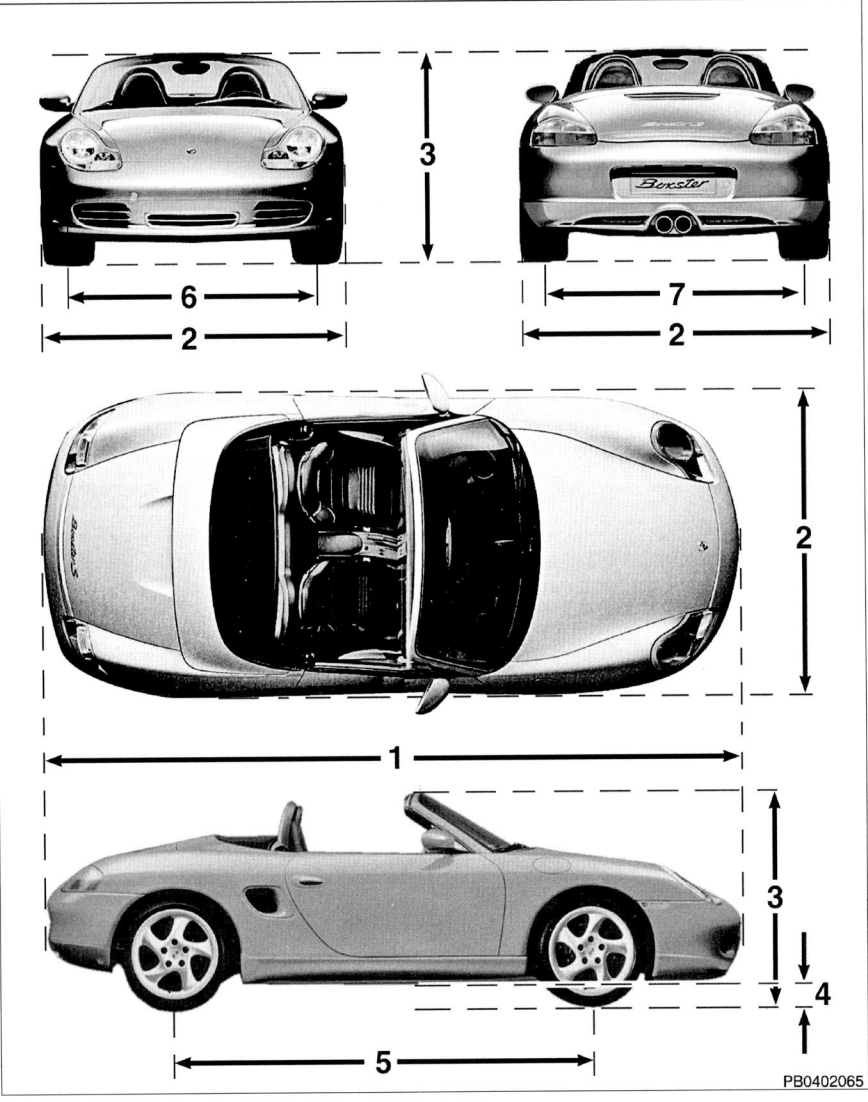

Body dimensions

1. **Length**
   - 4320 mm (170.1 in)
2. **Width**
   - 1780 mm (70.1 in)
3. **Height**
   - 1290 mm (50.8 in)
4. **Ground clearance**
   - 110 mm (4.3 in)
5. **Wheel base**
   - 2415 mm (95.1 in)
6. **Front track (17" wheels)**
   - Boxster 1465 mm (57.7 in)
   - Boxster S 1455 mm (57.3 in)
7. **Rear track (17" wheels)**
   - Boxster 1528 mm (60.2 in)
   - Boxster S 1514 mm (59.6 in)

NOTE—
- *Dimensions are based on 2002 models. Other models are similar.*

Boxster Familiarization 02-17
Body

Roll bar

◄ The roll bar is comprised of two protective hoops supported by a cross tube above the front edge of the convertible top compartment.

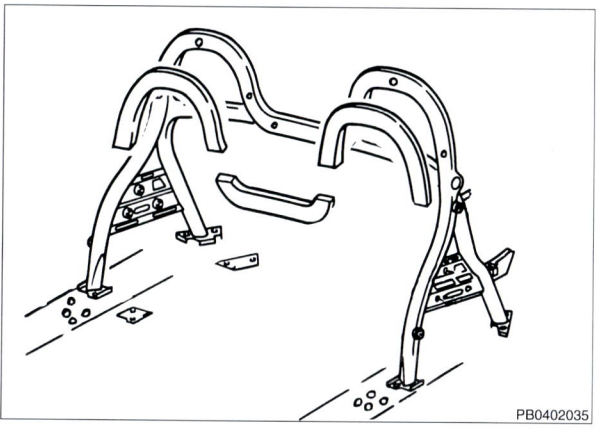

◄ The cross tube is supported at the base on solid frame member sections. The entire roll bar is a welded unit made of high tensile steel. The tube sections are made using an internal high pressure water forming technique.

Convertible top

The shape and design of the convertible top was developed and optimized in the wind tunnel with regards to wind noise, vibration, and deformation / inflation behavior. The convertible top is a three-layer structure. The top layer is fully synthetic for resistance to weathering. The middle layer is rubber to keep out water, and the inner layer is 100% cotton. The top can be raised or lowered in less than 12 seconds.

◄ Top-down air-draft was also minimized in the design, especially when the optional 3-piece wind stop is used.

Hardtop

◄ A hardtop is available as an optional extra. The light-weight factory hardtop is made of aluminum and weighs only 55 pounds (25 kg).

02-18 Boxster Familiarization

Body

Airbags

◁ Door-mounted airbags were gradually phased in during the first year of production (1997) and were standard in 1998 and later models.

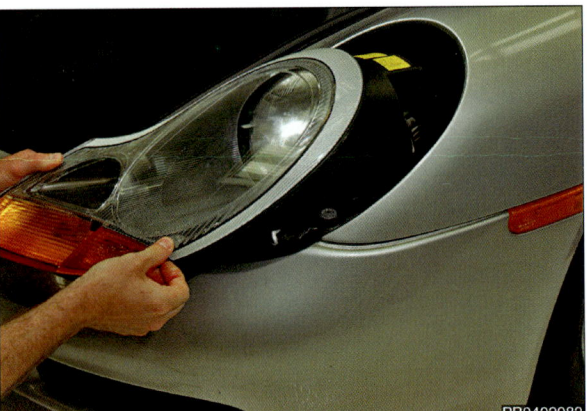

Headlights

◁ Aerodynamic halogen headlights were standard, with xenon lighting available as an option beginning in 1999. The headlights units are slide-in one-piece modules.

Instruments

◁ The instrument cluster is comprised of three round displays. The instruments use stepper motor measuring gauges for high accuracy.

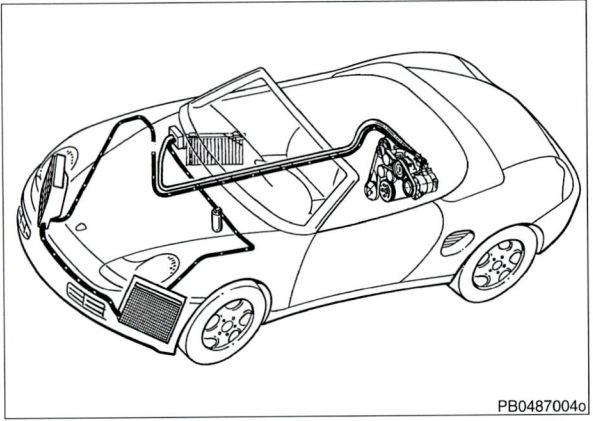

Climate control

◁ A fully automatic air-conditioning system is standard equipment on cars delivered to USA.

Boxster Familiarization 02-19
Model Year News

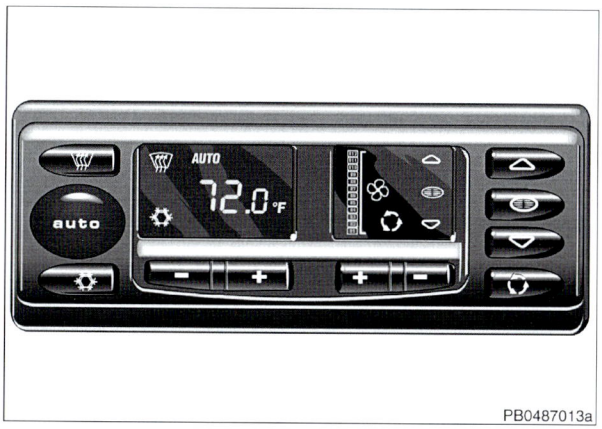

◁ The climate control system uses temperature sensors and automatically selects blower speeds to maintain cabin temperature.

MODEL YEAR NEWS

The 986 Boxster was produced for an eight year run. The information below details the model's annual product features and technical updates.

1998 model year

Fuel supply

◁ Redesigned fuel tank venting system, including fuel tank leakage diagnostics. New diagnostic components include:

- Fuel tank pressure sensor, called differential pressure sensor
- Electromagnetic carbon canister shut-off valve

This system checks the fuel tank and associated lines and hoses for hydrocarbon leaks into the atmosphere. If a leak is detected, the Check Engine light is turned on and a fault is set in DME fault memory. See **24 Fuel Injection** for more information on the fuel tank evaporative control system.

Wheels and tires

The 1998 Boxster can be fitted with 18 inch wheels owing to structural body modifications to the rear end.

Body

Rear body structural changes included the following redesigned components:

- Wheel well and coil spring mount
- Lower engine compartment bulkhead
- Rear wall crossmember
- Rear axle mount reinforcements.

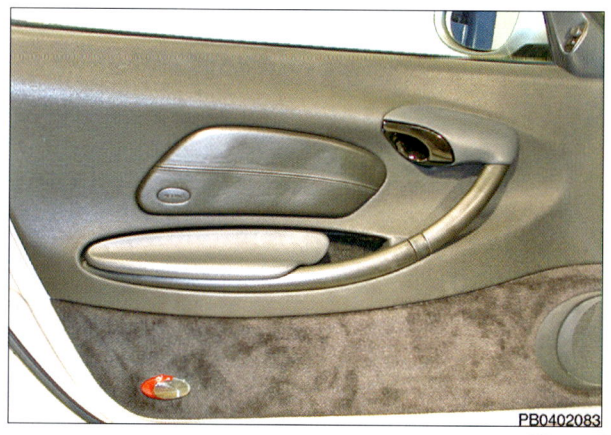

◁ Side (door-mounted) airbags with a volume of approx. 30 liters (1.1 cu.ft.) are designed to protect the chest cavity and the head. The door panel is redesigned to protect the pelvis.

Model Year News

Electrical system

Optional parking assistant aids the driver when backing into parking spots. The system warns the driver of obstacles through an audible gong.

◁ Four ultrasonic sensors **(arrows)** were fitted to the rear bumper. Whenever reverse gear is selected and the ignition is on, a brief gong is sounded to notify the driver that parking assistant is active

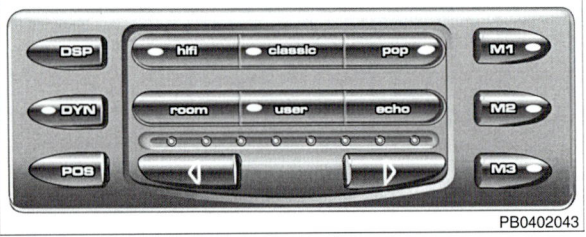

◁ Optional Digital Sound Processing (DSP) enables digital processing of speech and music in the amplifier. DSP allows for simulated acoustic spatial effects.

◁ Optional Porsche Communication Management (PCM) is a satellite supported (global positioning system or GPS) information and navigation system. PCM uses a digitized street map CD-ROM to provide navigational instructions (voice output, arrows and maps) to the driver. The GPS antenna is inconspicuously integrated in the dashboard. The PCM display is protected against theft by a device code. A PCM code and a NAVI code are needed before the system is operational.

1999 model year

Electrical system

◁ Optional Litronic (low beam xenon) headlights provide illumination that is approximately double that of the standard headlight and uses 30% less power. Other Litronic benefits:

- Increased illumination range
- Improved road edge lighting
- Improved illumination for safer cornering
- Improved color vision (higher color temperature imparts a daylight quality)

Litronic was only available as a package together with automatic dynamic headlight beam adjustment (AHBA) and headlight cleaning system. AHBA uses front and rear vehicle height sensors to detect changes in vehicle inclination, such as when accelerating or braking. Stepper motors in the headlight assemblies are used to compensate headlight beam aim.

Boxster Familiarization 02-21
Model Year News

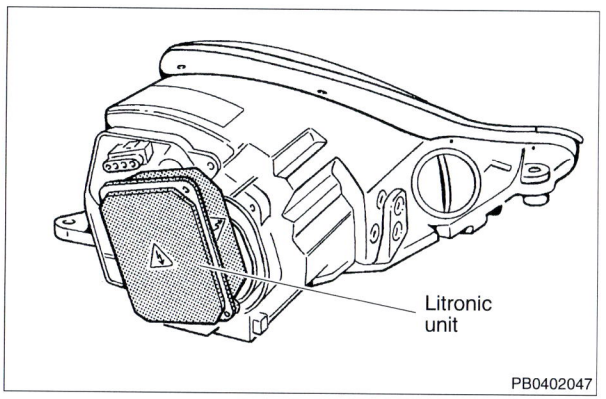

◁ The Litronic unit is mounted to the rear cover of the headlight assembly.

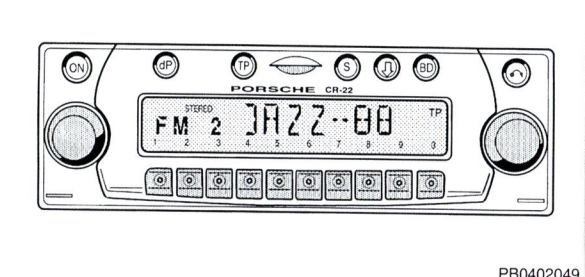

◁ New series of radios use a standard design with two rotary knobs and a double tuner concept for RDS diversity.

2000 model year

◁ New model introduced: Boxster S.

Engine

Boxster engine displacement was increased from 2.5 liters to 2.7 liters (engine code M96.22):

- Crankshaft stroke increased from 72 mm to 78 mm.
- To maintain compression ratio, pistons were shortened by 3 mm to 53 mm.

◁ The Boxster S engine (engine code M96.21) was an evolution of the 2.7 liter Boxster engine increased to 3.2 liter:

- Cylinder bore was increased 7.5 mm to 93 mm.

02-22 Boxster Familiarization

Model Year News

Engine management

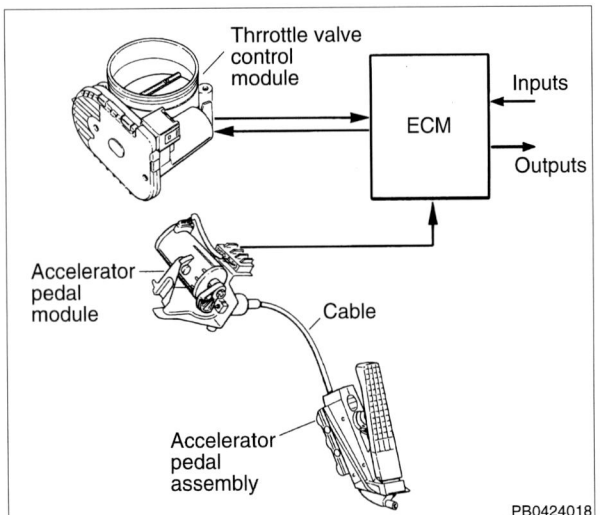

◁ New Bosch DME ME 7.2 system used in 2.7 liter Boxster and 3.2 liter Boxster S. The main difference with earlier DME M 5.2.2 is the electronic throttle system called E-gas. Instead of the accelerator pedal pulling a cable attached to the throttle valve, it pulls a short cable connected to a pedal module in the dashboard. The module uses a potentiometer to convert pedal travel and pedal speed into an electronic signal.

The DME control module processes the signal (driver's wish), sending it to an electric motor that operates the throttle valve. By providing electronic control over air intake at all engine speeds, the electronic throttle control enhances throttle response and helps reduce emissions. The system also eliminates the throttle linkage and a separate idle speed control valve.

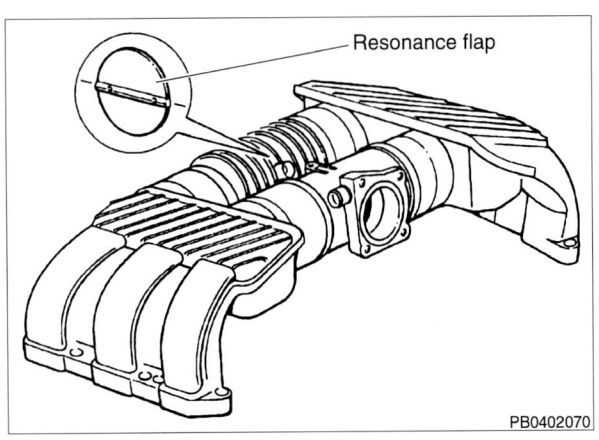

◁ The 2.7 liter and the 3.2 liter engines featured a new intake manifold design with a DME-controlled resonance flap. The resonance flap allows the length of the intake air tract to be optimized over a broader range of engine speed. The ME 7.2 control module activates a solenoid valve, which in turn activates a vacuum servo that open and close the resonance flap.

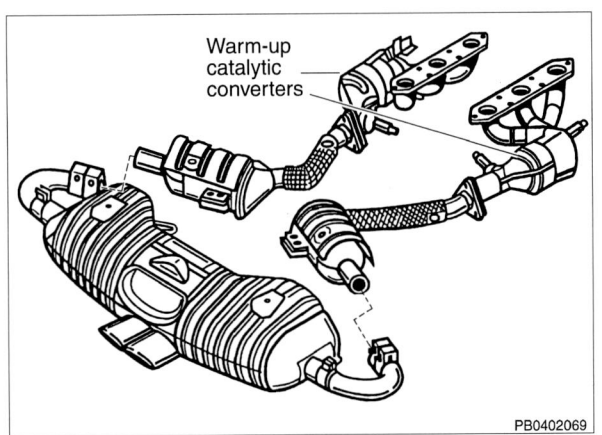

◁ 2000 and later US specification vehicles are equipped with warm-up catalytic converters. The converters are placed directly at the exhaust ports for quick warm up and faster emissions reduction.

Beginning in 2000, US specification cars are fitted with an ORVR (Onboard Refueling Vapor Recovering System) that directs gasoline fumes created during refueling into the evaporative emissions canister. This system included a larger evaporative canister volume (increased to 2 liters). The canister is located in the right front wheel well.

Boxster Familiarization 02-23
Model Year News

Transmission

A redesigned close-ratio 5-speed manual transmission (G86.01) was used to match the power and performance characteristics of the 2.7 liter engine.

◁ The Boxster S was available with a 6-speed manual transmission. The G86.20 gearbox was based on the unit used in the 911 Carrera.

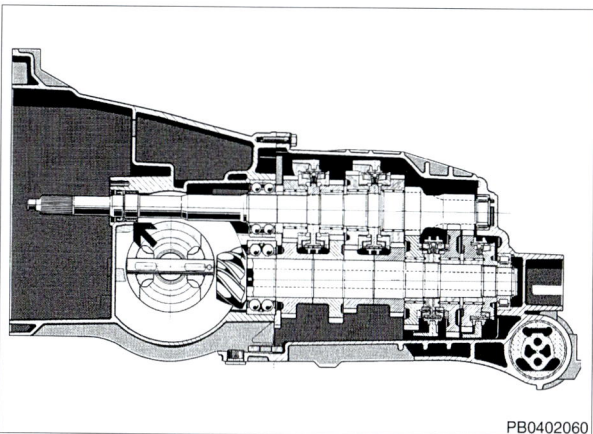

◁ 6th gear was configured to allow for maximum vehicle speed and the short 1st gear was designed to relieve the high load placed on the clutch during take off.

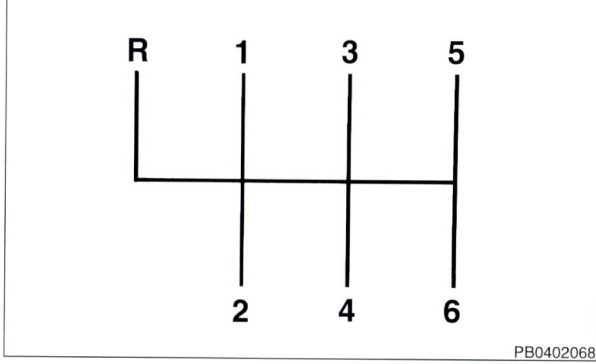

Suspension

The chassis and suspension were upgraded for the increased performance. Upgrades included reworked suspension springs, tubular sway bars, and shock absorbers for the different vehicle variants.

The rear suspension was modified for the increased weight and higher performance. The rear tie rod rubber bushings were stiffened and the friction of the ball joint increased for better straight line stability.

NOTE—
- *The modified rear tie rods can be retrofitted to earlier models. Replace tie rods in pairs.*

The Boxster S received a stronger rear wheel bearing carriers with larger wheel bearings and longer control arms for reduced bump steer.

Brakes

◁ Much of the brake system on the Boxster S was carried over from the 911 Carrera. The Boxster S brakes are recognizable by the red calipers and the larger cross-drilled rotors.

02-24 Boxster Familiarization

Model Year News

Additional center grille on Boxster S

Body

◁ The body of the Boxster S is nearly identical to the basic Boxster. Despite the installation of the additional center radiator behind the center grille of the front bumper and larger brake rotors and calipers, the volume of the front and rear trunks was unchanged.

The interior received a number of visual enhancements, as follows.

Previously unpainted interior plastic parts were painted in a matte black finish (painted plastic included parts on the instrument panel, center console, door trim panel, seats, floor and sill trim, and A-pillar windshield trim).

Previously laminated and all artificial leather surfaces were upgraded to a ribbed leather design.

Visible seams of black door pulls, hand brake lever, and other decorative seams were finished in grey.

The central groove seat covers were fitted with Alcantara.

Gear lever on the Boxster S 6-speed transmission was redesigned.

◁ The Boxster S received a 3-spoke sport steering wheel with a Porsche crest.

The gearshift pattern inlay, door handle, side airbag symbol, parking brake release knob and Tiptronic release button and shift gate received an aluminum finish.

Passenger seat manual height adjustment and illuminated vanity mirror became standard.

Table e identifies some of the noticeable interior differences between the Boxster and the Boxster S.

Table e. Overview of differences between Boxster and Boxster S interiors (2000 model year)		
Feature	Boxster	Boxster S
Interior door handle	Black	Aluminum look
Trunk openers	Black	Aluminum look
Gear lever top insert	5-speed aluminum look	6-speed aluminum look
Instrument rings	Black	Aluminum look
Door sill logo	not applicable	Boxster S
Center console logo	not applicable	Boxster S

An additional interior lining was used on the convertible top on the Boxster S model. This helped reduce wind noise and exterior noise. The main top bow was modified owing to this change. The convertible top drive motor microswitches were moved from the drive motor and the left B-pillar to the right side convertible top transmission.

The engine compartment venting (left rear side section) was visually modified. The front edge of the duct is more prominent and the opening between the two lower slats was concealed. This change applies to both Boxster and Boxster S models.

Boxster Familiarization 02-25

Model Year News

2001 model year

Engine

To reduce timing chain noise, the intermediate shaft drive chain was converted from a roller type chain to a tooth type chain. This change also required modifications to the crankshaft and intermediate shaft sprockets and corresponding chain tensioners.

Suspension

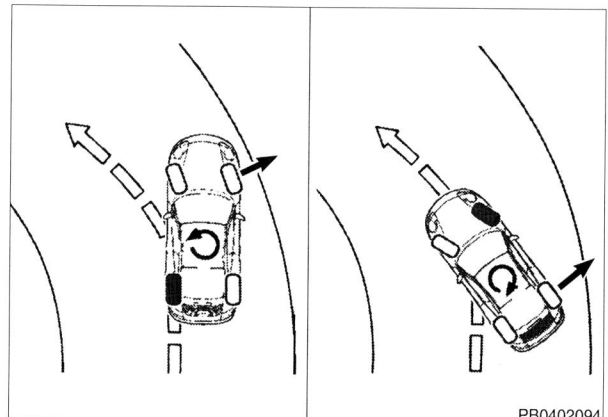

◁ Optional Porsche Stability Management system (PSM) replaced the optional traction control system for the Boxster. PSM, a system first introduced on the all-wheel drive 911 Carrera 4 model, gives the Boxster a high level of stability and dynamic safety on slippery roads.

WARNING—

- *While confident in the system's ability as a driving stability aid, Porsche cautions drivers that PSM cannot counteract the laws of physics, such as gravity and available friction.*

Body and interior

Interior changes for 2001 included dual cup holders and new light-emitting diode (LED) interior orientation lights. One LED provides subtle illumination of the cockpit and center console. An LED on the driver's side door handle illuminates the ignition lock and light switch, and an LED illuminates each door latch.

On the Boxster, a three-spoke steering wheel with color Porsche crest replaced the previous four-spoke wheel. The car keys also received a color Porsche crest.

Electrochromic interior mirror and outside driver's side mirror were available as an option.

The front and rear trunks used a thick black material and the roll bar cover used a softer padding. An additional interior lining was used on the convertible top on the Boxster model, as used on the 2000 Boxster S. This helped reduce wind noise and exterior noise. The main top bow was modified owing to the this change.

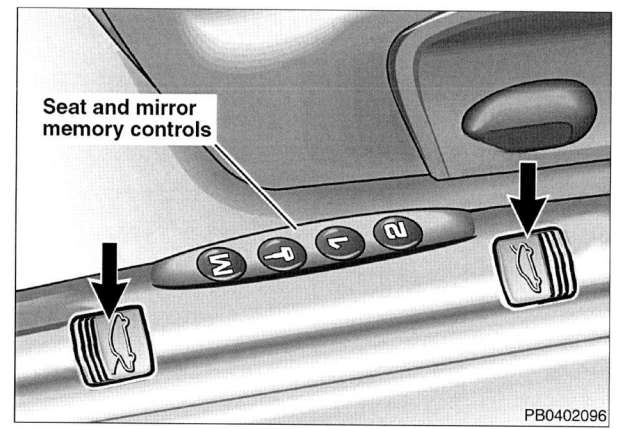

◁ The front and rear trunk lids received power release buttons **(arrows)** on the left door sill. The trunk lids can also be opened using the key remote control.

Individual key programming for seat and mirror control was added to the 2001 model. This feature allows the seat and mirror positions to be stored in the keyless entry control module for automatic adjustment when the vehicle is unlocked. Four color coded keys can be individually programmed.

NOTE—

- *The automatic seat adjustment can be cancelled by pressing the central locking button or any of the buttons for seat or memory seat adjustment.*

02-26 Boxster Familiarization

Model Year News

◁ An upgraded audio system was available as an option. The volume of the door loudspeaker housing was increased from 3 liters to approx. 5 liters.

The new sound system offered the following improvements:

- lower distortion (electrical and acoustical) in bass range
- increased overall volume
- increased bass sound pressure

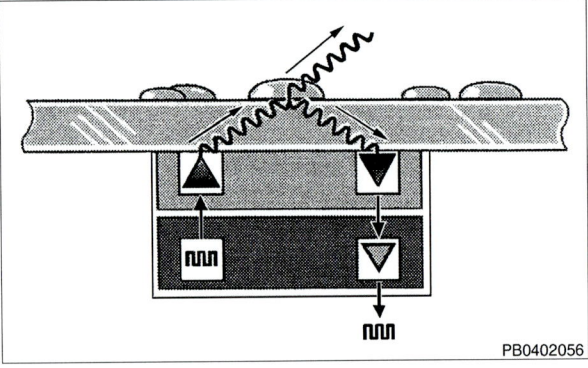

◁ An option for the windshield wipers was automatic rain sensing technology. The rain sensor measures the amount of precipitation (including snow) on the windshield. The wiper speed is automatically set based on the precipitation measured. Rain sensor technology is based on the principle of interrupted light on the windshield surface.

◁ The instrument cluster was equipped with diagnosis and dedicated fault memory.

As of January 2001, a redesigned Porsche Communication Management (PCM) computer was used. The enhanced system included a 4x CD-ROM for faster access time, a 12-channel GPS receiver (replaces 8 channel unit), a selective gyroscope function, a 72 MIPS processor (replaces 16 MIPS processor), flash memory with 4 MB instead of 2 MB, and a 16 MB memory instead of 2 MB.

NOTE—

- *To fit the new navigational system to an older car using the new software, consult an authorized Porsche service department.*

2002 model year

Engine management

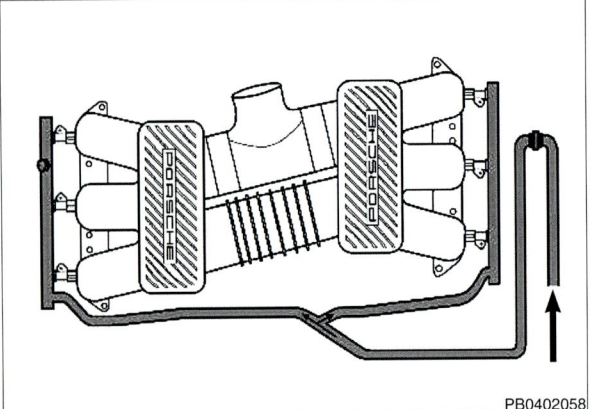

◁ The Bosch ME 7.2 system was modified to include the following:

- Non-return fuel supply system. This change helps to reduce hydrocarbon (HC) emissions.
- Mass air flow sensor less sensitive to dirt particles.
- Smaller, lighter fuel injectors (EV-6) with 4-hole spray disks.
- Planar (Bosch LSF) oxygen sensors. The LSF sensor provide more stable regulating characteristics, lower heating power demand, quick on-time and lower weight.

Boxster Familiarization 02-27

Model Year News

Transmission

The 6-speed Boxster S manual transmission was converted to a three bearing arrangement at the input shaft. The previous transmissions used a two bearing setup. This modification, introduced during the 2001 model year, was designed to reduce vibration.

Suspension

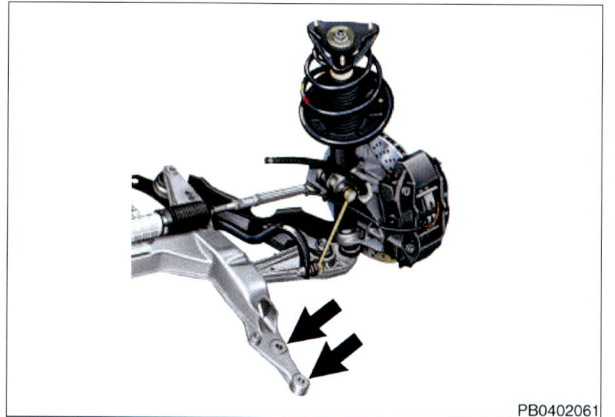

Front axle crossmember rear mounting points changed from one bolt per side to two bolts per side.

> **CAUTION—**
> - Early (single bolt mounting) and late (two bolt mounting) cross-members are not interchangeable.

The Porsche Stability Management (PSM 5.7) system was upgraded to include new type solenoid valves in the PSM hydraulic unit and active wheel speed sensor sensors (Hall effect). These active sensors can not only determine wheel speed, but also direction of rotation, wheel stop, and signal quality.

Body

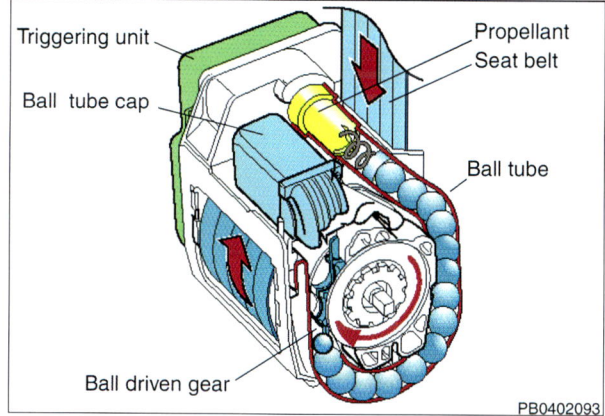

Seat belts were equipped with pyrotechnic (explosive charge) seat belt reels combined with belt force limiters. During a frontal collision, once the airbags are triggered, the seat belt reels are also triggered to automatically tension the seat belts. During a rear collision, only the belt tensioners are triggered. To help avoid bruising and internal injury to the passenger, the belt's pulling force is limited to a bearable level via the belt force limiter function.

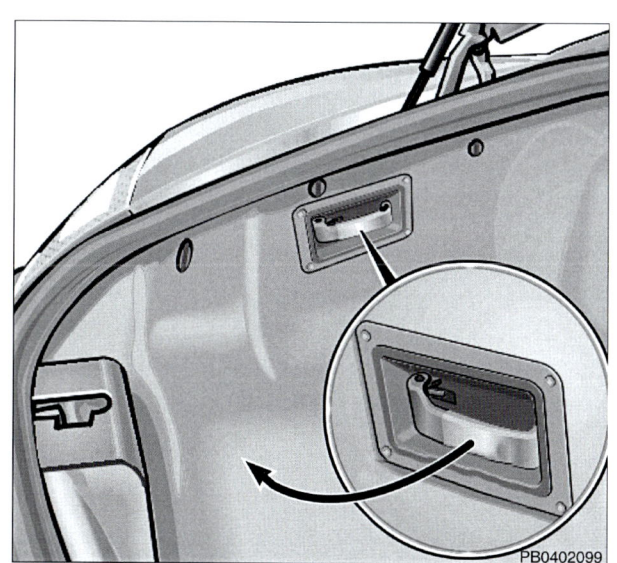

As required by law, 2002 and later models were fitted with an internal rear trunk release lever. This fluorescent lever allows a person trapped in the trunk to open the lid from the inside.

02-28 Boxster Familiarization

Model Year News

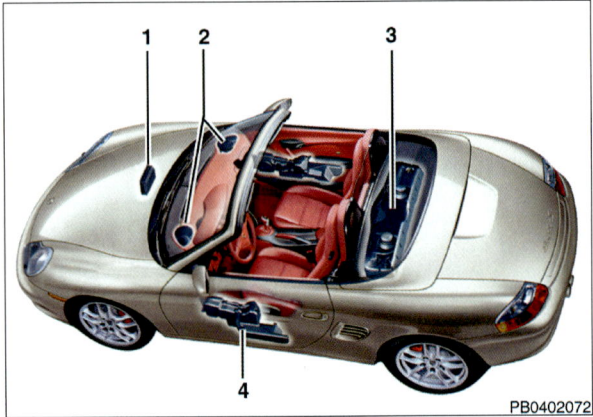

Electrical system

Interior switches were finished in matt black, as opposed to the polished look on pervious years. Black plastic trim bezels (e.g. radio, PCM display) also received the matt finish.

◁ The 2002 individual option program (I 680) offered the Bose® high end sound package, replacing the Digital Sound Processing (DSP) package introduced in model year 1998. This Bose system offered a remarkably higher sound quality, even with an open soft top.

1. Digital amplifier with 6-channel customized equalization
2. 3½" wide-range speaker and 2" tweeter on each side of dashboard
3. 11-liter tuned, ported base speaker enclosure with two 5¼" woofers and two 2½" wide-range speakers
4. 4½" speaker in customized enclosure in each door.

The vehicle alarm system received a functional enhancement. If the doors are opened manually with the key but the ignition is not switched on within 10 seconds, the alarm sounds. This change eliminated the function of lowering the door windows by holding the key in UNLOCK position.

2003 model year

Engine

◁ To reduce gas pressure in the crankcase, the bottom ends of the cylinder liners were arched **(arrows)**. Circulating crankcase gases help reduce power losses at higher engine rpms.

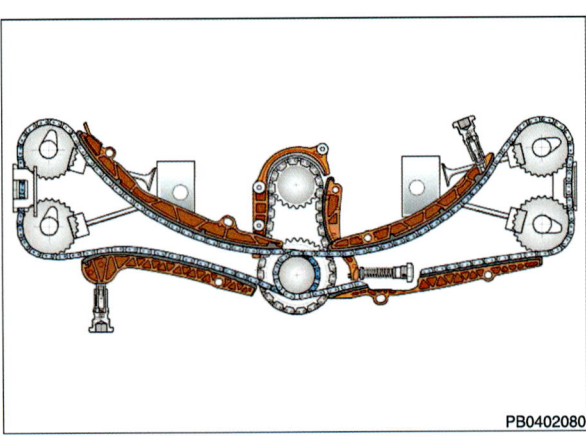

◁ A redesigned VarioCam system was introduced, eliminating the need for a drive chain between each pair of camshafts. The VarioCam system is controlled by the ME 7.8 engine management system.

Model Year News

Engine management

Boxster and Boxster S models were fitted with Bosch DME ME 7.8. This provided a noticeable increase in power and torque with an improved exhaust note. Some ME 7.8 features are as follows.

VarioCam system has continuously variable vane type adjuster on intake camshafts.

Exhaust emissions are reduced to meet National low emissions vehicle (LEV) standards.

Fuel consumption is reduced by approx. 2%.

Fuel loop is non-return type.

New style EV-6 fuel injectors are sequentially timed.

◄ Dual closed loop oxygen sensor control with Bosch LSF oxygen sensors:

1. Electrical connector
2. Protective sleeve
3. Planar sensor
4. Ceramic support
5. Sensor housing
6. Ceramic sealing package
7. Protective tube

The mass air flow sensor resists contamination.

Engine management components communicate over CAN-bus.

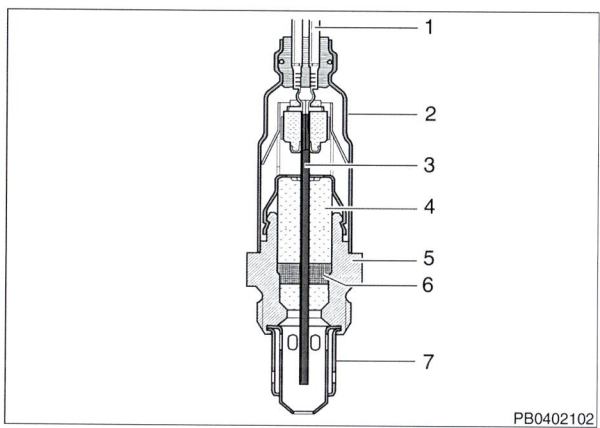

NOTE—

- *The EV-6 fuel injectors, Bosch LSF oxygen sensors and improved mass air flow sensor were introduced on 2002 models.*

Suspension

Only minor suspension changes were introduced for the 2003 model. Primarily, the front and rear springs were matched to the revised chassis dynamics. The springs were color coded for identification.

Body

◄ The convertible top features a single-pane safety glass rear window with defroster.

02-30 Boxster Familiarization

Model Year News

◁ Owing to this change the convertible top and convertible top mechanism were significantly reworked, including the addition of a center roof bow **(arrow)**.

◁ The front and rear bumper covers were redesigned. At the rear, new exhaust pipe tips and air extraction vents were integrated into the styling of the new rear end. The horizontal vents were added to help reduce muffler temperature owing to the increased performance and thermal load of the higher performance engine.

◁ To optimize engine air intake, the left side air duct was enlarged and the ducting to the air filter was modified. The side louvers, painted in body color, are located further outboard and a screen is mounted behind them. A water drain was integrated into the air duct to the air filter housing.

◁ The rear spoiler was redesigned to provide a more pronounced styling emphasis.

Boxster Familiarization 02-31
Model Year News

◁ A newly designed glove compartment with a storage capacity of 5 liters was provided. The storage compartment below the steering wheel was discontinued. Dual cup holders were standard equipment. To accommodate this change, the center vent was redesigned and also enlarged for improved ventilation.

◁ The steering wheel, dashboard and door handles were matched to the interior color.

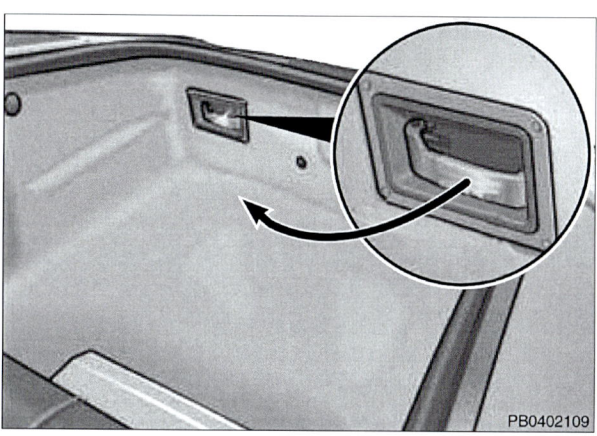

◁ An internal front trunk release lever was fitted.

Electrical system

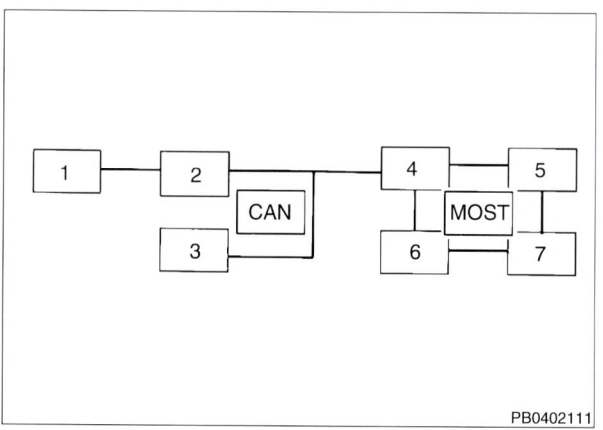

◁ MOST (Media Oriented System Transport) fiber optic bus was an addition to the CAN-bus system. MOST data transmission is used only for multimedia data.

1. Diagnostic test port
2. Instrument cluster
3. A/C system
4. Radio / PCM
5. Amplifier
6. Telephone
7. CD changer

02-32 Boxster Familiarization

Model Year News

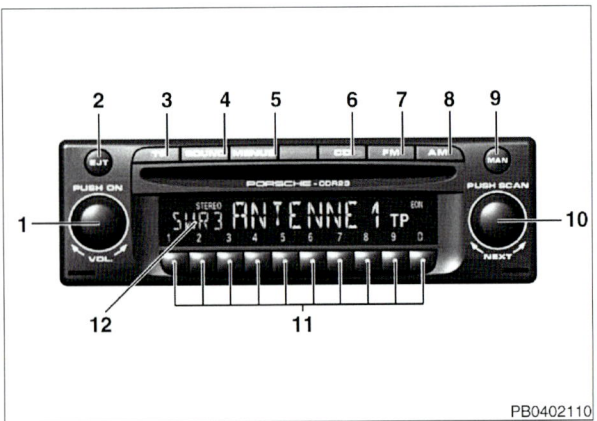

◁ The CDR 23 radio with diversity antenna is coded to the car and requires Porsche System Tester 2 (PST2) if replacement is required. A new CD changer is used with the CDR 23 radio. It features 6-second anti-slip capability and is slightly bigger than the unit it replaces.

1. Volume control
2. Eject button
3. Mute (traffic information button for ROW)
4. Sound button
5. Submenus button
6. CD button
7. FM button
8. AM button
9. Manual frequency
10. Right tuning / return button
11. Multifunction buttons
12. Display

◁ A second generation Porsche Communication Management (PCM) was installed, featuring the following enhancements:

- 2-tuner radio for improved reception
- CD-ROM drive that plays audio CDs
- 5.8 in. TFT color monitor in 16:9 resolution
- 12 button strip for improved functionality
- Detailed and easy to operate on-board computer with combined dot-matrix instrument cluster for basic audio and NAV menu display
- NAV display of complete route including present location and target with fast map setup

2004 model year

◁ The main feature of 2004 models was the special edition Boxster S. It was made available in early 2004 and only 1,953 (to mark the year of the early 550 Spyder) were made. Each special edition S came with a "50 years of the 550 Spyder" badge. The special edition Boxster S was finished in GT silver metallic, a color only found on the Carrera GT. The soft top was in cocoa.

The Boxster S acquired six more horsepower for a total of 266 bhp at 6,200 rpm. The Special Edition offers a slightly higher top speed and reaches 60 mph from a stop in just 5.7 seconds.

◁ The car received a special interior color scheme.

The car also received minor improvements including a redesigned shifter with a 15% shorter throw for the six-speed gearbox. Porsche Stability Management (PSM) was standard equipment. A sports suspension that lowered the car by 10 mm improved road holding and permitted higher lateral acceleration values.

The exhaust system used a specially styled, stainless-steel tailpipe for a distinctive sound.

The four-piston aluminum brake calipers of monobloc design used an aluminum paint finish, which were easily visible behind the larger 18-inch Carrera wheels.

OPTION CODES

Many Boxster features are packaged into factory-designated option codes, sometimes referred to as M-option codes because they are often listed in the wiring diagrams with a preceding M.

The option codes are listed on a sticker on the inside of the front trunk lid. **Table f** lists Boxster and Boxster S option codes.

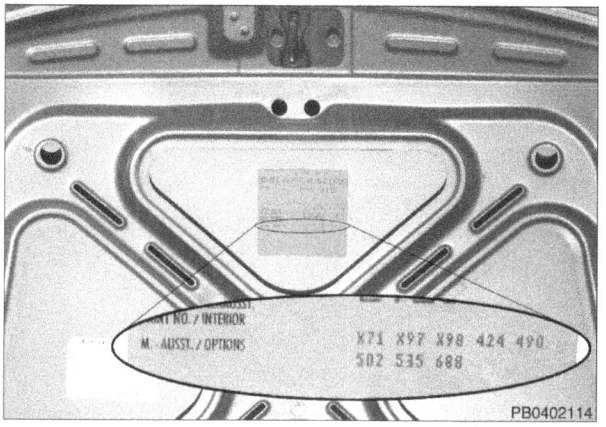

Table f. Option codes	
Code	Option
008	986 3.2 liter displacement
009	986 2.5 liter displacement
013	Sport Technik package
020	Speedometer with 2 scales (mph-kph)
024	Version for Greece
030	Sport chassis
034	Version for Italy
061	Version for Great Britain
062	Version for Sweden
063	Version for Luxembourg
064	Version for Netherlands
065	Version for Denmark
066	Version for Norway
067	Version for Finland
068	Version for Thailand
069	Version for other countries
071	Version for EU country
113	Version for Canada
114	Version for Taiwan
119	Version for Spain
124	Version for France
126	Control and indications in French
127	Control and indications in Swedish
130	Control and indications in English
139	Seat heating system, left seat

Table f. Option codes	
Code	Option
150	Without catalyst
193	Version for Japan
197	Battery with higher rating
198	Starter 1.7 KW
215	Version for Saudi Arabia
222	Traction control system
225	Version for Belgium
249	Tiptronic S
261	Outside mirror, plain, passenger side electrically heated and adjustable
265	Rain sensor
270	Outside mirror, plain, driver side electrically heated and adjustable
271	Outside mirror, aspherical, driver side electrically heated and adjustable
272	Outside mirror manually adjustable
273	Outside mirror electrically heated and adjustable
274	Vanity mirrors, Illuminated
277	Version for Switzerland
288	Headlight washers
329	Cassette radio (CR-210) with anti-theft coding and removable control unit. Standard radio (mandatory for USA)
330	Radio (Porsche CR-31)
340	Seat heating system, right seat
391	17" Wheels - Boxster design
396	17" Wheels - Boxster design
413	18" Turbo-look wheels
421	Cassette shelf in center console, front
424	CD shelf in center console, front
436	Three spoke steering wheel
437	Comfort seat, left
438	Comfort seat, right
441	Radio preparation
446	Wheel caps with colored Porsche crest
454	Cruise control
465	Rear foglight, left side
466	Rear foglight, right side
476	Porsche Stability Management (PSM)
479	Version for Australia
480	6-speed manual transmission
481	5-speed manual transmission
484	Version for USA
488	Inscriptions in German
490	Hi-fi sound system 6 speakers, amplifier
492	For left hand traffic
513	Lumbar support, passenger side
531	Electronic immobilizer (anti-theft)
534	Theft security system (ex. USA)
535	Remote controlled alarm system (USA)
536	Alarm siren and tilt sensor

Boxster Familiarization

Option Codes

Code	Option
537	Lumbar support, left seat
538	Lumbar support, right seat
539	Mechanical seat height adjustment, left
540	Mechanical seat height adjustment, right
551	Windstop
553	Version for USA, Canada
549	Roof transport system
550	Hard top in color of car body (including heated rear window)
551	Porsche windstop
562	Airbags, driver and passenger
563	Airbag, passenger
571	Active carbon cabin filter (for interior)
573	Air-conditioning system
580	Non-smoker package
586	Lumbar support, driver's side
601	Litronic headlights
614	Telephone preparation
618	Telephone
635	Parking Assistant
651	Power windows
659	On board computer
660	OBD II
661	Stricter emissions concept
662	PCM (information / navigation System)
664	On-board fuel vapor recovery system (ORVR)
680	Digital sound package
685	Radio (Porsche CDR-21)
688	CD-radio (Porsche CDR-210)
692	CD changer (Porsche CDC 3)
696	AM / FM radio with CD player
982	Soft look leather seats in interior color
999	Special Instructions (internal use only)
C02	US emissions
E51	Full leather interior package
E52	Small leather dashboard package
E53	Large mahogany dashboard package
E54	Small mahogany dashboard package
E55	Large carbon dashboard package
M6A	Black floor mat with Porsche lettering
M6F	Metropol Blue floor mat with Porsche lettering
M6H	Natural brown floor mat with Porsche lettering
M6J	Nephrite green floor mat with Porsche lettering
M6M	Boxster red floor mat with Porsche lettering
M6P	Graphite grey floor mat with Porsche lettering
M6S	Savanna beige floor mat with Porsche lettering
P01	Comfort package
P05	Cassette radio
P06	CD-radio
P09	Sport package
P10	Sport touring package
P11	Self-dimming inner and driver's side rear view mirrors and rain sensor
P14	Heated front seats package
P15	Power seats with driver's side and mirror memory
P16	Sport package
P17	Sport touring package
P18	Sport touring package
P37	Traction control with ABD
P38	Technic Sport Package including 17" wheels and traction control
P49	Digital Sound Processing (DSP)
P63	Sport package
P64	Sport touring package
P65	Sport touring package
P69	Sport Design package
P70	Sport Design package
P74	Litronic (xenon) headlight package
P77	Leather sport seats - manual
P78	Leather sport seats - manual length adjustment and electrical backrest adjustment
P84	Sport package (same as P63 except XRA 17" Sport Classic wheels)
P85	Sport Touring package (same as P64 except XRA 17" Sport Classic wheels)
P86	Sport Touring package (same as P65 except XRA 17" Sport Classic wheels)
TD4	Tourist delivery program
TT3	Tiptronic transmission 5 speed
VA3	Preparation / modification (tourist delivery vehicles only)
X21	Telephone console in leather
X26	4-spoke airbag steering wheel covered with leather
X45	Instrument dials painted interior color
X68	Convertible (tonneau cover) in color of roof
X69	Door sills in carbon embossed with model insignia.
X70	Stainless steel door entrance panel with the logo "Boxster"
X71	Instrument dials in aluminum look
X76	Flared rocker panels
X77	4-spoke airbag steering wheel covered with carbon / leather
X89	Wheel caps colored to sample, with colored crest
X99	Natural leather option
XAA	Boxster Aerokit: front spoiler, rear spoiler and rocker panel covers
XAB	Speedster rear in exterior Color
XD3	Rain sensor
XD9	Rims and caps painted in exterior color
XE3	Automatic dimmable rear view mirror
XJB	Rear center console painted in Arctic Silver
XKX	Lower part of instrument panel painted in silver
XLA	Boxster exhaust pipe

Boxster Familiarization 02-35
Option Codes

Table f. Option codes

Code	Option
XLF	Sport exhaust system
XME	Rear center console painted in exterior color
XMF	Front center console in interior leather
XMG	Side covers of center console rear in leather
XMH	Roll bar covers in leather
XMJ	Rear center console in carbon fiber
XMK	Roll bar painted exterior color
XML	Roll bar in chrome look
XMM	Front center console, including 1 compartment and CD holder in leather
XMN	Front center console, including 1 compartment and cassette holder in leather
XMP	Sun visors in leather
XMR	Sun visors in leather, additional reading light on passenger side
XMS	Roof control handle and ring in leather
XMT	Gear shift knob / parking brake grip in combination with mahogany / leather
XMU	Gear shift knob / parking brake grip in combination with mahogany / aluminum / leather
XMV	Tiptronic shifter / parking brake grip in mahogany / leather
XMW	Tiptronic shifter / parking brake grip in mahogany / aluminum / leather
XMX	Fresh air vents in the center painted
XMY	Roll bar painted silver
XMZ	Rear center console covered with leather
XN3	Airbag steering wheel combination of real wood / interior color of leather
XNC	Dashboard covered with leather in 2 colors of interior
XND	Lower part of instrument panel painted in color of exterior
XNE	4-spoke airbag steering wheel covered with mahogany / aluminum rivets / leather
XNF	Painted 3-part steering column (requires color)
XNG	Lower part of instrument panel covered with leather
XNH	Side fresh air vents and defroster covers (2-part) in leather
XNJ	Painted speaker covers and dashboard covered with 2 colors of leather, upper dashboard trim molding covered with leather
XNK	Painted speaker covers and dashboard covered with 2 colors of leather, upper dashboard trim molding and lower instrument panel covered with mahogany
XNL	Painted speaker covers and dashboard covered with 2 colors of leather, upper dashboard trim molding and lower instrument panel covered with carbon
XNM	Center air vent and heater controls / frame of air-conditioner carrier and switch covers left and right covered with leather
XNP	Center air vent and heater controls / frame of air-conditioner carrier and switch covers left and right covered with mahogany
XNR	Center air vent and heater controls / frame of air-conditioner carrier and switch covers left and right covered with carbon
XNS	Steering column casing in interior leather
XNT	4-spoke airbag steering wheel covered with mahogany / leather
XNU	Trim molding for upper dashboard covered with leather

Table f. Option codes

Code	Option
XNX	Trim molding for the upper dashboard covered with carbon
XNY	Trim molding for the upper dashboard covered with painted aluminum look
XNZ	Trim molding for upper dashboard covered with mahogany
XPA	3-spoke steering wheel in leather (includes airbag cover)
XPC	3-spoke steering wheel in dark wood and leather
XPD	3-spoke steering wheel in carbon and leather
XPE	3-spoke Sport steering wheel covered with mahogany / leather
XPN	Lower part of instrument panel covered with mahogany
XPP	Speaker covers (2) in control panel painted
XPY	Lower part of instrument panel covered with carbon
XRA	17" Sport Classic wheels / tires
XRB	18" Sport Classic wheels / tires
XRH	17" Dyno wheels / tires
XRK	18" Turbo Look wheels / tires
XRL	18" Sport Design wheel
XSA	Sport seats left and right, rear of backrest painted in color of exterior
XSC	Porsche crest stamped in headrest
XSD	Seat adjustment switches, hinge covers and release, covered with leather
XTC	Storage compartment, door opener cover and door mirror adjustment controls covered with leather
XTD	Storage compartment and door opener cover covered with mahogany, lower part of armrest and cover for mirror adjustment controls covered with leather
XTE	Storage compartment and door opener cover covered with carbon, mirror adjustment controls, inner door handle and grip covered with leather
XTF	Lower part of armrest including armrest clip painted in aluminum look, door handle in leather
XTG	Inner rocker panel covers covered with leather
XX1	Front floor mats embroidered "Porsche"
XX2	Foot space lighting (over center console) door pocket light
XXA	Hardtop painted in fading color tone
XY5	Tiptronic shift gate in leather (requires full leather interior)
XZB	Removable back pack between the seats
XZC	Luggage fastening kit for rear trunk
Y03	Gear shift knob / parking brake grip in combination of carbon / leather (X63 / X64)
Y05	Carbon / aluminum shift knob and brake handle
Y06	Gear shift knob and brake handle in combination of leather (interior color) and aluminum
Y08	Dark maple burled wood / aluminum shifter knob and brake handle
Y23	Tiptronic shifter and parking brake grip in Interior color leather and aluminum inserts
Y24	Tiptronic shifter / parking brake grip in carbon / leather (X48 / X64)
Y61	Carbon / aluminum Tiptronic shift selector and brake handle

Option Codes

Table f. Option codes	
Code	Option
Y63	Dark maple burled wood / aluminum on Tiptronic shift selector and brake handle
Z100	Alternate carpeting color - from standard colors
Z101	Porsche crest stamped in headrests
Z102	Seat belts in alternate color (Guards Red, Speed Yellow, or Riviera Blue)
Z103	Alternate seat stitching color
Z104	Seats in Boxster red or nephrite green leather only
Z105	Leather seats in alternate current model year color
Z106	Door panel stitching in alternate color

03 Service and Maintenance

GENERAL 03-2	Air filter (engine), replacing 03-28
BASIC SERVICE INFORMATION 30-2	Throttle valve, checking 03-28
Safety 03-2	UNDER CAR MAINTENANCE 03-29
Raising car safely using car jack. 03-2	Radiator air inlet duct, cleaning. 03-29
Working under car safely 03-3	Coolant hoses, checking condition 03-29
Non-reusable fasteners. 03-4	Spark plugs, replacing. 03-30
Tightening fasteners 03-4	Brake pads and rotors, checking. 03-30
Buying parts 03-5	Fuel filter, replacing (1997 - 2001 models) 03-31
Genuine Porsche parts 03-5	Fluid lines and hoses, checking 03-32
Non-returnable parts 03-5	Constant velocity (CV) joint boots, checking. . . 03-32
Information you need to know 03-6	Suspension and steering components,
Porsche service. 03-6	checking. 03-33
Tools 03-6	Exhaust system, checking. 03-33
MAINTENANCE TABLES 03-7	Underbody, checking 03-33
DIAGNOSIS SYSTEM 03-11	EXTERIOR AND INTERIOR
Fault memory, reading out 03-11	BODY MAINTENANCE 03-33
FLUIDS AND LUBRICANTS 03-11	Airbags, inspecting 03-33
Engine oil level, checking 03-11	Ventilation microfilter, replacing 03-34
Engine oil and filter, changing 03-12	Battery service 03-34
Coolant level, checking. 03-14	Battery notes 03-35
Brake fluid, checking. 03-15	Body exterior 03-35
Brake fluid, changing. 03-15	Clutch, checking 03-36
Power steering fluid, checking 03-16	Door check straps and hinges, lubricating 03-36
Manual transmission oil, checking 03-16	Door locks, trunk lid locks
Manual transmission oil, changing 03-17	and front lid safety hooks, checking. 03-36
Differential oil level, checking	Exterior lights, checking 03-37
(automatic transmission) 03-19	Headlights, adjusting. 03-38
Automatic transmission fluid (ATF),	Interior lights, checking 03-38
checking and topping off 03-19	Parking brake, checking and adjusting 03-38
Automatic transmission fluid (ATF), changing. . 03-21	Seat belts, checking 03-39
Windshield wipers and washer fluid,	Tire service 03-40
checking. 03-24	Test drive 03-41
ENGINE COMPARTMENT MAINTENANCE 03-24	**TABLES**
Convertible top service position 03-24	a. Bolt tightening torques–general 03-5
Top engine cover, removing and installing . . . 03-26	b. Minor maintenance tasks 03-8
Front engine cover, removing and installing . . 03-26	c. Major maintenance tasks 03-9
Drive belt, checking and replacing 03-27	d. Annual maintenance tasks 03-10
	e. Additional maintenance tasks. 03-10

03-2 Service and Maintenance

General

GENERAL

Carry out the maintenance work described in this repair group at the factory specified time or mileage interval. Following these intervals will help ensure safe and dependable operation. In addition, many of the maintenance procedures are necessary to maintain warranty protection.

NOTE—

- *Porsche is constantly updating their recommended maintenance procedures and requirements. The information contained here may not include updates or revisions made by Porsche since the publication of the documents supplied with the car. If there is any doubt about what procedures apply to a specific model or model year, or what intervals should be followed, remember that an authorized Porsche dealer has the latest maintenance information.*

BASIC SERVICE INFORMATION

Most of the necessary maintenance and minor repair described in the book can be done with ordinary tools. Below is some important information on how to work safely, a discussion of what tools will be needed and how to use them.

Safety

Although an automobile presents many hazards, common sense and good equipment can help ensure safety. Many accidents happen because of carelessness. Pay attention and stick to safety rules in this manual. Read the cautions and warnings given in **00 Warnings and Cautions** prior to working on the vehicle.

Raising car safely using car jack

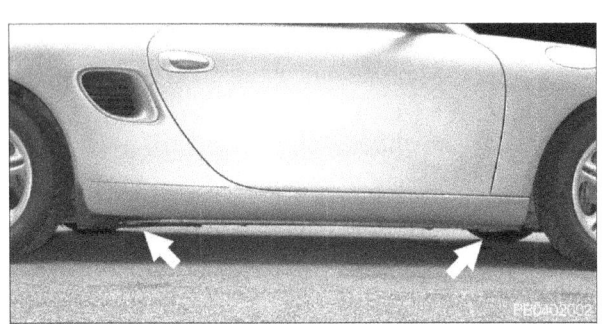

◁ Use the proper jacking points to raise the car safely and avoid damage. The jack supplied with the car can only be used at the front and rear side jacking lugs (**arrows**).

> *WARNING—*
> - *Do not work under a raised car unless it is solidly supported on jack stands that are intended for that purpose.*
> - *Watch the jack closely. Make sure it stays stable and does not shift or tilt. As the car is raised, the car may roll slightly and the jack may shift.*

- Park car on flat, level surface. Use chocks to block wheel opposite to one being raised.

> *WARNING—*
> - *Do not rely on the transmission or the parking brake to keep the car from rolling.*

- If changing a tire, loosen lug bolts before raising car.

◁ Remove jack from front trunk.
  - Remove large wing nut and washer at center of spare and tilt spare tire back.
  - Remove jack and jack handle from insert at center of spare.

Service and Maintenance 03-3

Basic Service Information

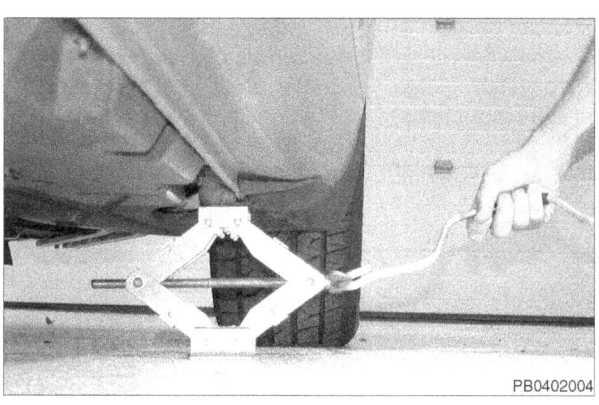

◄ Place jack into position, making sure raised part on jack fits into hole in jacking lug. Make sure jack is resting on flat, solid ground. Use a board or other support to provide a firm surface for jack, if necessary.

− Raise car slowly while constantly checking position of jack and car.

Working under car safely

The procedure given below describes how to raise the car using a floor jack and how to support it with jack stands.

> *WARNING—*
> - *Do not work under a raised car unless it is solidly supported on jack stands that are intended for that purpose.*
> - *Watch the jack closely. Make sure it stays stable and does not shift or tilt. As the car is raised, it may roll slightly and the jack may shift.*
> - *A jack is a temporary lifting device. Do not use to support the car while you are under it.*
> - *Do not use wood, concrete blocks, or bricks to support a car. Wood may split. Blocks or bricks, while strong, are not designed for that kind of load, and may break or collapse.*
> - *Use at least two jack stands to support car. Use automotive jack stands designed for supporting a car.*
> - *Place jack stands on a firm, solid surface. If necessary, use a flat board or similar solid object to provide a firm footing.*

> *CAUTION—*
> - *When raising the car using a floor jack or a hydraulic lift, carefully position the jack pad to prevent damaging the car body. If possible, use a rubber pad between the jack and the body.*

− Disconnect negative (−) cable from battery so that no one can start car. Let others know what you are doing.

> *CAUTION—*
> - *Prior to disconnecting the battery, read the battery disconnection cautions in **00 Warning and Cautions**.*

◄ To raise front corner, position floor jack under floor structural member (area within circle). Raise vehicle until jack stand can fit under jacking lug. If possible, place a rubber pad between body and jack stand.

- Lower car slowly until its weight is fully supported on jack stand. Watch to make sure jack stand does not tip or lean as car settles. Repeat for opposite front corner.

03-4 Service and Maintenance

Basic Service Information

◄ To raise rear of car, position floor jack under rear axle support in center of car (area within circle). Raise vehicle until jack stands can fit under both rear jacking lugs. If possible, place a rubber pad between body and jack stand.

– Lower car slowly until its weight is fully supported by jack stands. Watch to make sure jack stands do not tip or lean as car settles on them.

– Observe jacking precautions again when raising car to remove jack stands.

Non-reusable fasteners

Many fasteners used on the cars covered by this manual must be replaced with new ones once they are removed. These include but are not limited to: bolts, nuts (self-locking, nylock, etc.), cotter pins, studs, brake fittings, roll pins, clips and washers. Use genuine Porsche parts for this purpose.

Some bolts are designed to stretch during assembly and are permanently altered, rendering them unreliable once removed. These are known as torque-to-yield fasteners. Replace fasteners where instructed to do so. Failure to replace these fasteners could cause vehicle damage and personal injury. See an authorized Porsche dealer for applications and ordering information.

Tightening fasteners

It is good practice to tighten fasteners gradually and evenly to avoid misalignment or over-stressing any one portion of the component. For components sealed with gaskets, this method helps to ensure that the gasket will seal properly.

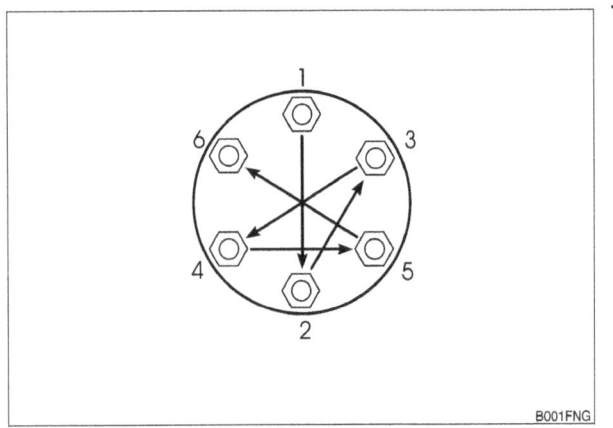

◄ Where there are several fasteners, tighten them in a sequence alternating between opposite sides of the component. Repeat the sequence until all the bolts are evenly tightened to the proper specification.

For some repairs a specific tightening sequence is necessary, or a particular order of assembly is required. Such special conditions are noted in the text, and the necessary sequence is described or illustrated. Where no specific torque is listed, use **Table a** as a general guide for tightening fasteners.

> *WARNING—*
> • **Table a** *is a general reference only. The values listed are not intended to be used as a substitute for torques specifically called out in the text.*

NOTE—
• *Metric bolt classes or grades are marked on the bolt head.*
• *Do not confuse wrench size with bolt diameter.*

Service and Maintenance 03-5
Basic Service Information

Table a. Bolt tightening torques–general (in Nm) (max. permissible)						
Bolt diameter	Bolt class (according to DIN 267)					
	5.6	5.8	6.8	8.8	10.9	12.9
M5	2.5	3.5	4.5	6	8	10
M6	4.5	6	7.5	10	14	17
M8	11	15	18	24	34	40
M10	23	30	36	47	66	79
M12	39	52	62	82	115	140
M14	62	82	98	130	180	220
M16	94	126	150	200	280	340
M18	130	174	210	280	390	470

Buying parts

Many of the maintenance and repair tasks in this manual call for the installation of new parts, or the use of new gaskets and other materials when reinstalling parts. In most cases, make sure needed parts are on hand before beginning the job. Read the introductory text and the complete procedure to determine which parts are needed.

For bigger jobs, partial disassembly and inspection are required to determine a complete parts list. Read the procedure carefully and, if necessary, make other arrangements to get the necessary parts while your car is disassembled.

Genuine Porsche parts

Genuine Porsche replacement parts from an authorized Porsche dealer are designed and manufactured to the same high standards as the original parts. They are guaranteed to fit and work as intended by the engineers who designed the car.

Many independent repair shops make a point of using genuine Porsche parts, even though they may at times be more expensive. They know the value of doing the job right with the right parts. Parts from other sources can be as good, particularly if manufactured by one of Porsches original equipment suppliers, but it is often difficult to know.

Porsche is constantly updating and improving their cars, often making improvements during a given model year. Porsche may recommend a newer, improved part as a replacement, and your authorized dealer's parts department will know about it and provide it. The Porsche parts organization is best equipped to deal with any Porsche parts needs.

Non-returnable parts

Some parts cannot be returned. The best example is electrical parts. Buy electrical parts carefully, and be as sure as possible that a replacement is needed, especially for expensive parts such as electronic control modules. It may be wise to let an authorized Porsche dealer or other qualified shop confirm your diagnosis before replacing a non-returnable part.

Service and Maintenance

Basic Service Information

Information you need to know

Model. When ordering parts it is important that you know the correct model designation for your car. Boxster models include the Boxster and Boxster S, and are known as model type 986.

Model year. This is not necessarily the same as date of manufacture or date of sale. A 1997 model may have been manufactured in late 1996, and perhaps not sold until early 1998. It is still a 1997 model. Model years covered by this manual are 1997 to 2004.

Date of manufacture. This information is may be necessary when ordering replacement parts or determining if any of the warranty recalls are applicable to your car. The label on the driver's door below the door latch specifies the month and year that the car was built.

 Vehicle identification number (VIN). This is a combination of letters and numbers that identify the particular car. The VIN appears on the state registration document and on the car itself. One location is in the lower left corner of the windshield. See also **01 Vehicle Identification and VIN Decoder**.

Engine code. Cars covered in this manual are powered by various 6-cylinder engines. For information on engine codes and engine applications, see **02 Boxster Familiarization**.

Transmission code. The transmission type with its identifying code may be important when buying clutch parts, seals, gaskets, and other transmission-related parts. For information on transmission codes and applications, see **02 Boxster Familiarization**.

Porsche service

Porsche dealers are uniquely qualified to provide service for Porsche cars. Their authorized relationship with the large Porsche service organization means that they are constantly receiving special tools and equipment, together with the latest and most accurate repair information.

The Porsche dealer's service technicians are highly trained and very capable. Authorized Porsche dealers are committed to supporting the Porsche product. On the other hand, there are many independent shops that specialize in Porsche service and are capable of doing high quality repair work. Checking with other Porsche owners for recommendations on service facilities is a good way to learn of reputable Porsche shops in your area.

Tools

Most maintenance can be accomplished with a small selection of the right tools. Tools range in quality from inexpensive junk, which may break at first use, to very expensive and well-made tools for the professional. The best tools for most do-it-yourself Porsche owners lie somewhere in between.

Many reputable tool manufacturers offer good quality, moderately priced tools with a lifetime guarantee. These are your best buy. They cost a little more, but they are good quality tools that will do what is

Service and Maintenance

Maintenance Tables

expected of them. Sears' Craftsman® line is one such source of good quality tools.

Some of the repairs covered in this manual require the use of special tools, such as a custom puller or specialized electrical test equipment. These special tools are called out in the text and can be purchased through an authorized Porsche dealer. As an alternative, some special tools mentioned may be purchased from the following tool manufacturers and/or distributors:

- Baum Tools Unlimited, Inc.
  P.O. Box 5867, Sarasota, FL 34277-5867
  800-848-6657
  http://www.baumtools.com
- Samstag Sales
  Carthage, Tennessee USA
  615-735-3388
  http://www.samstagsales.com
- Zelenda Machine and Tool Corp.
  65-60 Austin Street, Forest Hills, NY 11374-4695
  718-896-2288
  http://www.zelenda.com

MAINTENANCE TABLES

Except where noted, the maintenance items listed apply to all models and model years covered by this manual.

There are two mileage-based inspection requirements. These inspections alternate throughout the vehicle's maintenance history. If the last inspection interval was the **minor maintenance** (every 15,000 miles), the next inspection interval will be the **major maintenance** (every 30,000 miles), the next after that will be minor maintenance and so on.

Minor maintenance inspection tasks are listed in **Table b**. Major maintenance inspection tasks are listed in **Table c**. On vehicles driven fewer than 9,000 miles (15,000 km), perform an annual maintenance. See **Table d**. Additional maintenance items are listed in **Table e**.

Service and Maintenance

Maintenance Tables

Table b. Minor maintenance tasks
15,000, 45,000, 75,000, 105,000 miles, etc. (24,000, 72,000, 120,000, 168,000 km etc.)

Maintenance item	Tools required	New parts required	Warm engine required	Dealer service recommended	Additional repair information
Diagnosis system: Read out fault memory of all systems.	✻			✻	
Change oil (no filter).	✻	✻	✻		
Vehicle underside and engine compartment: Inspect for leaks (oils and fluids) and abrasion (chafed wires, lines and hoses).					
Power steering: Check fluid level.					
Coolant hoses: Check condition. Radiators and air inlets at front: Inspect for external contamination and blockage. Coolant: Check level and antifreeze protection.	✻				
Interior ventilation microfilter (particle filter): Replace.	✻	✻			
Brake hoses and lines: Visual inspection for damage, routing and corrosion. Brake fluid: Check brake fluid level.					
CV (drive axle) joints: Inspect boots for leaks and damage.					
Tires and spare wheel: Check condition and tire pressure.	✻				
Check door locks, lid locks and safety hooks of front lid to ensure proper function and secure latching.					
Vehicle lighting: Check lights for proper function. Headlights: Check adjustment. Horn: Check operation.					
Windshield wiper/washer system, headlight washer: Check fluid level and nozzle settings, pay attention to antifreeze protection in winter months.		✻			
Electrical equipment, warning and indicator lights: Check for proper function and operation. Check remote control key.					
Test drive: Check remote control key, front seat operation, foot and parking brakes, engine, clutch, steering, transmission, Parking Assistant, cruise control, heating, air-conditioning system and instruments.					

Service and Maintenance 03-9

Maintenance Tables

Table c. Major maintenance tasks
30,000, 60,000, 90,000, 120,000 miles, etc. (48,000, 96,000, 144,000, 192,000 km etc.)

Maintenance item	Tools required	New parts required	Warm engine required	Dealer service recommended	Additional repair information
Diagnosis system: Read out fault memory of all systems.	*			*	
Engine drive belt: Check condition.					
Change oil and filter.	*	*	*		
Vehicle underside and engine compartment: Inspect for leaks (oils and fluids) and abrasion (chafed wires, lines and hoses).					
Coolant hoses: Check condition. Radiators and air inlets at front: Inspect for external contamination and blockage. Coolant: Check level and antifreeze protection, check for leaks.	*				
Air filter: Replace filter element.	*	*			
Interior ventilation microfilter (particle filter): Replace.	*	*			
Fuel system: Inspect for damage, correct routing and secure fit of line connections.					
Power steering: Check fluid level.					
Brake system: Inspect brake pads and brake rotors for wear. Parking brake: Check parking brake lever travel.					
Brake hoses and lines: Inspect for damage, routing and corrosion. Check fluid level.					
Clutch: Check pedal end position.					
Throttle actuation: Check operation of throttle (full throttle position) using Porsche Service Tester 2 (PST2) (omitted as of model year 2002).					
Steering gear: Inspect rubber boots for damage. Tie rods: Inspect joints for play, inspect rubber boots for damage.					
CV (drive axle) joints: Inspect boots for leaks and damage.					
Exhaust system: Inspect for leaks or damage. Check hangers and suspension.					
Tires and spare wheel: Check condition and tire pressure.	*				
Check door locks, lid locks and safety hook of front lid to ensure proper function and secure latching. Check door and lid fasteners for proper tightening torques. Check inner trunk release (trunk entrapment), where applicable.					
Seat belts: Check operation and condition.					
Vehicle lighting: Check lights for proper function. Headlights: Check adjustment. Horn: Check operation.					
Windshield wiper/washer system, headlight washer: Check fluid level and nozzle settings, pay attention to antifreeze protection in winter months.		*			
Electrical equipment, warning and indicator lights: Check for proper function and operation.					
Test drive: Check remote control key, front seats, foot and parking brakes, engine, clutch, steering, transmission, Parking Assistant, cruise control, heating, air-conditioning system and instruments.					

03-10 Service and Maintenance

Maintenance Tables

Table d. Annual maintenance tasks

Maintenance item	Tools required	New parts required	Warm engine required	Dealer service recommended	Additional repair information
Diagnosis system: Read out fault memory of all systems.	*			*	
Vehicle underside and engine compartment: Inspect for leaks (oils and fluids) and abrasion (chafed wires, lines and hoses).					
Power steering: Check fluid level.					
Coolant hoses: Check condition. Radiators and air inlets at front: Inspect for external contamination and blockage. Coolant: Check level and antifreeze protection, check for leaks.	*				
Brake hoses and lines: Inspection for damage, routing and corrosion. Check fluid level.					
CV (drive axle) joints: Inspect boots for leaks and damage.					
Tires and spare wheel: Check condition and tire pressure.	*				
Door locks, lid locks, front lid safety hooks: Check proper function and secure latching. Check door and lid fasteners for proper tightening torques. Check inner trunk release (trunk entrapment), where applicable.					
Vehicle lighting: Check lights for proper function. Headlights: Check adjustment. Horn: Check operation.					
Windshield wiper/washer system, headlight washer: Check fluid level and nozzle settings, pay attention to antifreeze protection in winter months.		*			
Electrical equipment, warning and indicator lights: Check for proper function and operation.					
Test drive: Check remote control key, front seats, foot and parking brakes, engine, clutch, steering, transmission, Parking Assistant, cruise control, heating, air-conditioning system and instruments.					

Table e. Additional maintenance tasks

Maintenance item	Tools required	New parts required	Warm engine required	Dealer service recommended	Additional repair information
Engine drive belt: Replace every 60,000 miles.	*	*			
Fuel filter (1997 - 2001 models): Replace every 60,000 miles.	*	*			
Spark plugs (1997 - 2000 models): Replace every 30,000 miles. (2001 - 2004 models): Replace every 60,000 miles or 4 years at the latest.	*	*			28
Manual transmission fluid: Replace every 90,000 miles.	*	*	*		
Automatic transmission fluid: Replace ATF and filter every 90,000 miles.	*	*	*		
Differential oil: Replace every 90,000 miles.	*	*	*		
Brake fluid: Replace every 2 years.	*	*			47
Engine, transmission and running gear mounts: Inspect rubber mounts for damage after 4, 8, 10 years, then every 2 years.					
Airbag system: Inspect airbag system after 4, 8, 10 years, then every 2 years.	*	*		*	69

Service and Maintenance 03-11

Diagnosis System

Diagnosis System

Fault memory, reading out

Boxsters sold in the US market are equipped with On-Board Diagnostics II (OBD II). 2000 and later European-market Boxsters are equipped with European On-Board Diagnostics (EOBD). OBD II and EOBD systems monitor engine and automatic transmission functions to insure compliance with specified emission levels.

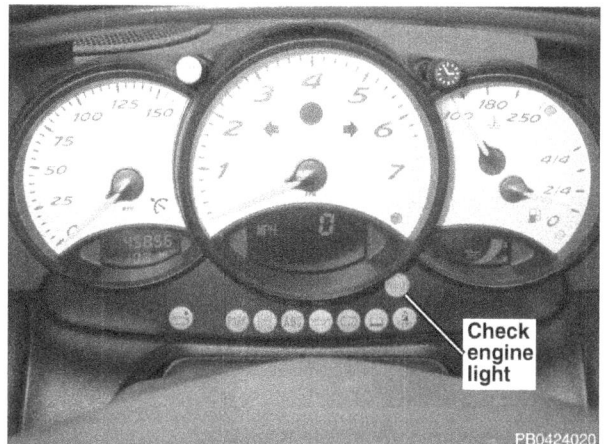

◀ Vehicle emission levels are continually monitored by the OBD II / EOBD system and problems are recognized and recorded. The Check Engine light (also known as malfunction indicator light or MIL) in the instrument cluster alerts the driver to the problem and the need to have the system checked for codes. These codes follow a standard format and are known as diagnostic trouble codes (DTCs).

Use a diagnostic scan tool to access fault codes. OBD data can be accessed using either the factory Porsche System Tester 2 (PST2) or a generic OBD II / EOBD scan tool.

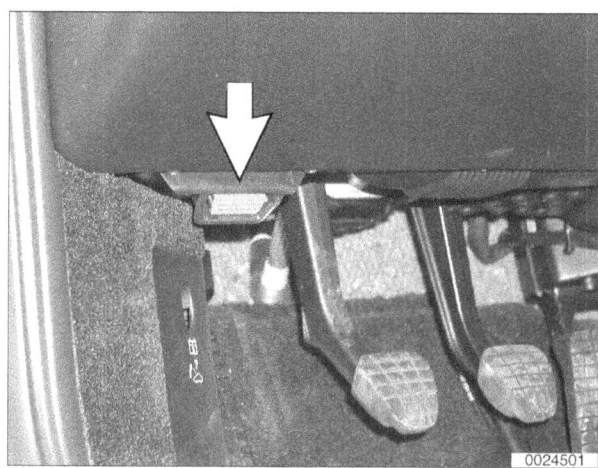

◀ The diagnostic scan tool is connected via the 16-pin diagnosis socket (**arrow**) located inside the vehicle below the left side of the dashboard.

NOTE—
- *Many automotive aftermarket OBD II / EOBD compatible diagnostic scan tools can also be used to retrieve DTCs.*
- *A list of DTCs can be found in* **24 Fuel Injection**.

Fluids and Lubricants

Engine oil level, checking

A small amount of oil consumption is normal. The rate of consumption depends on quality and viscosity of the oil and operating conditions. Regularly check the oil level gauge on the instrument panel. The oil level may also be checked using the dipstick in the rear trunk.

− After switching ignition OFF, wait at least 3 minutes to allow oil to flow back into oil pan.

◀ Remove dipstick (**arrow**) located at upper right of rear trunk. Wipe dipstick with clean cloth and fully reinsert into tube.

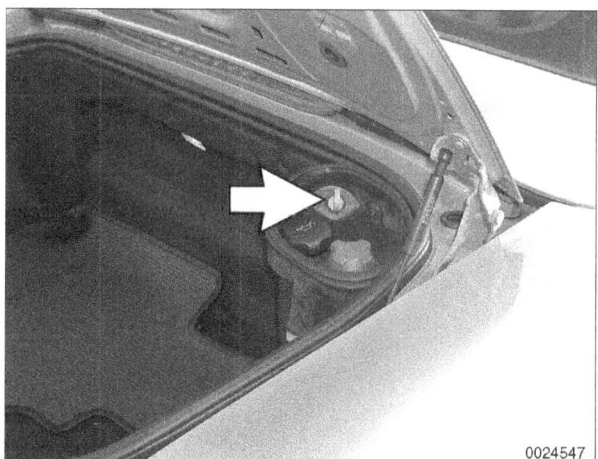

Fluids and Lubricants

◂ Remove dipstick again and read oil level. Make sure oil level is between MIN and MAX marks as shown by **arrows**.

> *CAUTION—*
> * *Risk of damage to the catalytic converter if oil level is too high.*

If necessary, add engine oil in ½ liter (½ quart) increments through oil fill tube.

Engine oil	
Oil specification: Synthetic	API SH or SJ ILSAC GF 3 Europe: ACEA A3
Oil viscosity and ambient temperatures: • Generally higher than 50°F (10°C) • Generally lower than 50°F (10°C)	10W-40, 15W-40, 15W-50 10W-40, 10W-30, 5W-30
Oil capacity: • Without filter change • With filter change	8.5 liters (9 US qt) 8.75 liters (9.25 US qt)
Recommended oil change interval (Porsche)	15,000 miles

> *CAUTION—*
> * *Lubricant specifications are subject to change. Consult your authorized dealer for the latest information regarding lubricant applications and specifications.*

Engine oil and filter, changing

Requirements for changing engine oil are:
* Engine oil warm
* Vehicle raised and level

> *WARNING—*
> * *Do not drain engine oil when engine is hot. Hot engine oil can scald.*
> * *Wear hand and face protection when draining warm engine oil.*

− Raise car and support safely.

> *WARNING—*
> * *Make sure the car is firmly supported on jack stands designed for the purpose. Place jack stands underneath structural chassis points, not under suspension parts.*

Service and Maintenance 03-13

Fluids and Lubricants

◄ Remove oil drain plug using 8 mm Allen bit. Drain engine oil into suitable container.

– Clean oil drain plug.

– After oil has stopped draining (allow at least 20 minutes), install oil drain plug with new sealing ring.

Tightening torque	
Oil drain plug to oil pan	50 Nm (37 ft-lb)

◄ Unscrew oil filter housing using Porsche special tool 9204 or equivalent cap or strap wrench.

NOTE—

- *The oil filter cap wrench, measuring 74 mm (2.9 in) across the flats of the wrench, is readily available from auto parts stores.*

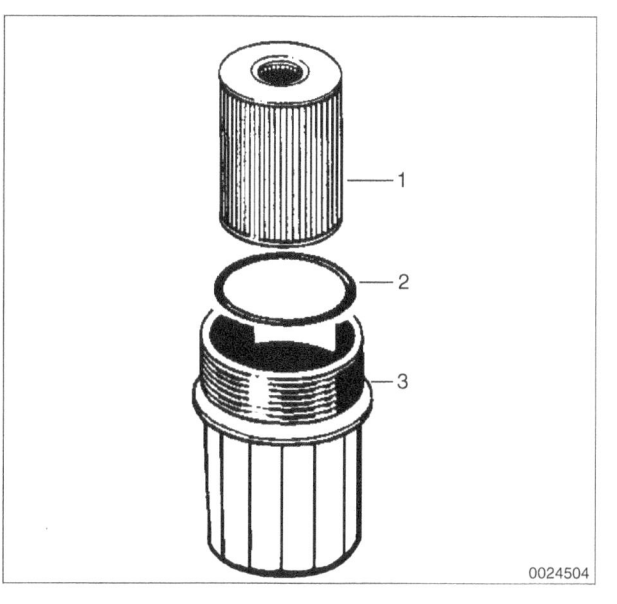

◄ Remove oil filter element (**1**) from housing (**3**).

– Install new filer element into housing. Using new O-ring (**2**), install housing and tighten.

Tightening torque	
Oil filter housing to crankcase	25 ± 1 Nm (19 ± 1 ft-lb)

Engine oil capacity	
Without filter change	8.5 liters (9 quarts)
With filter change	8.75 liters (9.25 quarts)

NOTE—

- *Engine oil capacity specifications listed above are approximate. Always check dipstick for proper oil level after oil change.*

Fluids and Lubricants

Coolant level, checking

The engine cooling system is filled with Porsche phosphate-free coolant. Do not mix or replace this coolant with other coolants. Use only original Porsche coolant when topping up.

> *WARNING—*
> - *Hot coolant can scald. Do not work on the cooling system until it is fully cool.*
> - *Do not open coolant fill cap on a hot engine.*
> - *Use extreme care when draining and disposing of coolant. Coolant is poisonous and lethal. Children and pets are attracted to it because of its sweet smell and taste. See a doctor or veterinarian immediately if any amount is ingested.*

NOTE—
- *Coolant draining, filling, and system bleeding procedures are in* **19 Cooling System**.

 Open rear trunk lid. Check coolant level through transparent window (**arrow**) of coolant expansion tank.
  - When engine is cold and car is level, check that coolant level is between MIN and MAX markings.

- If coolant level is low, top off as follows:
  - Cover coolant expansion tank cap with a shop towel and slowly unscrew cap.
  - Add mixture of phosphate-free coolant (antifreeze) and distilled water in equal parts.

> *CAUTION—*
> - *Be careful when topping up coolant not to spill any in trunk or on surface of vehicle.*

Coolant / distilled water mixture	Freeze protection
50% coolant / 50% water	down to -35°C (-31°F)
60% coolant / 40% water	down to -40°C (40°F)

NOTE—
- *Do not exceed a concentration of 60% coolant as this will decrease the freeze protection and cooling capacity.*

Service and Maintenance 03-15

Fluids and Lubricants

Brake fluid, checking

> **WARNING—**
> - Brake fluid is poisonous. DO NOT ingest brake fluid. Wash thoroughly with soap and water if brake fluid comes into contact with skin.

> **CAUTION—**
> - DO NOT let brake fluid come in contact with paint. Wash immediately with soap and water.
> - Brake fluid absorbs moisture from the air. Store in an airtight container.
> - DO NOT allow brake fluid to exceed the maximum level in the fluid reservoir.

For routine brake system maintenance:

- Maintain adequate level of brake fluid in reservoir.
- Replace brake fluid every 2 years.
- Check brake pads for wear.
- Check parking brake function.
- Inspect system for fluid leaks or other damage.

See **Under Car Maintenance** in this repair group for pad inspection and brake system inspection.

 Check that brake fluid level in reservoir (**arrow**) is between MAX and MIN marks.

> **NOTE—**
> - The brake fluid reservoir is also the reservoir for the hydraulically operated clutch.
> - A slight decrease in the brake fluid level due to brake pad wear is normal.

- Top off brake fluid as necessary so fluid level is at or below MAX mark.

Brake fluid specifications	
Porsche brake fluid	Super DOT 4

> **WARNING—**
> - Do not mix DOT 5 (silicone) brake fluid with DOT 4 brake fluid. Severe component corrosion results, leading to brake system failure.

Brake fluid, changing

Brake fluid readily absorbs moisture from the atmosphere. This moisture can cause brake system corrosion and adversely affect braking performance. Therefore, flush out old brake fluid and add new, fresh fluid at least every 2 years, regardless of mileage. See **47 Brakes–Hydraulic** for brake system bleeding procedures and brake fluid replacement.

Fluids and Lubricants

Power steering fluid, checking

Damage to the power-assisted steering is often the result of running the system while low on fluid. Grunt-like noises when the steering is turned lock-to-lock or foam formation in the fluid reservoir indicates low fluid level or that air bubbles are present in the fluid. Before topping up the reservoir, remedy any leaks on the suction side and replace faulty parts on the pressure side.

The reservoir is located in the engine compartment. There are two markings on the dipstick located on the reservoir cap:

- Cold marking = engine temp. approx. 20°C (68°F)
- Hot marking = engine temp. approx. 80°C (176°F)

Check power steering fluid level when engine is cold and not running.

– Open and remove top engine cover. See **Top engine cover, removing and installing** in this repair group.

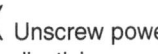 Unscrew power steering fluid reservoir cap (**arrow**). Wipe off dipstick.

– Close and then reopen cap. Check that fluid level is in shaded area below Cold marking.

- Cold = maximum engine temperature 20°C (68°F)

– Top up with Pentosin (CHF 11 S) power steering fluid as necessary. Do not overfill.

Power steering fluid	
Boxster, Boxster S	Pentosin CHF 11 S

CAUTION—
- *If coolant hoses come into contact with Pentosin, clean them thoroughly with water immediately.*
- *Replace any visibly swollen hoses.*

Manual transmission oil, checking

Requirement for checking manual transmission oil are:
- Transmission oil warm
- Vehicle raised and level
- 5-speed transmission: 17 mm Allen wrench
- 6-speed transmission: 10 mm Allen wrench
- 12-point anti-tamper wrench (Hazet special tool 2567-16)

– Raise vehicle and support safely. Make sure vehicle is level.

WARNING—
- *Make sure the car is firmly supported on jack stands designed for the purpose. Place jack stands underneath structural chassis points, not under suspension parts.*

Service and Maintenance 03-17
Fluids and Lubricants

◂ Remove transmission oil fill plug.
- 5-speed transmission: Use 17 mm Allen socket.
- 6-speed transmission: Use 10 mm Allen socket.

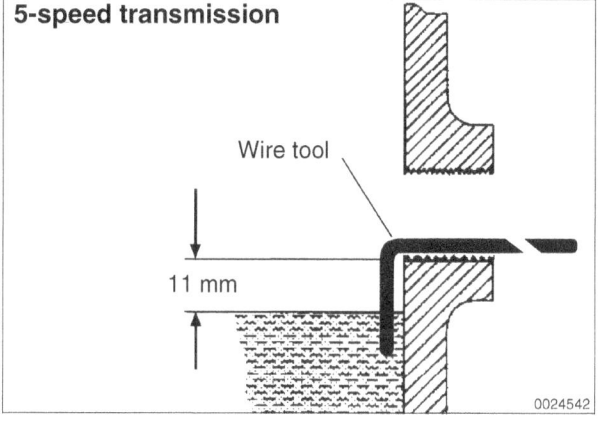

◂ 5-speed transmission: Check transmission oil level using a stiff wire hook inserted into fill hole.

Manual transmission oil level	
5-speed transmission	11 mm (7/16 in) below fill hole opening

− 6-speed transmission: Add fluid until it runs out of fill hole.

Manual transmission oil, changing

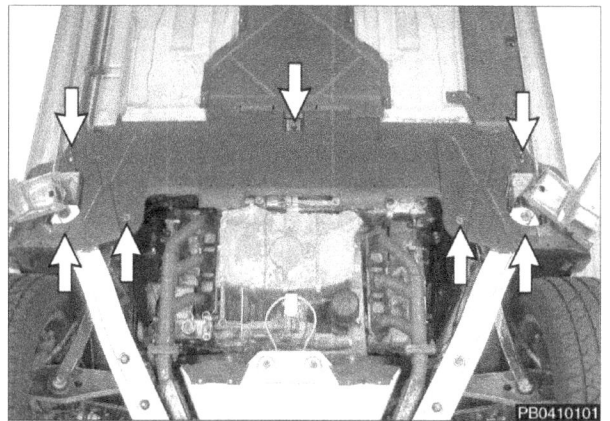

◂ Working beneath car, remove plastic shield under engine. **Arrows** indicate fasteners to be removed.

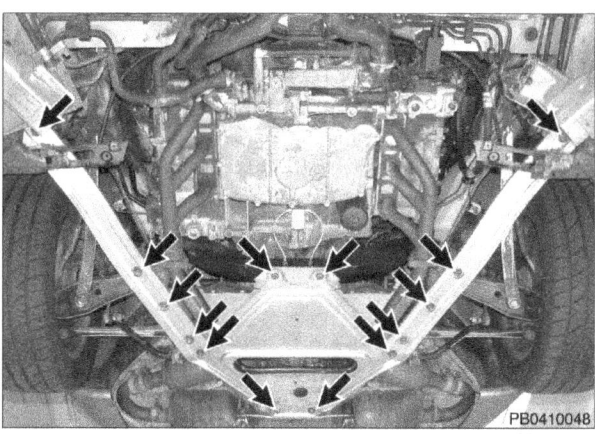

◂ Remove aluminum diagonal braces and chassis reinforcement plate (14 fasteners) (**arrows**).

03-18 Service and Maintenance

Fluids and Lubricants

◁ Remove transmission oil fill plug.
- 5-speed transmission: Use 17 mm Allen socket.
- 6-speed transmission: Use 10 mm Allen socket.

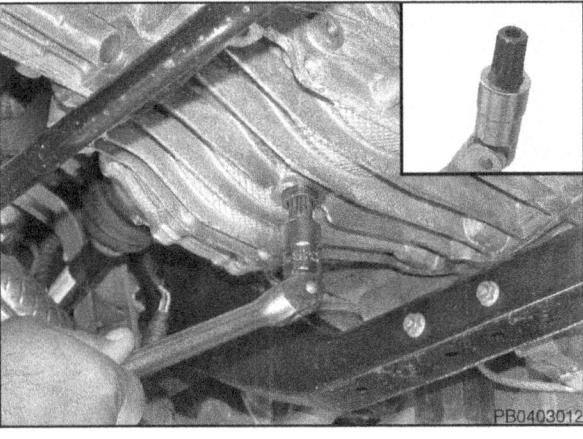

◁ Remove transmission oil drain plug:
- 5-speed transmission: Use Hazet 2567-16 or equivalent 16 mm triple-square anti-tamper wrench (**inset**).
- 6-speed transmission: Use 10 mm Allen socket.

NOTE—
- *Hazet tool 2567-16 is 12-point bit with a drilled center, ½-inch drive.*

- Drain oil into suitable container.
- Clean drain and fill plugs.
- When oil is fully drained, install oil drain plug.
- Fill transmission through fill hole.

Manual transmission oil capacity (Hypoid gear oil SAE 75W90)	
Boxster (5-speed)	2.25 liters (2.4 US qt)
Boxster S (6-speed)	2.8 liters (3.0 US qt)

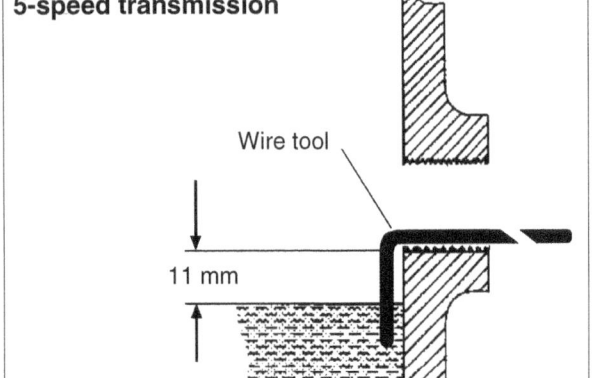

◁ 5-speed transmission: Check transmission oil level using a stiff wire hook inserted into fill hole.

Manual transmission oil level	
5-speed transmission	11 mm (⁷⁄₁₆ in) below fill hole opening

- 6-speed transmission: Add fluid until it runs out of fill hole.
- Install and tighten drain and fill plugs.

Tightening torques	
Drain plug to manual transmission	25 Nm (18 ft-lb)
Fill plug to manual transmission • M10 • M12	 30 Nm (22 ft-lb) 25 Nm (18 ft-lb)

- Replace reinforcement plate, diagonal braces and plastic undershield.

Service and Maintenance 03-19

Fluids and Lubricants

Tightening torques	
Reinforcement plate • To rear axle support (M10 bolt) • To support / suspension (M10 nut)	46 Nm (34 ft-lb) 65 Nm (48 ft-lb)
Diagonal brace to body / suspension	65 Nm (48 ft-lb)

Differential oil level, checking (automatic transmission)

NOTE—
* *Manual transmission: Differential oil and transmission oil is shared. See **Manual transmission oil, checking** and **Manual transmission oil, changing** in this repair group.*

– Raise vehicle and support safely. Make sure vehicle is level.

WARNING—
* *Make sure the car is firmly supported on jack stands designed for the purpose. Place jack stands underneath structural chassis points, not under suspension parts.*

– There is no differential oil drain plug. To drain oil, remove differential cover. To remove cover, remove drive axle and drive axle flange. See **42 Rear Suspension**.

◂ Unscrew differential oil fill plug (**arrow**).

– Check to see if oil level is up to lower edge of fill opening. If not, fill with specified fluid.

Automatic transmission differential oil capacity (Hypoid gear oil SAE 75W90)	
Boxster, Boxster S	0.8 liter (0.85 US quart)

– Replace sealing ring for fill plug and install plug.

Tightening torque	
Fill plug to differential housing	30 Nm (22 ft-lb)

Automatic transmission fluid (ATF), checking level and topping off

Check ATF level with the car raised and level, the engine running. Also, monitor ATF temperature using Porsche System Tester 2 (PST2) or an infrared thermometer (with laser pointer).

◂ If fluid needs to be added, use Porsche special tool VAG 1924 or equivalent with hooked filling tip to pump fluid into transmission.

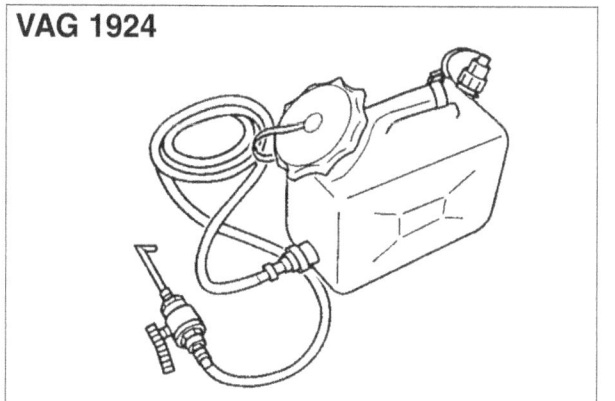
VAG 1924

Requirement for checking ATF are:

03-20 Service and Maintenance

Fluids and Lubricants

- Transmission not in limp-home mode: No fault codes present in Tiptronic control module.
- Vehicle raised and level.
- ATF temperature 30°C - 45°C (86°F - 113°F).
- ATF filler tool VAG 1924.

> **CAUTION—**
> - *Incorrect ATF temperature when filling ATF may cause the following:*
> *Temperature too cold = over-filling*
> *Temperature too hot = under-filling*

– Raise vehicle and support safely. Make sure vehicle is level.

> **WARNING—**
> - *Make sure the car is firmly supported on jack stands designed for the purpose. Place jack stands underneath structural chassis points, not under suspension parts.*

◄ Working underneath car, remove plastic under shield. **Arrows** indicate fasteners to be removed.

◄ Remove aluminum diagonal braces and chassis reinforcement plate (14 fasteners) (**arrows**).

– Make sure gear selector lever is in PARK and parking brake is firmly set. Start engine and let transmission fluid warm up to specified temperature.

ATF level checking temperature	
Engine running	30°C - 45°C (86°F - 113°F)

> **NOTE—**
> - *Use Porsche System Tester 2 (PST2) to determine ATF temperature. If the factory diagnostic tester is not available, use an infrared thermometer.*

– Make sure air-conditioner and heater are OFF.

– Place drain pan under transmission.

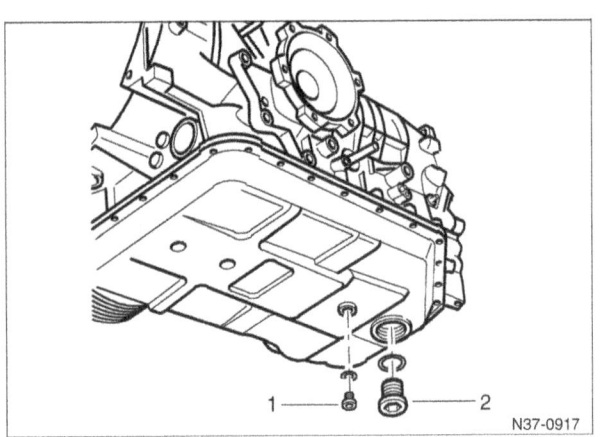

◄ With ATF at specified temperature (30°C - 45°C or 86°F - 113°F), unscrew ATF fill plug (**2**) using 17 mm Allen wrench.

> **WARNING—**
> - *Wear protective eye goggles.*

Service and Maintenance 03-21
Fluids and Lubricants

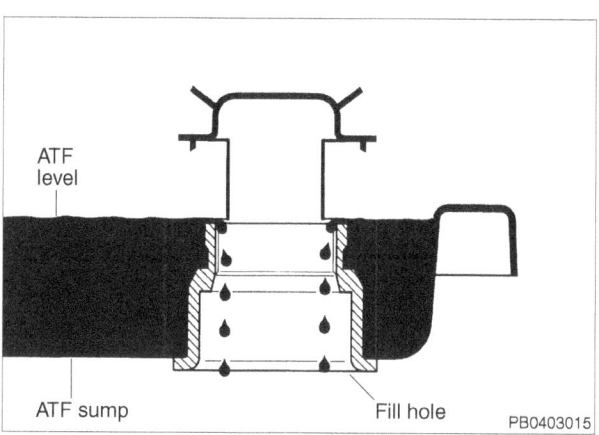

◂ If a small amount of fluid escapes from fill hole, ATF level is correct.

– If no fluid escapes from fill hole, top up ATF.

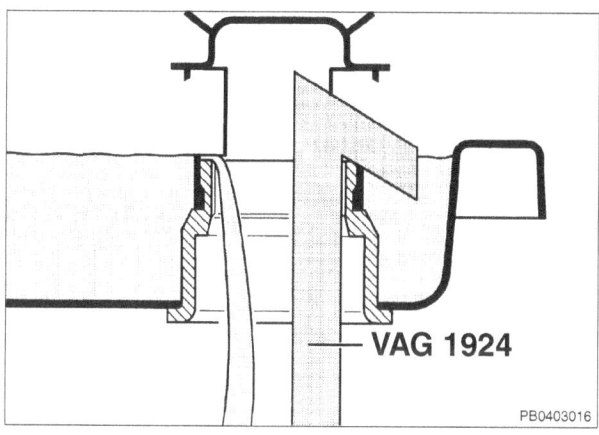

◂ Use Porsche special tool VAG 1924, or equivalent:
  • Insert filler tool into opening of ATF guard cap in fill hole.

  CAUTION—
  • *Do not press filler hook upwards more than necessary when inserting into fill plug opening in the well. It is possible to push the deflector cap off the top of the well.*

  • Fill with ATF until fluid runs out of fill hole.

ATF specification	
Boxster, Boxster S	ESSO LT 71141

– Install ATF fill plug with a new seal.

Tightening torque	
Fill plug to ATF sump	80 Nm (59 ft-lb)

WARNING—
• *To prevent hot fluid from escaping, install ATF fill plug before ATF temperature reaches 45°C (113°F).*

Automatic transmission fluid (ATF), changing

For ATF filter and ATF sump removal, see **37 Automatic Transmission**.

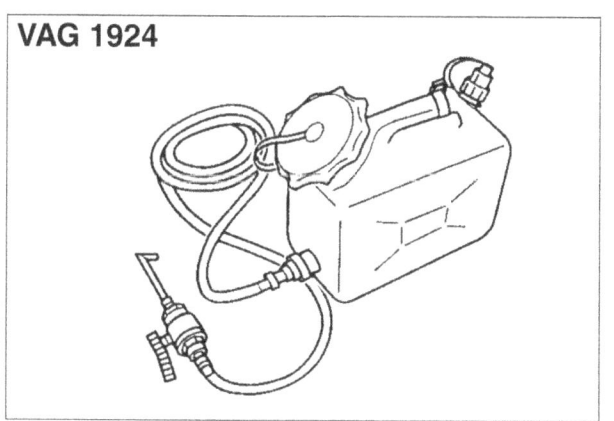

◂ Use Porsche special tool VAG 1924 or equivalent with hooked filling tip to pump fluid into transmission.

Requirement for changing ATF are:
• Transmission not in limp-home mode: No fault codes present in Tiptronic control module.
• Vehicle raised and level.
• ATF temperature 30°C - 45°C (86°F - 113°F).
• ATF filler tool VAG 1924.

CAUTION—
• *Incorrect ATF temperature when filling ATF may cause over-filling or under-filling.*

03-22 Service and Maintenance

Fluids and Lubricants

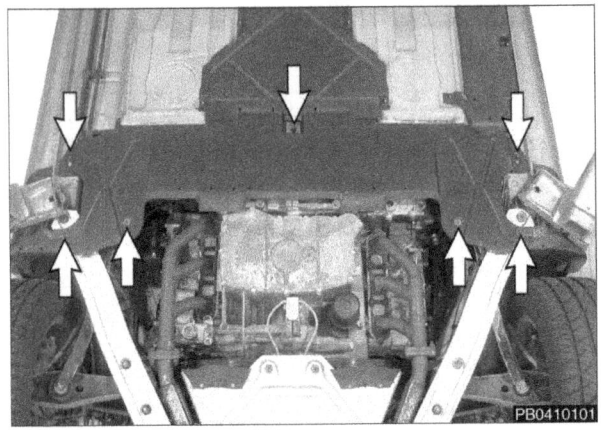

◀ Working beneath car, remove plastic shield under engine. **Arrows** indicate fasteners to be removed.

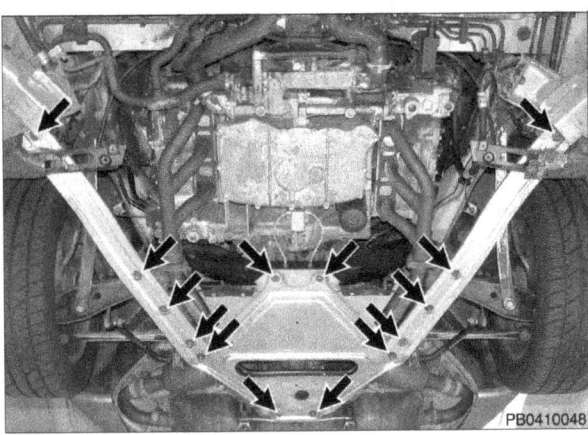

◀ Remove aluminum diagonal braces and chassis reinforcement plate (14 fasteners) (**arrows**).

– Place drain pan under transmission.

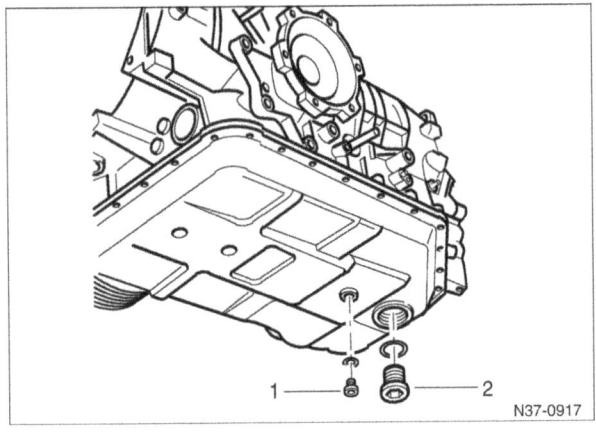

◀ Unscrew ATF drain plug (**1**) and drain ATF.

> **WARNING—**
> • Wear protective eye goggles.

> **CAUTION—**
> • Without ATF in transmission, do not start engine and do not tow vehicle.

– Replace sealing ring for ATF drain plug and reinstall plug.

Tightening torque	
Drain plug to ATF sump	40 Nm (30 ft-lb)

– Unscrew ATF fill plug (**2**) using 17 mm Allen wrench.

Service and Maintenance 03-23

Fluids and Lubricants

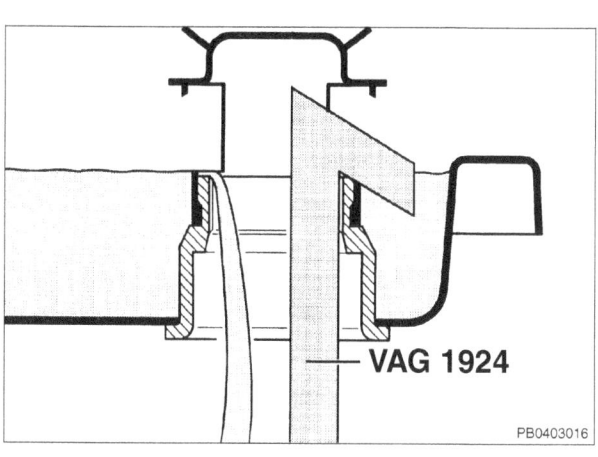

◀ Use Porsche special tool VAG 1924, or equivalent:
- Insert filler tool into opening of ATF guard cap in fill hole.

> **CAUTION—**
> - Do not press filler hook upwards more than necessary when inserting into fill plug opening in the well. It is possible to push the deflector cap off the top of the well.

- Fill with ATF until fluid runs out of fill hole.

ATF specification	
Boxster, Boxster S	ESSO LT 71141

ATF fill capacities	
Change fluid	3.5 liters (3.7 US qt)
Transmission capacity	9.0 liters (9.5 US qt)

– Start engine and allow to idle.

– With ATF at the specified temperature (30 - 45°C), top up until ATF emerges from fill hole.

– With brake pedal pressed, change through all selector lever positions, remaining in each position for approximately 10 seconds.

– Check ATF level again and top up if necessary. Make sure ATF temperature is 30°C - 45°C (86°F - 113°F) for final level check.

– Replace sealing ring for ATF fill plug and install and tighten plug.

Tightening torque	
Fill plug to ATF sump	80 Nm (59 ft-lb)

> **WARNING—**
> - To prevent hot fluid from escaping, install ATF fill plug before ATF temperature reaches 45°C (113°F).

03-24 Service and Maintenance

Engine Compartment Maintenance

Windshield wipers and washer fluid, checking

◄ Washer fluid reservoir filler (**arrow**) is located in rear of front trunk. Use washer fluid with correct temperature range for anticipated driving conditions.

Washer fluid capacity (approximate)	
Without headlight cleaning system	2.5 liters (2.6 US qt)
With headlight cleaning system	7 liters (7.4 US qt)

– Check spray pattern of front windshield washer nozzles. Adjust as necessary. See **92 Wipers and Washers**.

– Inspect windshield wiper blades, front and rear. Replace damaged or deteriorated parts. See **92 Wipers and Washers**.

NOTE—

* *In cases where cleaning of windshield and replacement of wiper blades fails to eliminate streaking conditions, thoroughly clean outside of windshield with a non-abrasive cleaner such as Bon Ami® or Soft Scrub® using a soft cloth and water. Rub until windshield is completely clean of all foreign material. Rinse windshield thoroughly.*

ENGINE COMPARTMENT MAINTENANCE

Convertible top service position

◄ Open convertible top approx. 4 in. (10 cm) from windshield frame.

– Remove ignition key to prevent convertible top operation.

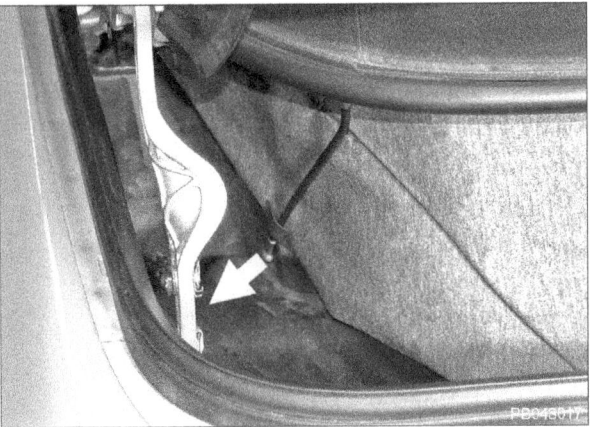

◄ Pull off lower ball head of convertible top tensioning cable (in direction of **arrow**) on left and right sides.

Service and Maintenance 03-25
Engine Compartment Maintenance

◄ Unclip and pull down fabric covering rod from two holders (**arrow**) at base of rear bulkhead.

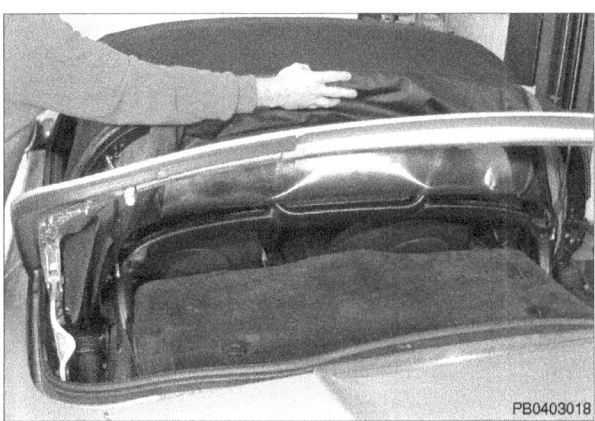

◄ Fold rear part of convertible top forward (**arrow**) until you feel it engage.

NOTE—

- *2003 - 2004 models: Support glass rear window by pulling retaining strap at base of window around left side of top and attaching it to centering pin at front of top.*

◄ Remove retaining spring clips (**arrow**) from convertible top compartment lid arms at left and right sides.

◄ Using an assistant, slide compartment cover up and forward until bore in arm aligns with bore in hinge bracket. Install drift (**arrow**) through bores at left and right sides.

– Assembly is reverse of removal.

Service and Maintenance

Engine Compartment Maintenance

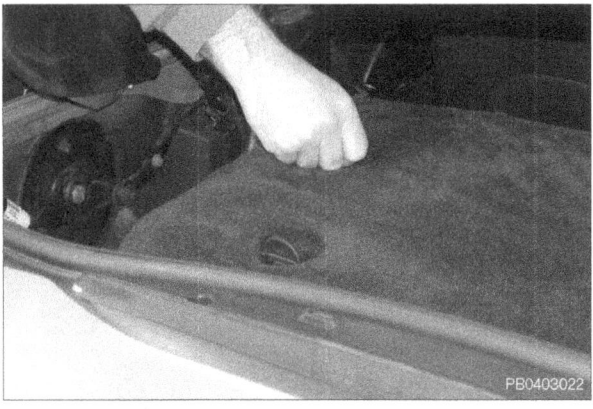

Top engine cover, removing and installing

- Place convertible top in service position. See **Convertible top service position** in this repair group.

◀ Loosen carpeted cover retaining locks and remove cover.

- In car with luggage pocket: Two retaining locks are inside pocket. In addition, loosen retaining lock on upper side of pocket.

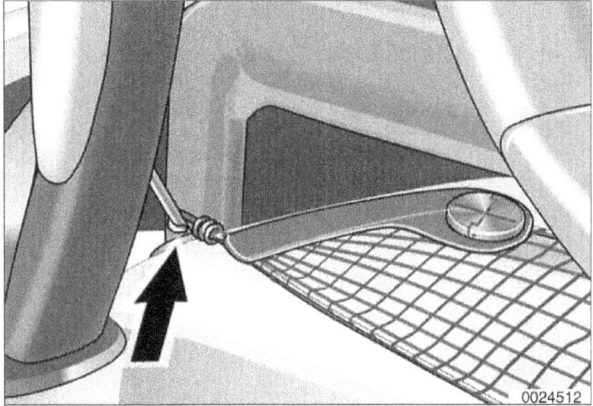

◀ In car with luggage net: Unhook retaining hook (**arrow**) on each side of supplemental safety bar.

◀ Loosen retaining locks (**arrows**) of top engine cover. Remove cover.

- Assembly is reverse of removal.

Front engine cover, removing and installing

- Remove left front seat. See **72 Seats**.

NOTE—
* *The front engine cover can be removed with the seat installed, but seat removal gains additional servicing access.*

- Remove carpeted cover at top engine cover. See **Top engine cover, removing and installing** in this repair group.

◀ Remove carpeted cover at front by unscrewing 4 threaded caps at top (**arrows**).

NOTE—
* *Use edge of coin to unscrew carpet retaining caps.*

Service and Maintenance 03-27

Engine Compartment Maintenance

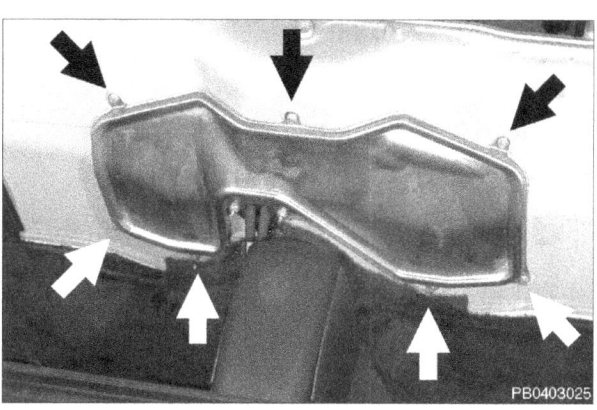

◁ Remove front engine cover bolts (**arrows**) and remove cover.

– Assembly is reverse of removal.

Drive belt, checking and replacing

– Remove front engine cover. See **Front engine cover, removing and installing** in this repair group.

– If reusing old belt, mark running direction on belt.

◁ Relieve belt tension by turning tensioner (wrench size 24 mm) clockwise while simultaneously removing belt from drive pulleys.

– Check drive belt condition. Replace belt if it is found to have any of these faults:
  • Subsurface cracks, core ruptures and cross-sectional breaks
  • Layer separation
  • Fraying of cord strands
  • Flank wear, fraying, brittleness, glassiness, cracks
  • Traces of oil and grease

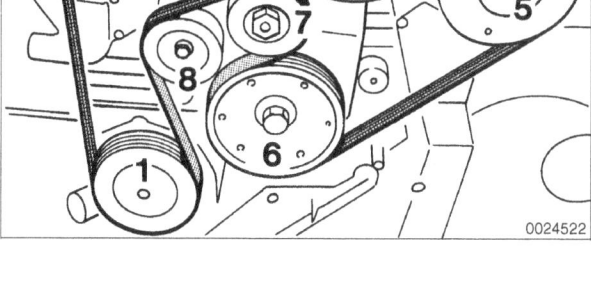

◁ Install ribbed belt over pulleys in following order:
  1. Coolant pump
  2. Alternator
  3. Upper idler
  4. Power steering pump
  5. Air-conditioning compressor
  6. Crankshaft
  7. Tensioner

– Use 24 mm wrench to turn tensioner (**7**) clockwise and simultaneously fit drive belt over lower idler pulley (**8**).

– Slowly release tensioner.

– Visually check that belt is correctly positioned on all drive pulleys. Run engine to confirm correct installation.

– Install front engine cover, rear wall carpet and front seat.

03-28 Service and Maintenance

Engine Compartment Maintenance

Air filter (engine), replacing

– Remove top engine cover. See **Top engine cover, removing and installing** in this repair group.

◄ Engine air filter is located on left side of engine compartment. Release spring clips (**arrows**) and pull out sliding drawer with filter element.

– Remove filter element.

– Clean air cleaner housing with a vacuum cleaner. Do not use compressed air.

> *CAUTION—*
> * *The mass air flow sensor can be damaged if compressed air is used to clean air cleaner housing.*

◄ Place new air filter element in filter tray so that it hangs downward in sliding filter tray.

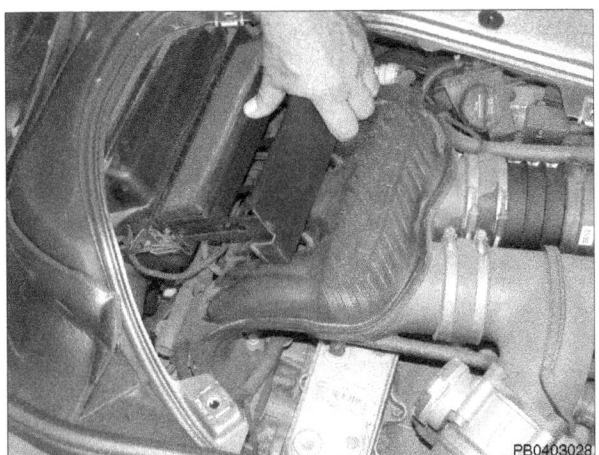

◄ Press in air filter element only using even pressure until it stops.

– Press in sliding filter tray until retaining springs engage.

Throttle valve, checking

– Check throttle operation, including full throttle position, using Porsche System Tester 2 (PST2) or equivalent scan tool.

Service and Maintenance 03-29

Under Car Maintenance

UNDER CAR MAINTENANCE

Many of the following procedures must be performed with the vehicle on a lift or safely positioned on jackstands.

> **WARNING—**
> - Never work under a raised car unless it is solidly supported on jack stands that are intended for that purpose.
> - When raising the car using a floor jack or hydraulic lift, carefully position the jack pad to prevent damaging the car body.
> - Watch the jack closely. Make sure it stays stable and does not shift or tilt. As the car is raised, it may roll slightly and the jack may shift.
> - A jack is a temporary lifting device. Do not use alone to support the car while you are under it.
> - Do not use wood, concrete blocks or bricks to support a car. Wood may split. Blocks and bricks, while strong, are not designed for that kind of load and may break or collapse.

Radiator air inlet duct, cleaning

◄ Clean cooling air inlet ducts on front of radiators with a vacuum cleaner nozzle inserted in front bumper grille openings.

− Boxster S: Also clean 3rd radiator in the center opening of front bumper.

Coolant hoses, checking condition

◄ Inspect hoses from below.
- Check that all connections are tight and dry. Coolant seepage indicates that hose clamp is loose, hose is damaged or connection is dirty or corroded. Dried coolant has a chalky appearance.
- Check condition of hoses by pinching them. Make sure hoses are firm and springy.

− Replace any hose that is cracked, soft and limp, or contaminated by oil or power steering fluid.

03-30 Service and Maintenance

Under Car Maintenance

Spark plugs, replacing

- Turn ignition off and remove key.
- Raise car and support safely.

> **WARNING—**
> - Make sure the car is firmly supported on jack stands designed for the purpose. Place jack stands underneath structural chassis points, not under suspension parts.

◂ Undo mounting bolts (**arrows**) for each ignition coil (3 coils on each side of engine).

- While twisting coil slightly, pull off coil and position it aside with connected cable.
- Remove spark plugs from cylinder head using socket wrench, extension and proper spark plug socket.
- Install in reverse order of removal.

Spark plugs	
2.5 liter engine	NGK BKR 6EK
	Bosch FR 7 LDC4
2.7 liter engine	Bosch FGR 6KQC
3.2 liter engine	Bosch FGR 7KQC
	Beru 14 FGR 6 KQU
	Beru 14 FGR 7KQU
Tightening torque	30 + 3 Nm (22 + 2 ft-lb)
Gap	1.6 mm + 0.05 mm

> **CAUTION—**
> - Spark plug applications are subject to change. Consult an authorized Porsche parts department or an automotive parts professional for the most up to date parts information.

Brake pads and rotors, checking

- Replace brake pads when brake pad warning indicator lights up, but no later than when there is a residual pad thickness of 2 mm (0.08 in). If brake pad wear is indicated by warning light, also replace warning contact (sender including wire and plug connection).

- To avoid replacing warning contacts, replace brake pads no later than when pad thickness is over 2.5 mm (0.1 in). Replace warning contacts if wire core is worn. However, if only plastic part of warning contact is worn, there is no need to replace it.

- With vehicle on ground, loosen but do not remove wheel lug bolts.

- Raise car and support safely. Remove wheel lug bolts and remove wheel.

> **WARNING—**
> - Make sure the car is firmly supported on jack stands designed for the purpose. Place jack stands underneath structural chassis points, not under suspension parts.

Service and Maintenance 03-31

Under Car Maintenance

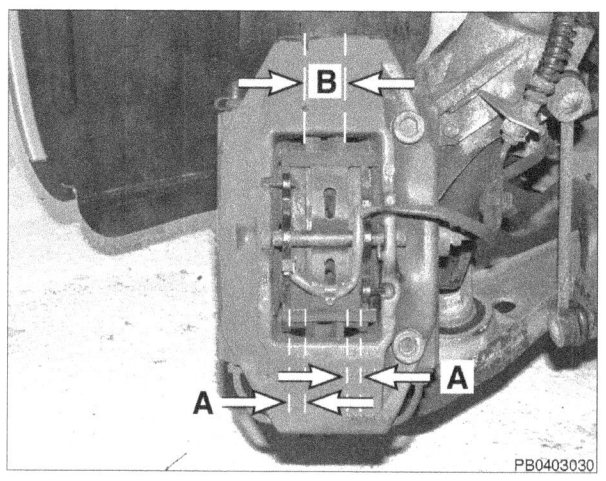

◄ Inspect brake pad thickness (**A**) and rotor thickness (**B**) for wear.

Brake wear specifications	
Brake pad wear limit (dimension **A**): • Pads only • Pads and wear indicator	2 mm (0.08 in) 2.5 mm (0.1 in)
Brake rotor wear limit (dimension **B**): • Front • Rear	22.6 mm (0.89 in) 18.6 mm (0.73 in)
Brake rotor machining limit (dimension **B**): • Front • Rear	22 mm (0.87 in) 18 mm (0.71 in)

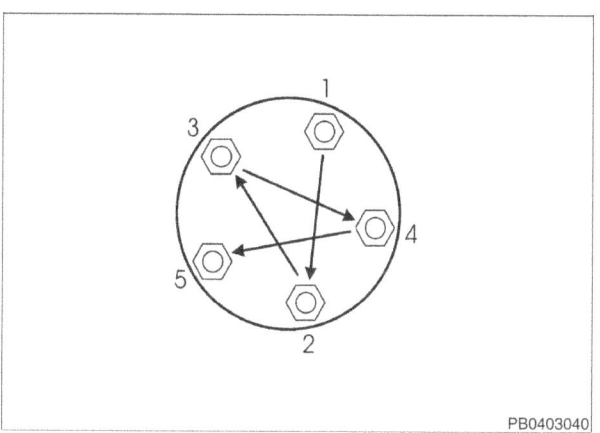

◄ Install wheel and tighten lug bolts by hand, evenly in a diagonal pattern.

– Repeat brake inspection procedure for each wheel.

– Lower vehicle to ground. Tighten lug bolts evenly to specification in a diagonal pattern.

Tightening torque	
Wheel to wheel hub	130 Nm (96 ft-lb)

Fuel filter, replacing (1997 - 2001 models)

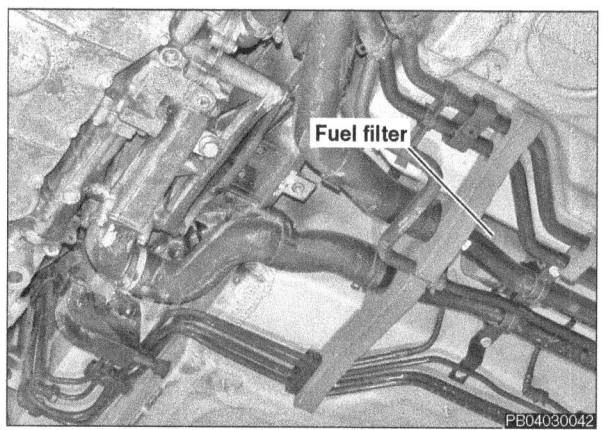

◄ 1997 - 2001 models: Fuel filter is located under center of car, approximately below seats, above engine coolant pipes. Remove large plastic underside protection cover to gain access to fuel filter.

2002 - 2004 models: Fuel filter is integrated with fuel pump and fuel level sender in fuel tank. It is specified by Porsche as "lifetime filter", owing to the reduced volume of fuel flowing through non-return system. No replacement interval is specified. See **24 Fuel Injection** for additional information.

– Loosen fuel filler cap.

– Raise car and support safely.

> **WARNING—**
> • Make sure car is firmly supported on jack stands designed for the purpose. Place jack stands underneath structural chassis points, not under suspension parts.

– Remove underside protection cover from below center of car.

– Remove ground wire from fuel filter (**A**).

03-32 Service and Maintenance

Under Car Maintenance

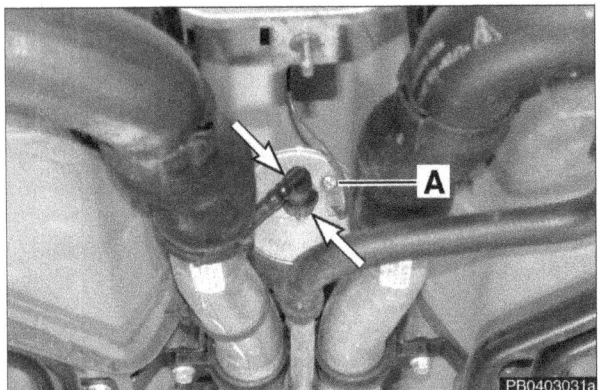

◄ Unclip fuel lines from filter by squeezing retaining collar on opposite sides (**arrows**). Be prepared to catch dripping fuel.

> **WARNING—**
> - *Gasoline is dangerous to your health. Wear suitable hand and skin protection when working on the fuel system. Do not breath fuel vapors. Always work in a well-ventilated area.*
> - *Fuel and fuel vapor is present when the fuel system is opened. Do not smoke or create sparks. Be aware of pilot lights in gas operated equipment (heating systems, water heaters, etc.). Have an approved fire extinguisher handy.*
> - *Gasoline fuel supply systems are designed to maintain system pressure after the engine is turned off. Fuel is expelled under pressure as fuel lines are disconnected. This can be a fire hazard, especially if the engine is warm. Always wrap a clean shop towel around fuel line fitting before loosening or disconnecting it.*

> **CAUTION—**
> - *Connect and disconnect fuel connections only in a straight line. Never use excessive force.*
> - *Protect fuel connections from dirt.*

- Undo fuel filter retaining strap and remove fuel filter toward rear.

- Replace fuel filter (noting direction of flow arrow on filter), and install in reverse order of removal. Reconnect ground wire.

> **WARNING—**
> - *Check that fuel line connections are properly locked in place after installation by pulling on them gently.*

Fluid lines and hoses, checking

- Check brake, fuel, power steering, coolant, oil and other fluid lines for leaks, damage, corrosion and loose connections.

Constant velocity (CV) joint boots, checking

Clean inner and outer CV joint (drive axle) boots prior to inspection.

- Raise car and support safely.

> **WARNING—**
> - *Make sure car is firmly supported on jack stands designed for the purpose. Place jack stands underneath structural chassis points, not under suspension parts.*

◄ Inspect inner and outer CV joint boots (**arrows**) on both drive axles. Press down on rubber boot with fingers to reveal hidden cracks.

- If a boot is found to have rips or holes, replace drive axle. See **42 Rear Suspension**.

Service and Maintenance 03-33

Exterior and Interior Body Maintenance

Suspension and steering components, checking

– Check ball joint dust boots for damage and correct seating. Check for play in ball joints.

– Check inner and outer tie rod end boots for damage and correct seating. Check for play by moving tie rods and wheels.

– Check attachment of ball joints and tie rod ends.

– If a rubber boot is found to have rips or holes, replace corresponding component. Dirt or moisture that has entered will destroy it. See **40 Front Suspension** or **42 Rear Suspension** for suspension component replacement.

– Check that suspension mountings (nuts, bolts, and fasteners) at front and rear are secure.

Exhaust system, checking

> *WARNING—*
> - *The exhaust system operates at extremely high temperatures. Do not touch the exhaust system while the engine is running.*
> - *Allow exhaust system to cool at least one hour before touching.*

– Inspect exhaust system for leaks or damage. Inspect exhaust system mounts. Replace faulty, missing or deteriorated parts.

Underbody, checking

Inspect underbody sealant for damage. Also check underbody, wheel housings and sill panels. Repair faults promptly.

EXTERIOR AND INTERIOR BODY MAINTENANCE

Airbags, inspecting

Boxster models are fitted with driver and passenger front airbags. 1998 and later models are fitted with side airbags in the doors as well.

> *WARNING—*
> - *Do not cover the padded airbag covers on the steering wheel and instrument panel nor affix any objects to them.*
> - *Do not apply chemical treatment to airbag unit covers. Clean with a dry or water moistened cloth only.*

– Inspect padded airbag unit covers on steering wheel and instrument panel and side airbag units for signs of external damage or excessive wear. Check with an authorized Porsche dealer if faults are found.

– Switch on ignition to perform a functional self-test of airbag warning light:
  - Airbag warning light illuminates for approx. 3 seconds.
  - If warning light does not light up, check bulb and power supply.

03-34 Service and Maintenance

Exterior and Interior Body Maintenance

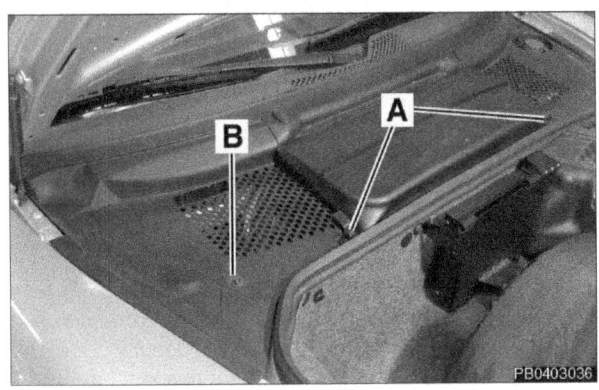

Ventilation microfilter, replacing

The passenger compartment particle air filter is located under the right cowl in the front trunk.

◂ Remove battery cover and right cowl cover:
- Twist left and right battery cover fasteners (**A**) 90° and lift off cover.
- Loosen cowl cover screw (**B**) and lift off cover.

NOTE—
- *The cowl cover is retained using an expanding rubber plug. If the expanding plug spins with the screw during removal, try pulling up on the cover while loosening the screw so that the plug is under tension.*

◂ Unclip (**A**) microfilter and remove (**B**).

– Insert new microfilter into housing. Check that filter is correctly seated, noting air flow direction on filter element.

Battery service

◂ The battery is under the center cowl in the front trunk. Under normal operating conditions, the battery is maintenance free.

Some batteries have a charge indicator (magic eye) on top that displays electrolyte level and charge condition.
- Green: Charge and electrolyte level are OK.
- Black: There is insufficient or no charge.
- Yellow or colorless: Electrolyte level is critically low. Top off with distilled water immediately.

NOTE—
- *If the battery age is in excess of 5 years and the charge indicators are colorless, replace battery.*
- *Air bubbles that occur normally during battery charging or during vehicle operation may adversely affect charge indicator reading. To obtain an accurate reading, gently tap the charge indicator with a screwdriver handle to displace air bubbles that have formed.*
- *Some batteries are equipped with sealing plugs with plastic foil. These batteries are considered maintenance free.*

– At high outside temperatures, check electrolyte level through translucent battery housing. Make sure electrolyte level is just above battery plates and separators.
- Remove filler caps to see battery plates. Electrolyte level should align with internal electrolyte level indicator (lip). This equates to external marking on battery case.

Service and Maintenance 03-35

Exterior and Interior Body Maintenance

- For a battery with removable electrolyte caps: If electrolyte level is low, replenish it by adding distilled water only. Distilled water prevents electrolyte impurities which cause self-discharging.

> **CAUTION—**
> - *DO NOT overfill battery. Overfilled batteries can boil over.*
> - *Too little electrolyte reduces the service life of the battery.*
> - *Extract excess electrolyte using a hydrometer.*
> - *Dispose of electrolyte (sulfuric acid) properly. Refer to local regulations pertaining to electrolyte disposal.*

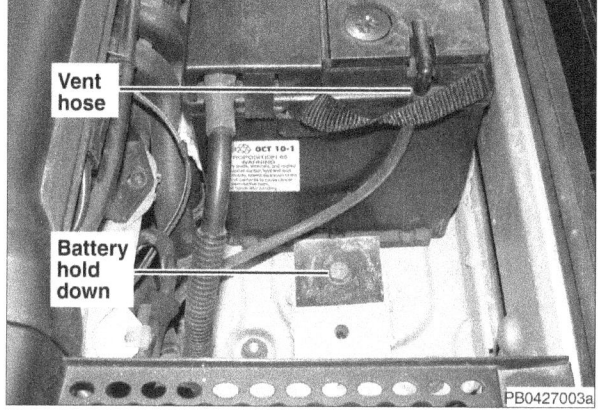

 During battery service, check the following:
- Battery securely bolted down.
- Battery cable clamping bolts tight.
- Battery vent hose correctly routed.
- Battery cell caps equipped with O-ring seals.

NOTE—
- *For battery testing, service and replacement, see* **27 Battery, Starter, Alternator**.
- *2001 and later cars: To open either of the electrically operated trunk lids on a car with a dead battery, see* **55 Trunk Lids**.

Battery notes

If the battery is disconnected and reconnected, the following systems or components may need to be reset or reintialized:
- Alarm siren
- Control module fault memories
- ECM adaptation
- Automatic transmission control module adaptation
- Power window motor limit position (standardization)
- Trip odometer and clock
- On-board computer
- Radio
- PCM

See **Battery disconnection notes** in **27 Battery, Alternator, Starter** for procedures necessary to reinitialize components.

Body exterior

Automobile finish is subject to deterioration from industrial fumes, corrosive road salt, acid rain and other damaging air-borne elements. Regular body finish care contributes to maintaining and preserving the exterior of your Porsche.

The best protection against environmental influences is frequent washing and waxing. How often this is required depends on the environment where the vehicle is used.

- Protect paint with a coat of hard wax at least twice a year.

- Check paint for chips and scratches. Touch up paint defects soon after they occur to prevent corrosion.

Service and Maintenance

Exterior and Interior Body Maintenance

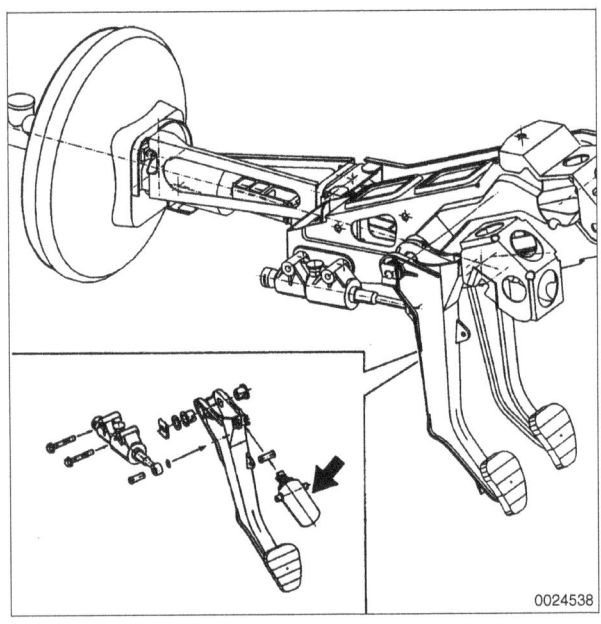

- After winter, thoroughly wash underside of vehicle.
- Keep exterior plastic and vinyl clean. Occasionally apply a colorless vinyl or leather preservative. Do not wax plastic or vinyl.

Clutch, checking

The clutch master cylinder has an internal stop. The clutch master cylinder push rod and clutch pedal are always pressed against the internal stop by the internal boost spring. Due to the automatic hydraulic adjustment of the clutch, it is not possible to check clutch play at the clutch pedal.

◀ For a quick functional check, pull pedal up toward driver seat. Make sure pedal does not move upward. If there is any play, check boost spring (**arrow**) or pedal assembly.

NOTE—

• *The clutch push rod and the boost spring are not adjustable.*

Proper clutch operation is dependent on:
• Correct bleeding of the clutch hydraulics. See **30 Clutch**.
• No leaks in hydraulic system.
• Pedal return to starting position.
• Proper installation position of pedals.

Door check straps and hinges, lubricating

◀ Lubricate door checks straps and door hinges periodically with a lubricant designed for door hinges. Clean old grease away before applying new lubricant.

Door locks, trunk lid locks and front lid safety hooks, checking

NOTE—

• *2001 and later cars: To open either of the electrically operated trunk lids on a car with a dead battery, see* **55 Trunk Lids***.*

- Check mounting bolt torques for door lock, front lid lock and rear lid lock, as well as retaining nuts for upper parts of locks of front and rear lids.

Tightening torque	
Door and lid lock fastening screws and retaining nuts	10 Nm (7 ft-lb)

Service and Maintenance 03-37

Exterior and Interior Body Maintenance

Door lock, inspection

– Check that door lock engages in two stages when doors are closed and disengages again when door handle is operated. Check operation inside and outside.

Front and rear trunk lid lock, inspection

– Check that lower lid locks engage by insertion of lock upper parts when lids (front and rear) are closed and disengage again when lid releases are pulled.

Front trunk lid safety hook, inspection

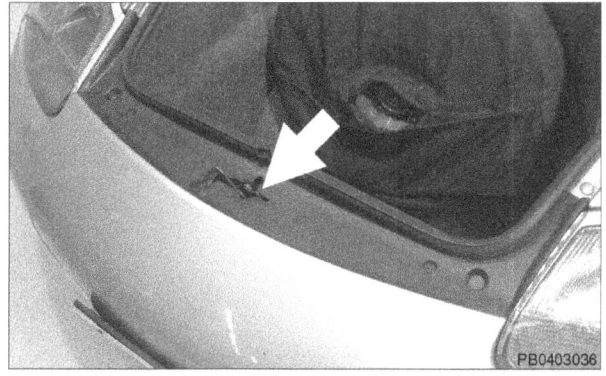

◂ Check that front trunk lid is held down by safety hook (**arrow**) after lid lock is opened. Check that safety hook engages in retaining plate at its lowest point.

– When lid is open, check that safety hook return spring loads hook toward base plate of lock upper part.

Emergency trunk release, checking (2002 and later models)

On 2002 and later models, an emergency trunk release lever was added to the interior of the rear trunk. On 2003 and later models, this feature was added to the front trunk.

> *WARNING—*
> - *If a person trapped in the front trunk unlocks the front lid when the vehicle is in motion, the locking hook of the lower part of lid lock opens completely at a speed of less than 3 kph (2 mph).*
> - *At a speed greater than 3 kph (2 mph), the locking hook of the lower part of lid lock does not unlock.*
> - *Stop the vehicle immediately if the trunk warning light in the instrument cluster illuminates.*

Checking this function requires a spare upper trunk lock (Porsche part no. 996 551 051 02) to simulate a closed trunk.

– Open rear trunk and engage upper trunk lock (spare part) into lower trunk lock.

– Operate entrapment lever and check that upper trunk lock automatically disengages from lower trunk lock.

Exterior lights, checking

– Check operation of headlights (high and low beam), marker lights, taillights, turn signal lights, brake lights, back-up lights and emergency flasher lights. See **94 Exterior Lights**.

Exterior and Interior Body Maintenance

— Use soapy water to clean exterior lights and plastic headlight lenses. Never use chemical cleaning agents. To avoid scratches, do not rub with dry or slightly damp cloths, paper towels or insect removal sponges.

Headlight washer nozzles are preset and do not need adjustment.

Light bulbs	Type, rating
Headlight, high and low beam	H7, 55 W
Front foglight	H7, 55 W
Rear foglight, turn signal light, back-up light, brake light	P 21 W
Taillight	R 5 W
Front parking light, side turn signal light, interior light	W 5 W
Third brake light	W 3 W
License plate light	C 5 W
Trunk light	K 10 W

NOTE—
- *Keep new bulbs clean and free from grease. Do not touch bulb surface with hands. Use a cloth or soft paper when replacing bulbs. Grease from fingerprints may lead to premature bulb failure.*
- *For bulb replacement, see* **94 Exterior Lights***.*

Headlights, adjusting

Headlight adjustment requires special equipment. See an authorized Porsche dealer or other qualified repair shop.

Interior lights, checking

— Check operation of indicator and instrument cluster warning bulbs. Check operation of interior cabin illumination bulbs. See **96 Interior Lights, Anti-theft**.

Parking brake, checking and adjusting

— Check to see if parking brake lever can be pulled up by more than 4 clicks without parking brake engaging. If so, adjust as follows.

— Loosen rear wheel lug bolts but do not remove.

— With parking brake disengaged, raise car and support safely. Remove rear wheels.

> *WARNING—*
> - *Make sure the car is stable and well supported at all times. Use a professional automotive lift or jack stands designed for the purpose. A floor jack is not adequate support.*

Service and Maintenance   03-39

Exterior and Interior Body Maintenance

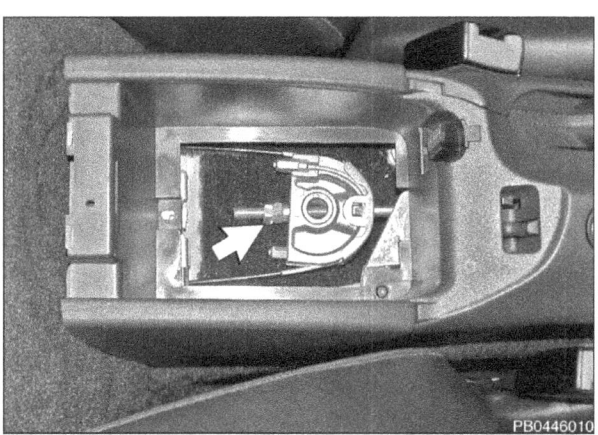

◀ Access parking brake turnbuckle. Open cover of center console storage compartment behind parking brake lever and remove rubber insert and bottom panel.

• Loosen adjusting nuts (**arrow**) on parking brake turnbuckle until cables are slack.

◀ Working at rear wheel, use screwdriver to advance parking brake adjuster by prying star wheel down through upper lug bolt bore until rotor can no longer be turned.

– Turn adjuster back 9 notches (5 notches until wheel can be turned freely, then another 4). Repeat procedure on opposite wheel.

– Pull up parking brake lever up by 2 clicks. Turn adjustment nut of turnbuckle until rear discs can be turned by hand with some resistance.

– Release parking brake lever and check that both wheels can be turned freely.

– Install parking brake lever trim panels and rear wheels.

– Lower vehicle to ground. Tighten lug bolts evenly to specification in a diagonal pattern.

Tightening torque	
Wheel to wheel hub	130 Nm (96 ft-lb)

Seat belts, checking

– Check that it is possible to pull out seat belt smoothly and evenly from belt retractor.

– Check that tongue of seat belt engages audibly in buckle.

– Check that abrupt pull on belt strap locks belt retractor.

– Inspect seat belt for damage. If cuts, fraying or seam tears are found, replace seat belt.

Exterior and Interior Body Maintenance

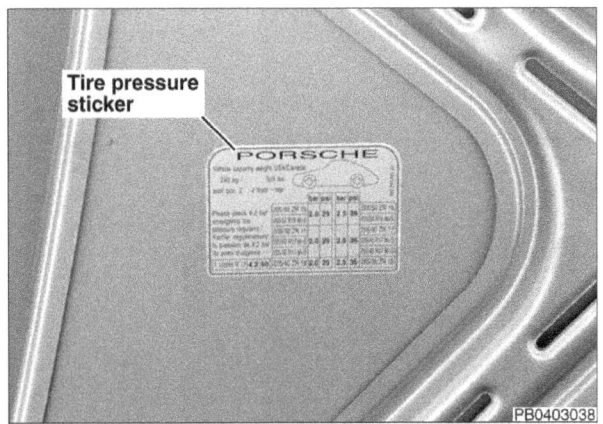

Tire service

◄ Check tire pressure on a regular basis when tires are cold. Refer to tire pressure sticker on underside of front trunk lid. Be sure to also check spare tire pressure.

– Check that all tires are the same type and tread pattern. Measure tread depth. If tread wear has exceeded minimum specification listed, replace tire.

Tire specifications	
Minimum tread depth	1.6 mm (1/16 in)

CAUTION—
- *Check wheel alignment after replacing tires to ensure maximum tire life.*

– Check tires for abnormal wear patterns. Tread wear pattern on front tires is an indication of whether the toe and camber settings need to be checked:
  - Feathering on tread: Incorrect toe.
  - Tread worn on one side: Incorrect camber.

NOTE—
- *When changing a wheel, use the threaded wheel centering stud (arrow) supplied with the spare tool kit.*

For more information on tires, see **44 Wheels, Tires and Alignment**.

Service and Maintenance

Exterior and Interior Body Maintenance

Test drive

- Upon completion of maintenance work, road test vehicle, checking the following:
  - Front seats
  - Brake pedal and parking brake free play:
    Brake pedal: max. ⅓ of overall pedal travel
    Parking brake lever: 4- 6 notches (clicks)
  - Brake operation and noise
  - ABS function
  - Engine
  - Steering play: Zero with engine running and vehicle on ground
  - Transmission
  - Parking Assistant (if equipped)
  - Cruise control
  - Traction control / PSM (if equipped)
  - Heating and air-conditioning system
  - Instruments
  - Radio operation, and reception
  - Vehicle handling and dynamics such as pulling to the side, cornering, vibrations in steering wheel and unusual noises

- Manual transmission car: Test the following:
  - Gearshift smooth operation
  - Clutch operation, smoothness and normal pedal pressure

- Automatic transmission car: Test the following:
  - Neutral safety switch: Make sure engine only starts in PARK or NEUTRAL.
  - Shift lock: Make sure shift lever does not move out of PARK or NEUTRAL unless foot brake is applied.
  - Kickdown: Depress accelerator pedal fully to floor while driving. Depending on vehicle and engine speed, upshift is either delayed or transmission shifts down into next lower gear.

10 Engine Removal and Installation

GENERAL 10-1	ENGINE AND TRANSMISSION MOUNTS 10-16
Warnings and Cautions.................. 10-1	Engine mount 10-16
ENGINE REMOVAL AND INSTALLATION...... 10-1	Transmission mount 10-17
Engine, removing 10-2	
Engine, installing........................ 10-11	

GENERAL

This repair group covers engine removal and installation.

The photos in this repair group illustrate work on a 1998 Boxster with 2.5 liter engine and manual transmission. The procedures are similar on other models.

Warnings and Cautions

WARNING—
- *Due to risk of personal injury, be sure the engine is cold before beginning the removal procedure.*

CAUTION—
- *Cover rear fender surfaces, painted areas and interior upholstery surrounding engine bay before beginning removal procedure.*
- *As an aid to installation, label all components, wires, and hoses before removing them.*
- *Do not reuse gaskets, O-rings or seals during reassembly.*

ENGINE REMOVAL AND INSTALLATION

To remove the engine, raise the car on an automotive lift and remove the transmission first. To do this, support the engine from above using a special engine support brace. Read the procedure through before starting the job.

NOTE—
- *Porsche recommends removing the engine and transmission as one unit, then separating the two.*

10-2 Engine Removal and Installation

Engine Removal and Installation

Engine, removing

◀ Move convertible top and convertible-top storage lid to service position. See **03 Service and Maintenance**.

– Remove left seat. See **72 Seats**.

– Open front and rear trunk lids.

– Disconnect negative (–) battery cable and cover battery terminal to keep cable from accidentally contacting terminal.

> **CAUTION—**
> - *Prior to disconnecting the battery, acquire the radio anti-theft code.*
> - *Read the battery disconnection cautions in **00 Warnings and Cautions**.*

– Remove carpeted panels from top and front engine access covers.

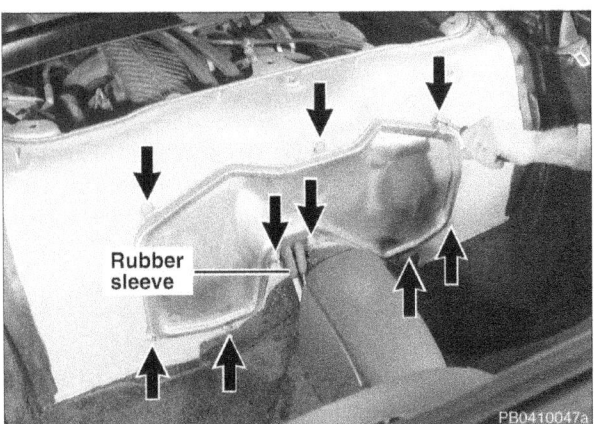

◀ Twist engine top cover hold-downs (**arrows**) and open cover.

◀ Undo engine access cover mounting fasteners at rear of passenger compartment (**arrows**) and lift out cover.

> **CAUTION—**
> - *Access cover edges are sharp.*

◀ Use 24 mm wrench to rotate engine drive belt tensioner clockwise. Slip belt off air-conditioning compressor (upper left pulley, upper right in photo).

> **NOTE—**
> - *If taking drive belt off, mark direction of rotation for reinstallation.*

Engine Removal and Installation 10-3

Engine Removal and Installation

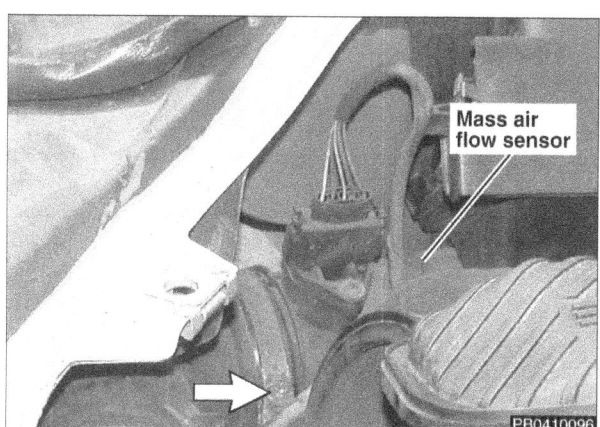

◁ Working at left rear of engine compartment:
- Detach mass air flow sensor harness connector behind air filter housing. Unhook harness from clip and place connector on engine.
- Loosen engine air intake duct clamp (**arrow**) at mass air flow sensor.

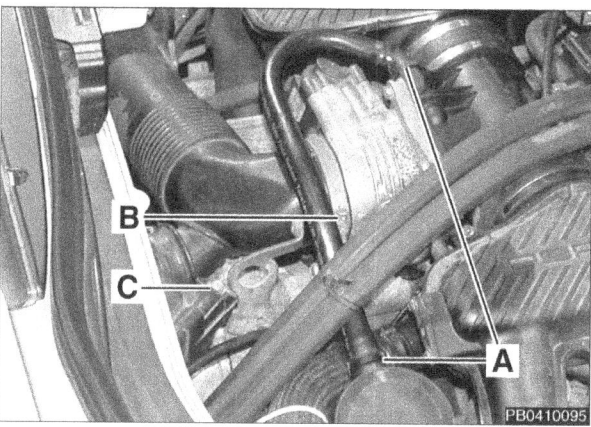

◁ Working on engine air intake ducts at rear of engine compartment:
- Squeeze plastic clamps (**A**) on crankcase vent duct to disconnect from oil separator and intake manifold. Remove duct.
- Loosen engine air intake duct clamp (**B**) at intake manifold.
- Remove air intake duct resonance chamber mounting bolt (**C**).
- Remove air intake duct from engine compartment.

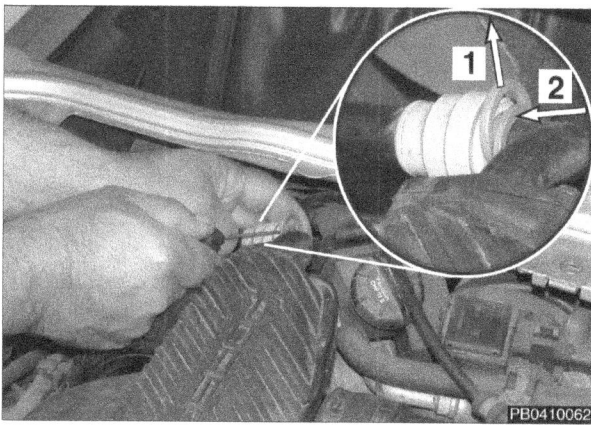

◁ Working at front of left intake manifold, remove brake booster vacuum line:
- Gently pry out grey plastic circlip (**1**).
- Press in plastic ring (**2**) and detach vacuum line from manifold fitting. Replace plastic circlip in groove to prevent loss.

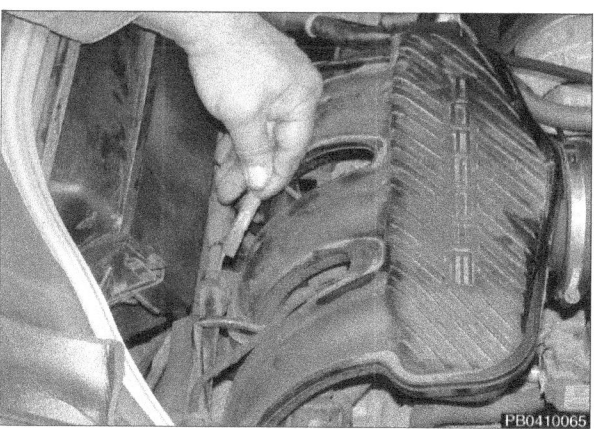

◁ Working at left front of engine, detach air-conditioning compressor clutch electrical connection:
- Cut wire tire holding connector to harness loom next to cylinder 4 and 5 intake runners.
- Pull connector apart.

10-4 Engine Removal and Installation

Engine Removal and Installation

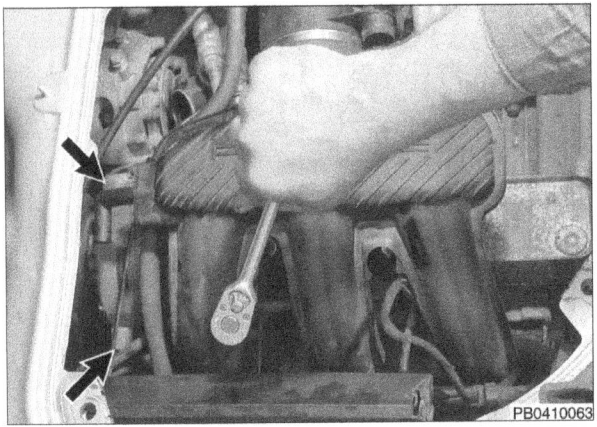

◄ Remove compressor mounting bolts (**arrows**) from above.

NOTE—
- Rear compressor mounting bolt is accessible between cylinders 4 and 5 intake runners.
- Compressor remains attached to refrigerant lines and stays in vehicle when the engine is removed. Remove compressor from engine when lowering engine.

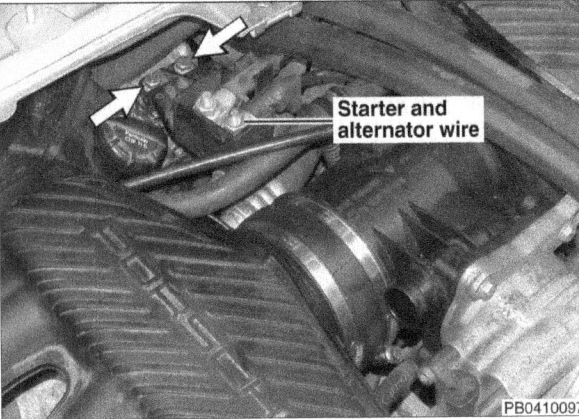

◄ Working at front center of engine:
- Flip open B+ junction box cover.
- Detach starter and alternator wire at B+ junction and set aside.
- Remove B+ junction mounting bolts (**arrows**) and move junction box aside.

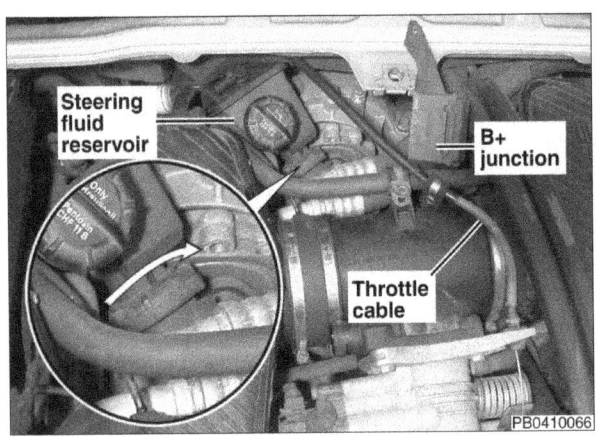

◄ Remove power steering reservoir:
- Loosen and move aside throttle cable, if applicable.
- Use clean syringe to siphon out Pentosin® fluid from power steering reservoir.
- Unscrew (**arrow**) bayonet lock at bottom of reservoir and lift off. Be prepared for dripping fluid.
- After removing reservoir, plug open steering fluid port.

CAUTION—
- Do not spill Pentosin® fluid. If coolant hoses come in contact with Pentosin® fluid, wash off immediately.

Engine Removal and Installation   10-5

Engine Removal and Installation

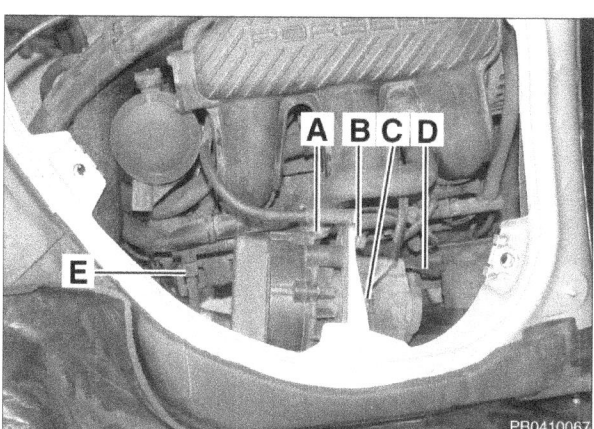

◀ Working at right side of engine compartment:
- Loosen secondary air pump hose clamp (**A**) and detach hose.
- Detach main engine ground cable lug (**B**) at body bracket.
- Disconnect secondary air pump (**C**) and engine compartment blower (**D**) harness connectors.
- Loosen gas cap, then disconnect fuel supply line (**E**) from right fuel rail.

> **WARNING—**
> - Before disconnecting the fuel hose, wrap a shop cloth around the hose to absorb any leaking fuel.
> - Fuel will be expelled under pressure. Be prepared to catch dripping fuel.
> - Do not smoke or work near heaters or other fire hazards.
> - Keep a fire extinguisher handy.
> - Immediately plug open fuel hoses.

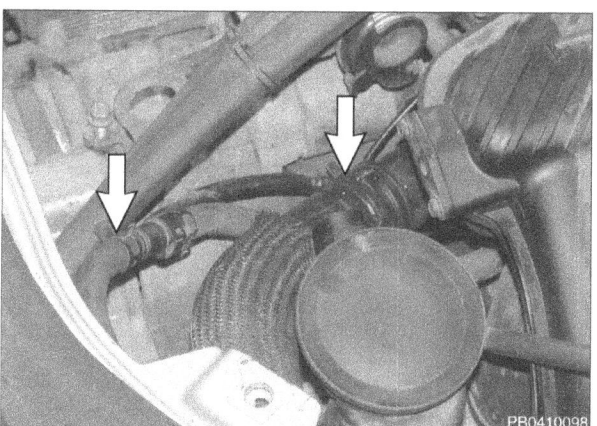

◀ Working at rear of engine compartment:
- Loosen oil filler neck and oil cooler coolant line clamps (**arrows**).
- Detach oil cooler line.

> **CAUTION—**
> - Be prepared to catch dripping coolant.

> **NOTE—**
> - Oil filler neck removal is covered later in this procedure.

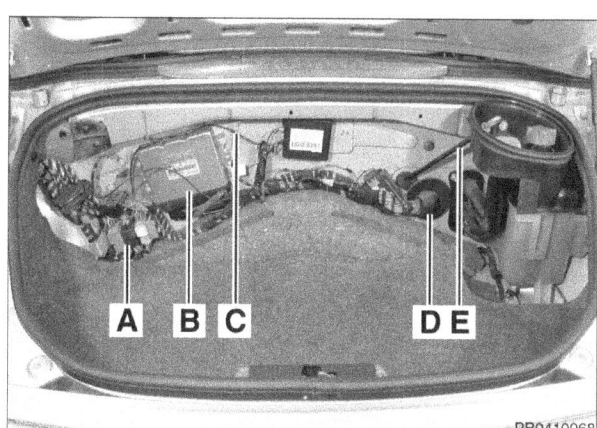

◀ Open rear trunk and remove trunk bulkhead trim panel to expose electrical harnesses and components:
- **A** Multiple harness connectors
- **B** Engine control module (ECM)
- **C** DME main ground
- **D** Engine harness grommet
- **E** Dipstick tube

◀ Detach harnesses from multiple connectors:
- Pull out locking clip (**arrow**).
- Pull harness plug out of connector.

10-6 Engine Removal and Installation

Engine Removal and Installation

◄ Detach ECM harness connector:
- Remove mounting nuts and remove ECM from bulkhead.
- Swing harness locking clip (**arrow**) away from body of module.
- Tilt harness plug to disconnect from module.
- Store module in safe place.

NOTE—
- *The ECM in some models may have harness connector(s) with different configuration.*

— Automatic transmission: Detach Tiptronic control module harness connector:
- Remove mounting nuts and remove control module from bulkhead.
- Detach harness connector from module.
- Store module in safe place.

— Disconnect DME and Tiptronic (if applicable) ground wire.

◄ Pull out dipstick. Use needle nose pliers to press together dipstick guide tube tabs and detach guide tube.

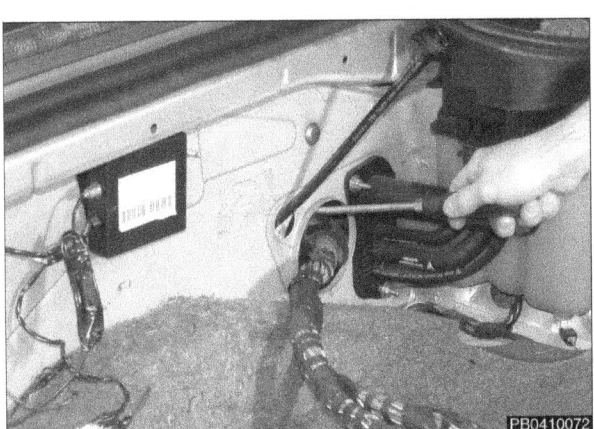

◄ Push dipstick guide and engine harness loom grommets into engine compartment.

CAUTION—
- *Take care to not pinch or kink electrical wires during disassembly or reassembly.*

◄ Place engine support brace (Porsche special tool 9591/1 or equivalent) across rear of engine bay and attach supporting chain to eye at rear of engine.

— Raise vehicle and support safely.

CAUTION—
- *Make sure the car is stable and well supported at all times. A floor jack is not adequate support.*

NOTE—
- *Make sure placement of lift pods or jack stands do not interfere with removal of rear diagonal brace fasteners or underbody splash shields.*

Engine Removal and Installation 10-7

Engine Removal and Installation

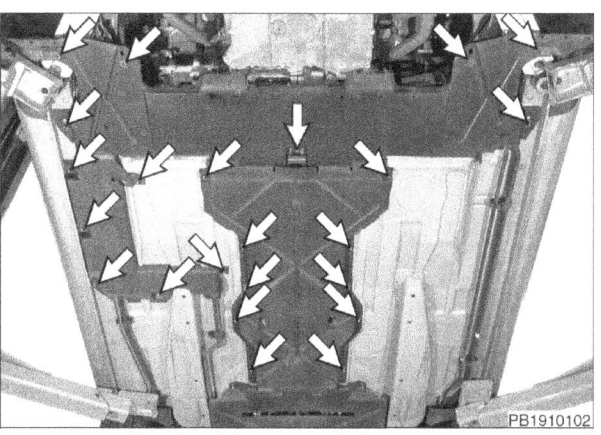

◀ Working underneath vehicle, remove plastic splash shield fasteners (**arrows**) from under engine, center tunnel and left rocker panel. Remove shields.

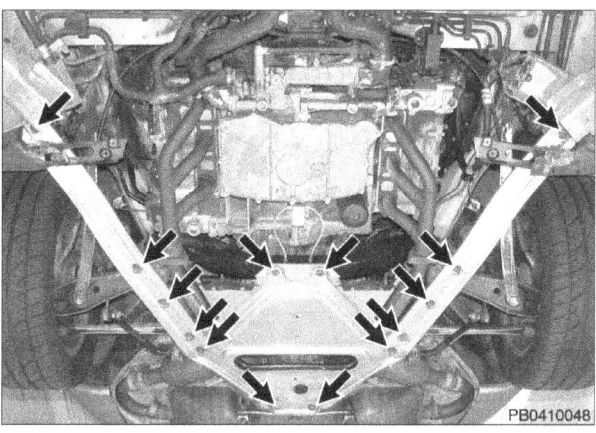

◀ Remove diagonal brace and reinforcement plate mounting fasteners (**arrows**). Set braces and plate aside.

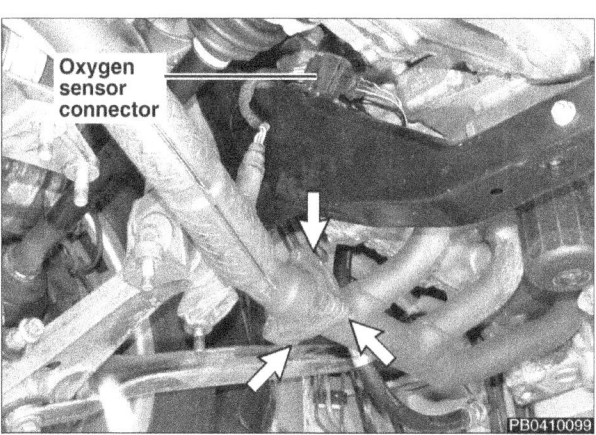

◀ On each side of engine, separate exhaust pipe from front exhaust manifold or catalytic converter at triangular flange by removing mounting fasteners (**arrows**).

– Separate oxygen sensor connectors at connecting pipes and also at rear of catalytic converters.

> *CAUTION—*
> * *Place protective sleeves over oxygen sensor connectors.*

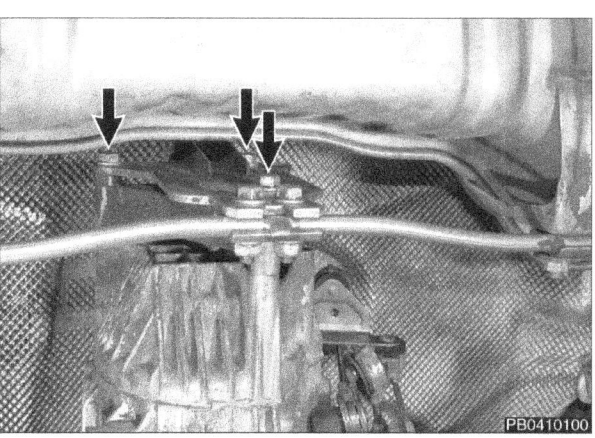

◀ Support exhaust system with tall jack stands. Working between muffler and transmission, remove exhaust bracket mounting bolts (**arrows**) from rear of transmission. With an assistant, carefully lower exhaust system, rocking it from side to side to clear the suspension.

> *CAUTION—*
> * *Store exhaust system in a safe place to prevent damage to oxygen sensors.*

– Remove transmission. See **34 Manual Transmission** or **37 Automatic Transmission**.

10-8 Engine Removal and Installation

Engine Removal and Installation

◀ Working underneath engine, detach passenger safety cable between engine and rear axle support. To avoid twisting cable, counterhold fastener with suitable wrench.

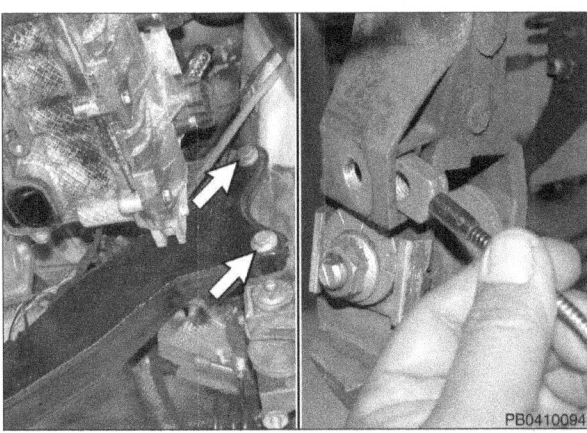

◀ Working at rear suspension arm mounts, remove rear axle support mounting bolts (**arrows**). Place support aside. Use magnet to remove trapped mounting nuts and store safely.

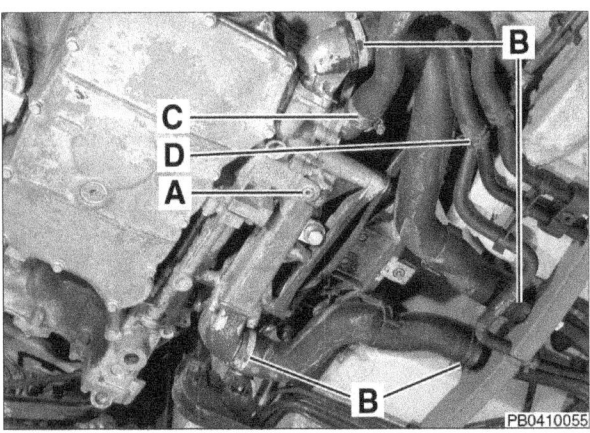

◀ Drain engine coolant:
- Loosen cap at coolant reservoir.
- Place 5-gallon pail under front of engine.
- Remove coolant manifold drain plug (**A**) at front of engine and drain engine coolant.
- Loosen radiator hose clamps (**B**). Remove radiator hoses to allow access to front engine mounting yoke. Plug openings.
- Loosen coolant filler hose clamp (**C**). Detach hose, allow coolant to drain. Plug openings.
- Loosen left heater hose clamp (**D**). Disconnect heater hose, then plug openings.

> *WARNING—*
> * Automotive antifreeze (coolant) is poisonous and lethal, specially to pets. Clean up spills immediately and rinse area with water. Dispose of coolant safely.*

◀ Working underneath left side of car:
- Remove small plastic splash shield.
- Disconnect power steering pressure line. Be sure to counterhold brass fitting while unscrewing pressure nut.

NOTE—
* Disconnect power steering return line as shown in the next 2 steps.*

Engine Removal and Installation   10-9

Engine Removal and Installation

◄ Pry steering line and brake vacuum hose retaining clip (**arrow**) from supporting bracket above left trailing arm.

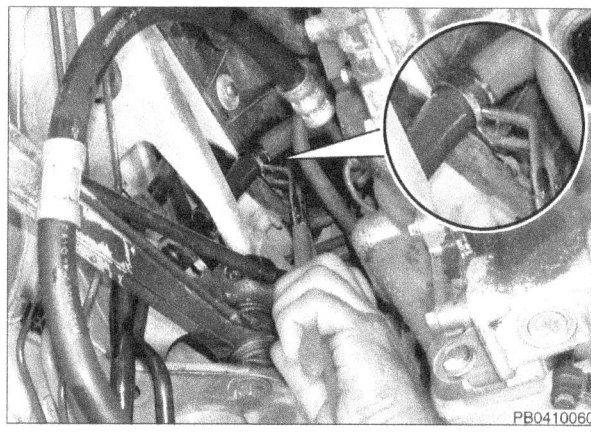

◄ Loosen power steering return line junction clamp (**inset**). Separate line at junction.

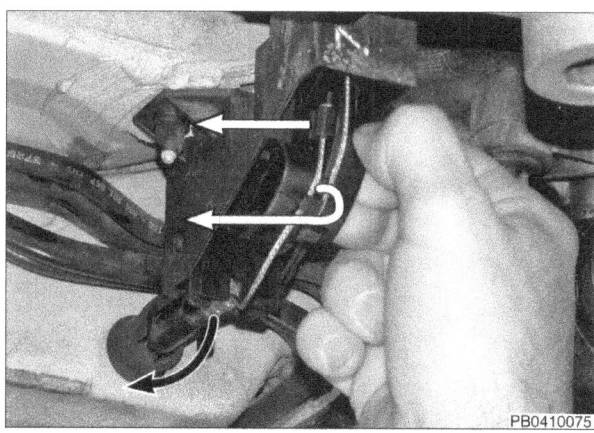

◄ Remove accelerator cable linkage cover. Unhook accelerator cable at bellcrank (**white arrows**) and detach cable sheath (**black arrow**) from linkage assembly.

◄ Remove 2 plastic nuts (**arrows**) to detach accelerator cable linkage assembly from chassis.

NOTE—
- *The accelerator linkage assembly remains with the engine during engine removal. Tie securely to engine to prevent damage.*

10-10 Engine Removal and Installation

Engine Removal and Installation

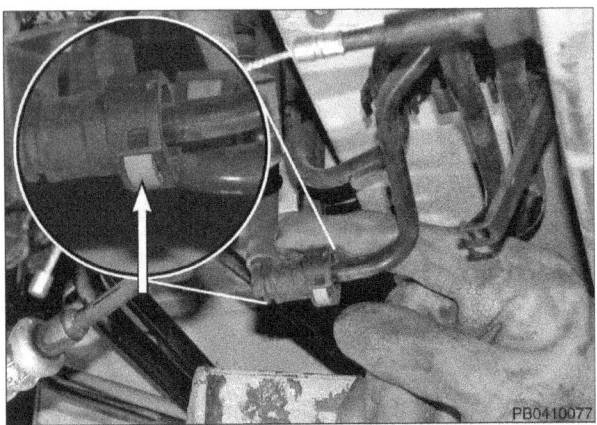

◀ Detach engine fuel return and fuel tank vent lines.
- Release lines from bracket on body.
- Separate by pressing gently on button (**arrow**).
- When separating fuel line, be prepared to catch dripping fuel.

> **WARNING—**
> - *Before disconnecting the fuel hose, wrap a shop cloth around the hose to absorb any leaking fuel.*
> - *Fuel will be expelled under pressure. Be prepared to catch dripping fuel.*
> - *Do not smoke or work near heaters or other fire hazards.*
> - *Keep a fire extinguisher handy.*
> - *Immediately plug open fuel hoses.*

◀ Working underneath right of engine, use large wooden pry bar to force oil filler neck (**arrow**) off engine oil tube.

NOTE—
- *If necessary, use hot air blower to heat hard plastic filler pipe and pull off oil tube.*

◀ Position support jack under engine. Alternatively, place heavy duty roll-away cart underneath engine assembly.

– Raise jack or lower car to relieve load on front engine yoke.

◀ Remove front engine yoke fasteners (**arrows**).

– Unhook engine support brace (Porsche special tool 9591/1) from top rear of engine and lower engine assembly or raise vehicle.

NOTE—
- *Before lowering engine or raising car fully, be sure to disengage A/C compressor. See next step.*

Engine Removal and Installation 10-11

Engine Removal and Installation

◁ Working inside left side passenger compartment, lift out air-conditioning compressor and set aside. Be sure to place blanket or other suitable protective pad underneath compressor.
- Guide engine wiring harness and dipstick guide carefully out of trunk bulkhead.
- Check for wiring, fuel hoses or mechanical parts that might become snagged as engine is lowered.

> **CAUTION—**
> - *If vehicle is to be placed on its wheels and moved, reinstall and torque rear axles, rear axle support and diagonal body braces.*

Engine, installing

While engine is out of vehicle, remove clutch and flywheel and renew as necessary. See **30 Clutch**.

◁ Place engine assembly under engine bay. Lower car or raise engine.

– Position engine in engine bay and raise carefully and in steps:
- Refit air-conditioning compressor as engine is raised.
- Thread engine wiring harness and dipstick guide tube through trunk bulkhead before final installation position is reached.
- Install engine yoke fasteners and tighten.

Tightening torque	
Engine yoke to body	65 Nm (48 ft-lb)

◁ Place engine support brace (Porsche special tool 9591/1 or equivalent) across rear of engine bay and attach supporting chain to eye at rear of engine.

– Remove engine support jack or roll-away cart.

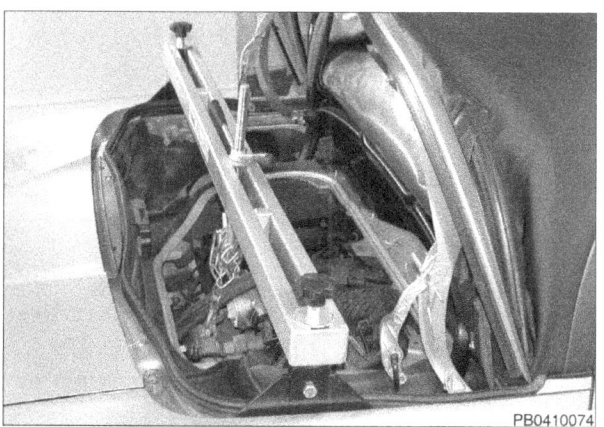

◁ Working underneath engine, reinstall rear axle support and torque bolts (**arrows**).

Tightening torque	
Rear axle support to rear suspension arm mounts	65 Nm (48 ft-lb)

– Reattach passenger safety cable between engine and rear axle support. Counterhold fastener with suitable wrench.

Tightening torque	
Passenger safety cable to rear suspension support	23 Nm (17 ft-lb)

Engine Removal and Installation

- Reinstall transmission. See **34 Manual Transmission** or **37 Automatic Transmission**.

- Reinstall rear suspension reinforcement plate support.

Tightening torque	
Reinforcement plate support to body	65 Nm (48 ft-lb)

- Reinstall exhaust system using new gaskets. See **26 Exhaust System**.

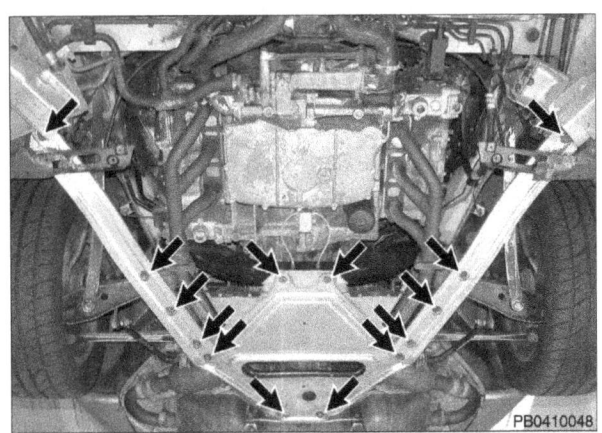

◀ Reinstall diagonal braces and rear suspension reinforcement plate. Torque fasteners (**arrows**).

Tightening torques	
Reinforcement plate to rear suspension arm mount:	
• Nuts	65 Nm (48 ft-lb)
• Bolts	46 Nm (34 ft-lb)
Diagonal brace to body and to rear suspension arm mounts	65 Nm (48 ft-lb)

- Working inside rear trunk, reconnect engine harness multiple connectors, engine control module (ECM) connector and Tiptronic (if applicable) control module connector. Reattach engine harness grounds to trunk bulkhead. Refit engine harness grommet in trunk bulkhead.

- Reattach oil dipstick guide tube. Refit guide tube grommet in trunk bulkhead.

- Reinstall trunk trim liner.

- Working at left front of engine, reinstall air-conditioning compressor mounting bolts and tighten.

Tightening torque	
A/C compressor to engine	28 Nm (21 ft-lb)

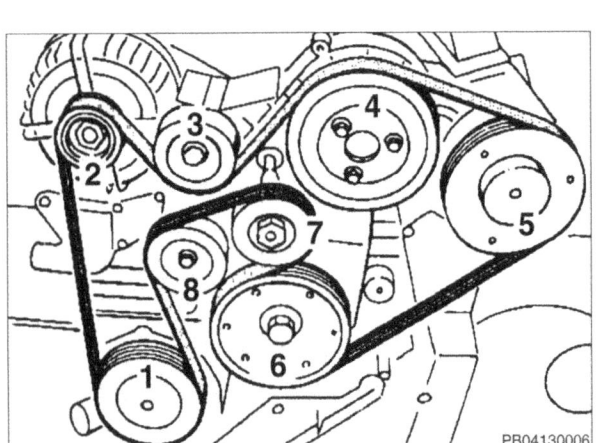

◀ Check engine drive belt for damage and replace if necessary. If reusing belt, fit in original direction of rotation. Place on pulleys in the following order:

1. Coolant pump
2. Alternator
3. Upper idler
4. Power steering pump
5. Air-conditioning compressor
6. Crankshaft pulley
7. Tensioner
8. Lower idler: Use 24 mm wrench to turn tensioner clockwise and fit drive belt.

NOTE—
- *After installation, check to make sure belt is correctly positioned on all pulleys.*

Engine Removal and Installation 10-13

Engine Removal and Installation

- Working at left side of engine compartment:
  - Reattach power brake booster vacuum line. Check that line is fully seated by pulling on it gently.
  - Check power steering fluid reservoir sealing O-rings and replace, if necessary. Reinstall reservoir.
  - Reattach starter and alternator wire to B+ junction.

Tightening torque	
Starter and alternator wire to B+ junction	10 Nm (7 ft-lb)

- Reinstall mass air flow sensor connector and engine air intake duct. Use new hose clamps as needed.

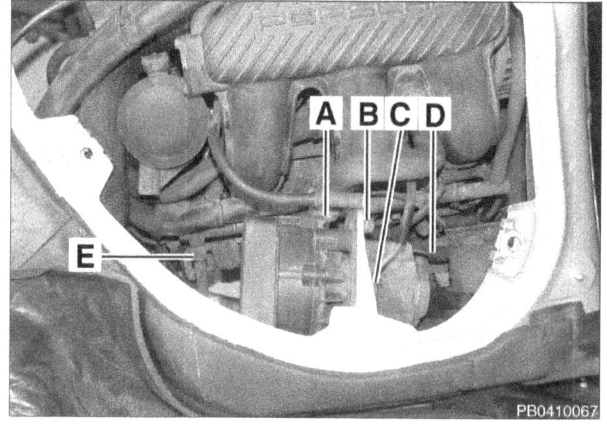

◄ Working at right side of engine compartment:
  - Reattach secondary air pump hose (**A**).
  - Reattach main engine ground cable lug (**B**) to body bracket.
  - Reconnect secondary air pump (**C**) and engine compartment blower (**D**) harness connectors.
  - Reconnect fuel supply line (**E**) to right fuel rail.

Tightening torque	
Fuel supply line to engine fuel rail	30 Nm (22 ft-lb)
Ground cable to engine	23 Nm (17 ft-lb)

- Reattach engine oil filler duct and oil cooler coolant line.

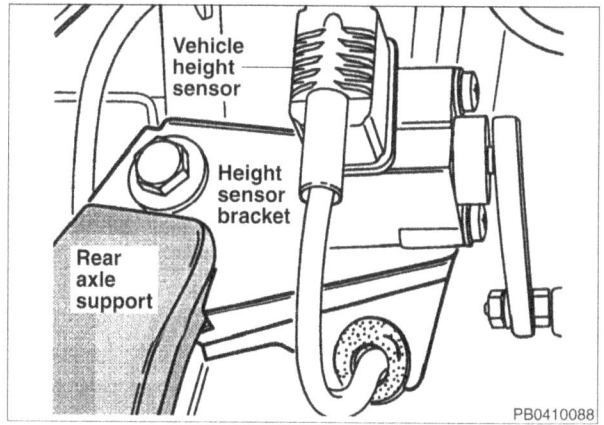

◄ Vehicle with automatic headlight beam adjustment (Litronic): Make sure bottom of vehicle height sensor bracket is in contact with rear axle support when support fasteners are tightened.

◄ Reattach accelerator cable linkage assembly to body. Refit accelerator cable to linkage bellcrank and reinstall cover.

Engine Removal and Installation

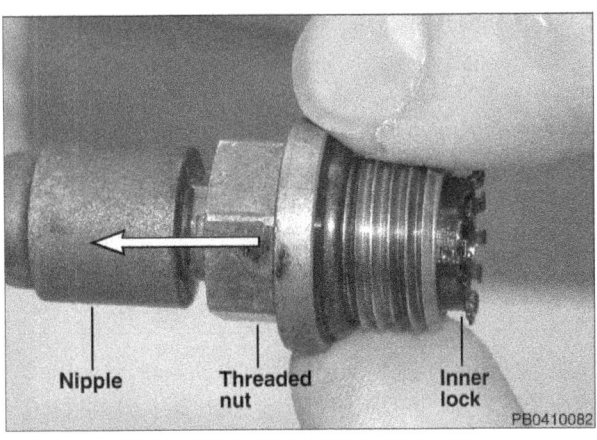

◀ Reconnect power steering fluid pressure line:
  • Hold pressure line from engine.
  • Press threaded nut toward nipple.

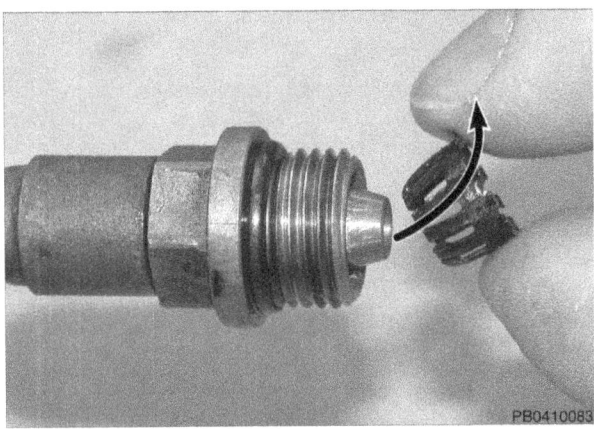

◀ Tilt inner lock and pull to remove from nipple.

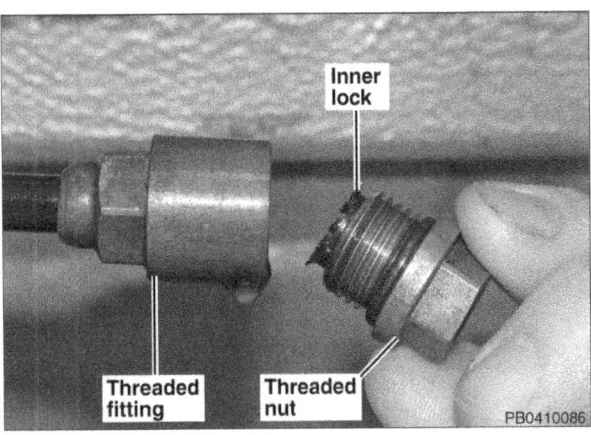

◀ Pull threaded nut off nipple and place inner lock loosely inside. Thread assembly into threaded fitting underneath car.

– Tighten assembly. Be sure to counterhold threaded fitting.

Tightening torque	
Steering pressure line to fitting (15 mm wrench size)	30 Nm (22 ft-lb)

– Press nipple at end of power steering pressure line into fitting assembly until it snaps in place.

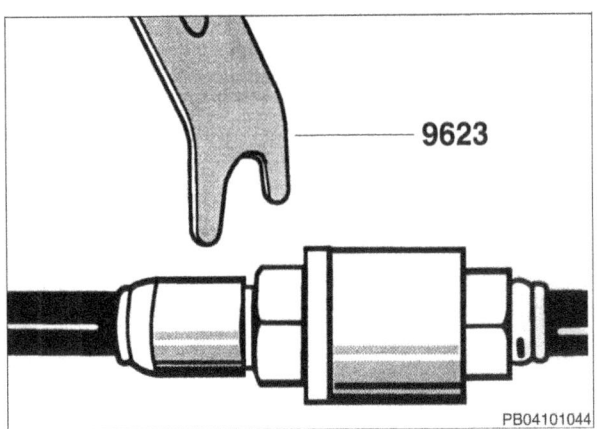

◀ Test fitting: Insert Porsche special tool 9623 in fitting slot and tilt slightly to make sure joint does not slide apart.

– Reconnect power steering return line and clamp properly.

– Replace brake booster line and power steering return line retaining clip.

– Reconnect fuel return line and fuel tank vent line. Make sure lines are not kinked.

WARNING—
• Check that fuel line connection is locked in place after installation by pulling on it gently.

Engine Removal and Installation 10-15

Engine Removal and Installation

- Reattach coolant hoses. Wet coolant hoses with water for ease of installation.

> **CAUTION—**
> - Do not apply lubricant to coolant hoses.

- Reinstall coolant drain plug with new seal.

Tightening torque	
Coolant drain plug to coolant manifold	10 - 15 Nm (7 - 11 ft-lb)

- Reinstall underside splash shields and right and left wheel housing trim panels.

◁ Reinstall front engine cover and torque bolts.

Tightening torque	
Engine access cover to body	10 Nm (7 ft-lb)

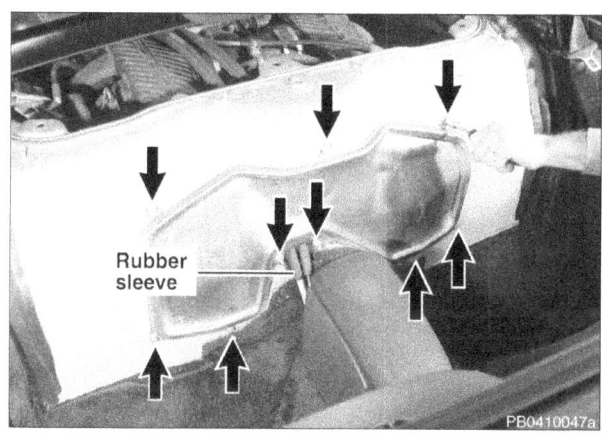

- Prior to reinstalling interior trim cover, make sure rubber sleeve over B+ lead and shift cables is located correctly.

- Remainder of installation is reverse of removal. Remember to:
  - Top up engine oil.
  - Top up and bleed cooling system. See **19 Cooling System**.
  - Top up power steering system with Pentosin® CHF 11 S fluid and bleed. See **48 Steering**.
  - Reinstall engine top cover and trim.

> **CAUTION—**
> - Do not overfill power steering reservoir. Do not spill Pentosin® fluid. If coolant hoses come in contact with Pentosin® fluid, wash off immediately.

- Reconnect negative battery terminal and tighten.

- Reinstall seat.

Tightening torque	
Seat to floor	65 Nm (48 ft-lb)

> **NOTE—**
> - Use Porsche System Tester 2 (PST2) to:
>   - Read out and clear DME and Tiptronic fault codes.
>   - Readjust headlight alignment and recalibrate Litronic control module.
> - See **27 Battery, Starter, Alternator** for **Battery disconnection notes**.

- Carry out test drive. After test drive, check that operating fluids are at correct level. Check engine and hoses for leaks.

10-16 Engine Removal and Installation

Engine and Transmission Mounts

ENGINE AND TRANSMISSION MOUNTS

The engine and transmission assembly are supported at three points:
- Rubber front engine mount
- Two hydraulic transmission mounts

Engine mount

◂ Front engine mount yoke is mounted to body via 4 fasteners (**arrows**).

◂ Front yoke is attached to engine mount via 2 long bolts (**arrows**).

◂ Engine mount is attached to engine via 3 bolts (**arrows**).

NOTE—
- *When reinstalling yoke to mount, locate rubber stops correctly.*

Tightening torques	
Engine mount to engine (M10)	46 Nm (34 ft-lb)
Engine yoke to body (M10)	65 Nm (48 ft-lb)
Engine yoke to engine mount (M10)	65 Nm (48 ft-lb)

Engine Removal and Installation 10-17

Engine and Transmission Mounts

Transmission mount

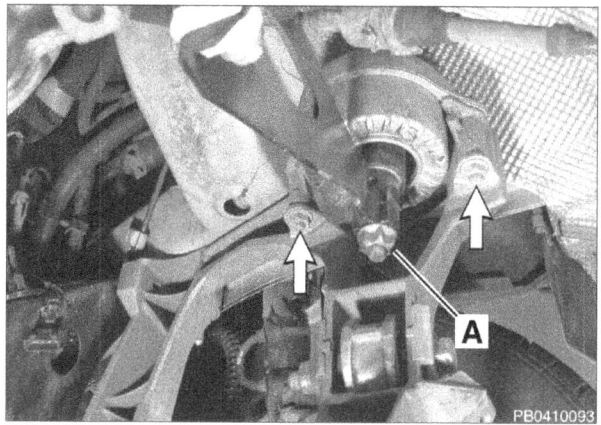

◄ Transmission hydraulic mount is attached to rear frame via 2 fasteners (**arrows**).

> *CAUTION—*
> - *Do not remove Torx nut (**A**) which attaches transmission mount to bracket. This destroys hydraulic mount.*

◄ Transmission mount bracket is attached to transmission via 2 bolts (**arrows**).

Tightening torques	
Transmission mount bracket to transmission • Boxster (M10 x 1.5) • Boxster S engine (M12 x 1.5)	65 Nm (48 ft-lb) 85 Nm (63 ft-lb)
Transmission mount to rear frame (M8)	33 Nm (24 ft-lb)

13 Engine Pulleys and Seals

GENERAL 13-1
CRANKSHAFT PULLEY 13-1
    Pulley system 13-1
    Idler and tensioner pulley components 13-2
    Crankshaft pulley, removing and installing 13-2

CRANKSHAFT SEALS 13-4
    Crankshaft pulley seal, removing and installing . 13-4
    Rear main seal, removing and installing 13-5

GENERAL

This repair group covers engine pulley removal and installation. Also covered is front and rear crankshaft seal replacement.

Engine drive belt service is covered in **03 Service and Maintenance**.

CRANKSHAFT PULLEY

Pulley system

 The multi-ribbed (serpentine) engine drive belt powers the air-conditioning compressor, alternator, coolant pump and power steering fluid pump. Idler and tensioner pulleys maintain correct belt routing and tension.

1. Alternator
2. Upper idler pulley
3. Tensioner pulley
4. Power steering pump
5. Air-conditioning compressor
6. Coolant pump
7. Lower idler pulley
8. Crankshaft

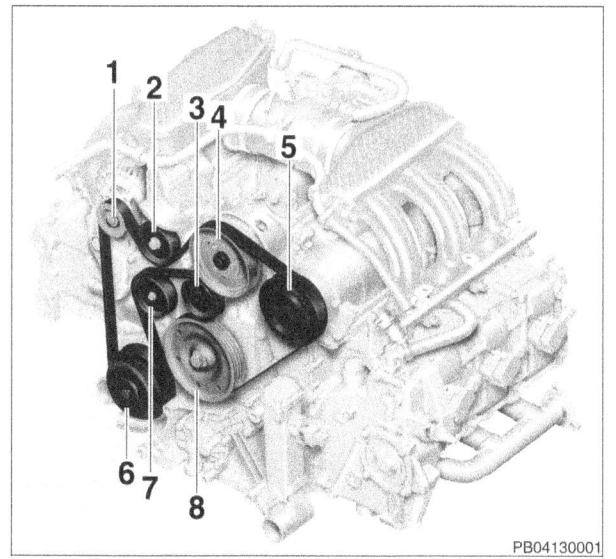

13-2 Engine Pulleys and Seals

Pulley system

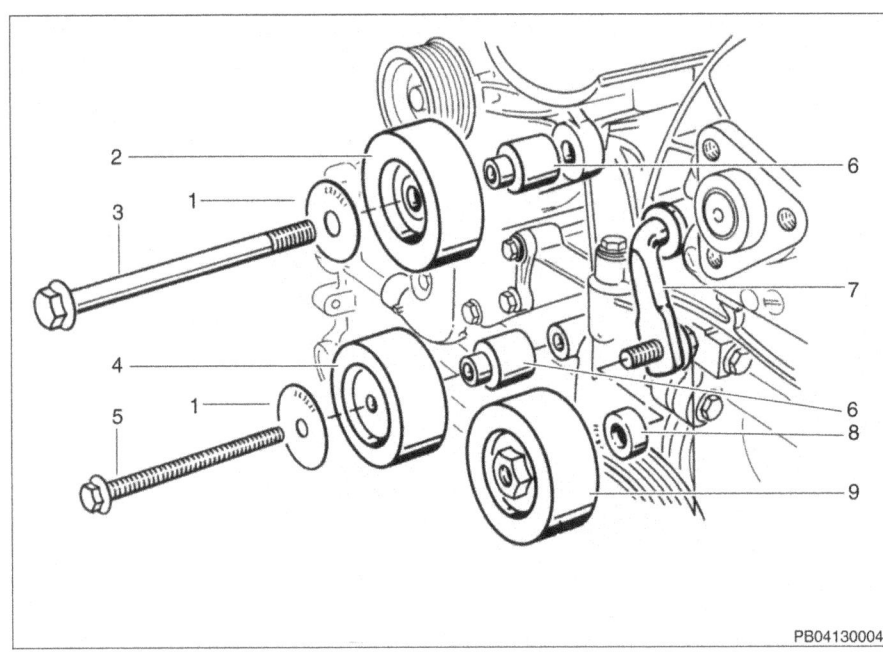

Idler and tensioner pulley components

1. **Cover plate**
   - dished side (with writing) faces bolt head
2. **Upper idler**
3. **M10 x 145 mm bolt**
   - tighten to 65 Nm (48 ft-lb)
4. **Lower idler**
5. **M8 x 55 mm bolt**
   - tighten to 23 Nm (17 ft-lb)
6. **Spacer pin**
7. **Tensioner spring arm**
8. **Spacer**
9. **Tensioner**

Crankshaft pulley, removing and installing

The crankshaft pulley is mounted to the crankshaft with a large amount of force. Be sure to have the Porsche special counterholding tool 9593 or equivalent available before starting the job.

- Remove left seat. See **72 Seats**.
- Remove trim cover behind seats from front engine access cover.

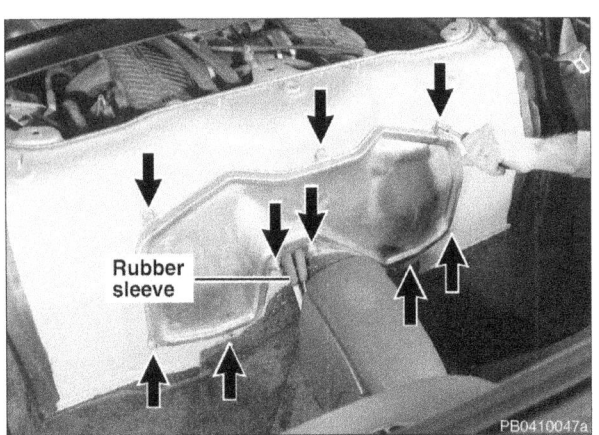

 Undo engine access cover mounting fasteners at rear of passenger compartment (**arrows**) and lift out cover.

> **CAUTION—**
> - Access cover edges are sharp.

 Use 24 mm wrench to rotate engine drive belt tensioner clockwise. Slip belt off pulleys.

> **NOTE—**
> - If reusing belt, mark direction of rotation before removing.

Engine Pulleys and Seals 13-3

Pulley system

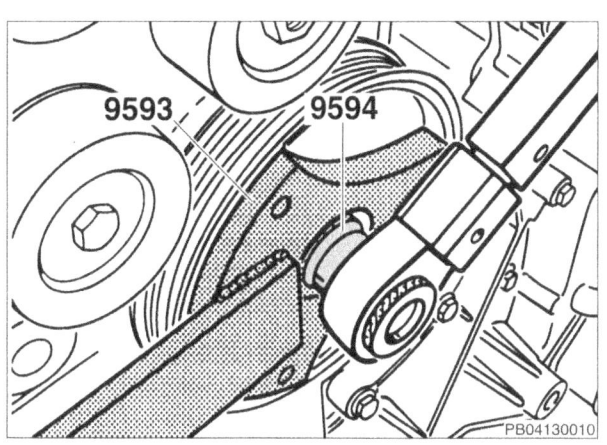

◀ Loosen crankshaft pulley:
- Counterhold pulley with Porsche special tool 9593 (spanner)
- Remove pulley bolt with short 24 mm socket (Porsche special tool 9594)

◀ Reinstall crankshaft pulley. Counterhold pulley and torque bolt in 2 stages.

Tightening torque	
Pulley to crankshaft (M16 x 1.5 x 60 bolt)	
• Stage 1	50 Nm (37 ft-lb)
• Stage 2	90° additional

– Inspect belt prior to installation and remove contaminants from grooves if necessary. Also inspect pulleys for contaminants in grooves and for smooth rotation. Replace components as necessary.

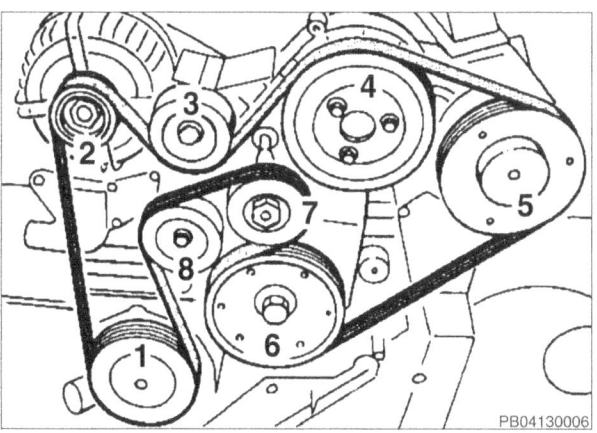

◀ Install drive belt on pulleys in the following order:
1. Coolant pump
2. Alternator
3. Upper idler
4. Power steering pump
5. Air-conditioning compressor
6. Crankshaft pulley
7. Tensioner
8. Lower idler: Use 24 mm wrench to turn tensioner clockwise and fit drive belt.

NOTE—
- *If reinstalling used belt, fit in original direction of rotation.*
- *After installation, check to make sure belt is correctly positioned on all pulleys.*

– Start engine and allow to idle for a few minutes. Shut engine OFF and recheck belt tension and location on pulleys.

– Refit engine access cover.

Tightening torque	
Engine access cover to body	10 Nm (7 ft-lb)

13-4 Engine Pulleys and Seals

Crankshaft Seals

CRANKSHAFT SEALS

Crankshaft pulley seal, removing and installing

- Disconnect negative (–) battery cable and cover battery terminal to keep cable from accidentally contacting terminal.

> **CAUTION—**
> - *Prior to disconnecting the battery, acquire the radio anti-theft code.*
> - *Read the battery disconnection cautions in **00 Warnings and Cautions**.*

- Remove crankshaft pulley. See **Crankshaft pulley, removing and installing** in this repair group.

◄ Use right angle drill to make 3 mm hole in face of crankshaft seal.

> **NOTE—**
> - *Drilling operation is illustrated with engine removed from vehicle.*

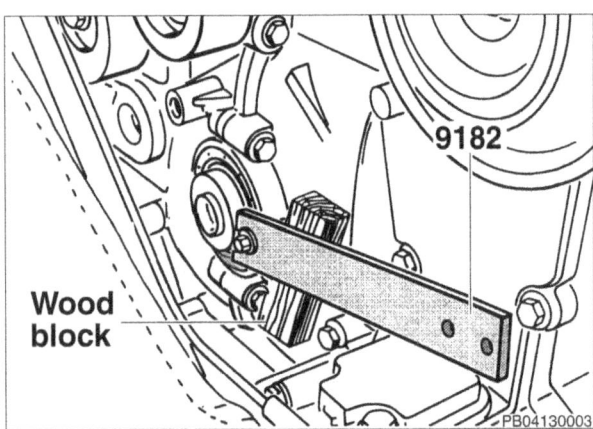

◄ Remove seal:
- Screw sheet metal screw with washer approx. 5 mm (¼ in) into drilled bore.
- Slide lever (Porsche special tool 9182 or equivalent) under washer.
- Support lever on small block of wood (approx. 2 x 1¼ x ¾ in) and pry out seal.

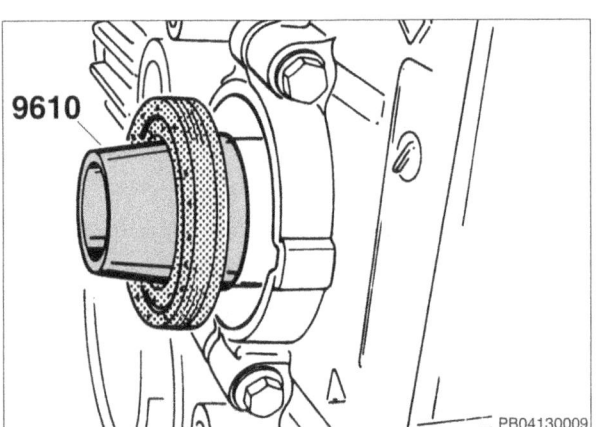

◄ To begin seal installation:
- Place assembly sleeve (part of Porsche special tool 9610) on crankshaft end.
- Thoroughly lubricate crankshaft end, assembly sleeve and sealing lip of seal.
- Coat outer edge of seal with Loctite® 574.
- Position seal on assembly sleeve by hand and slide on crankshaft.
- Remove assembly sleeve.

- Use installer part of special tool 9610 (installer cup and M16 x 1.5 x 60 mm bolt) to press in seal as far as possible.

- Reinstall crankshaft pulley and torque bolt in 2 stages.

Tightening torque	
Pulley to crankshaft (M16 x 1.5 x 60 mm bolt) • Stage 1 • Stage 2	50 Nm (37 ft-lb) 90° additional

Engine Pulleys and Seals 13-5

Crankshaft Seals

- Replace engine drive belt.

- Reconnect battery. Start engine and allow to idle for a few minutes. Shut engine OFF and recheck position and tension of belt.

- Refit engine access cover.

Tightening torque	
Engine access cover to body	10 Nm (7 ft-lb)

Rear main seal, removing and installing

Rear main seal problems are frequently the cause of oil leaks. Periodically inspect the seam between engine and transmission bellhousing for signs of fresh oil drips.

- Disconnect negative (–) battery cable and cover battery terminal to keep cable from accidentally contacting terminal.

> **CAUTION—**
> - Prior to disconnecting the battery, acquire the radio anti-theft code.
> - Read the battery disconnection cautions in **00 Warnings and Cautions**.

- Remove transmission. See **34 Manual Transmission** or **37 Automatic Transmission**.

- Manual transmission: Remove clutch. To avoid warping clutch pressure plate, loosen bolts in diagonal pattern and in several stages. See **30 Clutch**.

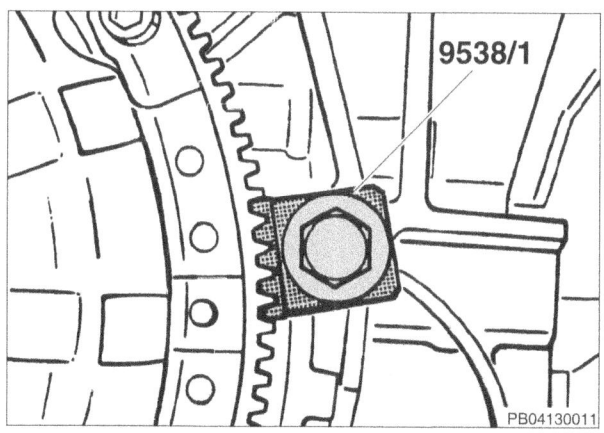

◂ Lock flywheel or drive plate using Porsche special tool 9538/1. Remove flywheel or drive plate mounting bolts. Remove and set aside flywheel or drive plate.

> **CAUTION—**
> - Do not damage the crankshaft flange when removing and installing flywheel or drive plate.
> - Do not use abrasive or polishing cloth or brush with metal bristles on sealing surface of crankshaft flange.

- Carefully pry out rear main seal.

◂ Alternatively, drill 3 mm hole in face of rear main seal.

Crankshaft Seals

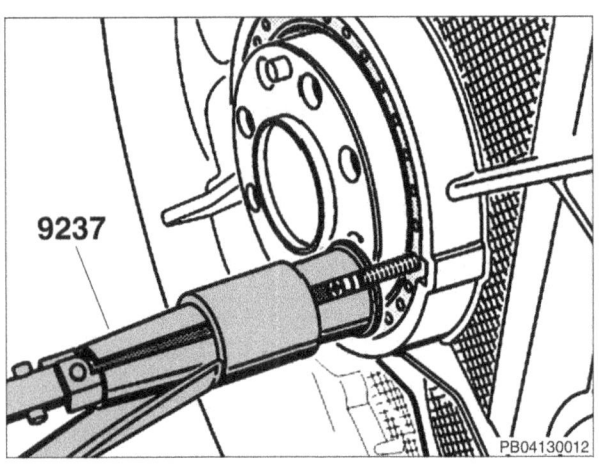

- Thread sheet metal screw with washer approx. 5 mm (¼ in) into drilled bore. Use puller (Porsche special tool 9237 or equivalent) to remove seal.

- Clean crankshaft flange and crankcase bore:
  - Use oilstone to remove small edges or burrs on crankshaft end chamfer.
  - Use clean lint-free cloth soaked in solvent naphtha or acetone to degrease end of crankshaft and crankcase bore. Blow dry using low air pressure, or allow to air dry.
  - Lightly oil crankcase bore.

> **CAUTION—**
> - Make sure crankshaft end flange is dry and thoroughly clean. Do not oil.

- Slide new rear main seal over crankshaft flange.

> **CAUTION—**
> - Do not oil rear main seal sealing lip.

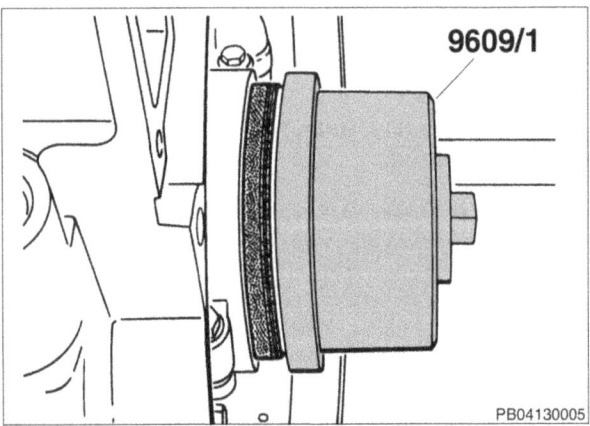

- Use Porsche special tool 9609/1 to press seal in.

> **CAUTION—**
> - Prior to use, make sure tool is thoroughly deburred and clean.

> **NOTE—**
> - The original rear main seal is installed to a depth of 11 mm. The new seal is pressed to a depth of 13 mm so that the sealing lip contacts new surface on the crankshaft.

- Replace 4 crankcase bolts (**arrows**) at rear of engine:
  - Remove bolts and discard. Clean oil residue out of threaded holes.
  - Install new bolts, Porsche part no. 999 385 004 09.

Tightening torque	
Crankcase bolt (M6)	10 Nm (7 ft-lb)

Engine Pulleys and Seals 13-7

Crankshaft Seals

◀ Replace 3 intermediate shaft flange bolts (**arrows**) one at a time:
- Remove one bolt and discard. Clean oil residue out of threaded hole.
- Install new bolt, Porsche part no. 900 385 275 09.
- Repeat for remaining 2 bolts.

Tightening torque	
Intermediate shaft cover to crankcase (M6)	10 Nm (7 ft-lb)

− Install flywheel and clutch or torque plate.

− Reinstall transmission. See **34 Manual Transmission** or **37 Automatic Transmission**.

Tightening torques	
Clutch pressure plate to flywheel (M8 x 16 mm bolt)	23 Nm (17 ft-lb)
Drive plate to crankshaft (M10 x 1 x 25 mm bolt) • Stage 1 • Stage 2	25 Nm (19 ft-lb) 90° additional
Flywheel to crankshaft (M10 x 1 x 50 mm bolt) • Stage 1 • Stage 2	25 Nm (19 ft-lb) 90° additional

15 Cylinder Head

GENERAL 15-1
  Warnings and Cautions.................... 15-1
CYLINDER HEAD COVER 15-2
  Cylinder head cover, removing
    (1997 - 2002 models)................... 15-2
  Cylinder head cover, installing
    (1997 - 2002 models)................... 15-5

SPARK PLUG WELL 15-8
  Spark plug well, resealing 15-8

GENERAL

This repair group covers removal and installation of cylinder head covers and spark plug wells.

The photos in this repair group illustrate work on a 1998 Boxster with 2.5 liter engine and manual transmission. The procedures are similar on other models.

For clarity, the procedures are illustrated on an engine removed from the car. It is possible to do this work on an engine in the car. However, the removal of additional parts, connectors or lines may be required for access to components.

Warnings and Cautions

WARNING—
- *To avoid personal injury, be sure the engine is cold before beginning any procedure in this repair group.*

CAUTION—
- *When working on internal engine components, maintain absolute cleanliness.*
- *To prevent damage to vehicle body or paint, use protective body covers.*
- *Lay removed engine parts on a clean surface and cover immediately. Even small dirt particles can block oil passages.*
- *Place matching marks on harness connectors, hardware and other components for ease of assembly.*

15-2 Cylinder Head

Cylinder Head Cover

CYLINDER HEAD COVER

> **CAUTION—**
> - *The procedures given here are not applicable to 2003 - 2004 Boxster models with DME ME 7.8.*

Cylinder head cover, removing (1997 - 2002 models)

− Disconnect negative (−) battery cable and cover battery terminal to keep cable from accidentally contacting terminal.

> **CAUTION—**
> - *Prior to disconnecting the battery, acquire the radio anti-theft code.*
> - *Read the battery disconnection cautions in* **00 Warnings and Cautions**.

− Raise rear of vehicle and support safely. Remove rear wheels.

> **CAUTION—**
> - *Make sure the car is stable and well supported at all times. Use a professional automotive lift or jack stands designed for the purpose. A floor jack is not adequate support.*

− Drain engine oil. See **03 Service and Maintenance**. After allowing oil to drip about 20 minutes, refit oil drain plug. Use new sealing O-ring.

Tightening torque	
Oil drain plug to oil pan	50 Nm (37 ft-lb)

◂ Working underneath engine, remove plastic splash shield fasteners (**arrows**) and remove shield.

◂ Remove splash shield bracket fastener (**arrow**). Take off bracket.

− 2000 or later model: Separate oxygen sensor connector underneath cylinder head cover. Detach harness connector bracket.

− 2000 or later model: Remove exhaust manifolds and catalytic converters. See **26 Exhaust System**.

− Right cylinder head cover, 2000 or later model: Remove plastic splash shields under cylinder head cover.

Cylinder Head 15-3

Cylinder Head Cover

◀ Left cylinder head cover: Unscrew VarioCam cover mounting fasteners (**arrows**) and remove power steering pressure line bracket. Pull pressure line aside.

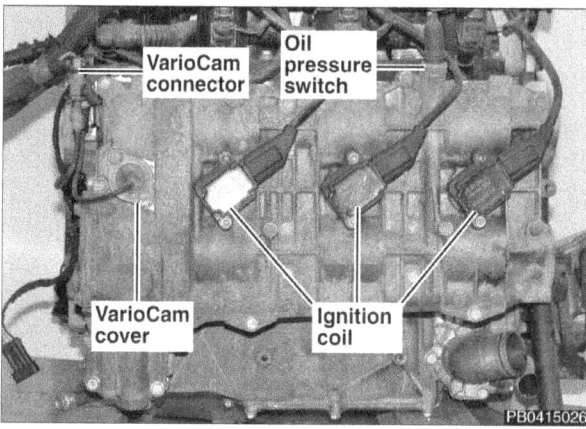

◀ Right cylinder head cover: Disconnect oil pressure switch connector.

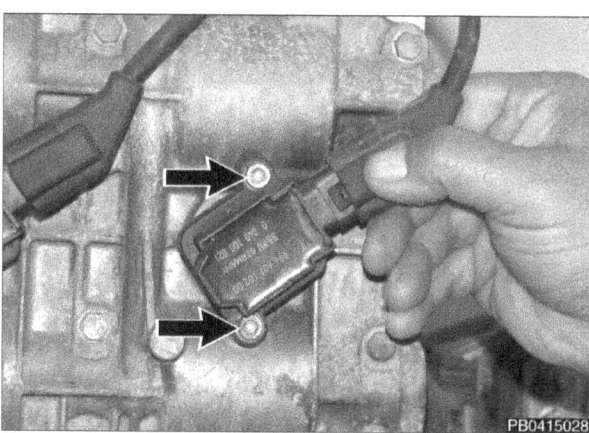

◀ Remove ignition coils:
- Squeeze tab to detach connector.
- Remove coil mounting Allen bolts (**arrows**).
- Pull coil off spark plug.

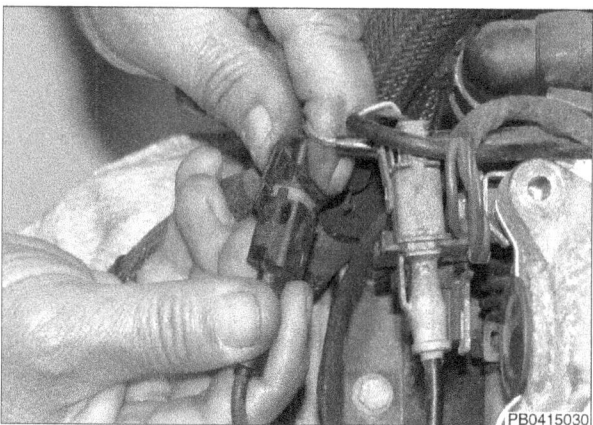

◀ Disconnect VarioCam harness connector.

15-4 Cylinder Head

Cylinder Head Cover

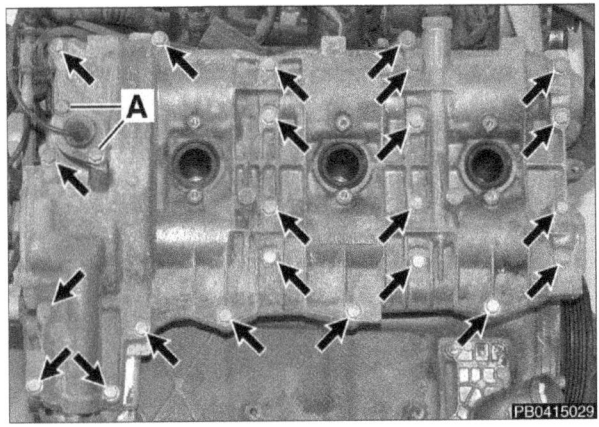

◀ Remove VarioCam cover mounting bolts (**A**).

– Loosen and remove cylinder head cover mounting bolts (**arrows**).

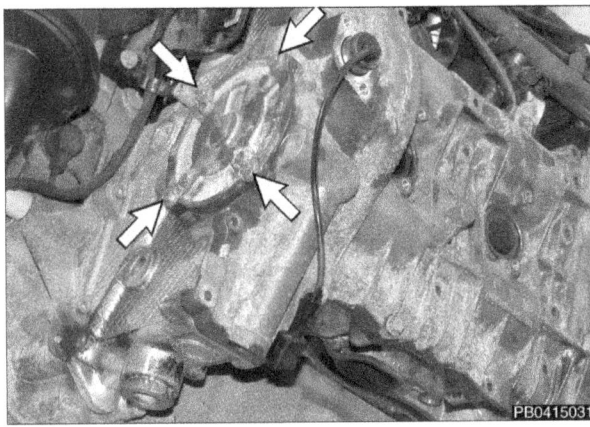

◀ Remove oil extraction pump mounting bolts (**arrows**). Remove pump and set aside.

◀ Use a pry bar to separate top of cylinder head cover from cylinder head.

> **CAUTION—**
> • Do not break the cast lugs on the cylinder head or cover.

– If necessary, loosen cylinder head cover by carefully tapping with plastic hammer. Be prepared to catch dripping oil.

– Right cylinder head cover, 2000 or later model: Remove coolant line mounting bracket bolt. Push coolant lines aside.

– Remove cylinder head cover by sliding over VarioCam solenoid.

Cylinder Head 15-5

Cylinder Head Cover

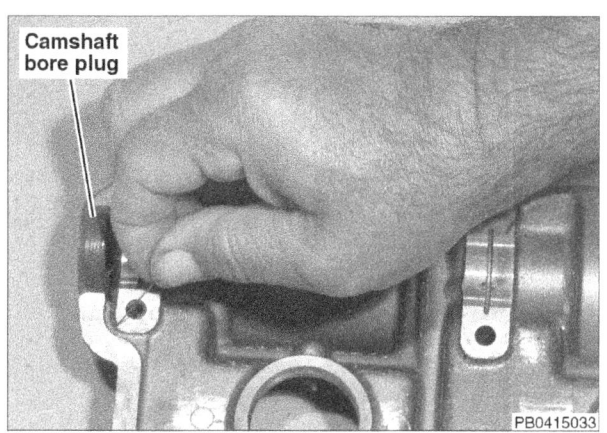

Cylinder head cover, installing (1997 - 2002 models)

◄ Clean cylinder head and cylinder head cover sealing surfaces:
- Remove old sealant residue thoroughly.
- Rub down surfaces with clean cloth soaked in degreasing agent (example: acetone, naphtha solvent).

> **CAUTION—**
> - Even small strands of dried sealant may cause a blockage in oil passageways.
> - To avoid damaging sealing surfaces, do not use sharp scraping tools.

– Remove camshaft bore plugs and spark plug wells from cover.

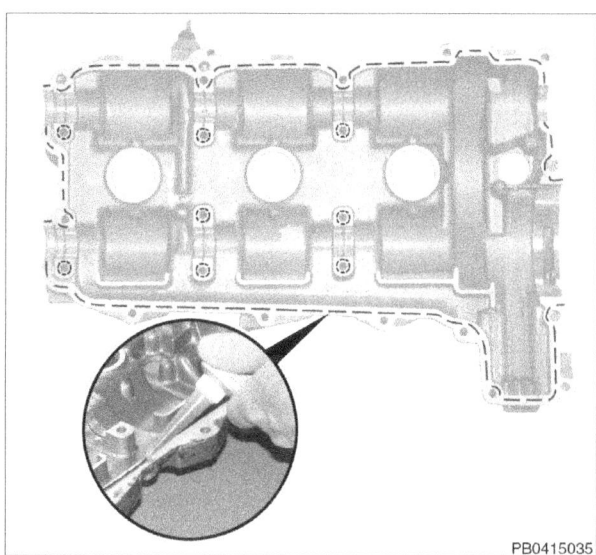

◄ Apply thin bead (approx. 1.5 mm thick) of silicone sealant to sealing surfaces and bearing saddle bolt bores (**dashed lines**) of cylinder head cover.

> **CAUTION—**
> - Make sure no sealant enters oil passages.
> - After sealant is applied, place cylinder head cover on cylinder head and torque bolts within 5 minutes.

> **NOTE—**
> - Use one of the following sealing products:
>   -3-Bond® 1209
>   -Loctite® 5900

– Place cylinder head cover on cylinder head. Install cover bolts and tighten finger tight.

> **NOTE—**
> - Right cylinder head: Push coolant lines aside when placing cover on cylinder head.

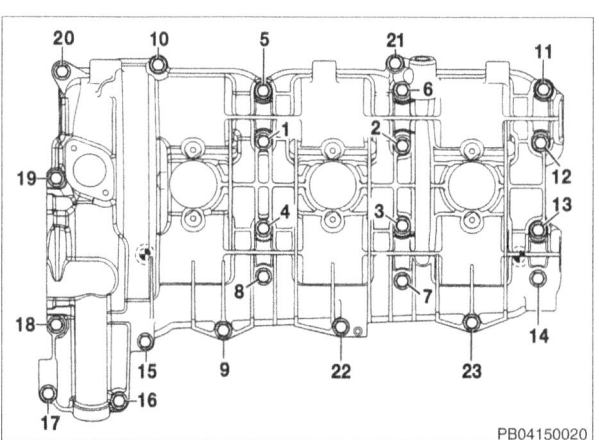

◄ Tighten cover bolts in sequence illustrated.

Tightening torque	
Cylinder head cover to cylinder head (M6 x 30 mm)	13 Nm (10 ft-lb)

Cylinder Head Cover

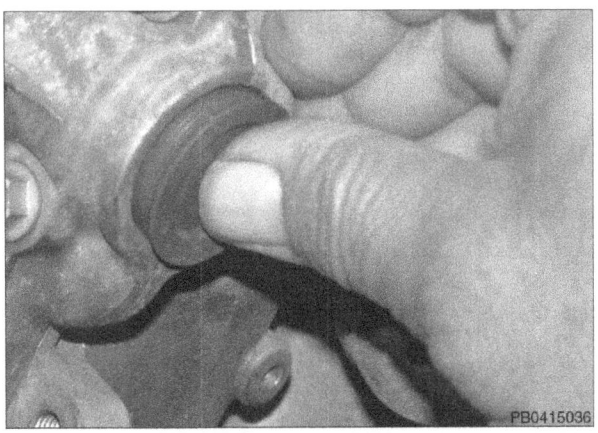

◀ If leaky, replace camshaft bore sealing plugs.
- If necessary, use plastic hammer to drive in plug.
- Do not use lubrication or sealant on plug.

◀ Inspect spark plug well O-ring seals (**arrows**) for damage and brittleness.

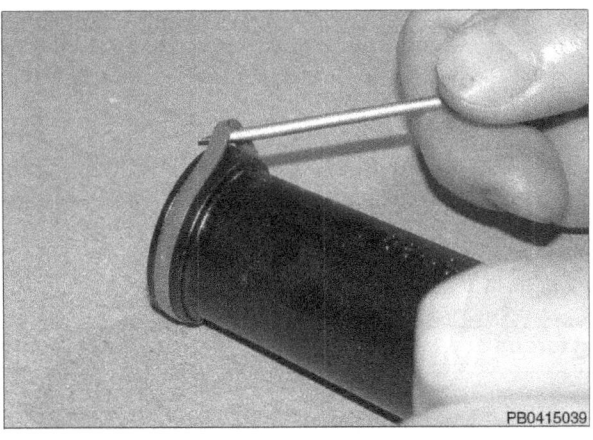

◀ If necessary, replace O-ring seals.

◀ Coat O-rings lightly with engine oil and press spark plug well into cylinder head cover firmly.

Cylinder Head 15-7
Cylinder Head Cover

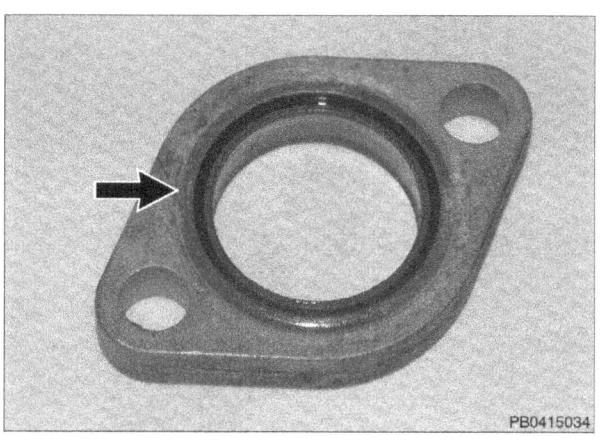

◄ Replace VarioCam cover seal (**arrow**). Install cover over VarioCam solenoid.

- Left cylinder head cover: Reattach power steering pressure line bracket using VarioCam cover mounting bolts.

Tightening torque	
VarioCam cover to cylinder head cover (M6 x 20 mm)	10 Nm (7 ft-lb)

– Reconnect VarioCam harness connector.

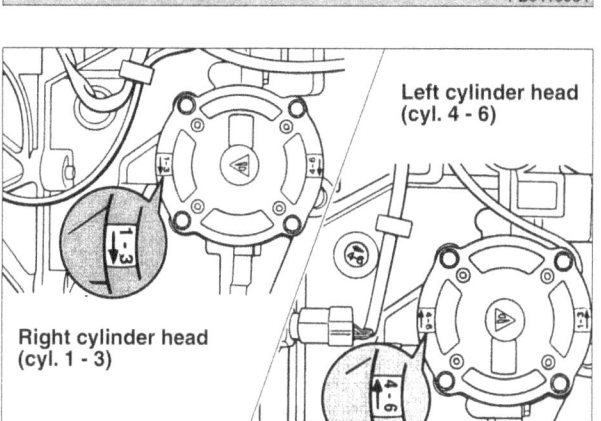

◄ Reinstall oil extraction pump:
- Clean off excess sealant at sealing surfaces.
- Oil new sealing O-ring.
- Use new self-locking bolts.

CAUTION—
- *Orient left and right oil extraction pumps correctly.*

Tightening torque	
Oil extraction pump to cylinder head cover (use new self-locking M6 x 20 mm bolts)	13 Nm (10 ft-lb)

– Remainder of installation is reverse of removal. Keep in mind the following:
- Replace exhaust manifold flange seals.
- Replace M8 x 28 mm exhaust manifold bolts. Replace other exhaust system hardware as necessary.

Tightening torques	
Exhaust manifold to cylinder head (use new M8 x 28 mm bolts)	23 Nm (17 ft-lb)
Ignition coil to cylinder head cover (Allen head M6 x 25 mm)	10 Nm (7 ft-lb)
Oxygen sensor harness bracket to cylinder head (M6)	10 Nm (7 ft-lb)
Plastic cover to right cylinder head (M6)	10 Nm (7 ft-lb)
Splash shield bracket to diagonal brace (M6)	10 Nm (7 ft-lb)

– Refill engine with oil.

– Start engine and check for oil and coolant leaks.

15-8 Cylinder Head

Spark Plug Well

SPARK PLUG WELL

Each spark plug is installed at the base of a spark plug well (oil protection tube), which is pressed into the cylinder head cover. Reseal spark plug wells in case of engine oil leak past the sealing O-rings.

Spark plug well, resealing

> *WARNING—*
> * *Due to risk of personal injury, be sure the engine is cold before beginning the procedure.*

- Raise rear of vehicle and support safely.

> *WARNING—*
> * *Make sure the car is stable and well supported at all times. Use a professional automotive lift or jack stands designed for the purpose. A floor jack is not adequate support.*

◀ Working underneath engine, remove plastic splash shield fasteners (**arrows**) and remove shield.

◀ Remove ignition coil:
- Squeeze tab to detach connector.
- Remove coil mounting Allen bolts (**arrows**).
- Pull coil off spark plug.

◀ Use expanding pliers to pull out spark plug well.

Cylinder Head 15-9

Spark Plug Well

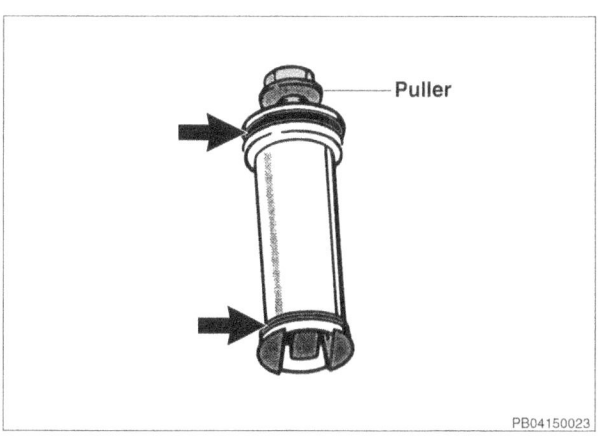

◄ In case of O-rings (**arrows**) adhering firmly to cylinder head cover, use an internal puller to work spark plug well out of bore.

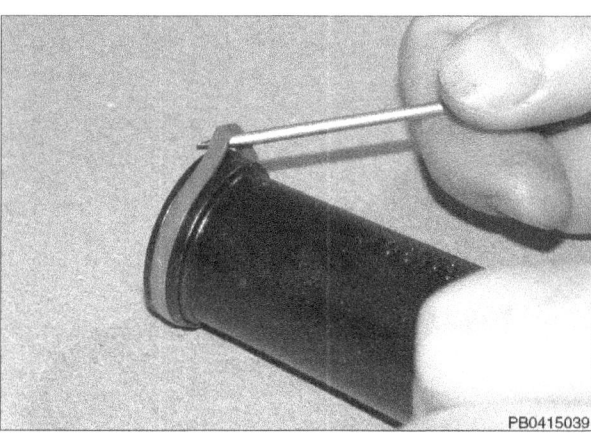

◄ Replace O-ring seals.

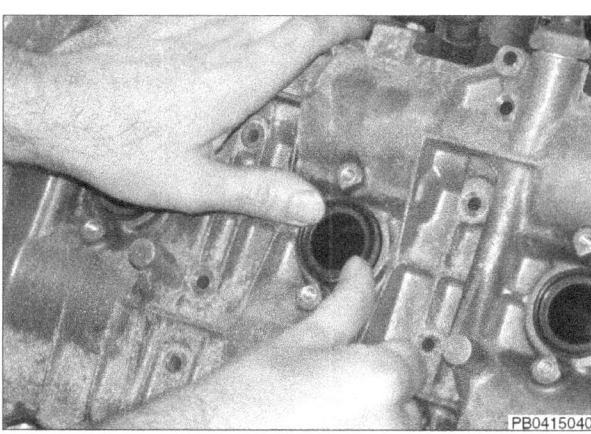

◄ Coat O-rings lightly with engine oil and press into cylinder head cover firmly.

– Reinstall ignition coil.

Tightening torque	
Ignition coil to cylinder head cover (Allen head M6 x 25 mm)	10 Nm (7 ft-lb)

17 Lubrication System

GENERAL 17-1	Engine oil cooler 17-2
LUBRICATION SYSTEM 17-1	Air-oil separator. 17-3
Engine oil 17-1	Oil level sensor 17-3
Lubrication system components overview 17-2	ENGINE OIL COOLER 17-4
Oil pumps 17-2	Engine oil cooler, removing and installing. 17-4
Oil filter 17-2	

GENERAL

This repair group covers the Boxster engine lubrication system.

LUBRICATION SYSTEM

To insure reliable engine oil supply, the Boxster engine has an integrated dry-sump lubrication system without an external oil tank. The oil sump is in the bottom of the crankcase.

Crankcase oil passages are cast. This eliminates the need to drill out oil passages in the crankcase.

Engine oil

Boxster oil characteristics	
Oil specification: Synthetic	API SH or SJ ILSAC GF 3 Europe: ACEA A3
Oil viscosity and ambient temperatures: • Generally higher than 50°F (10°C) • Generally lower than 50°F (10°C)	10W-40, 15W-40, 15W-50 10W-40, 10W-30, 5W-30
Oil capacity: • Without filter change • With filter change	8.5 liters (9 US qt) 8.75 liters (9.25 US qt)
Recommended oil change interval (Porsche)	15,000 miles

> *CAUTION—*
> * Lubricant specifications are subject to change. Consult your authorized dealer for the latest information regarding lubricant applications and specifications.*

17-2 Lubrication System

Lubrication System

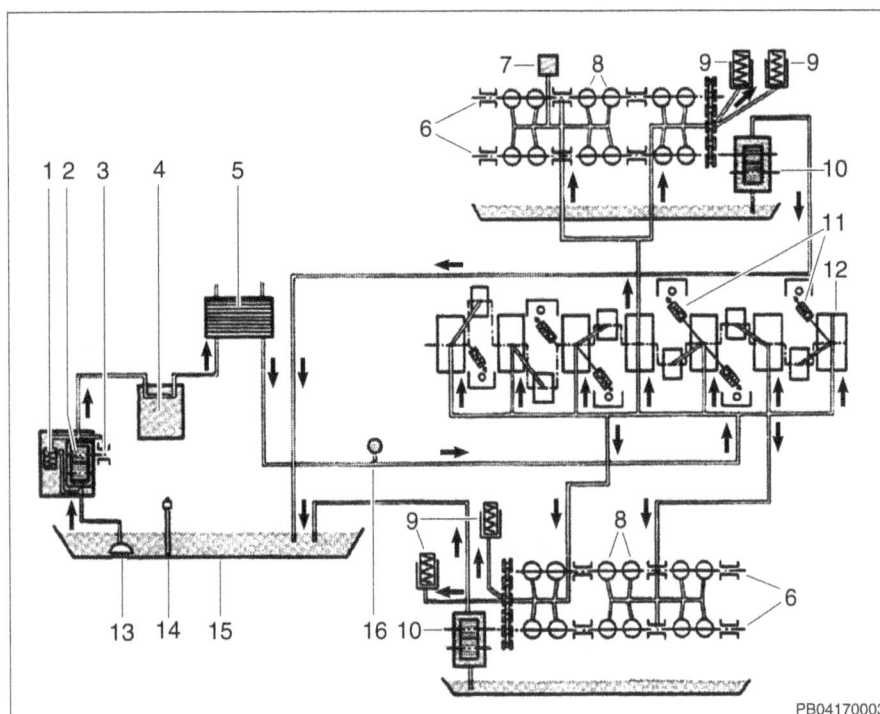

Lubrication system components overview

1. Oil pressure relief valve
2. Oil pump
3. Intermediate shaft
4. Oil filter
5. Engine oil cooler (oil-water heat exchanger)
6. Camshaft
7. Oil pressure sensor
8. Hydraulic valve lifter
9. Hydraulic chain tensioner
10. Oil return pump
11. Oil jet for piston-cooling
12. Crankshaft
13. Oil pump pickup
14. Oil level sensor
15. Oil sump
16. Oil temperature sensor

Oil pumps

The oil pump, driven by the intermediate shaft, draws oil out of the oil sump and distributes it to engine lubrication points. Oil baffles and an oil collection channel lead returning oil to the oil pump pickup.

Oil is pumped out of each cylinder head by a separate oil return pump and returned to the oil sump via an air-oil separator. Each return pump is driven by a camshaft.

Oil filter

The full flow oil filter is a replaceable pleated paper cartridge.

Porsche recommends that the oil filter be changed every other oil change, at approximately 30,000 miles.

Engine oil cooler

The Boxster engine is equipped with an oil-water heat exchanger above the crankcase and just below the left intake manifold.

Lubrication System 17-3

Lubrication System

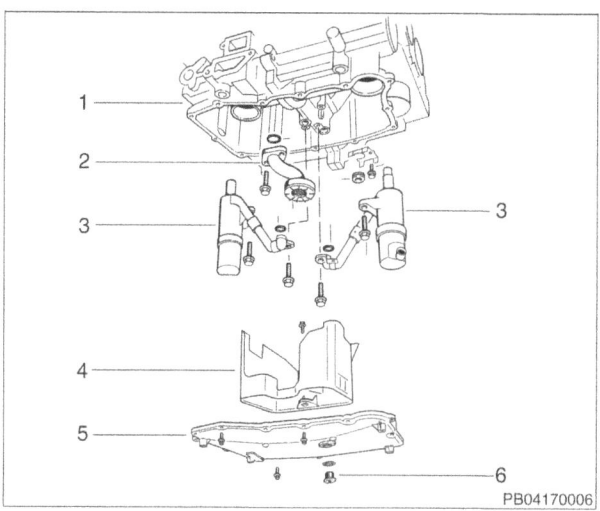

Air-oil separator

In the air-oil separator, the oil is centrifuged and defoamed before being returned to the sump. The supply of clear, defoamed oil to the hydraulic valve lifters ensures correct valve timing and clearance and maintains consistent engine performance and exhaust emissions.

1. Oil sump
2. Pump oil pickup
3. Air-oil separator
4. Baffle
5. Oil sump cover
6. Drain plug M18 x 1.5
   -tighten to 50 Nm (37 ft-lb)

Oil level sensor

Oil level information to the instrument cluster is provided by a sensor screwed into the oil sump.

1997 - 2000: There is a segmented oil level display in the instrument cluster right side gauge, next to the clock.

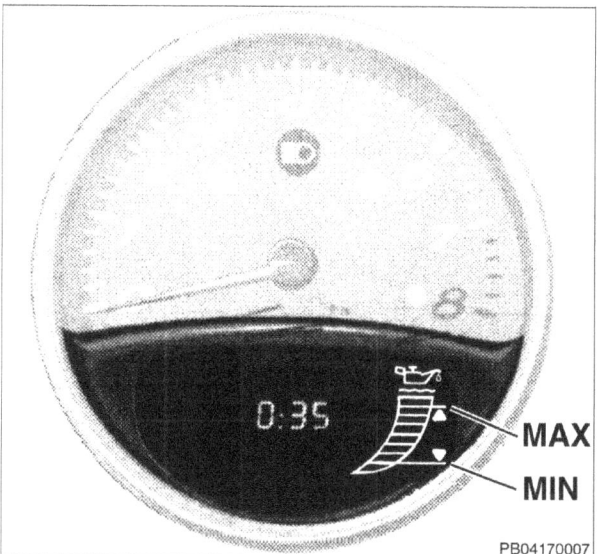

2001 - 2004: Oil level display is in the center gauge. It is turned on by the on-board computer when the ignition key is first turned ON.

The difference between MAX and MIN on the segmented display corresponds to approx. 1.5 liters (1.6 US qt).

If oil level drops below MIN, the bottom segment flashes. If there is a fault in the oil level detection system (example: broken wire, short to ground) the entire display flashes.

Correct oil level measurement is possible only with the vehicle horizontal. The time needed for the oil to settle for accurate measurement depends on engine temperature and how long the engine is OFF. When the ignition is switched OFF, the waiting time to detect correct oil level is shown on the clock as a decreasing interval in minutes and seconds. The segmented display also lights up in steps. For example, oil settling time at 80° (176°F) is approx. 30 seconds.

During refueling, oil level is measured automatically and displayed for approx. 1 minute after ignition is turned ON. The prerequisites of this feature are:

- Ignition OFF
- Engine warm
- Refueling of at least 12 liters (3 US gal)
- Refueling completed within 15 minutes

Lubrication System

Engine Oil Cooler

ENGINE OIL COOLER

The engine oil cooler (oil-water heat exchanger) is bolted to the top left rear of the crankcase, underneath the left intake manifold. Gain access to the cooler by removing the center intake manifold.

Engine oil cooler, removing and installing

> **WARNING**—
> • Due to risk of personal injury, be sure the engine is cold before beginning the removal procedure.

– Raise vehicle and support safely.

> **CAUTION**—
> • Make sure the car is stable and well supported at all times. A floor jack is not adequate support.

◀ Working underneath engine, remove plastic splash shield fasteners (**arrows**) and remove shield.

◀ Drain engine coolant:
  • Loosen cap at coolant reservoir in rear trunk.
  • Place 5-gallon pail under engine.
  • Remove coolant manifold drain plug (**arrow**) at front of engine and drain engine coolant.
  • Reinstall drain plug with new sealing O-ring.

> **WARNING**—
> • Automotive antifreeze (coolant) is poisonous and lethal, specially to pets. Clean up spills immediately and rinse area with water. Dispose of coolant safely.

Tightening torque	
Coolant drain plug to coolant manifold	10 - 15 Nm (7 - 11 ft-lb)

– Place convertible top and convertible top compartment lid in service position and remove engine compartment top cover. See **03 Service and Maintenance**.

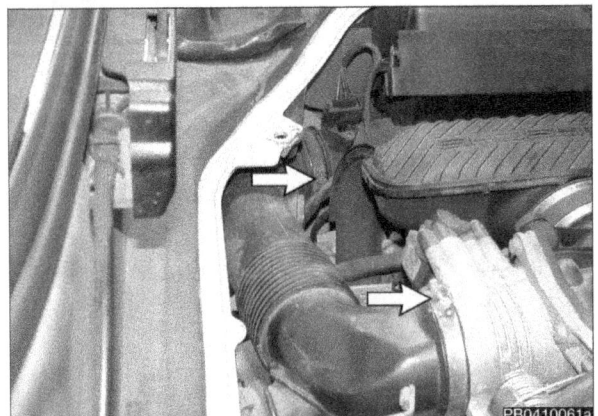

◀ Working at rear of engine compartment, loosen engine air intake duct clamps (**arrows**) and remove duct.

Lubrication System 17-5
Engine Oil Cooler

◄ Loosen throttle housing and center intake manifold and set aside:
 - Remove throttle cable clamp (**A**) if applicable.
 - Remove throttle housing hold down bolt (**B**).
 - Loosen center intake manifold hose clamps. Slide right manifold boot to left (**arrow**) and tilt manifold and throttle housing out and set aside. Leave vacuum hoses and cable attached.

– Loosen clamp and detach coolant line to oil cooler. Be prepared to catch dripping coolant.

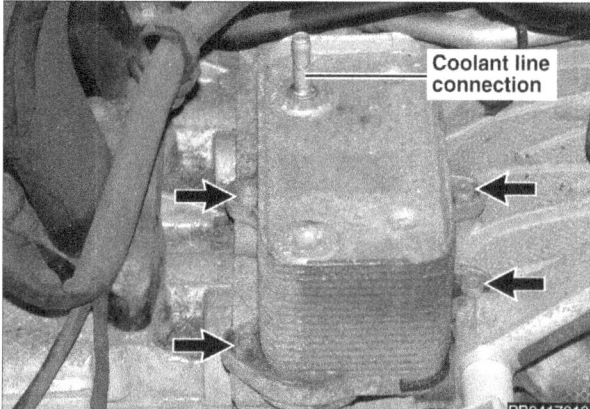

◄ Remove oil cooler mounting bolts (**arrows**). Lift oil cooler out of engine compartment. Use shop towels to catch dripping oil and coolant.

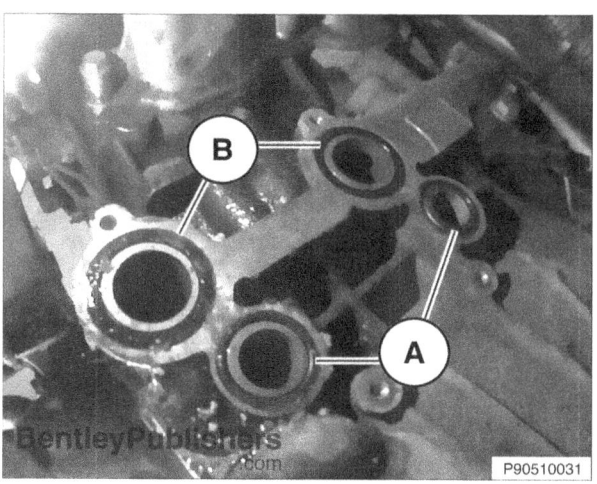

◄ Insert new sealing O-rings (**A** and **B**) on crankcase.

– Reinstall oil cooler.

Tightening torque	
Oil cooler to crankcase	10 Nm (7 ft-lb)

– Attach coolant line.

17-6 Lubrication System

Engine Oil Cooler

◄ Reinstall throttle housing and center intake manifold:
- Align manifold rubber boots with scribe marks on center manifold and tighten inner hose clamps.
- Shift center manifold between intake manifolds so that dimension **A** is identical at right and left.
- Tighten outer hose clamps.
- Tighten throttle housing hold down.
- Make sure vacuum line is seated in left rubber boot.

− Top up and bleed cooling system. See **17 Cooling System**.

− Fit engine compartment cover.

− Check engine oil and top up if necessary.

19 Cooling System

GENERAL 19-1
  Coolant and antifreeze 19-1
  Cooling system component overview 19-2
  Warnings and Cautions................... 19-3
SERVICE AND TROUBLESHOOTING 19-3
  Cooling system leak test................. 19-3
  Cooling system, flushing.................. 19-4
  Cooling system, draining, filling and bleeding... 19-5

COOLING SYSTEM
COMPONENT REPLACEMENT 19-7
  Coolant pump, removing and installing........ 19-7
  Coolant thermostat,
    removing and installing 19-9
  Coolant reservoir, removing and installing 19-10
RADIATORS AND ENGINE COOLING FANS .. 19-13
  Engine cooling fan, removing and installing ... 19-13
  Radiator, removing and installing 19-16

GENERAL

This repair group covers the engine cooling system.

For additional information see:

- **24 Fuel Injection** for engine coolant temperature (ECT) sensor replacement
- **80 Heating**
- **87 Air-conditioning**

Coolant and antifreeze

Cooling system capacity depends on a variety of options and configurations. If a repair procedure includes opening the sealed cooling system, make sure to fill and bleed system after repairs. See **Cooling system, draining, filling and bleeding** in this repair group.

Coolant capacity	
Year, model	Capacity
1997 - 2002 Boxster	17 liters (4.5 US gal)
2003 - 2004 Boxster: • Manual transmission • Automatic transmission	18 liters (4.76 US gal) 19 liters (5.02 US gal)
2003 - 2004 Boxster S: • Manual transmission • Automatic transmission	22 liters (5.81 US gal) 23 liters (6.08 US gal)

19-2 Cooling System

General

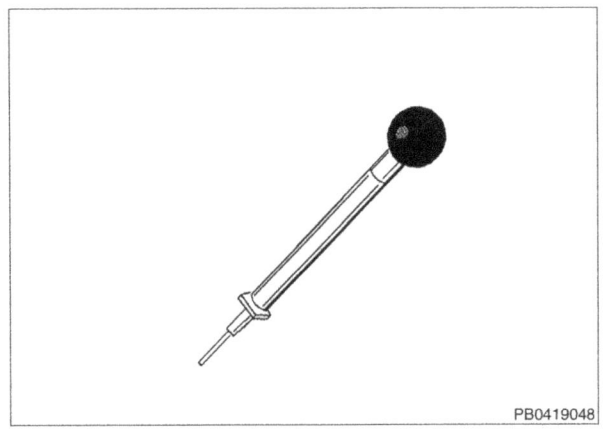

The Boxster engine cooling system is factory filled with a lifetime coolant mixture of 50% silicate-free antifreeze and 50% water. If topping up or replacing coolant, use fresh antifreeze mixture. In an emergency, use clear water.

◄ Use a coolant hydrometer to determine antifreeze concentration.

Coolant mixture recommendations	
Concentration	**Cold protection**
50% antifreeze	-35°C (-31°F)
60% antifreeze	-40°C (-40°F)

Do not use a higher concentration of antifreeze than a 60% mixture, as the heat transfer quality of the coolant decreases with higher antifreeze concentrations.

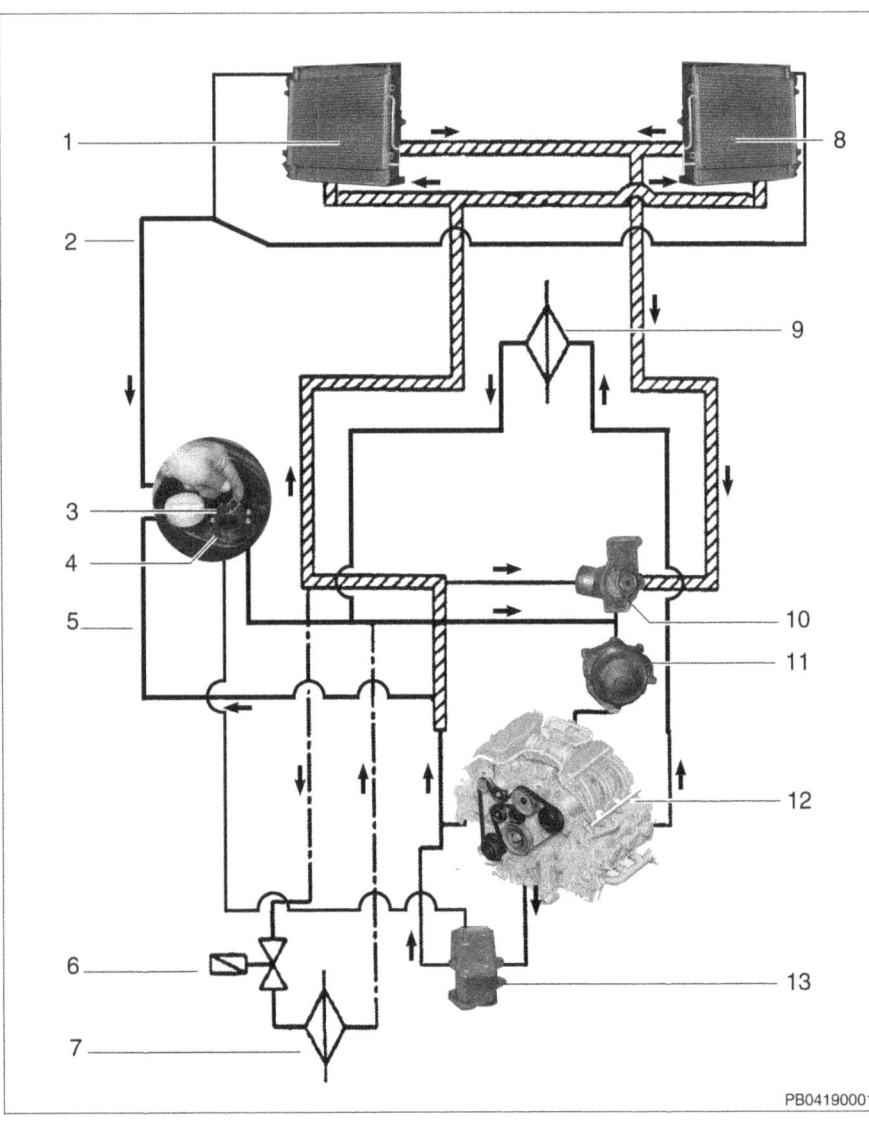

Cooling system components overview

1. Left radiator
2. Radiator vent line
3. Bleeder valve
4. Coolant expansion tank and filler
5. Engine vent line
6. ATF cooler electric shut-off valve
7. ATF cooler (heat exchanger)
8. Right radiator
9. Heater core
10. Coolant thermostat
11. Coolant pump
12. Crankcase
13. Engine oil cooler (heat exchanger)

Cooling System 19-3

Service and Troubleshooting

Warnings and Cautions

Observe the following warnings and cautions when working on the cooling system.

> **WARNING—**
> - *At normal operating temperature the cooling system is pressurized. Allow the engine to cool thoroughly (a minimum of one hour), then loosen the cooling system pressure cap slowly to allow safe release of pressure.*
> - *Releasing cooling system pressure lowers the boiling point of coolant and it may boil suddenly. Use heavy gloves and wear eye and face protection to guard against scalding.*
> - *Use extreme care when draining and disposing of engine coolant. Coolant is poisonous and lethal to humans and pets. Pets are attracted to coolant because of its sweet smell and taste. Seek medical attention immediately if coolant is ingested.*

> **CAUTION—**
> - *Avoid adding cold water to the coolant while the engine is hot or overheated. If it is necessary to add coolant to a hot system, do so only with the engine running and coolant pump turning.*
> - *To avoid excess silicate gel precipitation in the cooling system and loss of cooling capacity, use Porsche coolant or equivalent low silicate antifreeze.*
> - *Dispose of coolant in an environmentally safe manner.*
> - *If oil enters the cooling system, flush with cleaning agent.*
> - *When replacing cooling system components, replace spring-type hose clamps with the screw-type.*
> - *Prior to disconnecting the battery, read the battery disconnection cautions given in* **00 Warnings and Cautions**.

SERVICE AND TROUBLESHOOTING

Cooling system leak test

> **WARNING—**
> - *Due to risk of personal injury, be sure the engine is cold before beginning work on the cooling system.*

 Working at engine service tray in rear trunk, loosen coolant expansion reservoir cap (**arrow**) slowly to allow coolant pressure to vent. Remove cap.

19-4 Cooling System

Service and Troubleshooting

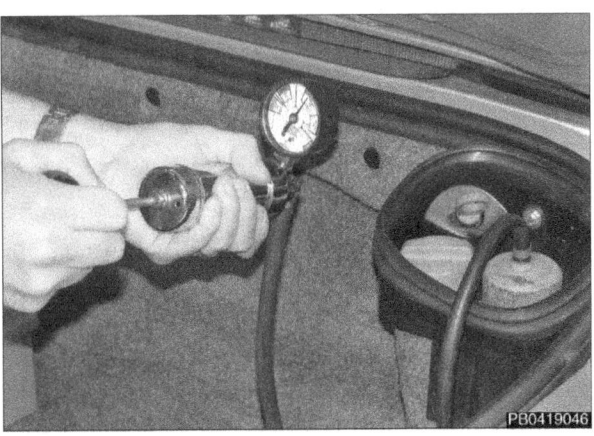

◂ Attach pressure tester and pump system to 1.3 bar (19 psi). If pressure drops, locate leak and repair.

– Check hoses, clamps and fittings for traces of dried coolant or drips.

– Check carpeted trim in rear trunk underneath coolant reservoir for signs of reservoir leakage.

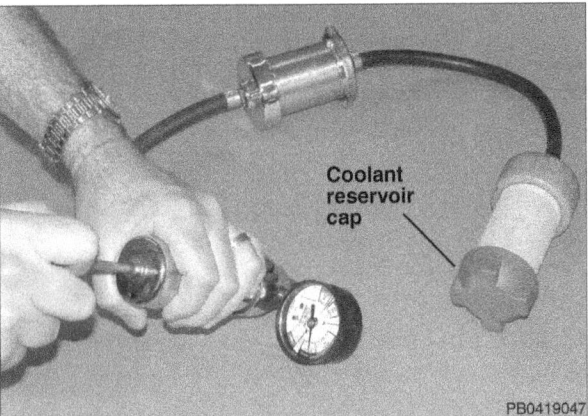

◂ Leak test pressure relief valve in coolant reservoir cap:
  • Screw cap, cap adapter and pressure tester together.
  • Pump up cap until pressure relief valve opens. Note opening pressure.
  • Allow pressure to vent until valve closes and pressure stabilizes.
  • Replace cap if pressure drops below specifications.

Cooling system pressure	
Cooling system	1.3 bar (19 psi)
Coolant reservoir cap • Pressure relief valve opens • Pressure relief valve closes	approx. 1.4 bar (20 psi) approx. 1 bar (14.5 psi)

– If the vehicle is equipped with a 2-step coolant cap (Porsche part no. 996 106 447 00), there is second pressure relief valve which opens at approx 1.8 bar (26 psi) if the pressure tester is pumped rapidly. If the valve does not open, replace the cap.

Cooling system, flushing

Porsche does not specify a periodic coolant change or flush interval.

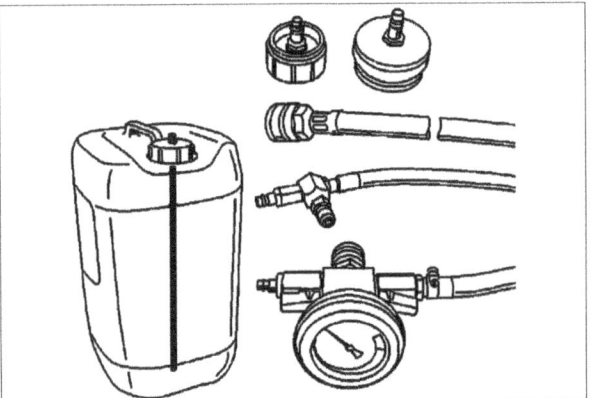

◂ Use cooling system flush kit (Porsche special tool set 9696 or equivalent) to flush cooling system. This equipment is used to pressurize, evacuate, store and recycle coolant in an environmentally responsible manner. If this equipment is not available, have this procedure performed at a Porsche dealer service department or other qualified Porsche service facility.

Cooling System   19-5

Service and Troubleshooting

Cooling system, draining, filling and bleeding

If the coolant needs to be drained and refilled during a repair, or the cooling system needs to be bled of air, use the following procedure.

> **WARNING—**
> - *Due to risk of personal injury, be sure the engine is cold before beginning work on the cooling system.*

- With ignition ON, turn heater fully ON to open heater valve.

◀ Working at engine service tray in rear trunk, loosen coolant expansion reservoir cap (**arrow**) slowly to allow coolant pressure to vent. Remove cap.

- Raise car and support safely.

> **WARNING—**
> - *Make sure the car is stable and well supported at all times. Use a professional automotive lift or jack stands designed for the purpose. A floor jack is not adequate support.*

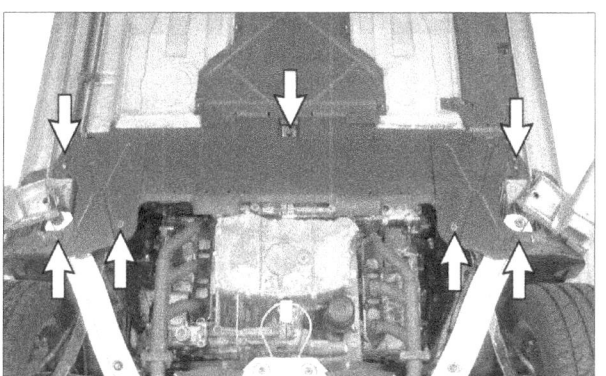

◀ Working underneath engine, remove plastic splash shield fasteners (**arrows**) and remove shield.

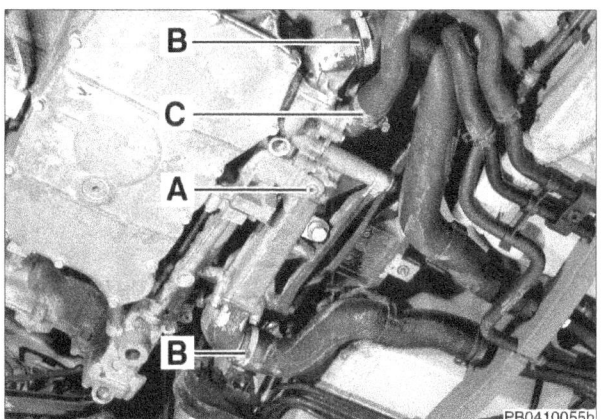

◀ Drain engine coolant:
- Place 5-gallon pail under engine.
- Remove coolant manifold drain plug (**A**) at front of engine and drain engine coolant.
- Loosen radiator hose clamps (**B**) and detach hoses from coolant manifold and thermostat housing.
- Loosen coolant filler hose clamp (**C**) and detach hose.

> **WARNING—**
> - *Coolant is poisonous and lethal to humans and pets. Clean up spills immediately and rinse area with water. Dispose of coolant safely.*

- Reinstall drain plug with new sealing O-ring.

Tightening torque	
Coolant drain plug to coolant manifold	10 - 15 Nm (7 - 11 ft-lb)

- Reattach coolant hoses and secure with new hose clamps.

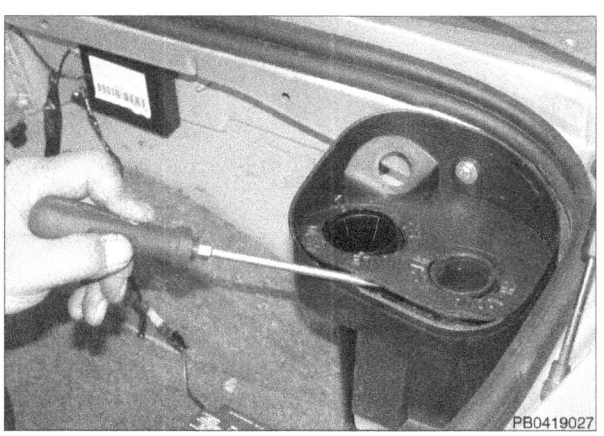

◀ Working at engine service tray in rear trunk, remove oil filler cap and lever off trim cover underneath filler necks. Replace oil filler cap.

19-6 Cooling System

Service and Troubleshooting

◀ Flip up coolant bleeder valve locking clip.

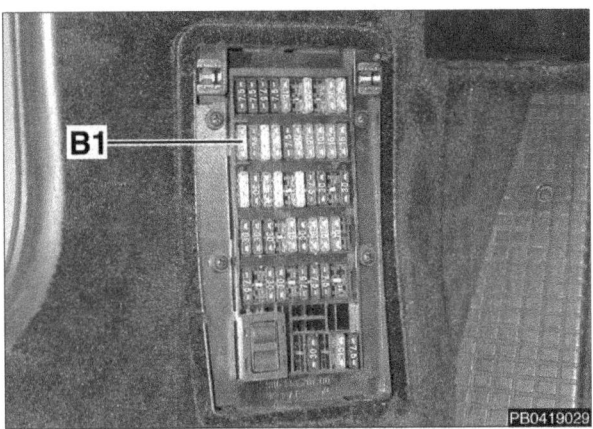

◀ Automatic transmission: Remove B1 fuse in left footwell fuse panel to disable ATF cooler shut-off valve.

> **CAUTION—**
> - For the following steps, protect the trunk trim by placing towels or a plastic sheet under the engine service tray to trap coolant.

- Fill with coolant up to bottom edge of filler neck.
- Run engine at idle. Top off coolant until no more coolant can be added when engine idle is raised sharply.

> **CAUTION—**
> - Make sure coolant temperature does not exceed 80°C (176°F).

- Replace coolant cap and tighten. Warm up engine for about 10 minutes at approx. 2500 rpm until coolant thermostat opens. To be sure engine is fully warmed up, check to see that radiator cooling fans in front wheel wells are ON.
- Allow engine to run an additional 5 minutes at 2500 rpm. Every 30 seconds briefly rev up engine to 5000 rpm.
- Open reservoir cap, relieving coolant pressure with care. Top off coolant, close cap and repeat process of intermittently revving engine to 5000 rpm for 5 minutes.
- Allow engine to idle until radiator cooling fans cycle ON and OFF once. Shut off engine and open reservoir cap, relieving coolant pressure with care.

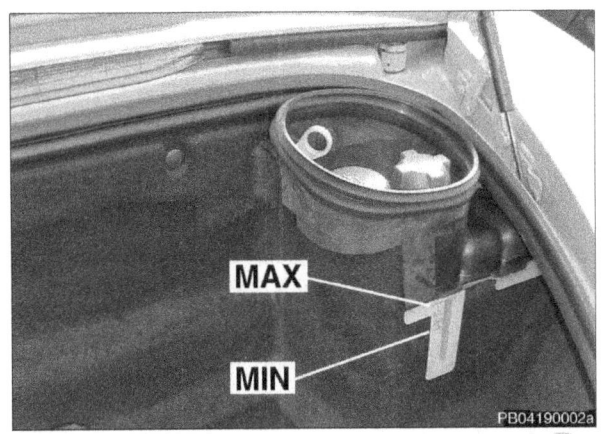

◀ Top off coolant until level reaches MAX.

- Flip down coolant bleeder valve locking clip.
- Replace filler neck trim cover and cap. Remove protective covers from trunk interior.
- Automatic transmission: Replace fuse B1.

Cooling System 19-7

Cooling System Component Replacement

COOLING SYSTEM COMPONENT REPLACEMENT

Coolant pump, removing and installing

> **WARNING—**
> - Due to risk of personal injury, be sure the engine is cold before beginning work on engine cooling components.

◀ Working at engine service tray in rear trunk, loosen coolant expansion reservoir cap slowly to allow coolant pressure to vent. Remove cap.

− Raise rear of car and support safely.

> **WARNING—**
> - Make sure the car is stable and well supported at all times. Use a professional automotive lift or jack stands designed for the purpose. A floor jack is not adequate support.

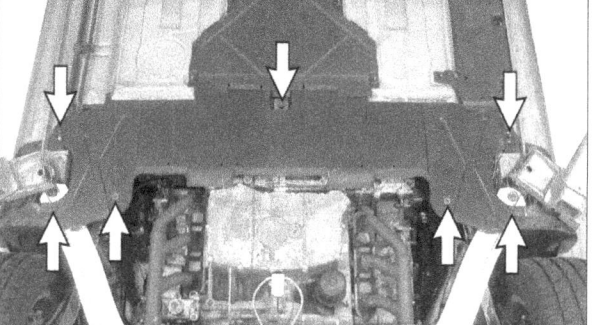

◀ Working underneath engine, remove plastic splash shield fasteners (**arrows**) and remove shield.

◀ To avoid draining the entire cooling system, use hose pinch tool to clamp shut radiator coolant hoses and coolant filler hose.

− Place 5-gallon pail under engine. Remove coolant manifold drain plug to drain crankcase coolant. See **Cooling system, draining, filling and bleeding** in this repair group.

> **WARNING—**
> - Coolant is poisonous and lethal to humans and pets. Clean up spills immediately and rinse area with water. Dispose of coolant safely.

− Remove right seat. See **72 Seats**.

− Remove carpeted panel from front engine access cover.

◀ Undo engine access cover mounting fasteners (**arrows**) behind seats and lift out cover.

> **CAUTION—**
> - Access cover edges are sharp.

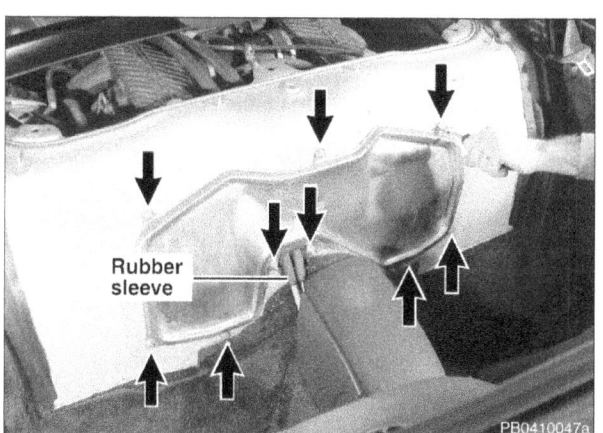

19-8 Cooling System

Cooling System Component Replacement

◀ Use 24 mm wrench to rotate engine drive belt tensioner clockwise. Slip belt off air-conditioning compressor (upper left pulley, upper right in photo).

NOTE—

* *If taking drive belt off, mark direction of rotation for reinstallation.*

◀ Remove coolant pump mounting bolts (**arrows**).
* Use flex-head 10 mm socket to access and remove leftmost bolt (rightmost in photo).
* Remove two lowest bolts from underneath car.

NOTE—

* *Engine shown removed from car for clarity.*

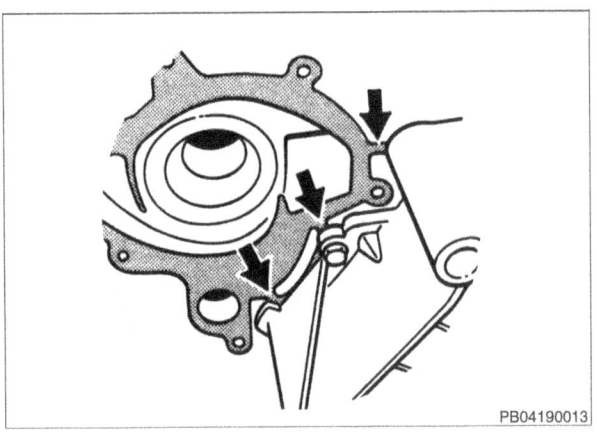

◀ Cut old metal gasket between coolant pump and coolant manifold at connecting webs (**arrows**).

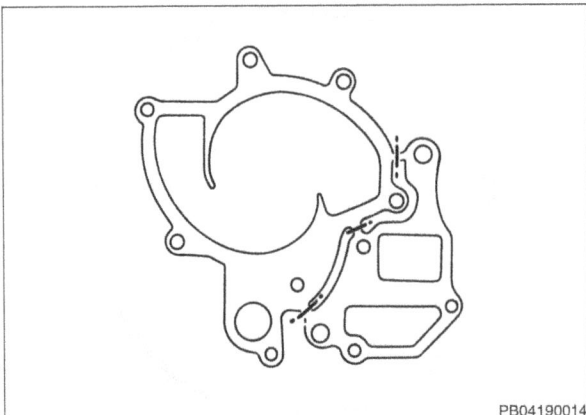

◀ Trim new coolant pump gasket along connecting webs (**dotted line**).

— Clean coolant pump sealing surface on crankcase.

— Install new pump and gasket and tighten mounting bolts.

Tightening torque	
Coolant pump to crankcase	10 Nm (7 ft-lb)

Cooling System 19-9

Cooling System Component Replacement

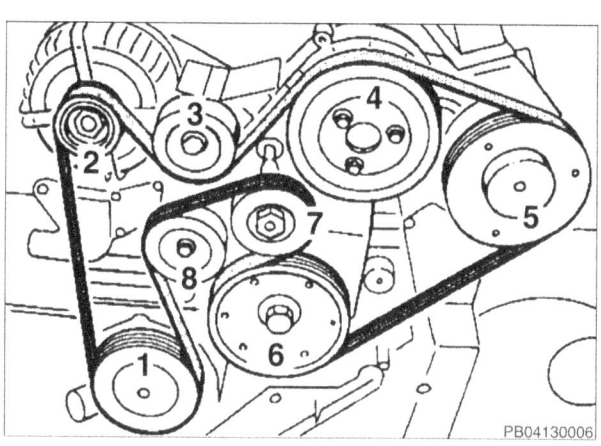

◂ Check engine drive belt for damage and replace if necessary. If reusing belt, fit in original direction of rotation. Place on pulleys as follows:

1. Coolant pump
2. Alternator
3. Upper idler
4. Power steering pump
5. Air-conditioning compressor
6. Crankshaft pulley
7. Tensioner: Use 24 mm wrench to turn tensioner clockwise.
8. Lower idler

– After installation, check to make sure belt is correctly positioned on all pulleys.

– Fill and bleed cooling system. See **Cooling system, draining, filling and bleeding** in this repair group. Run engine and check for coolant leaks.

– Reassemble engine access cover and reinstall seat.

Tightening torques	
Engine access cover to body	10 Nm (7 ft-lb)
Seat to floor	65 Nm (48 ft-lb)

Coolant thermostat, removing and installing

The coolant thermostat housing is on the right side front of the engine. Remove and install thermostat from underneath car.

> **WARNING—**
> - Due to risk of personal injury, be sure the engine is cold before beginning work on engine cooling components.

– Raise rear of car and support safely.

> **WARNING—**
> - Make sure the car is stable and well supported at all times. Use a professional automotive lift or jack stands designed for the purpose. A floor jack is not adequate support.

◂ Working underneath engine, remove plastic splash shield fasteners (**arrows**) and remove shield.

19-10 Cooling System

Cooling System Component Replacement

◂ To avoid draining the entire cooling system, use hose pinch tool to clamp shut radiator coolant hoses and coolant filler hose.

– Place 5-gallon pail under engine. Remove coolant manifold drain plug to drain crankcase coolant. See **Cooling system, draining, filling and bleeding** in this repair group.

> **WARNING—**
> • Coolant is poisonous and lethal to humans and pets. Clean up spills immediately and rinse area with water. Dispose of coolant safely.

– Loosen hose clamp and detach coolant hose from thermostat housing.

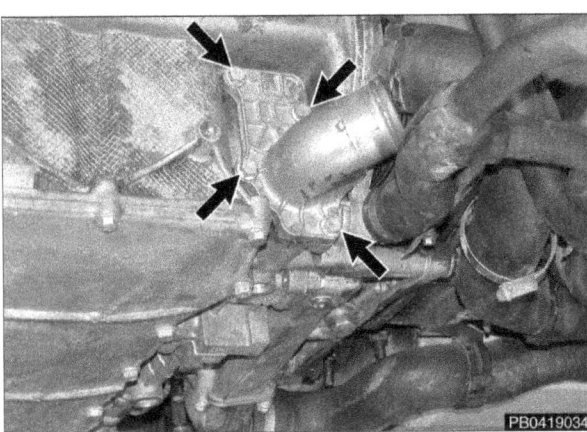

◂ Working under vehicle at right front underside of engine, remove thermostat housing mounting bolts (**arrows**). Remove thermostat housing and place on work bench.

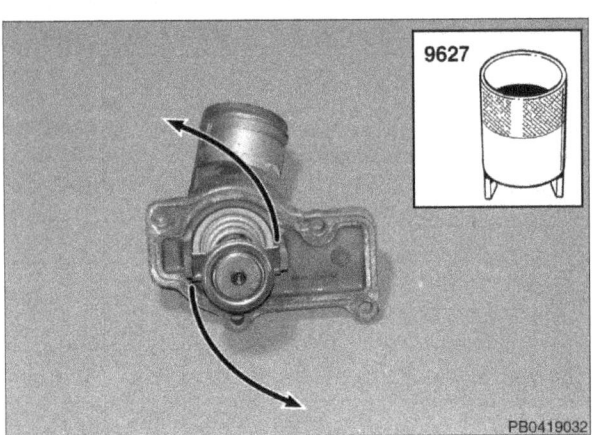

◂ Using Porsche special tool 9627 or equivalent, press down coolant thermostat, twist out of housing and install new.

– Reinstall thermostat housing with new gasket, reattach coolant hose, and fill and bleed cooling system. See **Cooling system, draining, filling and bleeding** in this repair group. Run engine and check for coolant leaks.

Tightening torque	
Coolant thermostat housing to crankcase	10 Nm (7 ft-lb)

Coolant reservoir, removing and installing

If rear trunk floor or carpeting is wet with coolant, remove coolant reservoir and inspect bottom. If leaks are found, replace coolant level sensor and gasket, coolant reservoir or both.

– Raise rear of car and support safely.

> **WARNING—**
> • Make sure the car is stable and well supported at all times. Use a professional automotive lift or jack stands designed for the purpose. A floor jack is not adequate support.

Cooling System 19-11

Cooling System Component Replacement

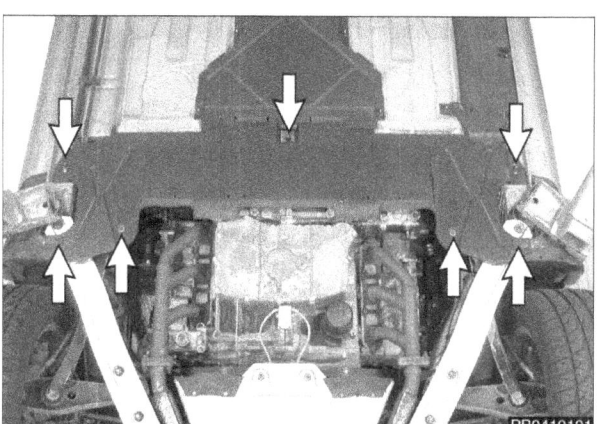

◁ Working underneath engine, remove plastic splash shield fasteners (**arrows**) and remove shield.

◁ To avoid draining the entire cooling system, use hose pinch tool to clamp shut radiator coolant hoses and coolant filler hose.

– Place 5-gallon pail under engine. Remove coolant manifold drain plug to drain crankcase coolant. See **Cooling system, draining, filling and bleeding** in this repair group.

> *WARNING—*
> * *Coolant is poisonous and lethal to humans and pets. Clean up spills immediately and rinse area with water. Dispose of coolant safely.*

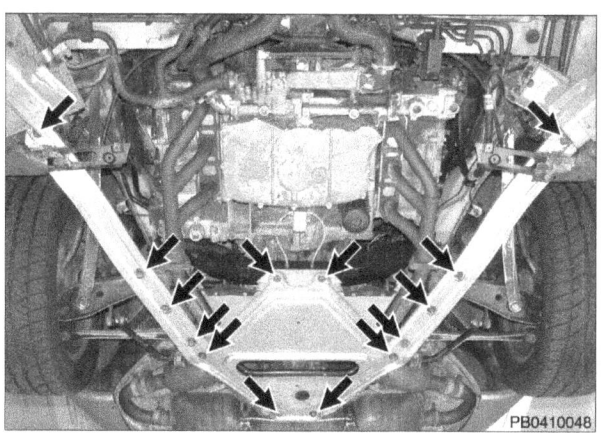

◁ Working underneath transmission, remove fasteners (**arrows**) for diagonal braces and rear suspension reinforcement plate. Set braces and plate aside.

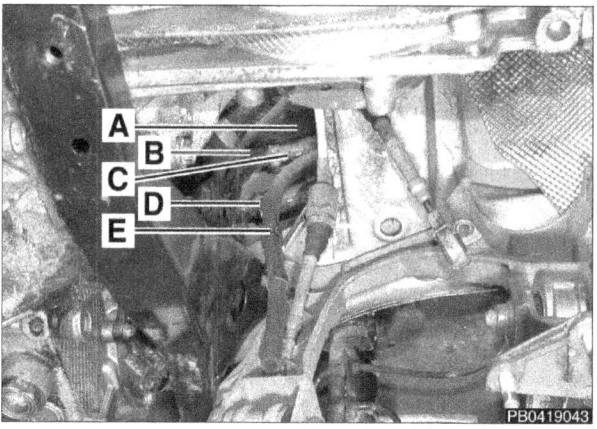

◁ Working underneath on right side of transmission, reach up and detach hoses at trunk bulkhead:
 • Oil filler hose (**A**)
 • Oil cooler coolant hose (**B**)
 • Radiator vent hose (**C**)
 • Coolant filler hose (**D**)
 • Coolant overflow hose (**E**)

19-12 Cooling System

Cooling System Component Replacement

◄ Open rear trunk and pull out dipstick. Use needle nose pliers to press together dipstick guide tube tabs and detach guide tube from engine service tray.

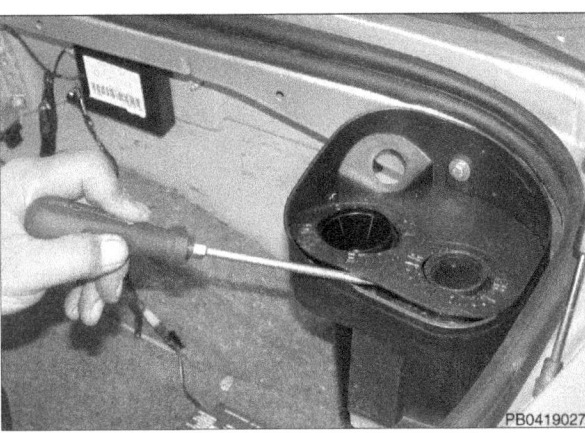

◄ Remove oil filler and coolant reservoir caps. Lever off trim cover underneath filler necks.

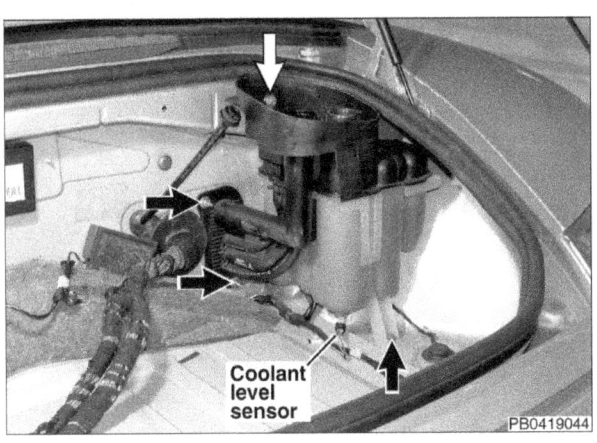

◄ Remove rear trunk floor carpet. Remove trunk trim along front bulkhead. Loosen and remove coolant reservoir mounting fasteners (**arrows**). Detach coolant level sensor harness connector.

– Check reservoir for leaks along seam and at coolant level sensor. Replace components as necessary.

– Installation is reverse of removal. Keep in mind the following:
  • Use new hose clamps on coolant lines.
  • Fill and bleed cooling system. See **Cooling system, draining, filling and bleeding** in this repair group. Run engine and check for coolant leaks.

Cooling System 19-13

Radiators and Engine Cooling Fans

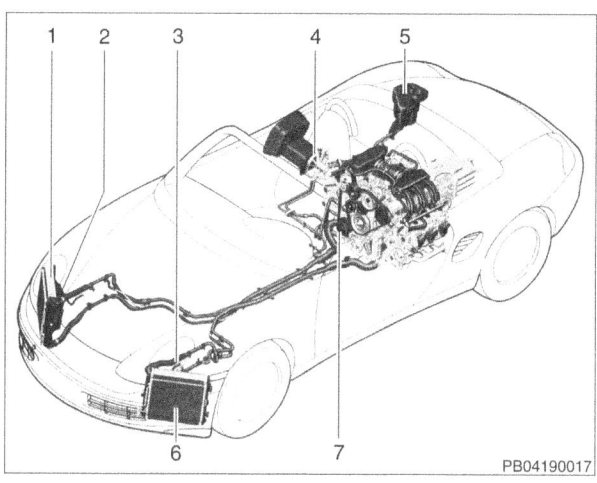

RADIATORS AND ENGINE COOLING FANS

◄ There are two radiators in the Boxster, one in each front wheel well behind the headlight. An air-conditioning condenser is mounted to the front of each radiator.

1. Right radiator and air-conditioning condenser
2. Right engine cooling fan
3. Left engine cooling fan
4. Engine compartment cooling fan (blower)
5. Coolant reservoir
6. Left radiator and air-conditioning condenser
7. Coolant pump

Boxster S is equipped with an additional center radiator in front.

The engine cooling fans operates at two different speeds, controlled by the engine control module (ECM):

- Low speed operation with coolant temperature above 96.75°C (206.2°F) or A/C switched ON
- High speed operation with coolant temperature above 102°C (215.6°F) or A/C pressure switch closed (refrigerant pressure above 16 bar / 232 psi)

Engine cooling fan, removing and installing

The procedures for removing the left or right engine cooling fan under the front fender are similar. Left fan removal in a 1998 Boxster is illustrated.

− Raise front of car and support safely.

> **WARNING—**
> • Make sure the car is stable and well supported at all times. Use a professional automotive lift or jack stands designed for the purpose. A floor jack is not adequate support.

− Remove front wheel.

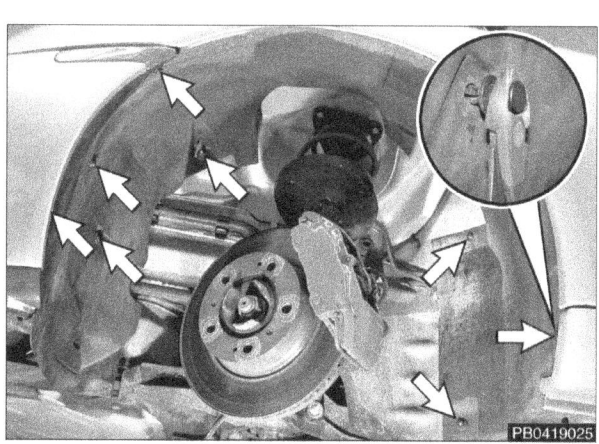

◄ Remove front wheel housing liner fasteners (**arrows**).
- To disengage plastic rivet, pull out rivet lock (**inset**).
- Remove plastic 10 mm nuts.

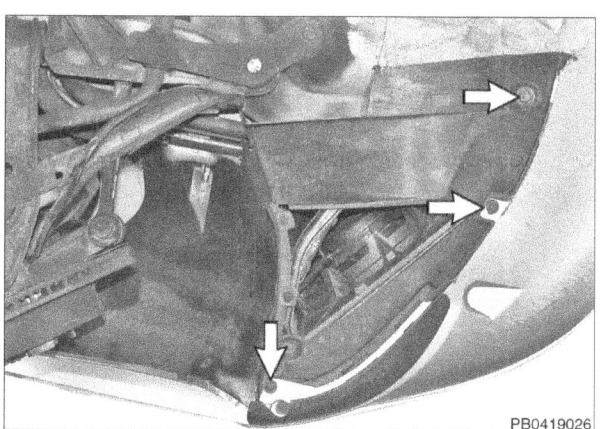

◄ Working underneath front fender, remove remainder of wheel housing liner fasteners (**arrows**). Lift off liner.

19-14 Cooling System

Radiators and Engine Cooling Fans

◀ Remove fender brace fasteners (**arrows**). Remove fender brace.

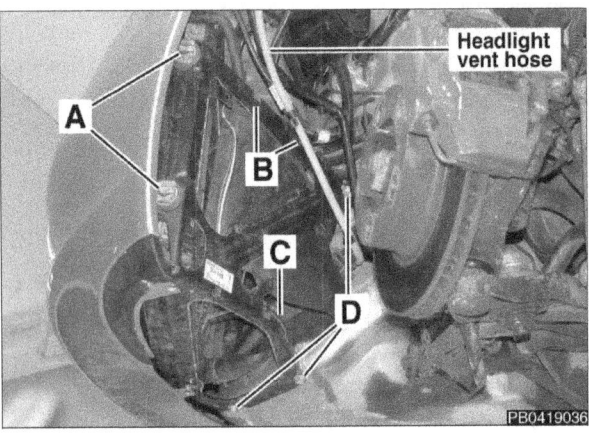

◀ Working at radiator support bracket:
- Remove radiator mounting circlips (**A**).
- Detach radiator vent hose clips (**B**) from bracket.
- Detach engine cooling fan resistor mounting clip (**C**) from bracket.
- Remove bracket mounting bolts (**D**) and carefully work bracket out of wheel well.

NOTE—
- *If necessary, detach headlight vent hose from headlight housing at top and move aside.*

◀ Working at cooling fan housing:
- Detach fan electrical connector (**A**).
- Remove fan housing mounting clips (**B**) and work fan off radiator.

◀ Place fan on work bench. Using small screwdriver, bend plastic clip to release electrical connector from fan housing.

Cooling System 19-15

Radiators and Engine Cooling Fans

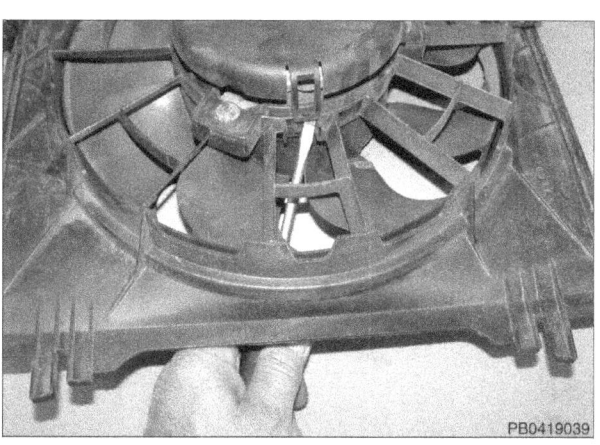

◄ Pry off outer fan cover.

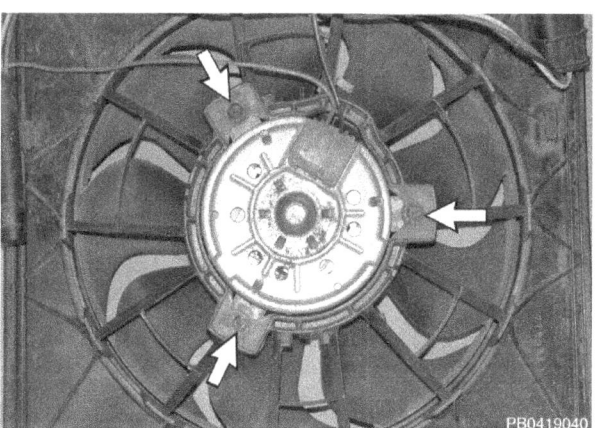

◄ Remove three mounting screws and detach fan from housing.

– Installation is reverse of removal.

Radiator, removing and installing

The procedures for removing the left or right radiator under the front fender are similar. Left radiator removal in a 1998 Boxster is illustrated.

> **WARNING—**
> * Due to risk of personal injury, be sure the engine is cold before beginning work on engine cooling components.

– Raise front of car and support safely.

> **WARNING—**
> * Make sure the car is stable and well supported at all times. Use a professional automotive lift or jack stands designed for the purpose. A floor jack is not adequate support.

– Remove front wheel, front wheel housing liner, radiator bracket and engine cooling fan. See **Engine cooling fan, removing and installing** in this repair group.

– Place 5-gallon pail under radiator coolant hoses. Pinch shut hoses, loosen clamps and detach hoses from radiator.

> **WARNING—**
> * Coolant is poisonous and lethal to humans and pets. Clean up spills immediately and rinse area with water. Dispose of coolant safely.

19-16 Cooling System

Radiators and Engine Cooling Fans

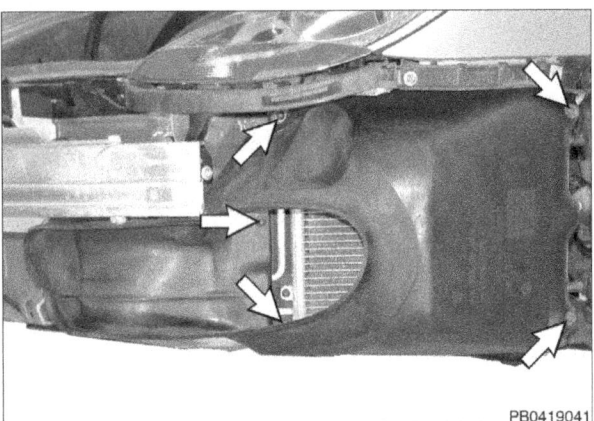

- Remove front bumper cover. See **63 Bumpers**.

◄ Remove radiator air duct mounting fasteners (**arrows**). Lift off duct.

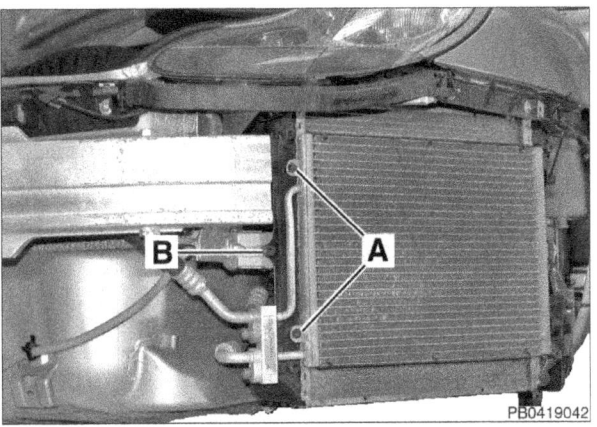

◄ Remove condenser mounting bolts (**A**). Pull condenser out of bracket at right and swing aside.

> *CAUTION—*
> * *Do not detach or damage refrigerant lines.*
> * *Use stiff wire to suspend condenser securely from bumper.*

- Slide radiator mounting pin (**B**) out of bracket. Remove radiator.

- Installation is reverse of removal. When finished, fill and bleed cooling system. See **Cooling system, draining, filling and bleeding** in this repair group. Run engine and check for coolant leaks.

20 Fuel Storage and Supply

GENERAL 20-1	FUEL TANK AND PUMP 20-5
System description 20-1	Fuel tank, draining. 20-5
Warnings and Cautions 20-2	Fuel pump and sending unit, removing
FUEL DELIVERY TESTS 20-3	and installing (1997 - 2001 models). 20-6
Relieving fuel system pressure 20-3	
Operating fuel pump for tests	
(1997 - 2001 models). 20-3	
Fuel pressure, checking	
(1997 - 2001 models). 20-4	
Fuel volume, checking (1997 - 2001 models) . . . 20-4	

GENERAL

This repair group covers service information for the fuel pump and related storage and supply components.

For fuel filter replacement see **03 Service and Maintenance**.

System description

The fuel pump delivers fuel at high pressure to the fuel injection system. During starting, the fuel pump runs as long as the ignition switch is in the start position and continues to run once the engine starts. If an electrical system fault interrupts power to the fuel pump, the engine will not run.

To diagnose a fuel related problem, begin by checking the fuel pump fuse, the fuel pump relay and DME main relay.

Fuel system fuse and relay locations	
Fuel pump fuse	Fuse panel in left footwell kick panel
Fuel pump relay	Relay panel 1 above fuse panel
DME main relay	Relay panel 2 in rear trunk

 1997 through 2001 Boxster and Boxster S models have an electric fuel pump mounted in the fuel tank in tandem with the fuel level sender. The fuel filter and pressure regulator are externally mounted.

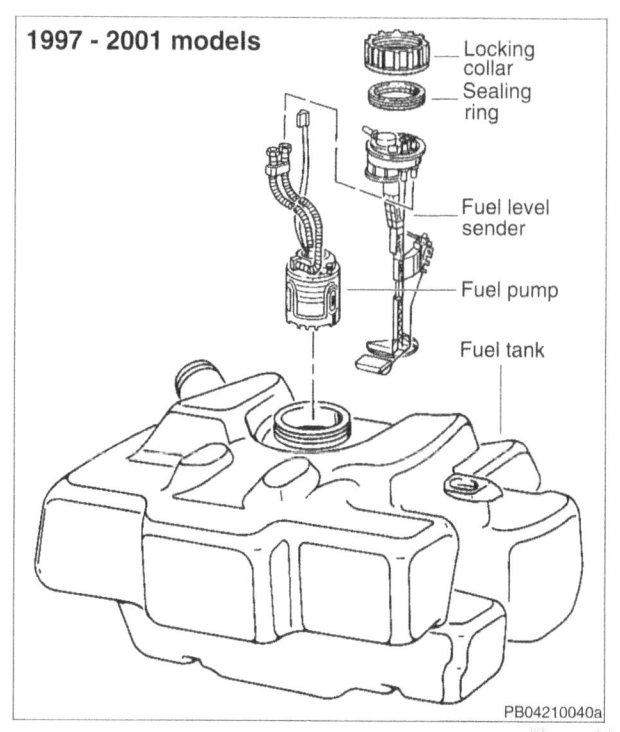

1997 - 2001 models
- Locking collar
- Sealing ring
- Fuel level sender
- Fuel pump
- Fuel tank

20-2 Fuel Storage and Supply

General

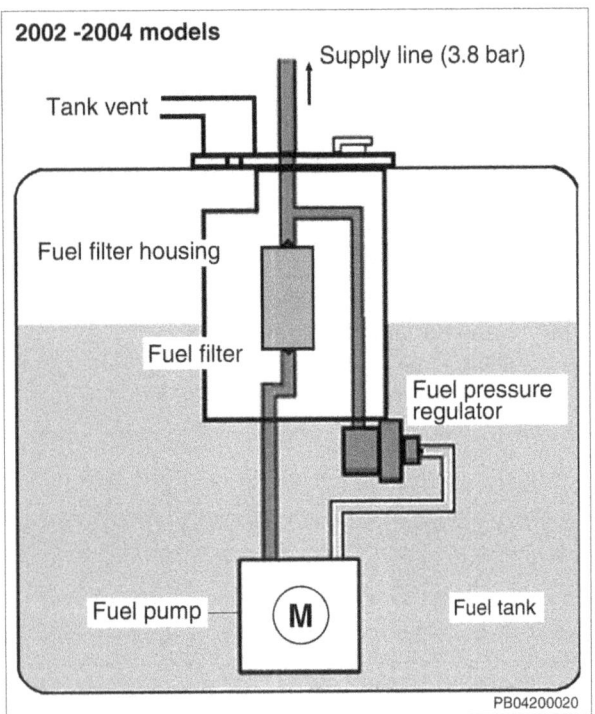

2002 - 2004 Boxster and Boxster S models use a maintenance-free non-return fuel system that incorporates the pump, level sender, fuel filter and pressure regulator within the fuel tank. These parts are not available separately.

NOTE—
- *On 2002 - 2004 models, the in-tank fuel filter is specified as a lifetime filter by Porsche. Due to the reduced volume of fuel flowing through the non-return system, no replacement interval is specified.*

Warnings and Cautions

Observe the following warnings and cautions when servicing the fuel system.

WARNING—
- *The fuel system is designed to retain pressure even when the ignition is OFF. When working with the fuel system, loosen the fuel lines slowly to allow residual fuel pressure to dissipate. Avoid spraying fuel. Use shop rags to capture leaking fuel.*
- *Before beginning work on the fuel system, place a fire extinguisher in the vicinity of the work area.*
- *Fuel is highly flammable. When working around fuel, do not disconnect wires that could cause electrical sparks. Do not smoke or work near heaters or other fire hazards.*
- *Unscrew the fuel tank cap to release pressure in the tank before working on the tank or lines.*
- *Do not use a work light with an incandescent bulb near fuel. Fuel may spray on the hot bulb causing a fire.*
- *Make sure the work area is properly ventilated.*

CAUTION—
- *Prior to disconnecting the battery, read the battery disconnection cautions in **00 Warnings and Cautions**.*
- *Before making any electrical tests with the ignition turned on, disable the ignition system. Be sure the battery is disconnected when replacing components.*
- *To prevent damage to the ignition system or other DME components, including the engine control module (ECM), always connect and disconnect wires and test equipment with the ignition OFF.*

Fuel Storage and Supply 20-3

Fuel Delivery Tests

CAUTION—
- *Cleanliness is essential when working with the fuel system. Thoroughly clean the fuel line unions before disconnecting any of the lines.*
- *Use only clean tools. Keep removed parts clean and sealed or covered with a clean, lint-free cloth, especially if completion of the repair is delayed.*
- *Do not move the car while the fuel system is open.*
- *Avoid using high pressure compressed air to blow out fuel lines and components. High pressure can rupture internal seals and gaskets.*
- *Always replace seals, O-rings and hose clamps.*

FUEL DELIVERY TESTS

This section describes fuel delivery tests for 1997 - 2001 Boxster and Boxster S models.

2002 - 2004 Boxster and Boxster S models are equipped with a non-return fuel system. Fuel delivery tests for non-return models require the Porsche System Tester 2 (PST2) and are beyond the scope of this manual.

Relieving system fuel pressure

The fuel system retains fuel pressure in the system when the engine is off. To prevent fuel from spraying on a hot engine, relieve system fuel pressure before disconnecting fuel lines. One method is to tightly wrap a shop towel around a fuel line fitting and loosen or disconnect the fitting.

Operating fuel pump for tests (1997 - 2001 models)

- To operate fuel pump for testing purposes without having to run engine, bypass fuel pump relay to power pump directly.

  Access relay panel 1 (above fuse panel, left footwell) and remove fuel pump relay (**arrow**). See **97 Fuses, relays, Component Locations**.

CAUTION—
- *Fuse and relay locations may vary. See* **97 Fuses, Relays, Component Locations** *for additional relay information.*

- Using a fused jumper wire, bridge terminals 30 and 87 (labeled 3 and 5 on the relay panel).
  - *Make jumper wire from 1.5 mm^2 (14 ga) wire and include an inline fuse holder with a 15 A fuse. To avoid relay panel damage from repeated connecting and disconnecting, also include a toggle switch in the jumper harness.*
  - *The fuel pump runs as soon as the jumper wire is attached.*

- Remove jumper harness after completing tests.

Fuel Delivery Tests

Fuel pressure, checking (1997 - 2001 models)

Fuel pressure directly influences fuel delivery. Use an accurate fuel pressure gauge for fuel pressure checks.

To operate the fuel pump for testing purposes, Follow one of two procedures:

- Bypass relay to power pump directly. See **Operating fuel pump for tests (1997 - 2001 models)** in this repair group.
- If the car is capable of idling, connect pressure gauge as described below and start engine. Shut off engine after reading fuel pressure.

– Remove top engine access cover. See **03 Service and Maintenance.**

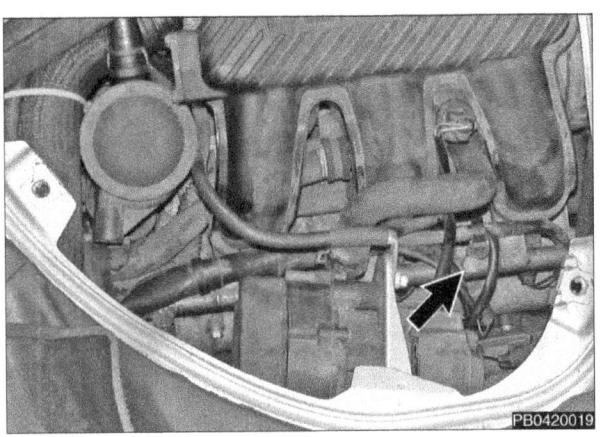

◀ Remove brass cap from fuel rail test connection fitting (**arrow**) and connect pressure gauge (Porsche special tool 378a / 9559).

> **WARNING—**
> - *Fuel will be expelled under pressure. Wrap a cloth around fuel line to absorb any leaking fuel.*

– Run fuel pump for several seconds and observe fuel pressure. See **Operating fuel pump for tests** in this repair group.

Nominal fuel pressure	
Engine off	3.8 ± 0.2 bar
Engine idling	3.3 ± 0.2 bar

– Remove fuel gauge from test connection and replace brass cap with new.

Tightening torque	
Brass cap to test connection	2.5 ± 0.5 Nm (1.8 ± 0.36 ft-lb)

Fuel volume, checking (1997 - 2001 models)

Fuel volume can be influenced by the condition of the fuel filter and battery. Make sure fuel filter is not clogged and battery is fully charged before beginning test. Use a suitable fuel safe container for fuel volume checks.

– Relieve fuel pressure. See **Relieving system fuel pressure** in this repair group.

– Raise vehicle.

> **WARNING—**
> - *Make sure the car is stable and well supported at all times. Use a professional automotive lift or jack stands designed for the purpose. A floor jack is not adequate support.*

– Remove rear underside panel to expose fuel lines.

Fuel Storage and Supply 20-5
Fuel Tank and Pump

◄ Working at left front of engine, locate fuel return line.

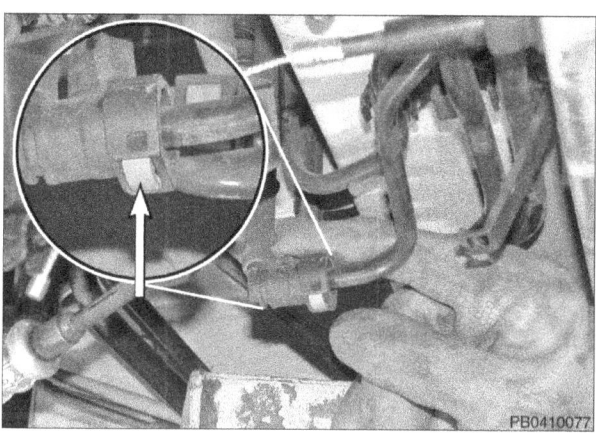

◄ Wrap a shop towel around fitting to absorb any leaking fuel. Separate fuel return line by gently pressing on button (**arrow**).

– Place and fuel return line in a measuring container.

– Bypass fuel pump relay. See **Operating fuel pump for tests** in this repair group. Run fuel pump for exactly 30 seconds and collect output.

NOTE—
- *Alternatively, connect Porsche System Tester 2 (PST 2) to operate fuel pump.*

– Minimum delivered fuel quantity is specified below.

Nominal fuel volume	
Engine off, fuel pump running	850 ml (0.9 US qt)

– Remove jumper harness (or PST 2) after completing tests.

– Reconnect fuel return line until audible click is heard. Make sure lines are connected properly by applying moderate pressure in an attempt to pull them apart.

FUEL TANK AND PUMP

Fuel tank, draining

Drain fuel tank into a safe storage unit using an approved fuel pumping device.

WARNING—
- *Before draining tank, be sure that all hot components, such as the exhaust system, are completely cooled down.*
- *Fuel may be spilled. Do not smoke or work near heaters or other fire hazards.*

– Disconnect negative (–) cable from battery.

CAUTION—
- *Prior to disconnecting the battery, read the battery disconnection cautions given in* **00 Warnings and Cautions**.

20-6 Fuel Storage and Supply

Fuel Tank and Pump

- Remove fuel tank filler cap.

- Slide suction hose into filler neck, twisting as necessary. Withdraw fuel into storage unit.

- Remove suction hose from tank filler neck carefully to avoid damaging filler neck baffle plate.

- After finishing repairs but before starting engine, fill fuel tank with at least 5 liters (1.5 gallons) of fuel.

> **CAUTION—**
> • Fuel pump is damaged if run without fuel.

Fuel pump and sending unit, removing and installing (1997 - 2001 models)

The fuel pump and fuel level sending unit are located in the fuel tank.

> **NOTE—**
> • 2002 - 2004 Boxster and Boxster S models are equipped with a maintenance-free non-return fuel system. Fuel pump assembly removal is not covered.

- Make sure ignition key is OFF and remove key.

◄ Working in front trunk under cowl, turn locking fasteners (**arrows**) 90° and remove battery cover.

- Remove negative battery lead first and cover with insulating tape. Remove positive lead.

> **CAUTION—**
> • Prior to disconnecting the battery, read the battery disconnection cautions in **00 Warnings and Cautions**.

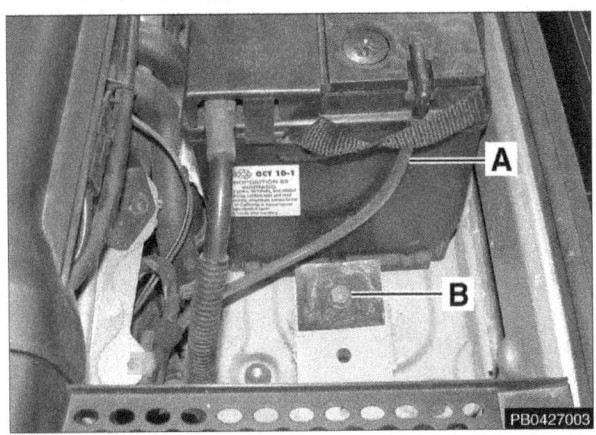

◄ Disconnect battery vent hose (**A**).

- Remove battery hold down fastener (**B**) and remove battery.

◄ Remove battery tray to expose top of fuel tank.

- Drain fuel tank. See **Fuel tank, draining** in this repair group.

Fuel Storage and Supply 20-7
Fuel Tank and Pump

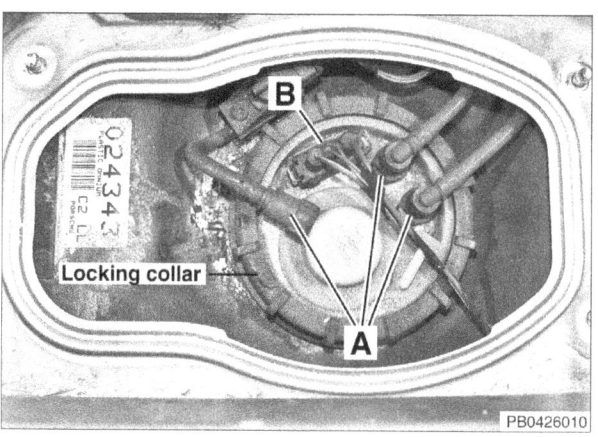

◄ Disconnect fuel lines (**A**) by pressing in on locking clips and pulling lines off fittings.

– Remove electrical connector (**B**) from top of tank by pressing in on retaining clips and pulling connector up.

– Use VW special tool 3217, or equivalent, to turn locking collar counter clockwise and remove locking collar.

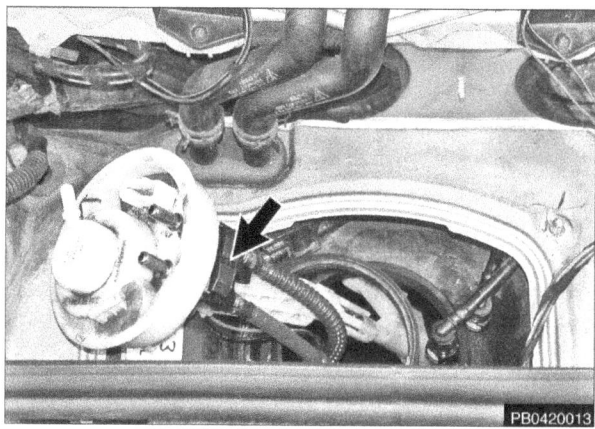

◄ Lift sending unit carefully from tank and release fuel lines by pressing locking tab (**arrow**) and simultaneously pulling both lines from their fittings.

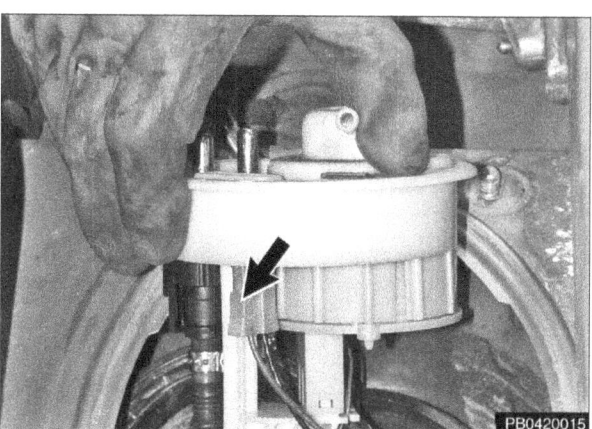

◄ Press locking tab (**arrow**) to release electrical connector from opposite side of sending unit.

– Remove sending unit.

CAUTION—
- *Sending unit is delicate. Make sure sending unit float arm is not bent when removing.*

– Siphon out remaining fuel.

– Using a fuel proof glove, reach down inside tank and firmly grasp fuel pump. Turn fuel pump counterclockwise approximately 15° to release from bayonet lock at base of tank.

– Remove fuel pump.

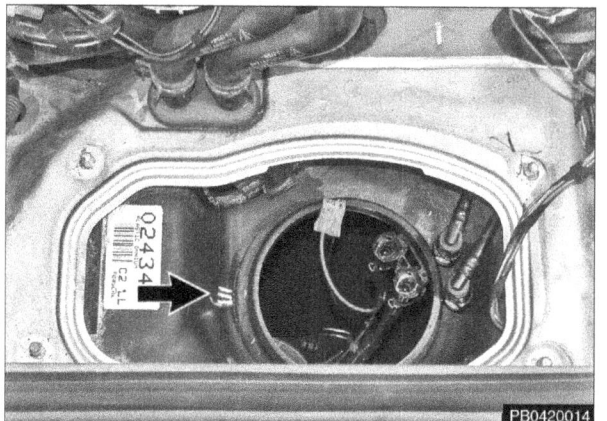

◄ To install, match left corner on flat side of fuel pump with sending unit position marks (**arrow**) on fuel tank.

– Lower fuel pump onto bayonet lock at bottom of tank.

– Turn fuel pump in bayonet lock clockwise as far as stop. Pull pump straight up to make sure it is seated properly.

– Attach electrical connector and fuel lines from fuel pump to sending unit. Fuel lines must snap audibly into place. Make sure lines are connected properly by applying moderate pressure in an attempt to pull them apart.

20-8 Fuel Storage and Supply

Fuel Tank and Pump

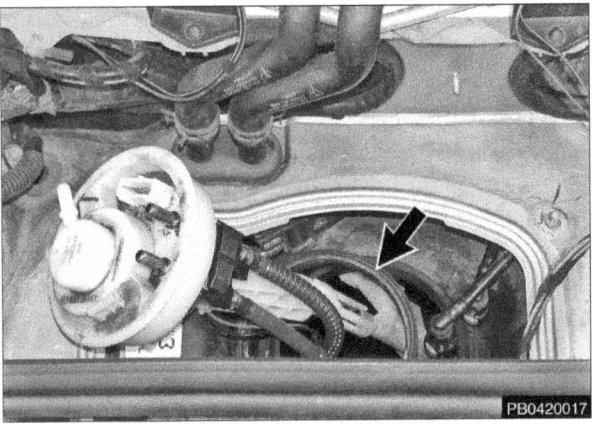

◂ Replace sealing ring (**arrow**) with new. Fit sealing ring into fuel tank opening.

− Lower sending unit into tank. Do not force. Make sure sending unit returns to its original position and sealing ring remains in place.

◂ Line up match marks on sending unit (molded triangle shape) and top of tank (molded dashes).

− Use VW special tool 3217 to tighten new locking collar to fuel tank.

Tightening torque	
Fuel level sending unit to fuel tank	70 Nm (52 ft-lb)

− Attach electrical connector and fuel lines to sending unit. Fuel lines must snap audibly into place. Make sure lines are connected properly by applying moderate pressure in an attempt to pull them apart.

NOTE—
* *The color coded green plug must be attached to the connection marked "V".*

− When reinstalling battery, first connect positive lead, then negative lead. Reattach ventilation hose.

Tightening torques	
Battery hold down to battery box	23 Nm (17 ft-lb)
Battery terminal clamp to battery post	5 Nm (4 ft-lb)

− After finishing repairs but before starting engine, fill fuel tank with at least 5 liters (1.5 gallons) of fuel.

CAUTION—
* *Fuel pump is damaged if run without fuel.*

NOTE—
* *It is common practice to replace the fuel filter when replacing the fuel pump. See* **03 Service and Maintenance**.

24 Fuel Injection

GENERAL ... 24-1
- Bosch M 5.2.2 engine management components (front of engine) ... 24-2
- Bosch M 5.2.2 engine management components (rear of engine) ... 24-2
- Engine management system overview ... 24-3
- Secondary air injection ... 24-4
- Secondary air injection components ... 24-4
- Evaporative control system ... 24-5
- Evaporative control system components ... 24-5
- DME adaptation (ME 7.2 / ME 7.8) ... 24-6
- Warnings and Cautions ... 24-6

DME COMPONENT TESTING ... 24-7
- FIO relay, testing ... 24-7
- DME main relay, testing ... 24-8
- Engine coolant temperature (ECT) sensor, testing ... 24-8

DME COMPONENT REPLACEMENT ... 24-9
- Engine coolant temperature (ECT) sensor, removing and installing ... 24-9
- Fuel injectors, removing and installing ... 24-10
- Mass air flow (MAF) sensor, removing ... 24-12
- Engine control module (ECM), removing ... 24-12

ECM PIN ASSIGNMENTS ... 24-12
- DME M 5.2.2 control module pin assignments ... 24-13
- DME ME 7.2 control module pin assignments ... 24-15
- DME ME 7.8 control module pin assignments ... 24-18

ON-BOARD DIAGNOSTICS ... 24-21
- Malfunction indicator light (MIL) ... 24-21
- Scan tool and scan tool display ... 24-22
- Diagnostic monitors ... 24-23
- Drive cycle ... 24-25
- Readiness codes ... 24-25
- Diagnostic trouble codes ... 24-26
- P-codes (DTCs) and Porsche fault codes ... 24-27

TABLES
a. Engine management systems ... 24-1
b. DME System M 5.2.2 DTCs ... 24-27
c. DME System ME 7.2 DTCs ... 24-34

GENERAL

This repair group covers service and repair for the various Bosch engine management systems used on the 1997 - 2004 Boxster models. Information about the on-board diagnostic system (OBD-II) is also provided. See also:

- **20 Fuel Storage and Supply**
- **28 Ignition System**
- **EWD Electrical Wiring Diagrams**

NOTE—

- *Fuel pressure testing and fuel pump repair information is covered in* **20 Fuel Storage and Supply***.*
- *For fuel filter replacement, see* **03 Service and Maintenance***.*

Table a. Engine management systems	
Model year: Engine code	**DME system**
1997 - 1999 Boxster 2.5 liter	Bosch M 5.2.2
2000 - 2002 Boxster 2.7 liter / Boxster S	Bosch ME 7.2
2003 - 2004 Boxster 2.7 liter / Boxster S	Bosch ME 7.8

24-2 Fuel Injection

General

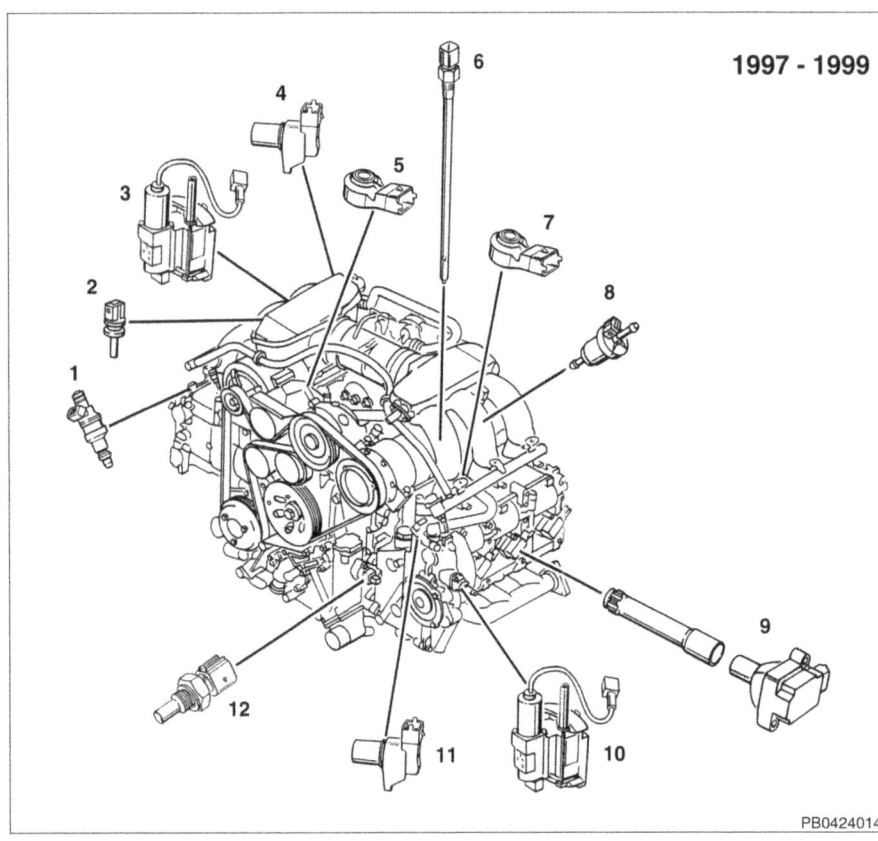

1997 - 1999

Bosch M 5.2.2 engine management components (front of engine)

1. Fuel injector
2. Engine compartment temperature sensor
3. Camshaft (VarioCam) adjuster
4. Camshaft position (CMP) sensor
5. Knock sensor
6. Oil temperature sender/oil level sensor
7. Knock sensor
8. Carbon canister (EVAP) purge valve
9. Ignition coil and spark plug connector
10. Camshaft (VarioCam) adjuster
11. Camshaft position sensor (CMP) sensor
12. Coolant temperature (ECT) sender

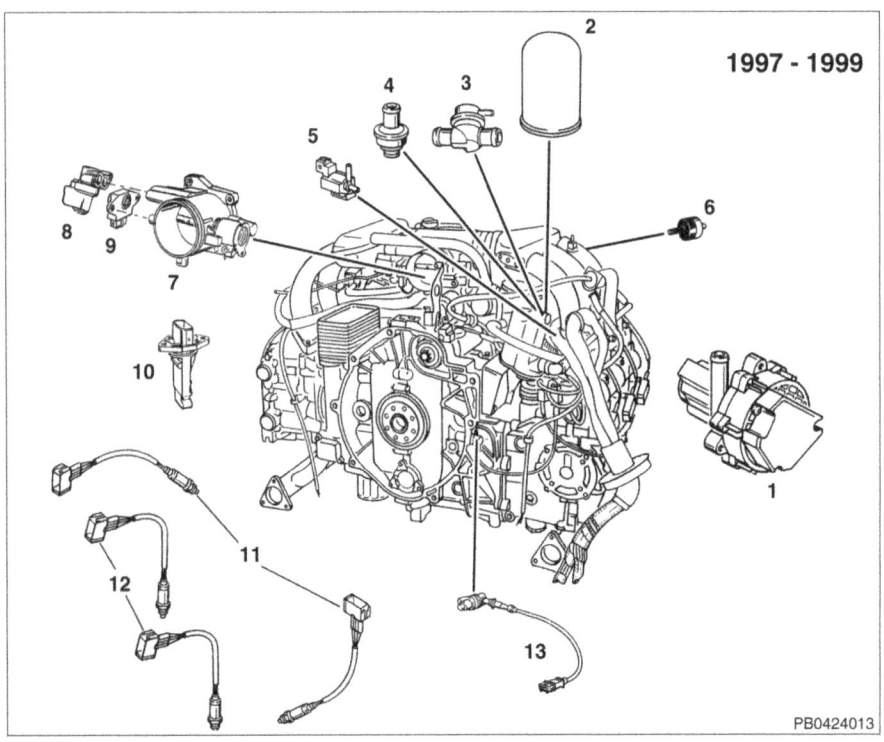

1997 - 1999

Bosch M 5.2.2 engine management components (rear of engine)

1. Secondary air pump
2. Vacuum reservoir
3. Fuel pressure regulator
4. Check valve
5. Secondary air vacuum valve
6. Check valve
7. Throttle housing
8. Idle air control valve
9. Throttle position sensor (TPS)
10. Hot film mass air flow sensor with intake air temperature sensor
11. Precatalyst oxygen sensors
12. Post-catalyst oxygen sensors
13. Crankshaft (speed/reference) position sensor

Fuel Injection 24-3

General

Engine management system overview

The DME (Digital Motor Electronics) engine management system controls fuel injection and ignition functions. It generates the injection signal, calculates the ignition angle, and provides adaptive control functions for various systems, including OBD (On Board Diagnostics) functions.

Bosch M 5.2.2 is installed on 1997 - 1999 Boxster models. The main features of the M 5.2.2 system are:

- Cylinder sequential fuel injection (separate fuel mixture control for each cylinder).
- Electronically mapped ignition system with individual coils at each cylinder (direct ignition).
- Adaptive heated oxygen sensor control, two sensors per catalytic converter.
- VarioCam (variable intake valve timing).
- Hot film mass air flow sensor with integral intake air temperature sensing.
- Idle air control using a twin-coil rotary adjuster.
- Adaptive throttle position sensor.
- Adaptive ignition knock control.
- Camshaft position sensors.
- Oil temperature measuring.
- Temperature sensor with analog instrument cluster display.
- DME-controlled radiator cooling fans.
- DME-controlled engine compartment blower.
- A/C compressor control.
- Check Engine light (malfunction indicator light or MIL) for emissions related and catalytic converter damaging faults.
- Secondary air injection.
- Evaporative emissions control with fuel tank leak detection
- Engine torque reduction (ignition angle and fuel injection intervention) for traction control.
- Active carbon canister venting
- Fuel pump disabling via SRS control module
- Idle speed adaptation for Tiptronic vehicles
- Engine torque control during Tiptronic gear shifting

 Bosch ME 7.2 is installed on 2000 - 2002 Boxster and Boxster S models. The major difference between ME 7.2 as compared to Bosch M 5.2.2 is the addition of the electronic throttle system.

NOTE—

- *On 2002 Boxster and Boxster S models, the ME 7.2 system was modified to include a non-return fuel supply system. This change helps to reduce hydrocarbon (HC) emissions. This technical upgrade also included a revised contaminant-resistant mass air flow sensor, smaller, lighter fuel injectors with 4-hole spray disks, and planar (Bosch LSF) oxygen sensors.*

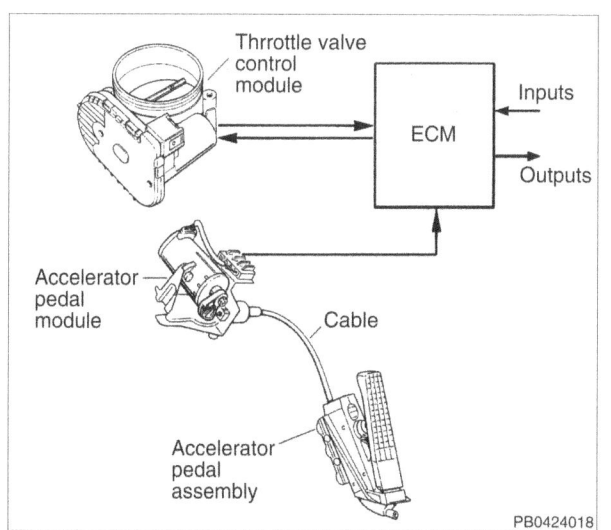

24-4 Fuel Injection

General

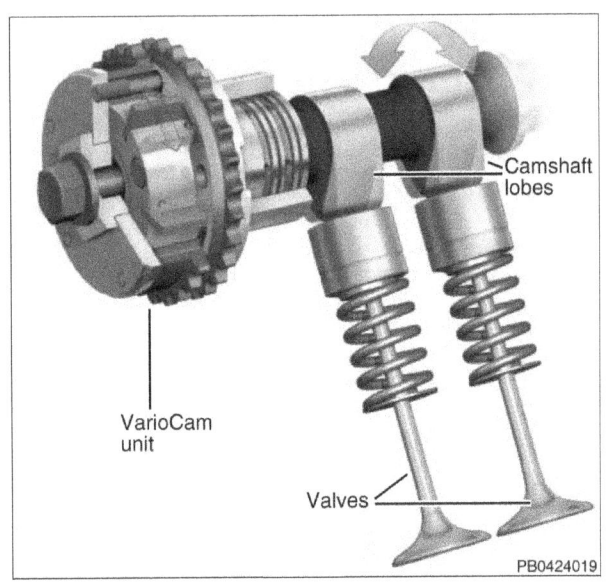

 Bosch ME 7.8 is installed on 2003 - 2004 Boxster and Boxster S models. The ME 7.8 system is a further refinement to the ME 7.2 system. The major technical upgrade to the ME 7.8 system is the VarioCam system with continuously variable vane-type adjuster on the intake camshaft.

Additional features include:

- Reduction in exhaust gas emissions (Low Emissions Vehicle - LEV compliant)
- Reduction in fuel consumption (approximately 2%)
- Fuel supply system with non-return flow (also introduced on 2002 ME 7.2)
- Sequential manifold injection with 4-hole fuel injectors (introduced on 2002 ME 7.2)
- Bosch LSF planar oxygen sensors (introduced on 2002 ME 7.2)
- Bosch HFM 5 hot film mass air flow sensor (introduced on 2002 ME 7.2)

Secondary air injection

The secondary air system pumps ambient air into the exhaust stream after a cold engine start to reduce the warm up time of the catalytic converters and to reduce HC and CO emissions. The engine control module (ECM) controls and monitors the secondary air injection system.

The injection pump draws in ambient air and supplies it to the air change-over valve. An electric solenoid, activated by the control module, switches vacuum to open the air change-over valve. Once opened, the air change-over valve allows air to be pumped into the exhaust stream via the cylinder heads.

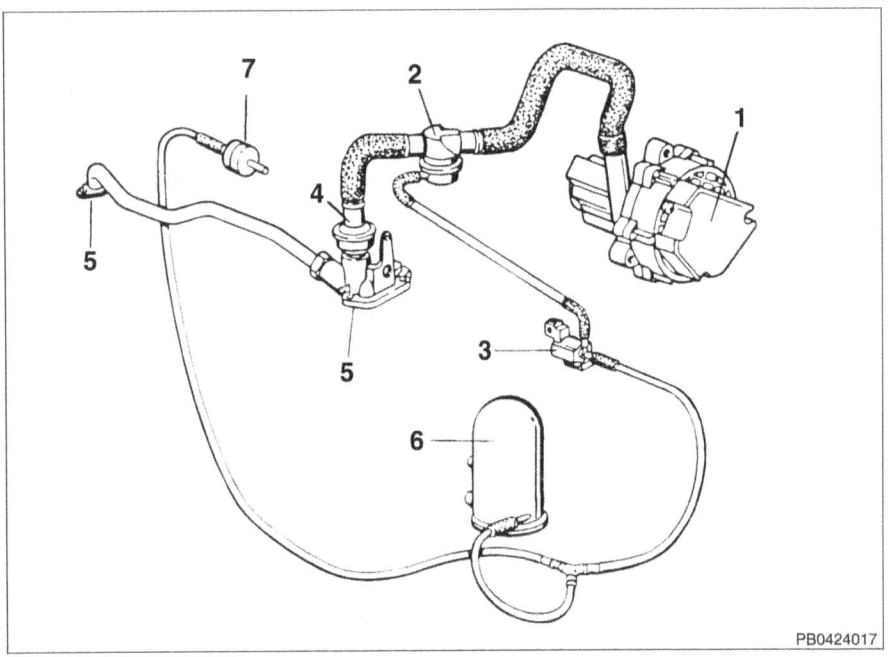

Secondary air injection components

1. Secondary air injection pump
2. Air change-over valve
3. Electric vacuum switching valve
4. Non-return valve
5. To cylinder heads
6. Vacuum reservoir
7. To vacuum source

Fuel Injection 24-5

General

Evaporative control system

The evaporative control system is designed to prevent fuel system evaporative losses from venting into the atmosphere. The evaporative system allows control and monitoring of evaporative losses by the on-board diagnostic (OBD II) software incorporated into the engine control module (ECM).

The system includes a carbon canister to store evaporated fuel, and plumbing to duct vapor from the fuel tank to the carbon canister and from the canister to the intake manifold. The carbon canister and associated valves are located in the rear of the right front fender, behind the fender lining.

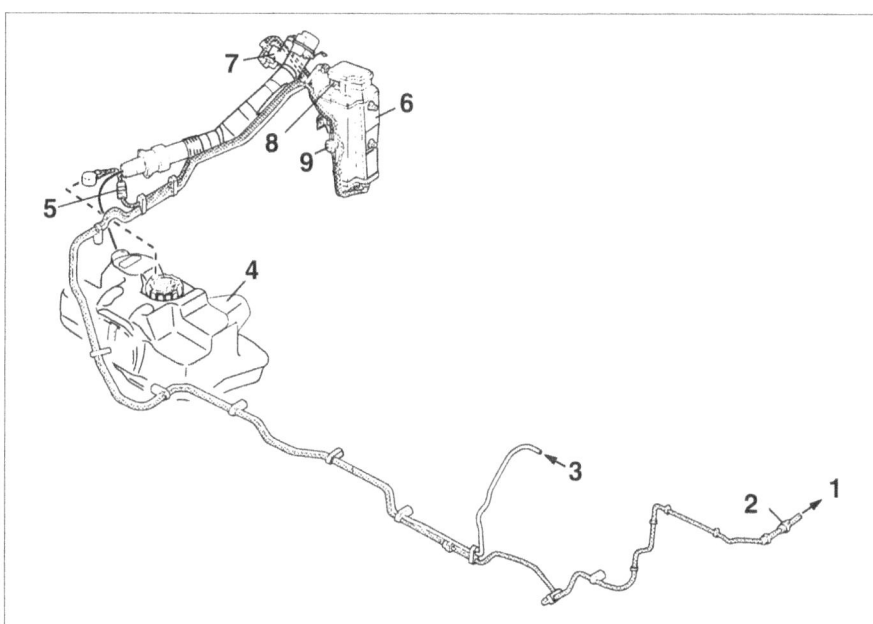

Evaporative control system components

1. **To intake manifold**
2. **Carbon canister (EVAP) purge valve**
   - On left side of engine, under intake manifold
3. **Purge air**
4. **Fuel tank**
5. **Differential pressure sensor**
   - For EVAP fuel tank leak detection
6. **Carbon canister**
   - In right front wheel well
   - To replace, disconnect vacuum lines and harness connector, remove mounting nut and pull canister from rear mounting grommets.
7. **Operating vacuum valve**
   - On top of fuel filler neck, in wheel well. Installing filler cap tightly opens valve.
8. **Carbon canister shutoff valve**
9. **Vacuum control valve**

General

DME adaptation (ME 7.2 / ME 7.8)

On ME 7.2 and ME 7.8 cars, the engine control module (ECM) software performs a learning and adaptation routine for the throttle unit if:

- The power supply to the ECM is interrupted.
- The ECM plugs are disconnected.
- A new ECM is installed.
- The throttle unit is replaced.
- The ECM is programmed.

– The following conditions must also be observed for adaptation to succeed:

- Vehicle is stationary.
- Battery positive voltage between 10 vdc and 16 vdc.
- Engine temperature between 5°C and 100°C (41°F and 212°F).
- Intake air temperature between 10°C and 100°C (50°F and 212°F).

– To perform DME adaptation:

1. Switch the ignition on for 1 minute without starting the engine. (Do not actuate accelerator pedal. Be sure pedal is in rest position, i.e. no carpet pressing on the pedal).
2. Switch off ignition for at least 10 seconds.

Warnings and Cautions.

WARNING—
- *When opening fuel lines, be prepared for fuel to be expelled under pressure (approx. 3 - 5 bar or 45 - 75 psi).*
- *Do not smoke or work near heaters or other fire hazards.*
- *Keep a fire extinguisher handy.*
- *Before disconnecting fuel hoses, wrap a cloth around fuel hoses to absorb any leaking fuel.*
- *Catch and dispose of escaped fuel.*
- *Plug all open fuel lines.*
- *Unscrew the fuel tank cap to release pressure in the tank before working on the tank or lines.*

CAUTION—
- *Cleanliness is essential when working with fuel circuit components. Thoroughly clean the unions before disconnecting fuel lines.*

Fuel Injection 24-7
DME Component Testing

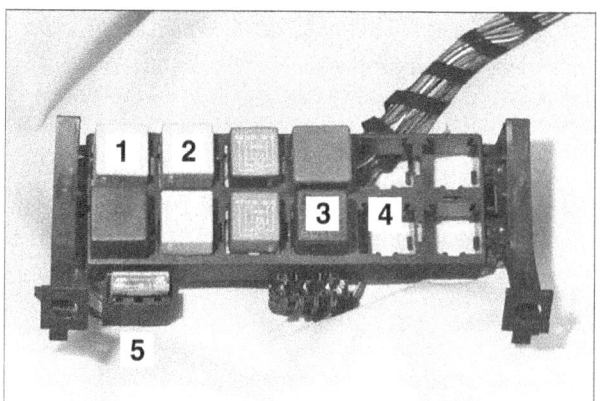

DME Component Testing

FIO relay, testing

◁ The FIO (fuel injector / ignition coil / oxygen sensor heater) relay is in relay panel 2 in the left side of the rear trunk.
1. DME main relay
2. FIO (fuel injector / ignition coil / oxygen sensor heater) relay
3. Secondary air pump relay (1998 and later)
4. Secondary air pump (1997)
5. Secondary air pump fuse (40 A)

> **CAUTION—**
> * Relay positions are subject to change and may vary from car to car. If questions arise, an authorized Porsche dealer is the best source for the most accurate and up-to-date information.

The FIO relay is energized via the DME main relay. Once energized, the FIO relay supplies positive (B+) power to the fuel injectors, the ignition coils, and the oxygen sensor heaters. If this relay is faulty, the engine does not run. The relay is located on relay panel 2 in the rear trunk, left side.

– Working in rear trunk, detach and fold back left side trim panel.

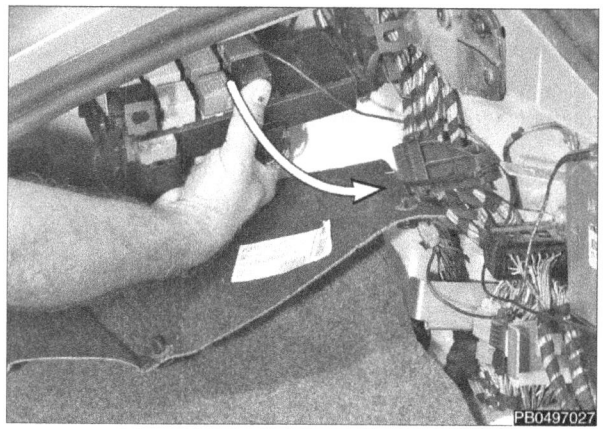

◁ Slide relay panel to right (**arrow**) to remove. Lower panel to trunk floor.

– With ignition off, remove FIO relay.

– Check for voltage at terminal 30 (red/white wire) on relay socket.
  • If battery voltage is present continue testing.

◁ If battery voltage is not present, check fuse 2 in row C (30 A) at main fuse panel. See **97 Fuses, Relays, Component Locations**.

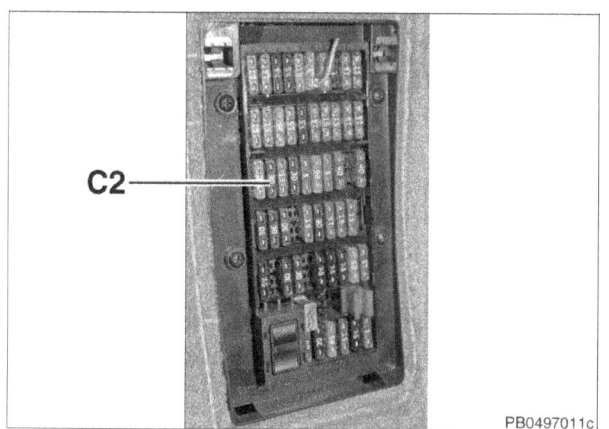

– Turn ignition on. Check for battery voltage at terminal 86 (red/blue wire).
  • If voltage is not present, check wiring. See **EWD Electrical Wiring Diagrams**.

– Turn ignition off.

– Check for continuity to ground at terminal 31 on relay block.
  • If ground is not present, check ground point 8 (GP8) in rear luggage compartment. See **97 Fuses, Relays, Component Locations**.

– Reinstall relay. Turn ignition on. Relay should audibly click. Working from underside of relay block, check for battery voltage at terminal 87 (large black wire).
  • If battery voltage is present, relay has energized and is functioning correctly.
  • If battery voltage is not present and all earlier tests are OK, relay is faulty. Replace.

24-8 Fuel Injection

DME Component Testing

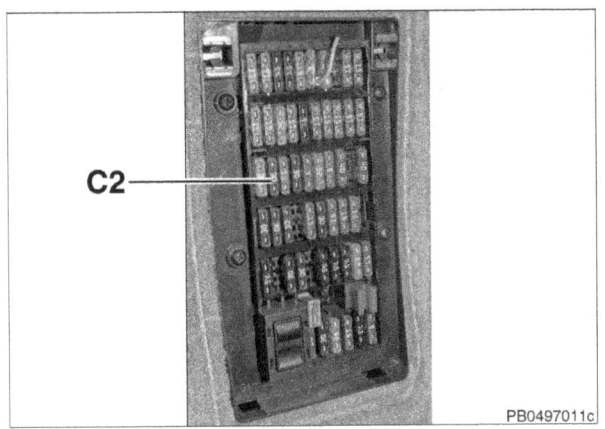

DME main relay, testing

The DME main relay is energized via the engine control module (ECM) and supplies battery positive (B+) power to FIO relay and many other engine management and emission system components. If this relay is faulty, the engine does not start.

The DME main relay is located on relay panel 2 in the rear trunk, left side. Access relay as described under **FIO relay, testing** in this repair group.

◄ With ignition off, remove DME main relay.

– Check for voltage at terminal 86 and terminal 30 (red/green wires) on relay socket.
  • If battery voltage is present continue testing.

◄ If battery voltage is not present, check fuse 1 in row C (25 amp) at main fuse panel.

– Turn ignition on.

– Check for continuity to ground at terminal 85 on relay block.
  • If ground is not present, signal from ECM is missing. Check wire between control module and relay. See **EWD Electrical Wiring Diagrams**.

– Turn ignition off.

– Reinstall relay. Turn ignition on. Relay should audibly click. Working from underside of relay block, check for battery voltage at terminal 87 (red/blue wire).
  • If battery voltage is present, relay has energized and is functioning correctly.
  • If battery voltage is not present and all earlier tests are OK, relay is faulty. Replace.

Engine coolant temperature (ECT) sensor, testing

◄ The engine coolant temperature (ECT) sensor (**inset**) is a negative temperature coefficient (NTC) resistor, which means that sensor resistance decreases as engine temperature increases. OBD II diagnostics monitors ECT sensor sensitivity and output.

Engine coolant temperature (ECT) sensor location	
Boxster, Boxster S engine	At coolant manifold, left front of engine, near cylinder 4

Fuel Injection 24-9

DME Component Replacement

- Check ECM reference voltage to sensor:
  - Disconnect harness connector from ECT sensor.
  - Turn ignition key ON.
  - Check for 5 volts between supply voltage (blue/white) wire of harness connector and ground.
  - Turn ignition key OFF.
  - If voltage is not present or incorrect, check wiring from ECM and check ECT sensor reference voltage output at ECM.

- Check ECT sensor resistance:
  - With harness connector disconnected, check resistance across sensor terminals. Compare test results to specified values.

Engine coolant temperature (ECT) sensor resistance	
Test temperatures	Resistance
0 °C (32 °F)	5 - 7 kΩ
20 °C (68 °F)	2 - 3 kΩ
60 °C (180 °F)	0.4 - 0.8 kΩ

DME COMPONENT REPLACEMENT

Engine coolant temperature (ECT) sensor, removing and installing

◀ The engine coolant temperature (ECT) sensor (**inset**) is a negative temperature coefficient (NTC) resistor, which means that sensor resistance decreases as engine temperature increases. OBD II diagnostics monitors ECT sensor sensitivity and output.

Engine coolant temperature (ECT) sensor location	
Boxster, Boxster S engine	At coolant manifold, left front of engine, near cylinder 4

Remove and install ECT sensor from underneath car.

> **WARNING—**
> - Due to risk of personal injury, be sure the engine is cold before beginning work on engine cooling components.

- Raise rear of car and support safely.

> **WARNING—**
> - Make sure the car is stable and well supported at all times. Use a professional automotive lift or jack stands designed for the purpose. A floor jack is not adequate support.

◀ Working underneath engine, remove plastic splash shield fasteners (**arrows**) and remove shield.

24-10 Fuel Injection

DME Component Replacement

◄ Working underneath engine, detach ECT sensor harness connector (**arrow**).

– Use 22 mm wrench to unscrew ECT sensor. Be prepared to catch dripping coolant.

– When reinstalling, check to make sure captive sealing ring is in place.

Tightening torque	
ECT sensor to coolant manifold	25 ± 5 Nm (18 ± 4 ft-lb)

– Check coolant level and top up if necessary.

Fuel injectors, removing and installing

NOTE—
* *The procedure given below applies specifically to 1997 through 2001 models. On 2002 and later cars (fitted with a non-return fuel rail), the procedure is similar.*

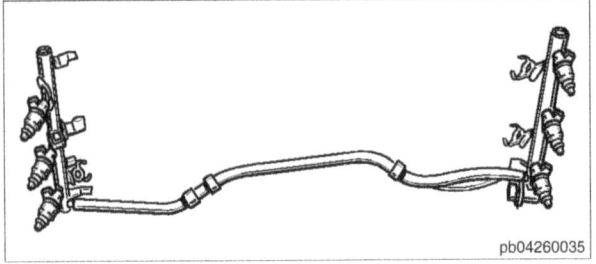

◄ Fuel rail and fuel injectors are removed as a complete unit.

– Move convertible top and convertible top storage lid to service position. See **03 Service and Maintenance**.

– Disconnect negative (-) battery lead and cover with insulating tape.

CAUTION—
* *Prior to disconnecting the battery, read the battery disconnection cautions in **00 Warnings and Cautions**.*

– Remove left seat. See **72 Seats**.

– Remove top and front engine access covers. See **03 Service and Maintenance**.

◄ Working at top of engine, remove clamp fastener (**arrow**) on fuel rail connecting line.

◄ Working at left front of engine, place a shop towel beneath fuel line at fuel pressure regulator to absorb any leaking fuel.

– Disconnect fuel line from left fuel rail. Use a wrench on each fitting to prevent damage to pressure regulator.

NOTE—
* *Engine shown removed for clarity.*

Fuel Injection 24-11

DME Component Replacement

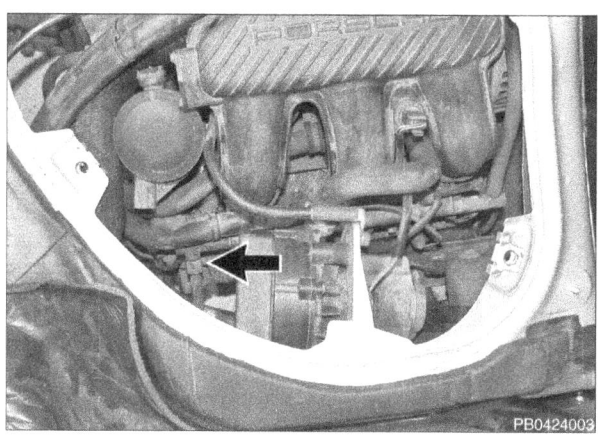

◄ Working at right rear of engine, place a shop towel beneath fuel line and the back of the right rail fitting to absorb any leaking fuel.

– Disconnect fuel line (**arrow**) from right fuel rail. Use a wrench on each fitting to prevent damage to lines.

◄ Remove fuel rail fastening bolts (**arrows**) on each side of fuel rail (right side shown).

– Release any wiring clips attached to fuel rail.

– Release electrical connectors at fuel injectors by releasing spring clips and pulling connectors straight off.

– Carefully work injectors out of intake bores.

– Lift fuel rail / injector assembly up and out of engine compartment.

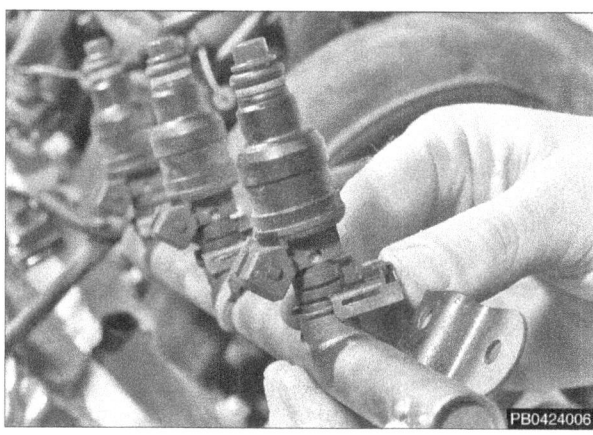

◄ With fuel rail removed, remove injector fastening clips. Carefully work injectors out of fuel rail bore.

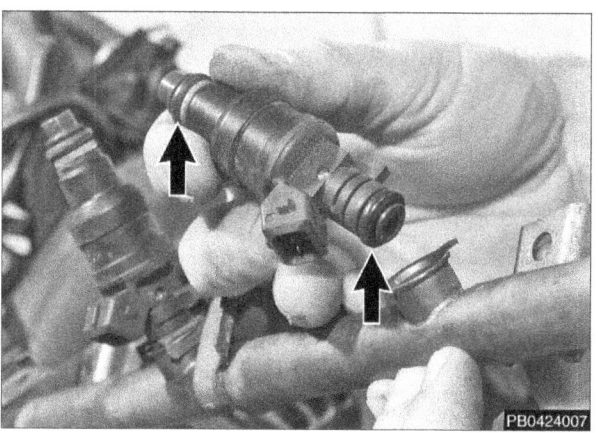

◄ Installation is reverse of removal. Remember to:
  • Replace sealing rings (**arrows**) with new.
  • Fully seat each injector in intake bore.
  • Attach electrical connectors.
  • Run engine and check fuel system for leaks.

24-12 Fuel Injection

ECM Pin Assignments

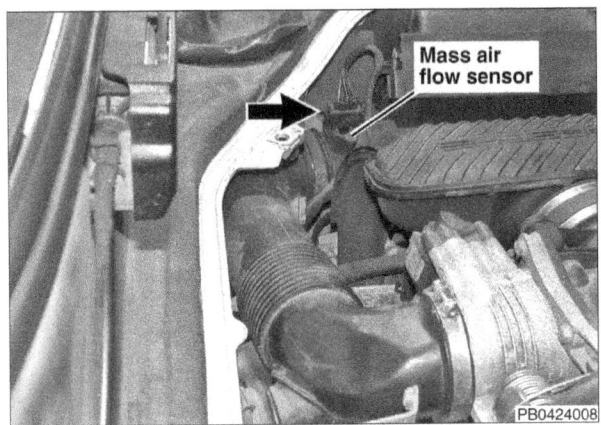

Mass air flow (MAF) sensor, removing

Access to the mass air flow (MAF) sensor is through the top engine access cover.

- Move convertible top and convertible top storage lid to service position. See **03 Service and Maintenance**.
- Remove top engine access cover. See **03 Service and Maintenance**.
◂ Release electrical connector at MAF sensor (**arrow**).
- Remove sensor using a Torx T 20 anti-tamper bit.
- Installation is reverse of removal. Make sure sensor seal is undamaged and correctly seated.

Tightening torque	
Mass air flow sensor to air intake housing	3 - 4 Nm (2 - 3 ft-lb)

Engine control module (ECM), removing

Access to the ECM is through the rear trunk.

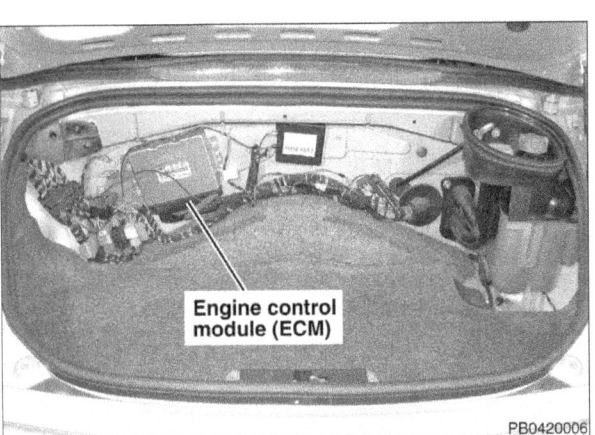

◂ Working in rear trunk, use a trim tool to remove retaining clips holding carpet to front of compartment to access ECM.

- Remove control module mounting bolts.

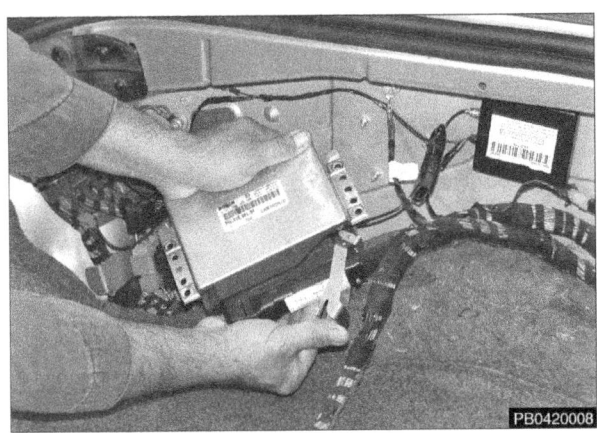

◂ Lift out control module and remove electrical connector(s) by sliding out retaining clip.

NOTE—
- *1998 Boxster single connector control module shown. Later models are equipped with multiple connector modules.*

- Remove ECM.

ECM PIN ASSIGNMENTS

Engine control module (ECM) pin assignments are given in the tables below. This information can be helpful when diagnosing faults to or from the control module.

Generally, absence of voltage or continuity means there is a wiring or connector problem. Test results with incorrect values do not necessarily mean that a component is faulty. Check for loose, broken or corroded connections and wiring before replacing

Fuel Injection 24-13

ECM Pin Assignments

components. For engine management system electrical schematics, see **EWD Electrical Wiring Diagrams**.

When making checks at the ECM, use a break-out box to allow tests to be made with the connector attached to the control module. This prevents damage to the small terminals in the connector. As an alternative, the connector housing can be separated so that electrical checks can be made from the back of the connector.

> *CAUTION—*
> - *Always wait at least one minute after turning off the ignition before removing the connector from the ECM. If the connector is removed before this time, residual power in the system relay may damage the control module.*
> - *Always connect or disconnect the control module connector and meter probes with the ignition off.*

DME M 5.2.2 control module pin assignments

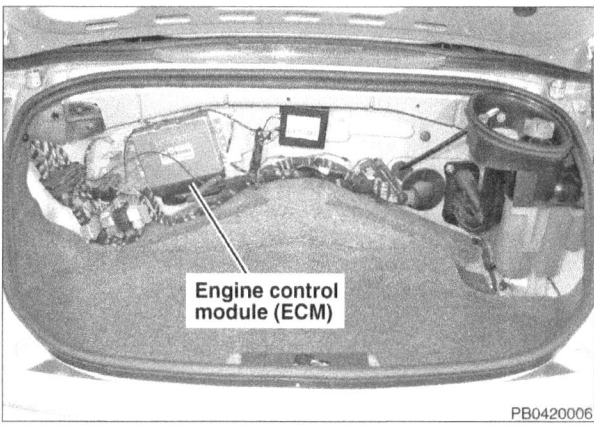

The M 5.2.2 ECM is located in the rear trunk. It has a single electrical connector with a total of 88 pins.

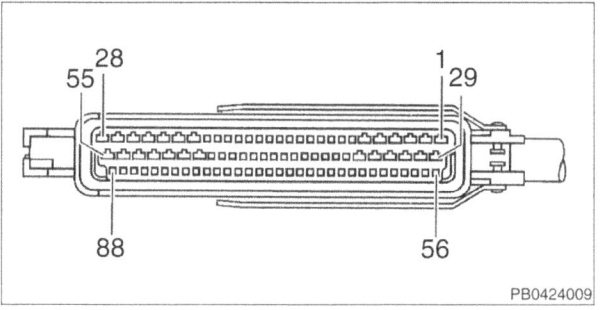

DME M 5.2.2 control module pin assignments Connector I (88-pin)	
1	Oxygen sensor heating (post-cat)
2	Idle air control valve winding (close)
3	Fuel injector 1
4	Fuel injector 3
5	Fuel injector 5
6	Ground, terminal 31
7	Carbon canister shutoff valve
8	Check engine lamp (MIL)
9	EX lamp
10	Vacant
11	Vacant
12	Electric fuel pump shutoff
13	Knock sensor 1
14	Pressure switch (A/C)
15	Intake air temperature sensor
16	Engine compartment temperature
17	Hot film mass air flow sensor
18	Signal, oxygen sensor 2 (pre-cat)
19	Signal, oxygen sensor 1 (pre-cat)
20	Ground, crankshaft sensor
21	Signal hall-effect sensor
22	Ignition coil 4
23	Ignition coil 5

ECM Pin Assignments

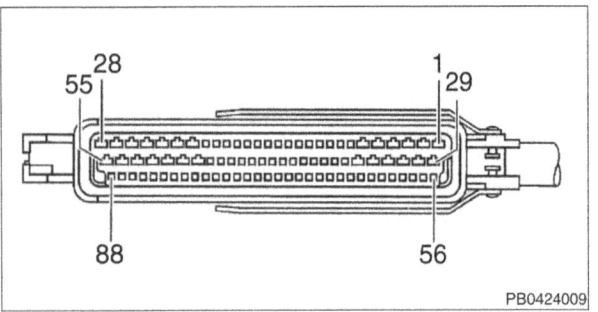

DME M 5.2.2 control module pin assignments Connector I (88-pin)

24	Ignition coil 6
25	VarioCam adjuster, cylinders 4 - 6
26	Terminal 30 (battery voltage at all times)
27	DME main relay
28	Electronics ground
29	Idle air control valve winding (open)
30	Oxygen sensor heating (pre-cat)
31	Fuel injector 2
32	Fuel injector 4
33	Fuel injector 6
34	Ground, sensors
35	Radiator fan
36	Radiator fan
37	Secondary air pump
38	Version coding
39	Indicator light, reserve
40	Vacant
41	Knock sensor 2
42	Automatic I/M test
43	Vacant
44	Signal, throttle position sensor
45	Ground, hot film mass air flow sensor
46	Ground, oxygen sensors
47	5 volt supply, hot film mass air flow sensor
48	Spec. engine torque, traction control
49	Ignition coil 1
50	Ignition coil 2
51	Ignition coil 3
52	VarioCam adjuster, cylinders 1 - 3

DME M 5.2.2 control module pin assignments Connector I (88-pin)

53	5 volt supply, throttle potentiometer
54	Voltage output to loads
55	Ignition ground
56	Terminal 15
57	Vacant
58	Engine torque, traction control
59	Vacant
60	Programming voltage
61	Fuel tank evaporative valve
62	Activation, A/C compressor on
63	Fuel pump relay
64	Vacant
65	Engine compartment cooling fan
66	Activation, start inhibit
67	Knock sensor signal
68	Start enable
69	A/C request
70	Vacant
71	Ground, knock sensors
72	Differential pressure sensor (fuel tank leak detection)
73	Oil temperature
74	Coolant temperature sensor
75	Vacant
76	Signal, oxygen sensor 2 (post-cat)
77	Signal, oxygen sensor 1 (post-cat)
78	Crankshaft sensor
79	Rpm sensor output, rear
80	Rpm signal
81	Vacant
82	Engine cooling fan monitoring
83	Fuel consumption display
84	Vacant
85	Data transfer - LOW
86	Data transfer - HIGH
87	Vacant
88	Diagnosis

Fuel Injection 24-15

ECM Pin Assignments

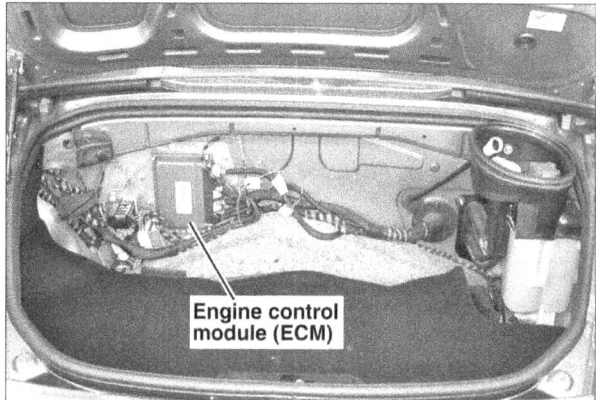

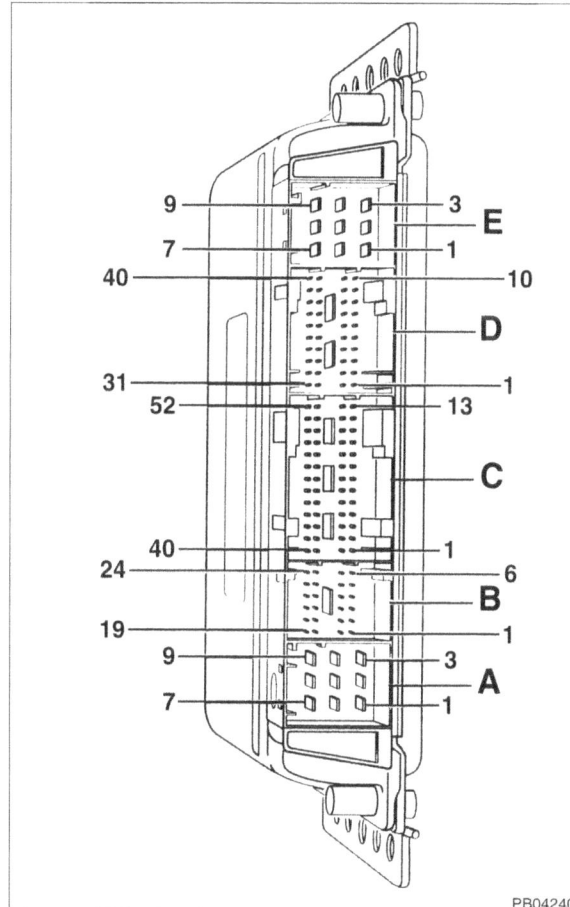

DME ME 7.2 control module pin assignments

 The ME 7.2 ECM is located in the rear trunk. It has 5 electrical connector with a total of 134 pins:

- Connector A: 9 pins
- Connector B: 24 pins
- Connector C: 52 pins
- Connector D: 40 pins
- Connector E: 9 pins

DME ME 7.2 control module pin assignments Connector A (9-pin)	
1	Terminal 15
2	Vacant
3	W-wire
4	Ground, electronic system
5	Ground, fuel injectors
6	Ground, output stages
7	Terminal 30
8	DME relay (terminal 87)
9	Vacant

DME ME 7.2 control module pin assignments Connector B (24-pin)	
1	Oxygen sensor heating 2 (post-cat)
2	Vacant
3	CAN low (Tiptronic)
4	CAN high (Tiptronic)
5	Vacant
6	Vacant
7	Oxygen sensor heating 1 (post-cat)
8	Ground, oxygen sensor 2 (post-cat)
9	Ground, oxygen sensor 1 (pre-cat)
10	Ground, oxygen sensor 2 (pre-cat)
11	Ground, oxygen sensor 1 (post-cat)
12	Automatic I/M testing
13	Oxygen sensor heating 2 (pre-cat)
14	Signal, oxygen sensor 2 (post-cat)
15	Signal, oxygen sensor 1 (pre-cat)
16	Signal, oxygen sensor 2 (pre-cat)
17	Signal, oxygen sensor 1 (post-cat)
18	Trigger, A/C compressor relay (terminal 85)

24-16 Fuel Injection

ECM Pin Assignments

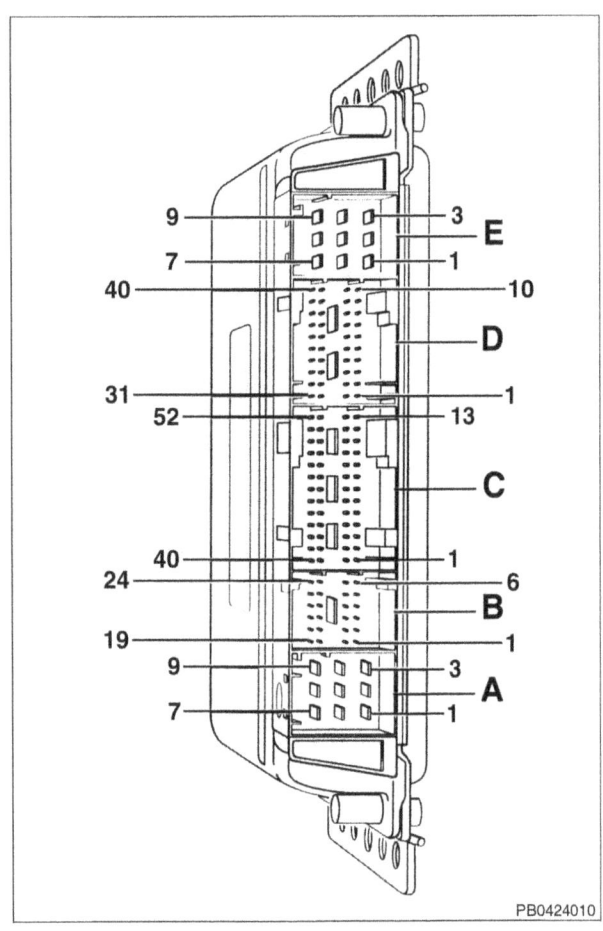

DME ME 7.2 control module pin assignments Connector B (24-pin)	
19	Oxygen sensor heating 1 (pre-cat)
20	Trigger, engine compartment fan (terminal 85)
21	Engine compartment temperature sensor
22	5-volt supply, mass air flow sensor
23	Trigger, DME relay (terminal 85)
24	Vacant

DME ME 7.2 control module pin assignments Connector C (52-pin)	
1	Knock sensor
2	Fuel injector, cylinder 5
3	EVAP canister purge valve
4	Variable-length manifold
5	Oil temperature sensor
6	Vacant
7	5-volt supply, camshaft position sensor, differential pressure sensor

DME ME 7.2 control module pin assignments Connector C (52-pin)	
8	Signal, throttle position sensor 2
9	Ground, mass air flow sensor
10	5-volt supply, throttle actuation
11	Trigger, secondary air pump relay (terminal 85)
12	Vacant
13	Start enable, Tiptronic (P + N)
14	Pilot light, engine compartment fan
15	Fuel injector, cylinder 3
16	Vacant
17	Ground, sensors
18	Vacant
19	Signal, camshaft position sensor 2
20	Signal, camshaft position sensor 1
21	Vacant
22	Engine coolant temperature sensor
23	Signal, mass air flow sensor
24	Signal, throttle position sensor 1
25	Ground, throttle position sensor 1 and 2
26	Vacant
27	Fuel injector, cylinder 4
28	Fuel injector, cylinder 6
29	Vacant
30	Secondary air valve
31	Vacant
32	Signal A, speed sensor
33	Vacant
34	Intake air temperature sensor
35	Vacant
36	Input, knock sensor 2
37	Ground, knock sensor 2
38	Vacant
39	Vacant
40	Fuel injector, cylinder 2
41	Fuel injector, cylinder 1
42	Throttle motor actuator
43	Throttle motor actuator
44	Vacant

Fuel Injection 24-17
ECM Pin Assignments

DME ME 7.2 control module pin assignments Connector C (52-pin)	
45	Shield, speed sensor
46	Signal B, speed sensor
47	Vacant
48	Vacant
49	Input, knock sensor 1
50	Ground, knock sensor 1
51	Vacant
52	Vacant

DME ME 7.2 control module pin assignments Connector D (40-pin)	
1	Clutch interlock switch
2	Vacant
3	Vacant
4	Coolant fan, stage 1
5	Variant coding (ground for Tiptronic)
6	A/C compressor requirement
7	Ground, pedal sensor 1
8	Signal, pedal sensor 1
9	5-volt supply, pedal sensor 1
10	Signal, fuel pump relay (terminal 85)
11	Dim information from instrument cluster
12	Ground, pedal sensor 2
13	Signal, pedal sensor 2
14	5-volt supply, pedal sensor 2
15	Ground, differential pressure sensor
16	Crash signal (airbag)
17	Speed signal output
18	Cruise control pilot light
19	Cruise control switch OFF
20	Coolant fan, stage 2
21	Signal, differential pressure sensor
22	Speed signal from ABS control module
23	Cruise control clutch switch
24	Stop light switch 1
25	Cruise control switch SET
26	Stop light switch 2
27	Cruise control switch RESUME

DME ME 7.2 control module pin assignments Connector D (40-pin)	
28	Stop light switch 2
29	Vacant
30	EVAP canister shutoff valve
31	Check engine (MIL)
32	Vacant
33	Fuel gauge
34	Fuel reserve pilot light
35	Vacant
36	CAN-high (PSM)
37	CAN-low (PSM)
38	Vacant
39	A/C pressure switch (medium)
40	Start-inhibit relay (terminal 85)

Connector E (9-pin)	
1	VarioCam, cylinders 1 - 3
2	Ignition coil 4
3	VarioCam, cylinders 4 - 6
4	Ignition coil 6
5	Ground
6	Ignition coil 1
7	Ignition coil 3
8	Ignition coil 2
9	Ignition coil 5

24-18 Fuel Injection

ECM Pin Assignments

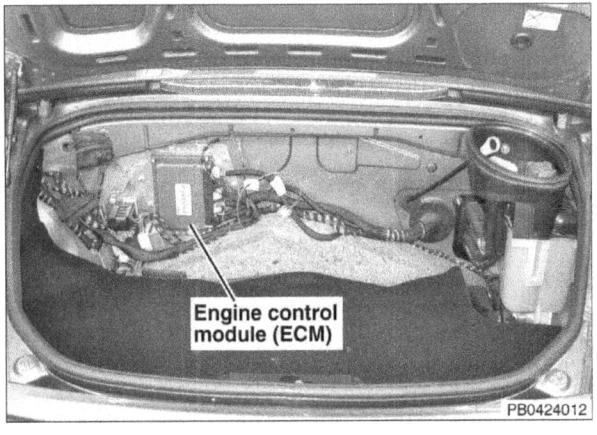

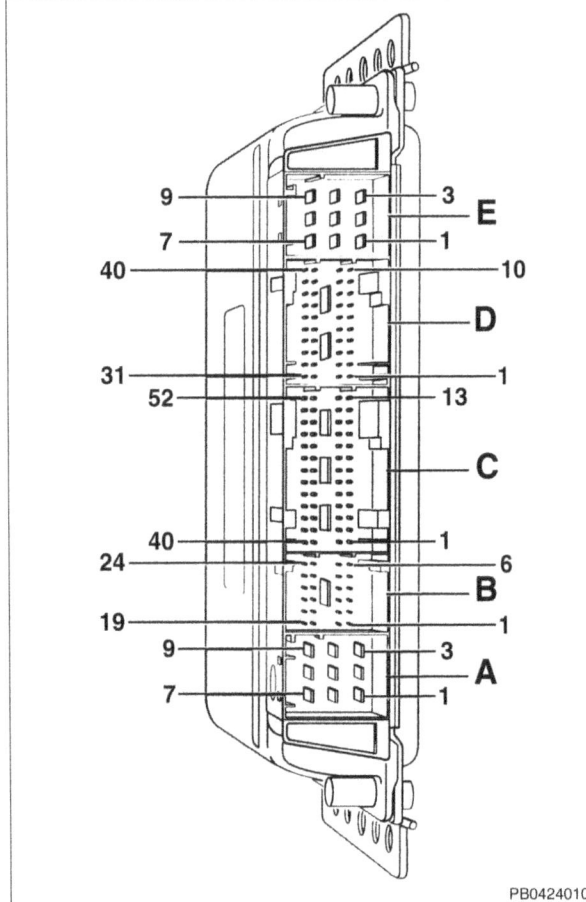

DME ME 7.8 control module pin assignments

The ME 7.8 ECM is located in the rear trunk. It has 5 electrical connector with a total of 134 pins:

- Connector A: 9 pins
- Connector B: 24 pins
- Connector C: 52 pins
- Connector D: 40 pins
- Connector E: 9 pins

DME ME 7.8 control module pin assignments Connector A (9-pin)	
1	Terminal 15
2	Terminal 30
3	W-wire
4	Ground, electronics
5	Ground, fuel injectors
6	Ground, output stages
7	Throttle motor actuator
8	DME relay (terminal 87)
9	Throttle motor actuator

DME ME 7.8 control module pin assignments Connector B (24-pin)	
1	Oxygen sensor heating 2 (post-cat)
2	Pump current regulator oxygen sensor 1 (pre-cat)
3	CAN-low (Tiptronic)
4	CAN-high (Tiptronic)
5	Pump current regulator oxygen sensor 1 (pre-cat)
6	Pump current regulator oxygen sensor 1 (pre-cat)
7	Oxygen sensor heating 1 (post-cat)
8	Ground, oxygen sensor 2 (post-cat)
9	Ground, oxygen sensor 1 (post-cat)
10	Ground, oxygen sensor 2 (post-cat)
11	Ground, oxygen sensor 1 (post-cat)
12	Vacant
13	Oxygen sensor heating 2 (pre-cat)
14	Signal, oxygen sensor 2 (post-cat)
15	Signal, oxygen sensor 1 (pre-cat)
16	Signal, oxygen sensor 2 (pre-cat)
17	Signal, oxygen sensor 1 (post-cat)
18	Vacant

Fuel Injection 24-19
ECM Pin Assignments

DME ME 7.8 control module pin assignments Connector B (24-pin)

19	Oxygen sensor heating 1 (pre-cat)
20	Vacant
21	Engine compartment temperature sensor
22	5-volt supply, mass air flow sensor
23	Vacant
24	Pump current regulator oxygen sensor 2 (pre-cat)

DME ME 7.8 control module pin assignments Connector C (52-pin)

1	VarioCam, cylinder 1 - 3
2	Fuel injector 5
3	EVAP canister purge valve
4	Frequency valve, charge pressure
5	Oil temperature sensor
6	Vacant
7	5v, camshaft position sensor, differential pressure sensor
8	Signal, throttle position sensor 2
9	Ground, mass air flow sensor
10	5-volt supply, throttle actuation
11	Trigger, secondary air pump relay (terminal 85)
12	Signal camshaft position sensor 1
13	Start enable, Tiptronic (P + N)
14	Secondary air valve
15	Fuel injector 3
16	Overrun recirculating air valve
17	Ground, sensors
18	Signal, camshaft position sensor 2
19	Charge control
20	Vacant
21	Vacant
22	Engine coolant temperature sensor
23	Signal, mass air flow sensor
24	Signal, throttle position sensor 1
25	Ground, throttle position sensor 1 and 2
26	VarioCam, cylinder 4 - 6
27	Fuel injector 4
28	Fuel injector 6

DME ME 7.8 control module pin assignments Connector C (52-pin)

29	Vacant
30	Vacant
31	Vacant
32	Ground
33	Vacant
34	Intake air temperature sensor
35	Vacant
36	Input, knock sensor 2
37	Ground, knock sensor 2
38	Vacant
39	Charge pressure sensor
40	Fuel injector, cylinder 2
41	Fuel injector, cylinder 1
42	Vacant
43	Vacant
44	Vacant
45	Signal A, speed sensor
46	---
47	Vacant
48	Vacant
49	Signal B, speed sensor
50	Ground, knock sensor 1
51	Vacant
52	Speed signal from ABS control module

DME ME 7.8 control module pin assignments Connector D (40-pin)

1	Clutch interlock switch
2	Vacant
3	Vacant
4	Coolant fan, stage 1
5	Vacant
6	Vacant
7	Ground, pedal sensor 1
8	Signal, pedal sensor 1
9	5-volt supply, pedal sensor 1
10	Signal, fuel pump relay (terminal 85)
11	Vacant

24-20 Fuel Injection

ECM Pin Assignments

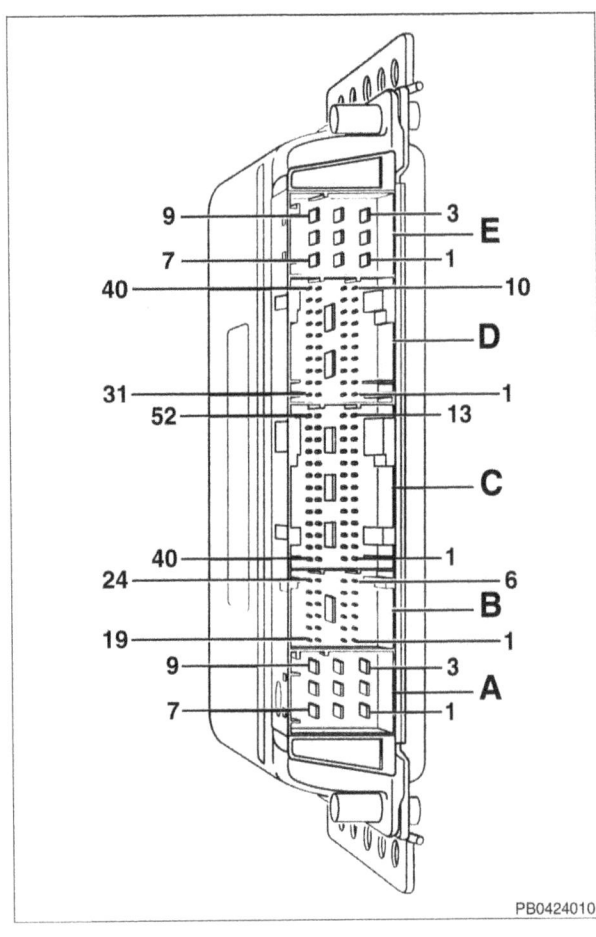

DME ME 7.8 control module pin assignments Connector D (40-pin)	
27	Trigger, A/C compressor fan relay (terminal 85)
28	Vacant
29	Knock sensor
30	EVAP canister shutoff valve
31	Coolant fan, stage 3
32	Vacant
33	Manual transmission, start enable
34	A/C medium pressure switch
35	Vacant
36	CAN-high (PSM and instrument cluster)
37	CAN-low (PSM and instrument cluster)
38	Vacant
39	Vacant
40	Tiptronic coolant shutoff valve

DME ME 7.8 control module pin assignments Connector E (9-pin)	
1	Ignition coil 6
2	Ignition coil 4
3	Ignition coil 2
4	Ignition coil 5
5	Ground
6	Ignition coil 1
7	VarioCam, cylinders 1 - 3
8	VarioCam, cylinders 4 - 6
9	Ignition coil 3

DME ME 7.8 control module pin assignments Connector D (40-pin)	
12	Ground, pedal sensor 2
13	Signal, pedal sensor 2
14	5-volt supply, pedal sensor 2
15	Ground, differential pressure sensor
16	Crash signal (airbag)
17	Speed signal output
18	Vacant
19	Vacant
20	Coolant fan, stage 2
21	Signal, differential pressure sensor
22	Vacant
23	Vacant
24	Automatic I/M testing
25	Trigger, engine compartment fan (terminal 85)
26	Trigger, DME relay (terminal 85)

Fuel Injection 24-21

On-Board Diagnostics

ON-BOARD DIAGNOSTICS

This section outlines the fundamentals and equipment requirements of the SAE (Society of Automotive Engineers) and CARB (California Air Resource Board) On-Board Diagnostics II standard as it applies to 1997 through 2004 Porsche Boxster and Boxster S models. Also given here is a listing of OBD II and Porsche-specific diagnostic trouble codes (DTCs).

OBD II standards have been applied to all passenger vehicles sold in the United States from model year 1996. Self diagnostic capabilities are incorporated into the engine control module (ECM) software to monitor components that can affect vehicle emissions. Each emission-influencing component is checked by a diagnostic routine to verify that it is functioning properly. If a problem or malfunction is detected, the OBD II system illuminates the malfunction indicator light (MIL) on the instrument panel.

The OBD II system also stores important information about the detected malfunction in the ECM so that a repair technician can find the problem and repair it accurately.

NOTE—

- Specialized OBD II scan tool equipment is needed to access the fault memory and OBD II data.
- The OBD II fault memory (including the Check Engine light) can only be reset using a special scan tool. Removing the connector from the ECM or disconnecting the battery does not erase the fault memory or turn the Check Engine light out.

Malfunction indicator light (MIL)

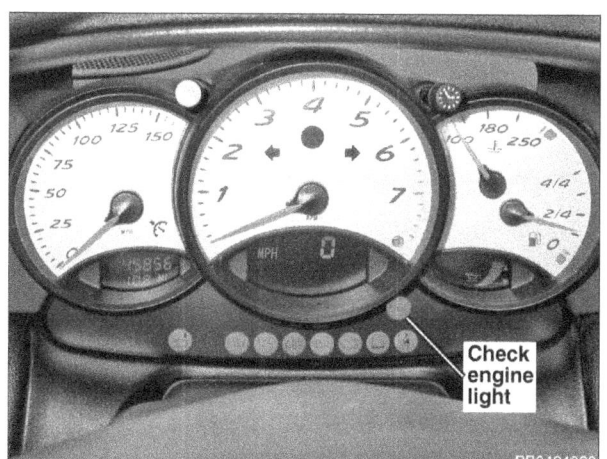

The OBD II system is designed to illuminate the malfunction indicator light (MIL), also called the Check Engine light, when emission levels exceed 1.5 times Federal standards.

Conditions that illuminate the MIL:

- Engine management system fault detected for two consecutive OBD II drive cycles (see **Drive cycle**).
- Catalyst damaging fault.
- Malfunction in component(s) monitoring emissions.
- Manufacturer-defined specifications exceeded.
- Implausible sensor input signal.
- Misfire.
- Leak detected in fuel tank evaporative system.
- No purge flow detected from purge valve / evaporative system.
- Failure of ECM to enter closed-loop operation within specified time.
- Engine management system enters "limp home" mode.

24-22 Fuel Injection

On-Board Diagnostics

Additional information on MIL:

- A fault code is stored in the ECM upon the first occurrence of a fault being identified.
- Two **complete** consecutive drive cycles with the fault present illuminate the MIL. The exception to the two-fault requirement is a catalyst-damaging fault which turns the light on immediately.
- If the second drive cycle is not completed and the fault not checked, the ECM counts the third drive cycle as the next consecutive drive cycle. The MIL is illuminated if the fault is still present.
- Once the MIL is illuminated, it remains illuminated unless the system is tested and found to be without faults through three complete consecutive drive cycles.
- Except for catalyst-damaging fault(s), the fault code is cleared from memory automatically if fault is not detected through **40** consecutive drive cycles.
- In case of a catalyst-damaging fault, the fault code is cleared from memory automatically if the fault is not detected through **80** consecutive drive cycles.

Scan tool and scan tool display

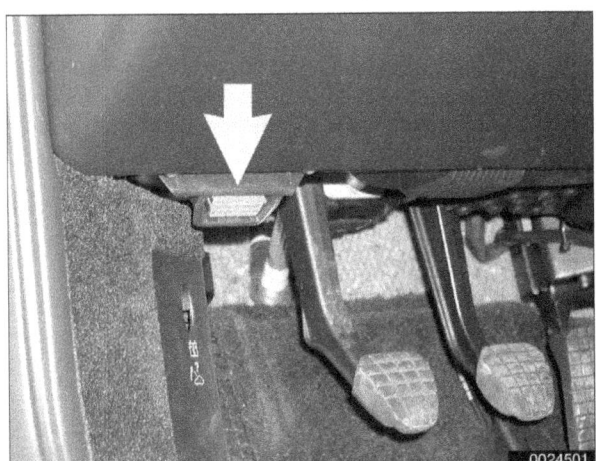

Owing to the advanced nature of OBD II adaptive strategies, all diagnostics need to start with a scan tool. Scan tools are connected to 16-pin OBD II diagnosis socket (**arrow**) under the left side of the dashboard.

OBD II standards mandate that the diagnosis socket be located within three (3) feet of the driver and must not require any tools to be exposed. The communication protocol used by Porsche is ISO 9141.

NOTE—

- *Professional diagnostic scan tools available at the time of this printing include the Porsche factory tool (PST2) and a small number of aftermarket Porsche-specific tools. The 'iScan' from Baum Tools Unlimited, is an example of a quality Porsche-specific scan tool.*
- *In addition to the professional line of scan tools, inexpensive 'generic' OBD II scan tool software programs and handheld units are readily available. These tools have limited capabilities, but they are nonetheless powerful diagnostic tools. These tools read live data streams and freeze frame data, turn off the Check Engine light, and much more.*

Diagnostic monitors

Diagnostic monitors are software routines which run tests and checks on specific systems, components, or functions.

A complete drive cycle (see **Drive cycle** in this repair group) is required for the tests to be valid. The diagnostic monitor signals the ECM of the loss or impairment of the signal or component and determines if a signal or sensor is faulty based on 3 conditions:

- Signal or component shorted to ground
- Signal or component shorted to B+
- Signal or component missing (open circuit)

The OBD II system must monitor *all* emission control systems that are on-board. Not all vehicles have a full complement of emission control systems. For example, a vehicle may not be equipped with secondary air system so naturally no secondary air readiness / function code would be present.

OBD II requires monitoring of the following:
- Oxygen sensors
- Catalysts
- Misfire
- Evaporative control system
- Secondary air system
- Fuel system

Monitoring these emissions related functions is done using DME input sensors and output actuators based on preprogrammed data sets. If the ECM cannot determine the environment or engine operating conditions due to missing or faulty signals it sets a fault code and, depending on conditions, illuminates the MIL.

Oxygen sensor monitoring: When drive conditions allow, response rate and switching time of each oxygen sensor is monitored. The oxygen sensor heater function is also monitored. The OBD II "diagnostic executive" knows the difference between upstream and downstream oxygen sensors and reads each one individually.

All oxygen sensors are monitored separately. In order for the oxygen sensor to be effectively monitored, the system must be in closed loop operation.

Catalyst monitoring: This strategy monitors the two heated oxygen sensors per bank of cylinders. It compares the oxygen content entering the catalytic converter to the oxygen leaving the converter.

The diagnostic executive knows that most of the oxygen should be used up during the oxidation phase and if it sees higher than programmed values, a fault is set and the MIL illuminates.

Misfire detection: This strategy monitors crankshaft speed fluctuations and determines if a misfire occurs by variations in speed between each crankshaft sensor trigger point. This strategy is so finely tuned that it can even determine the severity of the misfire.

Fuel Injection

On-Board Diagnostics

The diagnostic executive must determine if misfire is occurring, as well as other pertinent misfire information.

- Specific cylinder(s)
- Severity of the misfire event
- Emissions relevant or catalyst damaging

Misfire detection is an on-going monitoring process that is only disabled under certain limited conditions.

Secondary air injection monitoring: Secondary air injection is used to reduce HC and CO emissions during engine warm up. Immediately following a cold engine start (-10 to 40°C), fresh air/oxygen is pumped directly into the exhaust manifold. By injecting oxygen into the exhaust manifold, catalyst warm-up time is reduced.

System components:

- Electric air injection pump
- Pump relay
- Non-return valve
- Vacuum/vent valve
- Air injection pipes
- Vacuum reservoir

The secondary air system is monitored using the precatalyst oxygen sensors. Once the air pump is active and air is injected into the system, the signal at the oxygen sensor reflects a lean condition. If the oxygen sensor signal does not change, a fault is set identifying the faulty bank(s). If after completing the next cold start a fault is again present, the MIL illuminates.

Fuel system monitoring: This monitor receives high priority. It looks at the fuel delivery needed (long/short term fuel trim) for proper engine operation based on programmed data. If too much or not enough fuel is delivered over a predetermined time, a DTC is set and the MIL is turned on.

NOTE—
- *Fuel trim refers to adjustments to the base fuel schedule. Long-term fuel trim refers to gradual adjustments to the fuel calibration adjustment as compared to short term fuel trim. Long term fuel trim adjustments compensate for gradual changes that occur over time.*

Fuel system monitoring monitors the calculated injection time (ti) in relation to engine speed, load, and the precatalyst oxygen sensor(s) signals as a result of residual oxygen in the exhaust stream.

The diagnostic executive uses the precatalyst oxygen sensor signal as a correction factor for adjusting and optimizing the mixture pilot control under all engine operating conditions.

Evaporative system monitoring: This monitor checks the sealed integrity of the fuel storage system and related fuel lines.

This monitor has the ability to detect very small leaks anywhere in the system. A vacuum test is performed on the EVAP system on a continuous basis as the drive cycle allows. A differential pressure sensor is used to detect a leak in the system.

Fuel Injection 24-25

On-Board Diagnostics

Drive cycle

The purpose of the OBD II drive cycle is to run all of the emission-related on-board diagnostics over a broad range of driving conditions. When all diagnostics have been run, the system status (Inspection / Maintenance ready) "flags" are set to "Yes." See **Readiness codes** in this repair group.

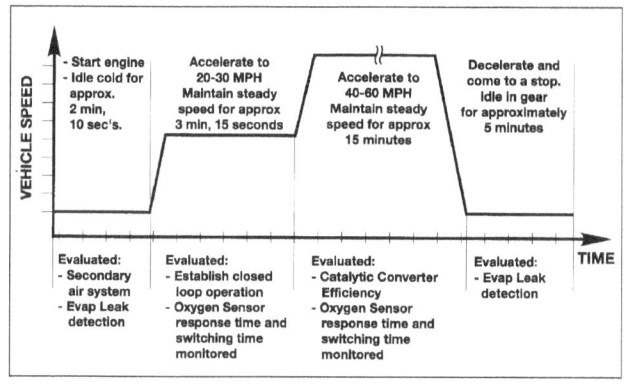

Shown is a the Federal Test Procedure (FTP) drive cycle. The OBD II drive cycle for the Boxster is similar.

A drive cycle is considered complete when all of the diagnostic monitors have run their tests without interruption. For a drive cycle to be initiated, the vehicle must be started cold and brought up to an oil temperature of 160°F, at least 40°F above its original starting temperature.

System status codes is set to "No" in the following cases:

- Battery or ECM is disconnected.
- DTCs have been erased after completion of repairs but drive cycle has not be completed.

Readiness codes

The inspection/maintenance (I/M) readiness codes are mandated as part of OBD II. The readiness code is stored after complete diagnostic monitoring of all components and systems has been carried out. The readiness code function was put into place to prevent manipulating an I/M emission test procedure by clearing faults codes or disconnecting the ECM or battery.

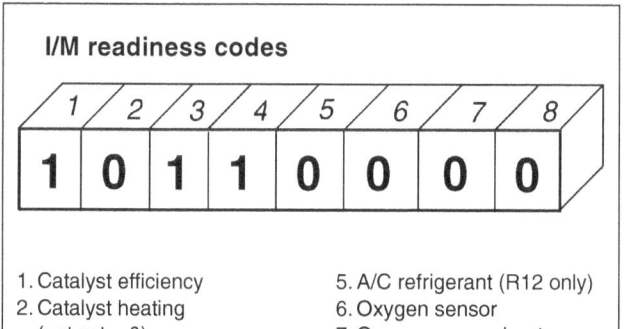

The readiness code can be displayed using an aftermarket scan tool. The code is binary: either "0" or "1". When all zeros are displayed, the system has established its readiness.

Diagnostic trouble codes (DTCs)

The SAE mandates a 5-digit diagnostic trouble code (DTC) standard. Emission related DTCs start with the letter "P" for power train and are commonly referred to as "P-codes". When the malfunction indicator light (MIL) is illuminated it indicates that a DTC has been stored:

- DTCs are stored as soon as they occur, whether or not the MIL illuminates.
- DTCs store and display a time stamp.
- DTCs record the current fault status.

DTC example: P 0 3 0 6

DTC digit interpretation	
1st digit P B C	powertrain body chassis
2nd digit 0 1	SAE Porsche
3rd digit (in P-codes) 0 1 2 3 4 5 6 7	total system air/fuel induction fuel injection ignition system or misfire auxiliary emission control vehicle speed & idle control ECM inputs/outputs transmission

- **P:** A powertrain problem
- **0**: SAE sanctioned or "generic"
- **3:** Related to an ignition system/misfire
- **06** Misfire has been detected at cylinder #6.

Freeze frame data: DTCs provide a "freeze frame" or snap-shot of a vehicle performance or emissions fault at the moment that the fault first occurs. This information is accessible through generic scan tools.

The freeze frame should contain, but is not limited to, the following data:

- Engine load (calculated)
- Engine RPM
- Short and long-term fuel trim
- Vehicle speed
- Coolant temperature
- Intake manifold pressure
- Open/closed loop operation
- Fuel pressure (if available)
- DTC

Fuel Injection 24-27

On-Board Diagnostics

P-codes (DTCs) and Porsche fault codes

Porsche fault codes (FC) expand on the SAE sanctioned DTCs and are accessible primarily through Porsche (or Porsche specific) diagnostic scan tools. Below is a listing of SAE P-codes, the corresponding Porsche fault codes, fault types and their meanings.

Table b. DME M 5.2.2 DTCs			
P-code	Porsche FC	Fault type	Explanation
P0102	115	Mass air flow sensor	Short to ground / below lower limit / lean mixture threshold
P0103	115	Mass air flow sensor	Short to B+ / above upper limit / rich mixture threshold
P0107	299	Ambient pressure sensor	Short to ground / below lower limit / lean mixture threshold
P0108	299	Ambient pressure sensor	Short to B+ / above upper limit / rich mixture threshold
P0112	124	Intake air temperature sensor	Short to ground / below lower limit / lean mixture threshold
P0113	124	Intake air temperature sensor	Short to B+ / above upper limit / rich mixture threshold
P0115	123	Engine temperature sensor	Signal implausible / implausible operating range / malfunction
P0117	123	Engine temperature sensor	Open circuit / no signal
P0117	123	Engine temperature sensor	Short to ground / below lower limit / lean mixture threshold
P0118	123	Engine temperature sensor	Short to B+ / above upper limit / rich mixture threshold
P0121	117	Throttle valve	Signal implausible / implausible operating range / malfunction
P0130	10	Precatalyst oxygen sensor, cyl. 1 - 3	Signal implausible / implausible operating range / malfunction
P0134	10	Precatalyst oxygen sensor, cyl. 1 - 3	Open circuit / no signal
P0132	10	Precatalyst oxygen sensor, cyl. 1 - 3	Short to B+ / above upper limit / rich mixture threshold
P0133	15	Oxygen sensor aging, cyl. 1 - 3	Short to ground / below lower limit / lean mixture threshold
P0133	15	Oxygen sensor aging, cyl. 1 - 3	Short to B+ / above upper limit / rich mixture threshold
P0136	12	Post-catalyst oxygen sensor, cyl. 1 - 3	Signal implausible / implausible operating range / malfunction
P0140	12	Post-catalyst oxygen sensor, cyl. 1 - 3	Open circuit / no signal
P0138	12	Post-catalyst oxygen sensor, cyl. 1 - 3	Short to B+ / above upper limit / rich mixture threshold
P0139	17	Post-catalyst oxygen sensor aging, cyl. 1 - 3	Signal implausible / implausible operating range / malfunction
P0139	17	Post-catalyst oxygen sensor aging, cyl. 1 - 3	Short to ground / below lower limit / lean mixture threshold
P0139	17	Post-catalyst oxygen sensor aging, cyl. 1 - 3	Short to B+ / above upper limit / rich mixture threshold
P0150	18	Precatalyst oxygen sensor, cyl. 1 - 3	Signal implausible / implausible operating range / malfunction
P0154	18	Precatalyst oxygen sensor, cyl. 1 - 3	Signal implausible /Open circuit / no signal
P0152	18	Precatalyst oxygen sensor, cyl. 1 - 3	Short to ground / below lower limit / lean mixture threshold
P0153	21	Oxygen sensor aging, cyl. 4 - 6	Short to ground / below lower limit / lean mixture threshold
P0153	21	Oxygen sensor aging, cyl. 4 - 6	Short to B+ / above upper limit / rich mixture threshold
P0156	20	Post-catalyst oxygen sensor, cyl. 4 - 6	Signal implausible / implausible operating range / malfunction
P0160	20	Post-catalyst oxygen sensor, cyl. 4 - 6	Signal implausible /Open circuit / no signal

Fuel Injection

On-Board Diagnostics

Table b. DME M 5.2.2 DTCs

P-code	Porsche FC	Fault type	Explanation
P0158	20	Post-catalyst oxygen sensor, cyl. 4 - 6	Short to B+ / above upper limit / rich mixture threshold
P0159	23	Post-catalyst oxygen sensor aging, cyl. 4 - 6	Signal implausible / implausible operating range / malfunction
P0159	23	Post-catalyst oxygen sensor aging, cyl. 4 - 6	Short to ground / below lower limit / lean mixture threshold
P0159	23	Post-catalyst oxygen sensor aging, cyl. 4 - 6	Short to B+ / above upper limit / rich mixture threshold
P0197	125	Oil temperature sensor	Short to ground / below lower limit / lean mixture threshold
P0198	125	Oil temperature sensor	Short to B+ / above upper limit / rich mixture threshold
P0300	62	Misfire detection (sum total)	Signal implausible / implausible operating range / malfunction
P0300	62	Misfire detection (sum total)	Short to ground / below lower limit / lean mixture threshold
P0300	62	Misfire detection (sum total)	Short to B+ / above upper limit / rich mixture threshold
P0301	508	Misfire, cylinder 1	Signal implausible / implausible operating range / malfunction
P0301	508	Misfire, cylinder 1	Short to ground / below lower limit / lean mixture threshold
P0301	508	Misfire, cylinder 1	Short to B+ / above upper limit / rich mixture threshold
P0302	509	Misfire, cylinder 2	Signal implausible / implausible operating range / malfunction
P0302	509	Misfire, cylinder 2	Short to ground / below lower limit / lean mixture threshold
P0302	509	Misfire, cylinder 2	Short to B+ / above upper limit / rich mixture threshold
P0303	510	Misfire, cylinder 3	Signal implausible / implausible operating range / malfunction
P0303	510	Misfire, cylinder 3	Short to ground / below lower limit / lean mixture threshold
P0303	510	Misfire, cylinder 3	Short to B+ / above upper limit / rich mixture threshold
P0304	511	Misfire, cylinder 4	Signal implausible / implausible operating range / malfunction
P0304	511	Misfire, cylinder 4	Short to ground / below lower limit / lean mixture threshold
P0304	511	Misfire, cylinder 4	Short to B+ / above upper limit / rich mixture threshold
P0305	512	Misfire, cylinder 5	Signal implausible / implausible operating range / malfunction
P0305	512	Misfire, cylinder 5	Short to ground / below lower limit / lean mixture threshold
P0305	512	Misfire, cylinder 5	Short to B+ / above upper limit / rich mixture threshold
P0306	513	Misfire, cylinder 6	Signal implausible / implausible operating range / malfunction
P0306	513	Misfire, cylinder 6	Short to ground / below lower limit / lean mixture threshold
P0306	513	Misfire, cylinder 6	Short to B+ / above upper limit / rich mixture threshold
P0327	210	Knock sensor 1	Short to ground / below lower limit / lean mixture threshold
P0328	210	Knock sensor 1	Short to B+ / above upper limit / rich mixture threshold
P0332	211	Knock sensor 2	Short to ground / below lower limit / lean mixture threshold
P0333	211	Knock sensor 2	Short to B+ / above upper limit / rich mixture threshold
P0336	110	Engine speed sensor signal	Signal implausible /Open circuit / no signal
P0341	112	Camshaft position sensor 1	Signal implausible / implausible operating range / malfunction
P0342	112	Camshaft position sensor 1	Short to ground / below lower limit / lean mixture threshold

Fuel Injection 24-29
On-Board Diagnostics

Table b. DME M 5.2.2 DTCs			
P-code	Porsche FC	Fault type	Explanation
P0343	112	Camshaft position sensor 1	Short to B+ / above upper limit / rich mixture threshold
P0413	85	Electric changeover valve	Signal implausible /Open circuit / no signal
P0414	85	Electric changeover valve	Short to ground / below lower limit / lean mixture threshold
P0414	85	Electric changeover valve	Short to B+ / above upper limit / rich mixture threshold
P0410	80	Secondary air injection system, cyl. 1 - 3	Signal implausible / implausible operating range / malfunction
P0410	80	Secondary air injection system, cyl. 1 - 3	Short to ground / below lower limit / lean mixture threshold
P0418	84	Secondary air injection pump	Signal implausible /Open circuit / no signal
P0418	84	Secondary air injection pump	Short to ground / below lower limit / lean mixture threshold
P0418	84	Secondary air injection pump	Short to B+ / above upper limit / rich mixture threshold
P0420	40	TWC conversion, cyl. 1 - 3	Short to B+ / above upper limit / rich mixture threshold
P0430	45	TWC conversion, cyl. 4 - 6	Short to B+ / above upper limit / rich mixture threshold
P0440	93	Fuel tank ventilation system	Short to B+ / above upper limit / rich mixture threshold
P0442	97	Fuel tank ventilation system (micro-leak)	Short to ground / below lower limit / lean mixture threshold
P0444	98	EVAP canister purge valve	Signal implausible /Open circuit / no signal
P0445	98	EVAP canister purge valve	Short to ground / below lower limit / lean mixture threshold
P0445	98	EVAP canister purge valve	Short to B+ / above upper limit / rich mixture threshold
P0446	95	EVAP canister shutoff valve (function)	Short to ground / below lower limit / lean mixture threshold
P0447	96	EVAP canister shutoff valve (output stage)	Signal implausible /Open circuit / no signal
P0448	96	EVAP canister shutoff valve (output stage)	Short to ground / below lower limit / lean mixture threshold
P0448	96	EVAP canister shutoff valve (output stage)	Short to B+ / above upper limit / rich mixture threshold
P0450	99	Tank pressure sensor	Signal implausible / implausible operating range / malfunction
P0452	99	Tank pressure sensor	Short to ground / below lower limit / lean mixture threshold
P0453	99	Tank pressure sensor	Short to B+ / above upper limit / rich mixture threshold
P0455	94	Fuel tank ventilation system (major leak)	Signal implausible / implausible operating range / malfunction
P0455	94	Fuel tank ventilation system (major leak)	Short to ground / below lower limit / lean mixture threshold
P0480	494	Fan output stage 1	Signal implausible /Open circuit / no signal
P0480	494	Fan output stage 1	Short to ground / below lower limit / lean mixture threshold
P0480	494	Fan output stage 1	Short to B+ / above upper limit / rich mixture threshold
P0481	495	Fan output stage 2	Signal implausible /Open circuit / no signal
P0481	495	Fan output stage 2	Short to ground / below lower limit / lean mixture threshold
P0481	495	Fan output stage 2	Short to B+ / above upper limit / rich mixture threshold
P0501	120	Vehicle speed	Signal implausible /Open circuit / no signal
P0506	32	Idle air control at stop	Short to ground / below lower limit / lean mixture threshold
P0507	32	Idle air control at stop	Short to B+ / above upper limit / rich mixture threshold
P0560	107	Voltage supply	Signal implausible / implausible operating range / malfunction
P0562	107	Voltage supply	Short to ground / below lower limit / lean mixture threshold

Fuel Injection

On-Board Diagnostics

Table b. DME M 5.2.2 DTCs

P-code	Porsche FC	Fault type	Explanation
P0563	107	Voltage supply	Short to B+ / above upper limit / rich mixture threshold
P0600	236	CAN time out, Tiptronic	Signal implausible /Open circuit / no signal
P0604	406	Control module faulty (RAM)	Signal implausible / implausible operating range / malfunction
P0605	405	Control module faulty (ROM)	Signal implausible / implausible operating range / malfunction
P0650	165	MIL (Check Engine)	Signal implausible /Open circuit / no signal
P0650	165	MIL (Check Engine)	Short to ground / below lower limit / lean mixture threshold
P0650	165	MIL (Check Engine)	Short to B+ / above upper limit / rich mixture threshold
P1115	13	Precatalyst oxygen sensor heating, cyl. 1 - 3	Signal implausible / implausible operating range / malfunction
P1115	13	Precatalyst oxygen sensor heating, cyl. 1 - 3	Signal implausible /Open circuit / no signal
P1115	13	Precatalyst oxygen sensor heating, cyl. 1 - 3	Short to ground / below lower limit / lean mixture threshold
P1115	13	Precatalyst oxygen sensor heating, cyl. 1 - 3	Short to B+ / above upper limit / rich mixture threshold
P1117	14	Precatalyst oxygen sensor heating, cyl. 1 - 3	Signal implausible / implausible operating range / malfunction
P1117	14	Precatalyst oxygen sensor heating, cyl. 1 - 3	Signal implausible /Open circuit / no signal
P1117	14	Precatalyst oxygen sensor heating, cyl. 1 - 3	Short to ground / below lower limit / lean mixture threshold
P1117	14	Precatalyst oxygen sensor heating, cyl. 1 - 3	Short to B+ / above upper limit / rich mixture threshold
P1118	4	Post-catalyst oxygen heating sensor, cyl. 4 - 6	Signal implausible / implausible operating range / malfunction
P1118	4	Post-catalyst oxygen sensor heating, cyl. 4 - 6	Signal implausible /Open circuit / no signal
P1118	4	Post-catalyst oxygen sensor heating, cyl. 4 - 6	Short to ground / below lower limit / lean mixture threshold
P1118	4	Post-catalyst oxygen sensor heating, cyl. 4 - 6	Short to B+ / above upper limit / rich mixture threshold
P1119	5	Precatalyst oxygen sensor heating, cyl. 4 - 6	Signal implausible / implausible operating range / malfunction
P1119	5	Precatalyst oxygen sensor heating, cyl. 4 - 6	Signal implausible /Open circuit / no signal
P1119	5	Precatalyst oxygen sensor heating, cyl. 4 - 6	Short to ground / below lower limit / lean mixture threshold
P1119	5	Precatalyst oxygen sensor heating, cyl. 4 - 6	Short to B+ / above upper limit / rich mixture threshold
P1121	430	Throttle position sensor 1	Signal implausible / implausible operating range / malfunction
P1121	430	Throttle position sensor 1	Short to ground / below lower limit / lean mixture threshold
P1121	430	Throttle position sensor 1	Short to B+ / above upper limit / rich mixture threshold

Fuel Injection 24-31
On-Board Diagnostics

Table b. DME M 5.2.2 DTCs

P-code	Porsche FC	Fault type	Explanation
P1122	431	Throttle position sensor 2	Signal implausible / implausible operating range / malfunction
P1122	431	Throttle position sensor 2	Short to ground / below lower limit / lean mixture threshold
P1122	431	Throttle position sensor 2	Short to B+ / above upper limit / rich mixture threshold
P1124	167	Fuel pump relay output stage	Signal implausible /Open circuit / no signal
P1124	167	Fuel pump relay output stage	Short to ground / below lower limit / lean mixture threshold
P1124	167	Fuel pump relay output stage	Short to B+ / above upper limit / rich mixture threshold
P1125	357	Oxygen sensing adaptation, upper load range, cyl. 1 - 3	Short to ground / below lower limit / lean mixture threshold
P1125	357	Oxygen sensing adaptation, upper load range, cyl. 1 - 3	Short to B+ / above upper limit / rich mixture threshold
P1126	356	Oxygen sensing adaptation, lower load range, cyl. 1 - 3	Short to ground / below lower limit / lean mixture threshold
P1126	356	Oxygen sensing adaptation, lower load range, cyl. 1 - 3	Short to B+ / above upper limit / rich mixture threshold
P1127	418	Oxygen sensing error by means of short test, cyl. 1 - 3	Short to ground / below lower limit / lean mixture threshold
P1127	418	Oxygen sensing error by means of short test, cyl. 1 - 3	Short to B+ / above upper limit / rich mixture threshold
P1128	360	Oxygen sensing adaptation, idle range, cyl. 1 - 3	Short to ground / below lower limit / lean mixture threshold
P1128	360	Oxygen sensing adaptation, idle range, cyl. 1 - 3	Short to B+ / above upper limit / rich mixture threshold
P1130	361	Oxygen sensing adaptation, idle range, cyl. 4 - 6	Short to ground / below lower limit / lean mixture threshold
P1130	361	Oxygen sensing adaptation, idle range, cyl. 4 - 6	Short to B+ / above upper limit / rich mixture threshold
P1132	359	Oxygen sensing adaptation, upper load range, cyl. 4 - 6	Short to ground / below lower limit / lean mixture threshold
P1132	359	Oxygen sensing adaptation, upper load range, cyl. 4 - 6	Short to B+ / above upper limit / rich mixture threshold
P1133	358	Oxygen sensing adaptation, lower load range, cyl. 4 - 6	Short to ground / below lower limit / lean mixture threshold
P1133	358	Oxygen sensing adaptation, lower load range, cyl. 4 - 6	Short to B+ / above upper limit / rich mixture threshold
P1134	419	Oxygen sensing error by means of short test, cyl. 4 - 6	Short to ground / below lower limit / lean mixture threshold
P1134	419	Oxygen sensing error by means of short test, cyl. 4 - 6	Short to B+ / above upper limit / rich mixture threshold
P1137	446	Clutch switch	Signal implausible / implausible operating range / malfunction
P1157	30	Engine compartment temperature	Short to ground / below lower limit / lean mixture threshold
P1158	30	Engine compartment temperature	Short to B+ / above upper limit / rich mixture threshold
P1237	150	Fuel injector, cylinder 1	Signal implausible /Open circuit / no signal

Fuel Injection

On-Board Diagnostics

Table b. DME M 5.2.2 DTCs

P-code	Porsche FC	Fault type	Explanation
P1225	150	Fuel injector, cylinder 1	Short to ground / below lower limit / lean mixture threshold
P1213	150	Fuel injector, cylinder 1	Short to B+ / above upper limit / rich mixture threshold
P1238	151	Fuel injector, cylinder 6	Signal implausible /Open circuit / no signal
P1226	151	Fuel injector, cylinder 6	Short to ground / below lower limit / lean mixture threshold
P1214	151	Fuel injector, cylinder 6	Short to B+ / above upper limit / rich mixture threshold
P1239	152	Fuel injector, cylinder 2	Signal implausible /Open circuit / no signal
P1227	152	Fuel injector, cylinder 2	Short to ground / below lower limit / lean mixture threshold
P1215	152	Fuel injector, cylinder 2	Short to B+ / above upper limit / rich mixture threshold
P1240	153	Fuel injector, cylinder 4	Signal implausible /Open circuit / no signal
P1228	153	Fuel injector, cylinder 4	Short to ground / below lower limit / lean mixture threshold
P1216	153	Fuel injector, cylinder 4	Short to B+ / above upper limit / rich mixture threshold
P1241	154	Fuel injector, cylinder 3	Signal implausible /Open circuit / no signal
P1229	154	Fuel injector, cylinder 3	Short to ground / below lower limit / lean mixture threshold
P1217	154	Fuel injector, cylinder 3	Short to B+ / above upper limit / rich mixture threshold
P1242	155	Fuel injector, cylinder 5	Signal implausible /Open circuit / no signal
P1230	155	Fuel injector, cylinder 5	Short to ground / below lower limit / lean mixture threshold
P1218	155	Fuel injector, cylinder 5	Short to B+ / above upper limit / rich mixture threshold
P1219	256	Accelerator pedal	Signal implausible / implausible operating range / malfunction
P1265	301	Airbag signal	Signal implausible / implausible operating range / malfunction
P1266	409	Fuel cutoff function monitor	Signal implausible / implausible operating range / malfunction
P1275	16	Precatalyst oxygen sensor aging, delay time, cyl. 1 - 3	Short to ground / below lower limit / lean mixture threshold
P1275	16	Precatalyst oxygen sensor aging, delay time, cyl. 1 - 3	Short to B+ / above upper limit / rich mixture threshold
P1276	22	Precatalyst oxygen sensor aging, delay time, cyl. 4 - 6	Short to ground / below lower limit / lean mixture threshold
P1276	22	Precatalyst oxygen sensor aging, delay time, cyl. 4 - 6	Short to B+ / above upper limit / rich mixture threshold
P1324	325	Position of camshaft in relation to crankshaft, cyl. 4 - 6	Short to ground / below lower limit / lean mixture threshold
P1324	325	Position of camshaft in relation to crankshaft, cyl. 4 - 6	Short to ground / below lower limit / lean mixture threshold
P1325	178	Camshaft adjustment, cyl. 4 - 6	Signal implausible / implausible operating range / malfunction
P1325	178	Camshaft adjustment, cyl. 4 - 6	Short to ground / below lower limit / lean mixture threshold
P1325	178	Camshaft adjustment, cyl. 4 - 6	Short to B+ / above upper limit / rich mixture threshold
P1340	322	Position of camshaft in relation to crankshaft, cyl. 1 - 3	Short to ground / below lower limit / lean mixture threshold
P1340	322	Position of camshaft in relation to crankshaft, cyl. 1 - 3	Short to B+ / above upper limit / rich mixture threshold

Fuel Injection 24-33
On-Board Diagnostics

Table b. DME M 5.2.2 DTCs

P-code	Porsche FC	Fault type	Explanation
P1341	174	Camshaft adjustment, cyl. 1 - 3	Signal implausible / implausible operating range / malfunction
P1341	174	Camshaft adjustment, cyl. 1 - 3	Short to ground / below lower limit / lean mixture threshold
P1341	174	Camshaft adjustment, cyl. 1 - 3	Short to B+ / above upper limit / rich mixture threshold
P1384	220	Knock control zero test	Signal implausible / implausible operating range / malfunction
P1385	221	Knock control offset	Signal implausible / implausible operating range / malfunction
P1386	222	Knock control test pulse	Signal implausible / implausible operating range / malfunction
P1397	113	Camshaft position sensor 2	Signal implausible / implausible operating range / malfunction
P1397	113	Camshaft position sensor 2	Short to ground / below lower limit / lean mixture threshold
P1397	113	Camshaft position sensor 2	Short to B+ / above upper limit / rich mixture threshold
P1411	208	Secondary air injection system, cyl. 4 - 6	Signal implausible / implausible operating range / malfunction
P1411	208	Secondary air injection system, cyl. 4 - 6	Short to ground / below lower limit / lean mixture threshold
P1455	170	A/C compressor control	Signal implausible /Open circuit / no signal
P1457	170	A/C compressor control	Short to ground / below lower limit / lean mixture threshold
P1456	170	A/C compressor control	Short to B+ / above upper limit / rich mixture threshold
P1501	403	Throttle jacking unit output stage	Signal implausible / implausible operating range / malfunction
P1502	412	Throttle jacking unit spring test	Short to B+ / above upper limit / rich mixture threshold
P1503	402	Throttle jacking unit position error	Signal implausible / implausible operating range / malfunction
P1504	410	Throttle jacking unit emergency air position	Signal implausible / implausible operating range / malfunction
P1505	404	Throttle jacking unit control range	Signal implausible /Open circuit / no signal
P1505	404	Throttle jacking unit control range	Short to ground / below lower limit / lean mixture threshold
P1505	404	Throttle jacking unit control range	Short to B+ / above upper limit / rich mixture threshold
P1506	413	Throttle jacking unit lower mechanical stop	Signal implausible / implausible operating range / malfunction
P1507	411	Throttle jacking unit gain adjustment	Signal implausible / implausible operating range / malfunction
P1508	408	Torque comparison function monitor	Signal implausible / implausible operating range / malfunction
P1509	429	Torque limiter	Short to B+ / above upper limit / rich mixture threshold
P1570	39	Immobilizer	Signal implausible / implausible operating range / malfunction
P1571	39	Immobilizer	Signal implausible /Open circuit / no signal
P1574	364	Stop light switch	Signal implausible / implausible operating range / malfunction
P1576	274	Cruise control standby lamp	Signal implausible / implausible operating range / malfunction
P1576	274	Cruise control standby lamp	Signal implausible /Open circuit / no signal
P1576	274	Cruise control standby lamp	Short to ground / below lower limit / lean mixture threshold
P1576	274	Cruise control standby lamp	Short to B+ / above upper limit / rich mixture threshold
P1577	427	Accelerator pedal potentiometer 1	Signal implausible / implausible operating range / malfunction
P1577	427	Accelerator pedal potentiometer 1	Short to ground / below lower limit / lean mixture threshold
P1577	427	Accelerator pedal potentiometer 1	Short to B+ / above upper limit / rich mixture threshold
P1578	428	Accelerator pedal potentiometer 2	Signal implausible / implausible operating range / malfunction

Table b. DME M 5.2.2 DTCs

P-code	Porsche FC	Fault type	Explanation
P1578	428	Accelerator pedal potentiometer 2	Short to ground / below lower limit / lean mixture threshold
P1578	428	Accelerator pedal potentiometer 2	Short to B+ / above upper limit / rich mixture threshold
P1579	111	Crankshaft position sensor	Signal implausible / implausible operating range / malfunction
P1600	216	CAN time out, PSM	Signal implausible /Open circuit / no signal
P1668	233	Start enable (output stage)	Signal implausible /Open circuit / no signal
P1668	233	Start enable (output stage)	Short to ground / below lower limit / lean mixture threshold
P1668	233	Start enable (output stage)	Short to B+ / above upper limit / rich mixture threshold
P1670	175	Electric change-over valve, variable-length manifold	Signal implausible /Open circuit / no signal
P1670	175	Electric change-over valve, variable-length manifold	Short to ground / below lower limit / lean mixture threshold
P1670	175	Electric change-over valve, variable-length manifold	Short to B+ / above upper limit / rich mixture threshold
P1671	407	Control module faulty (computer monitor: Reset)	Signal implausible / implausible operating range / malfunction

Table c. DME ME 7.2 DTCs

P-code	Porsche DTC	Fault type	Explanation
P0101	115	Mass air flow sensor	Signal implausible / implausible operating range / malfunction
P0102	115	Mass air flow sensor	Short to ground / below lower limit / lean mixture threshold
P0103	115	Mass air flow sensor	Short to B+ / above upper limit / rich mixture threshold
P0112	124	Intake air temperature sensor	Short to ground / below lower limit / lean mixture threshold
P0113	124	Intake air temperature sensor	Short to B+ / above upper limit / rich mixture threshold
P0115	123	Engine temperature sensor	Signal implausible / implausible operating range / malfunction
P0117	123	Engine temperature sensor	Short to ground / below lower limit / lean mixture threshold
P0118	123	Engine temperature sensor	Short to B+ / above upper limit / rich mixture threshold
P0123	117	Throttle position sensor	Signal implausible / implausible operating range / malfunction
P0130	10	Precatalyst oxygen sensor, cyl. 1 - 3	Signal implausible / implausible operating range / malfunction
P0134	10	Precatalyst oxygen sensor, cyl. 1 - 3	Signal implausible /Open circuit / no signal
P0131	10	Precatalyst oxygen sensor, cyl. 1 - 3	Short to ground / below lower limit / lean mixture threshold
P0132	10	Precatalyst oxygen sensor, cyl. 1 - 3	Short to B+ / above upper limit / rich mixture threshold
P0133	15	Precatalyst oxygen sensor aging, cyl. 1 - 3	Short to ground / below lower limit / lean mixture threshold
P0136	12	Post-catalyst oxygen sensor, cyl. 1 - 3	Signal implausible / implausible operating range / malfunction
P0140	12	Post-catalyst oxygen sensor, cyl. 1 - 3	Signal implausible /Open circuit / no signal
P0137	12	Post-catalyst oxygen sensor, cyl. 1 - 3	Short to ground / below lower limit / lean mixture threshold
P0138	12	Post-catalyst oxygen sensor, cyl. 1 - 3	Short to B+ / above upper limit / rich mixture threshold

Fuel Injection 24-35
On-Board Diagnostics

Table c. DME ME 7.2 DTCs

P-code	Porsche DTC	Fault type	Explanation
P0139	17	Post-catalyst oxygen sensor aging cylinders 1 - 3	Short to ground / below lower limit / lean mixture threshold
P0150	18	Precatalyst oxygen sensor cyl. 4 - 6	Signal implausible / implausible operating range / malfunction
P0154	18	Precatalyst oxygen sensor cyl. 4 - 6	Signal implausible / Open circuit / no signal
P0151	18	Precatalyst oxygen sensor cyl. 4 - 6	Short to ground / below lower limit / lean mixture threshold
P0152	18	Precatalyst oxygen sensor cyl. 4 - 6	Short to B+ / above upper limit / rich mixture threshold
P0153	21	Aging precatalyst oxygen sensor cyl. 4 - 6	Short to ground / below lower limit / lean mixture threshold
P0156	20	Post-catalyst oxygen sensor cyl. 4 - 6	Signal implausible / implausible operating range / malfunction
P0160	20	Post-catalyst oxygen sensor cyl. 4 - 6	Signal implausible / Open circuit / no signal
P0157	20	Post-catalyst oxygen sensor cyl. 4 - 6	Short to ground / below lower limit / lean mixture threshold
P0158	20	Post-catalyst oxygen sensor cyl. 4 - 6	Short to B+ / above upper limit / rich mixture threshold
P0159	23	Post-catalyst oxygen sensor aging cylinders 4 - 6	Short to ground / below lower limit / lean mixture threshold
P0197	125	Oil temperature sensor	Short to ground / below lower limit / lean mixture threshold
P0198	125	Oil temperature sensor	Short to B+ / above upper limit / rich mixture threshold
P0300	75	Misfire damaging to TWC	Signal implausible / implausible operating range / malfunction
P0301	63	Misfire, cylinder 1, catalyst damaging	Signal implausible / implausible operating range / malfunction
P0302	64	Misfire, cylinder 2, catalyst damaging	Signal implausible / implausible operating range / malfunction
P0303	65	Misfire, cylinder 3, catalyst damaging	Signal implausible / implausible operating range / malfunction
P0304	66	Misfire, cylinder 4, catalyst damaging	Signal implausible / implausible operating range / malfunction
P0305	67	Misfire, cylinder 5, catalyst damaging	Signal implausible / implausible operating range / malfunction
P0306	68	Misfire, cylinder 6, catalyst damaging	Signal implausible / implausible operating range / malfunction
P0336	111	Crankshaft position sensor	Signal implausible / implausible operating range / malfunction
P0341	112	Camshaft position sensor 1	Signal implausible / implausible operating range / malfunction
P0341	112	Camshaft position sensor 1	Short to ground / below lower limit / lean mixture threshold
P0341	112	Camshaft position sensor 1	Short to B+ / above upper limit / rich mixture threshold
P0410	80	Secondary air injection system cyl. 1 - 3	Signal implausible / implausible operating range / malfunction
P0420	40	TWC conversion cylinders 1 - 3	Short to ground / below lower limit / lean mixture threshold
P0430	45	TWC conversion cylinders 4 - 6	Short to ground / below lower limit / lean mixture threshold
P0440	93	Fuel tank ventilation system	Short to B+ / above upper limit / rich mixture threshold
P0441	93	Fuel tank ventilation system	Short to B+ / above upper limit / rich mixture threshold
P0444	98	EVAP canister purge valve	Signal implausible / Open circuit / no signal
P0445		EVAP canister purge valve	Short to ground / below lower limit / lean mixture threshold or Short to B+ / above upper limit / rich mixture threshold
P0446	95	EVAP canister shutoff valve (function)	Short to ground / below lower limit / lean mixture threshold
P0501	120	Vehicle speed	Signal implausible / implausible operating range / malfunction
P0506	32	Idle air control	Short to ground / below lower limit / lean mixture threshold

Fuel Injection

On-Board Diagnostics

Table c. DME ME 7.2 DTCs

P-code	Porsche DTC	Fault type	Explanation
P0507	32	Idle air control	Short to B+ / above upper limit / rich mixture threshold
P0600	236	CAN time out	Signal implausible /Open circuit / no signal
P0603	102	Control module faulty (external RAM)	Signal implausible / implausible operating range / malfunction
P0604	101	Control module faulty (internal RAM)	Signal implausible / implausible operating range / malfunction
P0605	103	Control module faulty (EPROM)	Signal implausible / implausible operating range / malfunction
P1115	13	Heating of oxygen sensor, cyl. 1 - 3	Short to ground / below lower limit / lean mixture threshold
P1102	13	Heating of oxygen sensor, cyl. 1 - 3	Short to B+ / above upper limit / rich mixture threshold
P1117	14	Heating of post-catalyst oxygen sensor, cyl. 1 - 3	Short to ground / below lower limit / lean mixture threshold
P1105	14	Heating of post-catalyst oxygen sensor, cyl. 1 - 3	Short to B+ / above upper limit / rich mixture threshold
P1119	5	Heating of precatalyst oxygen sensor, cyl. 4 - 6	Short to ground / below lower limit / lean mixture threshold
P1107	5	Heating of precatalyst oxygen sensor, cyl. 4 - 6	Short to B+ / above upper limit / rich mixture threshold
P1121	4	Heating of post-catalyst oxygen sensor, cyl. 4 - 6	Short to ground / below lower limit / lean mixture threshold
P1110	4	Heating of post-catalyst oxygen sensor, cyl. 4 - 6	Short to B+ / above upper limit / rich mixture threshold
P1123	27	Oxygen sensing, area 1 cylinders 1 -3	Short to ground / below lower limit / lean mixture threshold
P1124	27	Oxygen sensing, area 1 cylinders 1 -3	Short to B+ / above upper limit / rich mixture threshold
P1125	35	Oxygen sensing, area 1 cylinders 4 - 6	Short to ground / below lower limit / lean mixture threshold
P1126	35	Oxygen sensing, area 1 cylinders 4 - 6	Short to B+ / above upper limit / rich mixture threshold
P1127	26	Oxygen sensing, area 2 cylinders 1 - 3	Short to ground / below lower limit / lean mixture threshold
P1128	26	Oxygen sensing, area 2 cylinders 1 - 3	Short to B+ / above upper limit / rich mixture threshold
P1129	34	Oxygen sensing, area 2 cylinders 4 - 6	Short to ground / below lower limit / lean mixture threshold
P1130	34	Oxygen sensing, area 2 cylinders 4 - 6	Short to B+ / above upper limit / rich mixture threshold
P1140	121	Load sensing	Signal implausible / implausible operating range / malfunction
P1157	30	Engine compartment temperature sensor	Short to ground / below lower limit / lean mixture threshold
P1158	30	Engine compartment temperature sensor	Short to B+ / above upper limit / rich mixture threshold
P1237	150	Fuel injector, cylinder 1	Signal implausible /Open circuit / no signal
P1225	150	Fuel injector, cylinder 1	Short to ground / below lower limit / lean mixture threshold
P1213	150	Fuel injector, cylinder 1	Short to B+ / above upper limit / rich mixture threshold
P1238	151	Fuel injector, cylinder 2	Signal implausible /Open circuit / no signal
P1226	151	Fuel injector, cylinder 2	Short to ground / below lower limit / lean mixture threshold
P1214	151	Fuel injector, cylinder 2	Short to B+ / above upper limit / rich mixture threshold
P1239	152	Fuel injector, cylinder 3	Signal implausible /Open circuit / no signal
P1227	152	Fuel injector, cylinder 3	Short to ground / below lower limit / lean mixture threshold

Fuel Injection 24-37
On-Board Diagnostics

Table c. DME ME 7.2 DTCs

P-code	Porsche DTC	Fault type	Explanation
P1215	152	Fuel injector, cylinder 3	Short to B+ / above upper limit / rich mixture threshold
P1240	153	Fuel injector, cylinder 4	Signal implausible /Open circuit / no signal
P1228	153	Fuel injector, cylinder 4	Short to ground / below lower limit / lean mixture threshold
P1216	153	Fuel injector, cylinder 4	Short to B+ / above upper limit / rich mixture threshold
P1241	154	Fuel injector, cylinder 5	Signal implausible /Open circuit / no signal
P1229	154	Fuel injector, cylinder 5	Short to ground / below lower limit / lean mixture threshold
P1217	154	Fuel injector, cylinder 5	Short to B+ / above upper limit / rich mixture threshold
P1242	155	Fuel injector, cylinder 6	Signal implausible /Open circuit / no signal
P1230	155	Fuel injector, cylinder 6	Short to ground / below lower limit / lean mixture threshold
P1218	155	Fuel injector, cylinder 6	Short to B+ / above upper limit / rich mixture threshold
P1265	301	Airbag signal	Signal implausible / implausible operating range / malfunction
P1275	16	Aging precatalyst oxygen sensor cylinders 1 - 3	Short to ground / below lower limit / lean mixture threshold
P1276	22	Precatalyst oxygen sensor aging, cyl. 4 - 6	Short to ground / below lower limit / lean mixture threshold
P1313	50	Misfire, cylinder 1, emission relevant	Signal implausible / implausible operating range / malfunction
P1314	51	Misfire, cylinder 2, emission relevant	Signal implausible / implausible operating range / malfunction
P1315	52	Misfire, cylinder 3, emission relevant	Signal implausible / implausible operating range / malfunction
P1316	53	Misfire, cylinder 4, emission relevant	Signal implausible / implausible operating range / malfunction
P1317	54	Misfire, cylinder 5, emission relevant	Signal implausible / implausible operating range / malfunction
P1318	55	Misfire, cylinder 6, emission relevant	Signal implausible / implausible operating range / malfunction
P1319	62	Misfire, emission relevant	Signal implausible / implausible operating range / malfunction
P1324	325	Timing chain out of position, cyl. 4 - 6	Short to ground / below lower limit / lean mixture threshold
P1324	325	Timing chain out of position, cyl. 4 - 6	Short to B+ / above upper limit / rich mixture threshold
P1340	322	Timing chain out of position, cyl. 1 - 3	Short to ground / below lower limit / lean mixture threshold
P1340	322	Timing chain out of position, cyl. 1 - 3	Short to B+ / above upper limit / rich mixture threshold
P1384	212	Knock sensor 1	Signal implausible / implausible operating range / malfunction
P1385	213	Knock sensor 2	Signal implausible / implausible operating range / malfunction
P1386	222	Knock control test pulse	Signal implausible / implausible operating range / malfunction
P1397	113	Camshaft position sensor 2	Signal implausible / implausible operating range / malfunction
P1397	113	Camshaft position sensor 2	Short to ground / below lower limit / lean mixture threshold
P1397	113	Camshaft position sensor 2	Short to B+ / above upper limit / rich mixture threshold
P1411	208	Secondary air injection system cyl. 4 -6	Signal implausible / implausible operating range / malfunction
P1455	170	A/C compressor control	Signal implausible /Open circuit / no signal
P1457	170	A/C compressor control	Short to ground / below lower limit / lean mixture threshold
P1456	170	A/C compressor control	Short to B+ / above upper limit / rich mixture threshold
P1541	167	Fuel pump relay output stage	Signal implausible /Open circuit / no signal
P1501	167	Fuel pump relay output stage	Short to ground / below lower limit / lean mixture threshold

Fuel Injection

On-Board Diagnostics

Table c. DME ME 7.2 DTCs			
P-code	Porsche DTC	Fault type	Explanation
P1502	167	Fuel pump relay output stage	Short to B+ / above upper limit / rich mixture threshold
P1514	168	Idle air control valve opening coil	Signal implausible /Open circuit / no signal
P1513	168	Idle air control valve opening coil	Short to ground / below lower limit / lean mixture threshold
P1510	168	Idle air control valve opening coil	Short to B+ / above upper limit / rich mixture threshold
P1539	178	Camshaft adjustment, cyl. 4 - 6	Short to ground / below lower limit / lean mixture threshold
P1524	178	Camshaft adjustment, cyl. 4 - 6	Short to B+ / above upper limit / rich mixture threshold
P1530	174	Camshaft adjustment, cyl. 1 - 3	Short to B+ / above upper limit / rich mixture threshold
P1531	174	Camshaft adjustment, cyl. 1 - 3	Short to ground / below lower limit / lean mixture threshold
P1551	169	Idle air valve control closing coil	Signal implausible /Open circuit / no signal
P1552	169	Idle air valve control closing coil	Short to ground / below lower limit / lean mixture threshold
P1553	169	Idle air valve control closing coil	Short to B+ / above upper limit / rich mixture threshold
P1570	39	Immobilizer	Signal implausible / implausible operating range / malfunction
P1571	39	Immobilizer	Signal implausible /Open circuit / no signal
P1585	8	Misfire with empty fuel tank	Short to B+ / above upper limit / rich mixture threshold
P1600	107	Voltage supply	Signal implausible / implausible operating range / malfunction or Short to ground / below lower limit / lean mixture threshold
P1601	107	Voltage supply	Short to B+ / above upper limit / rich mixture threshold
P1602	108	Voltage supply	Signal implausible /Open circuit / no signal
P1640	105	Control module faulty (EPROM)	Signal implausible / implausible operating range / malfunction
P1671	251	ECM faulty	Signal implausible / implausible operating range / malfunction
P1673	253	Fan output stage	Signal implausible /Open circuit / no signal or Short to ground / below lower limit / lean mixture threshold Short to B+ / above upper limit / rich mixture threshold
P1689	104	Control module faulty (fault memory)	Signal implausible / implausible operating range / malfunction
P1691	165	MIL (Check Engine)	Signal implausible /Open circuit / no signal
P1692	165	MIL (Check Engine)	Short to ground / below lower limit / lean mixture threshold
P1693	165	MIL (Check Engine)	Short to B+ / above upper limit / rich mixture threshold
P1782	135	Engine engagement / nominal engine torque	Short to ground / below lower limit / lean mixture threshold

26 Exhaust System

GENERAL 26-1

EXHAUST SYSTEM COMPONENTS 26-1
  Muffler, removing and installing 26-2
  Catalytic converter, removing and installing 26-3

Oxygen sensors, removing and installing 26-4
Exhaust system diagram (1997-1999) 26-7
Exhaust system diagram (2000 - 2004) 26-8

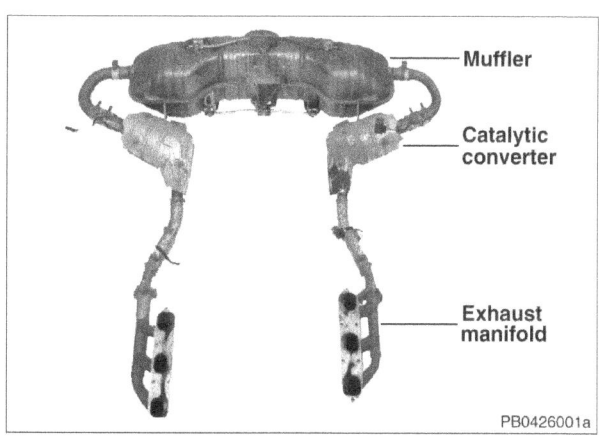

GENERAL

This repair group covers muffler, catalytic convertor and oxygen sensor replacement.

EXHAUST SYSTEM COMPONENTS

The exhaust system is designed to be maintenance free, although regular inspection is warranted due to the harsh operating conditions. Under normal conditions, catalytic converters do not require replacement unless they are damaged.

NOTE—
- *1997 - 1999 Boxster exhaust system shown. 2000 - 2004 models include 2 additional warm-up catalytic converters integral with the exhaust manifolds.*

Use new fasteners, clamps, mounts and gaskets when replacing exhaust components. A liberal application of penetrating oil to the exhaust system nuts and bolts in advance makes removal easier.

> *WARNING—*
> - *The exhaust system and catalytic converter operate at very high temperatures. Allow components to cool before servicing. Wear protective gloves to prevent burns. Do not use flammable chemicals near a hot catalytic converter.*
> - *Exhaust gases are colorless, odorless, and very toxic. Run the engine only in a well-ventilated area. Immediately repair any leaks in the exhaust system or structural damage to the car body that might allow exhaust gases to enter the passenger compartment.*
> - *Corroded exhaust system components crumble easily and often have exposed sharp edges. To avoid injury, wear eye protection and heavy gloves when working with exhaust parts.*

> *CAUTION—*
> - *Use care not to drag or bang the oxygen sensors. Oxygen sensors are easily ruined.*

Exhaust System Components

Muffler, removing and installing

Removal and installation procedures given here are similar for all models.

◄ Muffler connections:
1. Double clamp
2. Muffler clamp
3. Retaining bracket on catalytic converter
4. Retaining bracket on transmission
5. Muffler mounting bracket

– With exhaust system fully cold, raise and support car for access to exhaust system.

> **WARNING—**
> • Do not work under a lifted car unless it is solidly supported on jack stands designed for that purpose. Never work under a car that is supported solely by a jack.

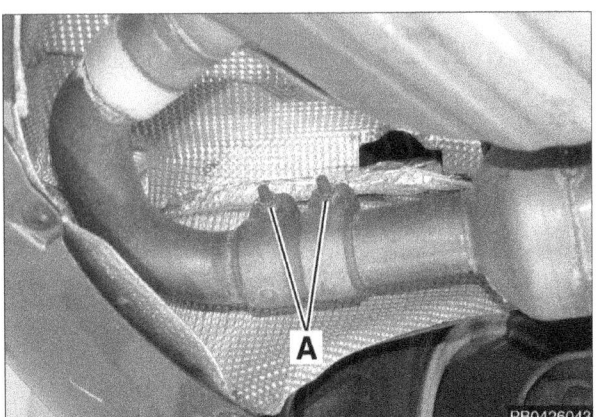

◄ Loosen double clamp fasteners (**A**) between exhaust pipes and catalytic converters. Slide clamp towards converter.

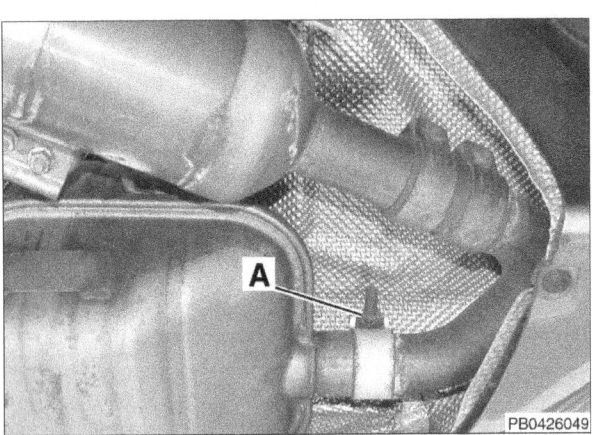

◄ Loosen clamp fasteners between exhaust pipes and muffler (**A**). Slide clamp towards exhaust pipe and remove pipe from muffler by pulling and twisting.

Exhaust System 26-3

Exhaust System Components

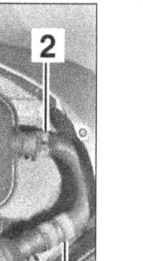

◂ Remove retaining brackets fasteners (**3**) on catalytic converters and on muffler holder (**4**).

NOTE—
- If necessary, remove bolts (**5**) on bracket and remove bracket to gain access to remaining fasteners.

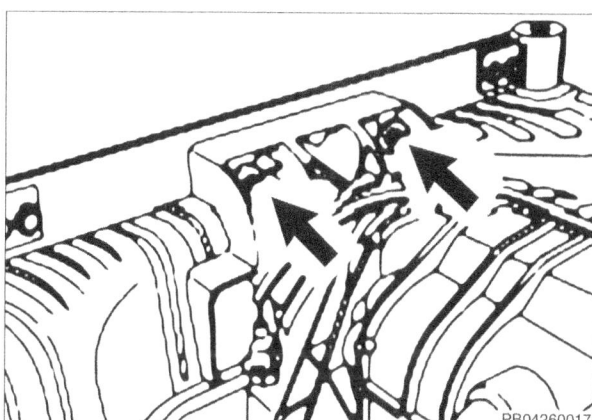

◂ Working above transmission, remove fasteners (**arrows**) on top of muffler bracket.

WARNING—
- Muffler is heavy. To avoid injury and to prevent damage to muffler or rear apron, use a transmission jack or an assistant to support and carefully lower muffler.

- With help from an assistant, guide studs out of top muffler bracket and tilt rear of muffler down past rear apron.

- Installation is reverse of removal. Remember to:
  - Use new fasteners and clamps.
  - Check for sufficient clearance between exhaust system and chassis.
  - Run engine and check exhaust system for leaks.

Catalytic converter, removing and installing

1997 - 1999 Boxster exhaust systems include 2 catalytic converters. 2000 - 2004 models are equipped with 2 additional warm-up catalytic converters integral with the exhaust headers.

Catalytic converter, removing and installing (all models)

- With exhaust system fully cold, raise and support car to access exhaust system.

WARNING—
- Do not work under a lifted car unless it is solidly supported on jack stands designed for that purpose. Never work under a car that is supported solely by a jack.

◂ Loosen double clamp fasteners between exhaust pipe and catalytic converter. Slide clamps towards converter.

- Disconnect oxygen sensor electrical connectors. See **Oxygen sensor, removing and installing** in this repair group.

- Remove retaining brackets fasteners (**3**) on catalytic converters and on muffler holder (**4**).

26-4 Exhaust System

Exhaust System Components

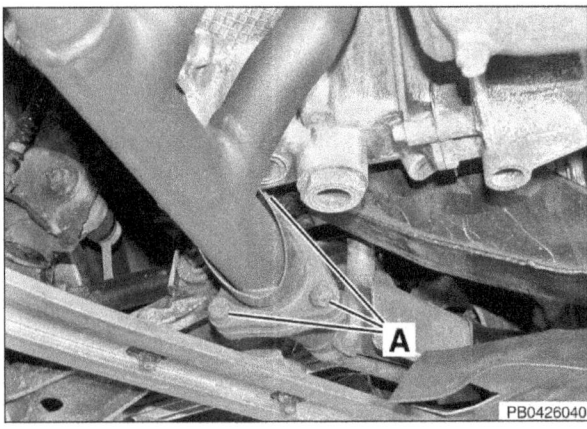

- Remove bolts from exhaust mounting flange (**A**).
- Tilt converter down and back to remove.
- Installation is reverse of removal. Remember to:
  • Check that converter flange sealing surfaces are not damaged.
  • Replace gaskets and bolts with new.
  • Reconnect oxygen sensors.
  • Run engine and check exhaust system for leaks.

Warm-up catalytic converter, removing and installing (2000 - 2004)

- With exhaust system fully cold, raise and support car to access exhaust system.

> **WARNING**—
> • Do not work under a lifted car unless it is solidly supported on jack stands designed for that purpose. Never work under a car that is supported solely by a jack.

- Disconnect oxygen sensor electrical connectors. See **Oxygen sensor, removing and installing** in this repair group.

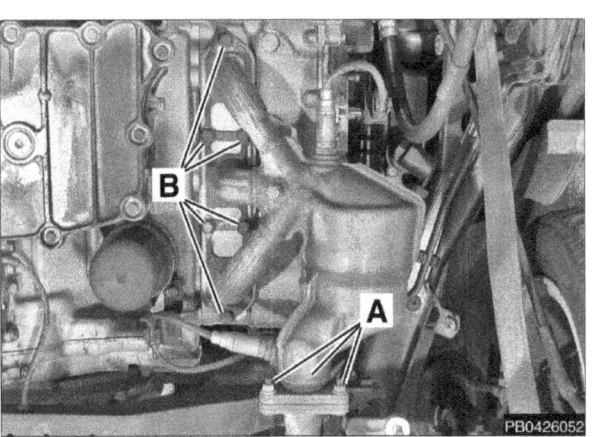

- Remove bolts from exhaust mounting flange (**A**).
- Remove exhaust manifold bolts from cylinder head (**B**) and remove converter.

> **CAUTION**—
> • Exhaust manifold bolts may break. Heat area surrounding manifold bolts before attempting to loosen.

- Installation is reverse of removal. Remember to:
  • Check that exhaust manifold and flange sealing surfaces are not damaged. Remove old gasket remnants.
  • Replace exhaust manifold and flange gaskets with new.
  • Replace all bolts with new.
  • Reconnect oxygen sensors.
  • Run engine and check exhaust system for leaks.

Oxygen sensor, removing

Oxygen sensors are located in front of (precatalyst) and behind (post-catalyst) each catalytic converter. Oxygen sensor replacement is similar for all models.

- With exhaust system fully cold, raise and support car to access exhaust system.

> **WARNING**—
> • Do not work under a lifted car unless it is solidly supported on jack stands designed for that purpose. Never work under a car that is supported solely by a jack.

Exhaust System 26-5

Exhaust System Components

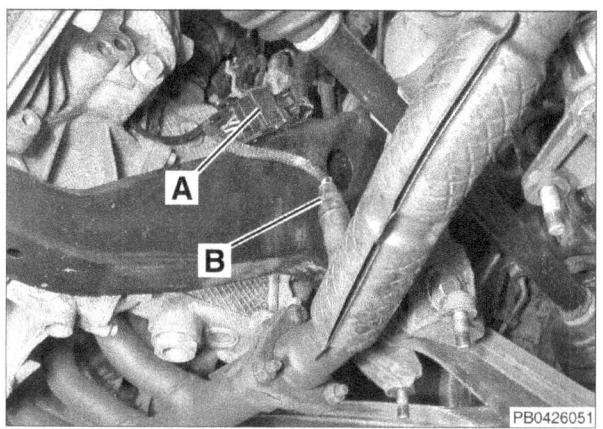

Oxygen sensors (1997 - 1999)

– Remove rear wheel.

◄ Remove electrical connector (**A**) from precatalyst oxygen sensor.

– Use an oxygen sensor socket or open end wrench to loosen sensor (**B**) and unscrew by hand.

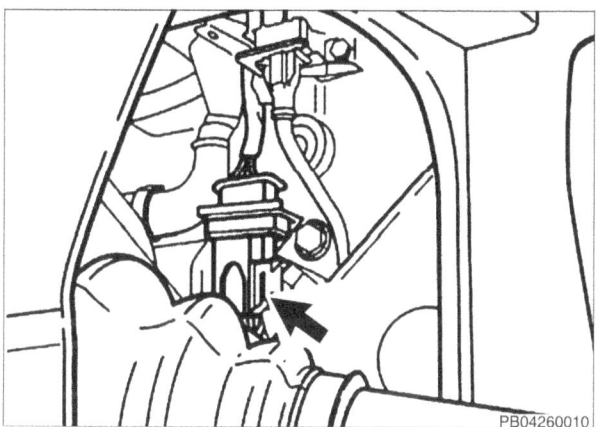

◄ Remove electrical connector from post-catalyst oxygen sensor (**arrow**).

– Use an oxygen sensor socket or open end wrench to loosen sensor and unscrew by hand.

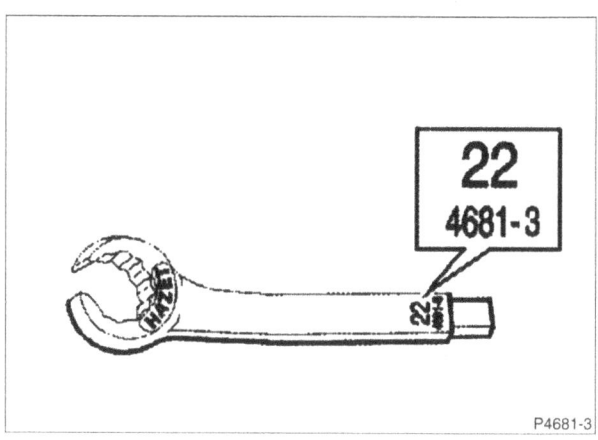

◄ Installation is reverse of removal, noting the following:
  • Oxygen sensors come precoated with antiseize paste.
  • Do not contaminate tip of sensor.
  • Use Porsche special tool 4681-3 or equivalent 22 mm open end crow foot wrench to tighten oxygen sensor to proper torque.
  • Attach harness connector.

Tightening torques	
Oxygen sensor to exhaust pipe:	
• Using Porsche special tool 4681-3	38 - 46 Nm (28 - 34 ft lb)
• Using generic tool	50 - 60 Nm (37 - 44 ft lb)

26-6 Exhaust System

Exhaust System Components

Oxygen sensors (2000 - 2004)

◀ Remove electrical connector (**A** or **B**) from post-catalyst oxygen sensor or precatalyst oxygen sensor.

◀ Use an oxygen sensor socket or open end wrench to loosen sensor (**A** or **B**) and unscrew by hand.

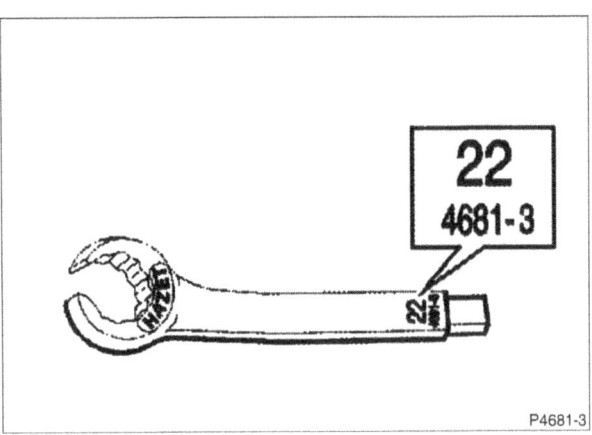

◀ Installation is reverse of removal, noting the following:
- Oxygen sensors come precoated with antiseize paste.
- Do not contaminate tip of sensor.
- Use Porsche special tool 4681-3 or equivalent 22 mm open end crow foot wrench to tighten oxygen sensor to proper torque.

Tightening torques	
Oxygen sensor to exhaust pipe:	
• Using Porsche special tool 4681-3	38 - 46 Nm (28 - 34 ft lb)
• Using generic tool	50 - 60 Nm (37 - 44 ft lb)

Exhaust System 26-7

Exhaust System Components

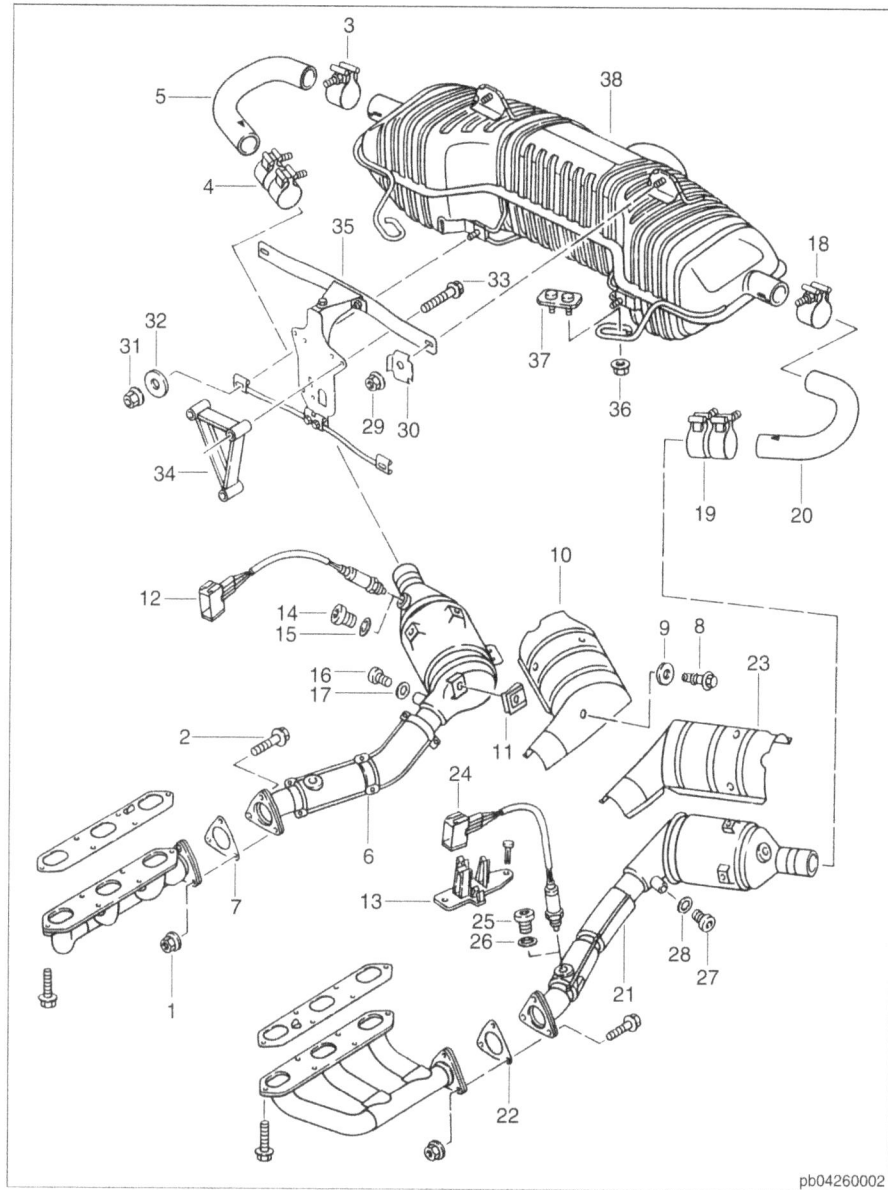

Exhaust system diagram (1997 - 1999)

1. M8 nut
2. M8 x 35 mm bolt
3. Clamp
4. Double clamp
5. Connection tube
6. Catalytic converter (cylinders 1 - 3)
7. Gasket
8. Heat shield bolt
9. Washer
10. Heat shield
11. Sheet metal nut
12. Oxygen sensor (OBD II vehicles)
13. Connector bracket
14. Plug (non-OBD II vehicles)
15. Sealing ring
16. Plug
17. Sealing ring
18. Clamp
19. Double clamp
20. Connection tube
21. Catalytic converter (cylinders 4 - 6)
22. Gasket
23. Heat shield
24. Oxygen sensor
25. Plug
26. Sealing ring
27. Plug
28. Sealing ring
29. M8 nut
30. Reinforcing plate
31. M8 nut
32. Washer
33. Bolt
    - M8 x 20 mm (automatic transmission)
    - M8 x 60 mm (manual transmission)
34. Bracket adapter
35. Bracket
36. M8 nut
37. Mounting plate
38. Muffler

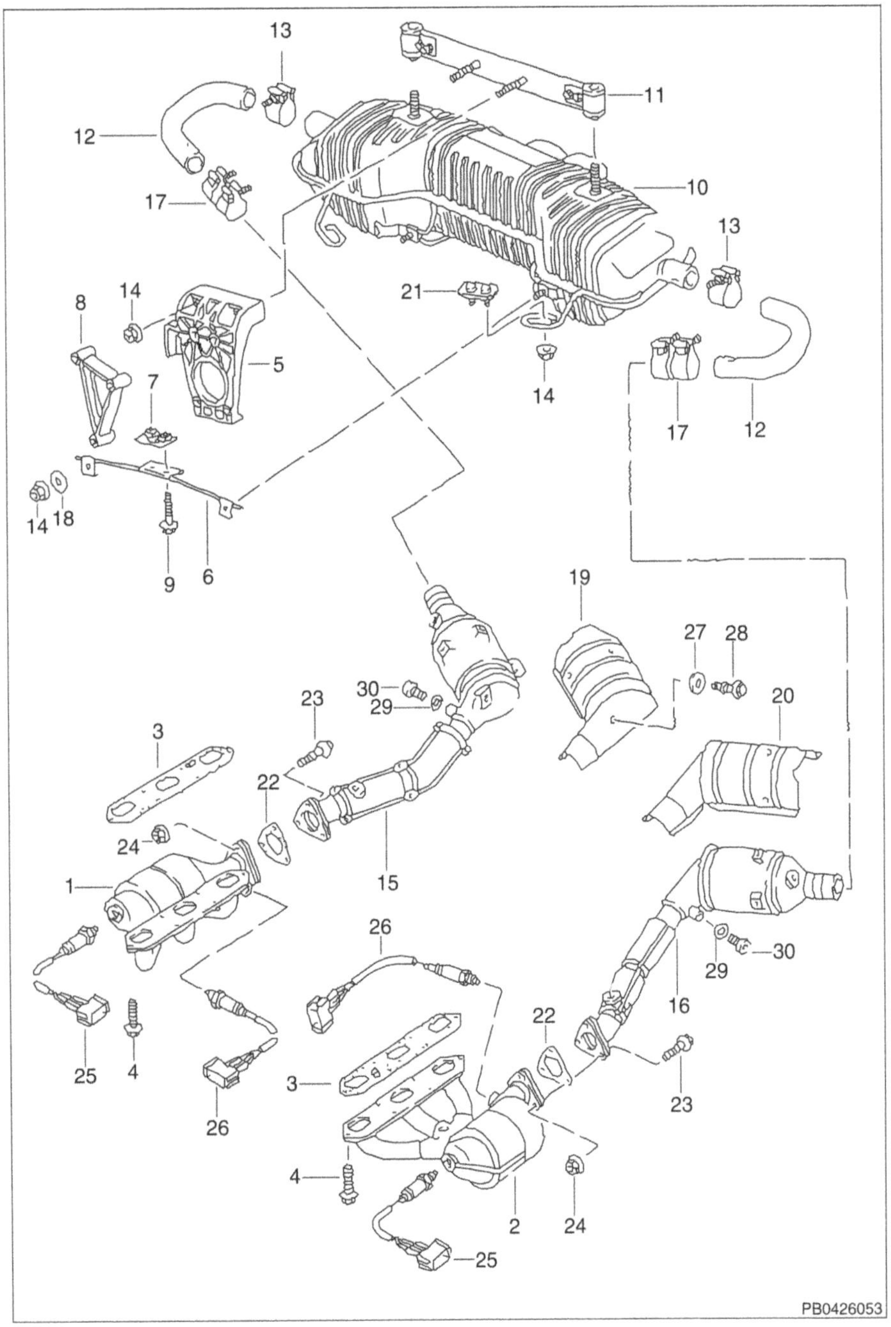

26-8 Exhaust System

Exhaust System Components

Exhaust system diagram (2000 - 2004)

1. Exhaust manifold - precatalytic converter
2. Exhaust manifold - precatalytic converter
3. Exhaust manifold gasket
4. M8 bolt
5. Support
6. Bracket
7. Retaining plate
8. Adapter
9. M8 bolt
10. Muffler
11. Mounting bracket
12. Connecting tube
13. Clamp
14. M8 nut
15. Catalytic converter
16. Catalytic converter
17. Double clamp
18. Washer
19. Heat shield
20. Heat shield
21. Mounting plate
22. Gasket
23. M8 bolt
24. M8 nut
25. Oxygen sensor (precatalyst)
26. Oxygen sensor (post-catalyst)
27. Sealing ring
28. Plug
29. Sealing ring
30. Plug

27 Battery, Starter, Alternator

GENERAL 27-1	STARTER 27-8
Battery disconnection notes 27-2	Starter troubleshooting 27-8
Troubleshooting 27-3	Starter, removing and installing 27-8
Warnings and Cautions 27-3	ALTERNATOR 27-10
BATTERY 27-4	Charging system quick check 27-10
Battery, removing and installing 27-4	Alternator, removing and installing 27-10
Battery charging 27-5	TABLES
Battery maintenance 27-5	a. Battery, starter and alternator troubleshooting 27-3
Battery testing 27-6	
Hydrometer test 27-6	b. Specific gravity of battery electrolyte at 27°C (80°F) 27-6
Battery open-circuit voltage test 27-6	
Battery load voltage test 27-7	c. Open-circuit voltage and battery charge. 27-7
	d. Battery load test - minimum voltage 27-7

GENERAL

This section covers battery, starter and alternator. Troubleshooting information for these components is found in **Table a**. For additional electrical troubleshooting information, see **97 Fuses, Relays, Component Locations**.

The alternator and starter are wired directly to the battery. To prevent accidental shorts that might blow a fuse or damage wires and electrical components, always disconnect the negative (-) battery lead before working on the electrical system.

Various versions of alternators, starters, and batteries have been used in Boxster models. Replace components according to the original equipment specification. When in doubt, consult an authorized Porsche parts department.

Battery disconnection notes

In addition to the general battery / power supply warnings and cautions in this repair group and in **00 Warnings and Cautions**, observe the following whenever the battery is disconnected or accidentally discharged.

- Control module faults may be erased. If possible, check fault memory using Porsche System Tester 2 (PST2) before disconnecting battery.

- After reconnecting battery and test-driving vehicle, "supply voltage" fault may be stored in various control modules. Erase using PST2.

General

— Idle speed fluctuations and minor driveability issues, including incorrect Tiptronic shifting characteristics, may be encountered after battery voltage is restored. Drive vehicle for several minutes to restore adaptation settings. On cars with DME ME 7.2 (2000 - 2002), carry out throttle adaptation as follows:

• Turn ignition ON for one minute prior to starting engine.

• Switch ignition OFF for at least 10 seconds, then start engine.

— Power window standardization: Reset power window limit positions once battery voltage is restored. With convertible top closed, raise window and keep power window switch in raise position for about 5 seconds.

— Trip odometer, clock and trip computer settings are lost. See owner's manual to reset.

— 1997 - 2000 models: Radio and PCM codes are lost and units do not function. Enter code found on radio or PCM code card, or use PST2 to retrieve code from DME control module.

NOTE—

• *2003 - 2004 models: Radio is electronically coded to the car. The code input function is no longer applicable. Note that a replacement radio requires coding using PST2.*

— On cars with PCM: Leave PCM on (with a panoramic view of the sky) for 20 minutes after successfully inputting PCM code.

Troubleshooting

Tests for individual electrical system components are described under component headings in this repair group. **Table a** gives some general troubleshooting ideas.

Table a. Battery, starter and alternator troubleshooting		
Symptom	**Probable cause**	**Corrective action**
Engine does not crank.	Fault in immobilizer system.	Try another ignition key. If problem persists, contact your authorized dealer.
	Faulty automatic transmission gear position switch.	Check, and if necessary, replace automatic transmission gear position switch. See **37 Automatic Transmission**.
Engine cranks slowly or not at all, solenoid clicks when starter is operated.	Battery cables loose, dirty or corroded.	See **03 Service and Maintenance** for battery service.
	Battery discharged.	Charge battery and test. Replace if necessary.
	Battery to body ground cable loose, dirty or corroded.	Inspect ground cable. Clean, tighten or replace if necessary.
	Poor connection at starter motor terminal 30.	Check connections, test for voltage at starter. Test for voltage at clutch switch or transmission range position switch. See **97 Fuses, Relays, Component Locations**.
	Starter motor or solenoid faulty.	Test starter.
Battery does not stay charged more than a few days.	Short circuit draining battery.	Test for excessive current drain with everything electrical off.
	Short driving trips and high electrical drain on charging system.	Evaluate driving style. Where possible, reduce electrical consumption when making short trips.
	Engine drive belt loose, worn, damaged.	See **03 Service and Maintenance** for belt service.
	Battery faulty.	Test battery and replace if necessary.
	Battery cables loose, dirty or corroded.	See **03 Service and Maintenance** for battery service.
	Alternator faulty.	Test alternator regulator.

Battery, Starter, Alternator 27-3

General

Warnings and Cautions

WARNING—
- *Wear goggles and rubber gloves when working around the battery or battery acid (electrolyte).*
- *Battery acid contains sulfuric acid and can cause skin irritation and burning. If acid is spilled on your skin or clothing, flush the area at once with large quantities of water. If electrolyte gets into your eyes, flush them with large quantities of clean water for several minutes and call a physician.*
- *When charging battery, make sure battery cover is off.*
- *Hydrogen gas given off by the battery during charging is explosive. Do not smoke. Keep open flames away from the top of the battery, and prevent electrical sparks by turning off the battery charger before connecting or disconnecting it.*

CAUTION—
- *Only use a digital multimeter when testing automotive electrical components.*
- *If a repair procedure specifies disconnecting the battery, follow the instruction for safety reasons.*
- *Prior to disconnecting the battery, read the battery disconnection cautions in* **00 Warnings and Cautions**.
- *Before disconnecting battery, switch ignition OFF. If the ignition is not turned off when the battery is disconnected, diagnostic troubles codes (DTCs) may be set in some electronic control modules.*
- *Disconnecting the battery cables may erase DTCs stored in ECM memory.*
- *Disconnecting the battery erases the radio code and radio presets. Note radio code and stored stations and restore them after connecting the battery.*
- *On-board computer and clock stored settings may be lost.*
- *Always disconnect the negative (–) battery cable first and reconnect it last. Cover the battery post with an insulating material whenever the cable is removed.*
- *Do not disconnect battery, alternator or starter wires while the engine is running.*
- *Never reverse the battery cables. Even a momentary wrong connection can damage the alternator and other electrical components.*
- *Do not depend on the color of insulation to tell battery positive and negative cables apart. Label cables before removing.*
- *Always allow a frozen battery to thaw before attempting to recharge it.*
- *If a quick charger is used to charge the battery, disconnect battery from vehicle electrical system and remove. This prevents damage to paintwork and upholstery.*

27-4 Battery, Starter, Alternator

Battery

BATTERY

◄ The battery is located in the front trunk under a protective cover (**arrow**) at the base of the windshield.

Under normal operating conditions, the battery is maintenance free. However, if battery is exposed to prolonged high temperatures or when servicing the battery, make sure that:

- Battery is secure and clamping bolt is tight.
- Battery vent hose is correctly routed.
- Cell caps are equipped with O-ring seals.
- Electrolyte level is just above battery plates or between minimum and maximum markings on battery case.
- Only distilled water is used to top-up batteries (this prevents impurities which cause self-discharging).
- Excess electrolyte is extracted using a hydrometer.
- Battery and electrolyte (sulfuric acid) are disposed of properly. Refer to local regulations for recycling/disposal information.

Battery, removing and installing

- Make sure ignition key is off and remove key. Observe battery disconnection warnings and cautions. See **Warnings and Cautions** in this repair group.

◄ Working in front trunk, turn locking fasteners 90° and remove battery cover.

◄ Remove negative battery lead first and cover with insulating tape. Remove positive lead.

Battery, Starter, Alternator 27-5

Battery

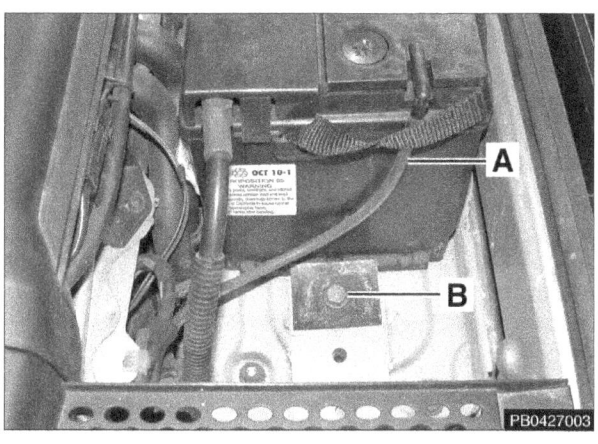

◄ Detach battery vent hose (**A**).
- Remove battery hold down fastener (**B**) and remove battery.

– When reinstalling, first connect positive lead, then negative lead. Reattach vent hose.

Tightening torques	
Battery hold down to battery box	23 Nm (17 ft-lb)
Battery terminal clamp to battery post	5 Nm (4 ft-lb)

Battery charging

Prolonged battery charging causes electrolyte evaporation to a level that can damage the battery. It is best to use a low-current charger (6 amperes or less) to prevent battery damage caused by overheating.

– Before charging battery, test battery as described in **Battery open-circuit voltage test** in this repair group.
- If voltage is 10 vdc or less, one or more cells may be faulty or battery may already be damaged.
- In this case, remove battery from vehicle to recharge, because escaping battery gases could damage interior equipment and trim.
- Remove battery cap covers and top up cells with distilled water. Leave caps off while charging.
- Attempt to revive faulty cells with low charging current.

Battery maintenance

Check battery acid level at least once a year. If necessary, top up with distilled water. Some "maintenance free" batteries are sealed. In that case, topping up is not required. See **03 Service and Maintenance** for more information.

Some electrical consumers require power supply from the battery even when the vehicle is not being driven. The battery also self-discharges. Battery self-discharging time is dependent on vehicle model and vehicle equipment.

If a long time passes without the battery being recharged, it will eventually result in concealed damage. This will in turn lead to early failure of the battery.

- If battery is connected to vehicle circuits: Recharge every 6 weeks.
- If battery is not connected to vehicle circuits: Recharge every 12 weeks.
- For vehicle in continual but light use, use a cigarette lighter trickle charger to maintain optimum battery charge. This type of charger, with electronic circuitry for controlling voltage and current, switches charge voltage off at 13.8 vdc and back on again at 12.6 vdc. Charge current is no greater than 5.5 amperes.

Battery

Battery testing

Battery testing determines the state of battery charge. On conventional or low-maintenance batteries, the most common method of testing the battery is to check the specific gravity of the electrolyte using a hydrometer. Before testing the battery, check that the cables are tight and free of corrosion.

NOTE—

- *In several battery tests given below, it is assumed that the battery cell caps are removed. This is not possible in some "maintenance-free" batteries.*

Hydrometer test

Before hydrometer testing, load the battery with 15 amperes for one minute. If the battery is installed in the vehicle, this can be done by turning on the headlights without the engine running. The state of battery charge based on specific gravity values is given in **Table b**.

The hydrometer indicates the specific gravity of the electrolyte. The more dense the concentration of sulfuric acid in the electrolyte, the higher the state of charge.

Electrolyte temperature affects hydrometer reading. Check the electrolyte temperature with a thermometer.

- Add 0.004 to the hydrometer reading for every 6°C (10°F) that the electrolyte is above 27°C (80°F).
- Subtract 0.004 from the reading for every 6°C (10°F) that the electrolyte is below 27°C (80°F).

Table b. Specific gravity of battery electrolyte at 27°C (80°F)	
Specific gravity	**State of charge**
1.265	Fully charged
1.225	75% charged
1.190	50% charged
1.155	25% charged
1.120	Fully discharged

- If electrolyte specific gravity is at or above 1.225, but battery lacks power for starting, determine battery service condition with a load voltage test. See **Battery load voltage test** in this repair group.

- If average specific gravity of battery cells is below 1.225, recharge battery.

- After charging several hours, if electrolyte specific gravity remains low or battery lacks starting power, replace battery.

Battery

Battery open-circuit voltage test

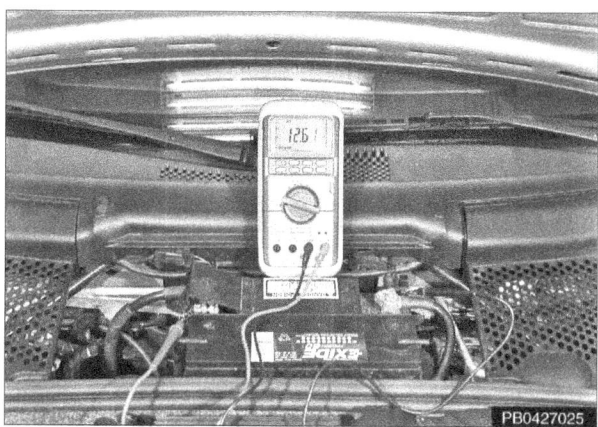

◄ Before testing, load battery with 15 amperes for one minute with battery load-tester or turn on headlights for about one minute without engine running. Connect digital voltmeter across battery terminals. Open-circuit voltage levels are given in **Table c**.

> **CAUTION—**
> - Only use a digital multimeter when testing automotive electrical components.

- If open-circuit voltage is OK but battery still lacks power for starting, perform load voltage test.

- If open-circuit voltage is below 12.4 vdc, recharge battery and retest.

Table c. Open-circuit voltage and battery charge	
Open-circuit voltage	State of charge
12.6 vdc or more	Fully charged
12.4 vdc	75% charged
12.2 vdc	50% charged
11.7 vdc or less	Fully discharged

Battery load voltage test

- Disconnect battery cables before making test.

> **CAUTION—**
> - Prior to disconnecting the battery, read the battery disconnection cautions in **00 Warnings and Cautions**.

- Using battery load tester, apply high resistive load (approx. 200 amps) to battery terminals for 15 seconds. Measure battery voltage.

> **NOTE—**
> - The battery should be fully charged for the most accurate results.

- Check results against data in **Table d**. Replace battery if voltage is below minimum.

Table d. Battery load test - minimum voltage	
Ambient temperature	Voltage*
27°C (80°F)	9.6 vdc
16°C (60°F)	9.5 vdc
4°C (40°F)	9.3 vdc
-7°C (20°F)	8.9 vdc
-18°C (0°F)	8.5 vdc
*Measured after applying a 200 amp load for 15 seconds	

Starter

The starter is located at the top of the engine under the throttle body.

Starter troubleshooting

On back of starter, the large wire at terminal 30 is direct battery voltage. The smaller wire at terminal 50 operates starter solenoid via ignition switch.

- If starter turns engine slowly when ignition is in START position:
  - Check battery state of charge.
  - Inspect starter wires, terminals and ground connections for good contact. In particular, make sure ground connections between battery, body and engine are completely clean and tight.
  - If no faults are found, starter may be faulty.

- If starter fails to operate:
  - Check electronic immobilizer. Try another ignition key. If no faults can be found, have the immobilizer system checked using Porsche scan tool equipment.
  - Check clutch pedal operated starter interlock switch or gear position switch (automatic transmission).

NOTE—
- *A factory-installed electronic immobilizer is used on Porsche cars. This system prevents operation of the starter if a specially coded ignition key is not used. See* **96 Interior Lights, Anti-theft**.
- *On automatic transmission cars, the transmission gear position switch signals the immobilizer to prevent the engine from starting in gear positions other than PARK or NEUTRAL. See* **37 Automatic Transmission**.
- *On manual transmission cars, a clutch switch is used to prevent the starter from operating unless the clutch pedal is pushed fully to the floor. See* **30 Clutch**.

- Check for battery voltage at terminal 50 of starter motor with key in START position.
  - If voltage is not present, check wiring between ignition switch and starter terminal. Check immobilizer system and other inputs that disrupt input to starter. See **EWD Electrical Wiring Diagrams**.
  - If voltage is present and no other visible wiring faults can be found, problem is most likely in starter motor.

Starter, removing and installing

- Disconnect negative battery lead and cover with insulating tape.

CAUTION—
- *Prior to disconnecting the battery, read the battery disconnection cautions in* **00 Warnings and Cautions**.

- Remove top and front engine access covers. See **03 Service and Maintenance**.

- Remove air duct between air cleaner housing and throttle body.

Battery, Starter, Alternator 27-9

Starter

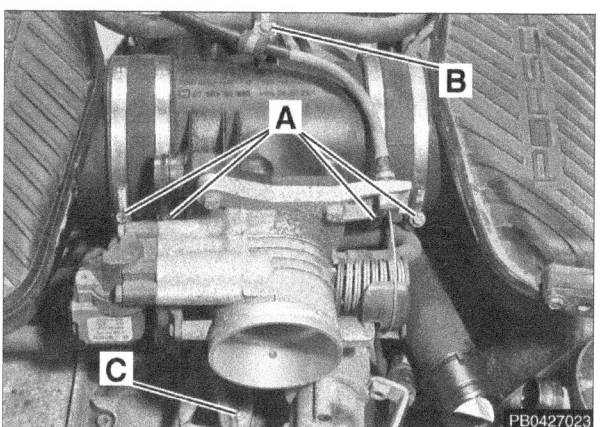

- Remove vent hose between intake distributor and oil separator.

◁ Loosen hose clamps on either side of intake distributor (**A**) and remove clamp (**B**).

- Release throttle body hold down (**C**) from crankcase. Swivel intake distributor along with throttle body up and around approximately 45° to remove throttle body.

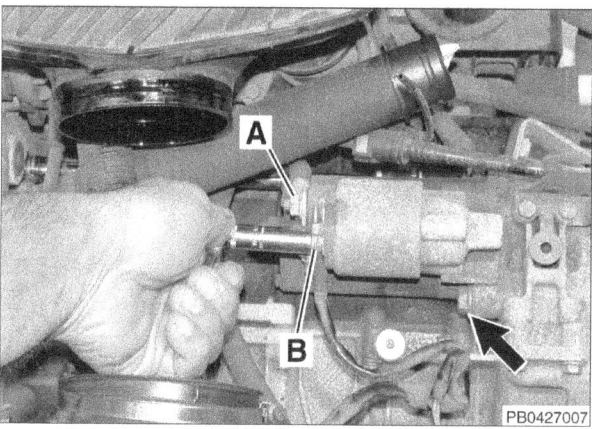

◁ Remove cables at terminal 30 (**A**) and terminal 50 (**B**) of starter solenoid.

- Remove upper bolt (**arrow**) holding starter to bellhousing.

◁ Remove lower starter bolt and ground strap. Insert a socket with long extension through gap (**arrow**) beneath upper idler pulley.

NOTE—

• *Drive belt removed for clarity.*

- Lift starter up and out of engine compartment.

- Installation is reverse of removal. Route starter wires in original location.

Tightening torques	
Starter to transmission (M10)	45 Nm (33 ft-lb)
Terminal 30 to starter (M8)	15 Nm (11 ft-lb)
Terminal 50 to starter (M6)	6.5 Nm (5 ft-lb)

27-10 Battery, Starter, Alternator

Alternator

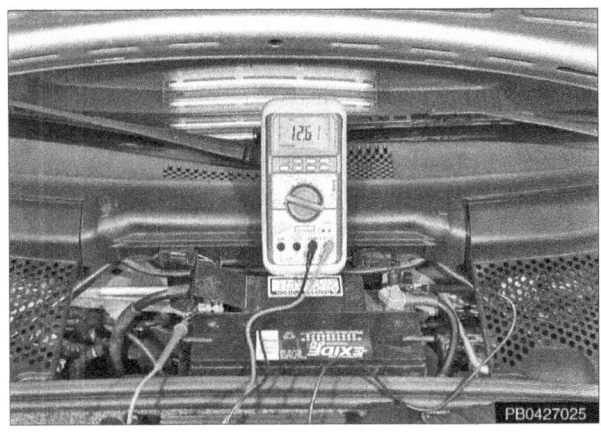

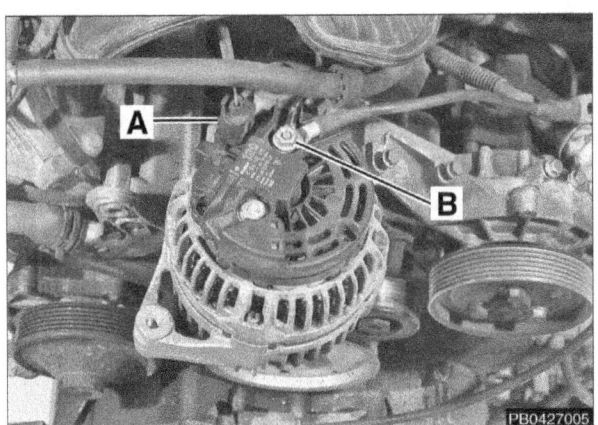

ALTERNATOR

Charging system quick check

◀ Use a digital multimeter to measure voltage across the battery terminals with key off and then again with engine running. Battery voltage should be about 12.6 volts with key off and between 13.5 and 14.5 volts with engine running.

NOTE—

- *If the voltage is higher than 14.8, the alternator is most likely faulty.*

Alternator, removing and installing

- Remove negative battery lead and cover with insulating tape.

- Move convertible top and convertible-top storage lid to service position. See **03 Service and Maintenance**.

- Remove right seat to aid access (optional). See **72 Seats**.

- Remove top and front engine access covers. See **03 Service and Maintenance**.

◀ Disconnect the B+ cables (**arrows**) to the alternator and starter at top of engine.

◀ Remove lower alternator bolt (**A**).

- To aid access, remove protective shield (**C**).

- Loosen bolt (**B**) by three turns. Using a soft-faced hammer, tap bolt to loosen bushing in alternator arm.

NOTE—

- *Tap bolt and bushing until alternator can be slid off bracket without resistance.*

- Lift alternator up and out of slotted bracket along with idler pulley.

- Turn alternator clockwise to clear brackets and guide it out from under intake.

◀ Gently pull alternator forward and release electrical connectors (**A**) and (**B**). Remove alternator.

- Installation is reverse of removal.

Tightening torques	
Alternator to engine block • M16 • M10 • M8	65 ± 5 Nm (48 ± 4 ft-lb) 45 Nm (33 ft-lb) 15 Nm (11 ft-lb)

28 Ignition System

GENERAL 28-1
   Engine management 28-1
   Ignition firing order 28-1
   Warnings and Cautions..................... 28-2

IGNITION SYSTEM COMPONENTS 28-2
   Disabling ignition system 28-2

Checking for spark 28-3
Camshaft position sensors,
   removing and installing 28-3
Crankshaft position sensor,
   removing and installing 28-4
Ignition coil, removing and installing 28-4
Knock sensors, removing and installing 28-5

GENERAL

This repair group covers service information for ignition system components. For related fuel and ignition information see:

- **20 Fuel Storage and Supply**
- **24 Fuel Injection**
- **EWD Electrical Wiring Diagrams**

NOTE—

- *For ignition switch replacement, see* **48 Steering**.

Engine management

Boxster engines use an advanced engine management system known as Digital Motor Electronics (DME). DME incorporates on-board diagnostics, fuel injection, ignition and other engine control functions.

Second generation On-Board Diagnostics (OBD II) is incorporated into the DME system. Using a Porsche-specific or OBD II electronic scan tool, it is possible to pinpoint ignition and other engine management problems.

Additional information about DME system diagnosis and DTCs is provided in **24 Fuel Injection** and **OBD On-Board Diagnostics**.

Ignition firing order

Each ignition coil is mounted above the corresponding spark plug. There is no distributor cap or ignition rotor. Cylinder 1 is at the right front of the engine, followed by cylinders 2 and 3. Cylinders 4, 5 and 6 are located on the left side of the engine, from front to back.

Ignition firing order	
Boxster, Boxster S	1 - 6 - 2 - 4 - 3 - 5

28-2 Ignition System

Ignition System Components

Warnings and Cautions

> **WARNING—**
> - Do not touch or disconnect any cables from the coils while the engine is running or being cranked by the starter.
> - The ignition system produces high voltages that can be fatal. Avoid contact with exposed terminals. Use extreme caution when working on a car with the ignition switched on or the engine running.
> - Connect and disconnect the DME system wiring and test equipment leads only when the ignition is OFF.
> - Before operating the starter without starting the engine (for example when making a compression test) always disable the ignition.

> **CAUTION—**
> - Prior to disconnecting the battery, read the battery disconnection cautions given at the front of this manual.
> - Do not attempt to disable the ignition by removing the coils from the spark plugs.
> - Connect or disconnect ignition system wires, multiple wire connectors, and ignition test equipment only while ignition is off. Switch multimeter functions or measurement ranges only with the test probes disconnected.
> - Do not disconnect the battery while the engine is running.
> - A high impedance digital multimeter should be used for all voltage and resistance tests. An LED test light should be used in place of an incandescent-type test light.
> - In general, make test connections only as specified by Porsche, as described in this manual, or as described by the instrument manufacturer.

IGNITION SYSTEM COMPONENTS

Disabling ignition system

The ignition system operates in a lethal voltage range and should be disabled any time engine service or repair work is being done that requires the ignition to be switched on.

The ignition system can be disabled by removing the engine control DME main relay.

> **NOTE—**
> - The DME main relay is also referred to as the ECM main relay.

− Working in rear trunk, detach and fold back left side trim panel.

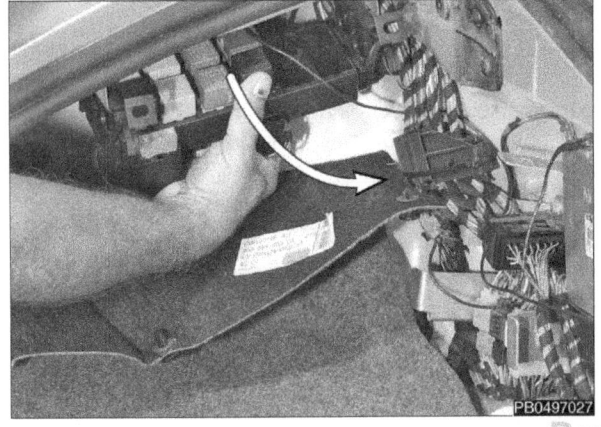

◂ Slide relay panel to right (**arrow**) to remove. Lower panel to trunk floor.

Ignition System 28-3

Ignition System Components

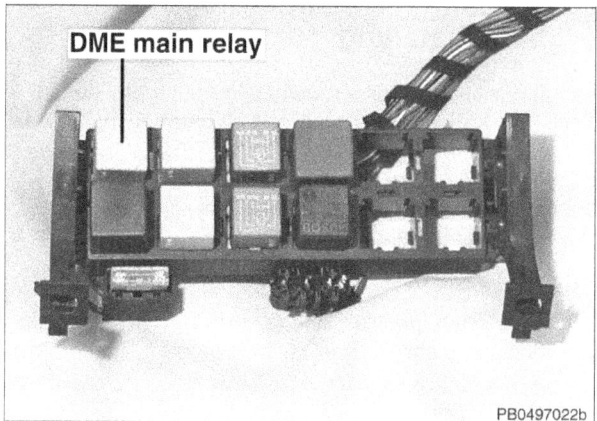

◀ With ignition off, remove DME main relay.

Checking for spark

CAUTION—
- If a spark test is done incorrectly, damage to the ECM or ignition coils may result.

NOTE—
- Spark plug replacement is covered in **03 Service and Maintenance**.

Checking for spark is difficult on engines with distributorless ignition systems.

Try removing the spark plugs and inspecting for differences between them. A poor-firing plug may be wet with fuel and/or black and sooty, but not always. If a coil is not operating, the ECM will electrically disable the fuel injector to that cylinder. The key is to look for differences between cylinders.

Camshaft position sensors, removing and installing

The camshaft position (CMP) sensors (Hall sensors) are used by the engine control module (ECM) for VarioCam control, fuel injection and knock control. There are two camshaft position sensors, one per cylinder bank.

− Make sure ignition is turned off.

− Remove top engine access cover. See **03 Service and Maintenance**.

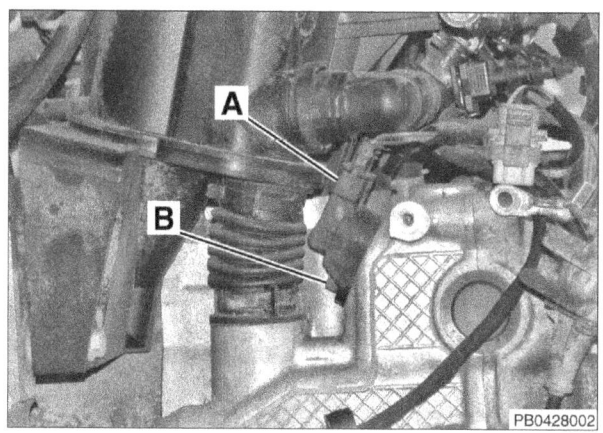

◀ Working at rear of right cylinder head (shown) or front of left cylinder head, disconnect electrical connector (**A**) from sensor.

− Remove bolt (**B**) and remove camshaft sensor.

− Installation is reverse of removal. Remember to:
  • Use new bolt and O-ring when installing sensor.
  • Route wire in installed position.
  • Check for ECM for fault codes.

Ignition System Components

Crankshaft position sensor, removing and installing

The crankshaft position sensor is mounted in right rear side of the cylinder block. The sensor reads a toothed pulse wheel mounted to the end of the crankshaft.

If the DME control module does not receive a signal from the crankshaft position sensor during cranking, the engine will not start.

If the OBD II system detects a catalyst damaging fault due to a malfunction in crankshaft position sensor, the Check Engine light (Malfunction Indicator Light MIL) will be illuminated.

– Remove right rear wheel.

– Working in right rear wheel well, remove electrical connector (**A**) from crankshaft sensor harness.

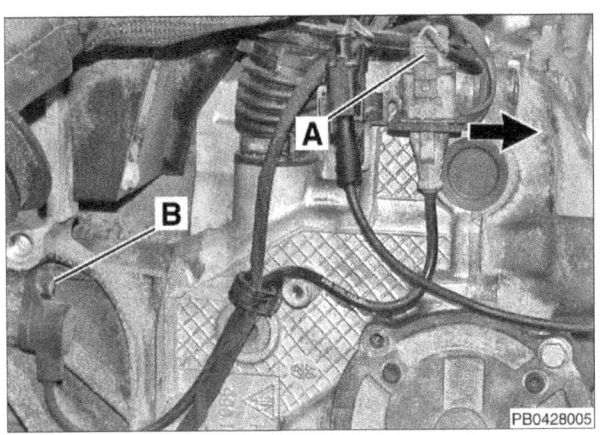

◄ Remove harness connector by sliding it out of bracket to the right (**arrow**).

- Remove bolt (**B**). Remove harness from cable clip and remove crankshaft sensor.

– Installation is reverse of removal. Make sure electrical connections are secure and that harness is routed in installed position.

Tightening torque	
Crankshaft sensor to engine block	10 ± 0.5 Nm
	7.5 ± 0.4 ft-lb

Ignition coil, removing and installing

Boxster engines use six individual ignition coils bolted directly to the cylinder head covers.

◄ Working at cylinder head cover, release electrical connector from ignition coil.

◄ Remove bolts (**A**). While turning slightly, pull coil up and out of spark plug well.

– Installation is reverse of removal. Make sure coil is fully seated on spark plug before replacing bolts and electrical connector.

> **CAUTION—**
> - Inspect ignition coil for evidence of arcing or other damage. Replace as needed.

Ignition System 28-5

Ignition System Components

Knock sensors, removing and installing

Boxster and Boxster S models use two piezo-electric knock sensors that monitor and control potentially damaging ignition knock through the ECM. Knock sensors function like microphones and are able to convert mechanical vibration (knock) into electrical signals. The ECM is programmed to react to frequencies that are characteristic of engine knock and adapt the ignition timing point accordingly.

Knock sensor 1 is mounted to the left side of the engine on top of the crankcase near cylinder 2. Knock sensor 2 is mounted to the right side of the engine on top of the crankcase near cylinder 5.

  Working on the top left of the engine, release electrical connector (**A**) and remove knock sensor mounting bolt (**B**).

> **CAUTION—**
> - Note installed angle of the knock sensor on crankcase before removing. Reinstall sensor in same position.

> **NOTE—**
> - Drive belt and A/C compressor shown removed for clarity.

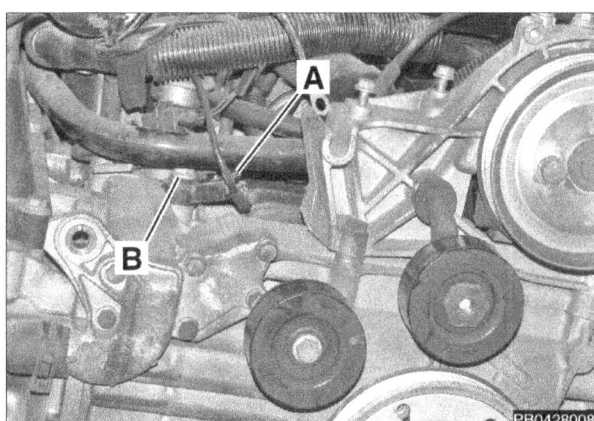

  Working on the top right of the engine, release electrical connector (**A**) and remove knock sensor mounting bolt (**B**).

> **CAUTION—**
> - Note installed angle of the knock sensor on crankcase before removing. Reinstall sensor in same position.

> **NOTE—**
> - Drive belt and alternator shown removed for clarity.

- Installation is reverse of removal. Remember to:
  - Clean knock sensor contact surface on crankcase and sensor.
  - Use a torque wrench when tightening sensor mounting bolt.

Tightening torque	
knock sensor to engine crankcase	20 ± 2 Nm 15 ± 1.5 ft-lb

30 Clutch

GENERAL 30-1
  Clutch system 30-1
  Dual-mass flywheel 30-1
  Troubleshooting 30-2
CLUTCH MECHANICAL 30-2
  Clutch assembly 30-2
  Clutch, replacing 30-3
  Clutch interlock switch, removing 30-5
CLUTCH HYDRAULIC SYSTEM 30-5
  Warnings and Cautions 30-5
  Clutch pedal and master cylinder 30-6
  Clutch hydraulic system, bleeding 30-6
  Clutch slave cylinder, removing and installing .. 30-7
  Clutch master cylinder, removing and installing . 30-7

GENERAL

This repair group covers replacement of clutch mechanical and clutch hydraulic components.

Special tools may be required for some of the procedures described in this repair group. Read the procedures through before beginning a job.

Mechanical clutch service requires transmission removal. See **34 Manual Transmission**.

Clutch system

The clutch mechanical system is made up of three major components: dual-mass flywheel, clutch disc and pressure plate.

The clutch is hydraulically operated by the master and slave cylinders. Clutch disc wear is automatically taken up by the pushrod travel of the slave cylinder, making periodic adjustment unnecessary.

Dual-mass flywheel

The dual-mass flywheel reduces harmonic disturbances and torsional vibrations transmitted to the driveline. The damper mechanism that is traditionally part of the clutch pressure plate is integrated into the dual-mas flywheel. This works to reduce the engine resonance point to a speed lower than engine idle, minimizing driveline vibrations.

The flywheel is split into two sections:
- Primary section: Bolted to crankshaft.
- Secondary section: Makes contact with clutch disc and secures pressure plate.

The primary section of the flywheel contains springs to isolate engine vibrations and a torque limiter to prevent engine torque spikes from exceeding engine and transmission component strength.

Clutch Mechanical

Troubleshooting

A soft or spongy feel to the clutch pedal, excessive pedal free-play, or grinding noises from the gears while shifting can all indicate problems with the clutch hydraulics. In these circumstances it is best to start with a clutch fluid flush, followed, if necessary, by replacement of the hydraulic parts.

Failure of the clutch to fully disengage (leading to difficult gear shifting) may be caused by one or more of the following:

- Leaky or faulty clutch slave cylinder or master cylinder.
- Inadequate travel or misalignment of the slave cylinder pushrod.
- Faulty or bent master cylinder pushrod.
- Binding clutch disc.
- Seized pilot bearing.

NOTE—

- *Inspect the bottom of the transmission bellhousing for oil. If oil is present, remove the flywheel to inspect the crankshaft rear main seal. See* **13 Engine Pulleys and Seals** *for more information.*

CLUTCH MECHANICAL

Clutch assembly

1. **Release lever**
2. **Throwout bearing**
3. **Retaining spring**
4. **M8 x 16 mm bolt**
   - Tighten to 23 Nm (17 ft-lb)
   - To avoid damaging pressure plate, bolts must be loosened and tightened in stages and in diagonally opposite sequence.
5. **Pressure plate**
6. **Clutch disc**
   - Install using special tool 3176 (clutch centering tool).
7. **M10 x 50 mm bolt**
   - Tighten to 25 Nm (19 ft-lb) + 120°
8. **Dual-mass flywheel**
   - Make sure that alignment pins are firmly seated.
   - Flywheel is not machinable. Replace if damaged.

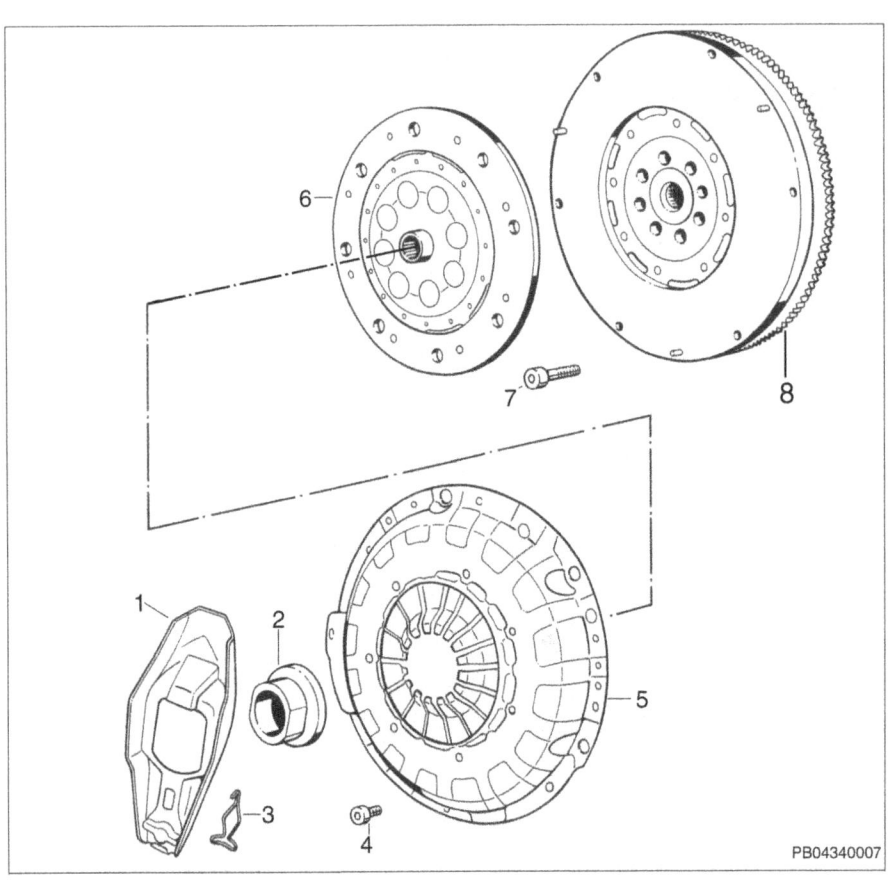

Clutch 30-3

Clutch Mechanical

Clutch, replacing

Clutch replacement requires the following:
- Lifting of vehicle
- Transmission removal
- Clutch disc centering tool

Read the procedure through before starting the job.

- Raise car and support safely.

> **WARNING—**
> - Make sure the car is stable and well supported at all times. Use a professional automotive lift or jack stands designed for the purpose. A floor jack is not adequate support.

- Remove transmission. See **34 Manual Transmission**.

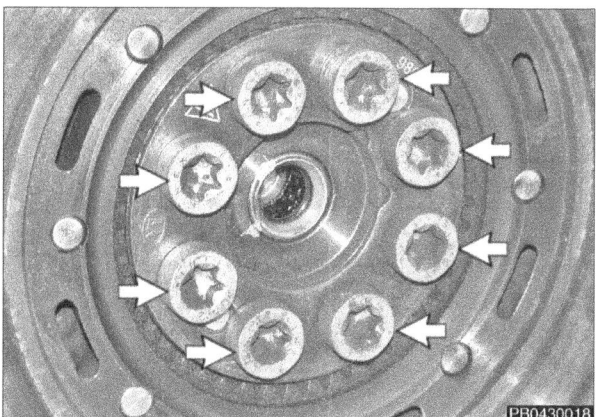

◂ Remove clutch pressure plate retaining bolts (**arrows**). Carefully pry pressure plate from flywheel alignment pins.

> **WARNING—**
> - Friction material in the clutch disc produces dangerous dust. Do not breathe in dust. Use water to wet down components and collect dripping mixture in a shop towel.

◂ Remove eight mounting bolts (**arrows**) and remove flywheel from crankshaft for inspection, pilot bearing replacement, or real main seal replacement.

> **NOTE—**
> - Oil residue, especially at the base of the bellhousing usually indicates a faulty crankshaft rear main seal. For information on seal replacement, see **13 Engine Pulleys and Seals**.

◂ With clutch pressure plate removed, inspect flywheel for signs or damage or scoring.
- If flywheel is scored or damaged, replace.
- If flywheel shows signs of mild glazing or blueing, clean surface using a drill with an abrasive pad.

> **NOTE—**
> - Due to construction details, the dual-mass flywheel cannot be machined or commercially resurfaced.

30-4 Clutch

Clutch Mechanical

◁ Clutch pilot bearing protrudes from rear of flywheel. Measure pilot bearing projection before driving bearing out from flywheel. Install new pilot bearing to appropriate depth.

◁ Install flywheel on crankshaft.
- Locate flywheel on crankshaft locating pin (**arrow**).

◁ Install flywheel with new retaining bolts using flywheel locking tool to counterhold flywheel. Torque bolts to specification in crisscross pattern and in several stages.

Tightening torque	
Flywheel to crankshaft	
• Stage 1	25 Nm (19 ft-lb)
• Stage 2	120° additional

◁ Use special tool to center clutch disc on clutch pressure plate and line up pressure plate locating pins. Mount to flywheel using new bolts. Torque bolts in crisscross pattern and in several stages.

Tightening torque	
Pressure plate to flywheel (use new bolts)	23 Nm (17 ft-lb)

- Remove clutch disc centering tool.

- Working in transmission bellhousing, remove throwout bearing from release lever and replace.

- Install transmission. See **34 Manual Transmission**.

Clutch 30-5

Clutch Hydraulic System

Clutch interlock switch, removing

The clutch interlock switch prevents starter operation until the clutch pedal is depressed.

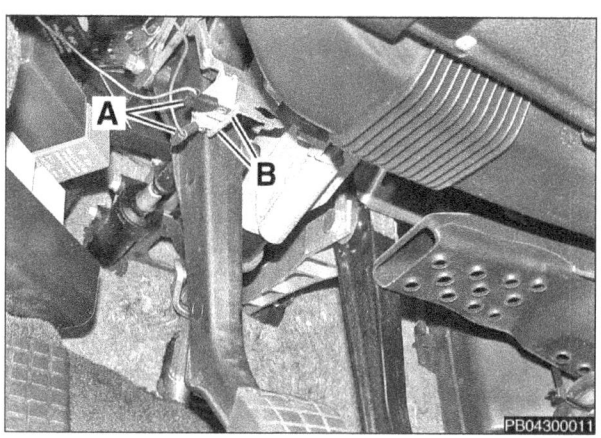

◄ Working in driver's side footwell, disconnect clutch switch electrical connectors (**A**). Remove mounting screws (**B**), and remove switch from support.

CLUTCH HYDRAULIC SYSTEM

◄ The hydraulic clutch system shares a common fluid reservoir with the brake system. The fluid reservoir (**arrow**) is located in the front trunk. See **Warnings and Cautions** in this repair group regarding brake fluid safety.

The clutch hydraulic system is not adjustable. The clutch slave cylinder automatically compensates for normal wear of the clutch disc.

The same problems that typically affect brake operation—air in the lines and moisture in the fluid—can also affect clutch operation. Change clutch and brake fluid periodically to minimize moisture absorption and corrosion in the hydraulic components.

Warnings and Cautions

> *WARNING—*
> - *Brake fluid is poisonous, highly corrosive, and dangerous to the environment. Dispose of old brake fluid properly.*
> - *Wear safety glasses and rubber gloves when working with brake fluid. Do not siphon brake fluid with your mouth.*

> *CAUTION—*
> - *Brake fluid is corrosive and dissolves paint. Immediately clean away any brake fluid spilled on painted surfaces and wash with water.*
> - *Always use brake fluid from a fresh, unopened container. Brake fluid absorbs moisture from the air. This can lead to corrosion problems in the clutch and brake hydraulic systems, and also lowers the brake fluid boiling point.*

30-6 Clutch

Clutch Hydraulic System

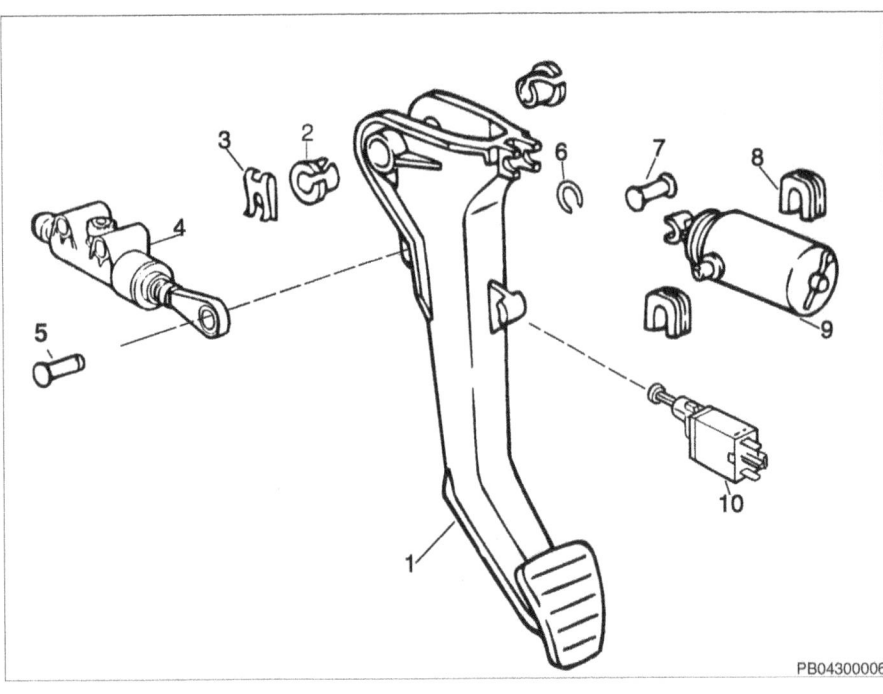

Clutch pedal and master cylinder

1. **Clutch pedal**
2. **Bushings**
   - Check for wear, replace if necessary
   - Lubricate with Optimol Optitemp LG 2, or equivalent grease
3. **Retaining clip**
   - Always replace
4. **Clutch master cylinder**
   - Fluid from brake fluid reservoir
5. **Pin**
6. **Retaining clip**
7. **Pin**
8. **Boost spring bearing**
   - Replace if necessary
   - Lubricate with Optimol Optitemp LG 2, or equivalent grease
9. **Boost spring with bracket**
   - Confirm proper spring diameter when replacing
10. **Clutch switch**

Clutch hydraulic system, bleeding

◄ This procedure given here requires a brake system pressure bleeder. The bleeder connects to the brake fluid reservoir and is filled with new brake fluid and pressurized. When the slave cylinder bleeder nipple is opened, new brake fluid is forced through the system.

NOTE—
- *Before the system is filled or bled, press clutch pedal to floor.*
- *Bleed the system until no more bubbles appear at the bleeder valve (use a collecting bottle with transparent hose).*
- *Use only new, Porsche brake fluid in hydraulic clutch system.*

Filling / bleeding

◄ Working at brake fluid reservoir in front trunk, fill reservoir up to MAX line with new brake fluid. Connect bleeder to fluid reservoir.

– Press pedal to floor.

– Switch on bleeding device until bleeding pressure reaches approx. 1.5 bar (22 psi).

CAUTION—
- *Do not exceed 2 bar (29 psi) pressure at the fluid reservoir when bleeding or flushing hydraulic system.*

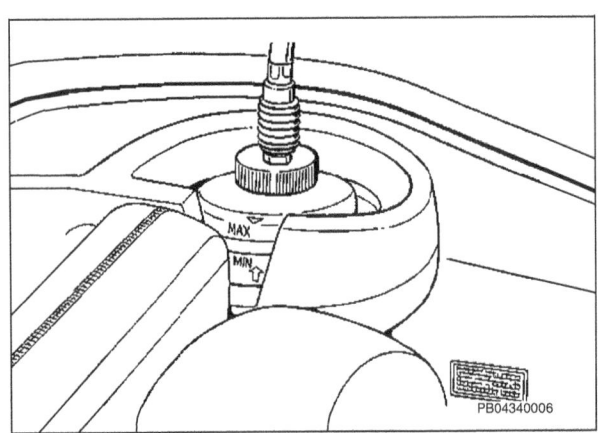

Clutch 30-7

Clutch Hydraulic System

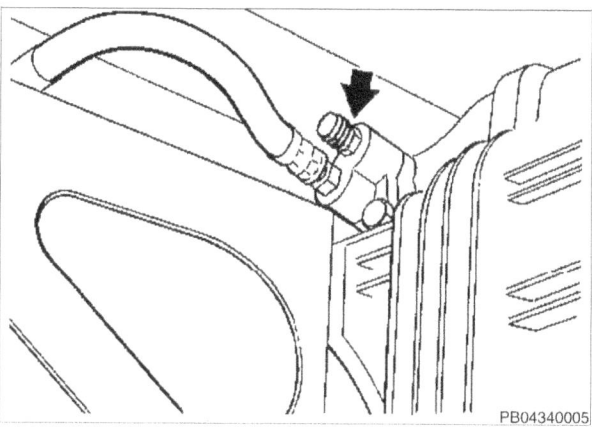

- Raise car and support safely.

> **WARNING—**
> - Make sure the car is stable and well supported at all times. Use a professional automotive lift or jack stands designed for the purpose. A floor jack is not adequate support.

◀ Working at upper left side of transmission, open bleeder valve on clutch slave cylinder until clear, bubble-free brake fluid emerges. Use a collecting bottle for precise examination of the emerging fluid for cleanliness and freedom from bubbles.

- Switch off and disconnect bleeding device. Correct brake fluid level if necessary.

Clutch slave cylinder, removing and installing

Removal of the clutch slave cylinder requires that the clutch hydraulic system be bled. See **Clutch hydraulic system, bleeding**.

◀ Remove clutch slave cylinder mounting bolt (**arrow**). Slide clutch slave cylinder towards rear of vehicle and remove from transmission. Suspend slave cylinder from chassis using mechanic's wire.

> **CAUTION—**
> - Clutch pedal must not be pressed once clutch slave cylinder has been removed from transmission. Hydraulic pressure could force actuating rod from slave cylinder assembly.

Tightening torque	
Clutch slave cylinder to transmission	23 Nm (17 ft-lb)

- Remove clutch slave cylinder hydraulic line from slave cylinder.

- Installation is reverse of removal, noting the following:
  - If clutch actuating rod needs to be pushed back into the slave cylinder during installation, open bleeder valve.
  - Bleed hydraulic system. See **Clutch hydraulic system, bleeding**.

Clutch master cylinder, removing and installing

Removal of the clutch master cylinder requires that the clutch hydraulic system be bled. See **Clutch hydraulic system, bleeding** in this repair group.

- Remove driver's seat. See **72 Seats**.

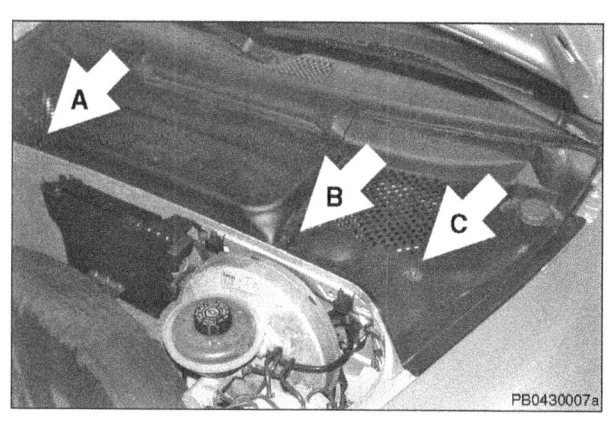

◀ Remove battery cover and left cowl cover:
  - Twist battery cover fasteners (**A** and **B**) 90° and lift cover up and straight out.
  - Loosen cowl cover screw (**C**) and lift cover up and straight out.

- Empty brake fluid reservoir using a syringe or other suction device, such as a turkey baster.

30-8 Clutch

Clutch Hydraulic System

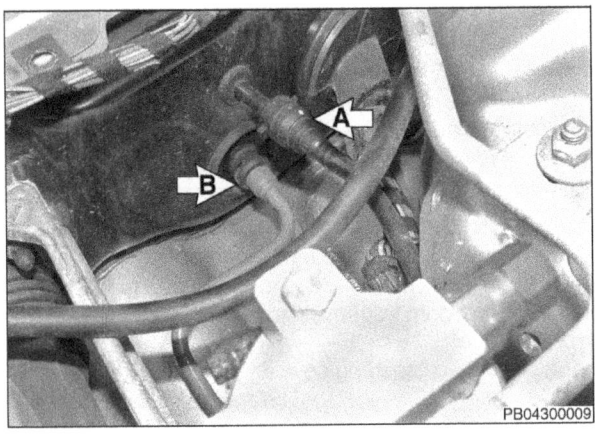

◀ Working in front trunk under left side trim panel at cowl, disconnect hydraulic fluid supply line (**A**) and line to slave cylinder (**B**).

- To remove supply line, press in on spring locks and pull line off.
- To remove line to slave cylinder, pry out locking clip and pull line off.

> **CAUTION—**
> - Brake fluid is corrosive and dissolves paint. Immediately clean away any brake fluid spilled on painted surfaces and wash with water.

NOTE—
- Place a shop rag around lines when disconnecting to soak up spilled fluid.

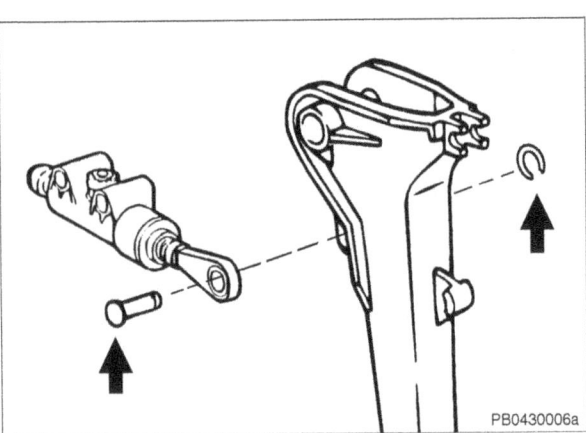

◀ Working under driver's side dashboard, remove clutch rod retaining pin circlip and remove pin from clutch pedal (**arrows**).

◀ Remove clutch master cylinder mounting bolts (**arrows**) and slide master cylinder down and out.

– Installation is reverse of removal. Once installation is complete, fill master fluid reservoir with fresh unopened brake fluid and bleed hydraulic system. See **Clutch hydraulic system, bleeding**.

32 Torque Converter

GENERAL 32-1

TORQUE CONVERTER 32-1
  Torque converter, removing and installing 32-1
  Torque converter seal, removing and installing . 32-2

GENERAL

The torque converter is a fluid coupling between the engine and transmission. The housing of the torque converter bolts to the driveplate of the engine, and turns at engine speed. As engine speed increases, fins inside the torque converter react with transmission fluid, causing the transmission to spin, moving the car.

TORQUE CONVERTER

Torque converter, removing and installing

- Remove transmission. See **37 Automatic transmission**.

- With transmission in horizontal position, carefully slide torque converter off of transmission input shaft.

  NOTE—

  • *Check the torque converter hub for traces of scoring. In the event of damage, the converter must be replaced.*

- Grease bearing pin of converter with Longtime® 3EP (part no. 000.043.024.00)

- Carefully guide torque converter onto transmission input shaft to first stop. Then turn converter until recess in converter hub engages impeller driver and converter slides inward.

  CAUTION—

  • *If torque converter is installed incorrectly, transmission pump can be destroyed when transmission is mounted to engine.*

Torque Converter

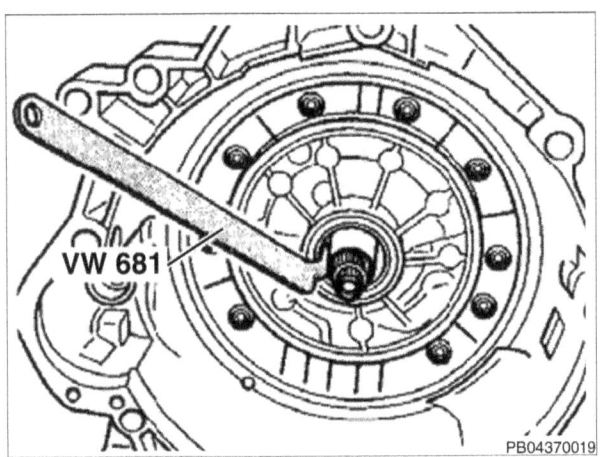

Torque converter seal, removing and installing

– Remove torque converter as described earlier.

◄ Remove torque converter seal with special tool VW 681. Place tool directly behind lip of seal so that mounting-face ring is not damaged.

NOTE—
* *Replace mounting face ring if damaged.*

– Coat circumference and lip of seal with light coating of Vaseline and drive seal flush with mounting face using Porsche special tool 3295.

34 Manual Transmission

GENERAL	34-1
Transmission applications	34-1
TRANSMISSION REMOVAL AND INSTALLATION	34-2
5-speed transmission, removing	34-2
5-speed transmission, installing	34-6
6-speed transmission, removing	34-8
6-speed transmission, installing	34-12
SHIFT KNOB	34-14
Manual transmission shift knob, removing	34-14

BACK-UP LIGHT SWITCH	34-14
Back-up light switch, removing (5-speed transmissions)	34-14
Back-up light switch, removing (6-speed transmission)	34-14
TABLES	
a. Manual transmission applications	34-1

GENERAL

This repair group covers external transmission service only, including removal and installation of the transmission. Internal repairs are not covered.

Transmission fluid service is covered in **03 Service and Maintenance**.

Transmission applications

Three manual transmissions were available in Porsche Boxster models, based on engine application.

Table a. Manual transmission applications			
Transmission type	5-speed G86.00 (Boxster 2.5 liter)	5-speed G86.01 (Boxster 2.7 liter)	6-speed G86.20 (Boxster S 3.2 liter)
Transmission code	CWA, DVY	EFD	n/a
1st gear	3.50	3.50	3.82
2nd gear	2.12	2.12	2.20
3rd gear	1.43	1.43	1.52
4th gear	1.03	1.09	1.22
5th gear	0.79	0.84	1.02
6th gear	-	-	0.89
Final drive ratio	3.89	3.56	3.44
Fluid type	75W90	75W90	75W90
Capacity	2.25 liter (2.37 US qt)	2.25 liter (2.37 US qt)	2.8 liter (2.95 US qt)

34-2 Manual Transmission

Transmission Removal and Installation

◀ The manual transmission can be identified by two code stampings, one located at the top of the bellhousing, the second located on the underside of the transmission.

Code stampings convey valuable information about the transmission. The first three letters indicate the transmission type. The second five number code is the date that the transmission was built: Day (05), month (11), year (7). The code given in the example on the left indicates that transmission was built on November 5, 1997.

TRANSMISSION REMOVAL AND INSTALLATION

5-speed transmission, removing

◀ With vehicle on ground, pry out wheel center cap from both rear wheels. Use round holes in cap to carefully pry out.

◀ Using a breaker bar with a 32 mm socket, loosen drive axle nut at left and right rear wheels.

– Raise car and support safely.

> **WARNING—**
> • Make sure the car is stable and well supported at all times. Use a professional automotive lift or jack stands designed for the purpose. A floor jack is not adequate support.

◀ Support engine using Porsche special tool 9591/1 or equivalent engine support bar.

Manual Transmission 34-3

Transmission Removal and Installation

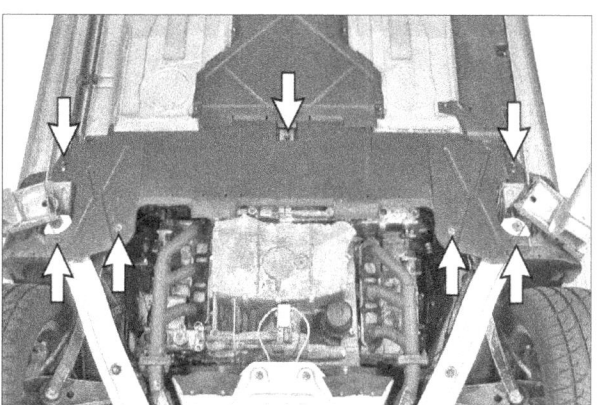

◂ Working beneath car, remove plastic under shield. **Arrows** indicate fasteners to be removed.

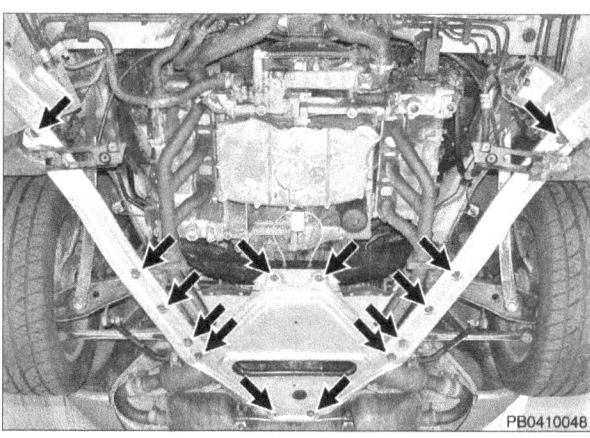

◂ Remove aluminum diagonal braces and chassis reinforcement plate (14 fasteners) (**arrows**).

◂ Remove rear sway bar from chassis. Disconnect sway bar at end links, and remove mounting brackets at bushings (**arrows**).

◂ Remove rear muffler.
- Remove band clamps (**arrows**) at rear muffler.
- Unbolt rear muffler support bracket at lower front and upper rear and remove muffler.

34-4 Manual Transmission

Transmission Removal and Installation

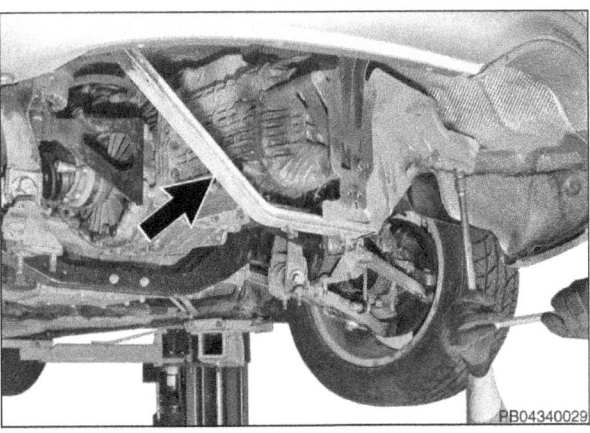

◀ Remove reinforcement plate support (**arrow**).

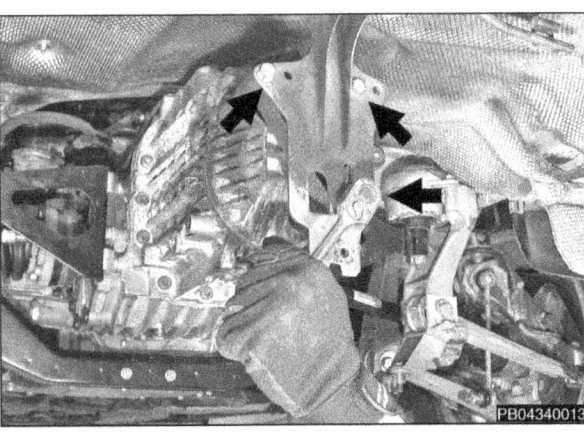

◀ Remove three retaining bolts (**arrows**) and remove muffler support bracket from rear of transmission housing.

◀ Working at transmission drive flanges, remove inner CV joint retaining bolts. Cover and protect removed inner CV joint ends from dirt and damage.

NOTE—
- *Firmly setting the parking brake and leaving transmission in gear makes removal of inner CV joint retaining bolts easier.*

– Remove drive axle nut at wheel bearing carrier.
  - Using a hammer and drift, knock outer CV joint free from hub and remove drive axle from vehicle.

CAUTION—
- *Rolling the vehicle on its wheels without the drive axles installed and properly torqued will damage rear wheel bearings.*

◀ Disconnect electrical harness connector at back-up light switch (**arrow**).

Manual Transmission   34-5
Transmission Removal and Installation

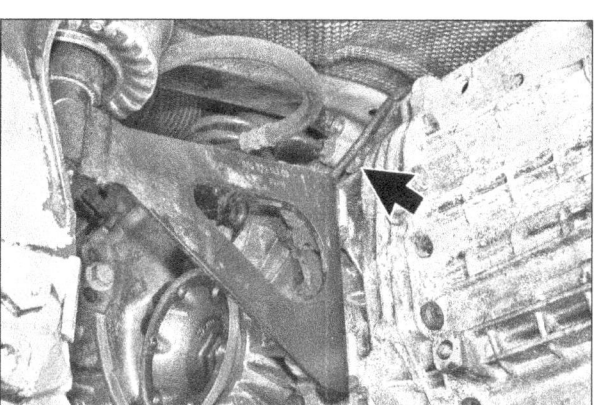

◂ Remove clutch slave cylinder mounting bolt (**arrow**). Slide clutch slave cylinder towards rear of vehicle and remove from transmission. Suspend slave cylinder from chassis using mechanic's wire.

> **CAUTION—**
> - Clutch pedal must not be pressed once clutch slave cylinder has been removed from transmission. Hydraulic pressure could force actuating rod from slave cylinder assembly.

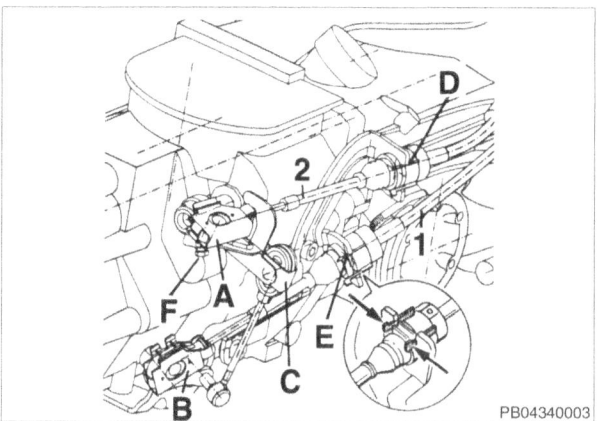

◂ Vehicles built before April 21, 1997: Disconnect shifter cables from transmission.
- Remove cotter pins for retaining clips (**A** and **B**). Shift lever must be removed to remove clip **A**.
- Remove retaining clip (**C**).
- Unclip upper joint rod.
- Loosen locking screw (**F**).
- Pull off retaining clips (**A** and **B**) and pull selector cables off ball head.
- Detach support bracket sleeves (**D** and **E**) at support bracket. Press tabs in direction of arrow and carefully pull support bracket sleeves out of support bracket.

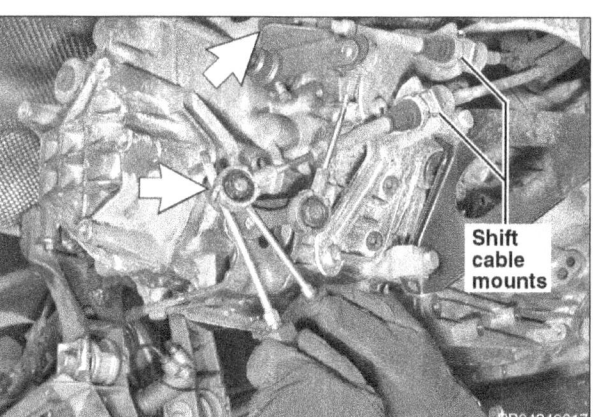

◂ Vehicles built after April 21, 1997: Disengage transmission shift cables (**arrows**) by prying cables from ball mounts. Unclip shift cables mounts at transmission. Pull shift cable ends up out of engine compartment and set aside.

− Secure transmission to transmission jack and support transmission.

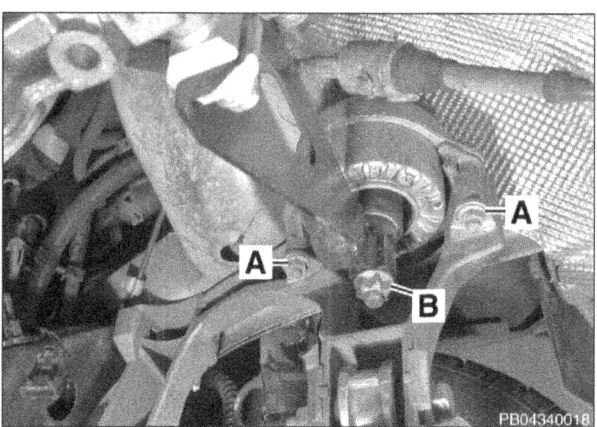

◂ Working at left and right transmission supports, remove transmission mount hardware (**A**). Remove support mounting hardware from transmission, and remove mount from chassis.

> **CAUTION—**
> - **DO NOT** remove nut **B**. Transmission mount can be damaged by removing nut.

34-6 Manual Transmission

Transmission Removal and Installation

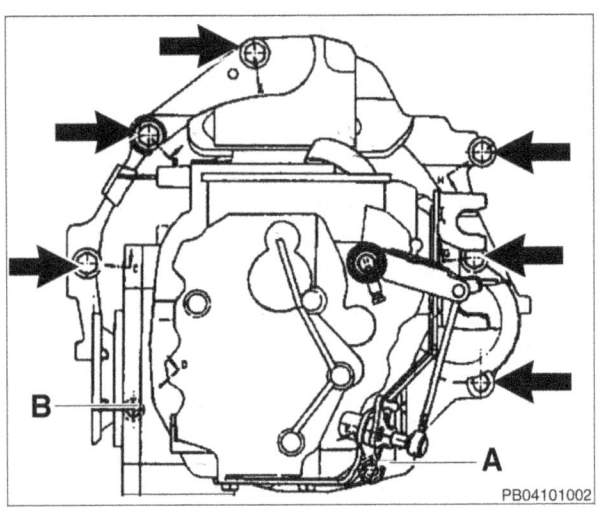

◂ Remove engine to transmission fastening bolts:
- Remove M10 nut (**A**).
- Remove M10 fastener (**B**) (10 mm triplesquare), if equipped.
- Remove 6 M12 bolts (**arrows**).

− Slide transmission backward and lower carefully.

NOTE—
- *For information on clutch disc and flywheel removal, see* **30 Clutch***.*

5-speed transmission, installing

◂ Clean splines of transmission input shaft and lightly lubricate with a thin coat of Olista Longtime 3 EP or equivalent grease.

− Ensure dowel sleeves are properly seated in transmission case.

− Carefully slide transmission into place, aligning transmission input shaft with clutch disc.

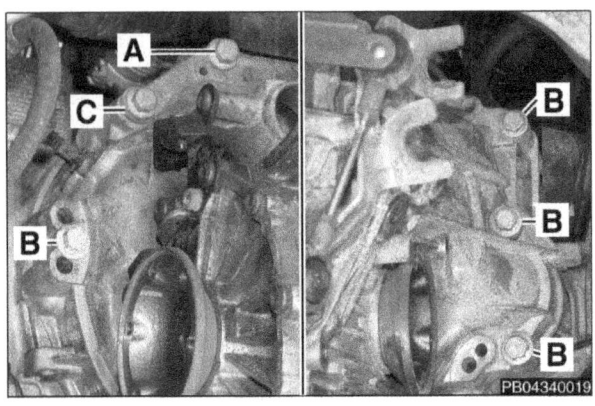

◂ With transmission bellhousing seated to engine, install engine to transmission mounting bolts.

Tightening torques	
A M 12 x 70 mm bolt	85 (63 ft-lb)
B M12 x 90 mm bolt	85 (63 ft-lb)
C M 12 x 100 mm bolt	85 (63 ft-lb)
M10 nut	45 (33 ft-lb)

− Install transmission supports with left and right transmission mounts.

Tightening torques	
Transmission mount to body (M8)	23 Nm (17 ft-lb)
Transmission support to transmission (M10)	65 Nm (48 ft-lb)

Manual Transmission 34-7

Transmission Removal and Installation

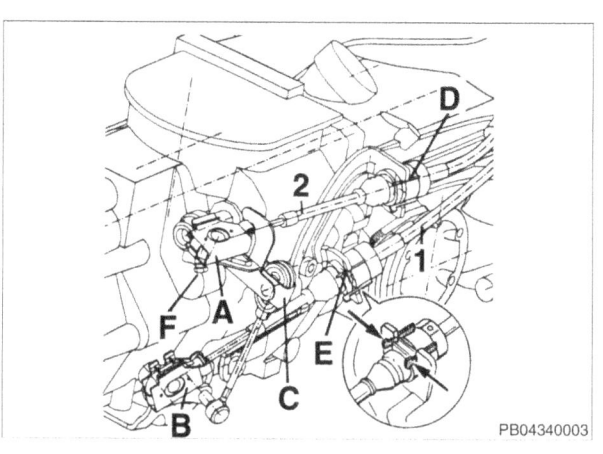

◄ Vehicles made before April 21, 1997: Connect shifter cables to transmission.

- Engage and lock support bracket sleeves (**D** and **E**) at bracket.
- Connect shift cables and push retaining clips (**A** and **B**) onto ball head.
- Tighten locking screw (**F**).
- Clip on upper joint rod.
- Install retaining clip (**C**).
- Install cotter pins for retaining clips (**A** and **B**).

– Vehicles made after April 21, 1997: Press shift cables into place and clip cables into retainers.

– Install clutch slave cylinder to transmission.

Tightening torque	
Clutch slave cylinder to transmission	23 Nm (17 ft-lb)

NOTE—

- *If clutch actuating rod needs to be pushed back into slave cylinder during installation, bleeder valve must be opened.*
- *If bleeder valve is opened, slave cylinder system must be bled. See* **30 Clutch**.

– Connect electrical harness connector for back-up light.

– Install drive axles. Slide outer CV joint into place at wheel hub and secure with axle nut. Install inner CV joints to transmission output flanges.

NOTE—

- *Due to the high torque of the drive axle nut, wait until the vehicle is reassembled and on the ground before tightening it to its final torque.*

Tightening torques	
Inner CV joint to output flange • M8 • M10	39 Nm (29 ft-lb) 81 Nm (60 ft-lb)
Drive axle to wheel hub	460 Nm (340 ft-lb)

– Install muffler support bracket to rear of transmission.

Tightening torque	
Muffler support bracket to transmission	65 Nm (48 ft-lb)

– Install reinforcement plate support.

Tightening torque	
Reinforcement plate support to body	65 Nm (48 ft-lb)

– Install rear muffler.

34-8 Manual Transmission

Transmission Removal and Installation

– Install rear sway bar.

Tightening torques	
Sway bar bracket to body	23 Nm (17 ft-lb)
End link to sway bar	50 Nm (37 ft-lb)

– Install diagonal braces and reinforcement plate.

Tightening torques	
Reinforcement plate to rear axle support (M10 bolt) to support / suspension (M10 nut)	46 Nm (34 ft-lb) 65 Nm (48 ft-lb)
Diagonal brace to body / suspension	65 Nm (48 ft-lb)

– Remainder of installation is reverse of removal, noting the following:
 • With vehicle reassembled and off supporting jacks, torque drive axles to specified torque.

Tightening torque	
Drive axle to wheel hub	460 Nm (340 ft-lb)

6-speed transmission, removing

◀ With vehicle on ground, pry out wheel center cap from both rear wheels. Use round holes in cap to carefully pry out.

◀ Using a breaker bar with a 32 mm socket, loosen drive axle hex nut at left and right wheels.

– Raise car, and safely support on a professional automotive lift, or jack stands.

> *WARNING—*
> • *Make sure the car is stable and well supported at all times. Use a professional automotive lift or jack stands designed for the purpose. A floor jack is not adequate support.*

Manual Transmission 34-9

Transmission Removal and Installation

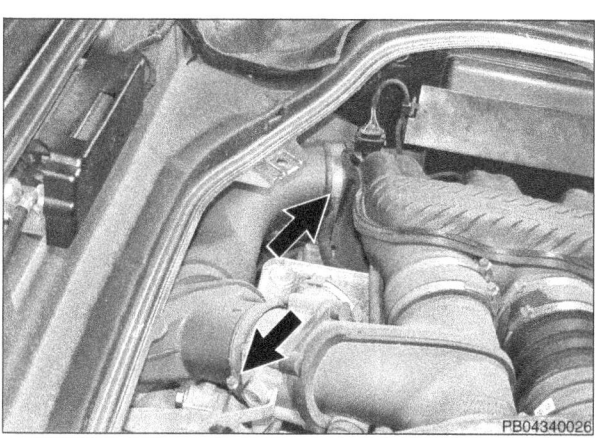

◄ Loosen clamps (**arrows**) and remove air intake duct.

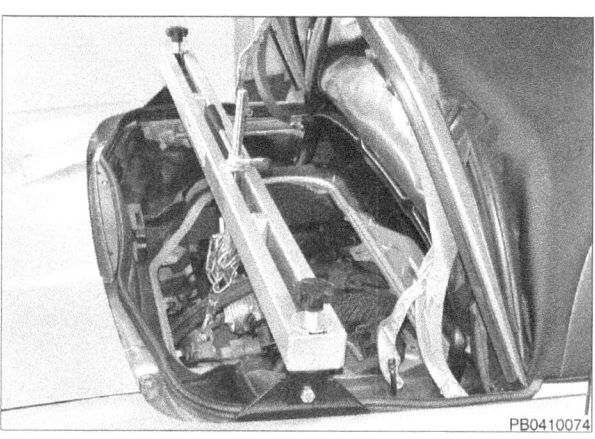

◄ Support engine using special tool 9591/1 or equivalent engine support bar.

◄ Working beneath car, remove plastic under shield. **Arrows** indicate fasteners to be removed.

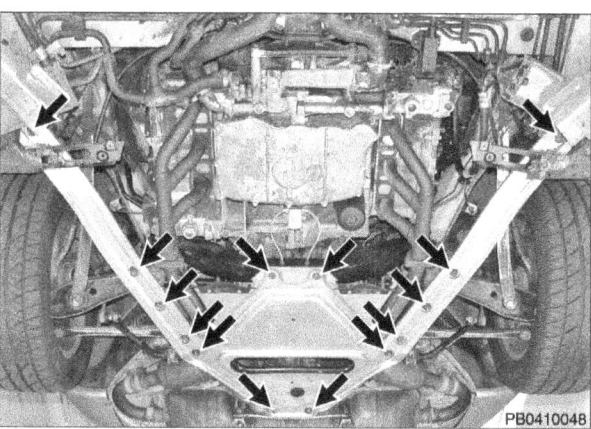

◄ Remove aluminum diagonal braces and chassis reinforcement plate (14 fasteners) (**arrows**).

34-10 Manual Transmission

Transmission Removal and Installation

◄ Remove sway bar from chassis. Disconnect sway bar at end links, and remove mounting brackets at bushings (**arrows**).

– Remove muffler with catalytic converters. See **26 Exhaust System**.

◄ Remove reinforcement plate support (**arrow**).

◄ Working on right side of transmission, remove shift linkage heat shield fasteners (**arrows**). Remove heat shield.

◄ Working at transmission drive flanges, remove inner CV joint retaining bolts.
- Cover and protect inner CV joint ends from dirt and damage.

NOTE —
- *Firmly setting the parking brake and leaving transmission in gear makes removal of inner CV joint retaining bolts easier.*

Manual Transmission 34-11

Transmission Removal and Installation

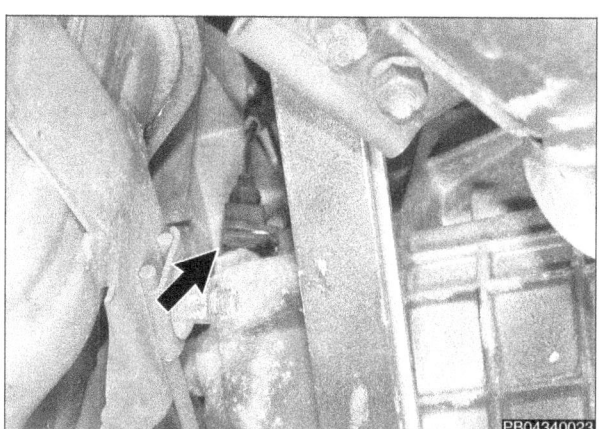

– Remove drive axle hex nut at wheel bearing carrier.
  • Using a hammer and drift, knock outer CV joint free from hub and remove drive axle from vehicle.

  CAUTION—
  • *Rolling the vehicle on its wheels without the drive axles installed and properly torqued will damage rear wheel bearings*

◄ Working at right side rear of transmission housing, disconnect electrical harness connection at back-up light switch (**arrow**).

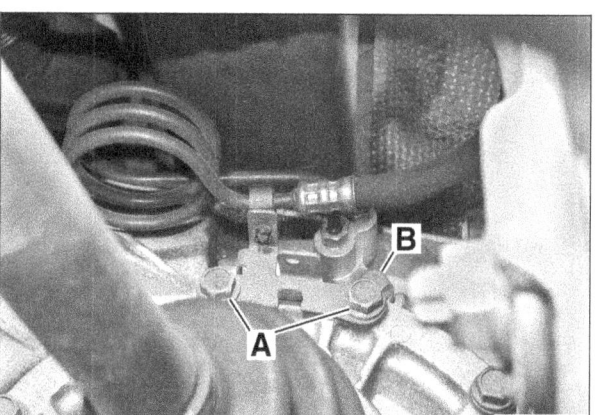

◄ Remove clutch fluid line retaining bracket bolts (**A**) and remove bracket from transmission. Remove clutch slave cylinder bolt (**B**) and remove slave cylinder from transmission. Suspend slave cylinder from chassis using mechanic's wire.

  CAUTION—
  • *Clutch pedal must not be pressed once clutch slave cylinder has been removed from transmission. Hydraulic pressure could force actuating rod from slave cylinder assembly.*

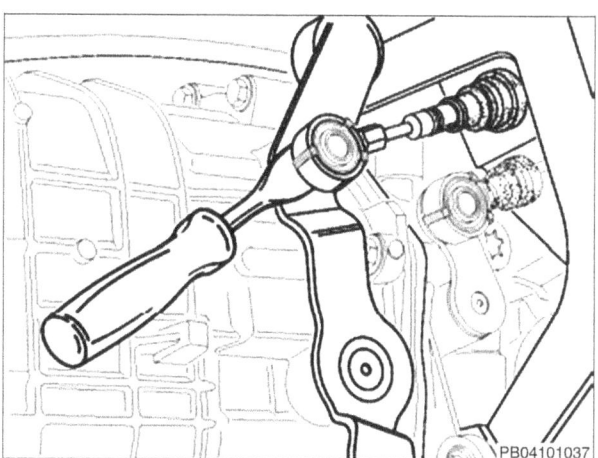

◄ Use prying tool to disengage transmission shift cables.

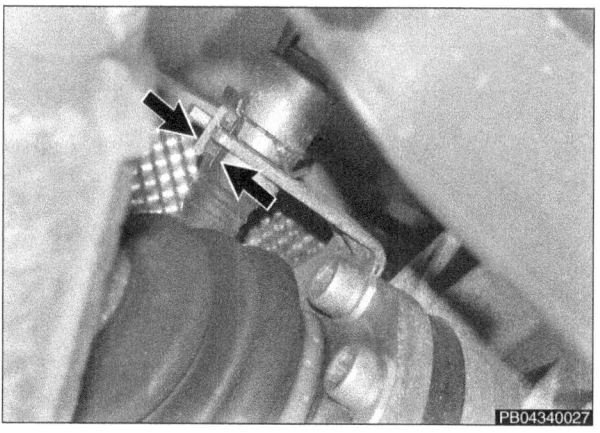

◄ Unclip shift cables at transmission mount:
  • Release shift cable support bracket sleeve by squeezing together tabs (**arrows**).
  • Pull shift cable ends out of engine compartment and set aside.

  NOTE—
  • *Make sure shift cable does not get kinked.*

– Secure transmission to transmission jack.

Manual Transmission

6-speed transmission, installing

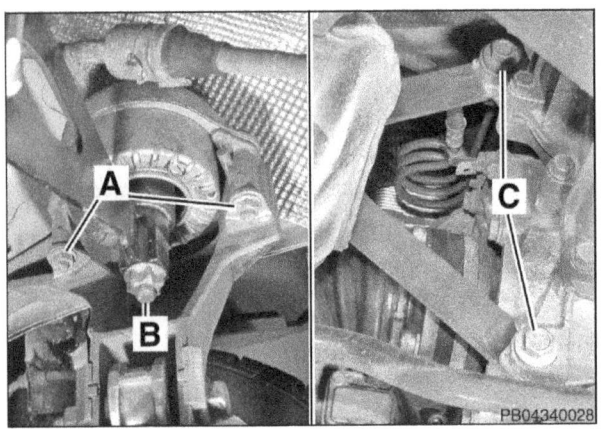

◄ Remove transmission mounts from body and transmission. Remove nuts (**A**) at right and left transmission mounts. Remove mount from transmission by removing bolts (**C**).

CAUTION—
* *Transmission mount can be damaged by releasing nut (**B**). Do not remove nut!*

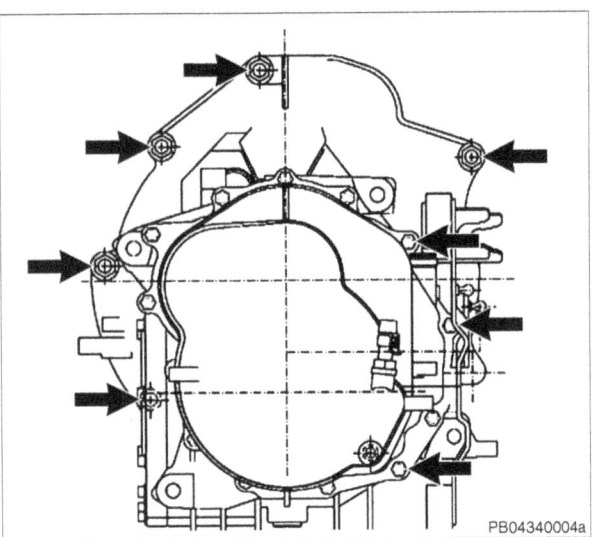

◄ Remove engine to transmission fastening bolts (**arrows**).

– Slide transmission backward and lower carefully.

6-speed transmission, installing

– Clean splines of transmission input shaft and lightly lubricate with a thin coat of Olista Longtime 3 EP (or equivalent grease).

– Ensure dowel sleeves are properly seated in crankcase.

– Carefully slide transmission into place, aligning transmission input shaft with clutch plate.

◄ With transmission bellhousing seated to engine, install engine to transmission mounting bolts.

No. Screw / nut	Tightening torque
1 M 12 x 70 mm	85 (63 ft-lb)
2 M12 x 100 mm	85 (63 ft-lb)
3 M12 x 100 mm	85 (63 ft-lb)
4 M10 x 50 mm (10 mm triplesquare)	45 (33 ft-lb)
5 M10 x 50 mm	45 (33 ft-lb)
6 M 12 x 50 mm	85 (63 ft-lb)
7 M 12 x 100 mm	85 (63 ft-lb)
8 M 12 x 100 mm	85 (63 ft-lb)

Manual Transmission 34-13

6-speed transmission, installing

— Install transmission supports with left and right transmission mounts.

Tightening torques	
Transmission mount (M8)	23 Nm (17 ft-lb)
Transmission mount (M10)	65 Nm (48 ft-lb)

— Press shift cables into place.

— Install clutch slave cylinder to transmission.

Tightening torque	
Clutch slave cylinder to transmission	23 Nm (17 ft-lb)

NOTE —
- *Bleeder valve must be opened in order for clutch actuating rod to be pushed back into the slave cylinder during cylinder installation. Connect a collection bottle with hose for this purpose.*

— Bleed clutch hydraulic system. See **30 Clutch**.

— Connect electrical harness connector for back-up light switch.

— Install drive axles. Slide outer CV joint into place at wheel hub and secure with axle nut. Install inner CV joints to transmission output flanges.

NOTE —
- *Due to the level of torque required to secure drive axle nut, wait until vehicle is reassembled before tightening axle nut to torque specification.*

Tightening torques	
Inner CV joint to output flange • M8 • M10	39 Nm (29 ft-lb) 81 Nm (60 ft-lb)
Drive axle to wheel hub	460 Nm (340 ft-lb)

— Install heat shield.

Tightening torque	
Heat shield to chassis	10 Nm (7.5 ft-lb)

— Install reinforcement plate support.

Tightening torque	
Reinforcement plate • To rear axle support (M10 bolt) • To support / suspension (M10 nut)	46 Nm (34 ft-lb) 65 Nm (48 ft-lb)

— Install muffler with catalytic converters. See **26 Exhaust System**.

34-14　Manual Transmission

Shift Knob

- Install diagonal braces and reinforcement plate.

Tightening torques	
Reinforcement plate • To rear axle support (M10 bolt) • To support / suspension (M10 nut)	46 Nm (34 ft-lb) 65 Nm (48 ft-lb)
Diagonal brace to body / suspension	65 Nm (48 ft-lb)

- Remainder of installation is reverse of removal, noting the following:
  - With vehicle reassembled and off supporting jacks, torque drive axles to specified torque.

Tightening torque	
Drive axle to wheel hub	460 Nm (340 ft-lb)

SHIFT KNOB

Manual transmission shift knob, removing

- Pry up shift lever boot at center console.
- Pull shift boot over shift knob, turning boot inside out.

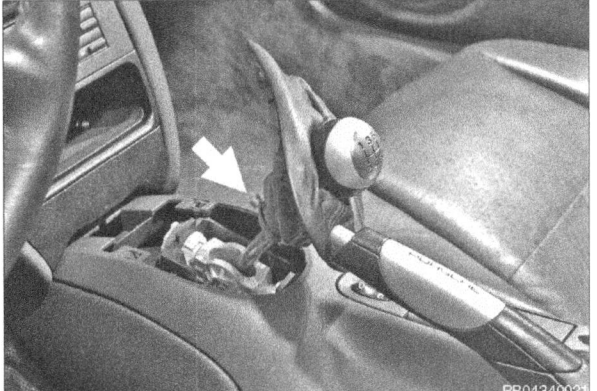

◀ Loosen 5 mm Allen shift knob retaining screw (**arrow**) and slide knob from shift lever.

BACK-UP LIGHT SWITCH

Back-up light switch, removing (5-speed transmission)

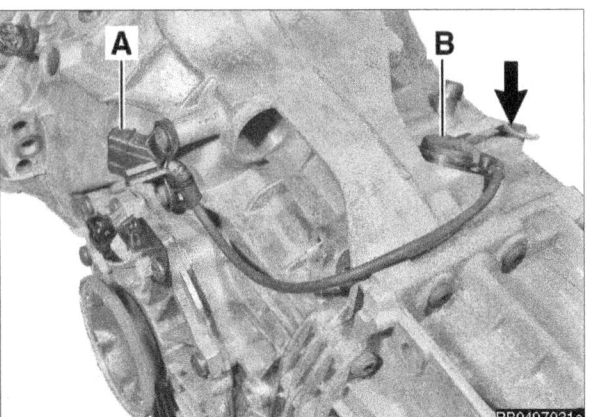

◀ The back-up light switch is mounted on the top right side of the transmission. To remove, disconnect electrical harness connector (**A**) and remove retaining bolt. Working at switch, loosen clamping bracket retaining bolt (**arrow**), slide bracket to side and pull switch (**B**) from transmission.

Back-up light switch, removing (6-speed transmission)

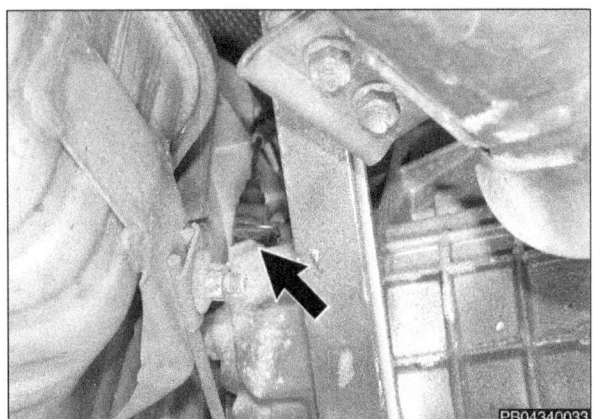

◀ The back-up light switch (**arrow**) is mounted on the right side of the rear of the transmission. To remove, disconnect electrical harness connector (**A**) and unscrew from transmission.

37 Automatic Transmission

GENERAL 37-1
  System description 37-1
  Automatic transmission electrical components .. 37-2
  Tiptronic control module pin assignments 37-3

AUTOMATIC TRANSMISSION SERVICE 37-4
  Automatic transmission fluid and filter,
    changing 37-4

AUTOMATIC TRANSMISSION REMOVAL
AND INSTALLATION 37-7
  Automatic transmission, removing and installing 37-7

AUTOMATIC TRANSMISSION
SHIFT ASSEMBLY 37-15
  Adjusting shift lever cable 37-15
  Automatic transmission shift knob,
    removing and installing 37-15
  Automatic transmission gear position switch,
    checking............................ 37-16
  Automatic transmission gear position switch,
    removing and installing 37-16

TABLES
a. Automatic transmission applications 37-1
b. Automatic transmission gear ratios 37-2

GENERAL

This repair group covers basic maintenance, removal and installation of the automatic transmission. Internal repairs to the automatic transmission are not covered.

NOTE—
- *ATF level checking is covered in* **03 Service and Maintenance**.

System description

The 5-speed automatic transmission used in the Porsche Boxster, model 5HP-19, is manufactured by ZF Getriebe, GmbH. See **Table a** for application information.

This transmission features a Tiptronic mode that allows the driver to select and shift through the forward gears using the gear shift lever like a manual transmission, without the need for clutch actuation.

Transmissions identification plates are located on the top of transmission near the transmission fluid cooler, and on underside of torque converter bellhousing. Porsche identifies transmissions using five digit codes. See **Table a**.

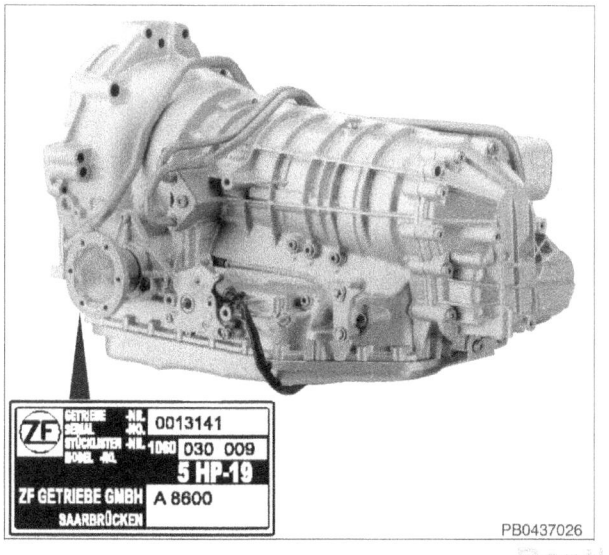

Table a. Automatic transmission applications	
Year, model	Transmission code
1997-1999 Boxster	A86.00
2000 - 2004 Boxster	A86.05
2000 - 2004 Boxster S	A86.20

37-2 Automatic Transmission

General

Table b. Automatic transmission gear ratios			
Gear	A86.00	A86.05	A86.20
1st	3.67	3.67	3.67
2nd	2.00	2.00	2.00
3rd	1.41	1.41	1.41
4th	1.00	1.00	1.00
5th	0.74	0.74	0.74
Reverse	4.10	4.10	4.10
Final drive ratio	3.36	3.33	3.09

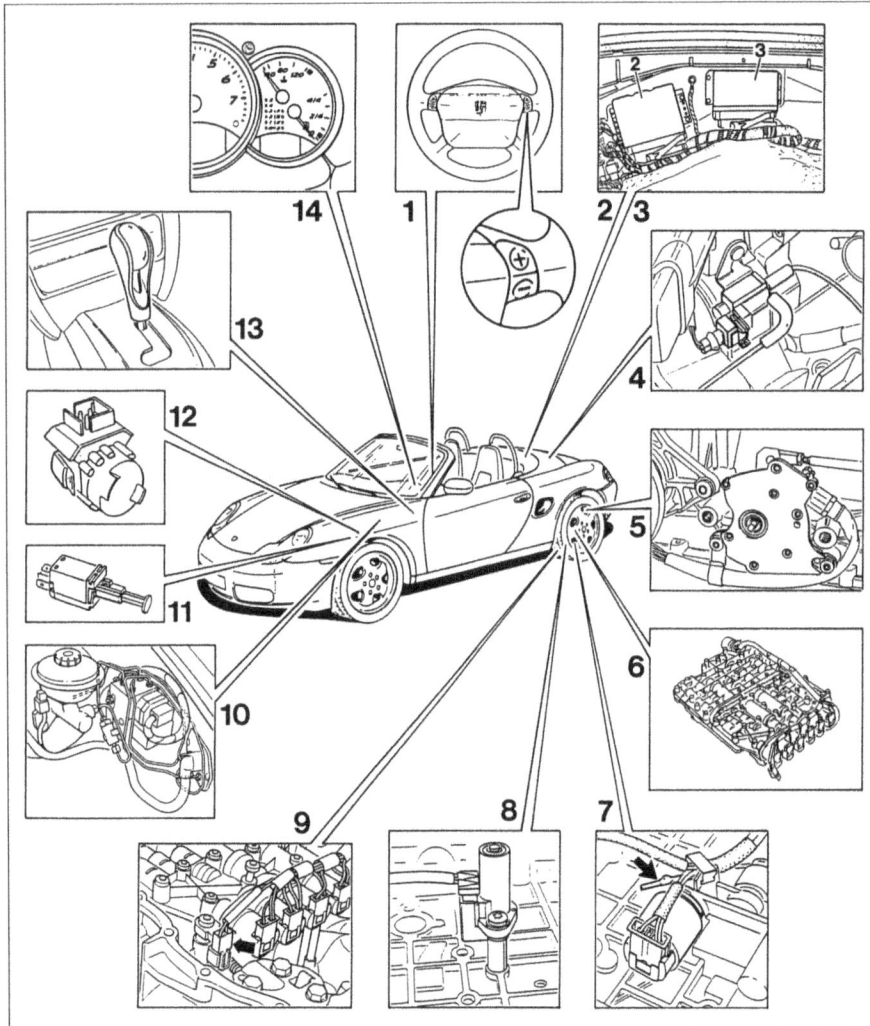

Automatic transmission electrical components

1. **Tiptronic steering wheel upshift and downshift switches**

2. **DME control module**
   - On front bulkhead, rear trunk

3. **Tiptronic control module (TCM)**
   - On front bulkhead, rear trunk

4. **Coolant change over valve**
   - Thermostatic control valve to open and close cooling circuit cooler for ATF

5. **Transmission gear position switch**
   - Communicates selector lever position to TCM
   - Controls back-up lights and disables starter when transmission gear is selected

6. **Transmission solenoid valves**
   - On transmission valve body

7. **ATF temperature sensor**
   - Integral with wiring harness. If damaged, entire harness must be replaced

8. **Transmission input speed sensor**

9. **Transmission output speed sensor**

10. **ABS control module**
    - Communicates ABS (wheel speed) and traction control information to TCM. When ABS or traction control is active, transmission control module selects special (mapped) program.

11. **Brake light switch**
    - Input to TCM to initiate downshifting and to activate shiftlock solenoid

12. **Kick-down switch**
    - Integral with accelerator pedal box

13. **Selector lever with manual switch**

14. **Instrument cluster**

Tiptronic control module pin assignments

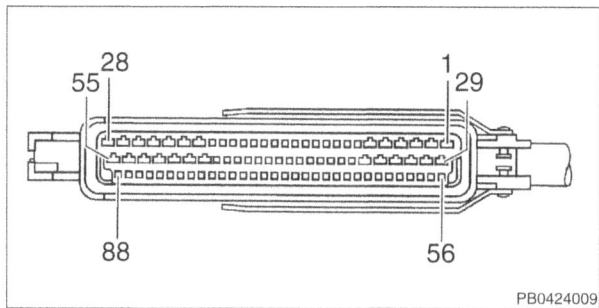

Connector I (88-pin)	
1	Pressure regulator 2
2	Shiftlock solenoid
3	Vacant
4	Pressure regulator 4
5	Pressure regulator 1
6	Ground
7	Vacant
8	Gear position switch, line W2
9	Gear position switch
10	Brake light
11	Vacant
12	Pin code 1
13	Manual program switch
14	Output shaft speed sensor, ground
15	Output shaft speed sensor, shielding
16	Input shaft speed sensor (+)
17	Vacant
18	Kickdown position, signal
19	Traction control active
20	Vacant
21	ATF temp. sensor, ground
22	ATF temp. sensor, signal
23	Input shaft speed sensor, shielding
24	Vacant
25	Display manual mode / Vacant
26	Terminal 30 (battery power)
27	Cruise control input
28	Ground

Connector I (88-pin)	
29	Pressure regulator 3
30	Solenoid valve 1
31	Vacant
32	Solenoid valve 3
33	Solenoid valve 2
34	Ground
35	Vacant
36	Gear position switch, line W3
37	Gear position switch, line W1
38	Front left wheel speed
39	Front right wheel speed
40	Vacant
41	Vacant
42	Output shaft speed sensor (+)
43	Vacant
44	Input shaft speed sensor, ground
45	Vacant
46	Upshift request
47	Downshift request
48	Pin code 2
49	Vacant
50	Vacant
51	Coolant shutoff valve
52	Pressure regulators, solenoid valves (+)
53	Pressure regulator, solenoid valves, shiftlock solenoid (+)
54	Terminal 15 (battery power, key on)
55	Terminal 15 (battery power, key on)
56 - 84	Vacant
85	CAN data transfer -low
86	CAN data transfer - high
87	Vacant
88	Diagnosis (K-line)

37-4 Automatic Transmission

Automatic Transmission Service

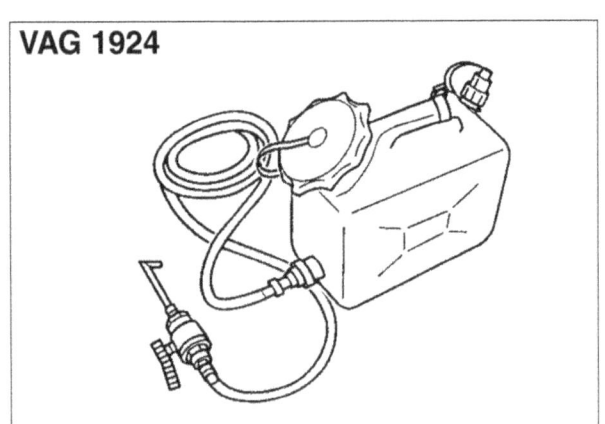

VAG 1924

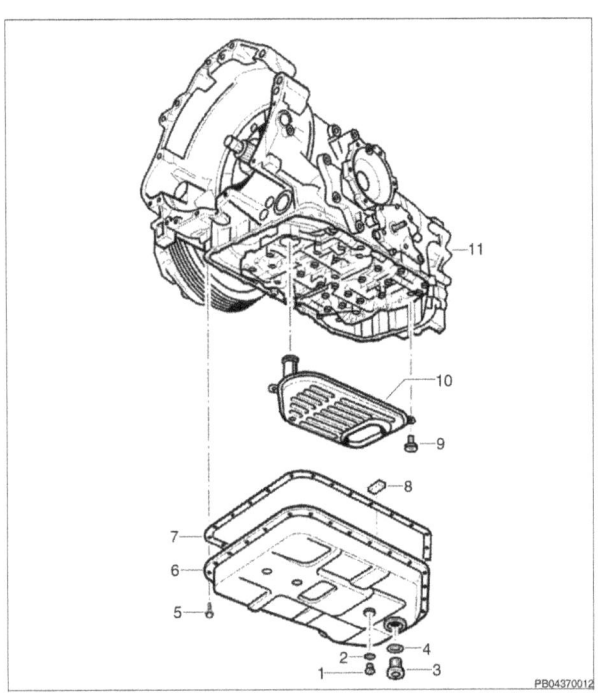

AUTOMATIC TRANSMISSION SERVICE

CAUTION—
- *ATF does not circulate unless the engine is running. When towing an automatic transmission vehicle, use a flat-bed truck or raise the rear wheels off the ground.*

Automatic transmission fluid and filter, changing

◄ Automatic transmission fluid changing requires a fluid pump, such as Porsche special tool VAG 1924, to pump fluid in through the filler hole in the bottom of the ATF sump filler hole.

NOTE—
- *Use only ATF approved by Porsche (part number 999.917.545.00), Esso LT 71141, or equivalent fluid.*

Test conditions:

- Transmission must not be in limp-home driving program, no faults present in transmission control module.
- ATF temperature between 30° and 40°C (85° and 100°F).
- Transmission in position "Park" and engine idling.
- Air conditioning and heater switched off.
- Vehicle horizontal.
- ATF filling system (Porsche special tool VAG 1924).

Automatic transmission pan and filter assembly

1. ATF drain plug
2. Sealing ring
3. ATF filler screw
4. Toroidal sealing ring
5. Screw (27)
6. ATF pan
7. Seal
8. Magnet (2)
9. Screw (2)
10. ATF filter
11. Transmission

Automatic transmission fluid capacity	
Fluid change quantity	3.5 liters (3.7 US qt)
Transmission fill quantity (including torque converter)	9 liters (9.5 US qt)

- Raise car, and safely support on a professional automotive lift, or jack stands.

WARNING—
- *Make sure the car is stable and well supported at all times. Use a professional automotive lift or jack stands designed for the purpose. A floor jack is not adequate support.*

Automatic Transmission 37-5
Automatic Transmission Service

◄ Working beneath car, remove plastic under shield. **Arrows** indicate fasteners to be removed.

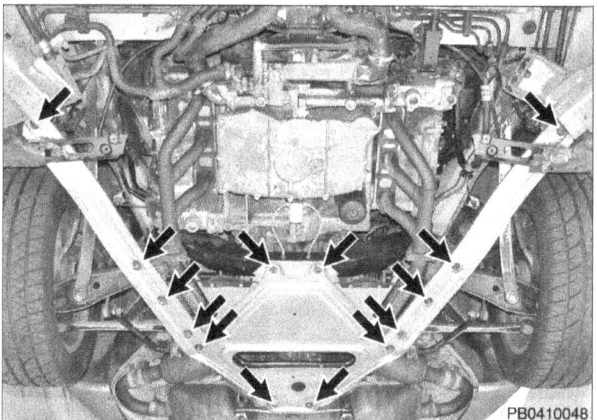

◄ Remove aluminum diagonal braces and chassis reinforcement plate (14 fasteners) (**arrows**).

– Place oil collection pan under transmission.

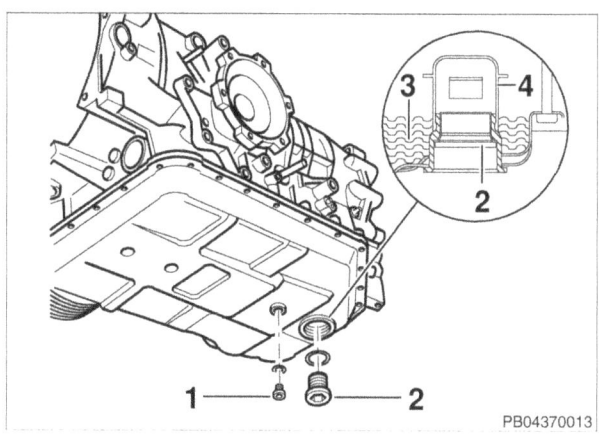

◄ Unscrew drain plug (**1**) and drain ATF.

CAUTION—
- Do not start engine or tow vehicle without ATF.

WARNING—
- Hot ATF can scald. Wear eye protection and protective clothing and gloves.

– Remove ATF pan by removing retaining bolts in a crosswise pattern.

– Remove ATF filter retaining screws and pull filter downwards.

– Insert new ATF filter and fasten with screws.

Tightening torque	
Transmission filter retaining screws	6 Nm (4.5 ft-lb)

NOTE—
- Apply a thin coat of petroleum jelly to gasket surface.

– Clean pan magnets and place in seams of ATF pan.

37-6 Automatic Transmission

Automatic Transmission Service

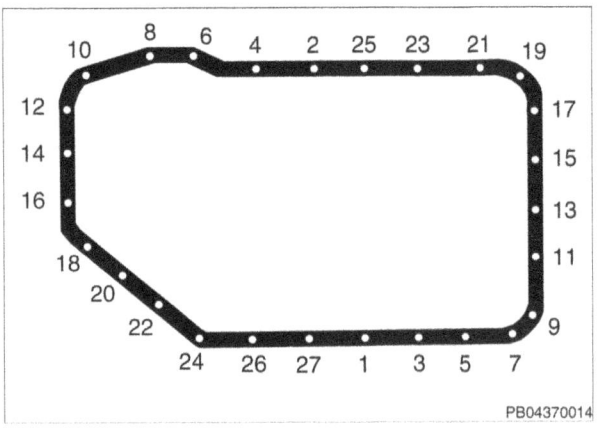

◄ Install ATF pan with new gasket. Tighten screws in sequence specified in illustration.

Tightening torque	
Transmission pan retaining screws	11 Nm (8 ft-lb)

– Replace sealing ring install ATF drain plug.

Tightening torque	
Transmission pan drain plug	40 Nm (29 ft-lb)

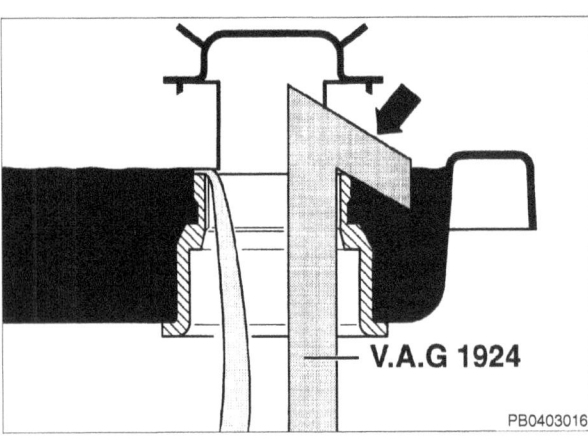

◄ Use special ATF filler tool VAG 1924, or equivalent, to fill with ATF until fluid runs out of fill hole.
  • Insert filler tool into opening of ATF guard cap in fill hole.

> **CAUTION—**
> • Do not press filler hook upwards more than necessary when inserting into fill plug opening in the well. It is possible to push the deflector cap off the top of the well.

– With transmission in "Park", start engine and allow to idle.

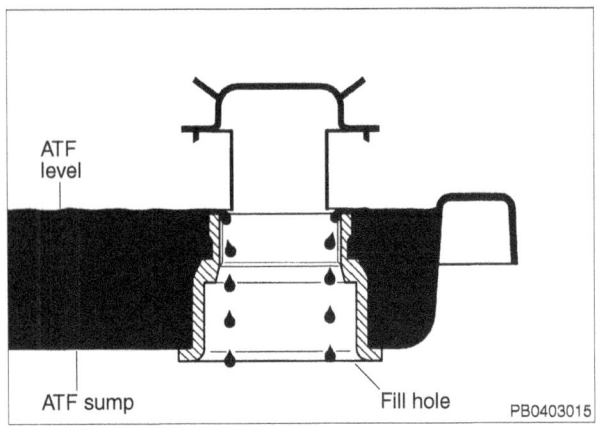

◄ With ATF at specified temperature (30 - 45°C), add fluid until it escapes from fill hole. Correct ATF level is indicated by arrow.

> **WARNING—**
> • Hot ATF can scald. Wear eye protection and protective clothing and gloves.

Tightening torque	
Filler plug to transmission	80 Nm (59 ft-lb)

– With brake pedal depressed, select all selector lever positions, holding each position for approximately 10 seconds.

– Check ATF fluid level again and top up if needed. See **03 Service and Maintenance**.

– Install diagonal braces and reinforcement plate.

Tightening torques	
Reinforcement plate:	
• To rear axle support (M10 bolt)	46 Nm (34 ft-lb)
• To support / suspension (M10 nut)	65 Nm (48 ft-lb)
Diagonal brace to body / suspension	65 Nm (48 ft-lb)

Automatic Transmission 37-7

Automatic Transmission Removal and Installation

AUTOMATIC TRANSMISSION REMOVAL AND INSTALLATION

Automatic transmission, removing and installing

Removing

◄ With vehicle on ground, pry out wheel center cap from both rear wheels. Use round holes in cap to carefully pry out.

◄ Using a breaker bar with a 32 mm socket, loosen drive axle hex nut at left and right wheels.

– Raise car, and safely support on a professional automotive lift, or jack stands.

> **WARNING—**
> • Make sure the car is stable and well supported at all times. Use a professional automotive lift or jack stands designed for the purpose. A floor jack is not adequate support.

– Move convertible top and convertible-top storage lid to service position. See **03 Service and Maintenance**.

– Disconnect battery and remove starter. See **27 Battery, Starter, Alternator**.

> **CAUTION—**
> • Prior to disconnecting the battery, read the battery cautions given in **001 General Warnings and Cautions**.

– Remove carpeted panels from top and front engine access covers.

◄ Twist engine top cover hold-downs (**arrows**) and open cover.

37-8 Automatic Transmission

Automatic Transmission Removal and Installation

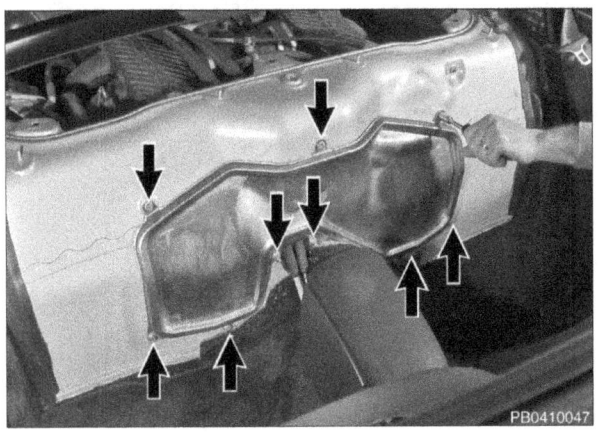

◀ Undo engine access cover mounting fasteners at rear of passenger compartment (**arrows**) and lift out cover.

> **CAUTION—**
> • Access cover edges are sharp.

− Using crankshaft pulley bolt, turn engine over so that one bore in pulley aligns with centring bore in crankcase.

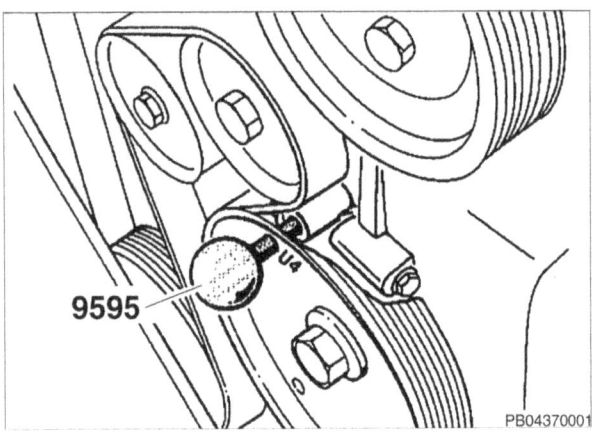

◀ Insert special tool 9595/1 (short pin) through crankcase pulley and into crankcase to lock crankshaft.

> **NOTE—**
> • If tool cannot be inserted, check that pulley is in the correct position on the crankshaft.

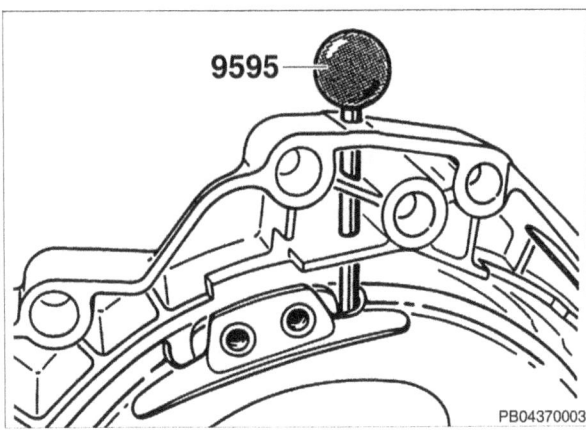

◀ Push special tool 9595/1 (long pin) through opening in transmission housing to secure torque converter from falling out.

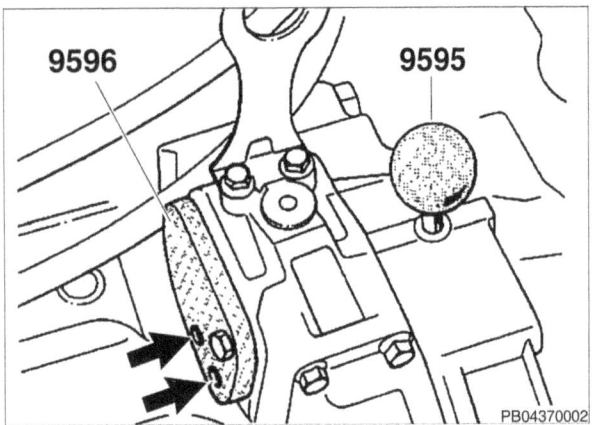

◀ Install special tool 9596 over starter opening at transmission bellhousing.

• Remove two torque converter bolts through openings (**arrows**) in special tool 9596 with 6mm Allen socket.

> **NOTE—**
> • This step is performed three times. Before each repetition, special tool 9595/1 (short pin) is removed from pulley, crankshaft turned 120° and tool is reinserted. All 6 bolts are removed this way.

Automatic Transmission 37-9

Automatic Transmission Removal and Installation

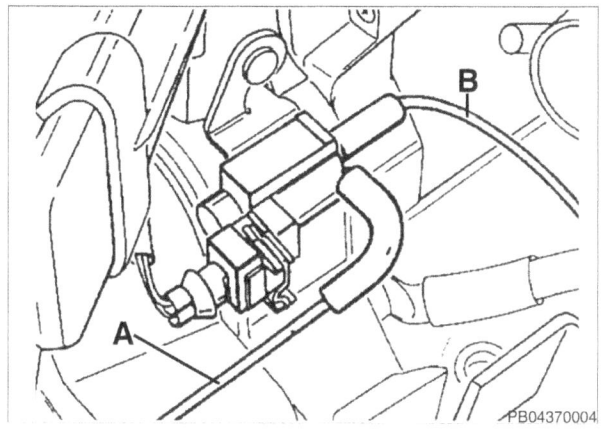

◄ Pull vacuum lines (**A** and **B**) off coolant change over valve.

− Remove vent line bracket from engine transport eyebolt.

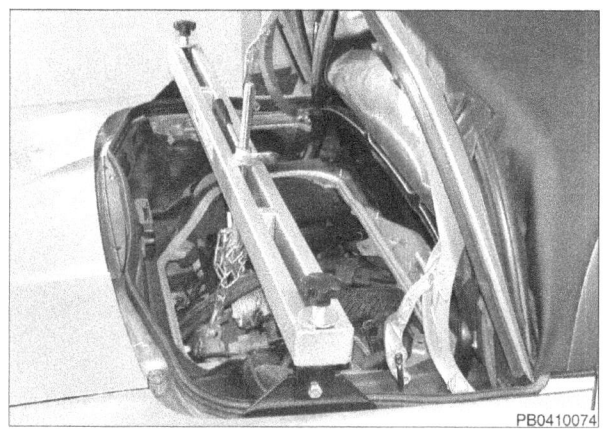

◄ Support engine using special tool 9591/1 or equivalent engine support brace.

◄ Working beneath car, remove plastic under shield. **Arrows** indicate fasteners to be removed.

◄ Remove aluminum diagonal braces and chassis reinforcement plate (14 fasteners) (**arrows**).

37-10　Automatic Transmission

Automatic Transmission Removal and Installation

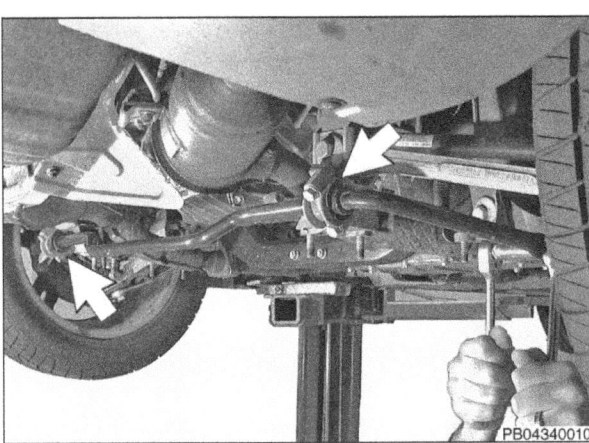

◀ Remove stabilizer bar from chassis. Disconnect stabilizer bar at end links, and remove mounting brackets at bushings (**arrows**).

− Remove muffler with catalytic converters. See **26 Exhaust System**.

◀ Remove reinforcement plate support (**arrow**).

NOTE—

• Manual transmission shown, automatic transmission procedure is similar.

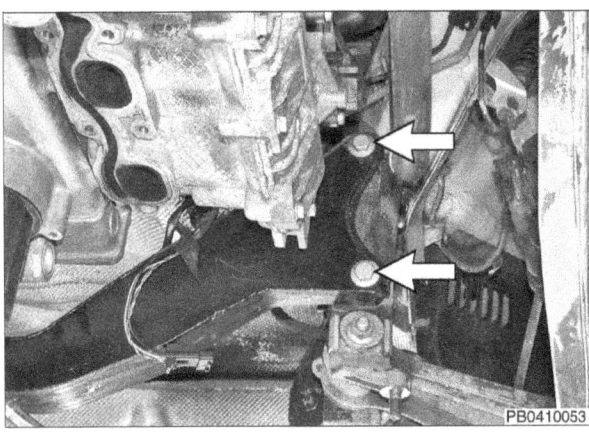

◀ Working at front of rear axle support, remove nuts holding passenger safety cable to rear axle support. Remove mounting bolts (**arrows**) on left and right side. Remove captive nuts from rear suspension arm mounts, and remove support from chassis.

◀ Working at transmission drive flanges, remove inner CV joint retaining bolts.

• Cover and protect inner CV joint ends from damage.
• Remove drive axle hex nut at wheel bearing carrier.
• Using a hammer and drift, knock outer CV joint free from hub and remove drive axle from vehicle.

CAUTION—

• Rolling vehicle on its wheels without drive axles installed and properly torqued will damage rear wheel bearings

Automatic Transmission   37-11

Automatic Transmission Removal and Installation

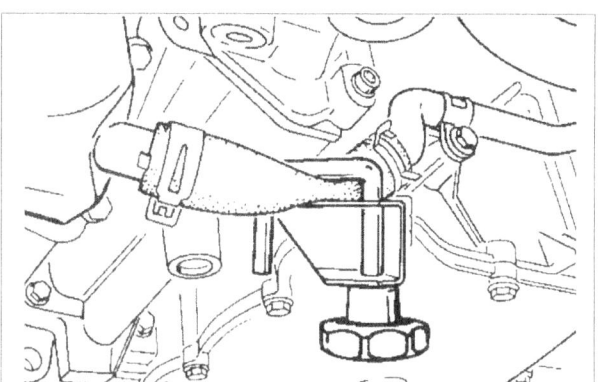

◀ Clamp off coolant supply hose with clamp and disconnect cooling line.

NOTE—

• *Collect coolant in a suitable drain pan.*

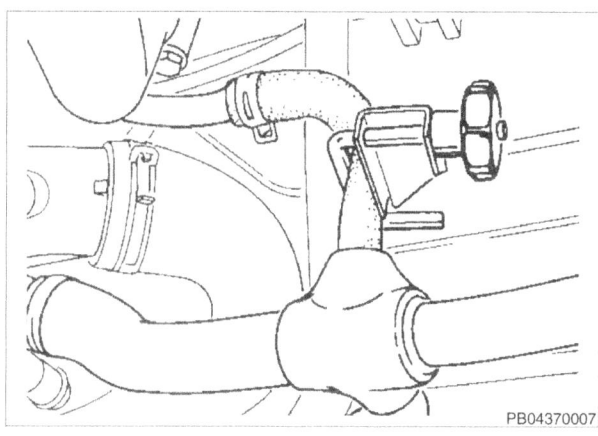

◀ Clamp off coolant return hose with clamp and disconnect cooling line.

NOTE—

• *Collect coolant in a suitable drain pan.*

− Disconnect plug connection for multi-function switch.

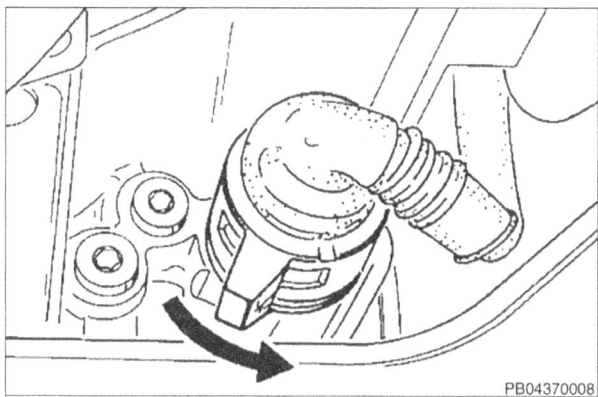

◀ Working at side of transmission, rotate Tiptronic harness connector counterclockwise and disconnect.

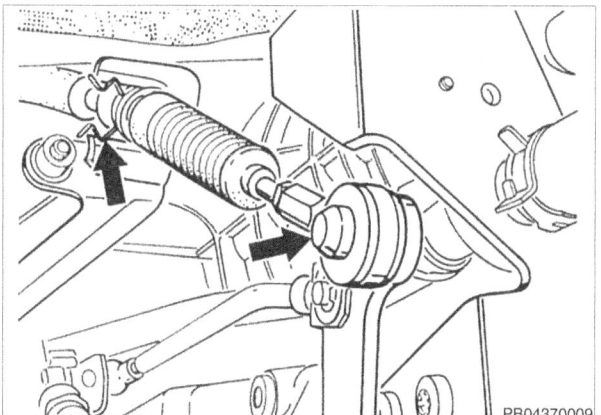

◀ Disconnect gear selector lever cable at transmission linkage and at support (**arrows**).

− Place transmission jack under transmission and secure using a strap.

37-12 Automatic Transmission

Automatic Transmission Removal and Installation

◀ Working at left and right transmission supports, remove transmission mount hardware (**A**). Remove support mounting hardware from transmission, and remove mount from chassis.

CAUTION—
* *Do not remove nut (**B**). Mount will be damaged.*

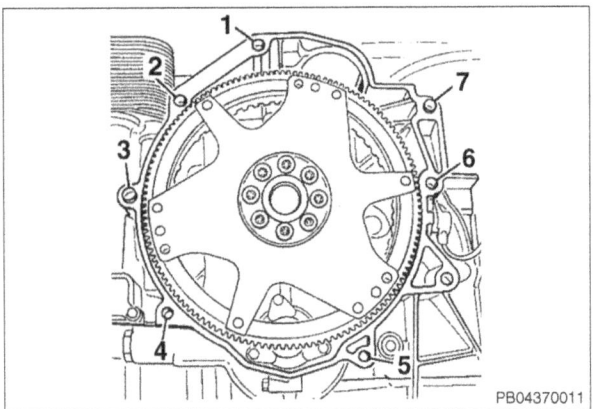

◀ Remove transmission mounting bolts in order specified in illustration.

– Pull transmission to rear and lower it carefully with a jack.

Installing

NOTE—
* *Before installing transmission, ensure that torque converter is locked in installation position with special tool 9595/1 (long pin).*

– Insert transmission carefully with transmission jack.

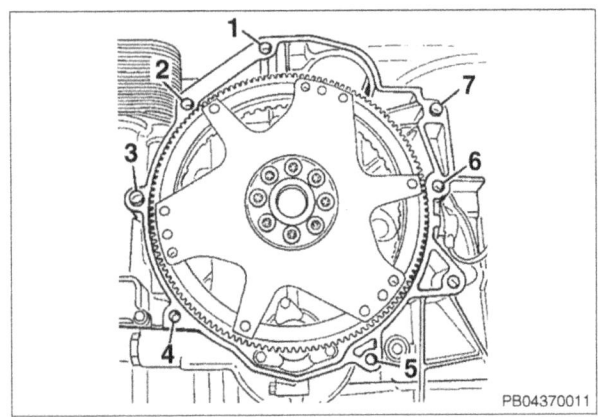

◀ Install transmission mounting bolts, torquing to specifications.

No. screw / nut tightening torque	
1 M 12 x 70 mm	85 (63 ft-lb)
2 M12 x 100 mm (with washer)	85 (63 ft-lb)
3 M12 x 90 mm	85 (63 ft-lb)
4 M10 x 35 mm (internal serration)	45 (33 ft-lb)
5 M10 mm (hexagon nut)	45 (33 ft-lb)
6 M 12 x 70 mm	85 (63 ft-lb)
7 M 12 x 90 mm	85 (63 ft-lb)

Automatic Transmission Removal and Installation

- Install right and left transmission mounts.

  NOTE—
  - *In order to install the right-hand transmission mount, the transmission must be pushed towards the left.*

Tightening torques	
Transmission mount • (M8) • (M10)	23 Nm (17 ft-lb) 65 Nm (48 ft-lb)

- Connect selector cable at lever and transmission

- Reconnect tiptronic harness connector. Turn clockwise to lock.

- Reconnect harness connector for multi-function switch.

- Reconnect coolant supply and return hoses.

- Working underneath engine, reinstall rear axle support and torque bolts (**arrows**).

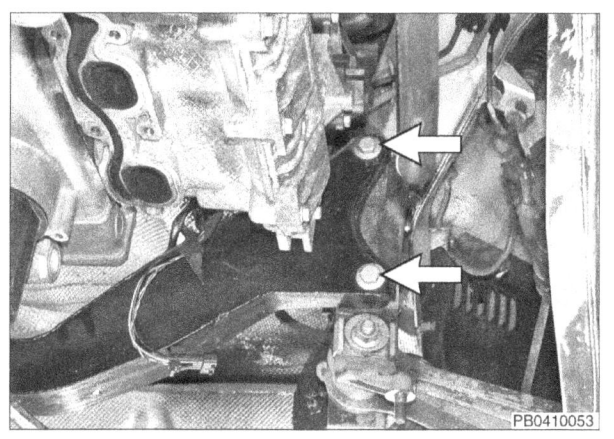

Tightening torque	
Rear axle support to rear suspension arm mounts	65 Nm (48 ft-lb)

- Reattach passenger safety cable between engine and rear axle support. Counterhold fastener with suitable wrench.

Tightening torque	
Passenger safety cable to rear suspension support	23 Nm (17 ft-lb)

- Install drive axles. Slide outer CV joint into place at wheel hub and secure with axle nut. Do not tighten axle nut at this time. Install inner CV joints to transmission output flanges.

  NOTE—
  - *Due to the high torque required to secure drive axle nut, wait until vehicle is reassembled and on the ground before tightening axle nut to specification.*

Tightening torques	
Inner CV joint to output flange M8 M10	 39 Nm (29 ft-lb) 81 Nm (60 ft-lb)

- Install muffler support bracket to rear of transmission.

Tightening torque	
Muffler support bracket to transmission	65 Nm (48 ft-lb)

37-14 Automatic Transmission

Automatic Transmission Removal and Installation

- Install reinforcement plate support.

Tightening torque	
Reinforcement plate support to body	65 Nm (48 ft-lb)

- Install rear muffler with catalytic converters. See **26 Exhaust System**.

- Install rear stabilizer bar.

Tightening torques	
Stabilizer bar bracket to body	23 Nm (17 ft-lb)
End link to stabilizer bar	50 Nm (37 ft-lb)

- Install diagonal braces and reinforcement plate.

Tightening torques	
Reinforcement plate: to rear axle support (M10 bolt) to support / suspension (M10 nut)	46 Nm (34 ft-lb) 65 Nm (48 ft-lb)
Diagonal brace to body / suspension	65 Nm (48 ft-lb)

- Screw vent line bracket to engine transport eyebolt.

Tightening torques	
Vent line to engine transport eyebolt	10 Nm (7.5 ft-lb)

- Secure torque converter to drive plate, installing bolts using special tool 9596 through starter opening.

Tightening torques	
Torque converter to drive plate	39 Nm (29 ft-lb)

NOTE—

- *This step is performed three times. Before each repetition, special tool 9595/1 (short) is removed from pulley, crankshaft turned 120° and tool reinserted. All 6 bolts are installed this way.*

- Remainder of installation is reverse of removal, noting the following:
  - Top off and bleed cooling system as necessary. See **19 Cooling System**.
  - With vehicle reassembled and off supporting jacks, torque drive axles to specified torque.

Tightening torque	
Drive axle to wheel hub	460 Nm (f0 ft-lb)

Automatic Transmission 37-15

Automatic Transmission Shift Assembly

AUTOMATIC TRANSMISSION SHIFT ASSEMBLY

Shift lever cable, adjusting

– Move shift lever to position "Park".

– Adjust cable length using barrel nut at transmission lever so that cable is free of tension.

– Check adjustment by selecting all gears and check that gear is indicated in instrument cluster. Also change gate from "D" to "M". A straight movement must be possible without catching.

Automatic transmission shift knob, removing and installing

– Turn ignition ON.

◀ Depress shift button, and insert appropriately sized Allen key into bore at base of button. Lever button off shift knob.

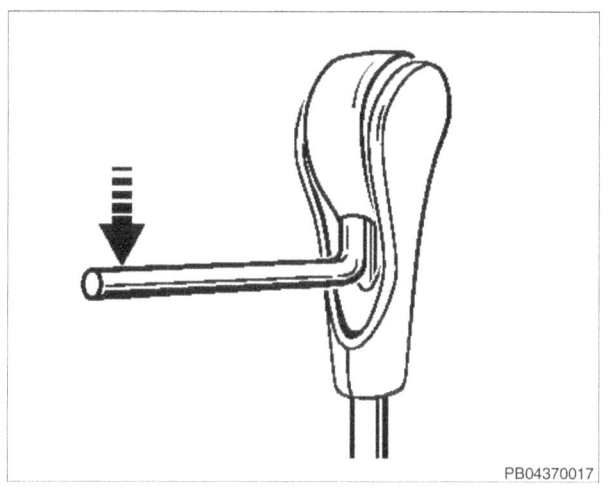

◀ Remove spring (**1**) and spring clip (**2**) and pull off shift knob.

– Assemble shift knob with spring clip, compression spring and shift button.

NOTE—
* Fit compression spring with small diameter facing the guide peg.

– Press complete shift knob onto shift lever until it bottoms out. Check function of shift button.

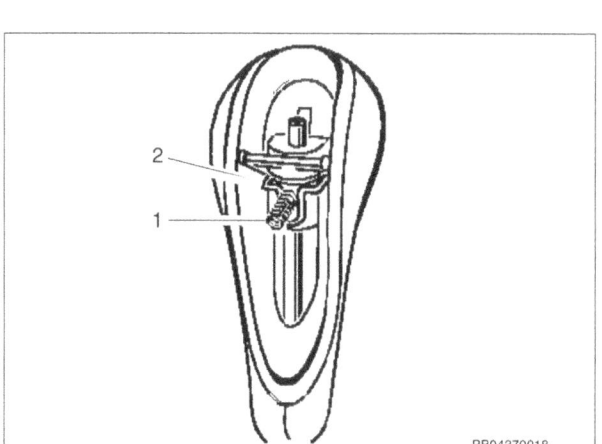

37-16 Automatic Transmission

Automatic Transmission Shift Assembly

Automatic transmission gear position switch, checking

The automatic transmission gear position switch functions are:
- Transmit selector lever position via 4 coding wires.
- Activate the back-up lights.
- Prevent starter operation in drive positions.

Gear position switch is located on the left side of the transmission housing. It is actuated by the selector lever via a Bowden cable.

NOTE—
- *The gear position switch is also known as the multi-function switch or the transmission range switch.*

- Make sure ignition is OFF.

- Access Tiptronic control module (TCM) in rear trunk and disconnect connector.

◂ Check for continuity to ground at TCM connector using table below.

NOTE—
- *When making the check below at the TCM connector, use extreme care not to bend or distort the small connectors.*

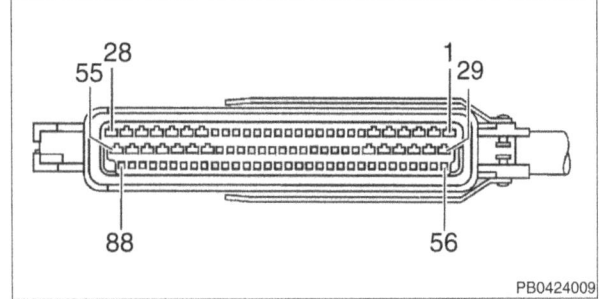

A/T gear position switch, continuity check	
Selector lever position	TCM terminal
P	terminal 36
R	terminal 8
N	terminal 8, 36, 37
D	terminal 9, 36, 37

- If any faults are found, make a continuity check on the harness between the TCM connector and the switch connector. Consult the wiring diagrams shown in **EWD Electrical Wiring Diagrams** for pinout information and wire colors.

Automatic transmission gear position switch, removing and installing

- Raise car, and safely support on a professional automotive lift, or jack stands.

WARNING—
- *Make sure the car is stable and well supported at all times. Use a professional automotive lift or jack stands designed for the purpose. A floor jack is not adequate support.*

- Move convertible top and convertible-top storage lid to service position. See **03 Service and Maintenance**.

- Remove carpeted panels from top access covers.

◂ Twist engine top cover hold-downs (**arrows**) and open and remove cover.

Automatic Transmission 37-17
Automatic Transmission Shift Assembly

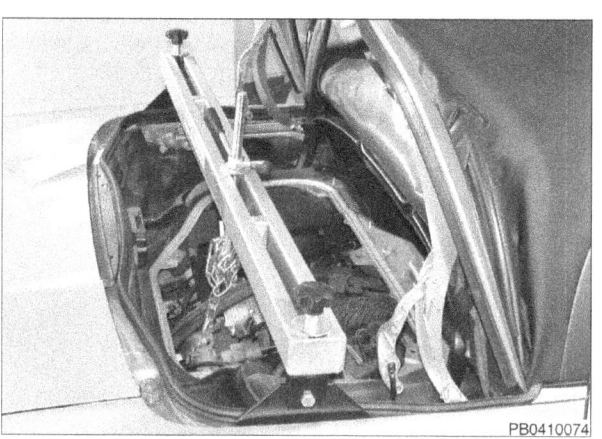

◄ Support engine using special tool 9591/1 or equivalent engine support brace.

◄ Working beneath car, remove plastic under shield. Arrows indicate fasteners to be removed.

◄ Remove aluminum diagonal braces and chassis reinforcement plate (14 fasteners).

– Remove left catalytic converters. See **26 Exhaust System**.

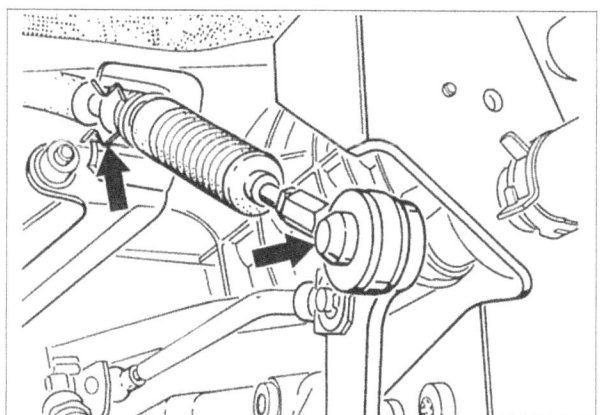

◄ Disconnect gear selector lever cable at transmission linkage and at support (**arrows**).

– Place jack under transmission.

37-18　Automatic Transmission

Automatic Transmission Shift Assembly

◀ Working at left transmission supports, remove transmission mount hardware (**A**). Remove support mounting hardware from transmission, and remove mount from chassis.

> **CAUTION—**
> - Do not remove nut (**B**). Mount will be damaged.

− Release harness connector at transmission gear position switch.

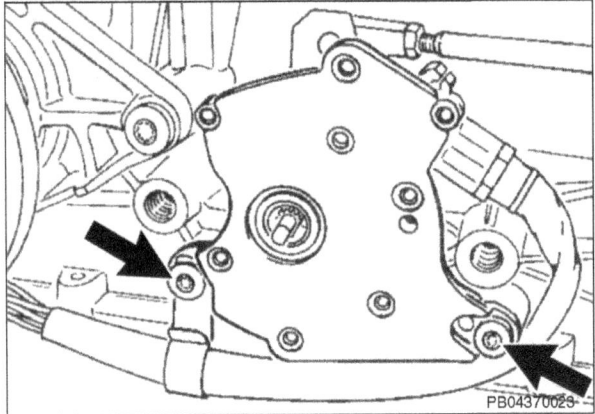

◀ Remove switch retaining screws (**arrows**) and pull switch from selector shaft.

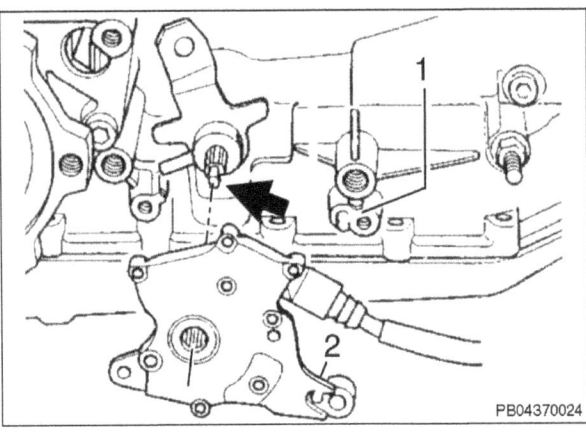

◀ Installation is reverse of removal, noting the following:
- Install switch to selector shaft so that flat area on splines in switch align to flat area on selector shaft (**arrow**).
- Turn switch until bore in switch (**2**) can be installed on dowel pin (**1**) on transmission.

> **CAUTION—**
> - *Place the switch on the selector shaft carefully to prevent damaging the switch contacts. Do not tilt or force the switch on during installation.*

Tightening torques	
Transmission mount (M8) (M10)	23 Nm (17 ft-lb) 65 Nm (48 ft-lb)
Reinforcement plate: to rear axle support (M10 bolt) to support / suspension (M10 nut)	46 Nm (34 ft-lb) 65 Nm (48 ft-lb)
Diagonal brace to body / suspension	65 Nm (48 ft-lb)

40 Front Suspension

GENERAL 40-1
　Warnings and Cautions 40-2
　Front suspension components 40-2
FRONT SUSPENSION 40-3
　Front control arms, replacing 40-3
　Sway bar link, replacing 40-4
FRONT STRUTS 40-5
　Front strut assembly 40-5

　Front struts, removing and installing 40-6
FRONT WHEEL BEARINGS 40-9
　Front wheel bearing carrier assembly 40-9
　Front wheel bearing, removing and installing ... 40-9

TABLES
a. Coil spring applications 40-6

GENERAL

This repair group covers front suspension service. Special press tools and procedures are required to disassemble and service many of the suspension components, including the front struts and wheel bearing assemblies. Read the procedure through before beginning a repair. For additional information, see:

- **44 Wheels, Tires and Alignment**.
- **48 Steering**.

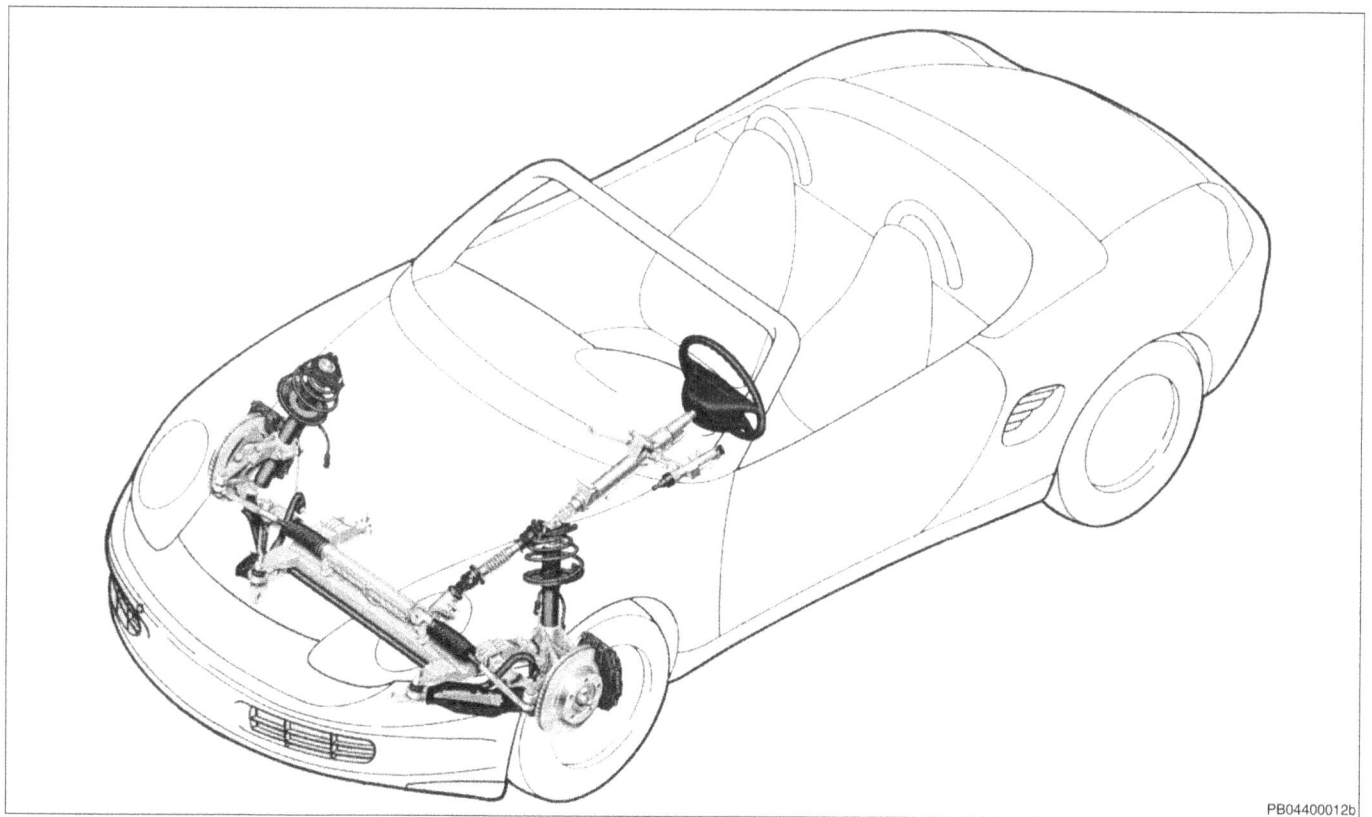

40-2 Front Suspension

General

Warnings and Cautions

> **WARNING—**
> - Do not reuse self-locking nuts or bolts. They are designed to be used only once and may fail if reused. Always replace self-locking fasteners any time they are loosened or removed.
> - Do not attempt to weld or straighten any suspension components. Replace damaged parts.

> **CAUTION—**
> - Due to the aluminum construction of suspension components, observe the following precautions:
> When replacing damaged suspension components, always inspect the condition of the control arms.
> Do not expose aluminum components to temperatures exceeding 80°C (176°F) unless noted.

Front suspension components

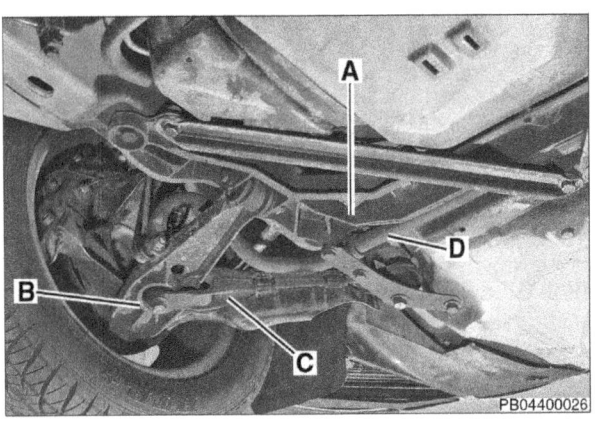

◄ The Boxster front suspension is comprised of an aluminum cross-member (**A**) with a transverse control arm (**B**) and a diagonal control arm (**C**) mounted to individual wheel bearing carriers at each side. The McPherson struts are twin-tube, gas pressurized. A tubular sway bar (**D**) links the transverse control arms to the cross-member.

The aluminum front cross-member supports the front control arms, rack and pinion steering rack and front sway bar. The cross-member is also an integral component of the front crash crumple zone.

> **NOTE—**
> - For the 2002 model year, Porsche changed the rear mounting points of the front axle cross-member from one bolt per side to two bolts per side.

> **CAUTION—**
> - Early (single bolt mounting) and late (two bolt mounting) cross-members are not interchangeable.

The aluminum control arm ball joints and bushings are not replaceable. If the control arm ball joints or bushings show signs of wear, the complete control arm must be replaced.

Standard suspension components do not allow for caster adjustment. In cases where caster adjustments are required, replacement control arms with adjustable caster bearings are available from Porsche.

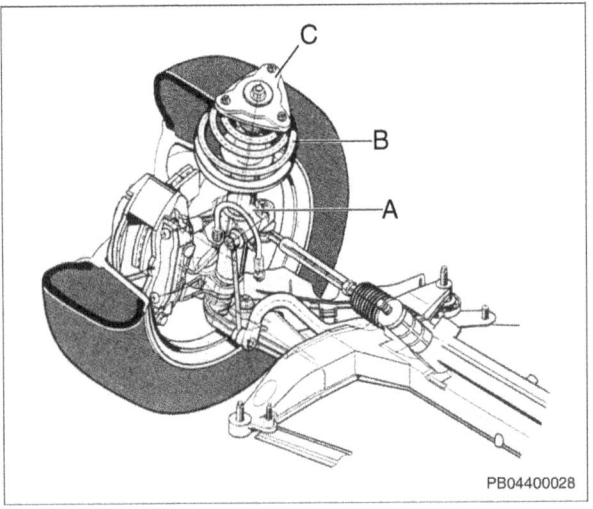

◄ The strut assembly (**A**) use twin-tube, gas pressure shock absorbers. The shock absorber body provides the lower perch for a conical spring (**B**). The top of the spring is retained by a strut mount and bearing (**C**), which attaches to the absorber piston rod.

Front Suspension

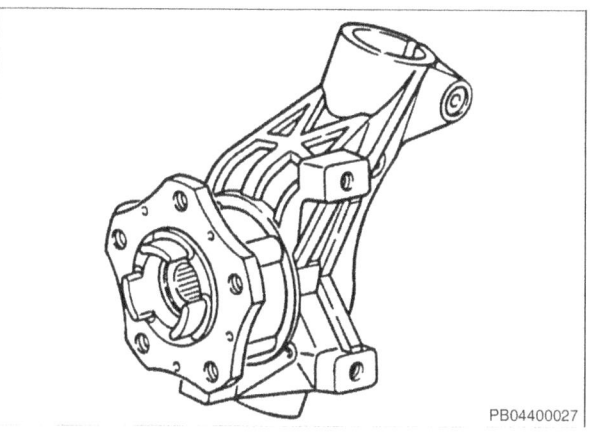

◁ The die-cast aluminum wheel bearing carrier supports the wheel bearing, wheel hub, brake caliper, strut assembly, tie rod, sway bar link, and ABS sensor.

FRONT SUSPENSION

NOTE—
- *The aluminum control arm ball joints and bushings are not replaceable. If the control arm ball joints or bushings show signs of wear, the complete control arm must be replaced.*

Front control arms, replacing

– Loosen front wheel lug bolts. Raise car, and safely support on a professional automotive lift, or jack stands. Remove front wheels.

> **WARNING—**
> - *Make sure the car is stable and well supported at all times. Use a professional automotive lift or jack stands designed for the purpose. A floor jack is not adequate support.*

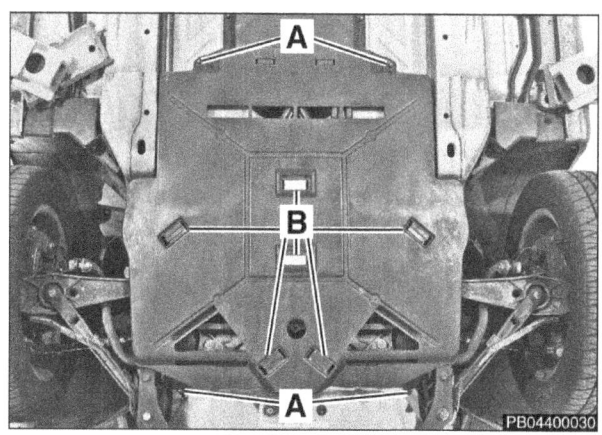

◁ Remove front plastic underbody shield:
- Remove plastic retaining nuts (**A**).
- Pry off six retaining clips (**B**).

NOTE—
- *Early cars used plastic retaining clips (B) that are easily damaged during removal. Later cars used spring steel clips that are easier to remove and install. Only the spring steel replacement clips are available as replacement parts.*

40-4 Front Suspension

Front Suspension

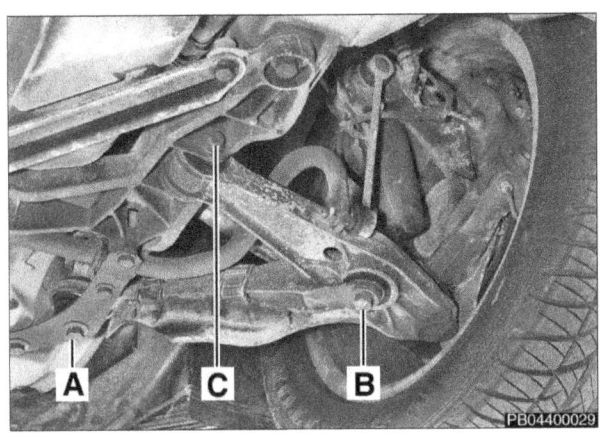

◀ Disconnect control arm retaining bolts at front of diagonal control arm (**A**), where diagonal control arm fastens to transverse control arm (**B**), and where transverse control arm fastens to crossmember (**C**).

– Separate transverse control arm from wheel bearing carrier by releasing ball joint. Loosen ball joint nut, counterholding ball joint studs with a T-40 Torx screwdriver. Press ball joint from wheel bearing carrier using special tool 9560.

– Installation is reverse of removal, noting the following:
  • Replace self locking nuts.
  • Retaining bolt at transverse control arm (**C**) must be torqued to specification with vehicle on the ground and suspension loaded with weight of vehicle.
  • If control arm with adjustable caster is being installed, reset wheel alignment. See **44 Wheels, Tires and Alignment**.

Tightening torques	
Control arms to cross-member	120 Nm (88 ft-lb)
Diagonal control arm to transverse control arm	160 (118 ft-lb)
Transverse Control arm to wheel bearing carrier	75 Nm (56 ft-lb)

Sway bar link, replacing

– Loosen front wheel lug bolts. Raise car, and safely support on a professional automotive lift, or jack stands. Remove front wheels.

> **WARNING—**
> • Make sure the car is stable and well supported at all times. Use a professional automotive lift or jack stands designed for the purpose. A floor jack is not adequate support.

◀ Disconnect sway bar link at sway bar.

– Working at top of wheel bearing carrier, remove retaining nut and remove clamping bolt/sway bar link from wheel bearing carrier.

> **NOTE—**
> • Clamping bolt/sway bar link secures strut assembly in wheel bearing carrier. Replace clamping bolt with similar diameter drift to prevent wheel bearing carrier from sliding off strut assembly.

– Installation is reverse of removal.

Tightening torques	
Strut assembly to wheel bearing carrier (at sway bar link mount)(M12)	85 Nm (63 ft-lb)
Sway bar link to sway bar (M10 x 1.5)	50 Nm (37 ft-lb)

Front Suspension 40-5

Front Struts

FRONT STRUTS

The strut assemblies use twin-tube, gas pressure shock absorbers. The shock absorber body provides the lower perch for a conical spring. The top of the spring is retained by a strut mount and bearing, which attach to the absorber piston rod.

Different strut/spring combinations are specified for different model variations. Please consult spring and strut applications before replacing components.

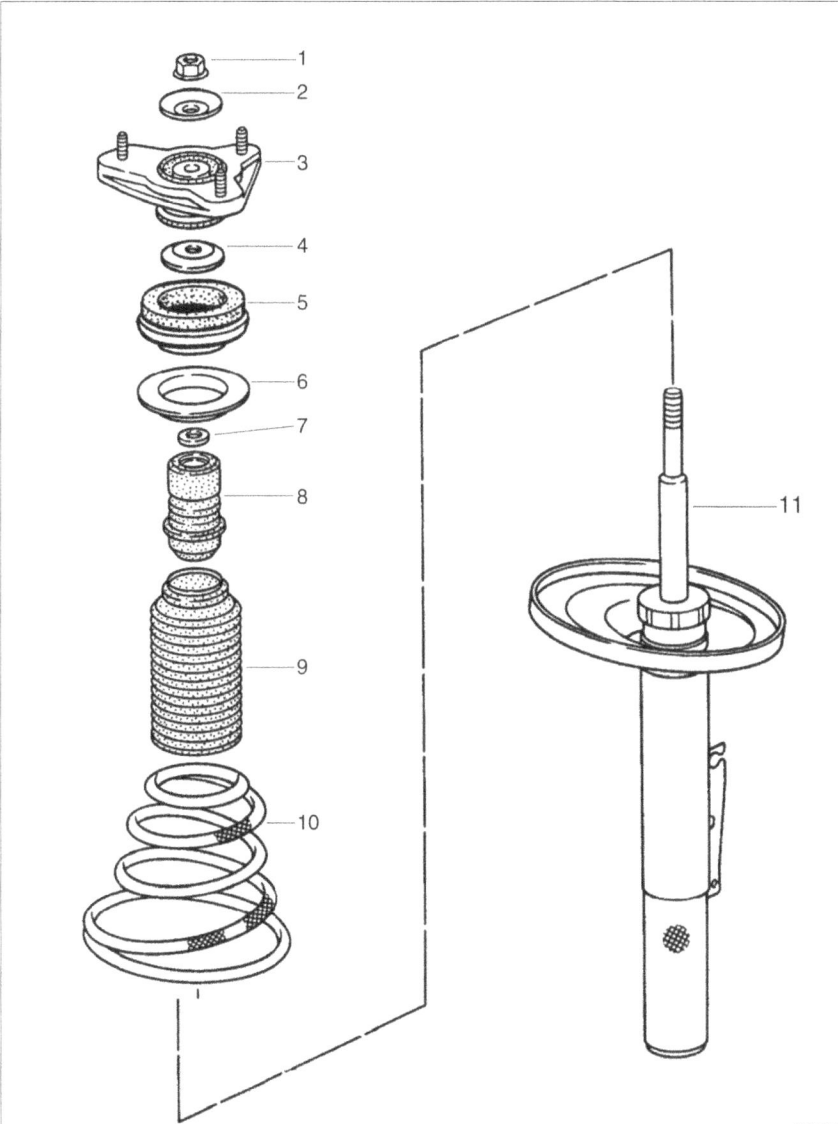

Front strut assembly

1. **Mounting nut (M14 x 1.5)**
   - Compress spring before removing nut.
   - Replace, tighten to 80 Nm (59 ft-lb)
2. **Stop plate**
   - Mount with dish facing upwards (numbers 2 and 4 are identical)
3. **Strut mount**
4. **Stop plate**
   - Mount with dish facing downwards (numbers 2 and 4 are identical)
5. **Strut bearing**
   - Make sure it is correctly seated in strut mount (number 3).
6. **Spring plate**
7. **Cup washer**
   - Mount in correct position
8. **Bump stop**
9. **Protective bellows**
10. **Coil spring**
    - Springs are color coded
11. **Shock absorber**

40-6 Front Suspension

Front Struts

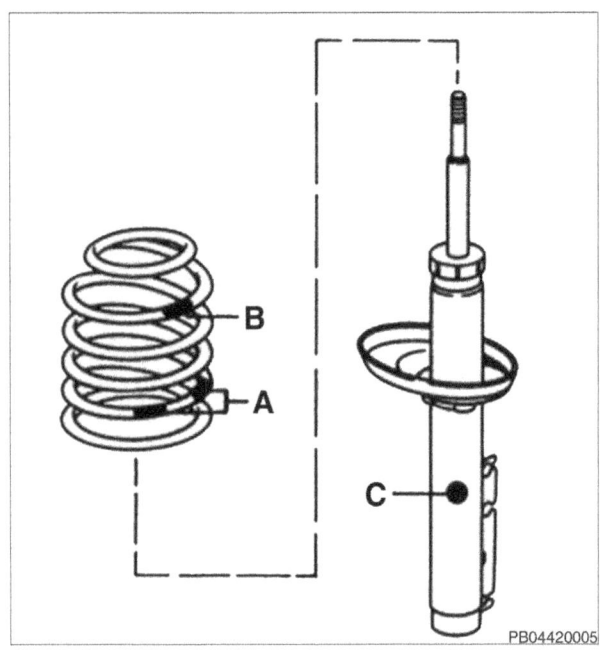

Strut component details

◂ Identifying marks for suspension packages

A indicates spring rates and lengths by line markings on the lower section of the spring. See **Table a**.

B indicates required coil plate thickness through color markings on upper section of coil spring (**B**).
White line = 3.0 mm spring plate
Green line = 6.5 mm spring plate

C indicates suspension package by color dot (**C**) on strut body.
Blue dot = standard suspension
Red dot = sport suspension

Table a. Coil spring applications		
Color	Year range	Model / options
blue / yellow	1997 - 2004	5-speed manual
blue / white	1997 - 2004	5-speed automatic
red / brown	1997 - 2002	5-speed manual / Sport
yellow / brown	2003 - 2004	5-speed manual / Sport
red / white	1997 - 2002	5-speed automatic / Sport
yellow / white	2003 - 2004	5-speed automatic / Sport
violet / yellow	2000 - 2004	S / 6-speed manual
violet / white	2000 - 2004	S / 5-speed automatic
orange / brown	2000 - 2004	S / 6-speed manual / Sport
orange / white	2000 - 2004	S / 5-speed automatic / Sport

Front struts, removing and installing

– Loosen front wheel lug bolts. Raise car, and safely support on a professional automotive lift, or jack stands. Remove front wheels.

> *WARNING—*
> • *Make sure the car is stable and well supported at all times. Use a professional automotive lift or jack stands designed for the purpose. A floor jack is not adequate support.*

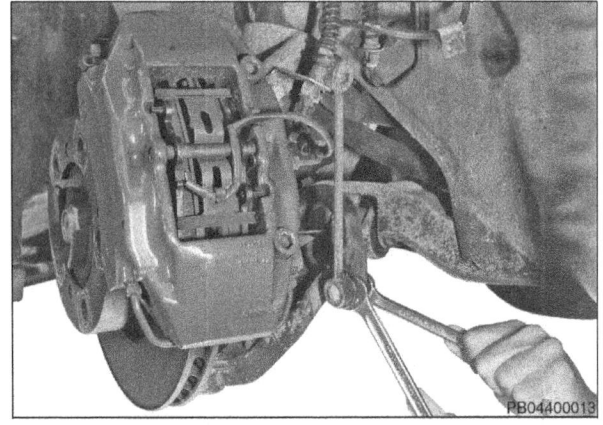

◂ Disconnect sway bar link at sway bar.

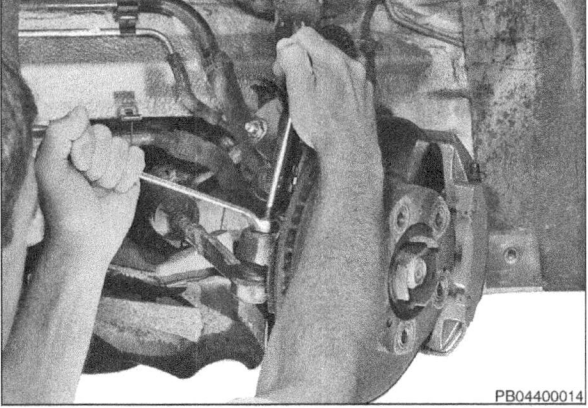

◂ Remove tie rod and control arms from wheel bearing carrier using special tool 9560 (ball joint press). When loosening fastening nuts, counterhold ball joint studs with a T-40 Torx screwdriver.

Front Suspension 40-7

Front Struts

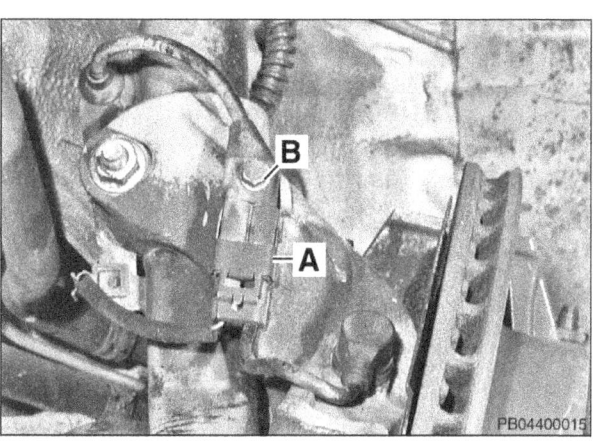

◀ Open clip (**A**) and disconnect wheel speed and brake wear electrical harness connectors at wheel bearing carrier and spring strut. Remove bolt (**B**) and remove electrical harness from strut body.

◀ Remove bolt (**arrow**) and remove brake line retaining bracket at strut. Remove brake caliper from wheel bearing carrier. See **46 Brakes—Mechanical**. Suspend brake caliper from chassis using mechanic's wire.

– Remove brake rotor retaining screws and remove front brake rotor from hub.

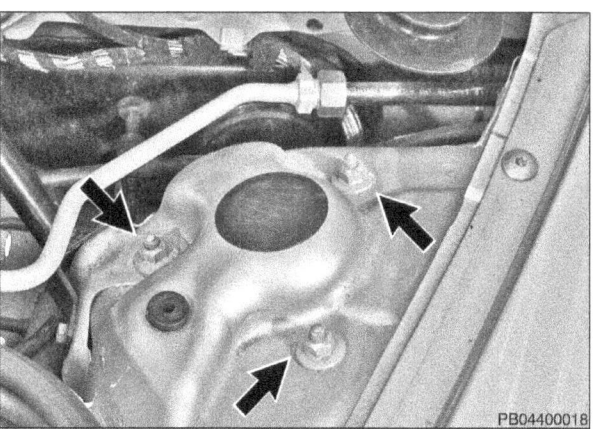

◀ Mark the location of the three strut mount nuts (**arrows**) in relation to body to indicate suspension alignment settings. Remove strut mount nuts. Remove front strut assembly from body with wheel bearing carrier.

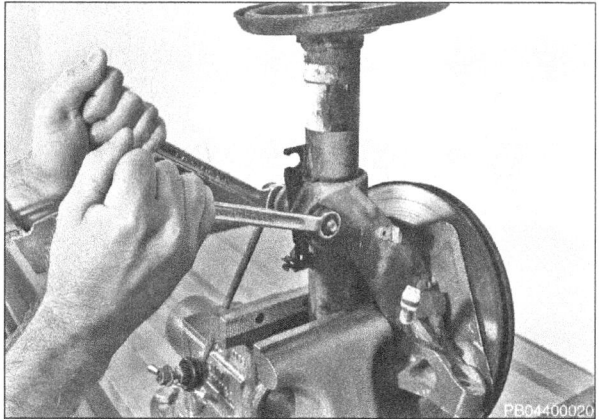

◀ With strut assembly secured in a bench vise, loosen clamped connection between wheel bearing carrier and strut assembly. Remove strut from wheel bearing carrier.

Front Struts

- Installation is reverse of removal, noting the following:
  - When installing strut assembly into vehicle, note the installed position of the strut mount. Arrow markings on strut mount must point to outside of vehicle. Use new fastening nuts. Before torquing nuts, adjust strut mount location in accordance to applied markings on chassis.
  - Replace brake caliper fastening bolts.
  - Do not grease fasteners in Dacroment finish (aluminum color).
  - Use correct tightening torques.
  - Check steering geometry values at front axle.

NOTE—
- *If work was performed or parts installed (replacement springs) that effect vehicle ride height, a complete vehicle alignment must be performed.*

Tightening torques	
Strut assembly to wheel bearing carrier (at sway bar link mount)(M12)	85 Nm (63 ft-lb)
Spring strut mount to body (M8)	33 Nm (24 ft-lb)
Brake caliper to wheel bearing carrier (M12)	85 Nm (63 ft-lb)
Brake line bracket to wheel bearing carrier (M6)	10 Nm (7.5 ft-lb)
Combination wire bracket to wheel bearing carrier (M6)	10 Nm (7.5 ft-lb)
Tie rod / control arm to wheel bearing carrier (ball joint, M12)	75 Nm (56 ft-lb)
Sway bar link to sway bar (M10 x 1.5)	50 Nm (37 ft-lb)
Wheel to wheel hub (M14)	130 Nm (96 ft-lb)

Front Suspension 40-9

Front Wheel Bearings

FRONT WHEEL BEARINGS

The Boxster uses front wheel bearing carriers and hubs that are interchangeable with rear wheel bearing carriers, and shared with other Porsche models Since the design allows the carriers to be used in either front axle of rear axle positions (on opposite sides), the bearing carriers use cartridge style wheel bearings, and have wheel hubs that are designed for constant velocity style drive axles.

With no front drive axles used in Boxster models, tension is maintained on front wheel bearings and hubs by a unique bolt that mimics the stub axle of a constant velocity drive axle. This bolt also incorporates the ABS sensor wheel.

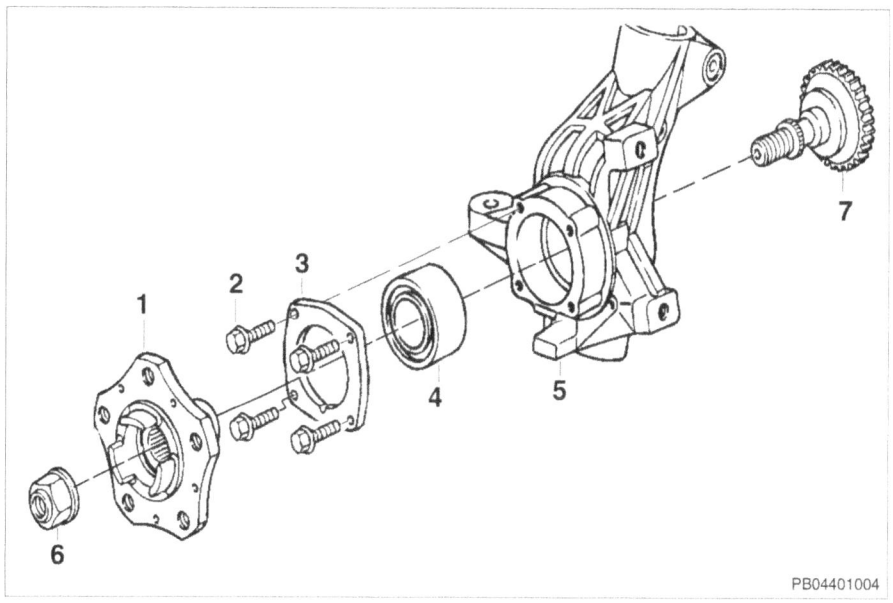

Front wheel bearing carrier assembly

1. **Wheel hub**
   - Press out using hydraulic press and appropriate press pieces.
   - Install using wheel centering device and pressure piece VW 415A.

2. **Bolt (M8 x 35 10.9)**
   - Tighten to 37 Nm (27 ft-lb)

3. **Retainer plate**

4. **Wheel bearing**
   - To remove and install, heat housing to 100°C (212°F), not exceeding 120°C (248°F).
   - Insert new bearing with bearing numbers facing wheel hub, using press tools 9247/4 and VW 415A.

5. **Wheel bearing housing**

6. **Hex nut (M22 x 1.5)**
   - Tighten to 460 Nm (340 ft-lb)

7. **Tension bolt**
   - Includes ABS pulse wheel.

Front wheel bearing, removing and installing

Cartridge style wheel bearings can be replaced either "on the car", using a mechanical press/puller set, or by removing the wheel bearing carrier and using a shop press. In order to illustrate details of both methods, this procedure demonstrates removing the wheel bearing carrier, and using a mechanical press/puller set to remove and install the wheel bearing.

 With vehicle on ground, pry out wheel center cap from front wheels. Use round holes in cap to carefully pry out.

40-10 Front Suspension

Front Wheel Bearings

◀ Using a breaker bar with a 32 mm socket, loosen hex nut at tension bolt using a breaker bar.

– Loosen front wheel lug bolts. Raise car, and safely support on a professional automotive lift, or jack stands. Remove front wheels.

> **WARNING—**
> • Make sure the car is stable and well supported at all times. Use a professional automotive lift or jack stands designed for the purpose. A floor jack is not adequate support.

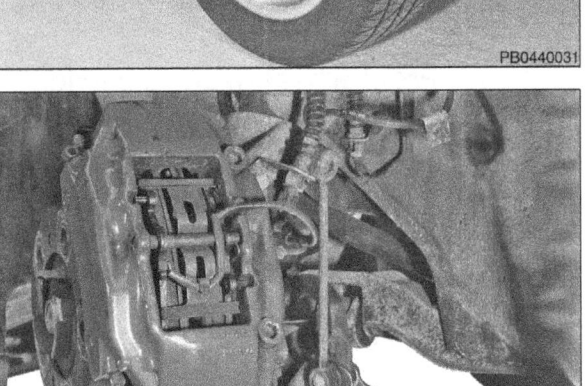

◀ Disconnect sway bar link at sway bar.

◀ Remove tie rod and control arms from wheel bearing carrier using special tool 9560 (ball joint press). When loosening fastening nuts, counterhold ball joint studs with a T-40 Torx screwdriver.

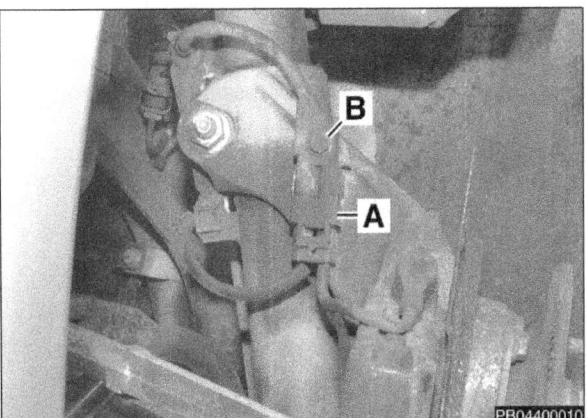

◀ Open clip (**A**) and disconnect wheel speed and brake wear electrical harness connectors at wheel bearing carrier and spring strut. Remove brake caliper and rotor. See **46 Brakes–Mechanical**.

– Loosen clamped connection between wheel bearing carrier and strut assembly. Remove wheel bearing carrier from strut.

– Remove hex nut and tension bolt from wheel bearing carrier.

Front Suspension 40-11

Front Wheel Bearings

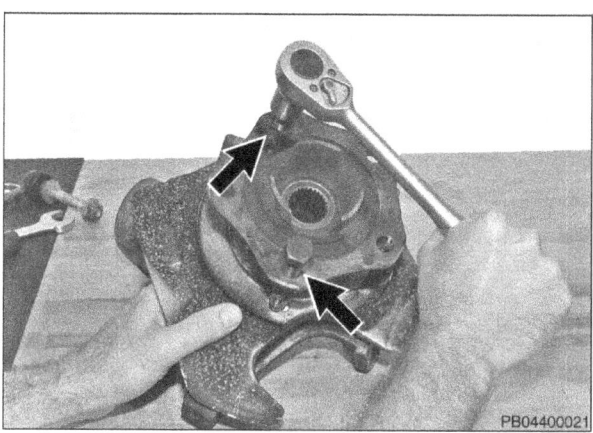

◀ Press wheel hub from carrier using bolts supplied in press/puller set (**arrows**) or suitable press tools.

NOTE—

- *Removing hub from wheel bearing carrier destroys wheel bearing.*

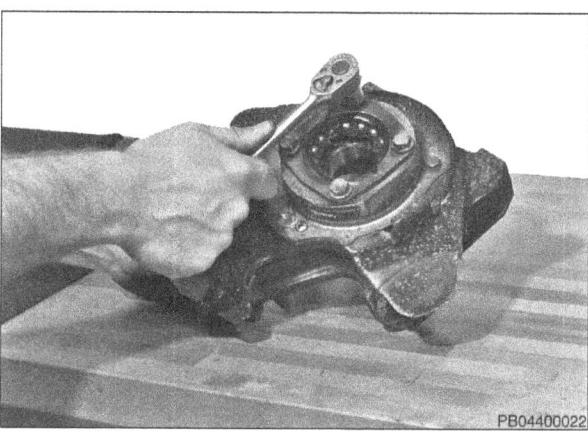

◀ Remove retainer plate and brake dust shield.

◀ Using threaded press tool and appropriate sized press plates, press wheel bearing from carrier.

NOTE—

- *To aid in wheel bearing removal, aluminum wheel bearing carrier can be heated to 100°C (212°F). Do not exceed 120°C (248°F).*

– Use threaded press tool and appropriate press plates to install new wheel bearing. During bearing installation, press on outer edge of wheel bearing only.
  - Heat bearing carrier to 100°C (212°F) if needed.
  - Install bearing with numbers facing toward wheel hub.

– Install brake dust shield and bearing retainer plate.

Tightening torque	
Brake dust shield to carrier	10 Nm (7.5 ft-lb)
Bearing retainer plate to bearing carrier	37 Nm (27 ft-lb)

◀ Remove outer bearing race from wheel hub using an appropriate puller.

NOTE—

- *If a suitable puller is not available, heat race using a torch and drive off using a brass drift.*

40-12　Front Suspension

Front Wheel Bearings

◀ Use appropriate sized press plates to support inner bearing race and press wheel hub into wheel bearing.

> **WARNING—**
> - *Failure to properly support inner bearing race while installing wheel hub damages bearing.*

- Installation of wheel bearing carrier is reverse of removal.

Tightening torques	
Strut assembly to wheel bearing carrier (at sway bar link mount)(M12)	85 Nm (63 ft-lb)
Spring strut mount to body (M8)	33 Nm (24 ft-lb)
Brake caliper to wheel bearing carrier (M12)	85 Nm (63 ft-lb)
Brake line bracket to wheel bearing carrier (M6)	10 Nm (7.5 ft-lb)
Combination wire bracket to wheel bearing carrier (M6)	10 Nm (7.5 ft-lb)
Tie rod / control arm to wheel bearing carrier (ball joints, M12)	75 Nm (56 ft-lb)
Sway bar link to sway bar (M10 x 1.5)	50 Nm (37 ft-lb)
Wheel to wheel hub (M14)	130 Nm (96 ft-lb)
Hex nut to tension bolt (M22 x 1.5)	460 Nm (340 ft-lb)

42 Rear Suspension

GENERAL ... 42-1
Warnings and Cautions ... 42-2
Rear suspension components ... 42-2

REAR STRUTS AND SWAY BAR ... 42-3
Rear struts, removing and installing ... 42-3
Rear strut assembly ... 42-6
Rear struts, disassembling and assembling ... 42-7
Rear sway bar, removing and installing ... 42-8

REAR CONTROL ARMS ... 42-9
Track arm, removing and installing ... 42-9
Control arm / trailing arm,
  removing and installing ... 42-10

REAR WHEEL BEARINGS AND DRIVE AXLES ... 42-11
Wheel bearing carrier assembly ... 42-11
Rear wheel bearing, replacing ... 42-11
Drive axle, removing and installing ... 42-15

TABLE
a. Rear coil spring applications ... 42-7

GENERAL

This repair group covers rear suspension service. Special press tools and procedures are required to disassemble and service the wheel bearing assemblies.

See also **44 Wheels, Tires and Alignment**.

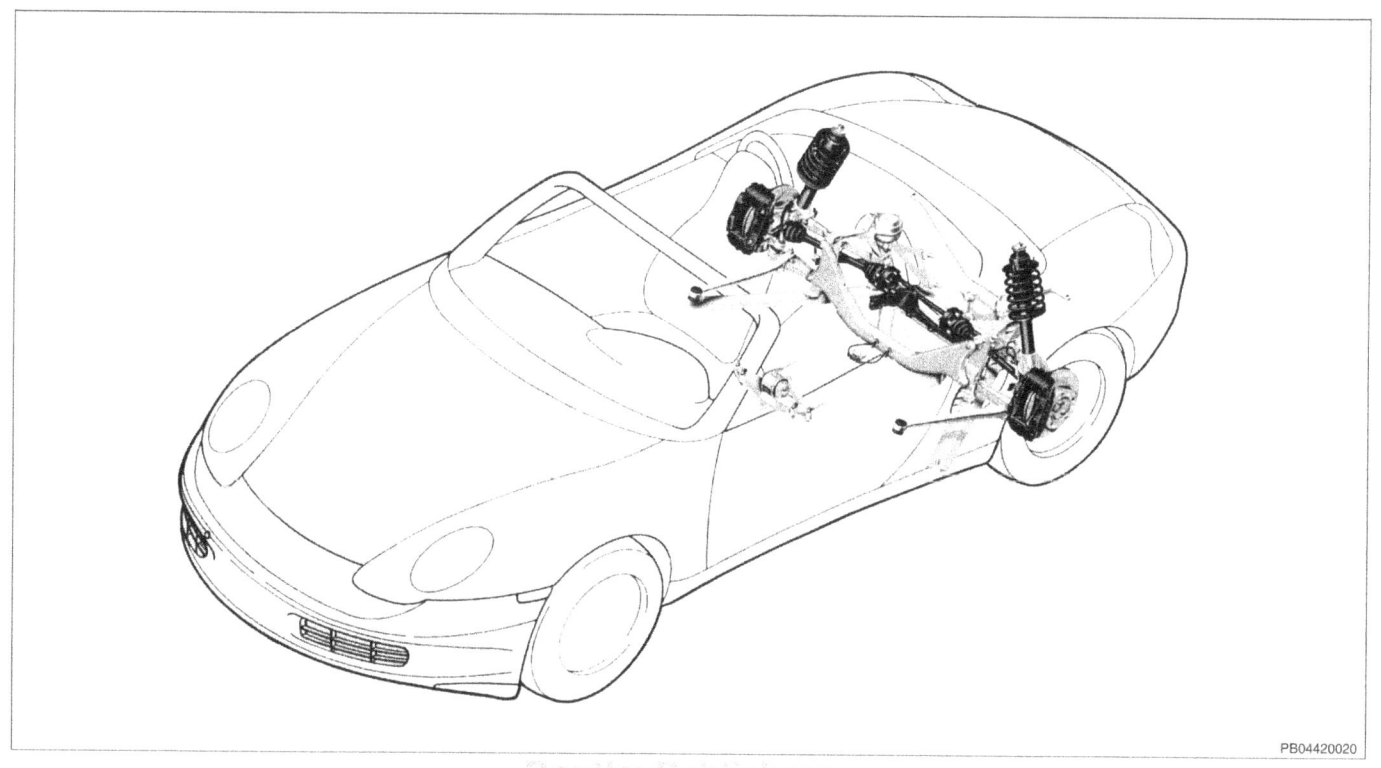

42-2 Rear Suspension

General

Warnings and Cautions

WARNING—
- Do not reuse self-locking nuts or bolts. They are designed to be used only once and may fail if reused. Always replace self-locking fasteners any time they are loosened or removed.
- Do not attempt to weld or straighten any suspension components. Replace damaged parts.

CAUTION—
- Due to the aluminum construction of suspension components, observe the following precautions:
  When replacing damaged suspension components, always check the condition of the control arms.
  Do not expose aluminum components to temperatures exceeding 80°C (176°F) unless noted.

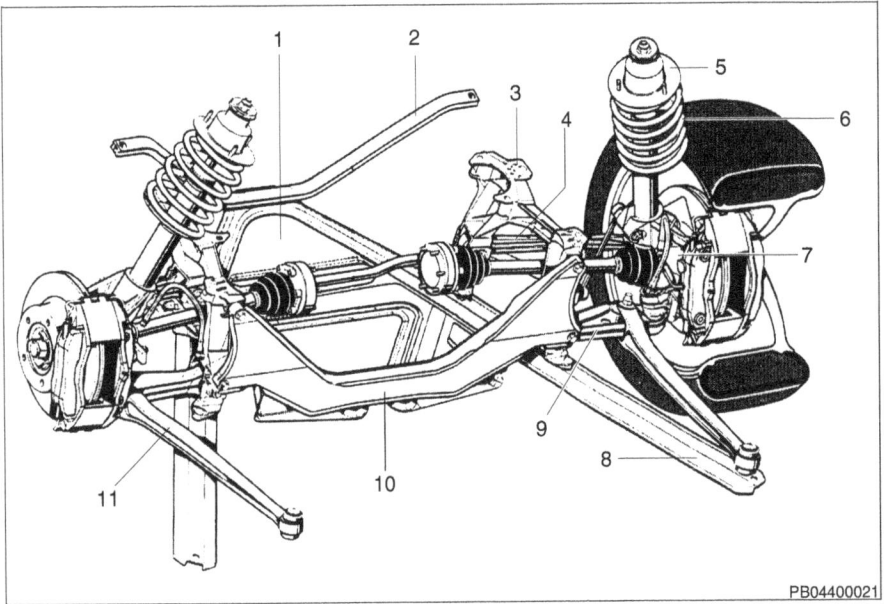

Rear suspension components

1. Chassis reinforcement plate
2. Reinforcement plate support
3. Rear suspension arm mount
4. Track arm
5. Strut bearing
6. Strut coil spring
7. Wheel bearing carrier
8. Diagonal brace
9. Control arm
10. Rear axle support
11. Trailing arm

The Boxster rear suspension uses multiple aluminum components resulting in an extremely light and rigid design.

The strut assemblies are comprised twin-tube, gas pressure shock absorbers. The shock absorber body provides the lower perch for a conical spring. The top of the spring is retained by the strut mount and bearing.

◀ Die-cast aluminum wheel bearing carriers support the wheel bearing, wheel hub, brake caliper, strut assembly, track arm, sway bar link, and ABS sensors.

The drive axles use traditional constant velocity joints at each end of the axle shaft. Individual CV joints, drive axle boots, and complete axles are available as replacement parts. Axles must be removed from the vehicle for service.

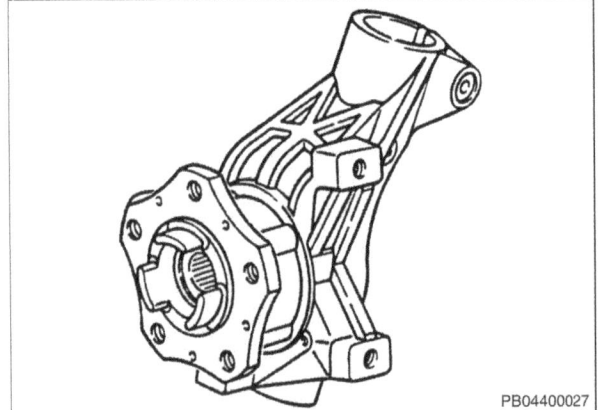

Rear Suspension 42-3

Rear Struts and Sway Bar

REAR STRUTS AND SWAY BAR

Rear struts, removing and installing

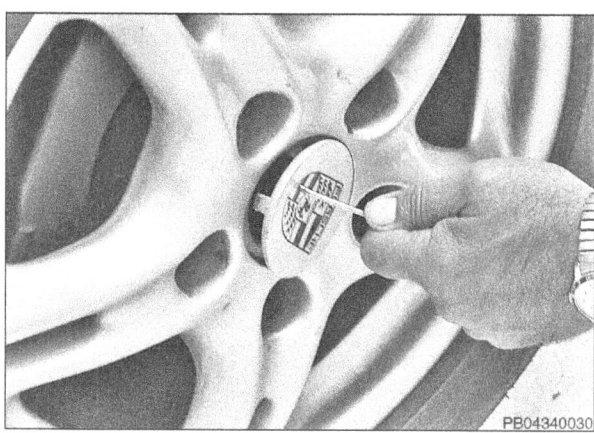

◀ With vehicle on ground, pry out wheel center cap from both rear wheels. Use round holes in cap to carefully pry out.

◀ Using a breaker bar with a 32 mm socket, loosen drive axle hex nut at left and right wheels.

– Loosen rear wheel lug bolts. Raise car and safely support on a professional automotive lift, or jack stands. Remove rear wheels.

> *WARNING—*
> * *Make sure the car is stable and well supported at all times. Use a professional automotive lift or jack stands designed for the purpose. A floor jack is not adequate support.*

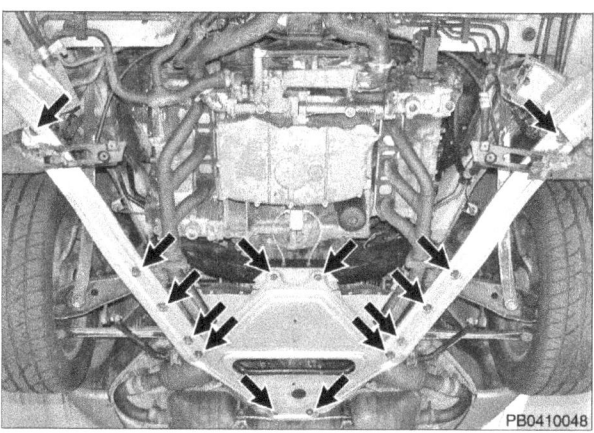

◀ Remove aluminum diagonal braces and chassis reinforcement plate (14 fasteners).

◀ Disconnect sway bar link at end of sway bar.

42-4 Rear Suspension

Rear Struts and Sway Bar

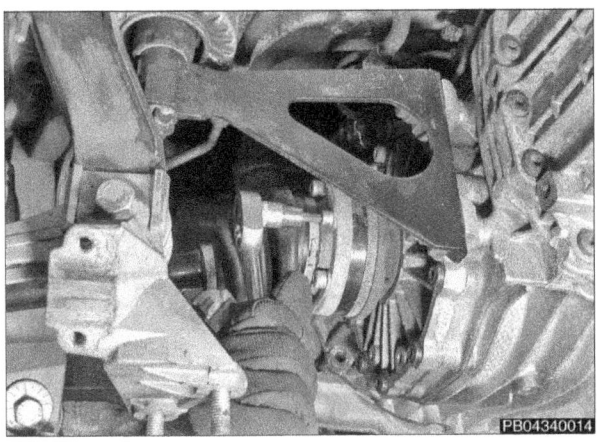

◄ Working at transmission drive flanges, remove inner CV joint retaining bolts and disconnect inner CV joint from transmission.

NOTE—
- *Leave transmission in gear and apply parking brake to make removal of inner CV joint retaining bolts easier.*

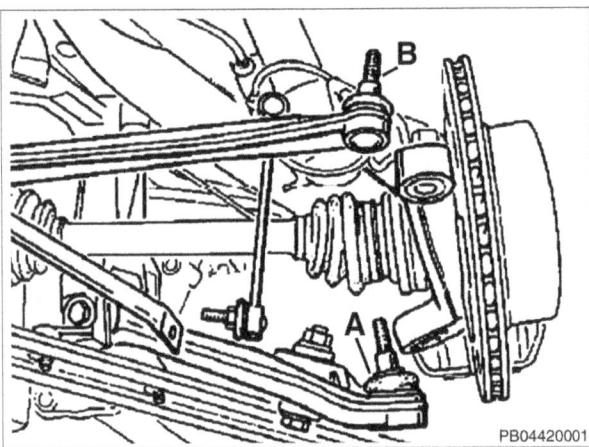

◄ Remove control arm (**A**) and track arm (**B**) from wheel bearing carrier. Counterhold ball joint studs using a T-40 Torx screwdriver when removing nuts. Press ball joints out using ball joint press tool (Porsche special tool 9560).

NOTE—
- *To free ball joints from wheel bearing carrier, pull down on control arms while swinging wheel bearing carrier outward.*

– Disconnect wheel speed sensor connectors at wheel bearing carrier and unclip electrical harness from strut assembly. Remove retaining bolt and remove electrical harness connector from wheel bearing carrier.

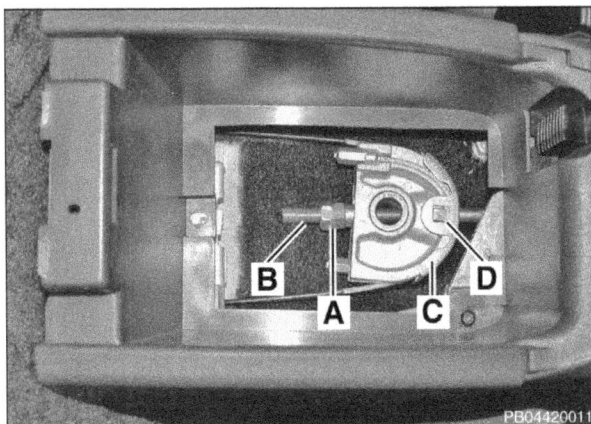

◄ Remove parking brake cable from turnbuckle at parking brake lever.
- Loosen lock nut and adjusting nut (**A**) on pull rod (**B**) and remove.
- Disengage tab washer (**C**) for cables from tab (**D**) on upper and lower sides.

NOTE—
- *To access turnbuckle, open the cover of the console compartment behind the parking brake lever and remove the rubber insert and bottom panel.*

– Remove applicable parking brake cable from body by pulling downward from guide.

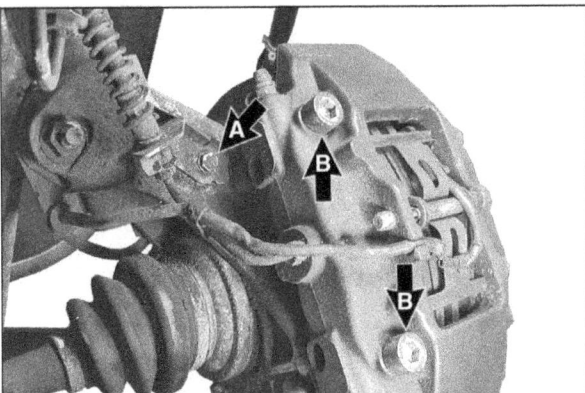

◄ Remove brake caliper.
- Remove brake line retaining bracket (**A**).
- Remove brake caliper retaining bolts (**B**) from wheel bearing carrier and suspend brake caliper from chassis using mechanic's wire. See **46 Brakes—Mechanical**.

– Remove brake rotor retaining screws and remove rear brake rotor from hub.

– Remove hex nut from drive axle and remove drive axle from wheel bearing carrier. If necessary, use a brass drift or suitable puller to free axle shaft from wheel bearing carrier.

Rear Suspension 42-5

Rear Struts and Sway Bar

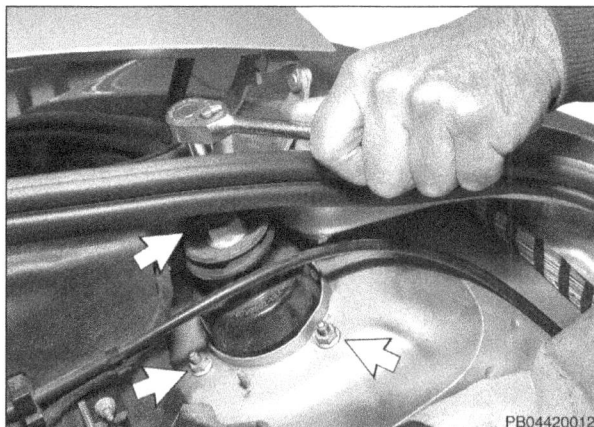

- Using a helper, remove drive axle by swinging strut assembly/wheel bearing carrier outward while bending drive axle out.

◀ Working at rear of engine compartment, peel back engine compartment liner. Open rear trunk lid, and remove rubber plug at at weather stripping channel.

- Mark location of strut mount nuts in relation to body to indicate suspension alignment settings.
- Remove strut mount nuts (**arrows**). Remove strut assembly from body complete with wheel bearing carrier.

NOTE—
- *Make sure ball joint and driveshaft boots are not damaged during removal and installation.*

- With assembly on work bench, loosen clamping bolt at wheel bearing carrier and remove from strut assembly. For rear strut service information, **Rear struts, disassembling and assembling**.

- Installation is reverse of removal, noting the following:
  - Tighten drive axle nut only with car on ground.
  - Replace brake caliper fastening bolts.
  - Do not lubricate Dacromet-finished (aluminum) coated fasteners.
  - Confirm correct strut alignment before tightening strut mount bolts.
  - Refit parking brake cables to turnbuckle at parking brake lever, adjust as necessary. See **42 Brakes–Mechanical**.
  - If new springs are installed, a rear wheel alignment should be carried out.

Tightening torques	
Strut mount to body (M8)	33 Nm (24 ft-lb)
Strut assembly to wheel bearing carrier (at sway bar link mount)(M12)	85 Nm (63 ft-lb)
Brake disc to wheel hub (M6)	10 Nm (7.5 ft-lb)
Brake line holder to wheel bearing carrier (M6)	10 Nm (7.5 ft-lb)
Brake caliper to wheel bearing carrier (M12 x 1.5)	85 Nm (63 ft-lb)
Electrical harness connector to wheel bearing carrier (M6)	10 Nm (7.5 ft-lb)
Control arms to wheel bearing carrier (ball joint)	75 Nm (56 ft-lb)
Inner CV joint to output flange M8 M10	39 Nm (29 ft-lb) 81 Nm (60 ft-lb)
Sway bar link to sway bar (M10 x 1.5)	50 Nm (37 ft-lb)
Reinforcement plate to rear axle support (M10 bolt) to support / suspension (M10 nut)	46 Nm (34 ft-lb) 65 Nm (48 ft-lb)
Drive axle nut to drive axle (use new nut)	440 Nm (325 ft-lb)
Diagonal brace to body / suspension	65 Nm (48 ft-lb)
Drive axle to wheel hub (use new nut)	460 Nm (340 ft-lb)

42-6 Rear Suspension

Rear Struts and Sway Bar

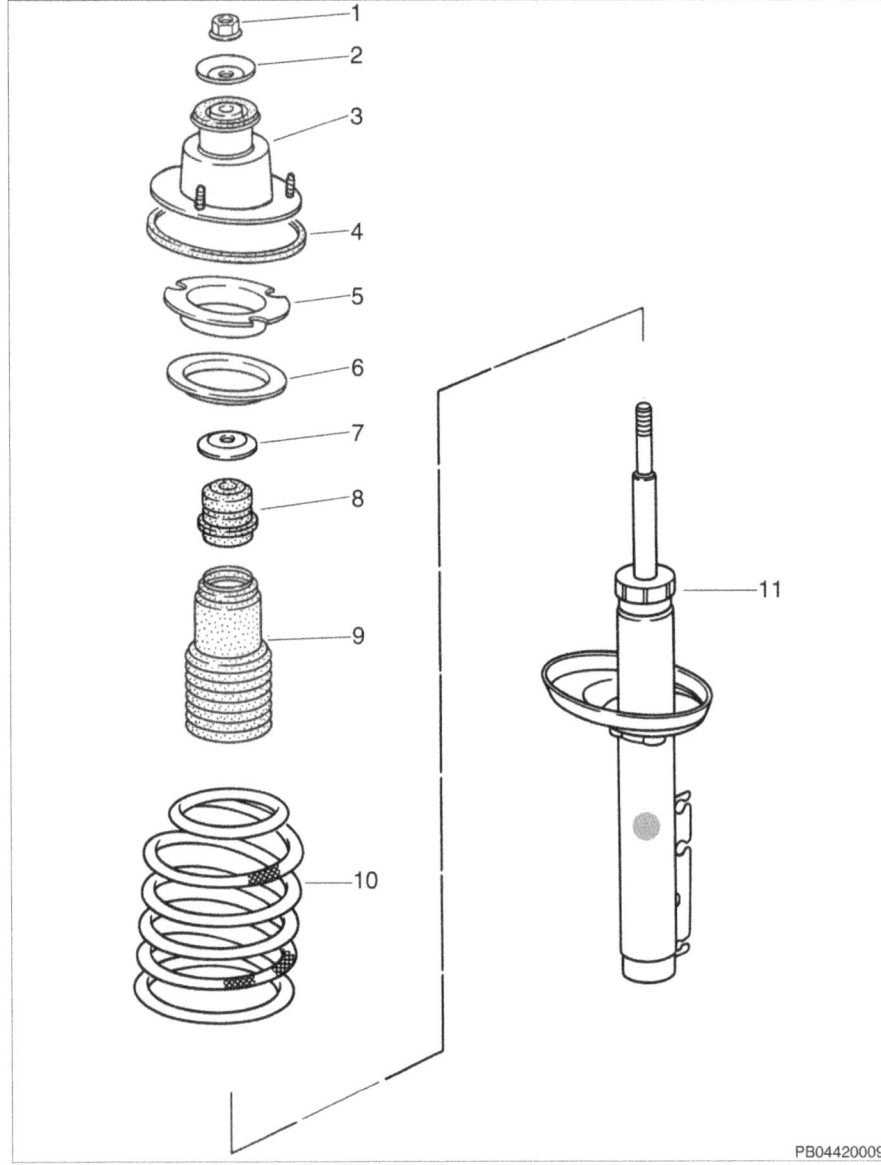

Rear strut assembly

1. **Fastening nut (M14 x 1.5)**
   - Tension coil spring before removing
   - Always replace
   - Tighten to 60 Nm (44 ft-lb)

2. **Stop plate**
   - Mount with dish facing upwards (numbers 2 and 7 are identical)

3. **Strut mount**
   - Remove with sealing ring (**4**)
   - Mounts for right and left sides are identical
   - Position mount in relation to spring before tightening fastening nut (**1**).

4. **Sealing ring**
   - Always replace

5. **Back-up ring**
   - Mount in correct position. Cutouts on backup ring should be aligned under stud heads of strut mount during reassembly.

6. **Spring plate**
   - Use correct part for spring used. See **Strut component details**.
   - Note correct position of back-up ring (**5**) before joining.

7. **Stop plate**
   - Mount with dish facing downwards (numbers 2 and 7 are identical)

8. **Bump stop**
   - Mount on dust shield (**9**).

9. **Dust shield**

10. **Coil spring**
    - For correct application see **Strut component details**.

11. **Strut**
    - For correct application see **Strut component details**.

Rear Suspension 42-7

Rear Struts and Sway Bar

Strut component details

Various strut/spring combinations are specified for different models and installed equipment. Consult the spring and strut application tables below when replacing components.

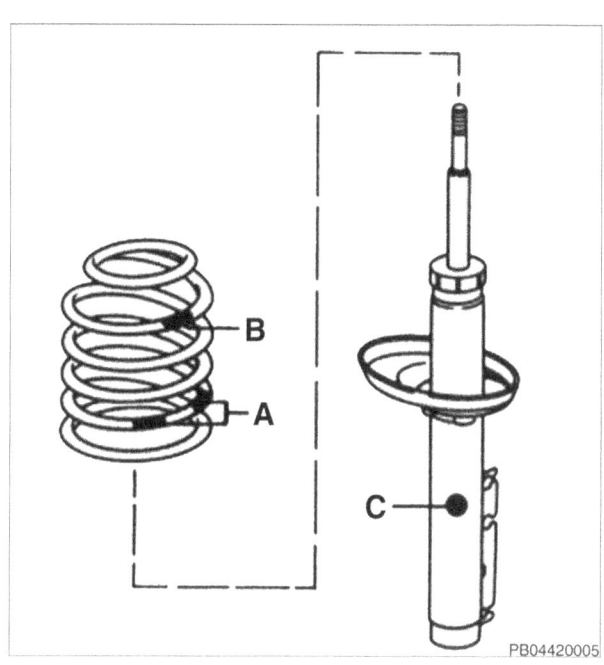

◄ **Identifying marks for suspension packages**

A indicates spring rates and lengths by line markings on the lower section of the spring. See **Table a.**

B indicates required coil plate thickness through color markings on upper section of coil spring (**B**).
White line = 3.0 mm spring plate
Green line = 6.5 mm spring plate

C indicates suspension package by color dot (**C**) on strut body.
Blue dot = standard suspension
Red dot = sport suspension

Table a. Rear coil spring applications		
Color	Year range	Model / options
blue / yellow	1997 - 2004	5-speed manual
blue / white	1997 - 2004	5-speed automatic
red / brown	1997 - 2002	5-speed manual / Sport
yellow / brown	2003 - 2004	5-speed manual / Sport
red / white	1997 - 2002	5-speed automatic / Sport
yellow / white	2003 - 2004	5-speed automatic / Sport
violet / yellow	2000 - 2004	S / 6-speed manual
violet / white	2000 - 2004	S / 5-speed automatic
orange / brown	2000 - 2004	S / 6-speed manual / Sport
orange / white	2000 - 2004	S / 5-speed automatic / Sport

Rear struts, disassembling and assembling

NOTE—
- *Always replace strut mount fastening nuts.*
- *Always replace coil springs in pairs.*
- *If coil springs are replaced, use of a different spring plate may be necessary.*

Disassembly

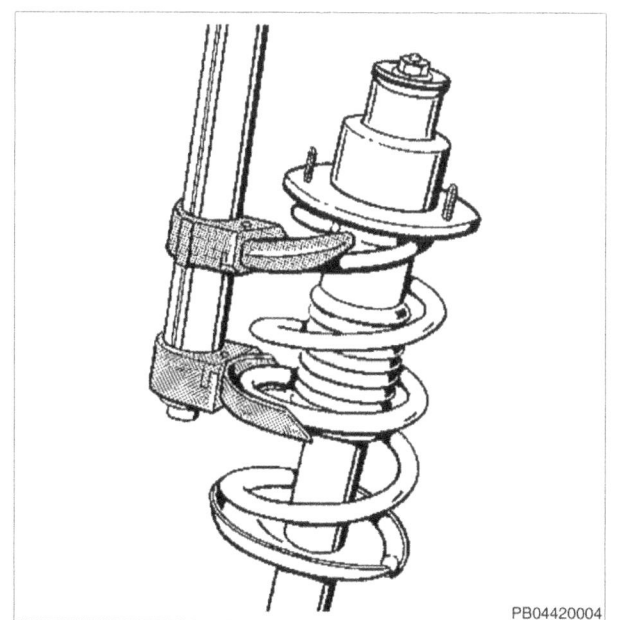

◄ To remove coil spring from strut assembly, compress coil spring using a coil spring compressor until strut mount can be turned freely.

– To remove strut mount, counter hold strut piston using 7mm hexagon socket key, and remove retaining nut (21 mm wrench).

> *CAUTION—*
> - *Never use an impact wrench to remove or tighten strut mount retaining hardware.*

– Remove all parts from strut. See **Rear strut assembly**.

42-8 Rear Suspension

Rear Struts and Sway Bar

Assembly

- Slide dust shield and bump stop into place on strut rod.
- Slide stop plate into correct position on strut rod.
- With coil spring tensioned, install spring so that spring end rests against stop in spring seat.
- Install spring plate into correct position on strut mount and fit to strut assembly. Screw new fastening nut onto strut rod, about 3 turns.
- Position strut mount so that downward angled fastening bolt aligns with plate on strut body.
- Counterholding strut rod with 7 mm hexagon socket key, tighten retaining nut.

Tightening torque	
Strut mount retaining nut	60 Nm (44 ft-lb)

CAUTION—
- *Never use an impact wrench to remove or tighten strut mount retaining hardware.*

Rear sway bar, removing and installing

- Raise car and support safely.

WARNING—
- *Make sure the car is stable and well supported at all times. Use a professional automotive lift or jack stands designed for the purpose. A floor jack is not adequate support.*

 Working beneath car, remove plastic under shield. **Arrows** indicate fasteners to be removed.

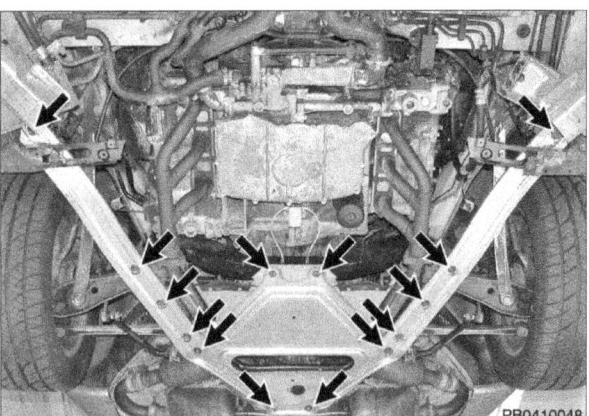

◀ Remove aluminum diagonal braces and chassis reinforcement plate (14 fasteners) (**arrows**).

Rear Suspension 42-9

Rear Control Arms

◄ Remove sway bar from chassis. Disconnect sway bar at end links, and remove mounting brackets at bushings (**arrows**).

– Install rear sway bar.

Tightening torques	
Sway bar bracket to body	23 Nm (17 ft-lb)
End link to sway bar	50 Nm (37 ft-lb)

– Install diagonal braces and reinforcement plate.

Tightening torques	
Reinforcement plate to rear axle support (M10 bolt) to support / suspension (M10 nut)	46 Nm (34 ft-lb) 65 Nm (48 ft-lb)
Diagonal brace to body / suspension	65 Nm (48 ft-lb)

REAR CONTROL ARMS

Track arm, removing and installing

NOTE—

* *In mid-1999, Porsche revised the track arm by changing the mount to a spherical shape and increasing the firmness of rubber bushings.*
* *Only revised track arms are available as replacement part. Revised track arms (part number 996.331.045.10) can be retrofitted to older models, but must only be done only in pairs.*

– Loosen rear wheel lug bolts. Raise car, and safely support on a professional automotive lift, or jack stands. Remove rear wheels.

WARNING—

* *Make sure the car is stable and well supported at all times. Use a professional automotive lift or jack stands designed for the purpose. A floor jack is not adequate support.*

– Counterhold track arm ball joint stud using a T-40 Torx screwdriver and remove nut.

◄ Press ball joint from wheel bearing carrier using a ball joint press tool (Porsche special tool 9560).

42-10 Rear Suspension

Rear Control Arms

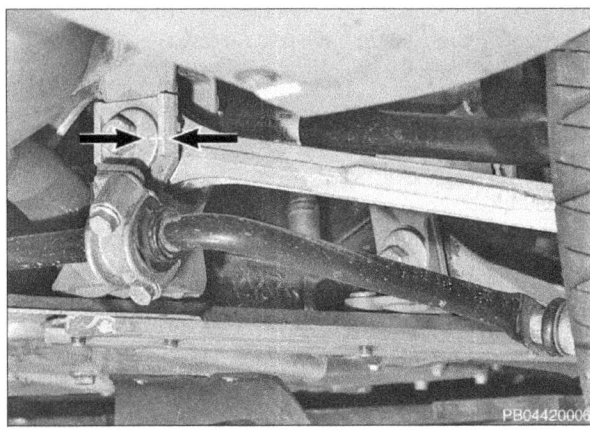

◂ Remove bolt and eccentric and pull out track arm. Before loosening track arm to frame bolt, mark position (**arrows**) of toe eccentric to maintain alignment.

– Installation is reverse of removal, noting the following:
  • Make sure that track arms are replaced in pairs only.
  • Replace self-locking fasteners.
  • Perform suspension alignment after installation. See **44 Wheels, Tires and Alignment**.

Tightening torques	
Track arm to • Wheel bearing carrier (M12 x 1.5) • Chassis (toe eccentric) (M12 x 1.5)	75 Nm (56 ft-lb) 100 Nm (75 ft-lb)
Wheel to wheel hub (M14 x 1.5)	130 Nm (96 ft-lb)

Control arm / trailing arm, removing and installing

– Loosen rear wheel lug bolts. Raise car, and safely support on a professional automotive lift, or jack stands. Remove rear wheels.

> **WARNING—**
> • Make sure the car is stable and well supported at all times. Use a professional automotive lift or jack stands designed for the purpose. A floor jack is not adequate support.

– On vehicles equipped with Xenon headlights (Litronic), remove height sensor strut rod from control arm.

– Remove ball joint nut from control arm, counterhold ball joint stud using a T-40 Torx screwdriver. Press ball joint from wheel bearing carrier using special tool 9560.

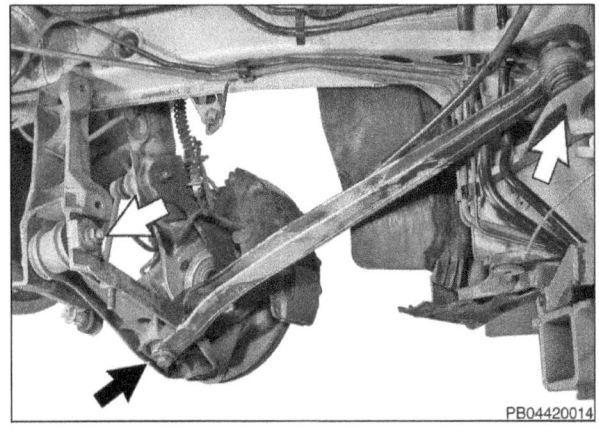

◂ Remove control arm / trailing arm retaining bolts (**arrows**), and remove control arm from chassis with trailing arm.

– Installation is reverse of removal, noting the following:
  • Replace fastening nuts.
  • Retaining bolt at control arm to body must be torqued to specification with vehicle on the ground and suspension loaded with weight of vehicle.
  • Perform suspension alignment after installation.

Tightening torques	
Control arm to • Body (M12 x 1.5) • Wheel bearing carrier (M12 x 1.5)	100 Nm (74 ft-lb) 75 Nm (56 ft-lb)
Trailing arm to • Control arm (M14 x 1.5) • Body (M14 x 1.5)	160 Nm (118 ft-lb) 160 Nm (118 ft-lb)
Wheel to wheel hub (M14 x 1.5)	130 Nm (96 ft-lb)

Rear Suspension 42-11

Rear Wheel Bearings and Drive Axles

REAR WHEEL BEARINGS AND DRIVE AXLES

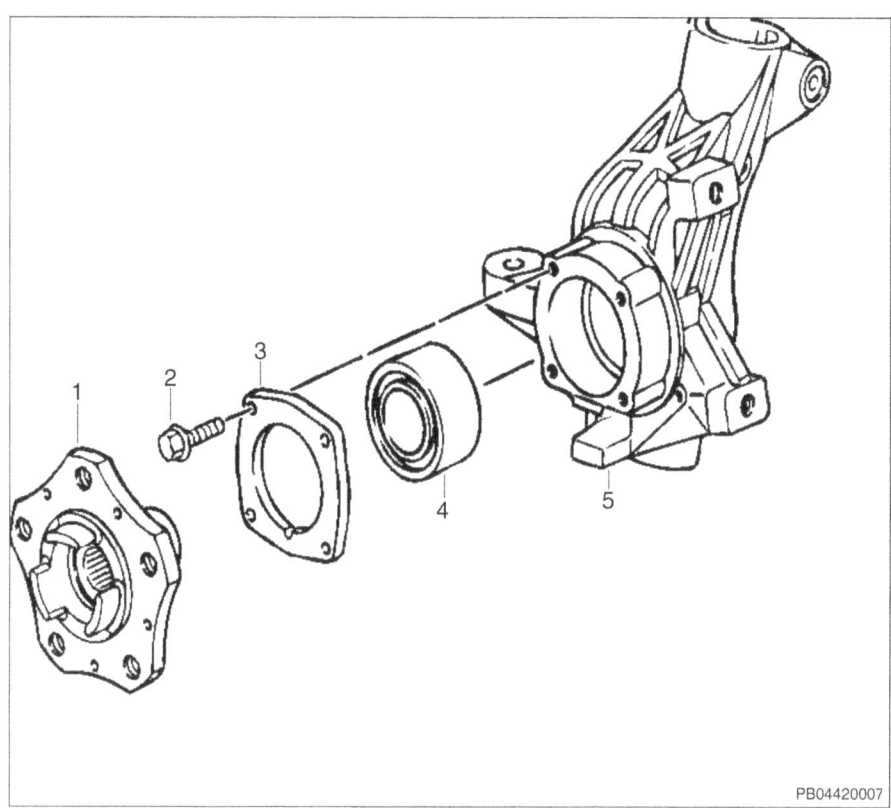

Rear wheel bearing carrier assembly

1. **Wheel hub**
   - Press out using hydraulic press and appropriate press pieces.
   - Install using wheel centering device and pressure piece VW 415A.

2. **Bolt (M8 x 35 10.9)**
   - Tighten to 37 Nm (27 ft-lb)

3. **Retainer plate**
   - Install with groove facing inward.

4. **Wheel bearing**
   - To remove and install, heat housing to 100°C (212°F), not exceeding 120°C (248°F).
   - Insert new bearing with bearing numbers facing wheel hub, using press tools 9247/4 and VW 415A.

5. **Wheel bearing carrier**

Rear wheel bearing, replacing

◄ With vehicle on ground, pry out wheel center cap from rear wheel. Use round holes in cap to carefully pry out.

◄ Using a breaker bar with a 32 mm socket, loosen drive axle hex nut.

– Raise car, and safely support on a professional automotive lift, or jack stands.

> **WARNING—**
> - Make sure the car is stable and well supported at all times. Use a professional automotive lift or jack stands designed for the purpose. A floor jack is not adequate support.

42-12 Rear Suspension

Rear Wheel Bearings and Drive Axles

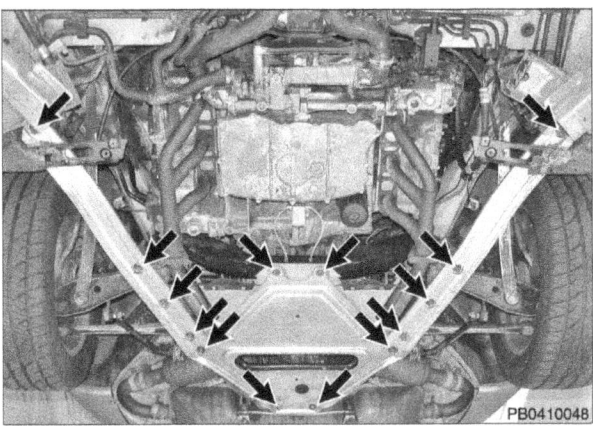

◀ Remove aluminum diagonal braces and chassis reinforcement plate (14 fasteners).

◀ Disconnect sway bar link at end of sway bar.

◀ Working at transmission drive flange, remove inner CV joint retaining bolts and disconnect inner CV joint from transmission.

• Cover and protect inner CV joint end from damage.

NOTE—

• *Leave transmission in gear and apply parking brake to make removal of inner CV joint retaining bolts easier.*

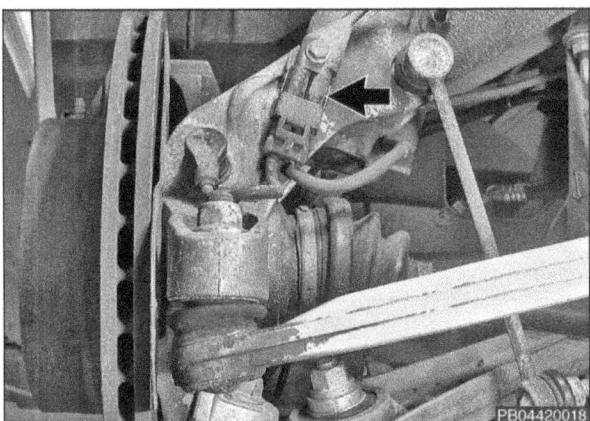

◀ Working at wheel bearing carrier, disconnect wheel speed /brake wear indicator electrical harness connector (**arrow**). Remove retaining bolt and remove connector from wheel bearing carrier.

Rear Suspension 42-13

Rear Wheel Bearings and Drive Axles

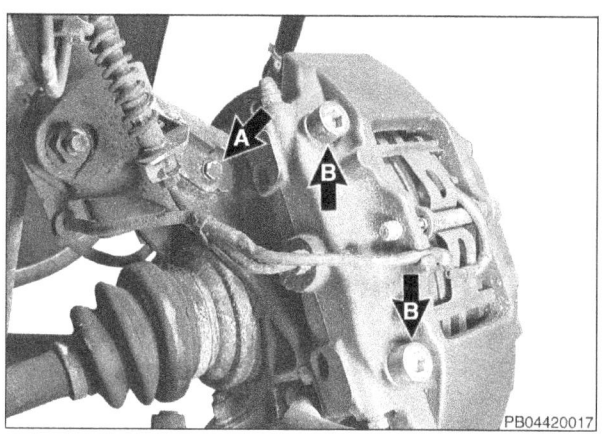

◀ Remove retaining bolt (**A**) and remove brake line retaining bracket at wheel bearing carrier. Remove brake caliper retaining bolts (**B**) from wheel bearing carrier and suspend brake caliper from chassis using mechanic's wire.

◀ Remove control arm from wheel bearing carrier using special tool 9560 (ball joint press). When loosening fastening nut, counterhold ball joint stud with T-40 Torx screwdriver.

NOTE—

- *To free ball joint from wheel bearing housing, pull down on control arms, while swinging wheel bearing carrier outward.*

— Remove counter-sunk screw and remove brake rotor. If necessary, tap rotor using a soft face hammer to release rotor from hub.

— Remove parking brake shoes and hardware. See **46 Brakes– Mechanical**.

— Loosen clamping bolt between wheel bearing carrier and strut assembly. Remove wheel bearing carrier from strut and remove bearing carrier with drive axle.

— Remove hex nut from drive axle and remove drive axle from wheel bearing carrier. If necessary, use a brass drift or suitable puller to remove axle shaft from wheel bearing carrier.

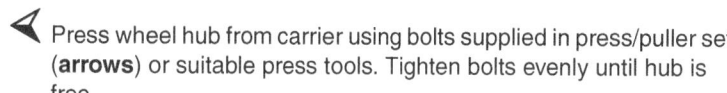

◀ Press wheel hub from carrier using bolts supplied in press/puller set (**arrows**) or suitable press tools. Tighten bolts evenly until hub is free.

NOTE—

- *Removing hub from wheel bearing carrier destroys wheel bearing.*

◀ Remove retainer plate and brake dust shield from wheel bearing carrier.

42-14 Rear Suspension

Rear Wheel Bearings and Drive Axles

◄ Using threaded press tool and appropriate sized press plates, press wheel bearing from carrier.

NOTE—
- *To aid in wheel bearing removal, aluminum wheel bearing carrier can be heated to 100°C (212°F). Do not exceed 120°C (248°F).*

− With wheel bearing removed, use threaded press tool and appropriate press plates to install new wheel bearing. During bearing installation, press on outer edge of wheel bearing only.
  - Heat bearing carrier to 100°C (212°F) if needed.
  - Install bearing with numbers facing toward wheel hub.

− Install brake dust shield and bearing retainer plate.

Tightening torque	
Brake dust shield to carrier	10 Nm (7.5 ft-lb)
Bearing retainer plate to bearing carrier	37 Nm (27 ft-lb)

◄ Remove outer bearing race from wheel hub using an appropriate puller.

NOTE—
- *If a suitable puller is not available, race can be heated and driven off using a brass drift.*

◄ Using appropriate sized press plates to support inner bearing race press wheel hub into wheel bearing.

WARNING—
- *Failure to properly support inner bearing race while installing wheel hub can lead to bearing damage.*

Rear Suspension 42-15

Rear Wheel Bearings and Drive Axles

– Installation of wheel bearing carrier is reverse of removal.

> **WARNING—**
> • Tighten drive axle nut only with car on ground.

Tightening torques	
Strut assembly to wheel bearing carrier (at sway bar link mount)(M12)	85 Nm (63 ft-lb)
Control arms to wheel bearing carrier (ball joints, M12)	75 Nm (56 ft-lb)
Brake disc to wheel hub (M6)	10 Nm (7.5 ft-lb)
Brake caliper to wheel bearing carrier (M12)	85 Nm (63 ft-lb)
Brake line bracket to wheel bearing carrier (M6)	10 Nm (7.5 ft-lb)
Combination wire bracket to wheel bearing carrier (M6)	10 Nm (7.5 ft-lb)
Inner CV joint to output flange • M8 • M10	39 Nm (29 ft-lb) 81 Nm (60 ft-lb)
Sway bar link to sway bar (M10 x 1.5)	50 Nm (37 ft-lb)
Reinforcement plate to rear axle support (M10 bolt) to support / suspension (M10 nut)	46 Nm (34 ft-lb) 65 Nm (48 ft-lb)
Diagonal brace to body / suspension	65 Nm (48 ft-lb)
Wheel to wheel hub (M14 x 1.5)	130 Nm (96 ft-lb)
Drive axle to wheel hub (use new nut)	460 Nm (340 ft-lb)

Drive axle, removing and installing

> **NOTE—**
> • A new drive axle was introduced for the 2003 model year. Up through 2002, drive axles used M8 bolts to secure the inner constant velocity joint. Beginning in 2003, a transitional style M8 bolt equipped with an M10 guide was used, allowing the later style joints (M10 bolt bores) to be bolted to earlier style drive flange (M8 bolt bores). Over the course of the 2003 model year, Porsche completed the transition to M10 bolts and drive flanges.

◀ With vehicle on ground, pry out wheel center cap from rear wheel. Use round holes in cap to carefully pry out.

42-16 Rear Suspension

Rear Wheel Bearings and Drive Axles

◀ Using a breaker bar with a 32 mm socket, loosen drive axle hex nut at rear wheel.

– Raise car, and safely support on a professional automotive lift, or jack stands.

WARNING—
- *Make sure the car is stable and well supported at all times. Use a professional automotive lift or jack stands designed for the purpose. A floor jack is not adequate support.*

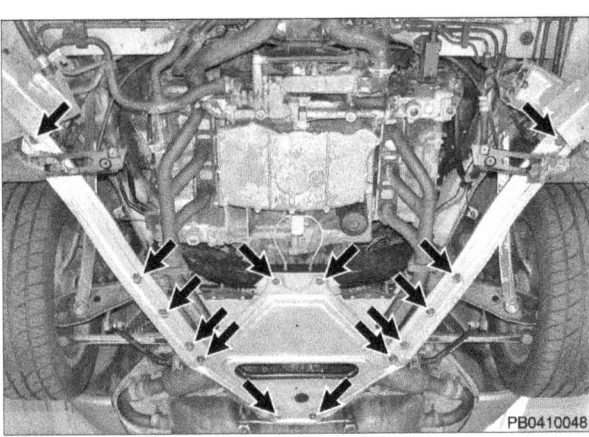

◀ Remove aluminum diagonal braces and chassis reinforcement plate (14 fasteners).

◀ Disconnect sway bar link at end of sway bar.

◀ Working at transmission drive flange, remove inner CV joint retaining bolts and disconnect inner CV joint from transmission.

NOTE—
- *Leave transmission in gear and apply parking brake to make removal of inner CV joint retaining bolts easier.*

Rear Suspension 42-17

Rear Wheel Bearings and Drive Axles

◂ Remove control arm and track arm from wheel bearing carrier using special tool 9560 (ball joint press). Counterhold ball joint stud with T-40 Torx screwdriver and loosen fastening nuts.

NOTE—

- *To free ball joint from wheel bearing housing, pull down on control arms, while swinging wheel bearing carrier outward.*

– With drive axle disengaged from wheel bearing carrier, remove drive axle from rear suspension.

– To install drive axle, fit to transmission flange and secure with two bolts. Lubricate splines of drive axle lightly with Optimoly HT grease. Swivel and pull wheel bearing carrier outwards. Bend outer CV joint and align with splines of wheel hub.

NOTE—

- *Use a helper when fitting the drive axle to the wheel bearing carrier.*

– Installation is reverse of removal, noting the following:
  - Tighten drive axle nut only with car on ground. Replace drive axle hex nut.
  - Inspect all parts before re-assembly. If drive axle CV boots are damaged, replace boots and repack CV joint grease before installing drive axle.

Tightening torques	
Control arms to wheel bearing carrier (ball joints, M12)	75 Nm (56 ft-lb)
Inner CV joint to output flange M8 M10	 39 Nm (29 ft-lb) 81 Nm (60 ft-lb)
Sway bar link to sway bar (M10 x 1.5)	50 Nm (37 ft-lb)
Sway bar to spring strut/wheel carrier (M12 x 1.5)	85 Nm (63 ft-lb)
Reinforcement plate to rear axle support (M10 bolt) to support / suspension (M10 nut)	 46 Nm (34 ft-lb) 65 Nm (48 ft-lb)
Diagonal brace to body / suspension	65 Nm (48 ft-lb)
Wheel to wheel hub (M14 x 1.5)	130 Nm (96 ft-lb)
Drive axle to wheel hub (use new nut)	460 Nm (340 ft-lb)
Reinforcement plate to rear axle support (M10 bolt) to support / suspension (M10 nut)	 46 Nm (34 ft-lb) 65 Nm (48 ft-lb)

44 Wheels, Tires and Alignment

WHEELS AND TIRES 44-1	
Approved wheel and tire combinations 44-1	
WHEEL ALIGNMENT 44-2	
Alignment adjustments 44-2	
Vehicle ride height. 44-3	

TABLES

a. Alignment specifications 44-3
b. Ride height 44-4

WHEELS AND TIRES

Approved wheel and tire combinations

Porsche has approved the following wheel and tire size combinations for Boxsters delivered from the factory. Porsche alloy wheels have been available in various styles.

Wheel and Tire sizes
16 Inch wheels Front 6 x 16 (ET 50) 205/55 ZR 16 Rear 7 x 16 (ET 40) 225/50 ZR 16
17 Inch wheels Front 7 x 17 (ET 55) 205/50 ZR 17 Rear 8.5 x 17 (ET 48/50) 255/40 ZR 17
18 Inch wheels Front 7.5 x 18 (ET 50) 225/40 ZR 18 Rear 9 x 18 (ET 52) 265/35 ZR 18 Front 9 x 18 (ET 52) 225/40 ZR 18 Rear 10 x 18 (ET 65) 265/35 ZR 18

NOTE—

- *Always replace tires in pairs (front and rear axles).*
- *Tires with high speed ratings (ZR, etc.) should be replaced after 6 years, regardless of tread condition.*
- *When balancing wheels, use adhesive weights, mounted to the inside of the wheel rim.*
- *Maximum radial runout for a bare rim is 0.7 mm.*
- *Maximum radial runout for a wheel and rim is 1.25 mm.*
- *Apply a light coating of Optimoly TA or equivalent aluminum paste anti-seize compound to lug bolt threads before installing.*
- *Lug bolts with damaged threads or heads should be replaced.*
- *Start lug bolts by hand and tighten evenly, observing specified pattern.*
- *18 inch wheels became available as an option for model year 1998, and standard equipment on Boxster S models.*

44-2 Wheels, Tires and Alignment

Wheel Alignment

Beginning in 1998, the following body components were modified to strengthen the rear of vehicles fitted with 18 inch wheels:

- Wheel well with spring mount.
- Lower engine compartment bulkhead.
- Rear-wall cross-member.
- Reinforcements for rear axle mount.

> *CAUTION—*
> - *Do not install 18 inch wheels on vehicles that have not received the factory reinforcements.*
> - *Lug bolts consist of two parts permanently joined together.*
> - *Never use an impact wrench to tighten bolts.*
> - *Only Porsche original equipment lug bolts should be used.*

Tightening torque	
Wheel to wheel hub	130 Nm (96 ft-lb)

WHEEL ALIGNMENT

Alignment adjustments

NOTE—
- *Alignment is measured with full tank of fuel, spare tire and emergency tools loaded, without driver.*

Front axle

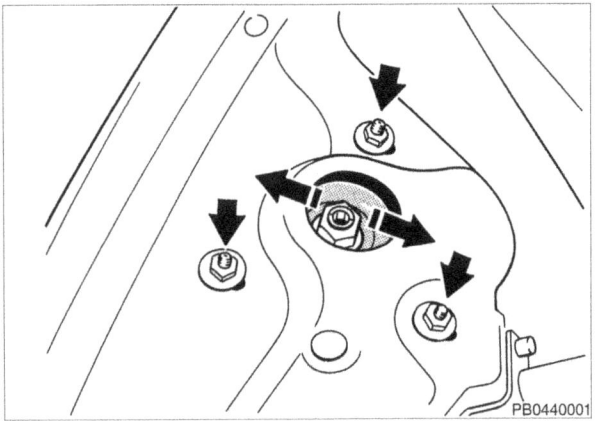

 Camber is adjusted by locating the front strut laterally.

- Toe is adjustable at the tie rods. After toe is adjusted, tighten tie rod lock nuts to 50 Nm (37 ft-lb).
- Caster is not an adjustable setting. Replacement transverse control arms with caster adjustment eccentrics are available from Porsche.

Rear axle

 Rear alignment is adjustable via eccentrics. Toe (**A**) and Camber (**B**).

Tightening torques	
Front strut mount to body	33 Nm (24 ft-lb)
Tie rod lock nuts	50 Nm (37 ft-lb)
Rear toe arm eccentric (**A**)	75 Nm (56 ft-lb)
Rear lower control arm eccentric (**B**)	100 Nm (74 ft-lb)

Wheels, Tires and Alignment 44-3

Wheel Alignment

Table a. Alignment specifications			
	USA (including Sport)	Rest of World (ROW)	ROW Sport
Front axle			
Toe unpressed (total)	+5' ± 5'	+5' ±5'	+5' ±5'
Toe difference angle (at 20° lock)	- 1° 20'± 30'	- 1° 50'± 30'	- 2° 20'± 30'
Camber with wheels forward max difference left to right	+5' ± 30' 20'	-10' ± 5' 20'	-15' ± 5' 20'
Caster max difference left to right	+5' ± 5' 20'	+5' ±5' 20'	+5' ±5' 20'
Rear axle			
Toe per wheel max difference left to right	+5' ± 5' 10'	+5' ±5' 10'	+5' ±5' 10'
Camber max difference left to right	- 1° 20'± 30' 20'	- 1° 20'± 30' 20'	- 1° 30'± 30' 20'

Vehicle ride height

Unlike previous generations of Porsche vehicles that used torsion bar suspensions, the ride height for a Boxster is not adjustable. Ride height can only be altered or changed by replacing coil springs and strut spring plates.

Coil spring height and rates are varied based upon differences in vehicle weight due to optional equipment (such as transmissions or an optional "Sport" package). Ride height soecifications are in **Table b**.

NOTE—
* *Vehicle ride height is measured with full tank of fuel, spare tire and emergency tools loaded, without driver.*

 Front axle ride height is measured from road surface to lower bolt head edge connecting tension strut to body.

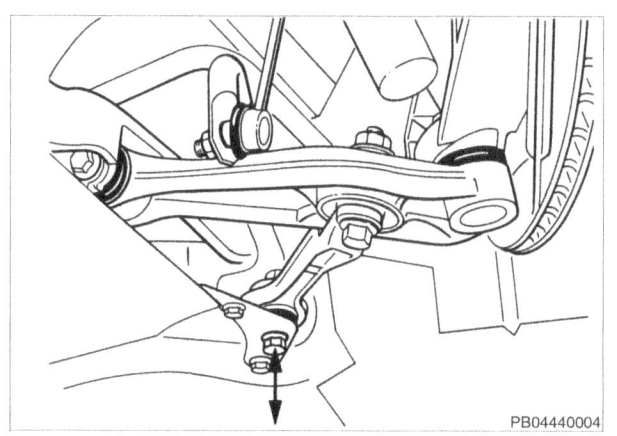

44-4 Wheels, Tires and Alignment

Wheel Alignment

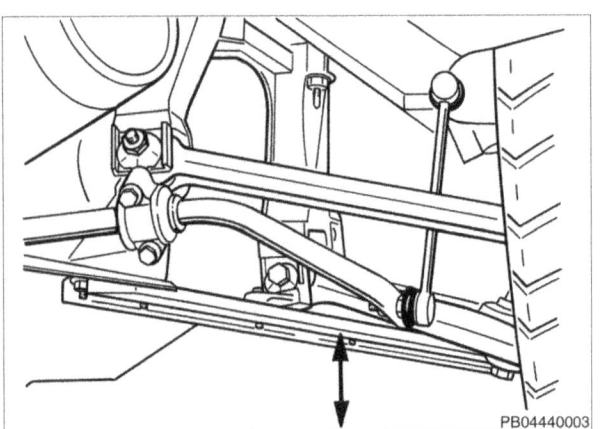

◄ Rear axle ride height is measured from road surface to lower edge of diagonal brace at control arm mount.

Table b. Ride height			
	USA (including Sport)	Rest of World (ROW)	ROW Sport
Front axle	152 mm ± 10	142 mm ± 10	132 mm ± 10
Rear axle	146 mm ± 10	146 mm ± 10	136 mm ± 10

45 Antilock Brakes (ABS)

GENERAL 45-1
  ABS system inspection 45-1
  Warnings and Cautions 45-1
ABS SYSTEMS 45-2
  System identification 45-2
  ABS 5.3 (3-channel system) 45-3
  ABS/TC 5.3 (4-channel system) 45-4

Porsche Stability Management
  (PSM, 4-channel system) 45-5
  Component descriptions 45-6
COMPONENT REPLACEMENT 45-7
  Wheel speed sensor, removing and installing ... 45-8
  Hydraulic unit / control module,
    removing and installing 45-8
  Precharge pump, removing and installing 45-10
  Rotational rate sensor, removing and installing .. 45-11

GENERAL

This repair group covers identification and replacement of components of the Antilock Brake (ABS) system.

See also the following repair groups:
- **46 Brakes—Mechanical**
- **47 Brakes—Hydraulic**

ABS system inspection

A visual inspection of the ABS system may help to locate system faults. If no visual faults can be found and the ABS light remains ON, have the system diagnosed by an authorized Porsche repair facility.

Carefully inspect the entire ABS wiring harness, particularly the pulse sensor harnesses and connectors near each wheel. Look for chafing or damage due to incorrectly routed wires.

Carefully remove the wheel speed sensors. See **Wheel speed sensor, removing and installing** in this repair group. Clean the sensor tips. Inspect the toothed pulse wheel at the wheel hubs. Check for missing, clogged or corroded teeth, or other damage that could alter the signal between the sensor tip and the pulse wheel.

Warnings and Cautions

> *WARNING—*
> - *Brake fluid is poisonous, corrosive and dangerous to the environment. Wear safety glasses and rubber gloves when working with brake fluid. Do not siphon brake fluid with your mouth. Dispose of brake fluid properly.*
> - *Do not reuse self-locking nuts, bolts or fasteners. They are designed to be used only once and may fail if reused. Replace with new self-locking fasteners.*

45-2 Antilock Brakes (ABS)

CAUTION—
- *Models covered by this manual require Porsche System Tester 2 to properly bleed the brake hydraulic system. See **47 Brakes–Hydraulic** for more information.*
- *Immediately clean brake fluid spilled on painted surfaces and wash with water, as brake fluid removes paint.*
- *Use new brake fluid from an unopened container. Brake fluid absorbs moisture from the air. This can lead to corrosion problems in the braking system, and also lowers the fluid boiling point.*
- *Plug open lines and brake fluid ports to prevent contamination.*
- *If carrying out electric welding work, be sure to disconnect electrical harness connector from electronic control module.*
- *Do not expose electronic control modules to high sustained heat, such as in a paint drying booth. Maximum heat exposure: 95°C (203°F) for short periods of time, 85°C (185°F) for long periods (approx. 2 hours)*

ABS SYSTEMS

The ABS system is designed to be maintenance free. There are no adjustments that can be made to the system. Repair and troubleshooting of the major ABS system components requires special test equipment and knowledge and should be done by an authorized Porsche dealer service department.

ABS is self-tested by the ABS control module each time the car is started. Once the test is complete, the ABS warning light turns OFF. If the light remains ON or comes ON at any time during driving, a system fault has occurred and the ABS system is electronically disabled. The conventional braking system remains fully functional.

System identification

Three ABS systems were available in the Porsche Boxster:
- ABS 5.3 is standard on Boxster models without traction control.
- ABS/TC 5.3 with Traction Control was optional from model years 1997 - 2000.
- Porsche Stability Management (PSM) was optional for model years 2001 - 2004. PSM added both traction control and dynamic stability control to the basic ABS system.

Models equipped with ABS/TC 5.3 and PSM can be identified by a system control switch (**arrow**) mounted on the upper left of the center console, above the radio.

Antilock Brakes (ABS) 45-3

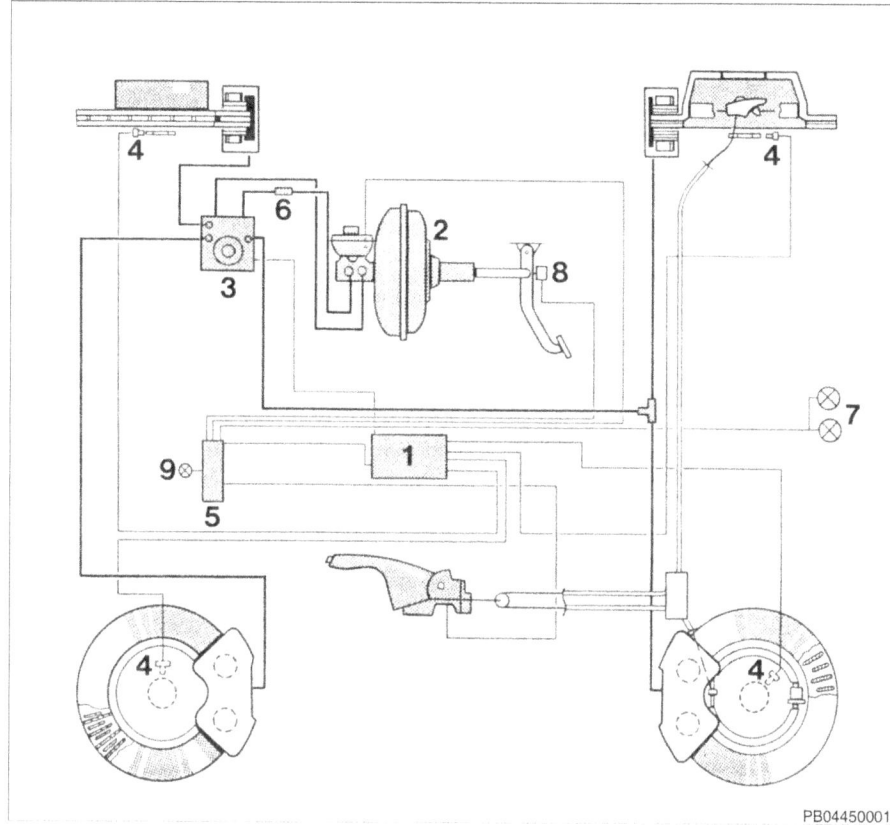

ABS 5.3
(3-channel system)

1. **ABS control unit**
   - Integrated with ABS hydraulic unit (3)
2. **Brake master cylinder with booster**
3. **ABS hydraulic unit**
   - Integrated with ABS control unit (1)
4. **ABS wheel speed sensors**
5. **Central information system**
6. **Brake proportioning valve**
7. **Brake light**
8. **Brake light switch**
9. **ABS warning light**

Electronically controlled ABS maintains vehicle stability and control during emergency braking by preventing wheel lock-up. ABS provides optimum deceleration and stability during adverse conditions by automatically adjusting brake system hydraulic pressure at each wheel to prevent wheel lock-up.

The system's main components are:
- Wheel speed (pulse) sensors
- ABS electronic control module
- Hydraulic unit

Wheel speed sensors continuously send wheel speed signals to the electronic control module. The control module compares these signals to determine, in fractions of a second, whether any of the wheels are about to lock. If any wheel is nearing a lock-up condition, the module signals the hydraulic unit to maintain or reduce pressure at the appropriate wheel(s). Pressure is modulated by electrically-operated solenoid valves in the hydraulic unit.

45-4 Antilock Brakes (ABS)

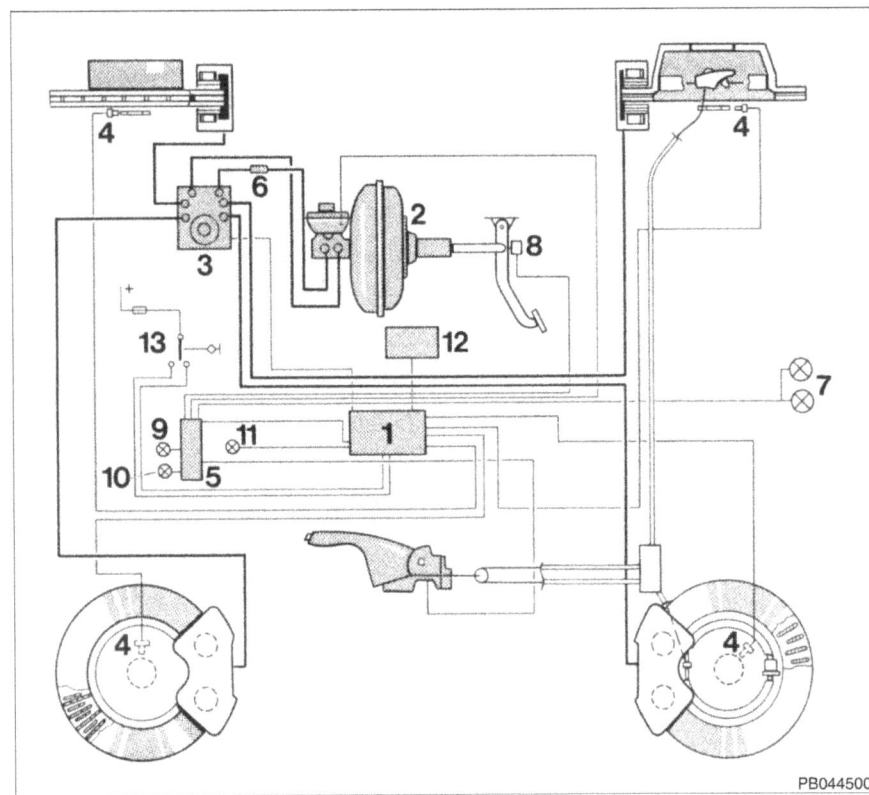

ABS/TC 5.3 (4-channel system)

1. **ABS/TC control module**
   - Integrated with ABS / TC hydraulic unit (3)
2. **Brake master cylinder with booster**
3. **ABS/TC hydraulic unit**
   - Integrated with ABS / TC control module (1)
4. **ABS wheel speed sensors**
5. **Central information system**
6. **Brake proportioning valve**
7. **Brake light**
8. **Brake light switch**
9. **ABS warning light**
10. **TC warning light (yellow)**
11. **TC function light (green)**
12. **DME control module**
13. **TC switch**

Available as an option through the 2000 model year, ABS/TC 5.3 is an extension of the basic Porsche ABS system with the added feature of traction control. The traction control system is made up of two sub-systems, Automatic Brake Differential (ABD) and Anti-Slip Regulation (ASR).

Automatic Brake Differential (ABD) functions similar to a locking or limited slip differential to help the vehicle retain traction. Using the ABS wheel speed sensors to detect wheel spin during acceleration, the ABS control module will pulse the hydraulic pressure line of the brake caliper of the spinning wheel to assist the vehicle in gaining traction.

If ABD intervention is not enough to regain traction, the ABS/TC system will employ Anti-Slip Regulation (ASR). ASR adjusts fuel injector cycles and retards ignition timing to reduce engine torque. When the Traction Control system is operational, the indicator lamp on instrument panel is illuminated.

The traction control function can be disabled by means of the "TC Off" switch located on the center console. If the TC system is switched off, the TC warning lamps in the instrument panel and the "TC Off" switch are illuminated.

NOTE—
- ABD remains active until vehicle speed reaches 62.5 miles per hour, even if the TC system is turned off.

Antilock Brakes (ABS) 45-5

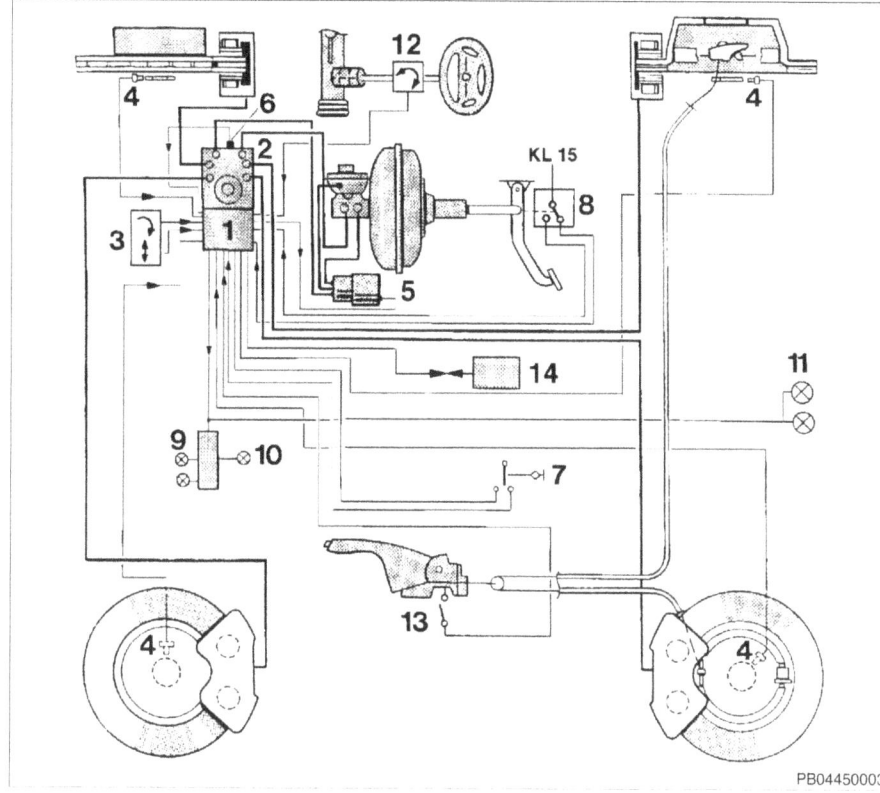

Porsche Stability Management (PSM, 4-channel system)

1. **PSM control module with relay**
   - Integrated with PSM hydraulic unit (2)
2. **Hydraulic unit**
   - Integrated with PSM control module
3. **Rotational rate sensor**
   - Measures lateral acceleration and yaw
4. **Engine speed sensor**
5. **Precharge pump**
6. **Pressure sensor**
7. **PSM switch**
8. **Brake light switch**
9. **Warning light**
10. **Information light**
11. **Brake light**
12. **Steering angle sensor**
13. **Hand brake switch**
14. **DME control module**

Available as an option beginning in model year 2001, PSM combines the functions of ABS/TC 5.3 and adds increased functions such as Electronic Brake force Distribution (EBD), Engine Drag Control (EDC) and Driving Dynamics Control (DDC).

PSM continually monitors driver inputs like throttle angle, steering angle, and brake pressure, as well as vehicle inputs such as engine speed, individual wheel speed and chassis rotational rate (pitch and yaw). Using these inputs, PSM looks for deviations in wheel speed as well as the driver's intended direction and speed. Should PSM detect deviations, it will activate the PSM system in order to maintain control of the vehicle.

PSM will activate brakes individually, balance braking force, and reduce vehicle torque in order to maintain safe vehicle control, all within a fraction of a second. PSM not only has the ability to close the vehicle throttle like ABS/TC, but it can also increase the throttle opening to apply more engine torque to maintain vehicle control.

PSM can be disabled by means of the "PSM Off" switch located on the center console. If the PSM system is switched off, the PSM warning lamps in the instrument panel and the "PSM Off" switch are illuminated.

NOTE—

- ABD system function remains active until vehicle speed reaches 62.5 miles per hour even if PSM system is turned off.
- Vehicle is still stabilized during braking even if PSM system is turned off.

45-6 Antilock Brakes (ABS)

Component descriptions

Brake light switch

A redundant switch sends signals to ABS system control module. If the ABS/TC or PSM control module receives a signal from this switch (brakes applied), the ASR function will be deactivated.

Brake pressure sensor

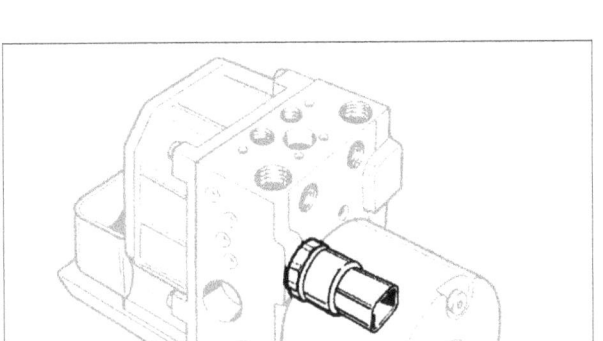

◀ The brake pressure sensor measures applied braking force in the ABS hydraulic unit. Braking force is used to calculate distributed wheel force and cornering stability force.

Parking brake switch

Application of the parking brake turns Electronic Drag Control (EDC) OFF.

Hydraulic unit / control module

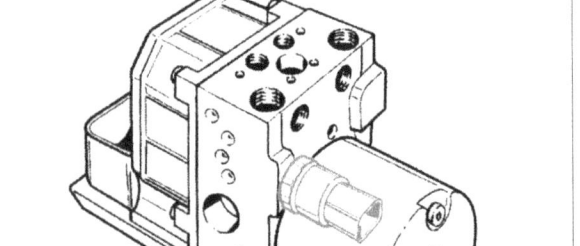

◀ Made up of the control unit, pump, solenoid valve relay, high speed electromagnetic valves and a pump unit, the hydraulic unit works to "pulse" braking force in ABS functions, as well as providing pressure for supplemental traction and stability functions.

Precharge pump (PSM only)

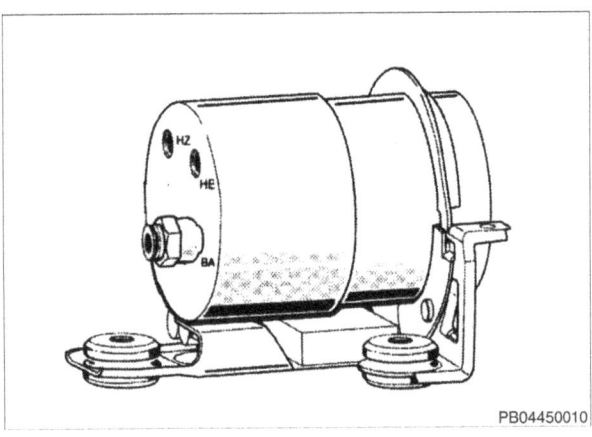

◀ PSM requires increased hydraulic system pressure and flow. Additional pressure is provided by a precharge pump added to the brake hydraulic system of PSM equipped vehicles.

Antilock Brakes (ABS) 45-7

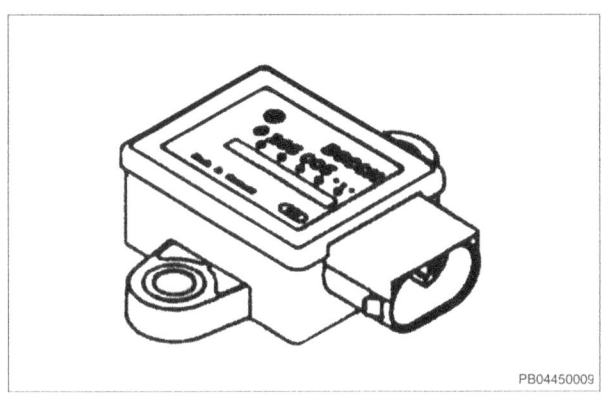

Rotational rate sensor (PSM only)

◄ The rotational rate sensor measures both lateral acceleration and rotational acceleration (yaw). This sensor is located under the center console.

Steering angle sensor (PSM only)

◄ Mounted on the steering column, the steering angle sensor measures the vehicle's intended direction of travel.

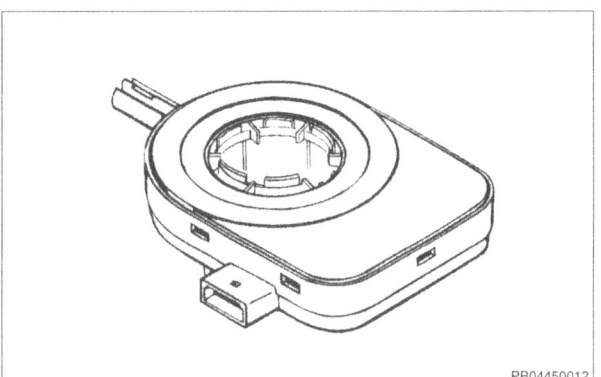

System switch (ABS/TC and PSM only)

Traction and dynamic drive control systems can be turned off via the center console mounted switch. While traction and stability functions will be turned off, basic ABS and ABD (Automatic Braking Differential) functions will be maintained.

Wheel speed sensors

◄ Located at the front and rear wheel bearing carriers, these inductive sensors read wheel speed via a pulse wheel mounted to the bearing hub.

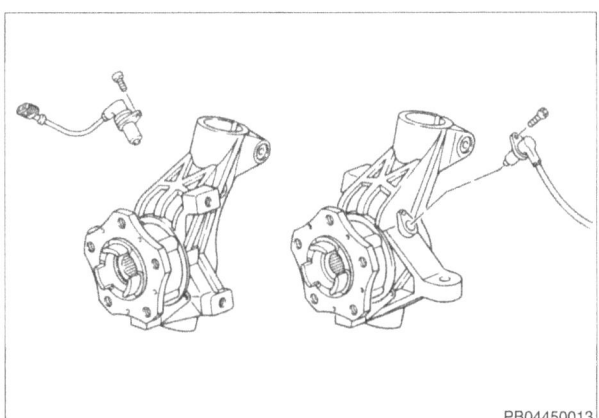

COMPONENT REPLACEMENT

Replacement of many components of the ABS and traction control systems require the use of Porsche System Tester 2 to clear system fault codes once the component has been replaced.

Replacement of the steering angle sensor requires coding and calibration using Porsche System Tester 2, and is beyond the scope of this manual. See an authorized Porsche repair facility for sensor replacement.

NOTE—

- *2003 - 2004: Steering angle sensor was redesigned. New sensor can be retrofit to earlier Boxster. Contact an authorized Porsche parts retailer for more information.*

45-8 Antilock Brakes (ABS)

Wheel speed sensor, removing and installing

CAUTION—
- *This procedure requires use of the Porsche System Tester 2 (PST2). Read the procedure through before beginning the job to determine the scope of the repair.*

NOTE—
- *Front and rear wheel speed sensors are interchangeable.*

- Loosen appropriate wheel lug bolts. Raise car and safely support on a professional automotive lift, or jack stands. Remove wheel.

WARNING—
- *Make sure the car is stable and well supported at all times. Use a professional automotive lift or jack stands designed for the purpose. A floor jack is not adequate support.*

 Open electrical harness connector (**A**) at strut assembly and remove connector for wheel speed sensor.
- Removing retaining bolt (**B**), and remove sensor from wheel bearing carrier.

- Before re-installation, note the following:
  - Check for chipping or contact at magnetic tip of sensor.
  - Coat sensor shaft and sensor bore with Molykote Longterm 2 or alternate lightweight grease.
  - If sensor was replaced due to ABS or traction control faults or accident damage, check system function using Porsche System Tester 2.

NOTE—
- *Do not allow grease on sensor tip.*

Tightening torque	
Wheel speed sensor to wheel bearing carrier	10Nm (7.5 ft-lb)

Hydraulic unit / control module, removing and installing

ABS and ABS/TC 5.3

CAUTION—
- *This procedure requires the Porsche System Tester 2. Read the procedure through before beginning the job to determine the scope of the repair.*

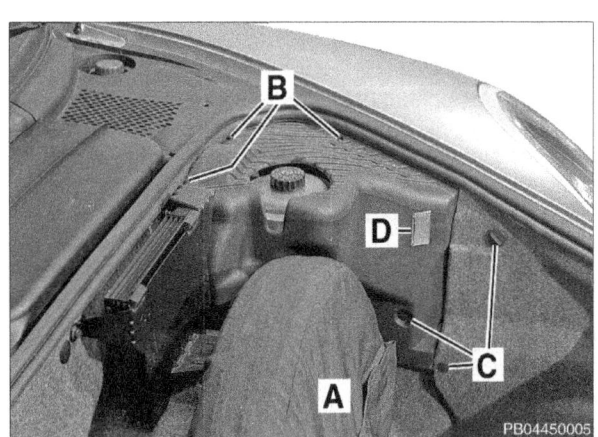

 Remove hydraulic unit / control module cover:
- Remove spare tire and tool kit (**A**) from front trunk.
- Remove screws (**B**), retaining hardware (**C**) and remove cover for hydraulic unit.
- With cover free, disconnect electrical harness connectors at trunk light (**D**).

Antilock Brakes (ABS)   45-9

◀ Drain brake fluid reservoir. Evacuate fluid using a fluid pump or baster.

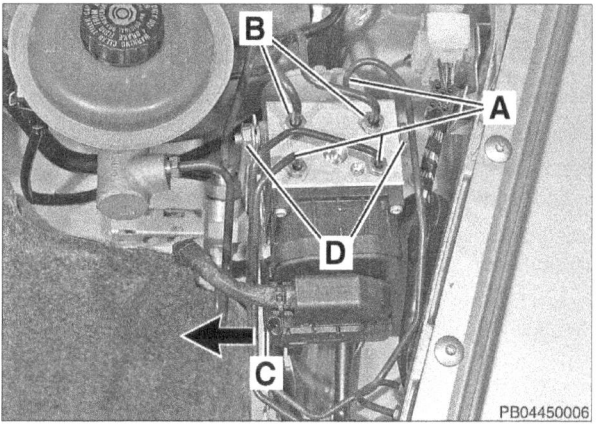

◀ To remove hydraulic unit / control module:
- Mark location of hydraulic lines in relation to hydraulic unit.
- Disconnect brake fluid distribution lines (**A**).
- Disconnect brake fluid supply lines (**B**).
- Slide electrical harness connector lock (**C**) out in direction of arrow, and remove connector from control module.
- Remove fastening nuts (**D**) and lift hydraulic unit / control module from bracket.

NOTE—
- *Seal hydraulic lines with suitable plugs.*

– Installation is reverse of removal, noting the following:
- Check brake lines and connections for leaks.
- Bleed brakes. See **47 Brakes—Hydraulic**.
- Using Porsche System Tester 2:
  Code control module for transmission type.
  Read and clear fault memory of ABS control module.
  Check electrical and hydraulic lines (brake lines) for incorrect assignment.
- Start engine. ABS-related lights in instrument cluster must go out when the engine is running.
- Road test brakes, performing at least one ABS control operation. Read ABS fault memory again. There must not be any stored faults.

PSM

CAUTION—
- *This procedure requires the Porsche System Tester 2. Read the procedure through before beginning the job to determine the scope of the repair.*

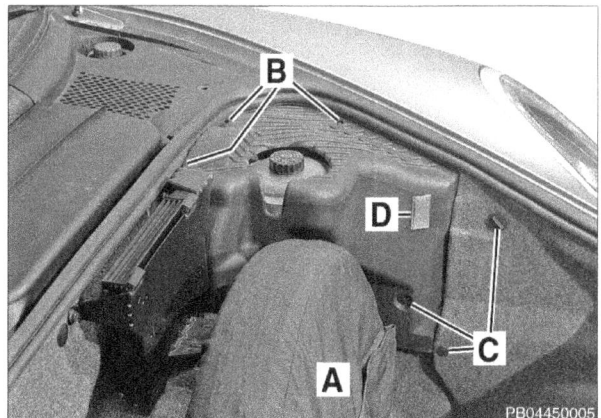

◀ Remove spare tire and tool kit (**A**) from front trunk. Remove screws (**B**), retaining hardware (**C**) and remove cover for hydraulic unit. With cover free, disconnect electrical harness connectors at trunk light (**D**).

45-10 Antilock Brakes (ABS)

◀ Drain brake fluid reservoir. Evacuate fluid using a fluid pump, or pump brake pedal to create pressure and "bleed" fluid through front and rear brake calipers.

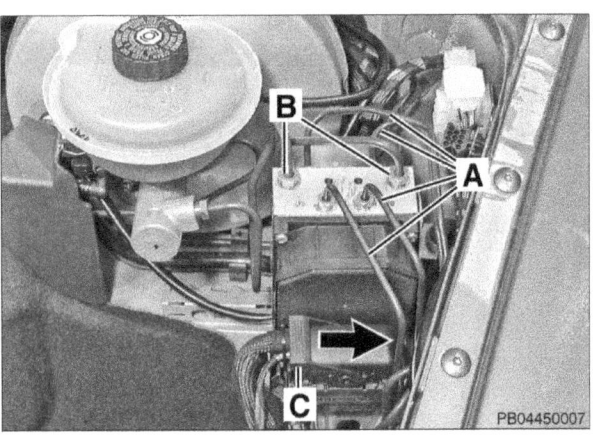

◀ To remove hydraulic unit / control module:
  • Mark location of hydraulic lines in relation to hydraulic unit.
  • Disconnect brake fluid distribution lines (**A**).
  • Disconnect brake fluid supply lines (**B**).
  • Lift electrical harness connector lock (**C**) out in direction of arrow, and remove connector from control module.
  • Disconnect electrical harness connector at pressure sensor at rear of hydraulic unit / control module.
  • Remove fastening nuts and lift hydraulic unit / control module from bracket.

NOTE—
• *Seal hydraulic lines with suitable plugs. Brake fluid on electrical contacts leads to corrosion and may cause faults in the system.*

− Installation is reverse of removal, noting the following:
  • Check brake lines and connections for leaks.
  • Bleed brakes. See **47 Brakes—Hydraulic**.
  • Using Porsche System Tester 2:
    - Code control module for transmission type.
    - Read and clear fault memory of PSM control module.
    - Check assignment of electrical and hydraulic lines.
  • Start the engine. PSM-related lights in instrument cluster must go out when engine is running.
  • Road test brakes, performing at least one ABS control operation. Read PSM fault memory again. There must not be any stored faults.

Precharge pump, removing and installing

> *CAUTION—*
> • *This procedure requires the Porsche System Tester 2. Please read the procedure through before beginning the job to determine the scope of the repair.*

The precharge pump is located on the opposite side of the front trunk from the brake master cylinder.

− Remove spare tire and tool kit from front trunk.

Antilock Brakes (ABS) 45-11

◀ Remove fastener (**arrow**) and peel back luggage compartment lining in the area around the pump.

- To remove precharge pump:
  - Disconnect hydraulic lines at rear of precharge pump. Plug lines and pump connections with suitable plugs.
  - Disengage electrical harness connector from side of pump.
  - Loosen retaining hardware at pump and remove pump from bracket.

> **WARNING—**
> - Fluid is expelled under pressure. Wrap a cloth around hydraulic line to absorb leaking fluid.

- Installation is reverse of removal, noting the following:
  - Check brake lines and connections for leaks.
  - Bleed brakes. See **47 Brakes—Hydraulic**.
  - If replacing hydraulic unit / control module in vehicles with traction control, code control module for transmission variation using Porsche System Tester 2, or have control module coded by a qualified Porsche repair facility.
  - Read out and erase PSM control module fault memory with Porsche System Tester 2 (PST 2). Additionally perform static test in PSM system.
  - Check electrical and hydraulic lines (brake lines) for incorrect assignment using the PST 2.
  - Start engine. PSM-related lights in instrument cluster must go out when engine is running.
  - Check brake function. Then perform a short test drive and perform at least one ABS control operation, taking road conditions into consideration. Read out PSM fault memory again.

Rotational rate sensor, removing and installing

> **CAUTION—**
> - This procedure requires use of the Porsche System Tester 2. Read the procedure through before beginning the job to determine the scope of the repair.
> - The rotational rate sensor is very sensitive to impact.

- Turn ignition OFF.

- Remove seats. See **72 Seats**.

- Remove center console. See **70 Trim—Interior**.

- Working under removed center console, remove rotational rate sensor retaining nuts. Remove sensor and disconnect electrical harness connector.

- Installation is reverse of removal, noting the following:
  - After installing sensor, clear PSM fault memory with PST2.
  - Start engine, confirm PSM warning lights are out with engine running. Read out PSM fault memory again.

46 Brakes–Mechanical

GENERAL 46-1	**PARKING BRAKE**........................ 46-5
Warnings and Cautions.................... 46-2	Parking brake components 46-5
BRAKE PADS AND ROTORS 46-2	Parking brake shoes, adjusting.............. 46-5
Brake pads, replacing 46-2	Parking brake shoes, replacing 46-6
Brake rotors, removing and installing 46-3	

GENERAL

This repair group covers replacement of brakes pads, brake rotors, and the parking brake assembly.

Also see the following repair groups:
- 45 Antilock Brakes (ABS)
- 47 Brakes—Hydraulic

Warnings and Cautions

> **WARNING—**
> - Brake friction materials such as brake linings or pads may contain asbestos fibers which can lead to illnesses. Do not create dust by grinding, sanding, or cleaning the pads with compressed air. Avoid breathing any asbestos fibers or dust.
> - Brake fluid is poisonous, highly corrosive and dangerous to the environment. Wear safety glasses and rubber gloves when working with brake fluid. Do not siphon brake fluid with your mouth. Immediately clean away any fluid spilled on painted surfaces and wash with water. Dispose of brake fluid properly.
> - Do not reuse self-locking nuts, or fasteners. They are designed to be used only once and may fail if reused. Always replace them with new self-locking nuts.

> **CAUTION—**
> - Always machine or replace brake rotors in pairs.
> - Always replace brake pads in sets (one set = one axle).
> - To prevent brake fluid overflow when caliper pistons are pushed back, siphon some fluid out of brake master cylinder reservoir.

46-2 Brakes–Mechanical

Brake Pads and Rotors

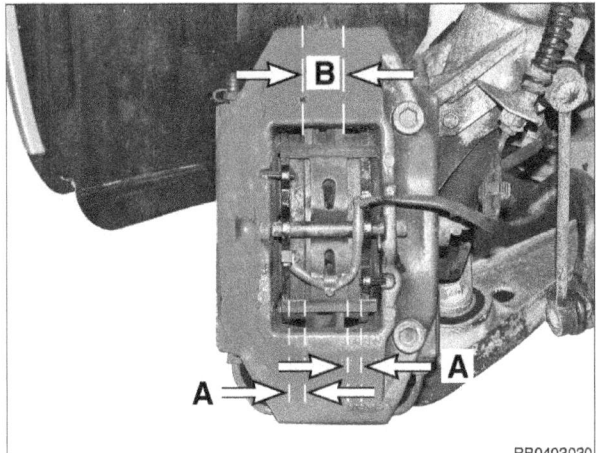

BRAKE PADS AND ROTORS

◀ Inspect brake pad thickness (**A**) and rotor thickness (**B**) for wear.

Brake wear specifications	
Approximate new brake pad thickness (dimension **A**): **Boxster** • Front • Rear **Boxster S** • Front • Rear	 11.0 mm (0.433 in) 11.0 mm (0.433 in) 12.0 mm (0.472 in) 10.5 mm (0.413 in)
Brake pad wear limit (dimension **A**): • Pads only (dimension **A**) • Pads and wear indicator	 2 mm (0.08 in) 2.5 mm (0.1 in)
New brake rotor (dimension **B**): **Boxster** • Front • Rear **Boxster S** • Front • Rear	 24 mm (0.944 in) 20 mm (0.787 in) 28 mm (1.024 in) 24 mm (0.866 in)
Brake rotor minimum thickness, after machining (dimension **B**): **Boxster** • Front • Rear **Boxster S** - *machining not permitted*	 22.6 mm (0.89 in) 18.6 mm (0.73 in)
Brake rotor wear limit (dimension **B**): **Boxster** • Front • Rear **Boxster S** • Front • Rear	 22 mm (0.866 in) 18 mm (0.709 in) 26 mm (1.023 in) 22 mm (0.866 in)

Brake pads, replacing

The following procedure applies to both front and rear brake pads. The brake calipers do not need to be removed in order to replace the brake pads.

– Loosen wheel lug bolts. Raise car, and support safely. Remove wheels.

> **WARNING—**
> • Make sure the car is stable and well supported at all times. Use a professional automotive lift or jack stands designed for the purpose. A floor jack is not adequate support.

◀ Remove brake pad wear indicator wiring harness (**arrows**) from caliper and remove wear indicators from brake pads.

> **NOTE—**
> • Replace pad wear indicator harness if wire is worn.

Brakes–Mechanical   46-3
Brake Pads and Rotors

◄ Remove clip (**arrow**) from retaining pin.

◄ Using a drift, drive retaining pin from brake caliper. Be prepared for spring clip (**arrow**) to spring off caliper.

− Pull brake pads from caliper using a brake pad puller.

◄ Press pistons back into calipers using a piston press tool or equivalent.

> *CAUTION—*
> * *Remove brake fluid reservoir cap before resetting pistons.*
> * *To prevent brake fluid overflow when caliper pistons are pushed back, siphon some fluid out of brake master cylinder reservoir.*

− Installation is reverse of removal, noting the following:
  * Fit new spreading spring clip, retaining pin, and retaining pin clip.
  * Firmly press brake pedal with vehicle stationary so that pads adjust to operating position.
  * Check brake fluid level.

NOTE—
* *New brake pads require approximately 150 miles to "break in" and attain their best friction and wear coefficient. During this period do not brake hard except in an emergency.*

Brake rotors, removing and installing

The Porsche Boxster uses a four wheel disc brake system with ventilated brake rotors at each wheel.

− Loosen lug bolts. Raise car, and support safely. Remove wheels.

46-4 Brakes–Mechanical

Brake Pads and Rotors

Brake rotor diameter	
Boxster	
• Front	298 mm (11.74 in)
• Rear	292 mm (11.50 in)
Boxster S	
• Front	318 mm (12.53 in)
• Rear	299 mm (11.78 in)

> **WARNING—**
> • Make sure the car is stable and well supported at all times. Use a professional automotive lift or jack stands designed for the purpose. A floor jack is not adequate support.

◁ Remove brake pad wear indicator wiring harness (**arrows**) from caliper and remove wear indicators from brake pads.

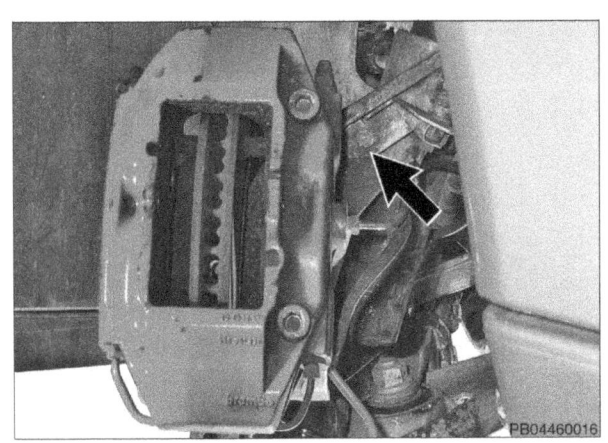

◁ Remove brake line bracket retaining bolt (**arrow**) and remove bracket from wheel bearing carrier.

> **NOTE—**
> • Brake pads do not need to be removed.

◁ For rear brake rotors, make sure parking brake lever is released. Working at rear wheel, retract parking brake adjuster by prying upwards on star wheel using a screwdriver through upper lug bolt bore.

◁ Using 10 mm Allen wrench, remove retaining bolts (**arrows**) and lift caliper from wheel bearing carrier. Suspend from chassis using mechanic's wire.

− Working at brake rotor, remove counter-sunk screws and remove rotor from hub. If necessary, tap screw heads using a drift to loosen screws and tap rotor using a soft face hammer to release rotor from hub.

Parking Brake

- Installation is reverse of removal, noting the following:
  - Clean the rotor mounting surface of wheel hub. Apply a thin coating of Optimoly TA or equivalent aluminum anti-seize compound.
  - Replace brake caliper mounting screws.
  - On rear brakes adjust parking brake shoes and cables. See **Parking brake shoes, adjusting**.

Tightening torques	
Brake rotor to hub	10 Nm (7.5 ft-lb)
Brake caliper to wheel bearing carrier	85 Nm (63 ft-lb)
Brake line bracket to wheel bearing carrier	10 Nm (7.5 ft-lb)
Wheel to wheel hub (M14 x 1.5)	130 Nm (96 ft-lb)

PARKING BRAKE

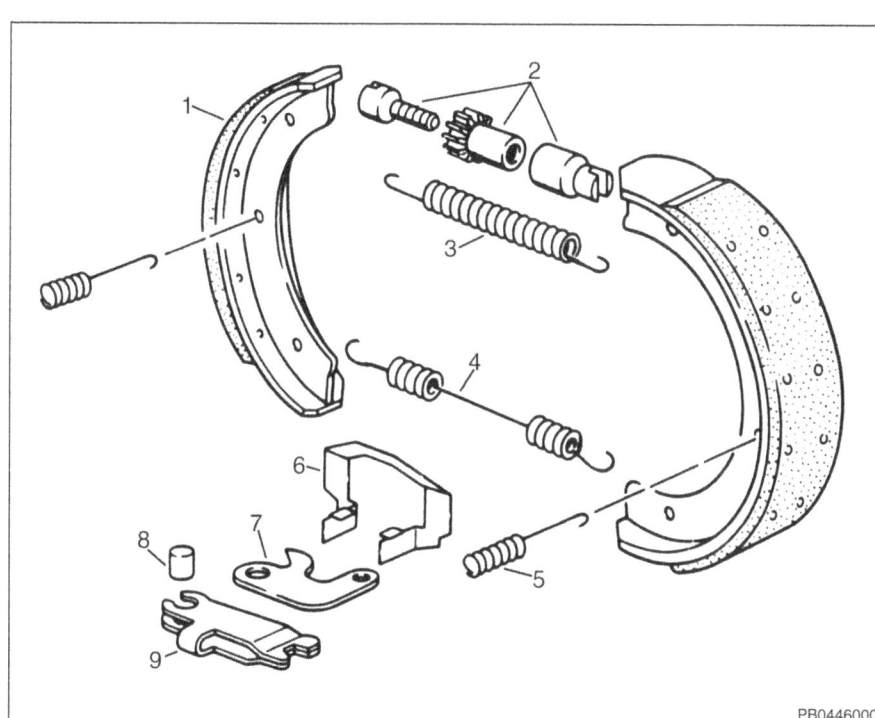

Parking brake components

1. Parking brake shoes
2. Star wheel adjuster
3. Return spring
4. Lower return spring
5. Compression spring
6. Supporting plate
7. Control lever
8. Screw
9. Clip

Parking brake shoes, adjusting

If the parking brake lever can be pulled up by more than 4 clicks without engaging, adjust parking brake.

- Loosen rear wheel lug bolts. Raise car, and support safely. Remove rear wheels

> **WARNING—**
> - Make sure the car is stable and well supported at all times. Use a professional automotive lift or jack stands designed for the purpose. A floor jack is not adequate support.

46-6 Brakes–Mechanical

Parking Brake

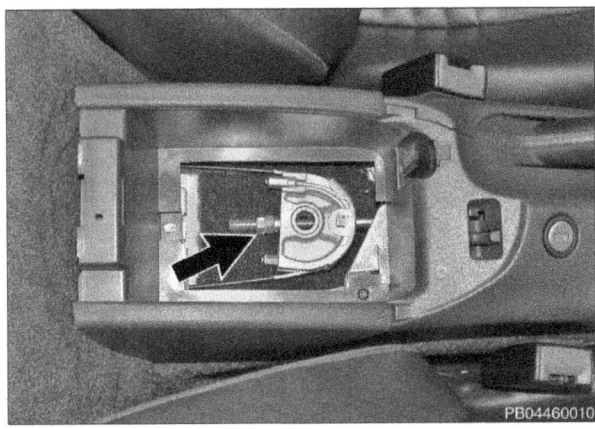

◄ Access parking brake turnbuckle. Open the cover of the console compartment behind the parking brake lever and remove the rubber insert and bottom panel.
  • Loosen adjusting nuts (**arrow**) on parking brake turnbuckle until cables are slack.

◄ Using a screwdriver, advance parking brake adjuster by prying star wheel down through upper lug bolt bore until rotor can not be turned. Next, turn adjuster back 9 notches (5 notches until wheel can be turned freely, then another 4). Repeat procedure on opposite wheel.

– Pull up parking brake lever by 2 clicks and turn adjusting nut of turnbuckle until both wheels can be turned with resistance.

– Release parking brake lever and check that wheels can be turned freely.

Tightening torque	
Wheel to wheel hub (M14 x 1.5)	130 Nm (96 ft-lb)

Parking brake shoes, replacing

– Loosen rear wheel lug bolts. Raise car, and support safely. Remove rear wheels.

> **WARNING—**
> • Make sure the car is stable and well supported at all times. Use a professional automotive lift or jack stands designed for the purpose. A floor jack is not adequate support.

◄ Remove brake pad wear indicator wiring harness (**arrows**) from brake pad and clip.

◄ Retract star wheel adjuster by prying upwards using a screwdriver through upper lug bolt bore.

Brakes–Mechanical 46-7

Parking Brake

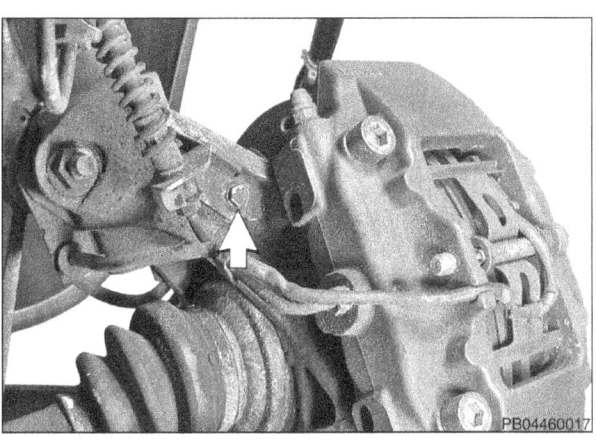

◁ Remove bolt (**arrow**) and remove brake line bracket from wheel bearing carrier.

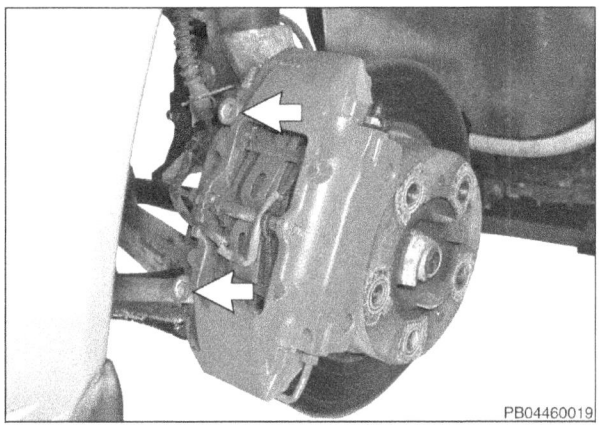

◁ Remove allen head retaining bolts (**arrows**) and lift caliper from wheel bearing carrier. Suspend from chassis using mechanic's wire.

– Working at brake rotor, remove counter-sunk screws and remove rotor from hub. If necessary, tap rotor using a soft face hammer to release rotor from hub.

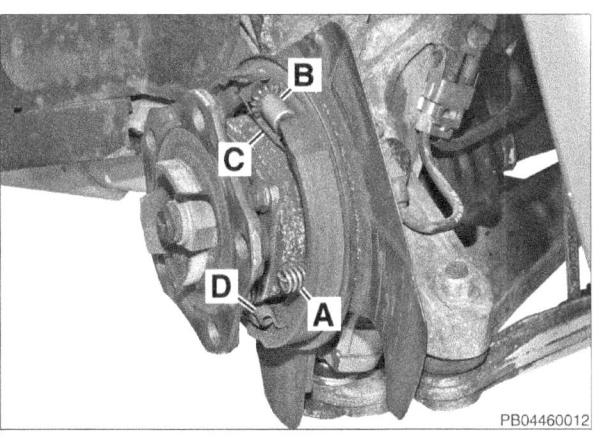

◁ Remove both compression springs (**A**), star wheel adjuster (**B**), and return spring (**C**). Remove parking brake shoes with lower return spring (**D**).

– Installation is reverse of removal, noting the following:
  - Grease adjuster, actuating lever pin, and brake shoe sliding surfaces.
  - Check that parking brake shoes, adjuster, return springs, compression springs and expanding lever are correctly positioned. Adjust as necessary.
  - Clean brake disc and hub mounting surface. Apply anti-seize compound to centering surface of wheel hub.
  - Adjust parking brake shoes. See **Parking brake shoes, adjusting**.
  - Fit brake caliper, replacing caliper mounting hardware.

Tightening torques	
Brake rotor to hub	10 Nm (7.5 ft-lb)
Brake caliper to wheel bearing carrier	85 Nm (63 ft-lb)
Brake line bracket to wheel bearing carrier	10 Nm (7.5 ft-lb)
Wheel to wheel hub (M14 x 1.5)	130 Nm (96 ft-lb)

47 Brakes–Hydraulic

GENERAL 47-1
   System description 47-1
   Troubleshooting 47-1
   Warnings and Cautions 47-2

BRAKE HYDRAULIC SYSTEM 47-2
   Replacing brake fluid 47-2
   Bleeding brakes 47-3
   Brake master cylinder, removing and installing.. 47-5
   Brake booster, removing and installing......... 47-6
   Brake caliper, removing and installing 47-6

GENERAL

This repair group covers brake bleeding, as well as replacement procedures for hydraulic brake master cylinder, vacuum brake booster and brake calipers.

Also see the following repair groups:
- **03 Service and Maintenance** for brake fluid level check.
- **45 Antilock Brakes (ABS)** for antilock brake system and traction control.
- **46 Brakes—Mechanical** for brake pads, rotors and parking brake assembly.

System description

Boxster models are equipped with power assisted four-wheel disc brakes. A vacuum brake booster provides power assistance to a dual-circuit master cylinder.

Troubleshooting

Brake performance is mainly affected by three things:
- Level and condition of brake fluid
- Brake system's ability to create and maintain hydraulic pressure
- Condition of friction components

Air in the brake fluid will make the brake pedal feel spongy during braking or will increase the brake pedal force required to stop. Fluid contaminated by moisture or dirt can corrode components. Inspect the brake fluid. If fluid is dirty or murky, or over a year old, fluid should be replaced.

Visually check hydraulic system starting at the master cylinder. To check the function of master cylinder hold brake pedal down hard with car stopped and engine running. The pedal should feel solid and stay solid. If pedal slowly falls to floor, either master cylinder is leaking internally, or fluid is escaping from other points. If no leaks can be found, the master cylinder is faulty and should be replaced. Check all brake fluid lines and couplings for leaks, kinks, chafing or corrosion.

Brake Hydraulic System

Check brake booster by pumping brake pedal approximately 10 times with engine off. Hold pedal down and start engine. Pedal should fall slightly. If not, check for any visible faults before suspecting a faulty brake booster, and check the one-way check-valve at the hose connection on the vacuum booster.

Warnings and Cautions

WARNING—
- *Brake fluid is poisonous, highly corrosive and dangerous to the environment. Wear safety glasses and rubber gloves when working with brake fluid. Do not siphon brake fluid with your mouth. Immediately clean away any fluid spilled on painted surfaces and wash with water. Dispose of brake fluid properly.*
- *Do not reuse self-locking nuts, bolts or fasteners. They are designed to be used only once and may fail if reused. Always replace them with new self-locking nuts.*
- *Do not mix DOT 5 (silicone) brake fluid with DOT 4 brake fluid as severe component corrosion will result. Such corrosion could lead to brake system failure.*

CAUTION—
- *Brake fluid is highly corrosive and dangerous to the environment. Dispose of it properly.*
- *When adding or replacing brake fluid always add new brake fluid from an unopened container.*
- *It is important to bleed the entire brake system whenever any part of the brake hydraulic system has been opened.*

BRAKE HYDRAULIC SYSTEM

Replacing brake fluid

Porsche recommends replacing the brake fluid at least once every two years. This is due to the fact that brake fluid readily absorbs moisture. Moisture in the brake fluid can adversely affect braking performance and may also damage the system, leading to costly repairs.

Replace the brake fluid using one of the procedures described later to expel the old fluid. Remove the filter/strainer from the brake fluid reservoir and clean it in new, unused brake fluid. Using new, unused brake fluid, pump at least 1 pint (500 cc) of brake fluid through each caliper to completely flush the system and expel the old fluid. Then refill the reservoir and bleed the brakes as described above. Use only Porsche-recommended brake fluid.

CAUTION—
- *Do not rely on flushing alone to clean a system contaminated with dirt or corrosion. The flushing procedure may actually force dirt in the lines into the calipers. To do the job thoroughly, diassemble the system must be disassembled and individually clean the parts. Use only brake fluid to flush the lines. Do not use alcohol since it encourages the accumulation of water in the system.*

Brakes–Hydraulic   47-3

Brake Hydraulic System

Brake fluid	
Porsche brake fluid	Super DOT 4

Bleeding brakes

◄ Brake bleeding on vehicles equipped with basic ABS 5.3 can be performed using a brake pressure bleeder, or though conventional methods.

> **CAUTION—**
> - Brake bleeding on vehicles equipped with ABS/TC 5.3 and PSM requires the Porsche System Tester 2 (PST 2) to activate the valves in the hydraulic unit for proper bleeding.

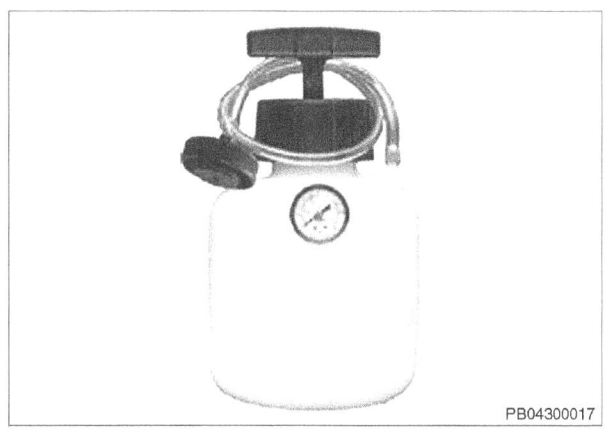

For information on ABS system identification, see **45 Antilock Brakes (ABS)**.

> **CAUTION—**
> - High mileage or older vehicles may experience brake master cylinder damage through aggressive pumping. For these vehicles, use a shorter pedal stroke and twice the number of strokes.

– Top off brake fluid reservoir with fresh brake fluid.

◄ Connect brake pressure bleeder to fluid reservoir. Fill bleeder fluid tank to sufficient level, and pressurize bleeder.

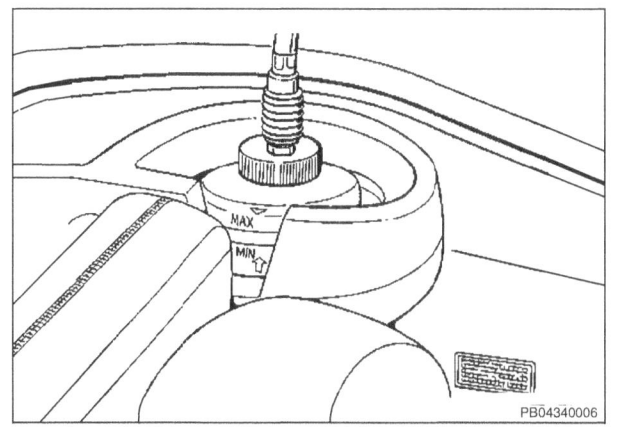

Brake bleeding pressure	
approximate pressure	1.0 - 2.0 bar (14 - 29 psi)

> **CAUTION—**
> - Do not exceed 2.0 bar (29 psi) at pressure bleeder.

– Bleed brakes using following sequence:
  1. Right rear
  2. Left rear
  3. Right front
  4. Left front

– Connect a transparent hose and collection bottle to bleeder valve. Open each bleeder valve until clear, bubble-free brake fluid appears.

– If brake master cylinder has been replaced, or brake hydraulic system required a significant volume of fluid, use the following steps to remove air bubbles from the primary circuit:
  - Open right rear bleeder valve, then fully depress brake pedal several times.
  - Hold pedal depressed for 2-3 seconds, then release slowly.
  - Repeat this procedure in sequence; rear left, front right, front left.

– If vehicle is equipped with basic ABS 5.3, disconnect bleeder and top off brake fluid. If vehicle is equipped with ABS/TC 5.3 or PSM, proceed to **additional brake bleeding procedures**.

Brake Hydraulic System

Additional brake bleeding procedure, vehicles equipped with ABS/TC 5.3

– Connect PST 2 (Porsche System Tester 2) to diagnostic socket.

– Switch ignition on. Using PST2 in ABS/TC function, select "Bleed" menu.

– Open right rear bleed valve with hose and collection bottle attached.

– Press start button on PST2.

NOTE—
* *Pressing the PST2 start button initiates specific functions in the hydraulic control unit (return pump and hydraulic solenoid valves are actuated).*

– Allow fluid to bleed from bleeder until clear and bubble-free.

NOTE—
* *Throughout bleeding process, fully depress the brake pedal at least 10 times.*

– Close right rear valve and immediately press "Stop" button of PST2.

– Switch ignition off and disconnect tester. Disconnect bleeding device and top off brake fluid as necessary.

Additional brake bleeding procedure, vehicles equipped with PSM

– Connect PST2 to diagnostic socket.

– Switch ignition on. Using PST2 in PSM function, select "Bleed" menu.

– Open right front bleed valve with hose and collection bottle attached.

NOTE—
* *The PST2 will switch off automatically if allowed to remain idle for extended periods.*

– Press start button on PST2 to initiate PSM booster pump.

– Bleed until brake fluid is clear and bubble-free.

NOTE—
* *Throughout bleeding process, fully depress the brake pedal at least 10 times.*

– Close right front bleeder valve and immediately press "Stop" button of PST2.

– Switch ignition off and disconnect tester. Disconnect bleeder and top off brake fluid as necessary.

Brakes–Hydraulic 47-5
Brake Hydraulic System

Brake master cylinder, removing and installing

– Working in front trunk, remove spare tire and emergency tool kit.

 Remove cover from master cylinder.
- Remove rubber trunk seal from edge of trunk compartment.
- Remove screws and plastic screw plug (**arrows**).
- Lift cover off master cylinder assembly, unplug trunk light harness connector, and remove cover from trunk.

 Drain brake fluid reservoir. Evacuate fluid using a fluid pump, or pump brake pedal to create pressure and and "bleed" fluid through front and rear brake calipers.

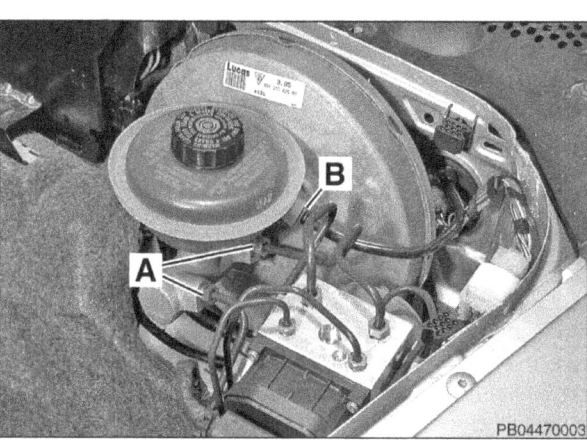 Disconnect brake lines (**A**) at the master cylinder. For vehicles equipped with manual transmissions, remove the clutch line (**B**) at reservoir. Cap brake lines and connections with suitable plugs.

 Remove fastening nuts (**arrows**) of the brake master cylinder. Remove master cylinder from fire wall.

– Installation is reverse of removal, noting the following:
- Top off brake fluid reservoir.
- First bleed the brakes as usual. See **Bleeding brakes**, then follow additional bleeding procedures as necessary.
- Bleed clutch system, if equipped. See **30 Clutch**.

Tightening torque	
Brake line to brake component	10-16 Nm (7-11 ft-lb)

47-6 Brakes–Hydraulic

Brake Hydraulic System

Brake booster, removing and installing

– Remove master cylinder. See **Brake master cylinder, removing and installing**.

– Remove brake hydraulic unit: See **45 Antilock Brakes (ABS)**.
  • Disconnect brake lines at hydraulic unit.
  • Seal brake lines and connections with suitable plugs.
  • On models equipped with PSM, disconnect electrical harness connector at pressure sensor.
  • Disconnect control module electrical harness connector at hydraulic unit.
  • Remove hydraulic unit from front trunk.

◄ Working under plastic trim panel in left side of battery compartment, loosen locking nut (**arrow**) and retaining nut and separate push rods from each other.

– Disconnect vacuum line at brake booster.

– Remove brake booster fastening bolts (Torx T-45). Remove brake booster from firewall.

– Installation is reverse of removal, noting the following:
  • Top off brake fluid reservoir.
  • Bleed brakes. See **Bleeding brakes**
  • If necessary, follow additional bleeding procedures. See **Additional brake bleeding procedure, vehicles equipped with ABS/TC 5.3** or **Additional brake bleeding procedure, vehicles equipped with PSM** in this repair group.
  • Bleed clutch system, if equipped. See **30 Clutch**.

Brake caliper, removing and installing

The following procedure applies to both front and rear brake calipers.

– Loosen wheel lug bolts. Raise car, and support safely. Remove wheels.

> **WARNING—**
> • Make sure the car is stable and well supported at all times. Use a professional automotive lift or jack stands designed for the purpose. A floor jack is not adequate support.

◄ Unclip and pull out brake pad wear indicator and harness (**arrows**) from brake pads.

> **NOTE—**
> • Replace pad wear indicator harness if wire is worn.

Brakes–Hydraulic 47-7

Brake Hydraulic System

◀ Remove clip (**arrow**) from retaining pin.

◀ Using a drift, drive retaining pin from brake caliper. Be prepared for spring clip (**arrow**) to spring off caliper.

– Pull brake pads from caliper using a brake pad puller.

◀ Press pistons back into calipers using a piston press tool or equivalent.

CAUTION—
- *Remove brake fluid reservoir cap before resetting pistons.*
- *To prevent brake fluid overflow when caliper pistons are pushed back, siphon some fluid out of brake master cylinder reservoir.*

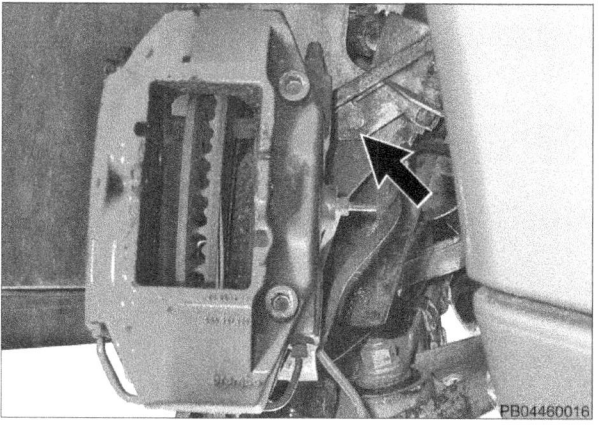

◀ Remove brake line bracket retaining bolt (**arrow**) and remove bracket from wheel bearing carrier.

– Using a flare nut wrench, loosen brake hydraulic line on back of caliper.

47-8 Brakes–Hydraulic

Brake Hydraulic System

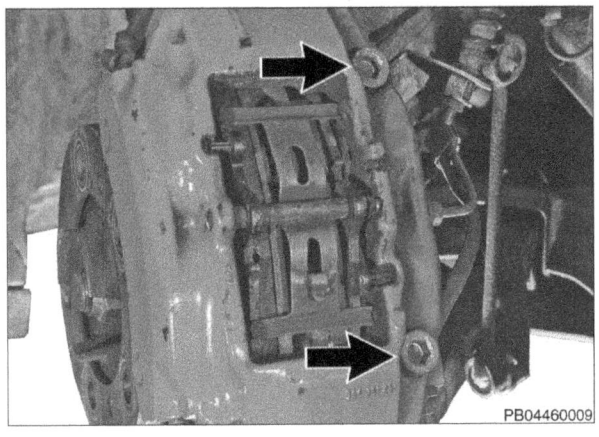

◀ Using a 10 mm allen wrench, remove retaining bolts (**arrows**) and lift caliper from wheel bearing carrier.

– Installation is reverse of removal, noting the following:
  • Fit new spreading spring clip, retaining pin, and retaining pin clip.
  • Firmly press brake pedal with vehicle stationary so that pads adjust to operating position.
  • Bleed brakes. See **Bleeding brakes**.
  • Check brake fluid level.

Tightening torque	
Brake line to brake component	10-16 Nm (7-11 ft-lb)

48 Steering

GENERAL 48-1	**TIE RODS AND STEERING RACK** 48-6
Steering system 48-1	Tie rods, removing and installing 48-6
Warnings and Cautions 48-2	Steering rack, removing and installing 48-7
STEERING WHEEL AND COLUMN 48-2	**POWER STEERING** 48-9
Steering wheel, removing and installing 48-2	Power steering fluid, checking 48-9
Steering column stalk switch,	Power steering fluid, bleeding 48-10
removing and installing 48-3	Power steering pump, removing and installing . 48-10
Ignition electrical switch, removing and installing 48-5	
Ignition lock cylinder, removing and installing .. 48-5	

GENERAL

This repair group covers servicing of the steering system, including steering rack, power steering pump, and tie rod replacement, as well as steering wheel removal, ignition lock and switch replacement and steering stalk switch replacement.

See also the following repair groups:
- **40 Front Suspension**
- **44 Wheels, Tires and Alignment**

Steering system

The Porsche Boxster is equipped with power rack and pinion steering. The steering rack is bolted to the front crossmember. The steering wheel and column are connected to the steering rack by a short shaft and a universal joint. There are no provisions for lubrication of the steering rack or linkage.

Wear and excessive play or clearance anywhere in the steering system will cause sloppy, loose steering. Before checking for any steering problems, check tires, tire inflation pressures and front suspension components. Inspect rubber rack boots and tie-rod end boots for tears or damage and replace if necessary. On cars with high mileage, tie rod ends are prone to wear. In general, the steering system is serviced by replacement of worn parts.

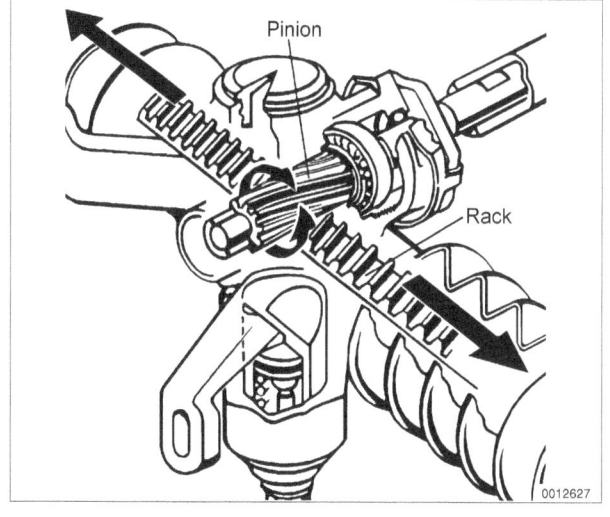

Steering specifications	
Steering turns (lock to lock)	2.98
Turning circle diameter	10.9 m (35.8 ft.)

48-2 Steering

Steering Wheel and Column

Warnings and Cautions

> **WARNING—**
> - The Boxster is equipped with an airbag mounted in the steering wheel. The airbag is an explosive device and should be treated with extreme caution. Follow the airbag removal procedure as outlined in **69 Seatbelts, Airbags.**
> - Do not reuse self-locking nuts. They are designed to be used only once and may fail if reused. Replace with new locking nuts.
> - Do not install bolts and nuts coated with undercoating wax, as correct tightening torque cannot be assured. Clean threads with solvent before installation, or install new parts.
> - Do not attempt to weld or straighten any steering components. Replace damaged parts.

> **CAUTION—**
> - Do not overfill power steering reservoir. Do not spill Pentosin fluid. If coolant hoses come into contact with Pentosin, clean them thoroughly with water immediately.
> - Replace any visibly swollen hoses.

STEERING WHEEL AND COLUMN

Steering wheel, removing and installing

- Disconnect negative (–) battery cable.

> **CAUTION—**
> - Prior to disconnecting the battery, read the battery disconnection cautions in **00 Warnings and Cautions**.
> - Be sure to have the radio anti-theft code before disconnecting the battery.

- Remove driver airbag. See **69 Seat Belts, Airbags**.

> **WARNING—**
> - Improper handling of the airbag can cause serious injury. Store the airbag with the horn pad facing up. If stored facing down, accidental deployment can propel it violently into the air, causing injury.

◄ Working at steering wheel center:
  - Detach electrical connectors (**A**).
  - Remove steering wheel mounting nut (**B**).

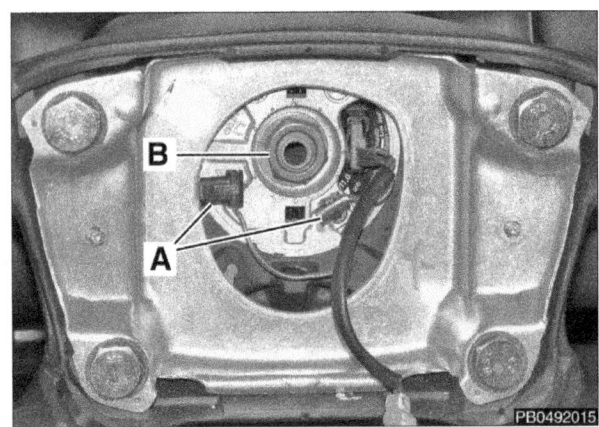

Steering 48-3

Steering Wheel and Column

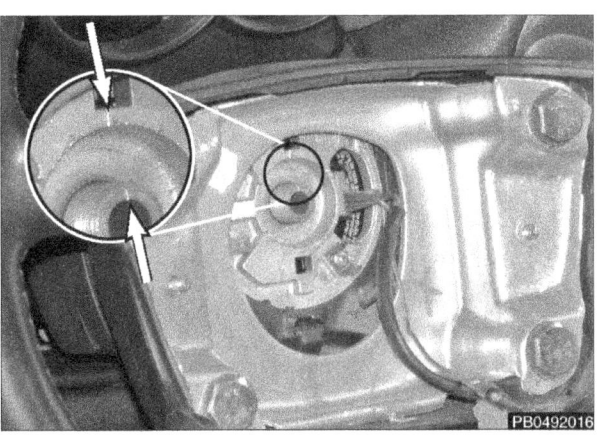

◂ Turn wheels to straight ahead position and mark position of steering wheel splines in relation to shaft splines (**arrows**). Remove steering wheel with front wheels in straight ahead position.

NOTE—
- *To prevent rotational damage to the airbag contact spiral spring, the spiral spring locks when the steering wheel is removed. The contact spring unlocks automatically when the steering wheel is refitted.*

◂ Prior to installing steering wheel, make sure airbag contact spiral spring alignment **arrows** line up for steering wheel center position.

– Fit steering wheel using previously made spline alignment marks and make sure upper spokes of wheel are horizontal. Install and tighten steering wheel mounting nut.

CAUTION—
- *Do not pinch airbag contact spring and horn harness.*

Tightening torque	
Steering wheel to steering shaft	46 Nm (34 ft-lb)

– Installation is reverse of removal, noting the following:
- Reattach airbag and horn harness connectors.
- Remount airbag using new mounting bolts. See **69 Seat Belts, Airbags.**

CAUTION—
- *Use care not to pinch horn and airbag harness when mounting airbag.*

Tightening torque	
Airbag to steering wheel	10 Nm (7 ft-lb)

Steering column stalk switch, removing and installing

The steering column stalk switch may include the following four switch stalks:
- Cruise control switch
- Windshield wiper/washer switch
- On-board computer
- Turn signal / high beam switch

– Remove steering wheel. See **Steering wheel, removing and installing**.

– Extend steering column as far as possible.

48-4 Steering

Steering Wheel and Column

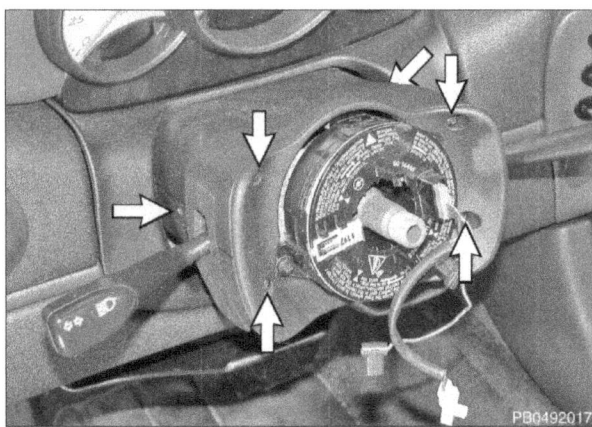

◀ Remove steering column cover fasteners (**arrows**). Remove covers.

◀ Remove airbag contact spiral spring mounting screws (**arrows**). Slide contact spiral spring off steering wheel shaft and place aside without separating harness connectors.

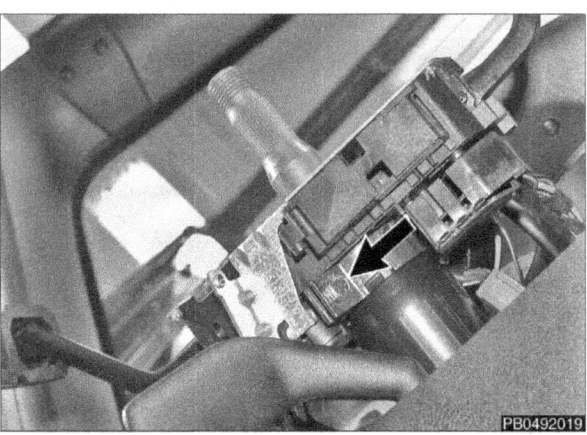

◀ Working below steering column:
- Loosen stalk switch assembly pinch collar bolt (M5) (**arrow**).
- Slide switch assembly off steering column, disconnecting harness connectors as necessary.

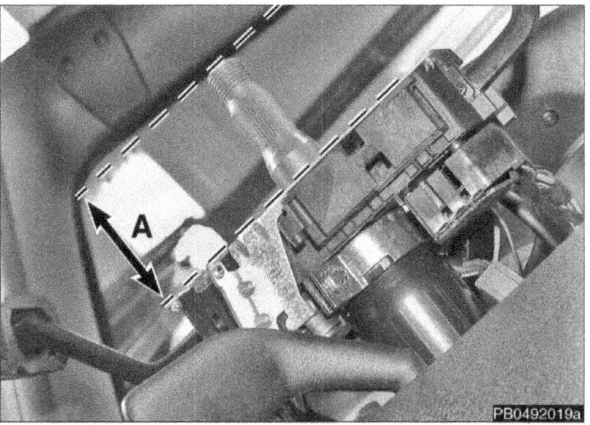

◀ When reinstalling, position switch assembly at correct depth on steering column before tightening pinch collar bolt.

Stalk switch assembly position	
Dimension **A** (tip of steering shaft to sheet metal cover of switch assembly)	55 ± 0.5 mm (2.16 ± 0.02 in)

- Install steering column covers and airbag contact spring. Make sure wiring harnesses do not get pinched.

- Reinstall steering wheel and airbag. See **Steering wheel, removing and installing**.

Steering 48-5

Steering Wheel and Column

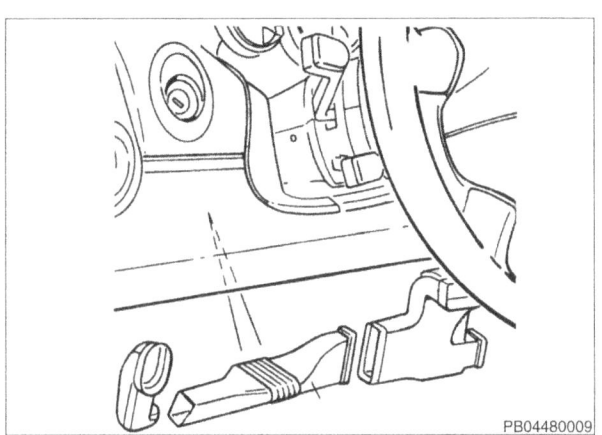

Ignition electrical switch, removing and installing

– Disconnect negative battery cable.

> **CAUTION—**
> - Prior to disconnecting the battery, acquire the radio anti-theft code.
> - Read the battery disconnection cautions in **00 Warnings and Cautions**.

– Turn ignition key to position O (off) and remove.

◄ Remove air duct under driver's side of dashboard.

– Disconnect electrical harness connector from back of ignition switch.

◄ Remove locking paint from switch retaining screws (**arrows**). Slightly loosen both screws using a short screwdriver, and pull ignition switch from housing.

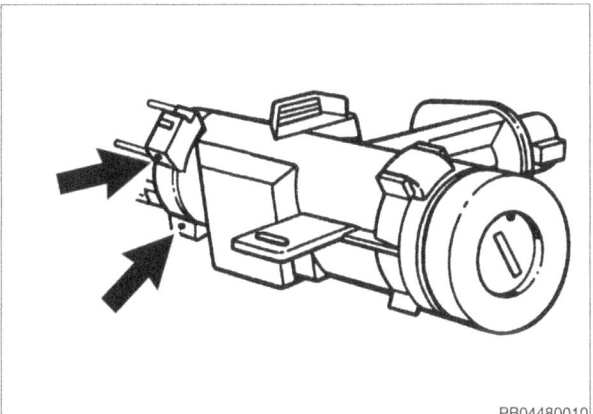

– Installation is reverse of removal, noting the following:
  - Move ignition switch to O (off) position.
  - Insert switch into housing. Tighten fastening screws using a short screwdriver and secure screws with locking paint.
  - Connect electrical harness connectors to switch and fit air duct.
  - Connect negative battery cable and test switch.

Ignition lock cylinder, removing and installing

– Disconnect negative battery cable.

> **CAUTION—**
> - Prior to disconnecting the battery, acquire the radio anti-theft code.
> - Read the battery disconnection cautions in **00 Warnings and Cautions**.

◄ Remove trim cover (rosette) at ignition lock cylinder.

48-6 Steering

Tie Rods and Steering Rack

◂ Insert ignition key and turn to position 1 (ignition ON). Insert straightened paper clip into hole at top right of key slot (**arrow**) as far as possible. This releases lock cylinder. Pull lock cylinder out of steering lock housing.

◂ Disconnect immobilizer induction coil electrical harness connector (**arrow**) from ignition cylinder barrel.

- Installation is reverse of removal, noting the following:
  - Place ignition lock in position 1 (ignition ON), and lock cylinder barrel pawl with paper clip.
  - Reconnect immobilizer induction coil electrical harness connector and push cylinder into housing. Turn key, if necessary, until lock cylinder is in fully. Remove paper clip.
  - Connect negative battery cable and test switch.

NOTE—
- *If a new ignition cylinder with key and transponder is used, code new key using Porsche System Tester 2 (PST2). See an authorized Porsche service facility.*

TIE RODS AND STEERING RACK

Tie rods, removing and installing

- Loosen front wheel lug bolts. Raise car, and support safely. Remove front wheels

> **WARNING—**
> - *Make sure the car is stable and well supported at all times. Use a professional automotive lift or jack stands designed for the purpose. A floor jack is not adequate support.*

◂ Remove tie rod and control arm from wheel bearing carrier using special tool 9560 (ball joint press). When loosening fastening nut, counterhold ball joint stud with a T-40 Torx screwdriver.

NOTE—
- *Brake rotor and caliper shown removed for clarity.*

Steering 48-7

Tie Rods and Steering Rack

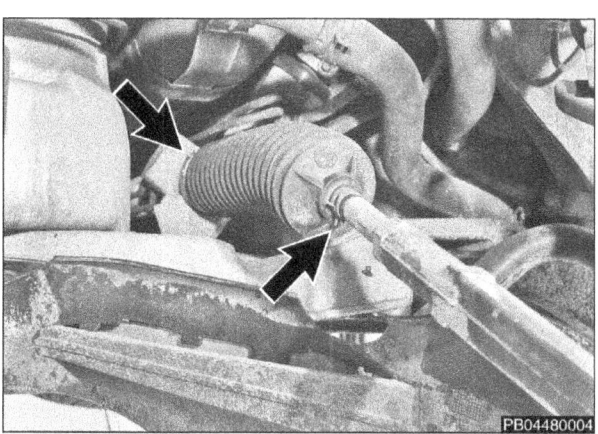

◄ Release retaining clamps (**arrows**) at each end of steering tie rod boot. Slide boot towards end of tie rod to expose steering rack.

– Using an appropriate wrench, release tie rod from steering rack.

◄ Installation is reverse of removal. Replace inner tie rod boot clamp. Crimp boot clamp (**arrow**) using crimping pliers or appropriate nippers.

Tightening torques	
Tie rod to steering rack	80 Nm (59 ft-lb)
Tie rod to wheel bearing carrier	50 Nm (37 ft-lb)

Steering rack, removing and installing

– Loosen front wheel lug bolts. Raise car, and support safely. Remove front wheels

> **WARNING—**
> - Make sure the car is stable and well supported at all times. Use a professional automotive lift or jack stands designed for the purpose. A floor jack is not adequate support.

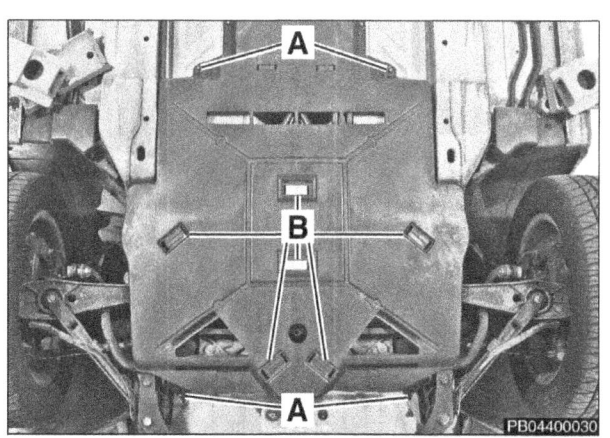

◄ Remove front plastic underbody shield:
- Remove plastic retaining nuts (**A**).
- Pry off six retaining clips (**B**).

> **NOTE—**
> - Early cars used plastic retaining clips (**B**) that are easily damaged during removal. Later cars used spring steel clips that are easier to remove and install. Only the spring steel replacement clips are available as replacement parts.

48-8 Steering

Tie Rods and Steering Rack

◀ With wheels pointing straight ahead, disconnect steering rack universal joint at column at left side of front suspension. Remove clamping bolt (**arrow**) and slide joint upwards.

NOTE—

* Be sure front wheels are in the straight ahead position before removing the u-joint clamping bolt. If this is not done, the steering wheel will have to be removed and the airbag contact spiral spring will have to be put in the center position after installing the steering rack.

◀ Remove tie rod end nut and press tie rod end from wheel bearing carrier. When loosening nut, counterhold tie rod end stud using T-40 Torx screwdriver. Press out tie rod end using appropriate press tool.

NOTE—

* Brake caliper and rotor shown removed for clarity

◀ Working at driver's side of steering rack, loosen hex head bolt (**arrow**) and remove fluid lines retaining plate. Disconnect lines.

NOTE—

* When the fluid lines are disconnected, power steering fluid will leak out. Catch power steering fluid in a suitable container. Once all fluid has drained, seal openings in rack and line ends with appropriate caps and plugs to prevent system contamination.

◀ Remove steering rack fastening bolts (**arrows**) on underside of crossmember.

– Turn steering wheel to left to extend steering gear to the left side (right on right hand drive vehicles), and remove rack from vehicle.

* If necessary, remove tie rods from steering rack before removing rack.

Steering 48-9

Power Steering

- Installation is reverse of removal, noting the following:
  - Replace rack mounting fasteners.
  - With rack fully extended, grease with VW steering gear grease AOF 063 000 04. Also grease outside of steering gear pinion shaft at teeth of universal joint.
  - With steering rack, steering wheel, and airbag contact unit in correct position, push shaft onto universal joint. For ease of assembly, put steering rack fastening bolts in place.
  - After fitting hydraulic lines to steering rack, fill reservoir with Pentosin CHF 11S power steering fluid and bleed system. See **Power steering fluid, bleeding**.

Tightening torques	
Steering rack to crossmember	65 Nm (48 ft-lb)
Tie rod to steering rack	80 Nm (59 ft-lb)
Hydraulic lines to steering rack	20 Nm (15 ft-lb)
Tie rod to wheel bearing carrier	50 Nm (37 ft-lb)
Universal joint to steering rack	23 Nm (17 ft-lb)

POWER STEERING

Power steering fluid, checking

Damage to the power-assisted steering is often the result of running the system while low on oil. Insufficient oil level will damage the servo pump as a result of the high pressure in the hydraulic circuit.

Grunt-like noises when the steering is locked or foam formation in the reservoir indicates a low oil level and/or that air has been sucked in. Before topping up the reservoir, remedy any leaks on the suction side and replace any faulty parts on the pressure side.

The reservoir is located in the engine compartment. There are two markings on the dipstick located on the reservoir cap. One surface is marked "Cold" and the other is marked "Hot".
- Cold = engine temp. approx. 20°C (68°F)
- Hot = engine temp. approx. 80° C (176°F)

Check power steering fluid level when the engine is cold and not running.

- Open and remove top engine cover. See **03 Service and Maintenance**.

Unscrew power steering fluid reservoir cap (**arrow**).

- Wipe off dipstick.

48-10 Steering

Power Steering

- Close and then reopen cap. Recheck level on dipstick. Fluid level should be in shaded area below "Cold" marking.
  - Cold = maximum engine temperature 20°C (68°F)

- Top up with Pentosin (CHF 11 S) power steering fluid as necessary. Do not overfill.

> **CAUTION—**
> - *If coolant hoses come into contact with Pentosin, clean them thoroughly with water immediately.*
> - *Replace any visibly swollen hoses.*

Power steering fluid, bleeding

After the installation of new steering components or after a substantial loss of fluid, bleed power steering system.

- Briefly start engine a few times and, when it fires, quickly switch it off again. This procedure causes fluid level in reservoir to fall rapidly. Continuously check reservoir level using this procedure: Make sure reservoir is not sucked dry.

> **CAUTION—**
> - *Do not overfill power steering reservoir. Do not spill Pentosin fluid. If coolant hoses come in contact with Pentosin fluid, wash off immediately.*

- If power steering fluid level does not fall when engine is turned over, start engine and let run at idle.

- Turn steering wheel steadily and quickly from stop to stop several times to allow air to escape. Do not turn steering wheel with more force than necessary (to prevent unnecessary build-up of pressure).

- Observe fluid level during procedure. If level continues to fall, add fluid until level remains constant and no more air bubbles stop rising.

> **NOTE—**
> - *If difference in fluid level between engine off and engine running differs by more than 10 mm, hydraulic fluid contains too much air.*

- Top off power steering fluid level with engine at idle, without operating steering wheel.

Power steering pump, removing and installing

- Raise car and support safely.

> **WARNING—**
> - *Make sure the car is stable and well supported at all times. Use a professional automotive lift or jack stands designed for the purpose. A floor jack is not adequate support.*

- Move convertible top and convertible-top storage lid to service position. See **03 Service and Maintenance**.

- Remove left seat. See **72 Seats**.

Steering 48-11

Power Steering

- Disconnect negative (–) battery cable.

  CAUTION—
  - *Prior to disconnecting the battery, acquire the radio anti-theft code.*
  - *Read the battery disconnection cautions in* **00 Warnings and Cautions**.

- Remove carpeted panels from top and front engine access covers.

◄ Twist engine top cover hold-downs (**arrows**) and open and remove cover.

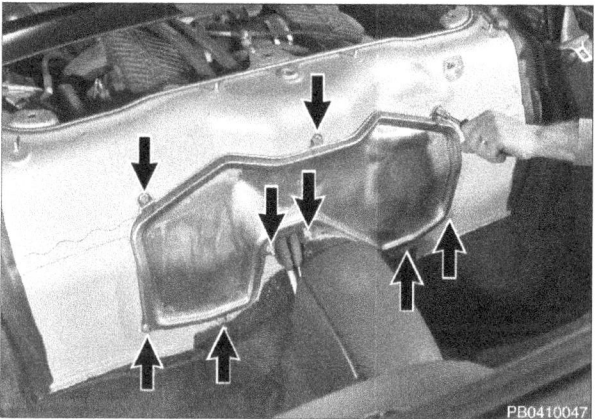

◄ Undo engine access cover mounting fasteners at rear of passenger compartment (**arrows**) and lift cover off.

  CAUTION—
  - *Access cover edges may be sharp.*

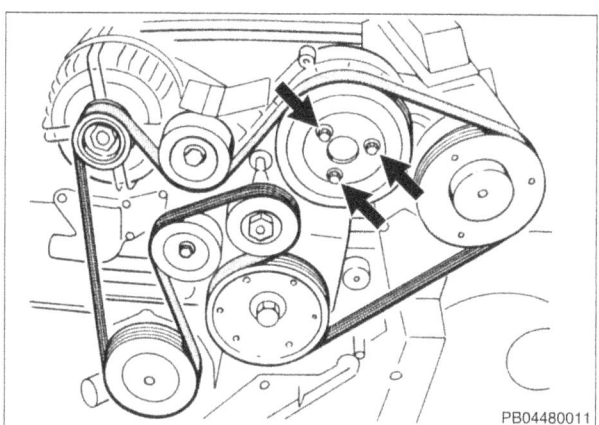

◄ Remove power steering pump drive pulley mounting bolts (**arrows**).

◄ Use 24 mm wrench to turn engine drive belt tensioner clockwise. Slip belt off air-conditioning compressor (upper left pulley, upper right in photo). Remove power steering pump pulley.

  NOTE—
  - *If taking drive belt off, mark direction of rotation for reinstallation.*

48-12 Steering

Power Steering

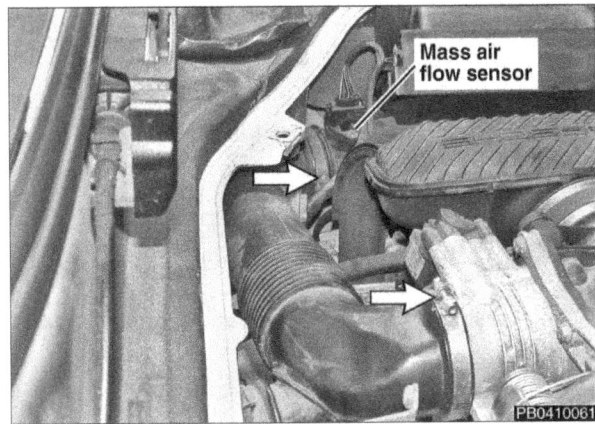

◀ Working at rear of engine compartment:
  - Detach mass air flow sensor harness connector at left behind air filter housing. Unhook harness from clip and place connector on engine.
  - Loosen engine air intake duct clamps (**arrows**) and remove duct.

◀ Remove throttle housing and center intake manifold and set aside:
  - Remove throttle cable clamp (**A**) if applicable.
  - Remove throttle housing hold down bolt (**B**).
  - Loosen center intake manifold hose clamps. Slide right manifold boot to left (**arrow**) and tilt manifold and throttle housing out and set aside. Leave vacuum hoses and cable attached.

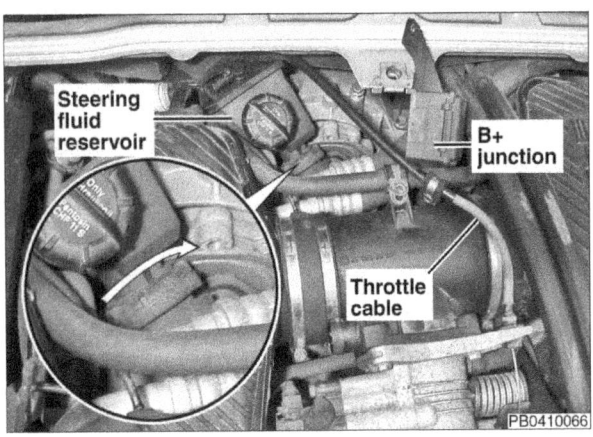

◀ Remove power steering fluid reservoir:
  - Loosen and move aside throttle cable, if applicable.
  - Use clean syringe to siphon out Pentosin fluid from power steering reservoir.
  - Unscrew (**arrow**) bayonet lock at bottom of reservoir and lift off. Be prepared for dripping fluid.
  - After removing reservoir, plug open steering fluid port.

> **CAUTION—**
> - *Do not spill Pentosin fluid. If coolant hoses come in contact with Pentosin fluid, wash off immediately.*

- Remove underbody cover.

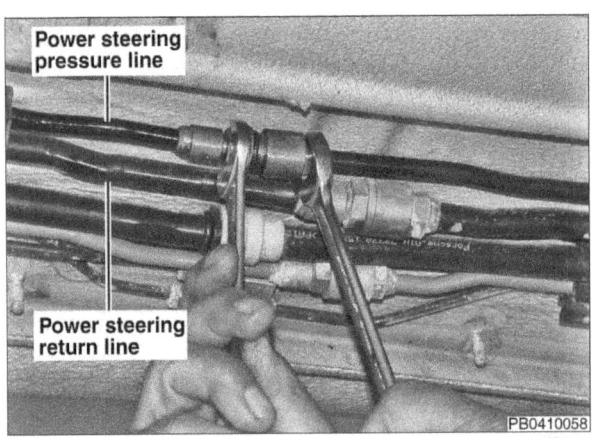

◀ Working underneath left side of car, disconnect power steering pressure and return lines. Be sure to counterhold brass fitting while unscrewing pressure nut.

Steering 48-13

Power Steering

◀ Remove air-conditioning compressor and set aside. See **87 Air-conditioning** for more information. Be sure to place blanket or other suitable protective pad underneath compressor.

− Remove lower air-conditioning compressor mounting bracket from engine, freeing power steering supply line bracket from engine.

− Detach steering return line from tank. Working at rear of pump, push red locking ring forward and pull out line. Use plastic tools to press unlocking ring. Carefully protect line against dirt and scratches.

− Disconnect supply line at power steering pump. Use a 17 mm wrench to remove fitting while counterholding fitting at power steering pump using a 24 mm wrench.

− Remove three hex-head bolts (M8x12) from front of power steering pump. Remove retaining bolt at lower left rear of pump and remove hydraulic pump to the rear.

> **CAUTION—**
> - If coolant hoses come into contact with Pentosin, clean immediately.

− Installation is reverse of removal, noting the following:
  - Reconnect power steering lines as shown below.
  - Reconnect power steering return line and clamp properly.
  - Replace brake booster line and power steering line retaining clip.
  - Reinstall A/C compressor.
  - Reinstall drive belt as shown below.
  - Check power steering fluid reservoir sealing O-rings and replace, if necessary. Reinstall reservoir.
  - Bleed power steering system once re-assembly is complete. See **Power steering fluid, bleeding** in this repair group.

Reconnecting power steering lines

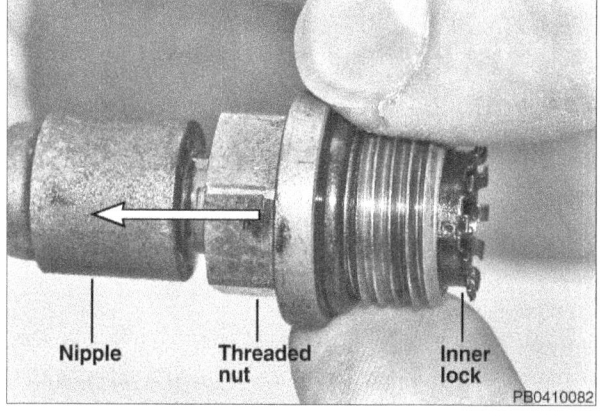

◀ Reconnect power steering fluid pressure line:
  - Hold pressure line from engine.
  - Press threaded nut toward nipple.

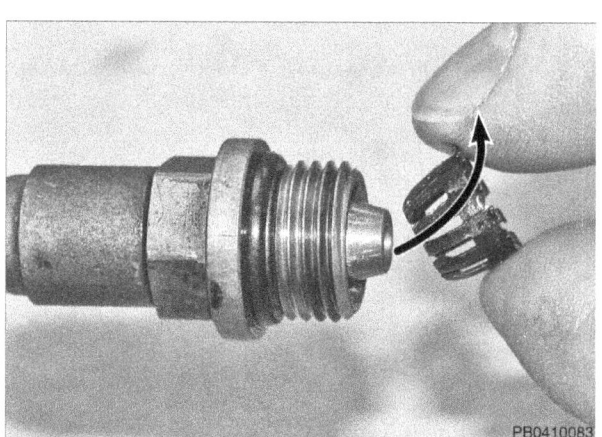

◀ Tilt inner lock and pull to remove from nipple.

48-14 Steering

Power Steering

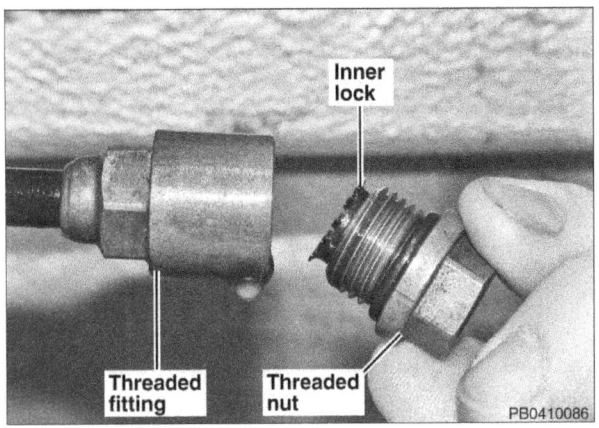

◀ Pull threaded nut off nipple and place inner lock loosely inside. Thread assembly into threaded fitting underneath car.

– Tighten assembly. Be sure to counterhold threaded fitting.

Tightening torque	
Steering pressure line to fitting (15 mm wrench size)	30 Nm (22 (ft-lb)

– Press nipple at end of power steering pressure line into fitting assembly until it snaps in place.

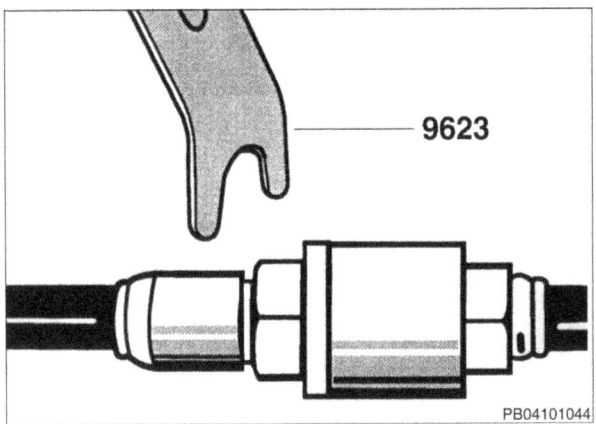

◀ Check fitting: Insert Porsche special tool 9623 in fitting slot and tilt slightly to make sure joint does not slide apart.

Drive belt routing

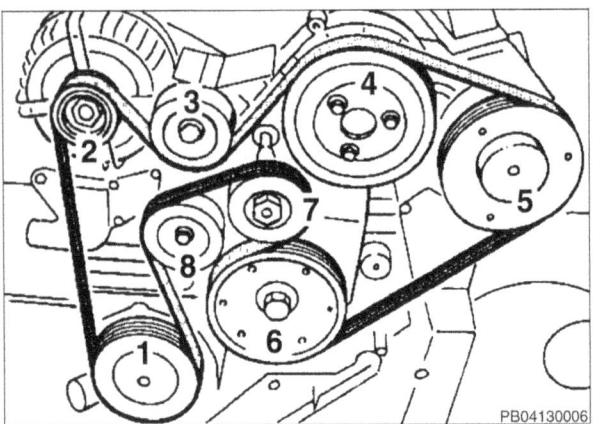

◀ Check engine drive belt for damage and replace if necessary. If reusing belt, fit in original direction of rotation. Place on pulleys in the following order:

1. Coolant pump
2. Alternator
3. Upper idler
4. Power steering pump
5. Air-conditioning compressor
6. Crankshaft pulley
7. Tensioner
   Use 24 mm wrench turn tensioner clockwise to fit drive belt
8. Lower idler

NOTE—
* *After installation, check to make sure belt is correctly positioned on all pulleys.*

50 Fenders, Drains

GENERAL 50-1
FRONT FENDERS 50-1
  Front wheel housing liner,
    removing and installing 50-1
  Front fender components 50-2
Fuel tank filler flap, removing and installing 50-2
Front fender, removing and installing 50-3
WATER DRAINS 50-4
  Front water drains, cleaning 50-4
  Rear water drains, cleaning 50-4

GENERAL

This repair group covers service procedures for the front fenders as well as water drain locating and cleaning. Also see:
- **55 Trunk Lids**
- **57 Doors and Locks**
- **63 Bumpers**
- **66 Exterior Equipment**

FRONT FENDERS

Front wheel housing liner, removing and installing

NOTE—
- *The procedure given here is based on 1998 Boxster.*

– Raise front of car and support safely. Remove front wheel.

WARNING—
- *Make sure the car is stable and well supported at all times. A floor jack is not adequate support.*

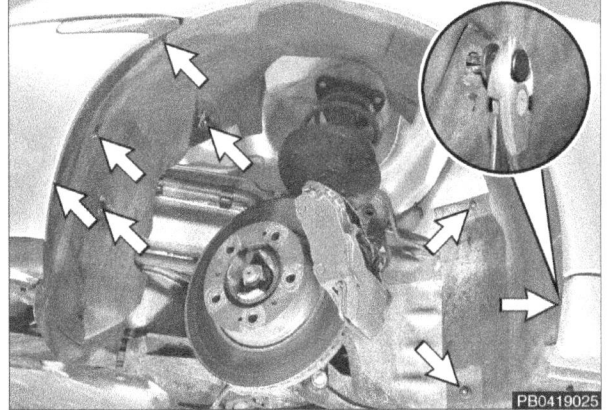

◀ Remove front wheel housing liner fasteners (**arrows**).
  - To disengage plastic rivet, pull out rivet lock (**inset**).
  - Remove plastic 10 mm nuts.

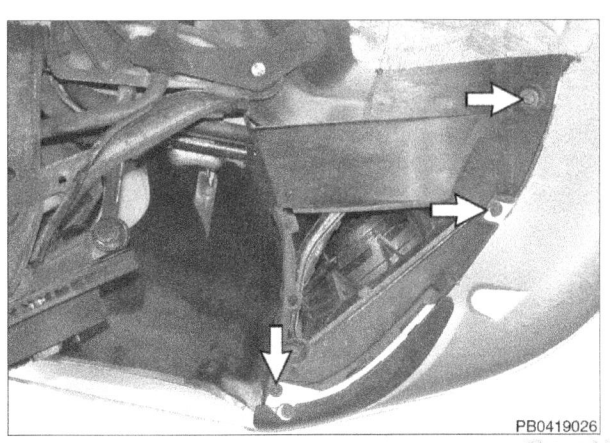

◀ Working underneath front fender, remove remainder of wheel housing liner fasteners (**arrows**). Lift off liner.

– Installation is reverse of removal. Replace any broken fasteners.

50-2 Fenders, Drains

Front Fenders

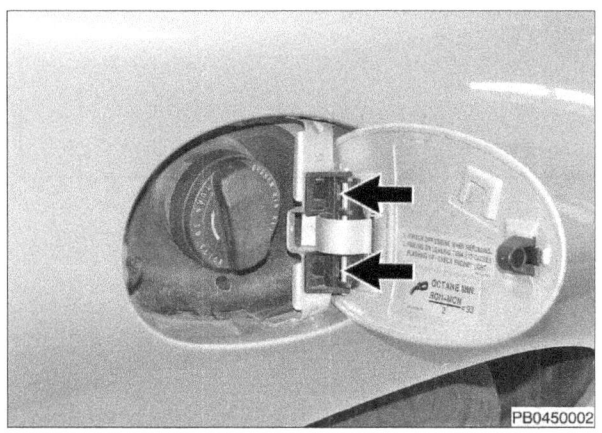

Fuel tank filler flap, removing and installing

◀ Insert a thin bladed screwdriver into each side of filler flap hinge (**arrows**) to release locking tabs and carefully lift flap off hinge.

– If replacing hinge, release locking tabs holding hinge to fender and remove hinge.

– To install, push hinge into fender until locking tabs engage and then push filler flap locking tabs into hinge.

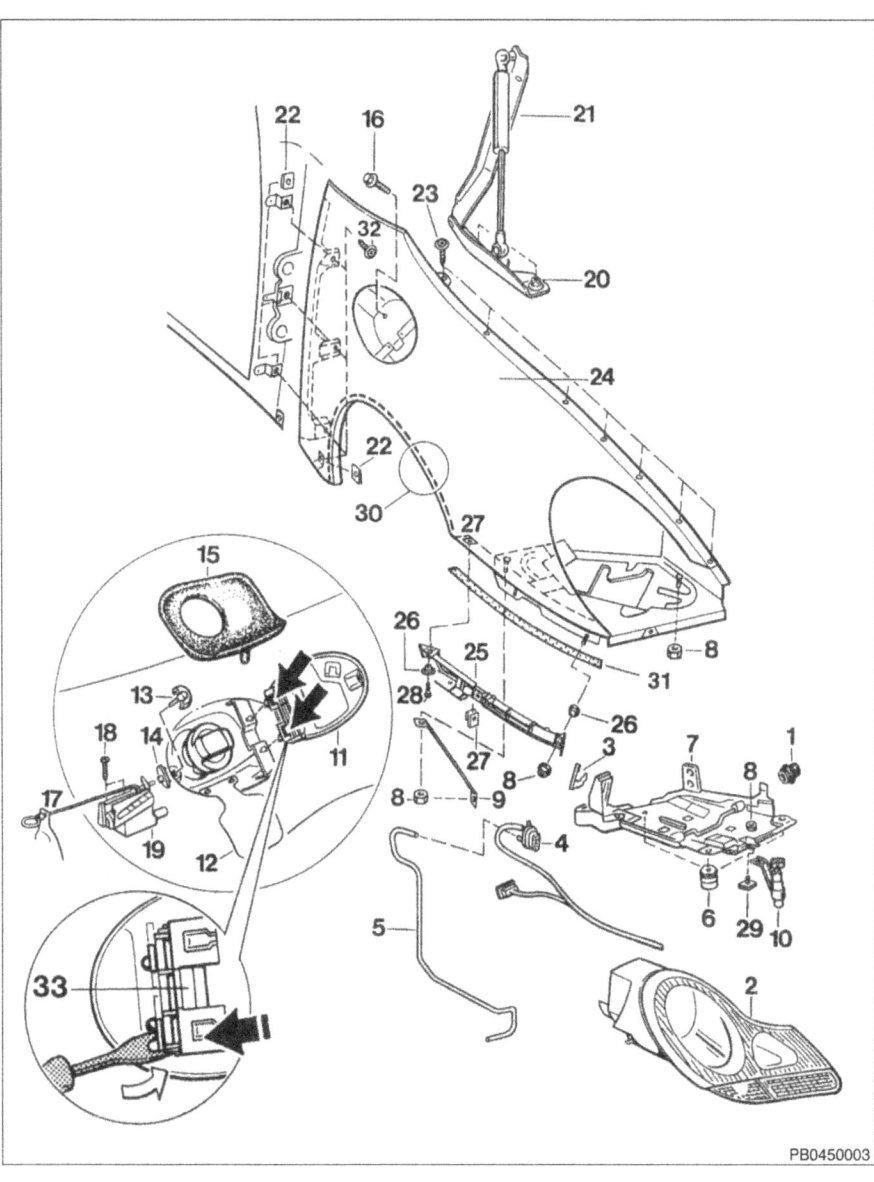

Front fender components

1. Rubber sleeve
2. Headlight
   - See **94 Exterior Lights** for headlight removal
3. Sliding piece
4. Plug connection
5. Vent hose
6. Fastening element
7. Mounting plate
8. Collar nut M6
9. Strut
10. Headlight washer nozzle
11. Filler flap
12. Protective tabs
13. Rubber sleeve
14. Guide ring
15. Sleeve
16. M6 bolt
17. Emergency cable

Fenders, Drains 50-3

Front Fenders

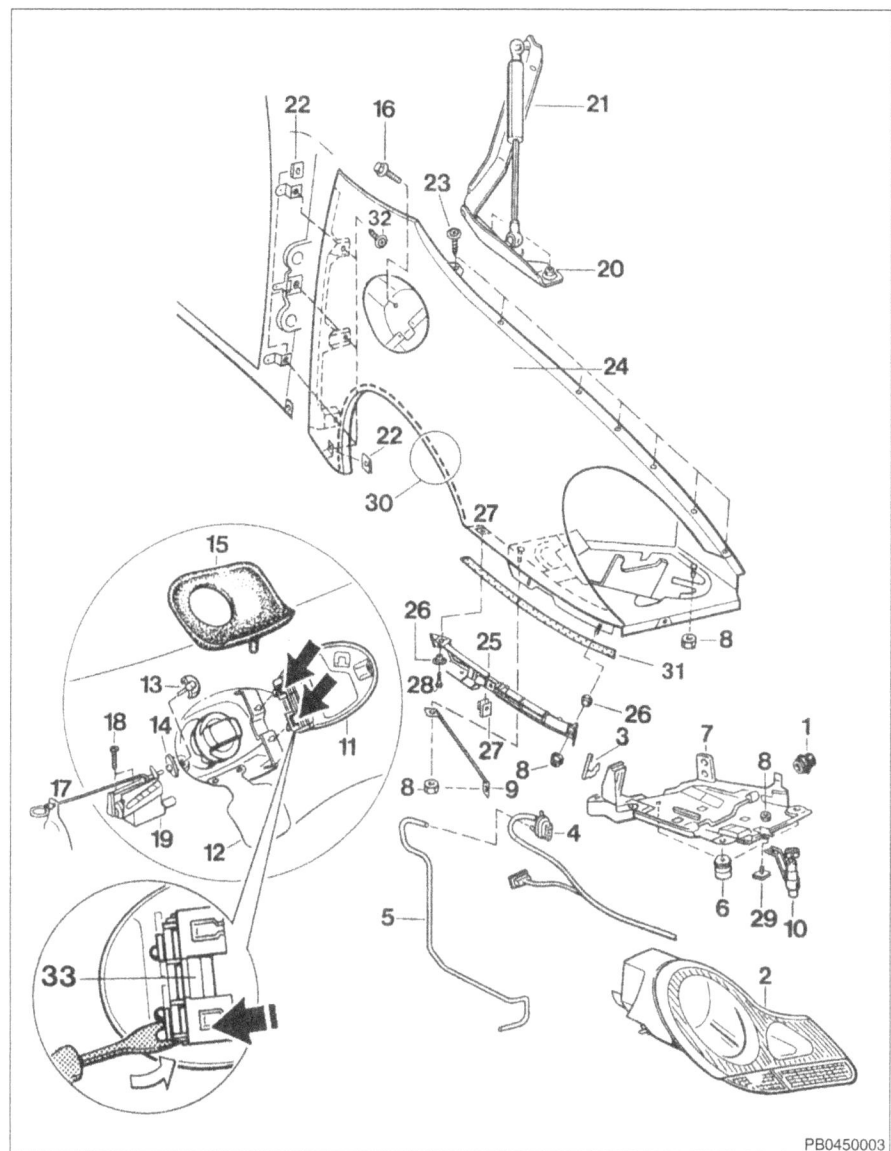

18. Sheet metal screw
19. Fuel filler flap servo motor
20. M6 bolt
21. Hinge
22. Nut
23. Sheet metal screw
24. Fender
    - When removing, first pull away and down at A-pillar and remove toward front
25. Rail
26. Spacer
27. Nuts
28. Sheet metal screw
29. Sliding piece
30. PVC seam seal
    - Replace on inside seam of wheel arch
31. Adhesive tape
32. Sheet metal screw
33. Plastic hinge

Front fender, removing and installing

− Raise car and support safely. Remove front wheel.

WARNING—
• Make sure the car is stable and well supported at all times. Use a professional automotive lift or jack stands designed for the purpose. A floor jack is not adequate support.

− Remove front wheel housing liner. See **Front wheel housing liner, removing and installing** in this repair group.

− Right fender: Remove tank filler flap. See **Fuel tank filler flap, removing and installing** in this repair group.

− Right fender: Remove carbon canister. See **Evaporative control system components** in **24 Fuel Injection**.

− Left fender: Remove windshield washer tank. See **92 Wipers and Washers**.

Water Drains

- Remove headlight and headlight mounting plate. See **94 Exterior Lights**.
- Remove front trunk lid hinge at fender. See **55 Trunk Lids**.
- For remainder of removal procedure, see diagram **Front fender components**, earlier in this repair group.

WATER DRAINS

The main water drains are located in the front cowl and convertible top / engine cover area. Additional drains are located at the fuel filler neck and in the A/C and heater housing. Drains can become clogged with leaves, pine needles and other debris. Clogged drains can cause damage to the body and allow water to leak into the passenger compartment. Periodically check water drains for blockage and clean as needed.

Front water drains, cleaning

- Working under front hood, remove battery cover and left and right cowl covers to access drains.

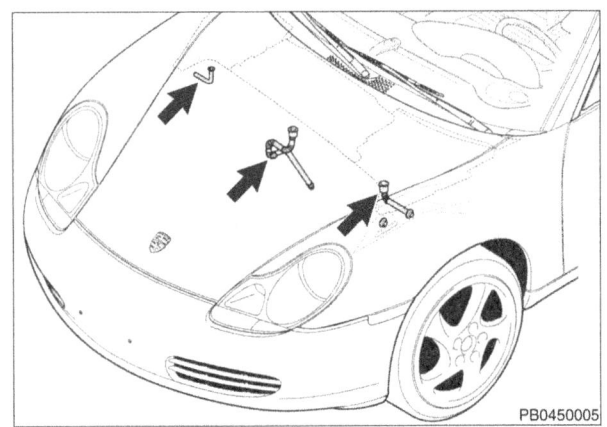

- Locate water drains (**arrows**) in center, left and right of cowl area.
- Use vacuum cleaner to remove debris.
- Carefully blow out water drain using compressed air.

NOTE—

- *To avoid water damage, make sure water drains are not pushed out of position.*

Rear water drains, cleaning

- Place convertible top in service position. See **03 Service and Maintenance**.

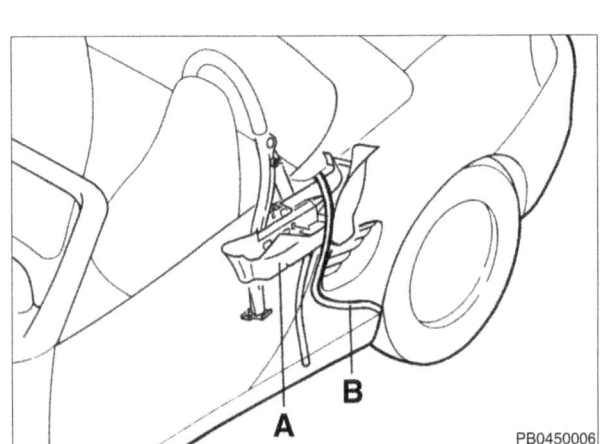

- Locate water drains in water collection tray (**A**) and water channel (**B**).
- Use vacuum cleaner to remove debris.
- Carefully blow out water drain using compressed air.

NOTE—

- *To avoid water damage, make sure water drains are not pushed out of position.*

55 Trunk Lids

GENERAL 55-1
  Emergency trunk lid release
    (2001 - 2004 models) 55-1
FRONT TRUNK LID 55-2
  Front trunk lid components 55-2
  Front trunk lid struts, removing and installing ... 55-3
  Front trunk lid, removing and installing 55-3
  Front trunk lid, aligning 55-3

REAR TRUNK LID 55-4
  Rear trunk lid components 55-4
  Rear trunk lid struts, removing and installing ... 55-4
  Rear trunk lid, removing and installing 55-5
  Rear trunk lid, aligning 55-5
TRUNK LID RELEASE 55-6
  Front and rear trunk lid release, servicing 55-6

GENERAL

This repair group covers service procedures for the front and rear trunk lid and trunk lid release mechanism.

Emergency trunk lid release (2001 - 2004 models)

If the battery is discharged on 2001 and later models, use a separate, external battery to open the trunk and charge the vehicle battery.

NOTE—

- *1997 - 2000 models: Cable operated trunk release*
- *2001 - 2004 models: Electrically powered trunk release*

◄ Working in left footwell:
  • Pull open main fuse panel cover in kick panel.
  • Pull out red emergency circuit contact (**inset**) about 3 cm (1¼ in).
  • Attach positive jumper cable clamp.

− Attach second jumper to vehicle ground such as left door striker.

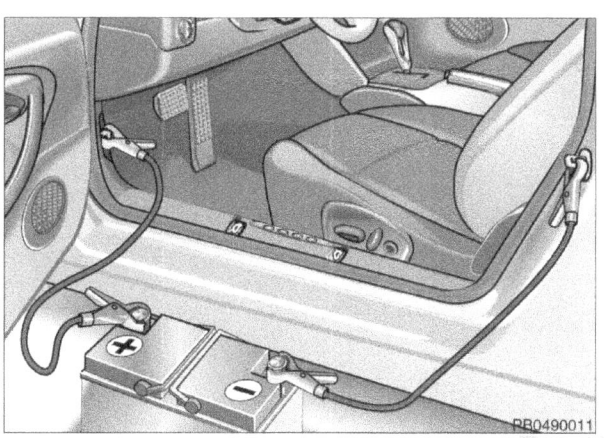

◄ Attach jumper cables to separate battery.

− Operate trunk lid release.

− Disconnect jumper cables and press emergency circuit contact back into fuse panel.

NOTE—

- *If the vehicle was locked, the alarm may sound when jumper cables are connected. To deactivate, lock and unlock the driver's door using the key.*
- *The power supply to the trunk lid release goes through a small relay located behind the fuse panel.*

Front Trunk Lid

FRONT TRUNK LID

CAUTION—
- *Do not switch on windshield wipers with the front trunk lid raised. As a precaution, remove wiper motor fuse. See* **97 Fuses, Relays, Electrical Component Locations**.
- *Trunk lid is heavy. Before removing trunk lid supports, have an assistant support the lid.*
- *The body is painted at the factory after assembly. Realignment of body panels may expose unpainted metal. Paint all exposed metal surfaces once work is complete.*

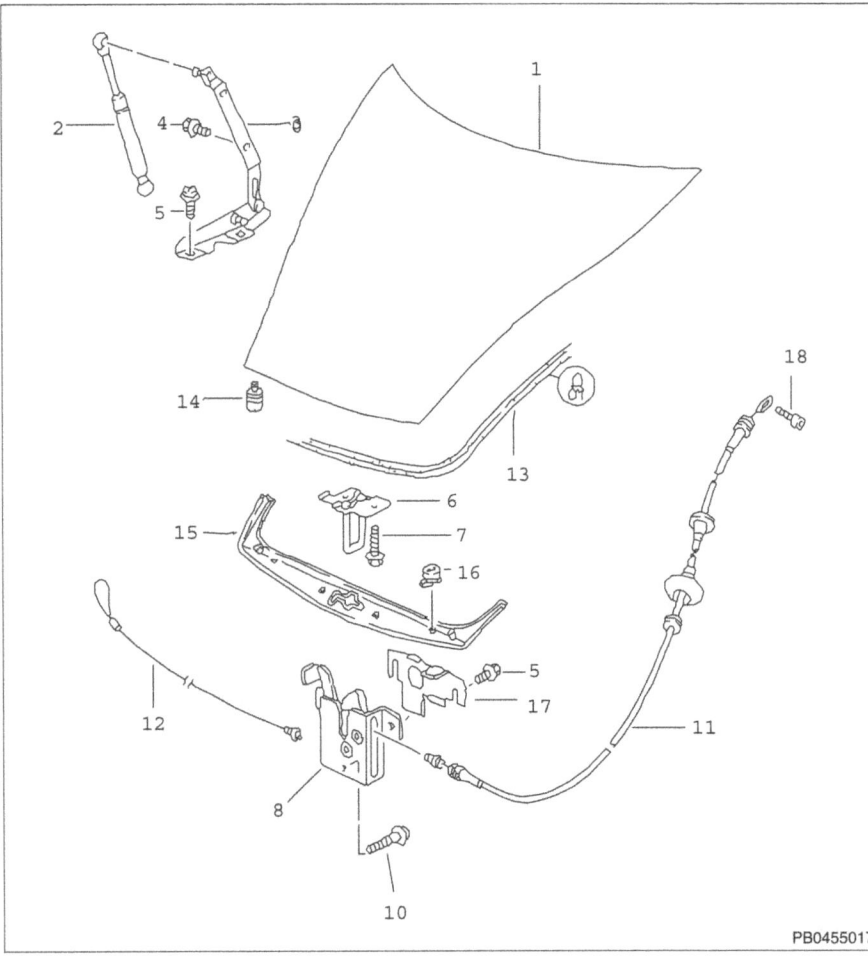

Front trunk lid components

1. Trunk lid
2. Trunk lid strut
3. Hinge
4. Bolt M6
5. Bolt M6
6. Striker
7. Bolt M6
8. Latch, mechanical (1997 - 2000)
9. Latch, electrical (2001 - 2004) (not shown)
10. Bolt M6
11. Bowden cable (1997 - 2000)
12. Emergency release cable
    - Located in right front fender
13. Trunk gasket
14. Bump stop
15. Trim
16. Trim fastener
17. Latch plate
18. Bolt M4

Trunk Lids 55-3

Front Trunk Lid

Front trunk lid struts, removing and installing

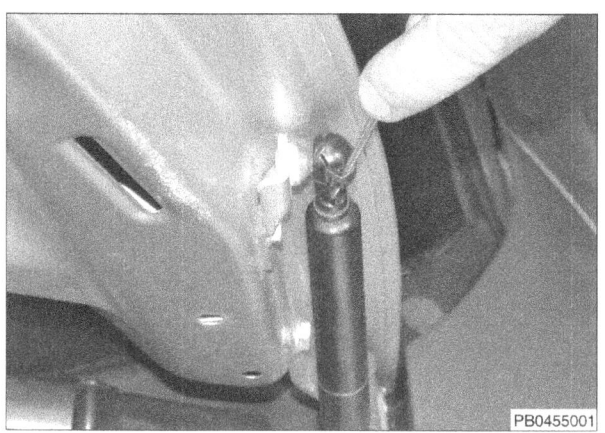

◁ Release top of gas strut from trunk lid by prying spring clip and removing strut from ball joint.

– Remove bottom of gas strut from body in a similar fashion.

– Repeat procedure for other side while an assistant supports trunk lid.

– To install, press strut socket onto ball joint until spring clip locks in place.

Front trunk lid, removing and installing

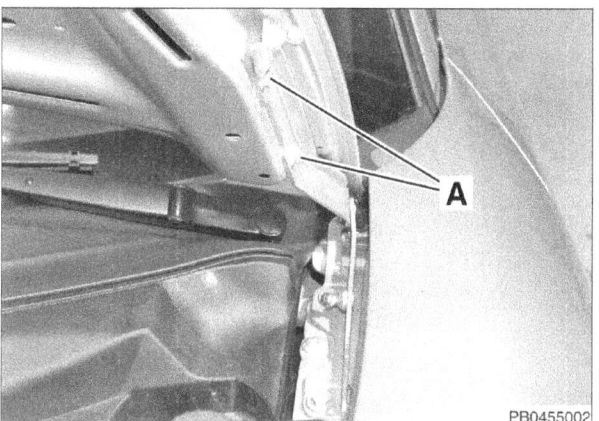

– Remove gas struts from trunk lid. See **Front trunk lid struts, removing and installing** in this repair group.

◁ With an assistant supporting trunk lid, remove hinge attachment bolts (**A**) from left and right sides and remove trunk lid.

– Installation is reverse of removal. Remember to:
  • Tighten hinges in previously installed position to maintain trunk lid alignment.
  • Press strut sockets onto ball joints until spring clips lock in place.

Tightening torque	
Trunk lid hinge to front trunk lid	10 Nm (7.5 ft-lb)

Front trunk lid, aligning

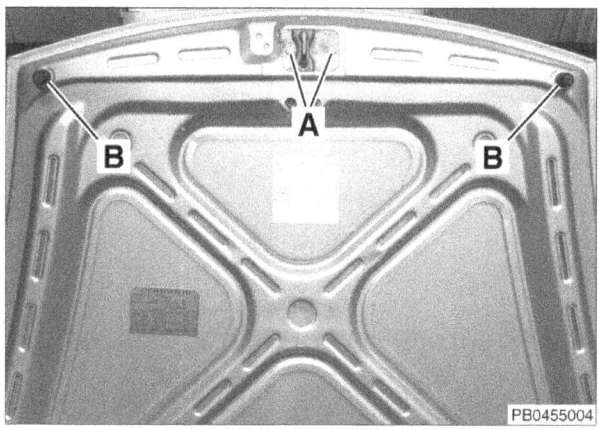

If trunk lid was removed, make sure hinges are aligned as close to the original mounting points as possible. Use touch up paint on any exposed surfaces.

◁ In addition to the hinge attaching bolts, front trunk lid alignment aids include the lid striker and rubber bump stops:
  • Adjust lid striker by loosening attaching bolts (**A**) and moving it, if necessary.
  • Adjust rubber bump stops (**B**) in or out, if necessary.

– Before shutting trunk lid, check the striker-to-lock alignment. Adjust striker as needed.

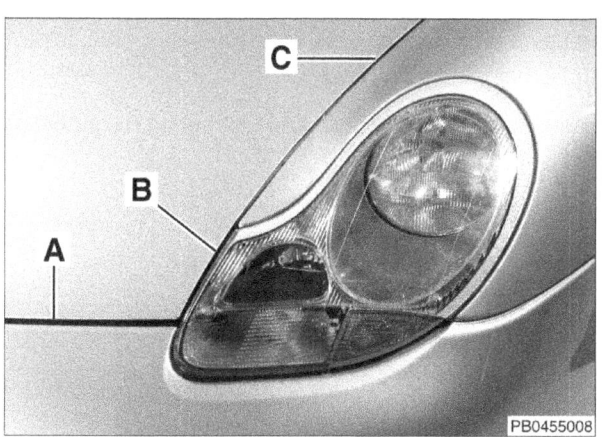

◁ Close trunk lid and check alignment. Align lid so that gap dimensions match the table below.

Front trunk lid gap dimensions	
Front trunk lid to front bumper (**A**)	5.8 ± 1.0 mm (0.23 ± 0.04 in)
Front trunk lid to headlight assembly (**B**)	4.5 ± 1.0 mm (0.18 ± 0.04 in)
Front trunk lid to front fender (**C**)	4 ± 0.5 mm (0.16 ± 0.02 in)

55-4 Trunk Lids

Rear Trunk Lid

REAR TRUNK LID

CAUTION—
- *The trunk lid is heavy. Before removing trunk lid supports, have an assistant support the lid.*
- *The body is painted at the factory after assembly. Realignment of body panels may expose unpainted metal. Paint all exposed metal surfaces once work is complete.*

Rear trunk lid components

1. Trunk lid
2. Trunk lid strut
3. Hinge
4. Bolt M6
5. Bolt M6
6. Striker
7. Latch
   - Mechanical (1997 - 2000)
   - Electrical (2001 - 2004) (not shown)
8. Latch plate
9. Bolt M6
10. Bowden cable (1997 - 2000)
11. Emergency release cable
    - Located in left rear of trunk
12. Locking clip
13. Trunk gasket
14. Bump stop
15. Pull rod
16. Bolt M4

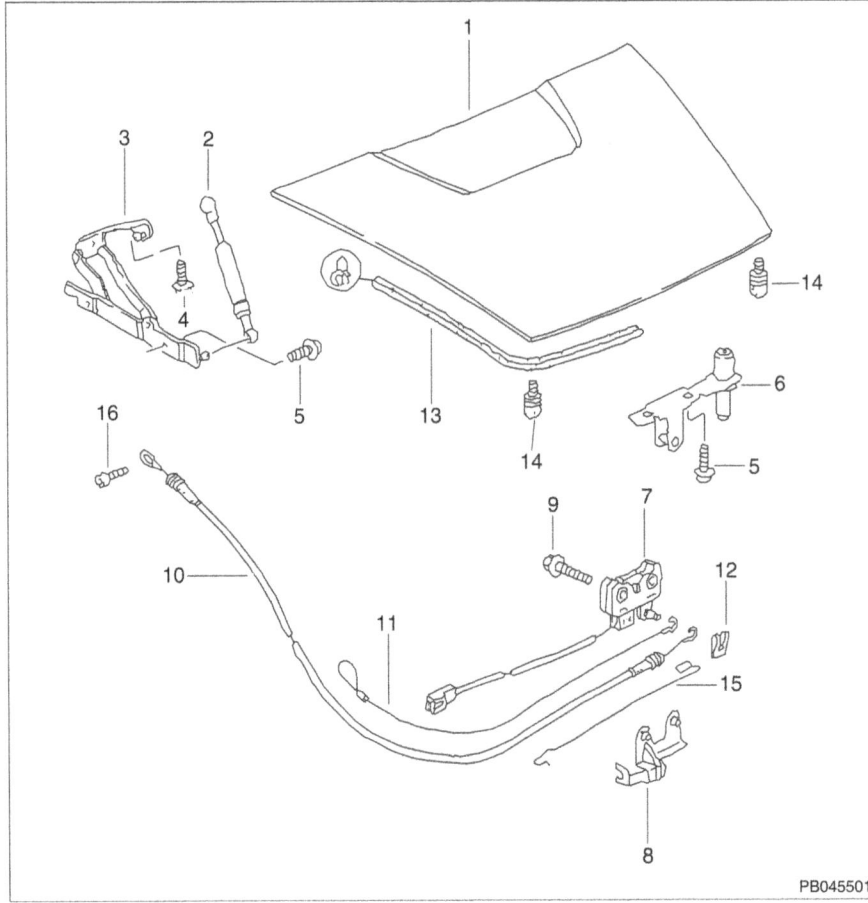

Rear trunk lid struts, removing and installing

◁ Release top and bottom of gas strut from trunk lid by prying spring clips (**arrows**) and removing strut ends from ball joints.

– Repeat procedure for other side while an assistant supports trunk lid.

– To install, press strut socket onto ball joint until spring clip locks in place.

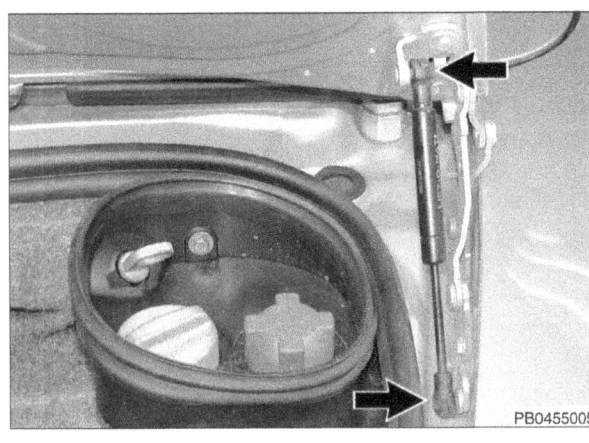

Trunk Lids 55-5

Rear Trunk Lid

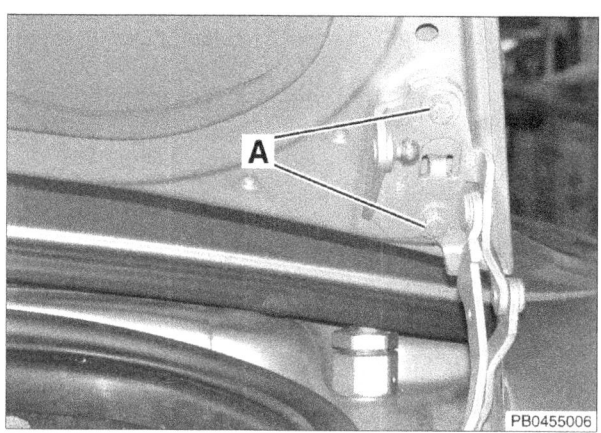

Rear trunk lid, removing and installing

- Remove top struts from trunk lid. See **Rear trunk lid struts, removing and installing** in this repair group.

◀ With an assistant supporting trunk lid, remove hinge attachment bolts (**A**) from left and right sides and remove lid.

- Installation is reverse of removal. Remember to:
  - Tighten hinges in previously installed position to maintain lid alignment.
  - Press strut sockets onto ball joints until spring clips lock in place.

Tightening torque	
Trunk lid hinge to rear trunk lid	10 Nm (7.5 ft-lb)

Rear trunk lid, aligning

If trunk lid was removed, make sure hinges are aligned as close to the original mounting points as possible. Use touch up paint on any exposed surfaces.

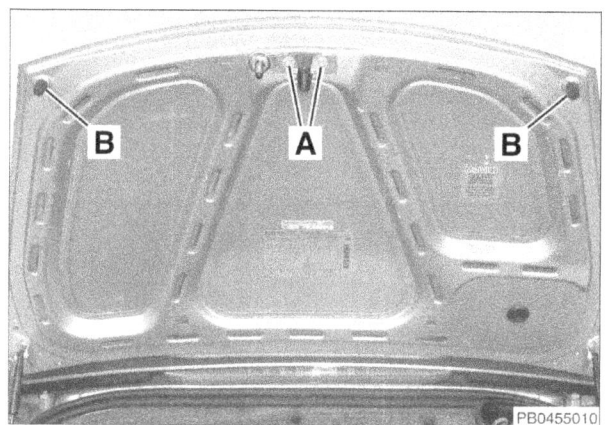

◀ In addition to hinge attaching bolts, rear trunk lid alignment aids include the lid striker and rubber bump stops:
- Adjust lid striker by loosening attaching bolts (**A**) and moving it, if necessary.
- Adjust rubber bump stops (**B**) in or out, if necessary.

- Before shutting trunk lid, check striker-to-lock alignment. Adjust striker as needed.

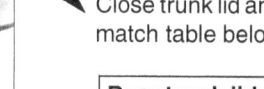

◀ Close trunk lid and check alignment. Align lid so that gap dimensions match table below.

Rear trunk lid gap dimensions	
Rear trunk lid to spoiler (**A**)	5 ± 1.0 mm 0.20 ± 0.04 in
Rear trunk lid to fender (**B**)	4.5 ± 1.0 mm 0.18 ± 0.04 in
Rear trunk lid to convertible top cover (**C**)	6 ± 1.0 mm 0.24 ± 0.04 in

55-6 Trunk Lids

Trunk Lid Release

TRUNK LID RELEASE

Front and rear trunk lid release, servicing

− Disconnect negative (−) cable from battery.

> **CAUTION—**
> • Prior to disconnecting the battery, read the battery disconnection cautions given in **00 Warnings and Cautions**.

− Remove driver's seat to access release mechanism. See **72 Seats** for more information.

◄ Pry off plastic covers and loosen allen bolts (**arrows**) 4 or 5 turns.

− Pry cover upwards using a plastic trim tool.

> **NOTE—**
> • 1997 - 2000 mechanical trunk lid release shown. 2001 - 2004 models use an electrical release. Disassembly is similar.

◄ Remove bolts (**arrows**) to remove front or rear Bowden cable from housing, as necessary.

> **NOTE—**
> • 2001 - 2004 models: No cable is used. Release electrical connectors to service interior switches.

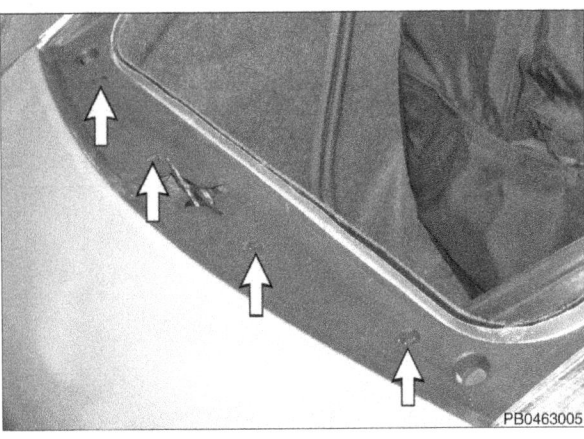

◄ Working in front trunk, turn trim fasteners 90° and remove trim.

Trunk Lids 55-7
Trunk Lid Release

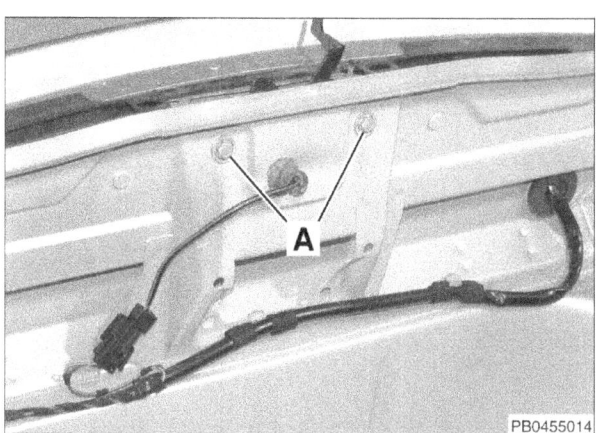

◂ Lift off trunk lid gasket and fold down front trunk trim to expose trunk lid latch mechanism. Remove bolts (**A**) and push latch out through top opening.

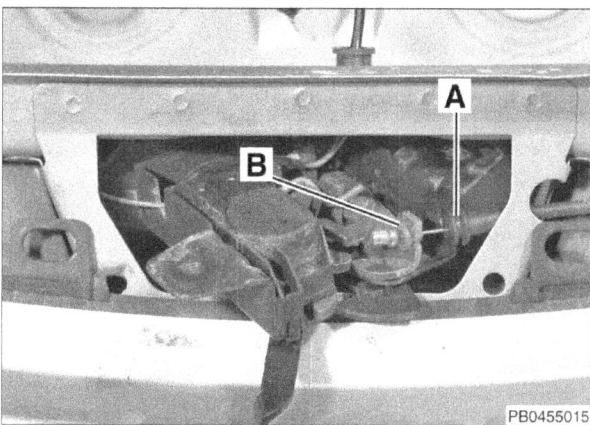

◂ Release Bowden cable housing (**A**) and cable end (**B**) from latch mechanism to service cable.

NOTE—
- *2001-2004 models: No cable is used. Release electrical connectors to service trunk lid latch.*

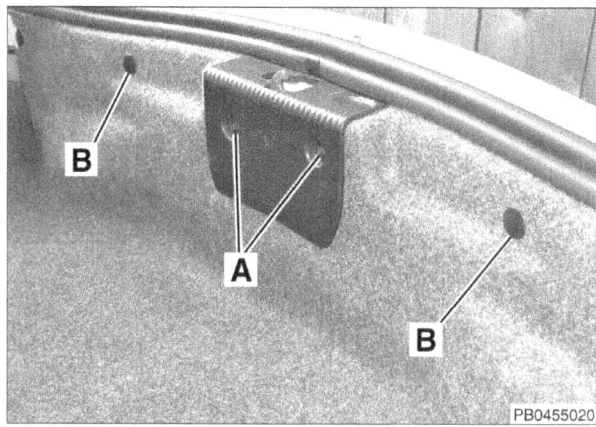

◂ Working in rear trunk, remove screws (**A**) and remove latch cover.

– Remove trim fasteners (**B**) and fold down rear trim to gain access to latch mechanism.

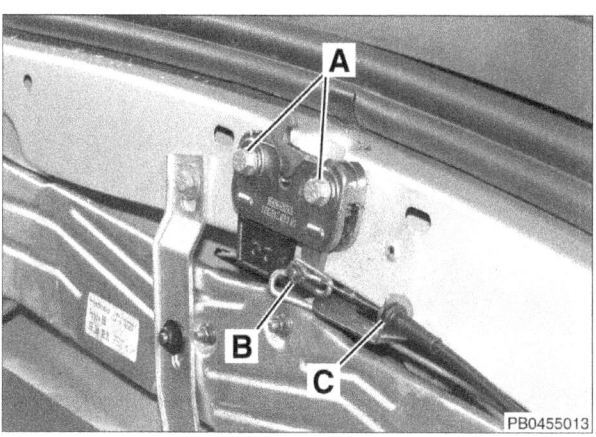

– Remove latch bolts (**A**) to service latch.

– Remove fastening clip (**B**) and cable housing (**C**) to service Bowden cable.

NOTE—
- *2001-2004 models: No cable is used. Release electrical connectors to service trunk lid latch.*

57 Doors and Locks

GENERAL 57-1

DOORS 57-1
   Door, removing and installing 57-1
   Door, adjusting 57-2

DOOR TRIM PANELS 57-3
   Door trim panel, removing and installing 57-3

DOOR LOCK REPLACEMENT 57-6
   Door striker, replacing 57-6
   Door handle and lock cylinder,
     removing and installing 57-6
   Door lock, removing and installing 57-7

GENERAL

This repair group covers removal and installation of the doors, door trim panels and door lock components. For related information see:

- **64 Door Windows** for power door windows and door glass removal.
- **96 Interior Lights, Anti-theft** for central locking and anti-theft system theory and operation.

> *WARNING—*
> - *Boxster models are equipped with side impact airbags installed in the doors. Read the warnings and cautions in **69 Seat Belts, Airbags** prior to starting work on door.*

DOORS

Door, removing and installing

— Disconnect negative (-) cable from battery.

> *WARNING—*
> - *Wait 1 minute after disconnecting battery before working in vicinity of airbag.*

> *CAUTION—*
> - *Prior to disconnecting the battery, read the battery disconnection cautions given in **00 Warnings and Cautions**.*

 Remove fasteners (**A**) on upper and lower door hinge pins.

> *NOTE—*
> - *2001 - 2004 models: counterhold top of hinge pin while removing lower fastener.*

Doors

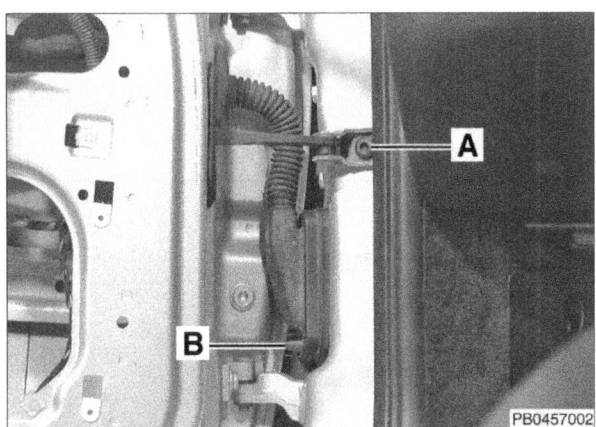

◄ Remove bolt (**A**) on door check.

> **CAUTION—**
> • With door check disconnected, door can hyper-extend, damaging door and body panels.

– Remove fastener (**B**) on wiring harness connector. Lift connector up and out of door post.

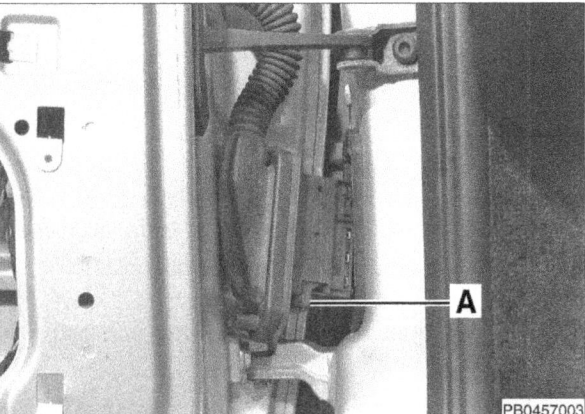

◄ Pull down on locking lever (**A**) and separate harness connector halves.

– Lift door up and off hinge pins to remove.

– Installation is reverse of removal. Remember to:
  • Replace door hinge pin fasteners with new.
  • Make sure wiring harness locking lever is fully seated.
  • Lock connector in door post opening by pushing down.

Tightening torques	
Door check to body	23 Nm (17 ft-lb)
Door hinge fastener to hinge pin (1997 - 2000)	10 Nm (7.5 ft-lb)
Door hinge fastener to hinge pin (2001 - 2004)	13 Nm (9.5 ft-lb)
Wiring harness connector to body	2.5 Nm (2 ft-lb)

Door, adjusting

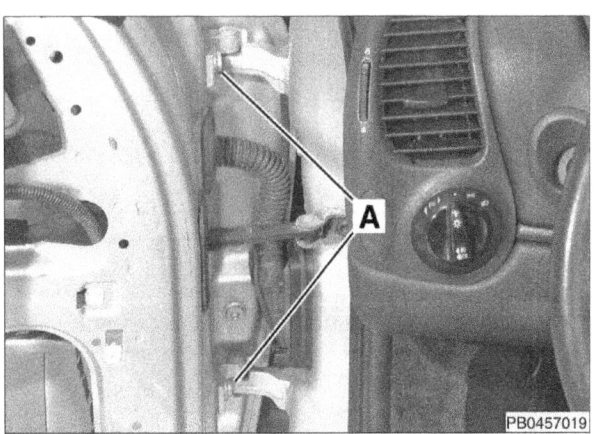

◄ Loosen door hinge bolts (**A**) to adjust door. If necessary, remove centering sleeves to increase adjustment range.

NOTE—
• *The body is painted at the factory after assembly. Realignment of the body panels may expose unpainted metal. Paint all exposed surfaces once work is complete.*

Doors and Locks 57-3

Door Trim Panels

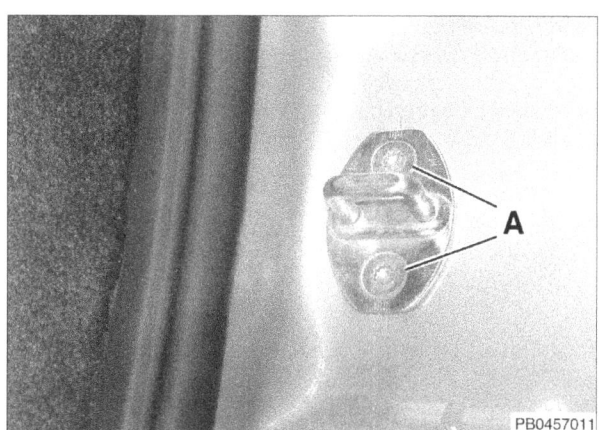

◄ Loosen striker bolts (**A**) to adjust striker.

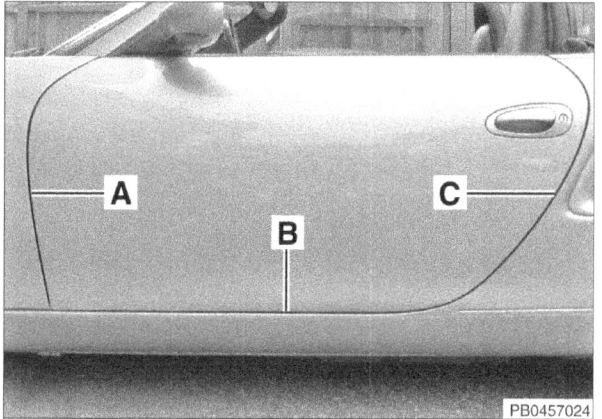

◄ Close door and check that gap dimensions match those given below.

Door gap dimensions	
Door to front fender (**A**)	5.0 + 1.0 mm (0.2 +0.04 in)
Door to sill (**B**)	4.4 +1.0 -0.5 mm (0.17 +0.04 -0.02 in)
Door to rear fender (**C**)	4.0 +1.0 -0.5 mm (0.16 +0.04 -0.02 in)

DOOR TRIM PANELS

Door trim panel, removing and installing

— Disconnect negative (-) cable from battery.

> **WARNING—**
> - Wait 1 minute after disconnecting battery before working in vicinity of airbag.

> **CAUTION—**
> - Prior to disconnecting the battery, read the battery disconnection cautions given in **00 Warnings and Cautions**.

> *NOTE—*
> - Trim panel plastic clips are easily damaged. Have extra clips on hand before beginning procedure.

◄ Working with the door open, use a plastic trim tool to pry up on back edge of inner door handle trim while sliding trim (**arrow**) toward front of car.

57-4 Doors and Locks

Door Trim Panels

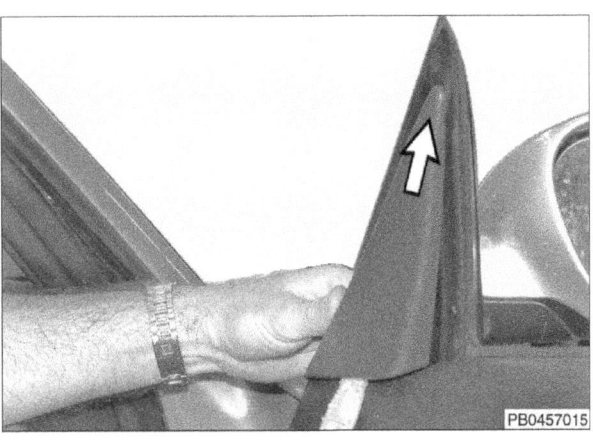

◄ Use a plastic trim tool to pry up on lower edge of window trim panel while sliding trim upward (**arrow**).

– Pry small trim covers off airbag cover and inner door pull to expose fasteners.

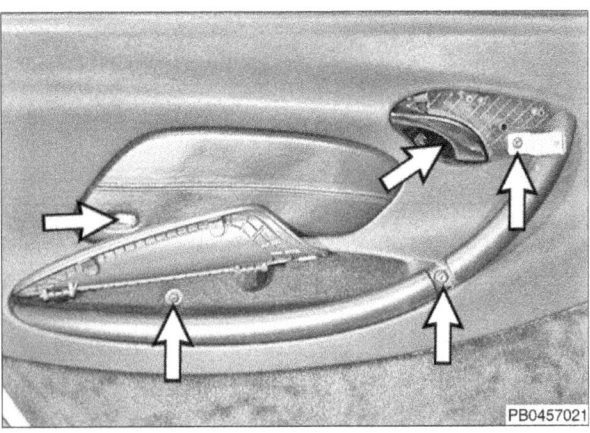

◄ Remove fasteners (**arrows**) on door pull and door trim panel.

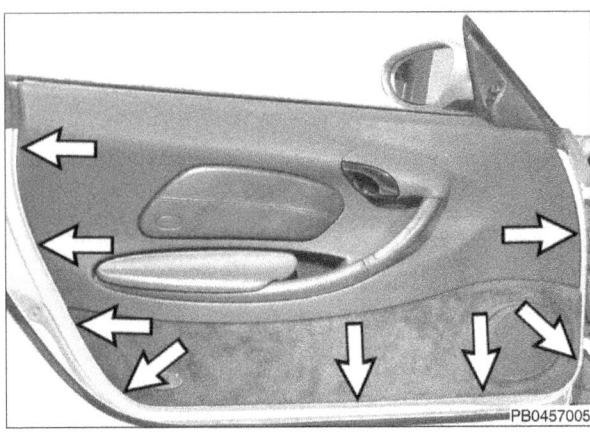

◄ Using a plastic trim tool, pry out door panel trim clips (**arrows**).

– Lift door panel up and off window sill.

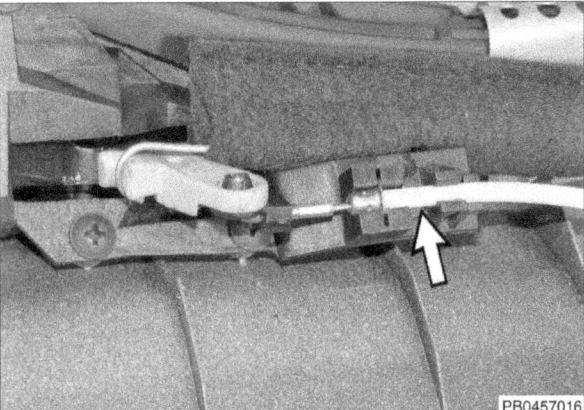

◄ Working at back of panel, cut plastic tie and release lock actuator cable from holder (**arrow**).

Doors and Locks 57-5

Door Trim Panels

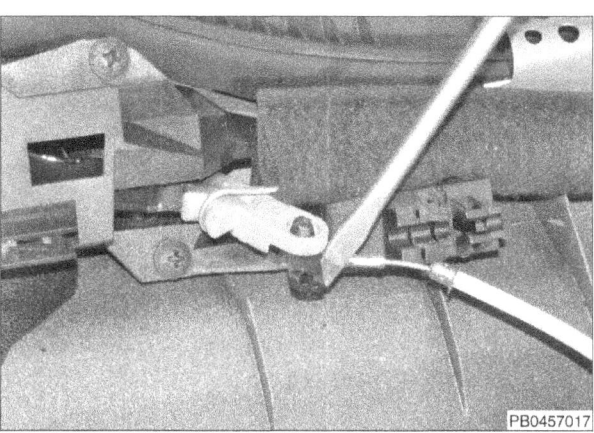

◁ Release locking clip at end of actuator cable and remove cable from inner door handle.

– Release electrical connectors at back of inner door handle and at lower courtesy light.

– Remove door trim panel.

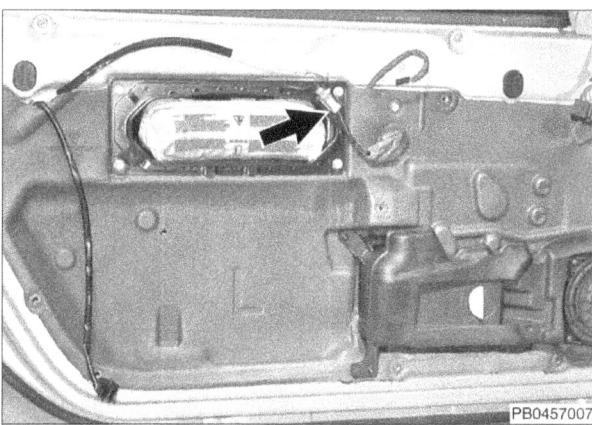

◁ If further door disassembly is required, release side airbag electrical connector (**arrow**).

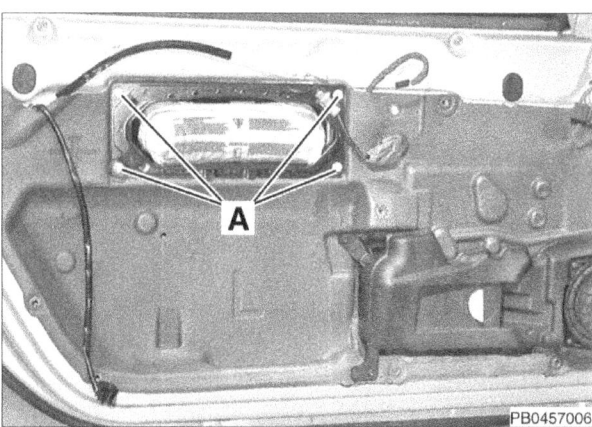

◁ Remove airbag fasteners (**A**) and remove airbag.

> **WARNING—**
> - *To avoid injury, store airbag in a safe place with airbag facing up.*

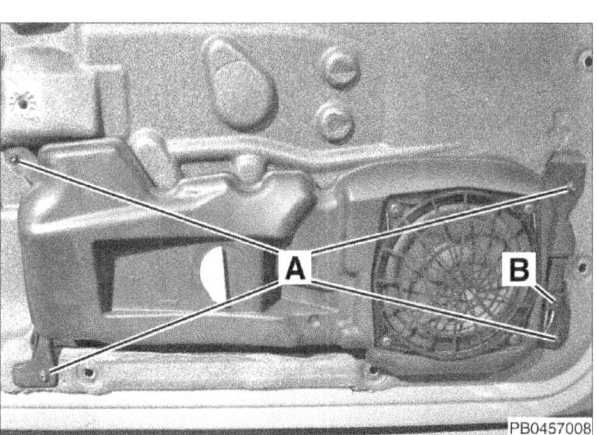

◁ Remove speaker enclosure fasteners (**A**) and release electrical connector (**B**) to remove speaker enclosure.

57-6 Doors and Locks

Door Lock Replacement

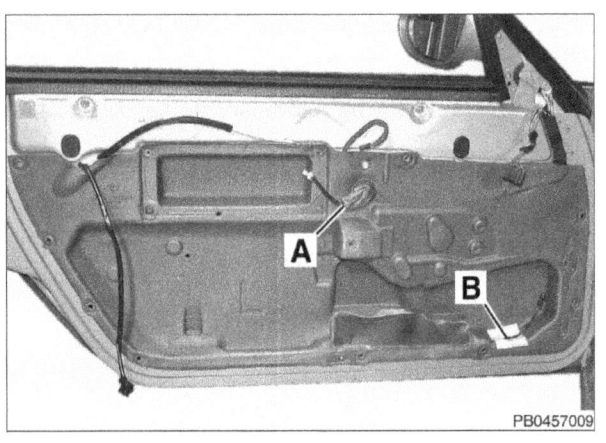

◀ Carefully peel insulating liner off door. Guide air bag wiring (**A**) and speaker wiring (**B**) out of liner.

- Remainder of installation is reverse of removal. Remember to:
  - Reinstall insulating liner in original position. Do not block bolt and trim clip holes.
  - Make sure all electrical connectors are securely fastened.
  - Replace plastic tie holding door lock actuator cable to bracket on back of inner door handle.
  - Replace missing or broken trim clips.

DOOR LOCK REPLACEMENT

Door striker, replacing

For door alignment information see **Door, adjusting** in this repair group.

- Mark position of door striker.

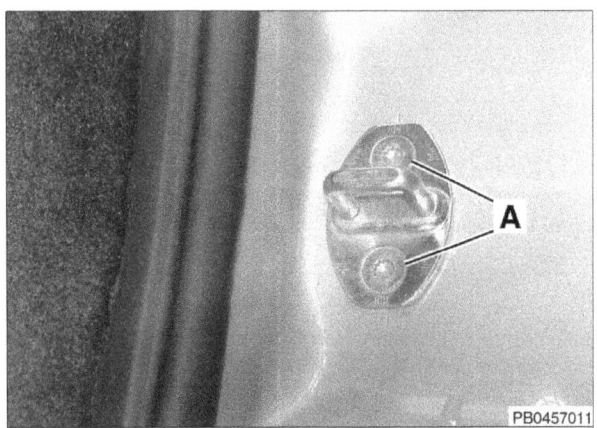

◀ Remove fasteners (**A**) and remove striker.

- Replace striker in previously installed position to maintain door adjustment.

Door handle and lock cylinder, removing and installing

- Remove door trim panel and insulating liner. See **Door trim panel, removing and installing** in this repair group.

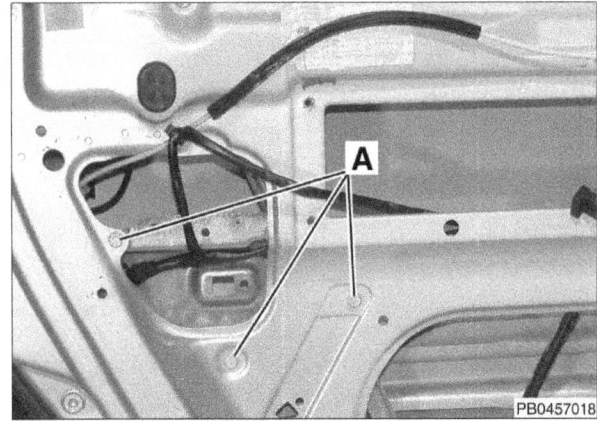

◀ Remove fasteners (**A**) holding cover plate to door skin to gain access to door handle assembly.

- Release door handle electrical connector.

Doors and Locks 57-7

Door Lock Replacement

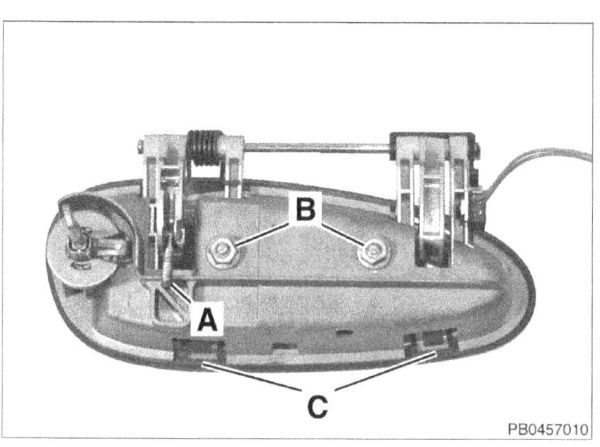

◄ Working inside door cavity, release door handle actuating rod (**A**) from door lock mechanism.

– Remove door handle fasteners (**B**) from lock cylinder bracket.

– Push up on locking tabs (**C**) and tilt door handle out of lock cylinder bracket and door.

NOTE—

• *Door handle shown removed for clarity.*

– With door handle removed, release lock cylinder arm from door lock mechanism. Remove lock cylinder along with lock cylinder bracket out of cavity.

– Remove lock cylinder retaining clip and remove cylinder from bracket.

– Installation is reverse of removal. Remember to:
  • Engage lock cylinder arm in door lock mechanism.
  • Reattach electrical connector and door handle actuating rod.
  • Test handle and lock function before attaching door panel.

Tightening torque	
Door handle to door lock bracket	10 Nm (7.5 ft-lb)

Door lock, removing and replacing

– Remove door trim panel and insulating liner. See **Door trim panel, removing and installing** in this repair group.

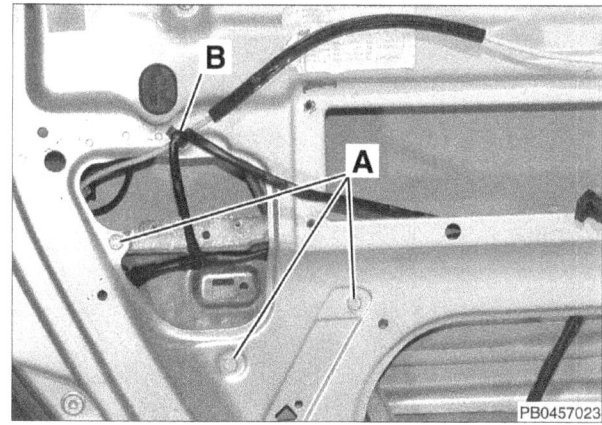

◄ Remove fasteners (**A**) holding cover plate to door skin to gain access to door handle assembly and door lock.

– Remove plastic clip (**B**) holding lock actuator cable.

– Working inside door cavity, release door lock electrical connector.

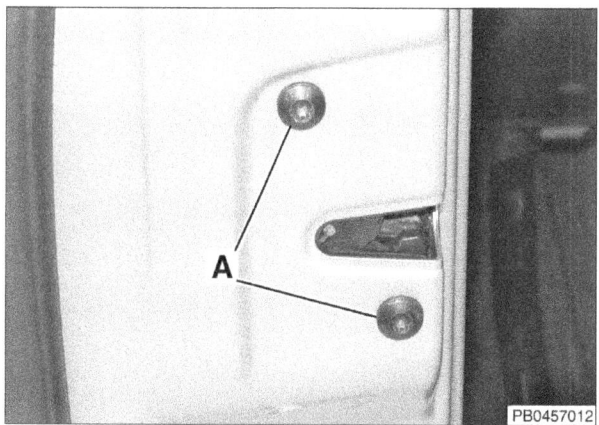

◄ Remove bolts (**A**) holding door lock to door. Remove door lock with actuating motor out of cavity.

57-8 Doors and Locks

Door Lock Replacement

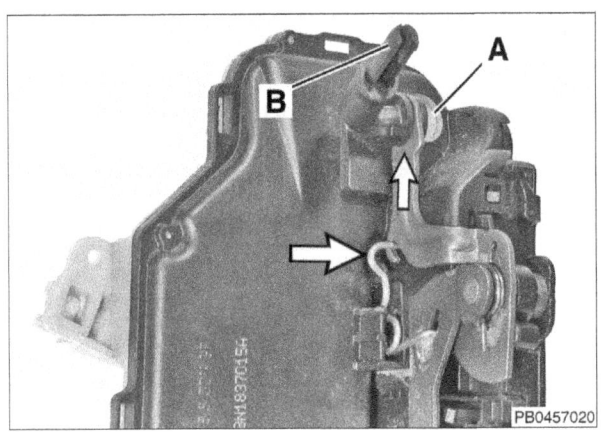

◀ Before reinstalling door lock, lift actuating lever and affix guide spring to spring hook (**arrows**).

> **CAUTION—**
> - Spring must be engaged in spring hook before installing door lock. Failure to do so can result in lock and alarm malfunction.

− Remainder of installation is reverse of removal. Remember to:
  - Guide lock cylinder arm into place on door lock (**A**).
  - Attach door handle actuating rod to door lock lever (**B**).
  - Attach servo motor electrical connector.

NOTE—
- Guide spring is released the first time handle is used. An initial increase in effort to operate door handle is normal.

Tightening torque	
Door lock to door	20 ± 2 Nm (15 ± 1.5 ft-lb)

61 Convertible Top

GENERAL ... 61-1
Convertible top components ... 61-1
Convertible soft top ... 61-2
Convertible top operation ... 61-2
Convertible top, opening and closing ... 61-3
Convertible top emergency closing ... 61-4
Troubleshooting ... 61-5
Warnings and Cautions ... 61-6

CONVERTIBLE TOP DRIVE SERVICE ... 61-7
Convertible top drive motor,
  removing and installing ... 61-7
Convertible top transmission,
  removing and installing ... 61-8

TABLES
a. Convertible top microswitch status ("open") ... 61-6
b. Convertible top microswitch status ("close") ... 61-6

GENERAL

This repair group covers convertible top normal and emergency operation. Troubleshooting, component locations and mechanical repair procedures are also provided. Convertible soft top repairs are not included in this manual.

See also:

- **03 Service and Maintenance** for convertible top service position.
- **EWD Electrical Wiring Diagrams** for circuit details.

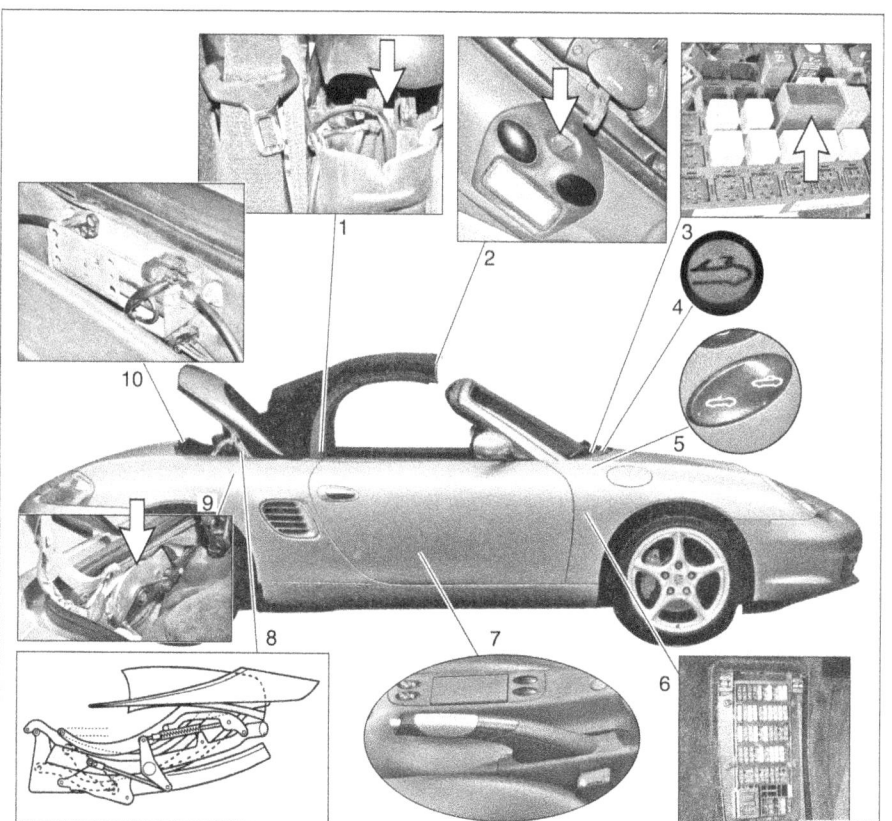

Convertible top components

1. **B-pillar microswitch (1997 - 1999)**
   - Behind left B-pillar trim
2. **Top lock microswitch**
3. **Convertible top control module**
   - Relay panel 1
4. **Convertible top warning light in instrument cluster**
5. **Convertible top toggle switch**
6. **Fuse panel**
   - In left footwell
   - Fuses B6 and D3
7. **Parking brake switch**
8. **Convertible top mechanism**
9. **Convertible top transmission**
   - Behind each seat in side panel
   - 2000 - 2004: Right transmission contains microswitch
10. **Convertible top drive motor and compartment lid microswitch**

61-2 Convertible Top

General

Convertible soft top

The convertible soft top is one piece which folds in a Z-shape. The soft top has a 3-layer structure:

- Top layer: Synthetic fiber ensures resistance to weathering.
- Middle layer: Rubber prevents leakage.
- Bottom layer: Cotton fabric provides a soft feel for the top interior.

The special weave and bonding of the layers ensures high quality and appearance.

There are two versions of the rear window:

- 1997 - 2002: Soft plastic rear window is glued and stitched into position. It is not replaceable separately.
- 2003 - 2004: Integrated tempered glass rear window is equipped with defroster electrical grid.

Convertible top operation

The convertible top is actuated by a single motor which drives two transmissions (one on each side of vehicle) via drive cables.

- Each transmission operates a drive lever (**9**).
- Each drive lever uses one link (**8**) to operate the convertible top frame.
- The drive lever uses another link (**10**) to operate the convertible top compartment lid or cover.

1. Convertible top frame
2. Main bow
3. Push bar
4. B-pillar arm
5. Drive rod
6. Tension bow pivot
7. Support bracket
8. Convertible top drive operating link (red-tipped)
9. Transmission drive lever
10. Convertible top compartment lid operating link (black-tipped)
11. Hinge lever
12. Guide arm
13. Convertible top compartment lid
14. Tension bow

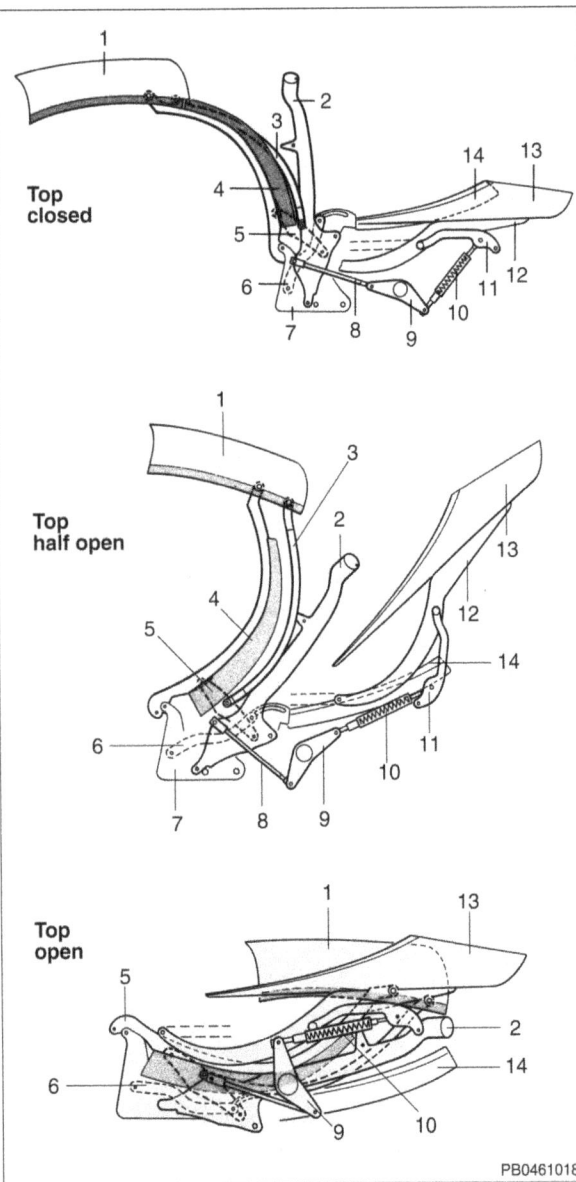

Convertible Top 61-3

General

Convertible top, opening and closing

Opening

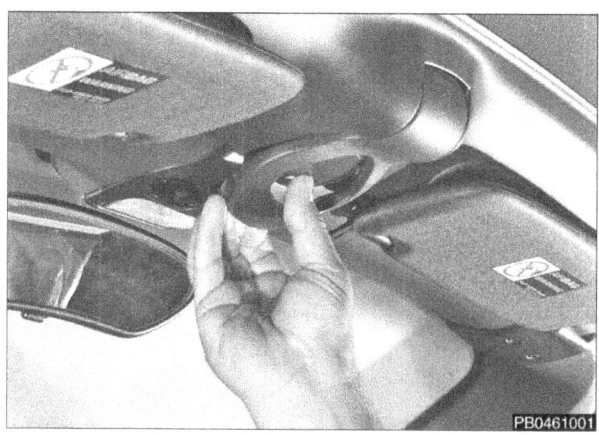

◄ With vehicle stationary, parking brake engaged and ignition ON, press top locking lever center button and pull lever down and back.

NOTE—
* *If windows are up, they automatically lower a few inches.*

◄ Press convertible top rocker switch (**inset**) in center dashboard.

◄ Keep switch pressed until convertible top warning light (**inset**) in instrument cluster goes OFF.

Closing

CAUTION—
* *Make sure rear shelf storage compartment is solidly locked down and does not interfere with convertible top closing.*
* *Make sure sun visors and vanity mirrors are folded down and out of the way.*

– With vehicle stationary, parking brake engaged and ignition ON, press convertible top rocker switch.

NOTE—
* *If windows are up, they automatically lower a few inches.*

– With top partially closed, release rocker switch. Then make sure top locking lever is in unlocked position.

◄ Press rocker switch again and keep pressed until convertible top warning light in instrument cluster goes OFF. If necessary, assist top by pulling on handle recesses in front frame. Lock convertible top against top of windshield.

61-4 Convertible Top

General

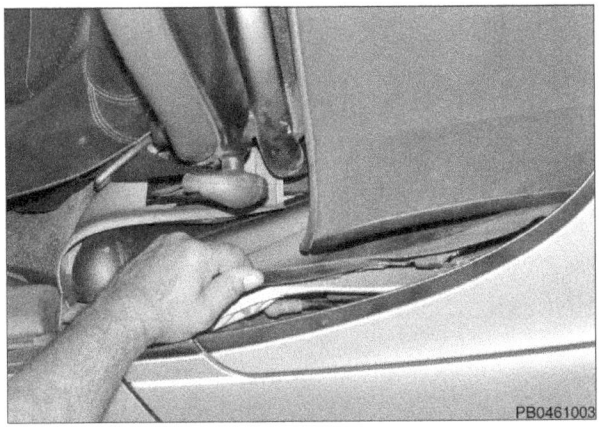

Convertible top emergency closing

In case of electrical failure, use the following procedure to close the convertible top manually.

> **WARNING—**
> • Perform the emergency top closing with great care. Serious personal injury may result if the top or lid are not held firmly in place when links are disconnected.

- Turn engine OFF and remove ignition key.

◄ Working behind seat, remove plastic convertible top link covers.

◄ Use a large screwdriver to pry convertible top compartment lid (black-tipped) links off top drive levers on each side.

> **CAUTION—**
> • Work carefully to prevent damage to body paint work.
> • If convertible top compartment lid is stuck half-way up, make sure an assistant supports lid when link on the other side is pried off.

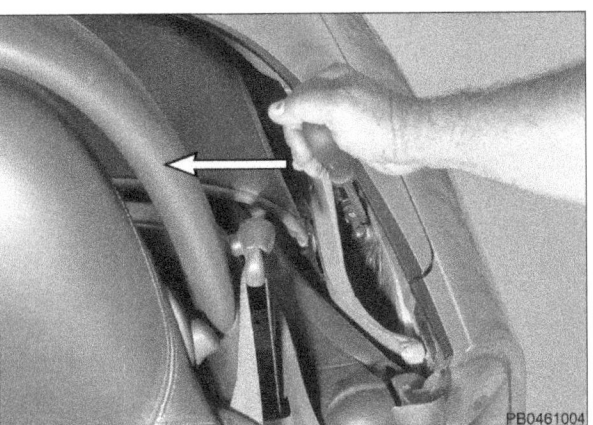

◄ Lift and tilt compartment lid backward and rest in open position.

◄ Use a large screwdriver to pry red-tipped links off convertible top drive levers on each side.

> **CAUTION—**
> • If convertible top is stuck half-way up, make sure an assistant supports the top when the link on the other side is pried off.

- Lift convertible top using both hands and lock against top of windshield.

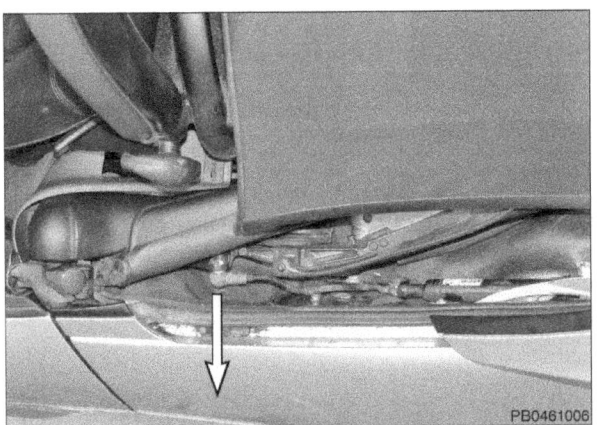

Convertible Top 61-5

General

- Manually close convertible top compartment lid.

> **CAUTION—**
> - *Convertible top compartment lid is not secured shut under these circumstances. Drive at moderate speeds (approx. 45 mph) until the top mechanism is repaired.*

Troubleshooting

When troubleshooting convertible top malfunctions, keep in mind the following:

- When attempting to open or close top, make sure vehicle is stationary and parking brake is set.
- Make sure ignition switch is ON.

Power supply

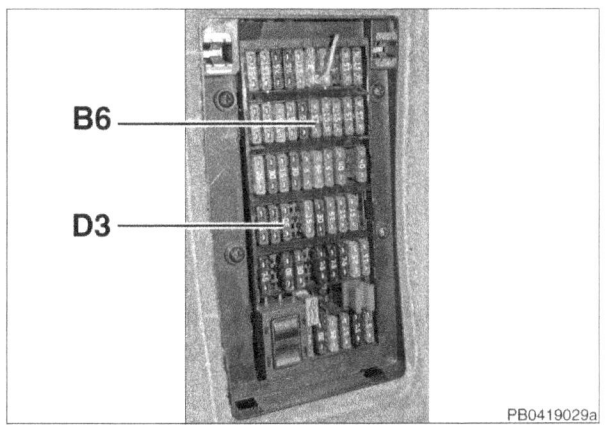

◀ **Fuse**. Remove main fuse panel cover at left footwell and check convertible top fuses:

- **B6**: Convertible top control module
- **D3**: Convertible top drive motor

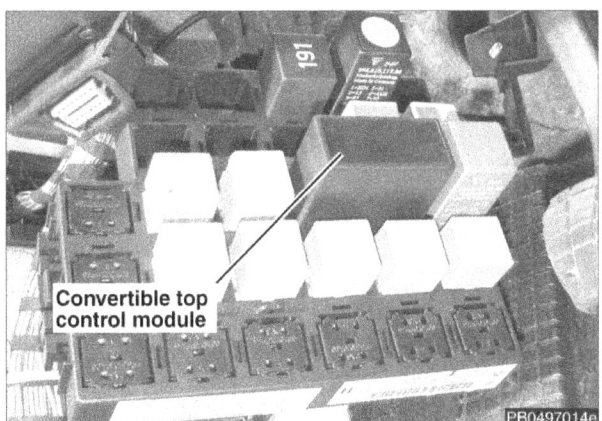

◀ **Control module**. Lower relay panel 1 behind left side of dashboard. See **97 Fuses, Relays, Component Locations** for access information.

Linkage

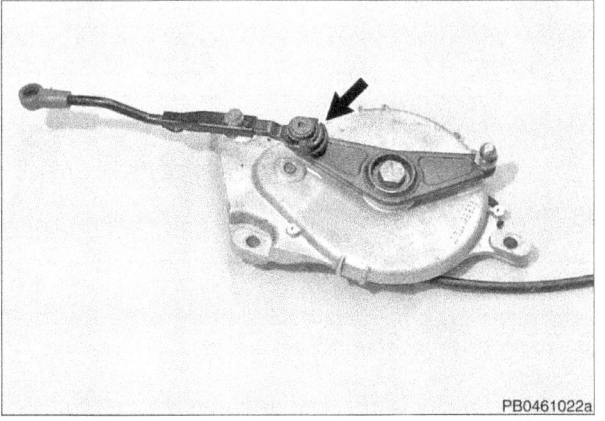

- Check pins, cables and linkages for wear or excessive play.

◀ **Rubber bushing (arrow)** at connection of convertible top link to transmission driver lever (Porsche part no. 986 561 881 00) is especially subject to wear.

61-6 Convertible Top

General

Microswitches

- Check convertible top microswitches for continuity. See **Table a** and **Table b** for microswitch continuity information.

◀ **Top lock microswitch** (**arrow**) at top center of windshield frame. If top is unlocked, microswitch is grounded.

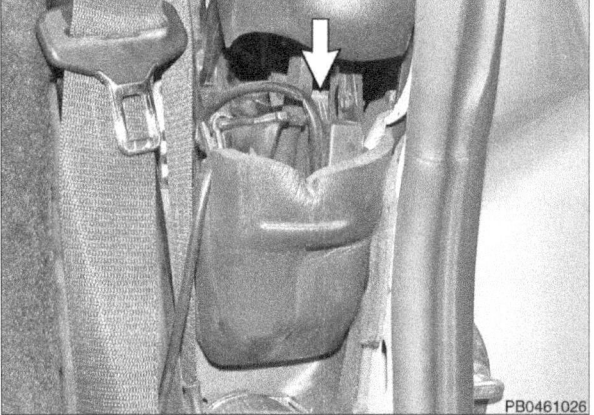

◀ **B-pillar microswitch (1997- 1999)** (**arrow**) underneath left B-pillar trim. For access information see **70 Interior Trim**.

◀ **Compartment lid microswitch** (**arrow**) at convertible top drive motor, center rear of convertible top compartment.

Table a. Convertible top microswitch status (convertible top switch in "open" mode)		
Convertible top position	B-pillar microswitch	Compartment lid microswitch
Top closed	Ground	Open
Compartment lid open	Ground	Ground
Top opening	Open	Ground
Top open	Open	Open

Table b. Convertible top microswitch status (convertible top switch in "close" mode)		
Convertible top position	B-pillar microswitch	Compartment lid microswitch
Top open	Open	Open
Compartment lid open	Open	Ground
Top closing	Ground	Ground
Top closed	Ground	Open

Warnings and Cautions

WARNING—
- *Do not operate the convertible top while performing repairs. Open convertible top to facilitate access and remove ignition key before proceeding with repairs.*

Convertible Top 61-7
Convertible Top Drive Service

CAUTION—
- Open or close the convertible top with the vehicle level and stationary and the parking brake engaged.
- Do not operate the top with the vehicle jacked up.
- Do not operate the top unless there is sufficient clearance above the vehicle.
- To prevent damage to top from mold or chafing, make sure the top is completely clean and dry before opening.
- To prevent the rear window from scratching, wash with clean water and dry before opening the top.
- When possible, park the vehicle in the shade. Exposure to sun damages textiles, rubber and colors in the convertible top.
- Do not attempt to drive the vehicle unless the top is completely open or closed.
- Do not modify the convertible top control system to bypass the built-in safety features.
- Do not open the top with temperatures below freezing. Rear window damage may result.
- When opening the top, be sure to help the plastic rear window to crease evenly along its length as it folds. A thick towel folded and placed in the crease helps protect the plastic from damage.
- If the top is open for an extended time, be sure to assist it during the closing procedure. The top has to go through an initial stretching before it will close without assistance.

CONVERTIBLE TOP DRIVE SERVICE

Convertible top drive motor, removing and installing

- Close convertible top half-way so that convertible top lid is at its highest point. Remove ignition key.

- Pull off drive motor plastic cover.

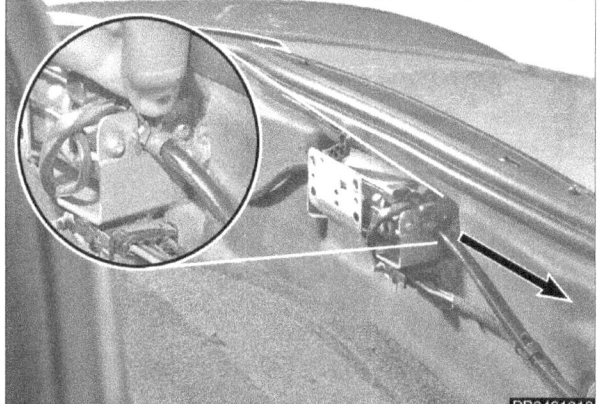

◄ Detach transmission cable on each side of motor:
- Remove cable retaining clip (**inset**).
- Pull cable straight out of motor (**arrow**).

CAUTION—
- Support convertible top lid while both cables are disconnected from motor.

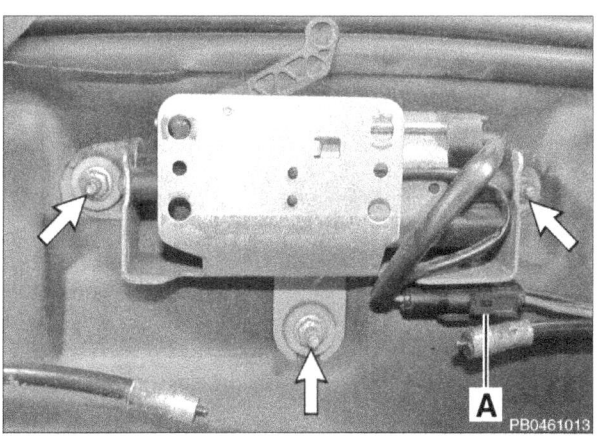

◄ Remove motor mounting nuts (**arrows**) and detach electrical connector (**A**).

- Slide motor off mounting studs.

- Installation is reverse of removal. Keep in mind the following:
  - Inspect motor mounting insulators and replace if necessary.
  - When reconnecting left and right transmission cables, make sure convertible top and lid are square to body.

61-8 Convertible Top

Convertible Top Drive Service

Convertible top transmission, removing and installing

The convertible top drive motor uses 2 cables to drive 2 gear-reduction transmissions, one on each side of the car, in the top compartment. The transmissions are connected to the convertible top and the compartment lid via link rods.

◀ Open convertible top approx. 4 in. (10 cm) from windshield frame.

− Remove ignition key to prevent convertible top operation.

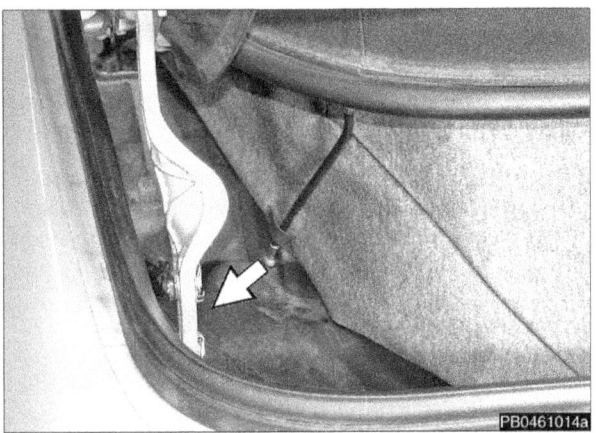

◀ Pull off lower ball head of convertible top tensioning cable in direction of **arrow** on left and right sides.

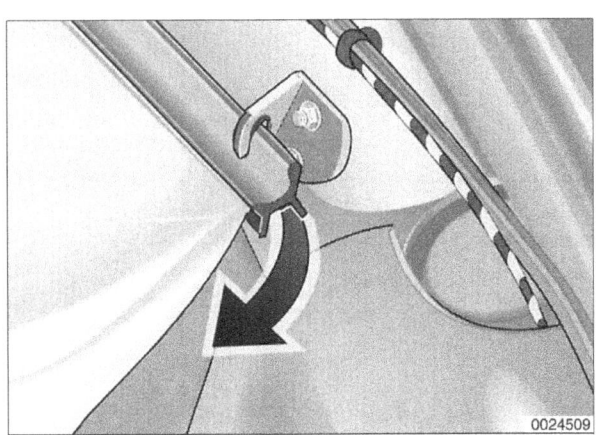

◀ Unclip and pull down fabric covering from two holders (**arrow**) at base of rear bulkhead.

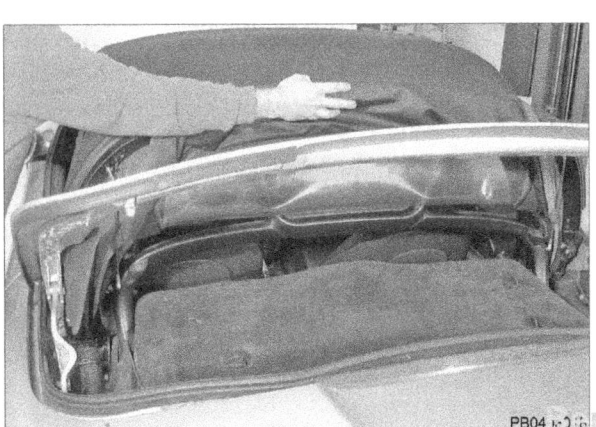

◀ Fold rear part of convertible top forward (**arrow**) until you feel it engage.

NOTE—
- *2003 - 2004 models, support glass rear window by pulling retaining strap at base of window around left side of top and attaching it to centering pin at front of top.*

Convertible Top 61-9

Convertible Top Drive Service

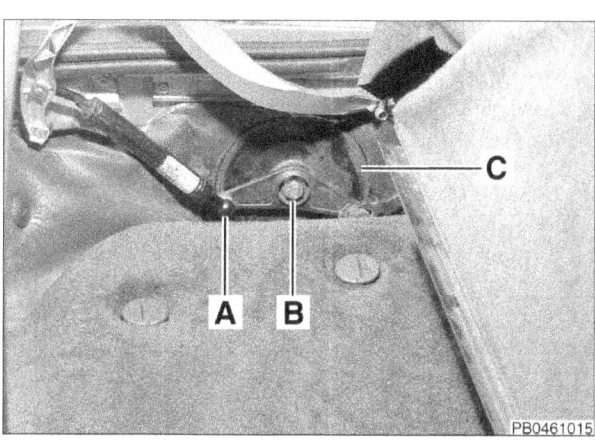

◀ Working at convertible top transmission:
- Use pry bar or large screwdriver to detach convertible top lid link ball socket (**A**) from transmission drive lever.
- Remove drive lever mounting bolt (**B**).
- Detach drive lever from transmission and fold forward.
- Remove transmission plastic cover (**C**).

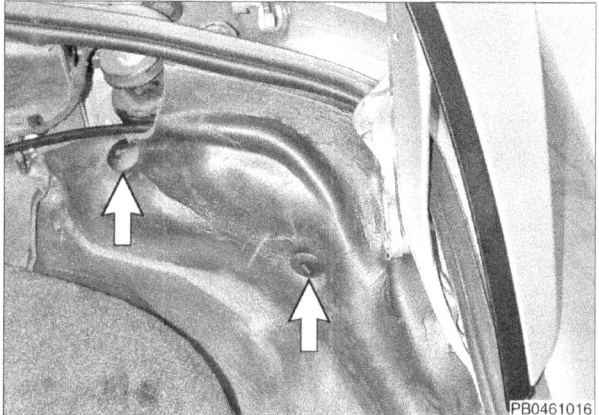

◀ Working in convertible top cavity, remove rubber insulation plastic retainers (**arrows**). Fold rubber insulation away from transmission.

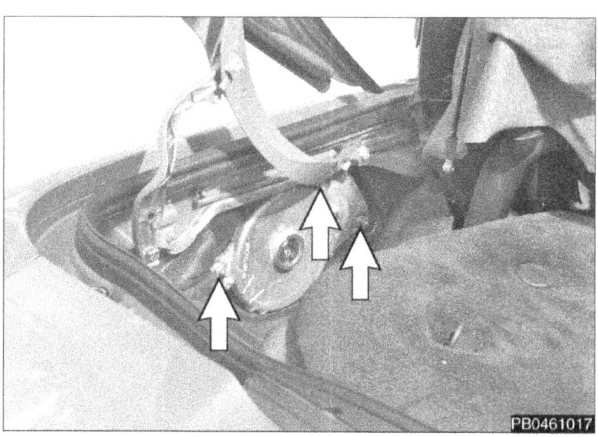

◀ Remove transmission mounting bolts (**arrow**). Lift out transmission and cable together.

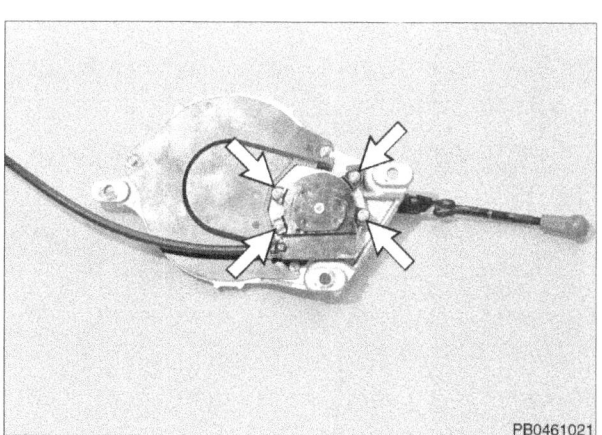

◀ Separate cable from transmission. Loosen transfer gear cover screws (**arrows**) and remove cable.

Convertible Top Drive Service

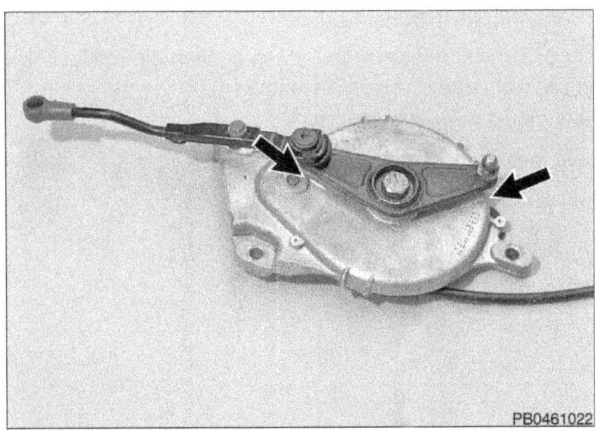

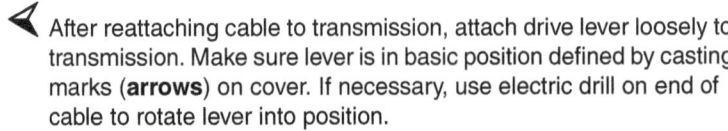 After reattaching cable to transmission, attach drive lever loosely to transmission. Make sure lever is in basic position defined by casting marks (**arrows**) on cover. If necessary, use electric drill on end of cable to rotate lever into position.

– Remove lever, then reinstall transmission, transmission plastic cover and rubber insulation. Reinstall lever.

Tightening torques	
Drive lever to transmission (M12)	80 Nm (59 ft-lb)
Transmission to side panel (M8)	22 Nm (16 ft-lb)

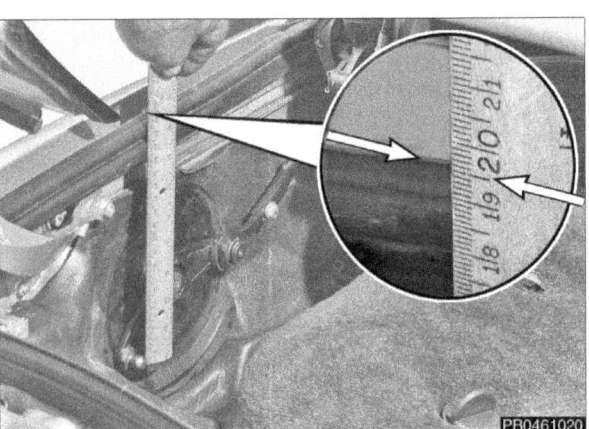

◂ Measure distance from compartment lid link pivot at drive lever to top of convertible top compartment lid gasket.

Convertible top in closed position	
Convertible top link pivot to lid gasket	195 ± 0.5 mm (7.68 ± 0.02 in)

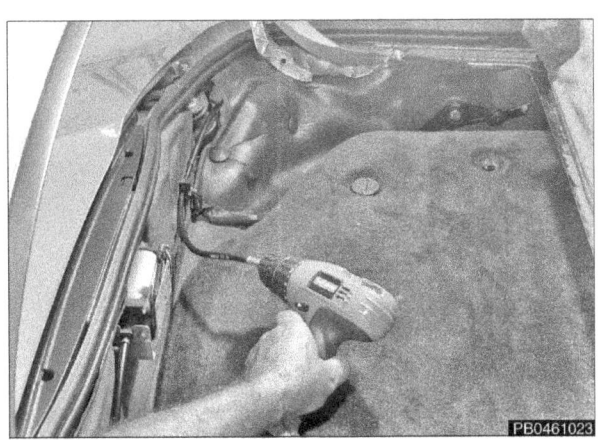

◂ If necessary, reposition arm using electric drill on end of cable.

– Once measurements are equal on both sides, reconnect cables to motor and reattach top cover (black-tipped) links to drive levers. Test top for correct operation.

63 Bumpers

GENERAL 63-1
  Bumper construction 63-1
  Bumper impact absorber 63-2
  Bumper covers 63-2

FRONT BUMPER 63-2
  Front bumper components (Boxster) 63-2
  Front bumper cover, removing and installing ... 63-3
  Front bumper, removing and installing ... 63-5

REAR BUMPER 63-5
  Rear bumper components 63-5
  Rear bumper cover, removing and installing ... 63-6
  Rear bumper, removing and installing ... 63-8

GENERAL

This repair group covers front and rear bumper repairs.

Parking Assistant ultrasonic emitters and detectors, where equipped, are located in the rear bumper. See **91 Radio and Communication**.

Bumper construction

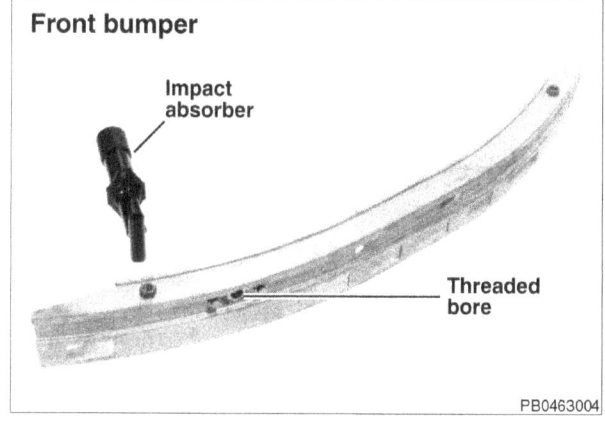

 The front and rear bumpers are made of extruded aluminum with a thickness of 5 mm (0.2 in). The bumpers are attached to the body via telescoping impact absorbers, one on each side.

A threaded bore on the right side allows a towing lug to be screwed into the front bumper.

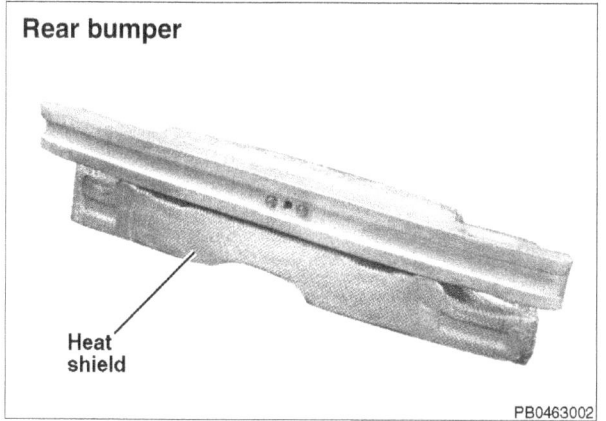

 The threaded towing lug bore in the rear bumper is in the center, behind the license plate.

A heat shield is attached below the bumper to protect the bumper and plastic bumper cover from the heat of the exhaust system.

63-2 Bumpers

Front Bumper

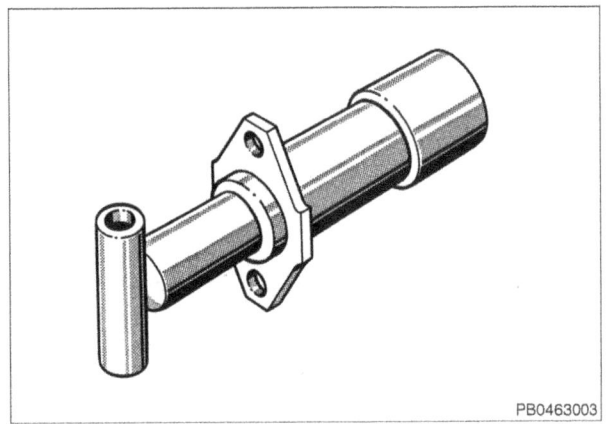

Bumper impact absorber

◀ The bumper mounting impact absorbers are filled with silicone elastomer. When pushed in, the elastomer absorbs energy, then releases by rebounding the absorber to its original length.

Bumper covers

The bumper covers are made of recyclable polypropylene-ethylene-propylene-diene-monomer (PPEPDM) plastic and are painted to match as the body.

In the front, air grilles (2 on the Boxster, 3 on the Boxster S) duct air to engine cooling radiators and A/C condensers. The side turn signals are attached to the sides of the front bumper cover.

In the rear, as an option, the bumper cover carries 4 Parking Assistant emitters.

FRONT BUMPER

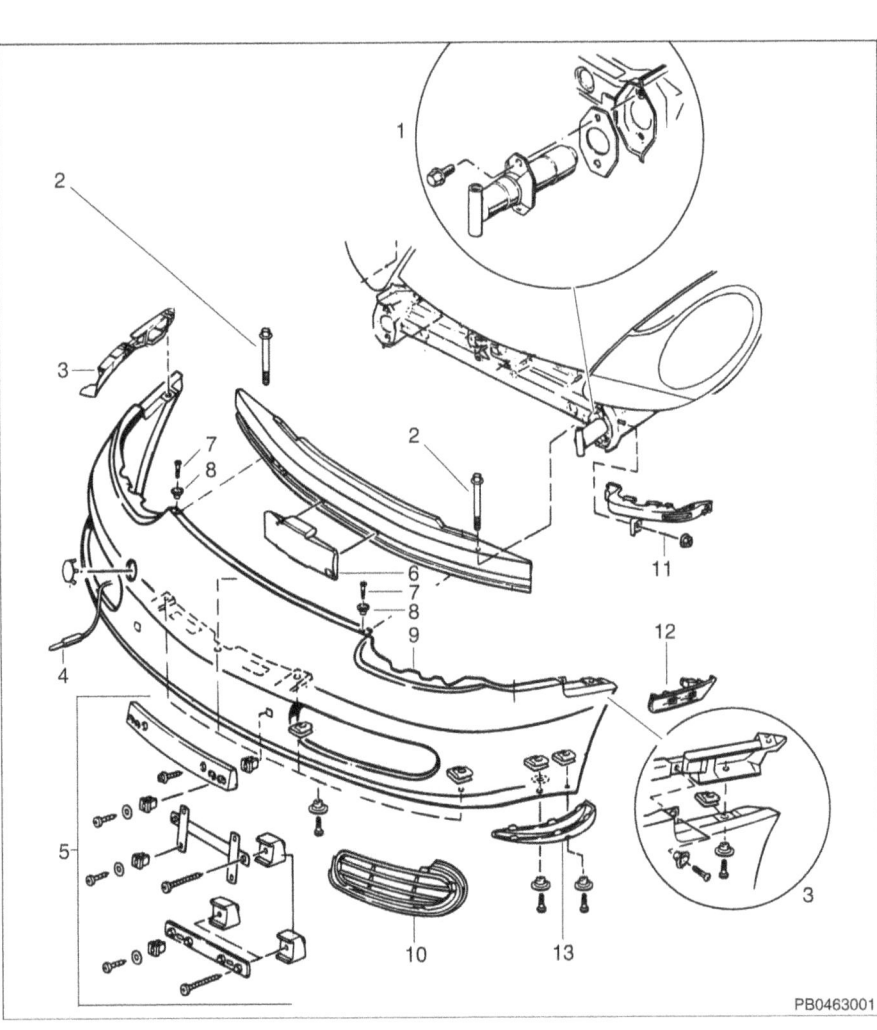

Front bumper components (Boxster)

1. **Impact absorber assembly**
   - M8 bolt, tighten to 23 Nm (17 ft-lb)
2. **Bumper mounting bolt**
   - M12 x 1.5 x 100 bolt tighten to 85 Nm (63 ft-lb)
3. **Side support**
4. **Outside temperature sensor**
5. **Front license plate bracket**
6. **Foam insulator**
7. **Sheet metal screw**
8. **Spacer sleeve**
9. **Bumper cover**
10. Grille
11. Retainer
12. Side turn signal light
13. Trim panel

Bumpers 63-3

Front Bumper

Front bumper cover, removing and installing

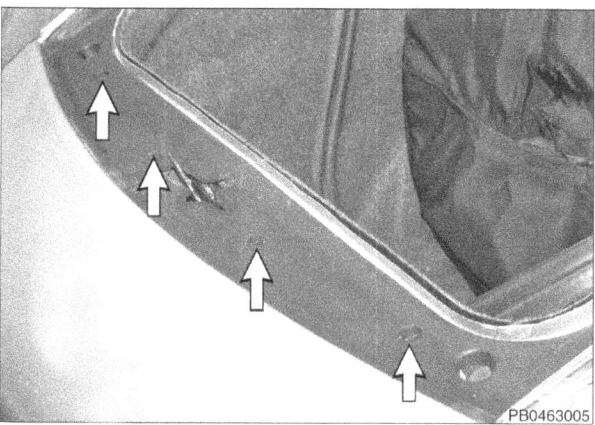

◄ Open front trunk and peel off trunk edge rubber trim. Use screwdriver to twist trunk lock cover plastic retainers (**arrows**) 90° to release cover. Remove cover.

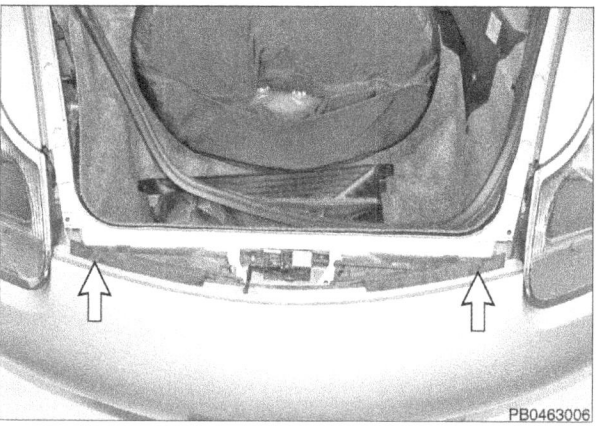

◄ Remove bumper cover mounting sheet metal screws (**arrows**).

– Raise front of car and support safely.

> **WARNING—**
> • Make sure the car is stable and well supported at all times. Use a professional automotive lift or jack stands designed for the purpose. A floor jack is not adequate support.

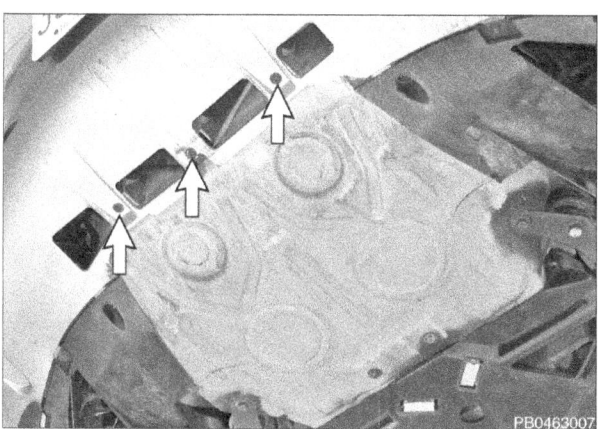

◄ Working underneath front, remove bumper cover mounting screws (**arrows**).

– Remove front wheels.

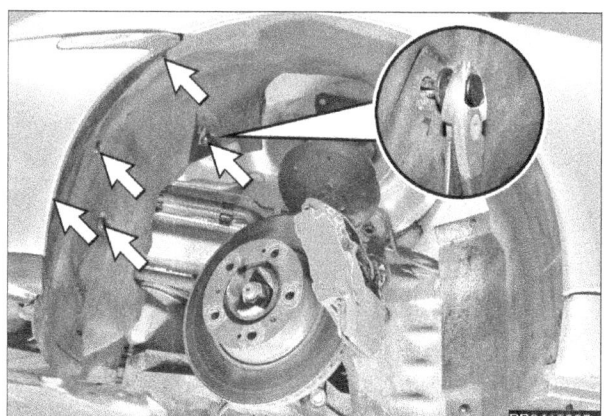

◄ Working in wheel housing underneath fender, remove wheel housing liner fasteners (**arrows**) at front of liner.
  • To disengage plastic rivet, pull out rivet lock (**inset**).
  • Remove plastic 10 mm nuts.

63-4 Bumpers

Front Bumper

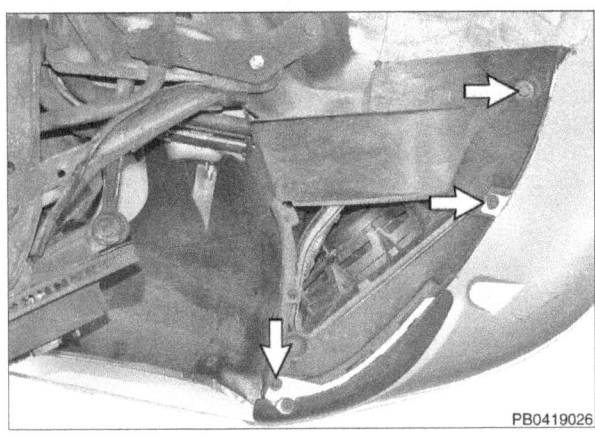

◄ Working underneath front of fender, remove remainder of wheel housing liner fasteners (**arrows**). Pull liner back sufficiently to gain access to side turn signal mounting.

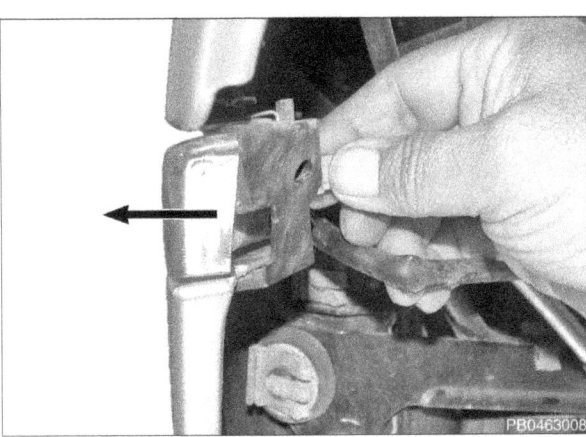

◄ Squeeze side turn signal retaining spring to disengage light socket and lens from fender. Detach electrical connector and set light aside.

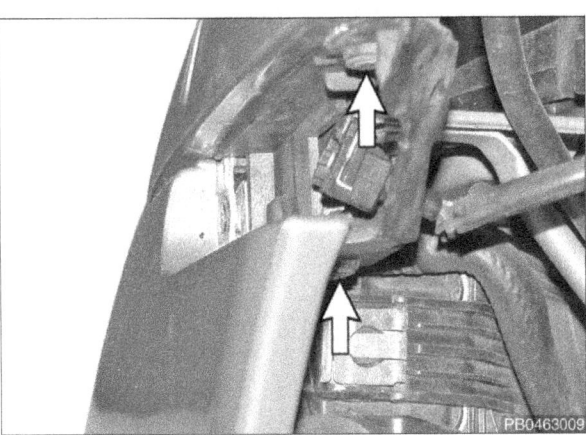

◄ Working at side turn signal opening, remove bumper cover side mounting screws (**arrows**).

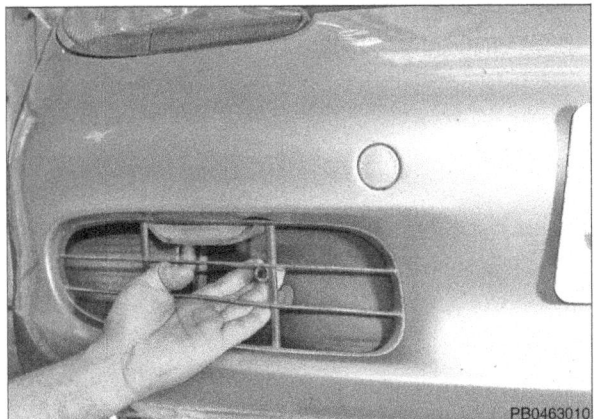

◄ Working through right side grille, press out outside temperature sensor and disconnect.

– With assistant, lift off bumper cover carefully, watching to make sure no wires, hoses or other components become snagged.

– Installation is reverse of removal.

Bumpers 63-5

Rear Bumper

Front bumper, removing and installing

– Raise front of car and support safely.

> **WARNING—**
> * Make sure the car is stable and well supported at all times. Use a professional automotive lift or jack stands designed for the purpose. A floor jack is not adequate support.

– Remove front bumper cover. See **Front bumper cover, removing and installing** in this repair group.

– Remove bumper mounting bolts (**A**). Lift bumper off impact absorbers.

– If necessary, remove impact absorber mounting bolts (**B**) and lift out absorber.

– Installation is reverse of removal.

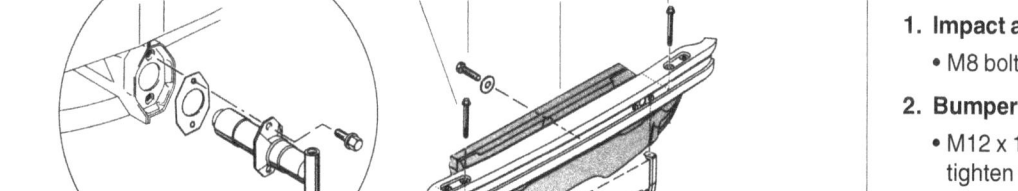

Tightening torques	
Bumper to impact absorber (M12 x 1.5 x 100 bolt)	85 Nm (63 ft-lb)
Impact absorber to body (M8 x 30)	23 Nm (17 ft-lb)

REAR BUMPER

Rear bumper components

1. **Impact absorber assembly**
   * M8 bolt, tighten to 23 Nm (17 ft-lb)
2. **Bumper mounting bolt**
   * M12 x 1.5 x 100 bolt tighten to 85 Nm (63 ft-lb)
3. **M5 bolt**
4. **Bumper and heat shield assembly**
5. **Rear wheel housing trim**
6. **Sheet metal screw**
7. **Reinforcing bracket**
8. **Rear license plate bracket**
9. **Support**
10. **License plate light harness**
11. **Sheet metal screw**
12. **Bumper cover top securing strip**
13. **License plate light assemblies**
14. **Rear bumper cover**
15. **Impact cushion**
16. **M8 bolt**

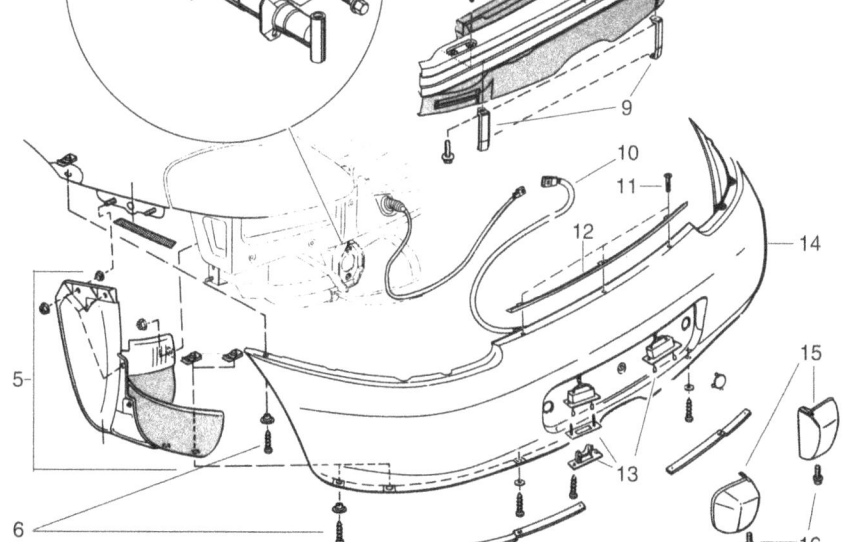

63-6 Bumpers

Rear Bumper

Rear bumper cover, removing and installing

– Turn ignition ON and raise rear spoiler. Use rear spoiler switch in main fuse panel, left footwell. Turn ignition OFF.

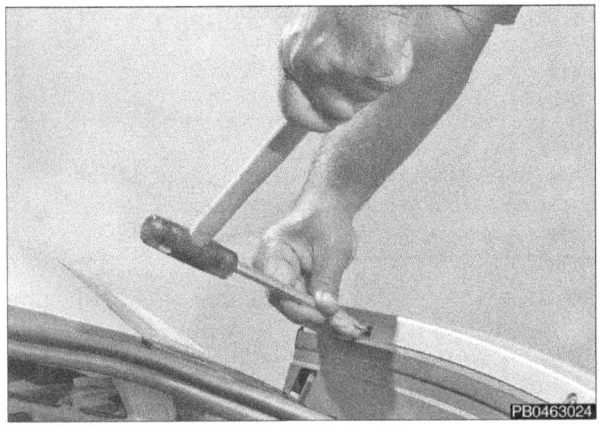

◀ Use thin pin punch and light hammer to drive in plastic rivet pins in spoiler cover.

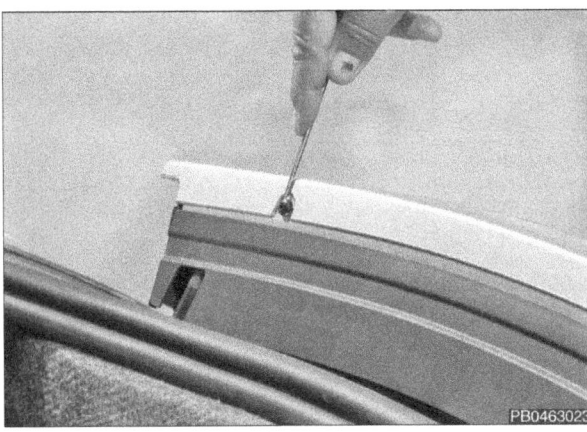

◀ Gently pry out spoiler cover plastic rivets.

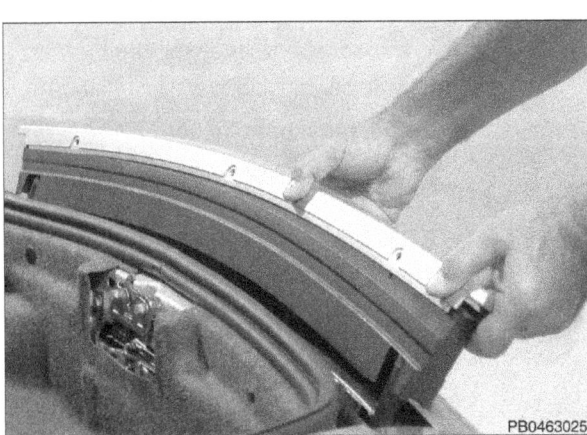

◀ Tilt spoiler cover backward and off spoiler.

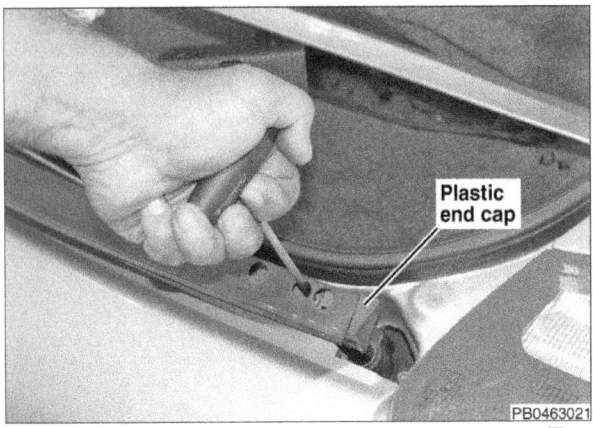

◀ Use 5 mm Allen wrench to remove spoiler top rail mounting bolts.

NOTE—
- *With top rail in hand, remove plastic end cap and shake out plastic rivet expanding pins that fell inside in previous step.*

Bumpers 63-7

Rear Bumper

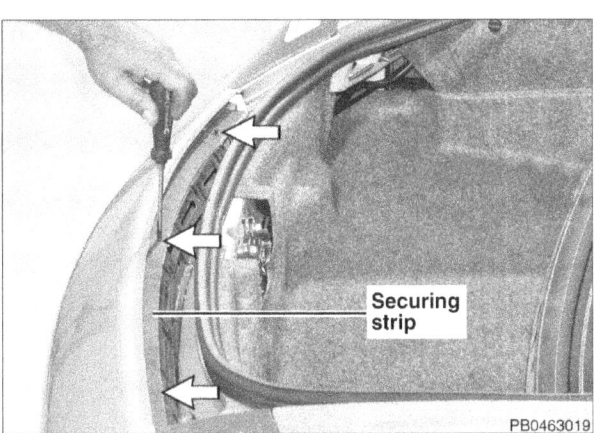

◄ Remove top bumper cover mounting screws (**arrows**). Remove securing strip.

– Raise rear of car and support safely.

> **WARNING—**
> - Make sure the car is stable and well supported at all times. Use a professional automotive lift or jack stands designed for the purpose. A floor jack is not adequate support.

◄ Remove bumper cover screws in rear of wheel housing.

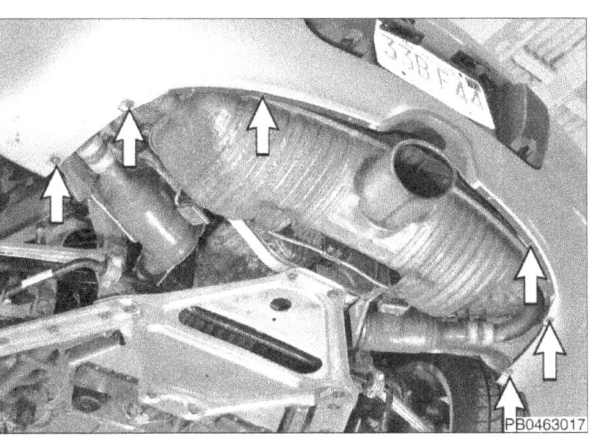

◄ Working underneath rear of car, remove 6 bumper mounting screws (**arrows**).

◄ Remove license plate light socket mounting screws (**arrows**). Gently pry sockets out of rear bumper cover.

63-8 Bumpers

Rear Bumper

◁ Pull off harness connectors off license plate sockets and set sockets aside. Push harnesses up into bumper holes.

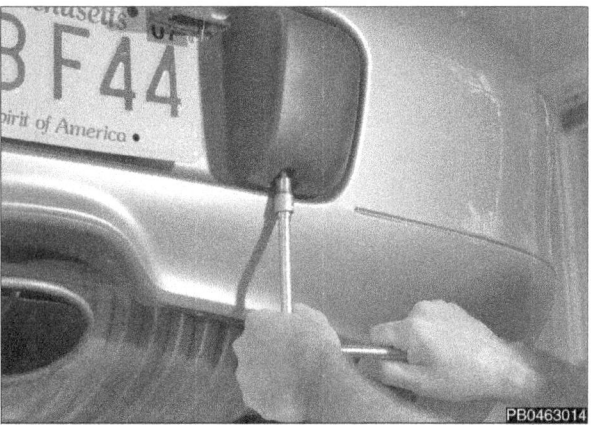

◁ Use 6 mm Allen socket to remove impact cushion mounting bolts. Unhook impact cushions and set aside.

◁ With assistant, lift off bumper cover carefully, watching to make sure no wires or other components become snagged.

– When installing, be sure to thread license plate light harnesses in bumper cover holes before cover is fully fitted in place.

Rear bumper, removing and installing

– Raise rear of car and support safely.

> **WARNING—**
> • Make sure the car is stable and well supported at all times. Use a professional automotive lift or jack stands designed for the purpose. A floor jack is not adequate support.

– Remove rear bumper cover. See **Rear bumper cover, removing and installing** in this repair group.

Bumpers 63-9

Rear Bumper

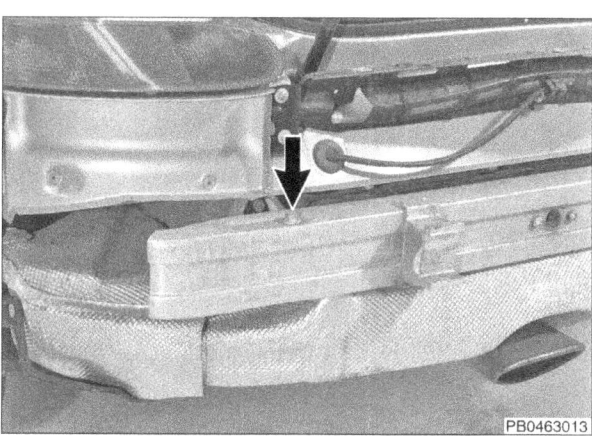

◄ Remove bumper mounting bolts (**arrow**). Lift bumper and attached heat shield off impact absorbers.

◄ If necessary, remove impact absorber mounting bolts (**arrows**) and lift out absorber.

– Installation is reverse of removal.

Tightening torques	
Bumper to impact absorber (M12 x 1.5 x 100 bolt)	85 Nm (63 ft-lb)
Impact absorber to body (M8 x 30)	23 Nm (17 ft-lb)

64 Door Windows

GENERAL 64-1
   Electric window control 64-1
   Window lifting components 64-2
   Window height adjustment 64-2
   Window drop-down feature 64-2
   Warnings and Cautions 64-3
POWER WINDOW CONTROLS 64-3
   Power window system fuse 64-3
   Power window switch, removing and installing .. 64-3

WINDOW REPAIRS 64-4
   Window glass, removing and installing 64-4
   Window position, adjusting 64-5
WINDOW REGULATOR REPAIRS 64-9
   Window motor standardization 64-9
   Window motor, removing and installing 64-9
   Window regulator, removing and installing 64-10

TABLE

a.   Window adjustment summary 64-8

GENERAL

This repair group covers door glass and power window regulator repairs.

Windshield and rear window replacement is beyond the scope of this manual.

Electric window control

The electric windows can be operated under the following conditions:

- Ignition ON, door open or closed.
- Ignition switched OFF but ignition key not removed from steering lock.
- Ignition switched OFF until a door is opened for the first time.

 Electrically powered windows have the following functions:

- **One-touch function.** Window toggle switch raises or lowers the window without the button having to be kept depressed. Window can be stopped at intermediate positions by briefly pressing the button.
- **Pinch protection.** If the window encounters resistance while lifting, the motor is reversed.
- **Drop-down feature.** When the door is opened, the window drops slightly to clear the door seal. When the door is closed, the window is raised fully up into the seal.
- **Convertible top operation.** When the convertible top is opened or closed, The windows drop 100 - 150 mm (4 - 6 in) to reduce the load placed on the window-to-top seals. After the top is fully opened or closed, raise windows using the toggle switches.

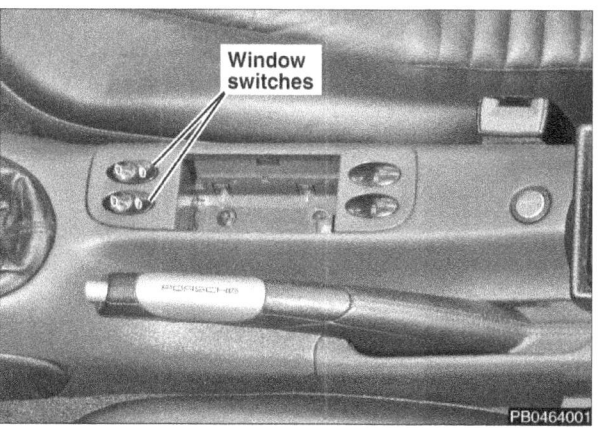

Door Windows

General

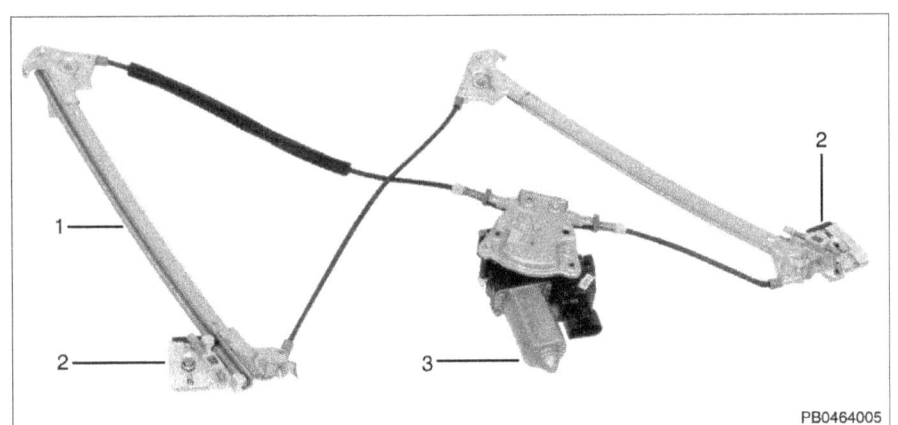

Window lifting components

1. Window regulator rail
2. Window driver with clamping jaws
3. Window motor with integrated control module

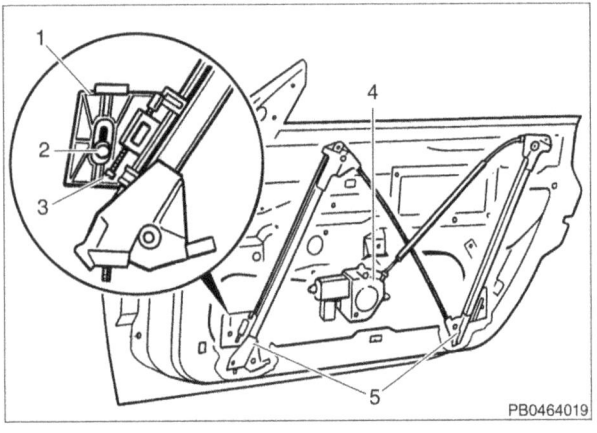

Window height adjustment

The window drivers allows window height adjustment. Adjusters are accessible from underneath the door. See **Window position, adjusting**.

1. Clamping jaw
2. Window mounting screw
3. Height adjuster screw (E6 Torx)
4. Window motor
5. Window driver

Window drop-down feature

When the door is opened, the window is automatically lowered approx. 13 mm (½ in) to clear the door seal. When the door is closed, the window is raised fully up into the seal. This gives the frameless door window improved sealing and reduces wind noise.

Any time the window or window regulator is loosened or removed, adjust window pretension and position carefully to:

- Prevent glass breakage.
- Avoid high wind noise.
- Avoid water leaks.

Door Windows 64-3

Power Window Controls

Warnings and Cautions

> **WARNING—**
> - Boxster models are equipped with side impact airbags installed in the doors. Read the warnings and cautions in **69 Seat Belts, Airbags** prior to starting work on window.
> - When connecting or disconnecting the side-impact airbag harness connector, make sure the battery is disconnected.
> - The airbags system contains a back-up power supply within the SRS control module. Allow a 1 minute discharge period after disconnecting the battery cable.
> - If the airbag warning light on the dashboard illuminates after completion of window repairs, the airbags will not deploy in case of an accident. Be sure to have the SRS system diagnosed and repaired immediately.
> - When working on door window, disconnect the harness connector to the window regulator to prevent pinching fingers in the moving window mechanism.
> - Wear hand and eye protection when working with broken glass.
> - If a window is broken, vacuum all of the glass bits out of the door cavity. Use a blunt screwdriver to clean out any remaining glass pieces from the window guide rails.

> **CAUTION—**
> - Peel off and reglue inner door insulator (vapor barrier) very carefully. If damaged in any way, replace.

POWER WINDOW CONTROLS

Power window system fuse

◀ Window system 30 A fuse is in position D1 in main fuse panel, behind left footwell cover.

Power window switch, removing and installing

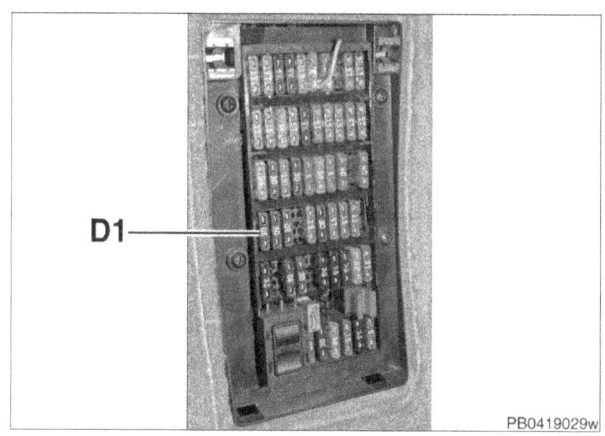

◀ Remove center console ashtray. Then remove two plastic Torx screws (**arrows**).

64-4 Door Windows

Window Repairs

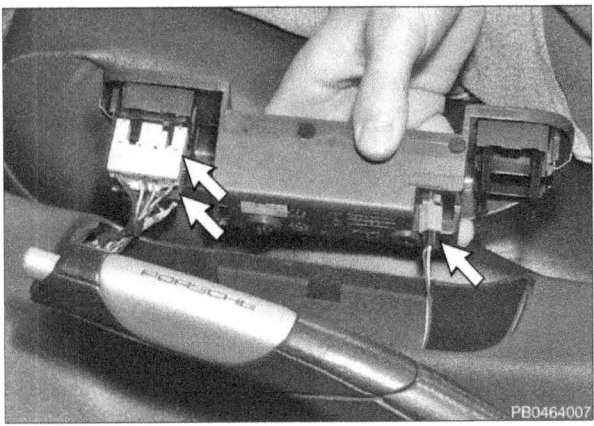

◀ Gently pry out window switch assembly. Disconnect harness connectors (**arrows**).

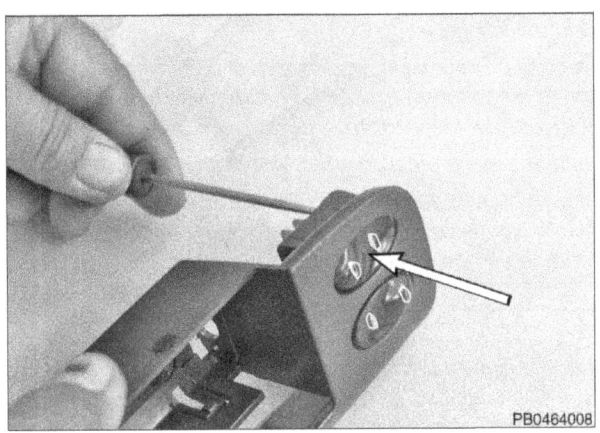

◀ Use thin prying tool to open switch holder sufficiently to press switch out of holder in direction of **arrow**.

– Installation is reverse of removal.

WINDOW REPAIRS

Window glass, removing and installing

– Raise window to top.

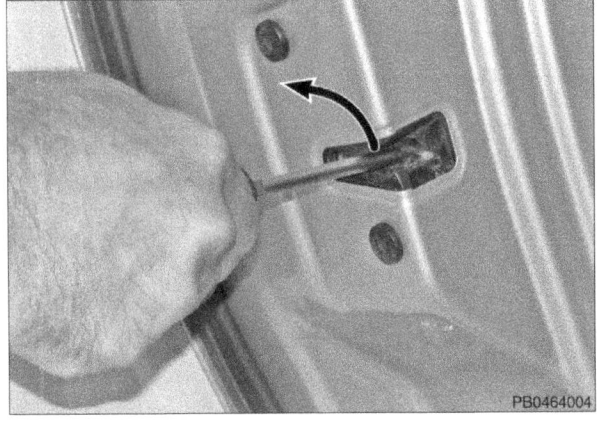

◀ With door open, use screwdriver to force door rotary latch in locked direction (**arrow**). This defeats window drop-down feature and forces window to stay at top of travel.

– Turn ignition OFF and remove ignition key.

– Remove door trim panel. See **57 Doors and Locks**. Do not remove door speaker, door-mounted airbag nor insulating liner (vapor barrier).

◀ Remove rubber plugs from oval bores near top of door. Use Torx driver to loosen window mounting screws (**arrow**) through oval bores (**inset**).

> **CAUTION—**
> • Do not remove Torx screws.

Door Windows 64-5
Window Repairs

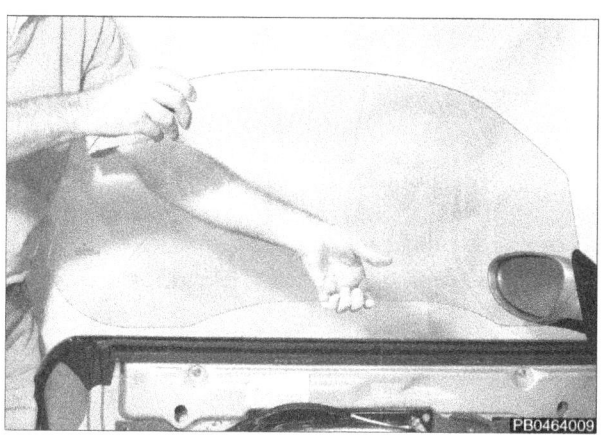

◄ Carefully lift window off clamping jaws and out of tracks.

– When installing, carefully slide window between rubber trim. Push down firmly to engage in clamping jaws. Tighten window mounting Torx screws.

Tightening torque	
Window driver clamping jaw to window	9 Nm (7 ft-lb)

– Adjust window position. See **Window position, adjusting** in this repair group.

> **CAUTION—**
> - Before closing door, pull up on door handle and make sure door lock rotary latch opens and window drops down slightly.

Window position, adjusting

Window position adjustment is summarized in **Table a**.

Window fore-aft position

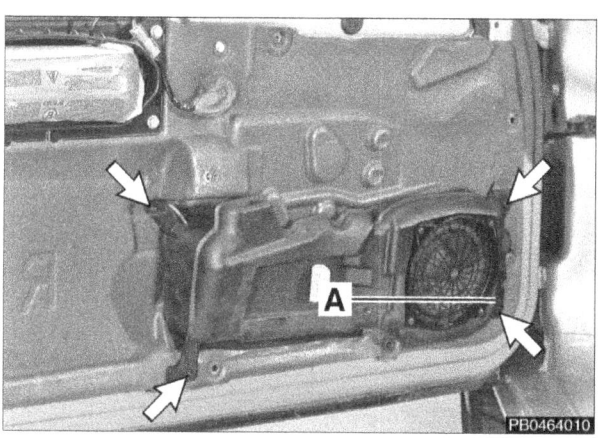

◄ Detach door speaker electrical harness (**A**) and remove speaker mounting screws (**arrows**). Remove speaker.

– Carefully fold back door insulating liner (vapor barrier).

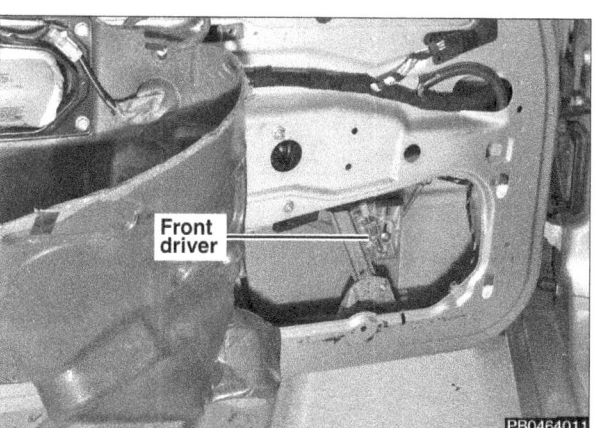

◄ Turn ignition ON and lower window until front window regulator driver is visible.

64-6 Door Windows

Window Repairs

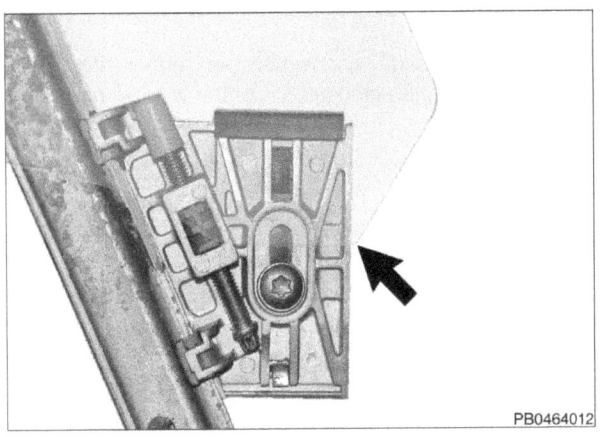

◁ Loosen Torx screws and slide window fore or aft in clamping jaws for correct position: Forward lower corner of glass (**arrow**) lines up with edge of front driver.

NOTE—
- *If it is necessary to raise window fully to access the clamping jaw screw for the rear driver, be sure to have the door rotary latch in closed position so that window can rise to top of travel.*

Tightening torque	
Window driver clamping jaw to window	9 Nm (7 ft-lb)

Window angle, adjusting

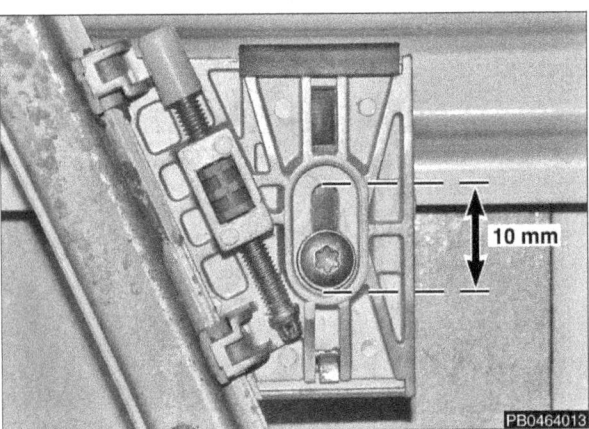

◁ Use slots in window driver clamping jaws to adjust window angle.
- Adjustment range: 10 mm (0.4 in).
- Make sure base height of window is set equally in front and rear drivers.

Tightening torque	
Window driver clamping jaw to window	9 Nm (7 ft-lb)

Window height, adjusting

– With convertible top closed, raise window to top.

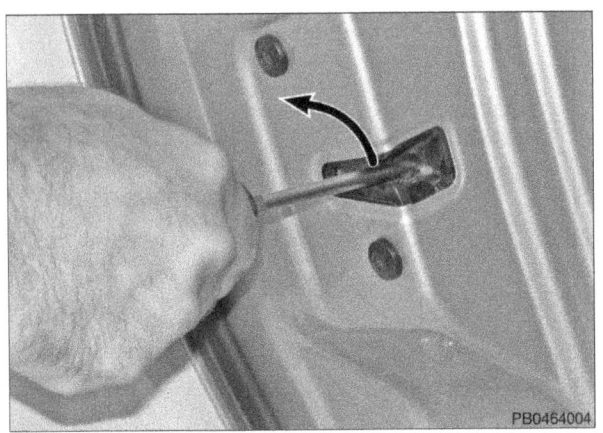

◁ With door open, use screwdriver to force door rotary latch in locked direction (**arrow**). This defeats window drop-down feature and forces window to stay at top of travel.

– Shut door gently until top of glass contacts outside of convertible top seal.

CAUTION—
- *Do not push door to latch.*

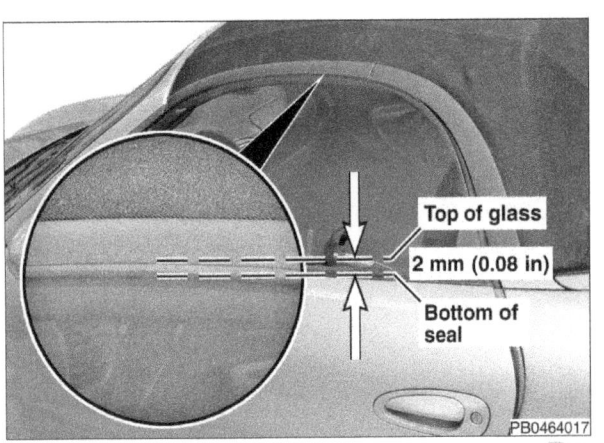

◁ Check top of window for correct height: 2 mm (0.08 in) above lower edge of seal.

Door Windows 64-7

Window Repairs

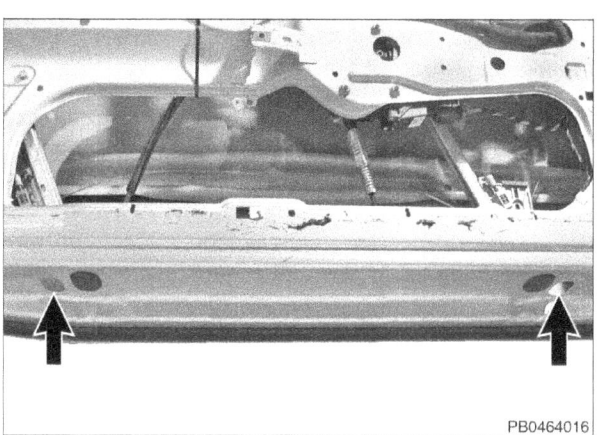

◂ If window height adjustment is needed, remove outboard rubber plugs at door bottom.

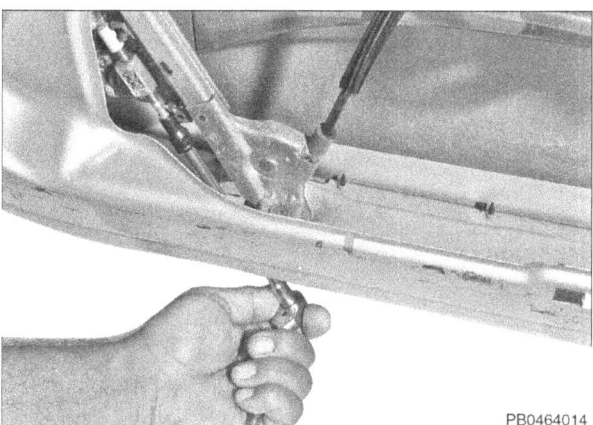

◂ Adjust window height by reaching through bores at door bottom to turn E6 Torx screws in window regulator drivers.
  - Adjustment range: Approx. 15 mm (0.6 in).
  - Adjust window height so that it projects evenly approx. 2 mm (0.08 in) above lower edge of convertible top sealing strip.
- To test adjustment:
  - Pull up on door handle and make sure door lock rotary latch opens and window drops down.
  - Close door. Window rises into convertible top seal.

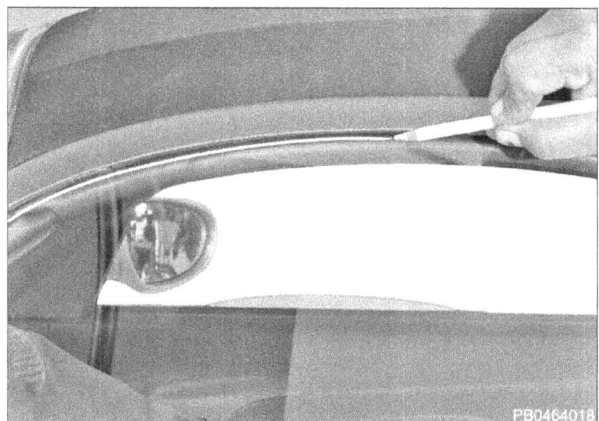

◂ Use wax marker to mark door window along seal.

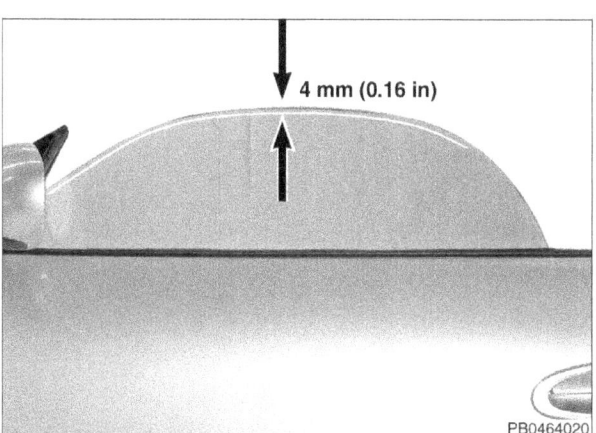

◂ Open door. Window drops down approx. 13 mm (½ in). Measure distance of wax line below window edge: Make sure it is uniformly approx 4 mm (0.16 in).

64-8 Door Windows

Window Repairs

Window sealing force, adjusting

Set window rake angle for best sealing.

◀ Clamp a sheet of paper between door window and convertible top seal. Correct sealing force: Paper very difficult to pull out or rips while pulling out.

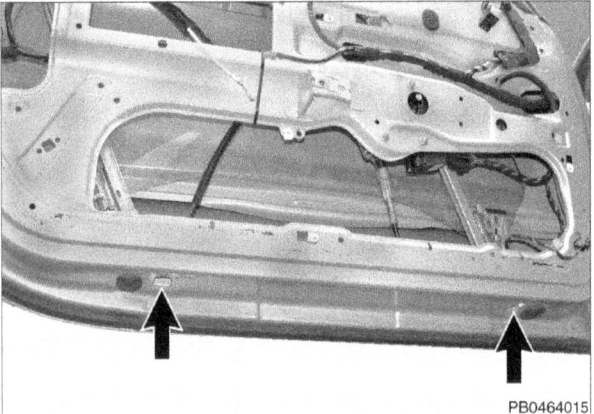

◀ If sealing force is incorrect, adjust window rake by sliding lower window rail mountings in or out:

- Remove inboard rubber plugs (**arrows**) at door bottom.
- Loosen M6 window rail mounting nuts and slide rails in or out.
- Adjustment range: Approx. 10 mm (0.4 in).
- Tighten M6 nuts.

Tightening torque	
Window rail to door	10 Nm (8 ft-lb)

– Recheck sealing force with sheet of paper, as before.

Table a. Window adjustment summary		
Parameter	Specification	Procedure
Window fore-aft position	Front lower corner of window lined up with edge of front driver.	Loosen window mounting screws, slide window fore or aft.
Window angle	Equal height at each clamping jaw.	Loosen window mounting screws, slide clamping jaw up or down.
Window height	Maximum height of window top 2 mm (0.4 in) above lower edge of convertible top seal.	Turn adjusters at window driver (E6 Torx).
Window sealing force (rake)	Check clamping force on sheet of paper.	Loosen lower window rail fasteners. Slide rails in or out.

Door Windows 64-9

Window Regulator Repairs

WINDOW REGULATOR REPAIRS

Window motor standardization

NOTE—
* *Power window standardization is also referred to as initialization.*

If the battery is disconnected, the power window regulator loses its reference point for closed position (limit position). Therefore, standardize door windows whenever power to the window motors is interrupted:

 With convertible top fully closed, raise window and keep power window switch in raise position for about 5 seconds.

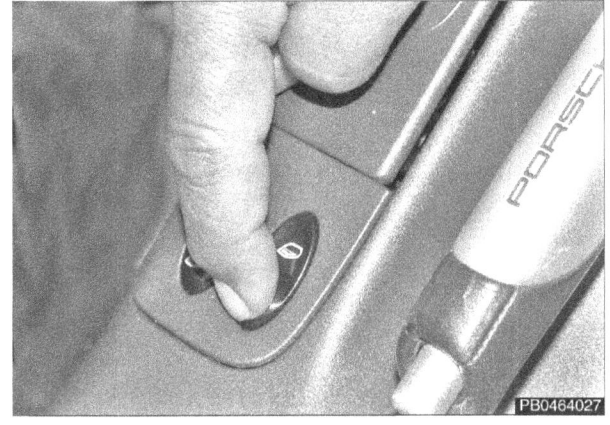

Window motor, removing and installing

— Close window.

— Disconnect negative (-) cable from battery.

> *WARNING—*
> * *Wait 1 minute after disconnecting battery before working in vicinity of airbag.*

> *CAUTION—*
> * *Prior to disconnecting the battery, read the battery disconnection cautions given in **00 Warnings and Cautions**.*

— Open door and remove inner door panel, door-mounted airbag, door speaker and insulating liner (vapor barrier). See **57 Doors and Locks**.

> *WARNING—*
> * *To avoid injury, store airbag in a safe place with airbag facing up.*

 Remove window motor mounting nuts (**arrows**).

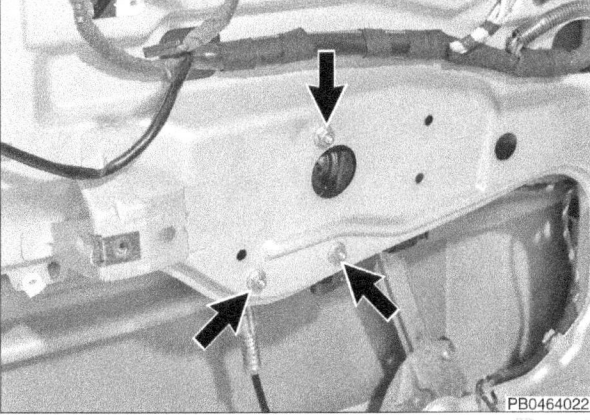

Door Windows

Window Regulator Repairs

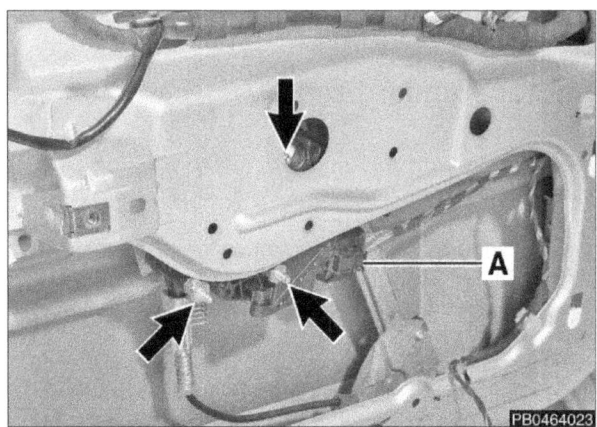

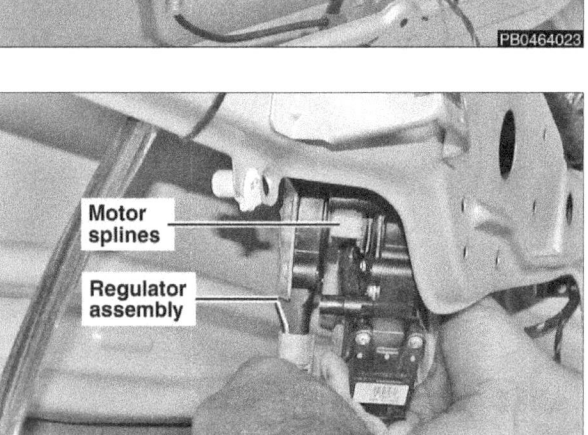

◄ Detach window motor:
  • Push motor into door to free mounting studs from bores in door.
  • Detach motor harness connector (**A**).
  • Remove 3 studs (**arrows**) mounting motor to window regulator assembly.

◄ Working inside door cavity, carefully work motor splines out of regulator assembly and remove motor.

− Installation is reverse of removal.

Tightening torques	
Window motor to door	10 Nm (8 ft-lb)
Window motor to regulator assembly	10 Nm (8 ft-lb)

− Standardize window motor after reconnecting battery. See **Window motor standardization** in this repair group.

Window regulator, removing and installing

− Close window.

− Disconnect negative (-) cable from battery.

> **WARNING—**
> • Wait 1 minute after disconnecting battery before working in vicinity of airbag.

> **CAUTION—**
> • Prior to disconnecting the battery, read the battery disconnection cautions given in **00 Warnings and Cautions**.

− Open door and remove inner door panel, door-mounted airbag, door speaker and insulating liner (vapor barrier). See **57 Doors and Locks**.

> **WARNING—**
> • To avoid injury, store airbag in a safe place with airbag facing up.

− Remove window. See **Window glass, removing and installing** in this repair group.

Door Windows 64-11

Window Regulator Repairs

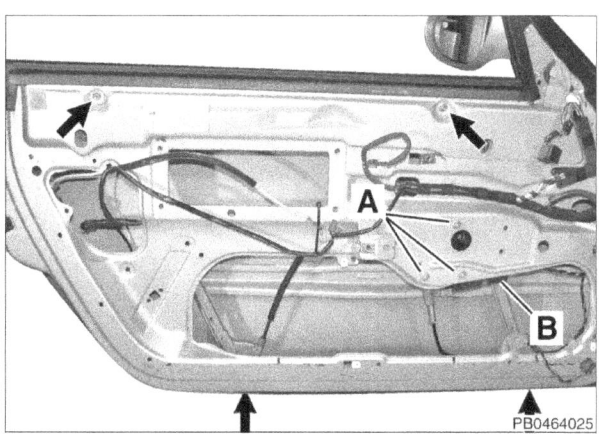

◄ Remove window regulator mounting nuts (**arrows**).
- Upper nuts mounts tops of window rails.
- Lower window rails fasteners are accessible through bores in door bottom. Remove 2 rubber plugs.
- Remove window motor mounting nuts (**A**).
- Disconnect window motor harness connector (**B**).

– Carefully work assembly out of door cavity through speaker opening.

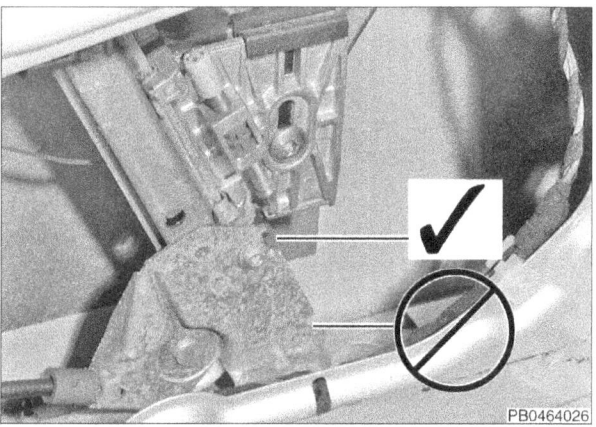

◄ If installing new regulator, make sure lower forward window driver limit stop is in top notch of forward regulator base.

– Installation is reverse of removal. Adjust window position. See **Window position, adjusting** in this repair group.

Tightening torques	
Window driver clamping jaw to window	9 Nm (7 ft-lb)
Window motor to door	10 Nm (8 ft-lb)
Window rail to door	10 Nm (8 ft-lb)

– Standardize window motor after reconnecting battery. See **Window motor standardization** in this repair group.

66 Exterior Equipment

GENERAL 66-1
OUTSIDE REAR VIEW MIRRORS 66-1
  Outside mirror, removing and installing 66-1
  Mirror glass, replacing 66-2
  Mirror heater, checking 66-3
REAR SPOILER 66-3
  Rear spoiler, removing and installing 66-3
  Rear spoiler motor, removing and installing 66-5

ENGINE COMPARTMENT VENTS 66-6
  Engine compartment vent,
    removing and installing 66-6
EMBLEMS 66-6
  Front emblem, removing and installing 66-6
  Rear emblem, removing and installing 66-7

GENERAL

This repair group includes repair information for the outside rear view mirrors, rear spoiler and engine compartment vents.

> *CAUTION—*
> * *Mirror and trim components should be at room temperature or above before removal to reduce the risk of breakage.*

OUTSIDE REAR VIEW MIRRORS

Mirror components available separately from an authorized Porsche dealer include mirror glass and outside plastic housing.

Outside mirror, removing and installing

 Working on inside of door, use a plastic trim tool to pry up on lower edge of window trim panel while sliding trim upward.

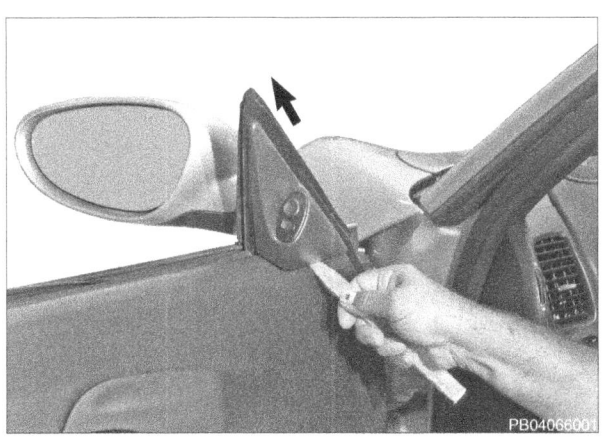

66-2 Exterior Equipment

Outside Rear View Mirrors

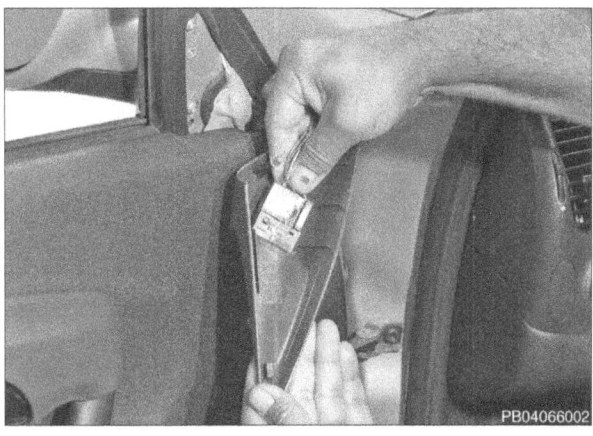

◄ Release electrical connector locking tabs and remove electrical connector from switch.

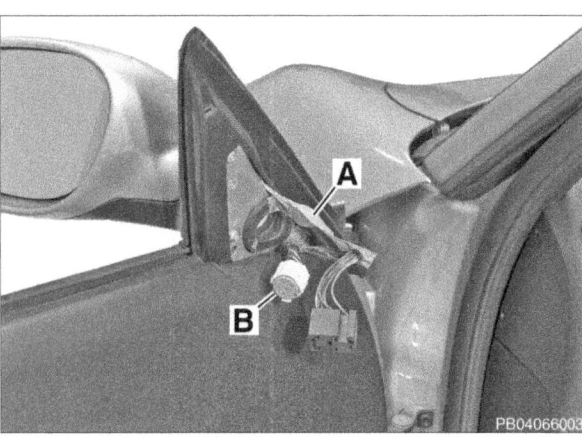

◄ Remove sealing tape (**A**) from connector grommet. Release electrical connector (**B**).

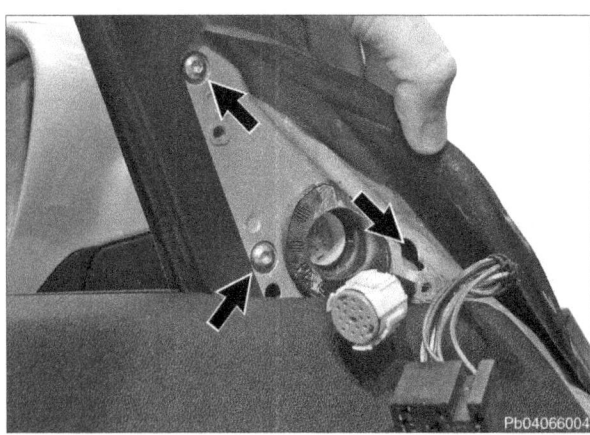

◄ Remove mirror fasteners (**arrows**) and remove outside mirror.

– Installation is reverse of removal. Make sure:
  • Sealing tape is in place.
  • Electrical connectors are properly fastened.
  • Interior trim is hooked on at top and pressed down into place.

Mirror glass, replacing

CAUTION—
• *Mirror glass replacement procedure can damage glass.*

◄ Use a plastic prying tool to carefully lever mirror glass out of metal clips in mirror housing.

– Remove electrical connectors for mirror heater.

– Installation is reverse of removal.

Exterior Equipment 66-3

Rear Spoiler

Mirror heater, checking

Perform mirror heater check with mirror at room temperature.

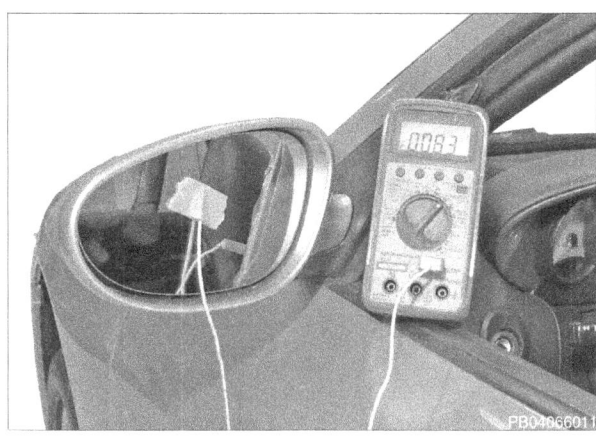

◄ Using a thermal sensor, measure temperature at center of mirror.

– Turn mirror heater ON and wait 1 minute.

– Measure mirror temperature again. Mirror temperature should increase by approximately 2 degrees.

– If no increase is measured, diagnose fault using wiring diagrams. See **EWD Electrical Wiring Diagrams.**

REAR SPOILER

The rear spoiler, standard on Boxster models, aids vehicle stability at high speeds by increasing aerodynamic downforce. The spoiler deploys automatically when vehicle speed reaches approximately 75 mph / 120 kph and retracts at approximately 50 mph / 80 kph. The spoiler can also be deployed manually using the rocker switch found on the main fuse panel in the left foot well.

> *WARNING—*
> * Danger of accident: If rear spoiler does not extend, high speed driving stability is impaired.

Rear spoiler, removing and installing

– Turn ignition ON and raise rear spoiler. Use rear spoiler switch in main fuse panel, left footwell. Turn ignition OFF.

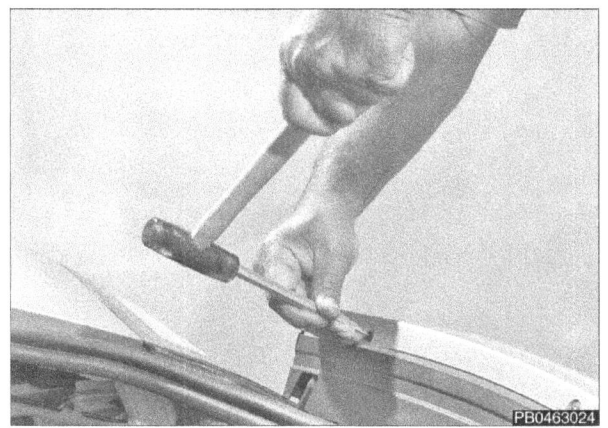

◄ Use thin pin punch and light hammer to drive in plastic rivet pins in spoiler cover.

> *NOTE—*
> * 2003 - 2004 models: Use a screwdriver through holes in lower part of spoiler cover to release locking tabs and lift spoiler cover off.

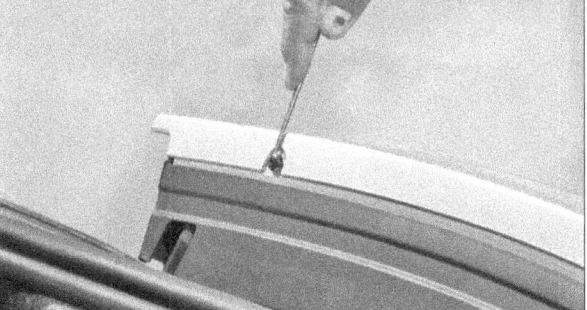

◄ Gently pry out spoiler cover plastic rivets.

66-4 Exterior Equipment

Rear Spoiler

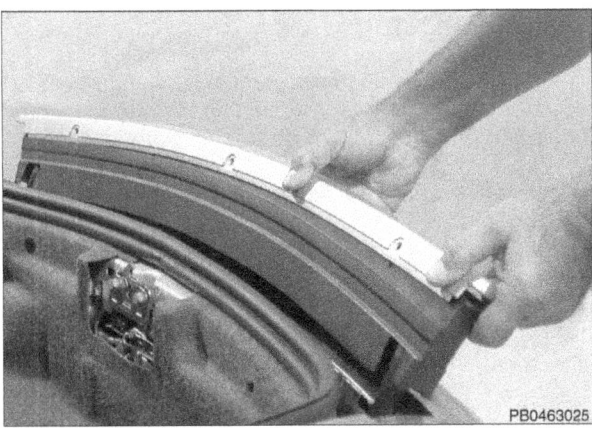

◀ Tilt spoiler cover backward and off spoiler.

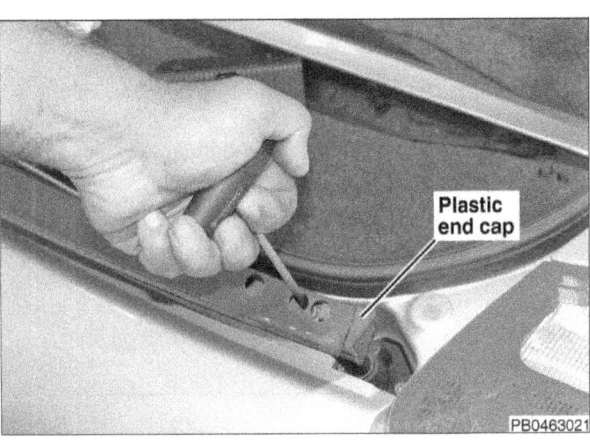

◀ Use 5 mm Allen wrench to remove spoiler top rail mounting bolts.

NOTE—
- 1997 - 2002 models: With top rail in hand, remove plastic end cap and shake out plastic rivet expanding pins that fell inside in previous step.

– Remove rubber trunk seal, trunk liner, and plywood floor from trunk.

– Turn ignition ON and lower rear spoiler. Use rear spoiler switch in main fuse panel, left footwell. Turn ignition OFF.

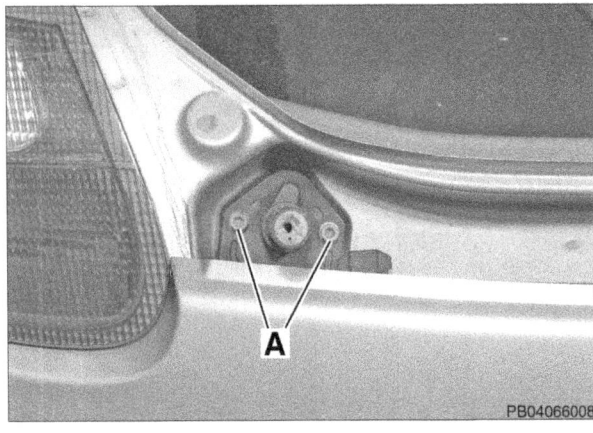

◀ Using a 5 mm Allen wrench, remove spoiler piston mounting plate screws (**A**) at left and right. Remove mounting plates.

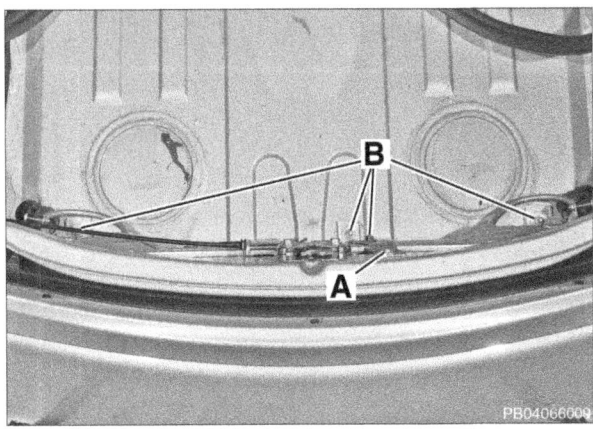

◀ Remove rear spoiler mechanism mounting hardware bolt (**A**) and nuts (**B**).

Exterior Equipment 66-5

Rear Spoiler

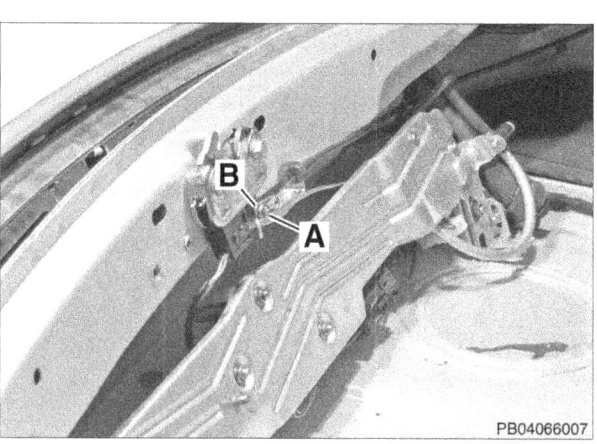

◀ Remove retaining clip (**A**). Slide Bowden cable and lock actuating rod (**B**) from trunk lid lock assembly.

– Slide lower end of rear spoiler assembly out and release lift pistons from vehicle body.

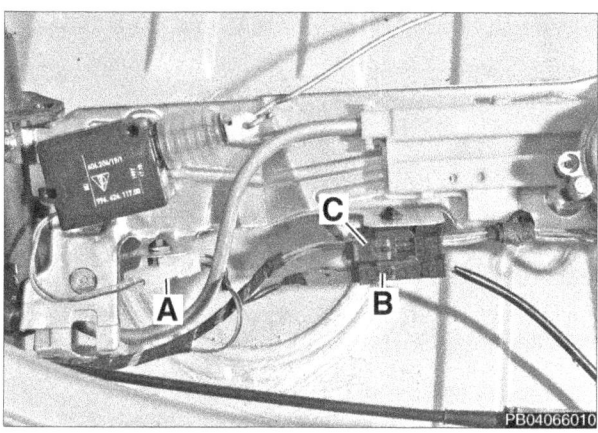

◀ Working inside rear trunk, disconnect electrical harness connectors for trunk lock actuator (**A**), trunk lock switch (**B**) and spoiler motor (**C**).

NOTE—
* Spoiler motor and trunk lid lock connectors are joined together and attached to spoiler housing.
* Connectors may be separated from spoiler housing by rotating connectors 45° counterclockwise.
* Connectors can be separated from each other by depressing tab located at center between two connectors and sliding trunk lid lock connector rearward.

– Installation is reverse of removal.

Spoiler gaps	
Spoiler to rear trunk	5.1 - 0.5 mm (0.2 - 0.02 in)
Spoiler to rear apron	3.1 mm (0.12 in)

Rear spoiler motor, removing and installing

– Remove rear spoiler assembly, see **Rear spoiler, removing and installing**.

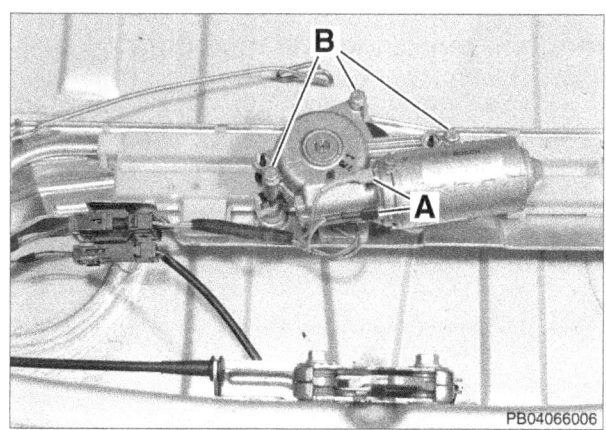

◀ Disconnect rear spoiler motor electrical harness connectors (**A**). Remove spoiler motor fasteners (**B**) and remove motor.

– Remainder of installation is reverse of removal. Make sure:
* Transmission spur gear and axle are in place.
* Wiring harness connectors are firmly attached.

66-6 Exterior Equipment

Engine Compartment Vents

ENGINE COMPARTMENT VENTS

Boxster models are equipped with matching engine compartment vents on the front edge of each rear fender. The left side vent supplies combustion air to the engine intake. The right side vent supplies air to the engine compartment cooling fan (blower).

Redesigned engine compartment vents were introduced on 2003 models. Removal and installation procedures are similar.

Engine compartment vent, removing and installing

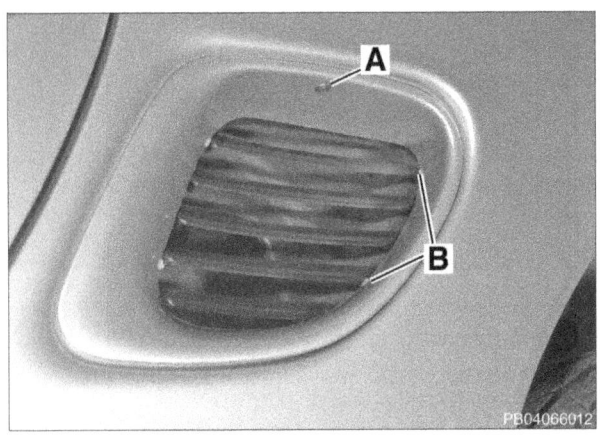

◄ Remove screw (**A**).

- 1997 to 2002 models: Using a plastic pry tool or narrow tipped screwdriver, pry back clips (**B**) and pull vent from quarter panel.
- 2003 and 2004 models: Push vent diagonally and upward to remove.

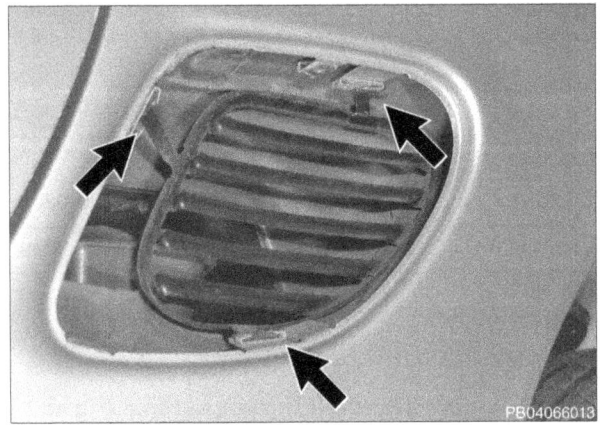

◄ Remove vent grill from retaining clips at body (**arrows**).

– Installation is reverse of removal.

NOTE—

- *2003 and 2004 models require that grill and vent be installed together.*

EMBLEMS

Front emblem, removing and installing

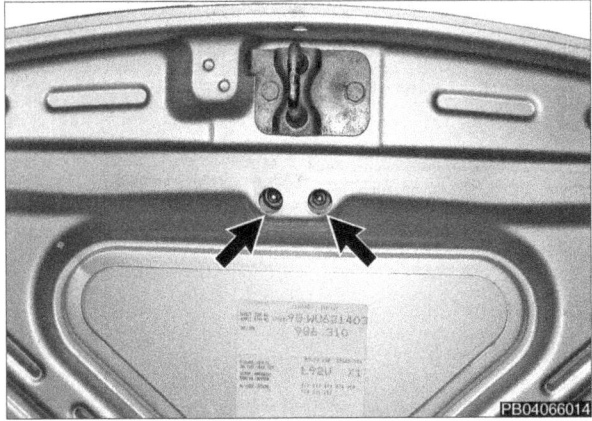

◄ With front trunk open, remove front emblem by removing speed nuts (**arrows**) located inside front trunk lid.

Exterior Equipment 66-7

Emblems

Rear emblem, removing and installing

- Heat emblem and rear trunk lid with a hair dryer.

◂ Using dental floss or monofiliment fishing line, gently separate emblem from trunk lid by sliding line under emblem using a sawing motion.

- Clean residual adhesive from emblem and trunk lid using an appropriate solvent.

- Emblem fastens to trunk lid using double-sided trim tape or similar adhesive.

69 Seat Belts, Airbags

GENERAL 69-1	**AIRBAG REPAIRS** 69-8
Seat belt system 69-1	Airbag control module, removing and installing . 69-8
Airbag system components................. 69-2	Side impact crash sensor,
Airbag deployment 69-2	removing and installing 69-9
Passenger airbag 69-3	Driver airbag, removing and installing 69-10
Smart airbags 69-3	Driver airbag contact spring,
Airbag control module 69-3	removing and installing 69-11
Airbag replacement 69-4	Passenger airbag, removing and installing ... 69-13
Warnings and Cautions 69-4	Side impact (door-mounted) airbag,
	removing and installing 69-14
SEAT BELT REPAIRS 69-5	**TABLE**
Seat belt reel, removing and installing 69-5	
Seat belt buckle, removing and installing 69-7	a. Airbag component replacement
	(Porsche recommendations) 69-4

GENERAL

This repair group covers seat belt and airbag component replacement.

Be sure to have airbag system diagnosis and component testing carried out by a Porsche service technician or other qualified personnel.

NOTE—

- *Airbags are known as the Supplemental Restraint System (SRS). The occupant safety system (seat belts and airbags) is called the Multiple Restraint System (MRS).*
- *Special test equipment is required to retrieve airbag fault codes, diagnose system faults and reset/turn off the airbag warning light. The warning light remains ON until problems are corrected and the fault memory cleared.*

Seat belt system

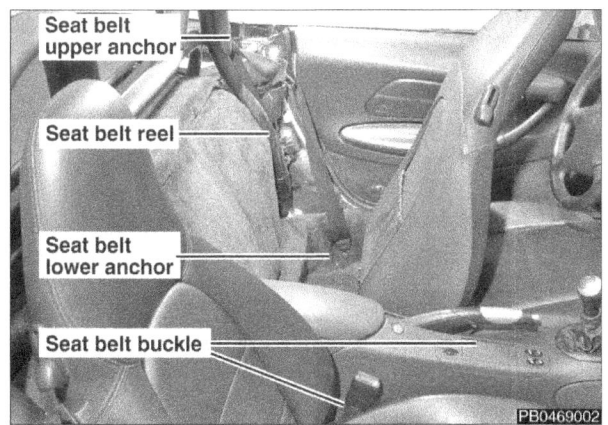

Boxster seat belts use a 3-point inertia reel system for driver and passenger. The belt reel is behind B-pillar trim, bolted to the roll-bar assembly. The lower belt anchor is at the base of the B-pillar while the upper belt anchor is also bolted to the roll-bar. The belt buckle or lock is bolted to the inner seat base.

NOTE—

- *Photo illustrates driver seat belt components with B-pillar trim removed for clarity.*

69-2 Seat Belts, Airbags

General

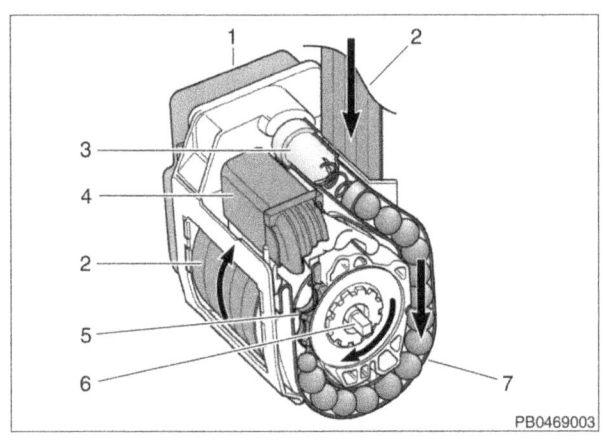

Beginning with 2002 models, the seat belt reel features a pyrotechnic (explosive charge) tensioning device triggered by the airbag system.

1. Triggering unit
2. Belt
3. Propellant (explosive charge)
4. Ball tube cap
5. Ball-driven gear
6. Belt take-up shaft
7. Ball tube

The propellant, once triggered, forces the balls in the tube to rotate the gear attached to the belt take-up shaft, thus tensioning the belt by up to 200 mm (7.8 in) in 10 ms. A belt-force limiter prevents bruising and internal injuries to the passenger due to excessive tensioning of the belt.

During a frontal collision, the belt tensioner is triggered together with the front airbags. During a rear collision, only the belt tensioner is triggered.

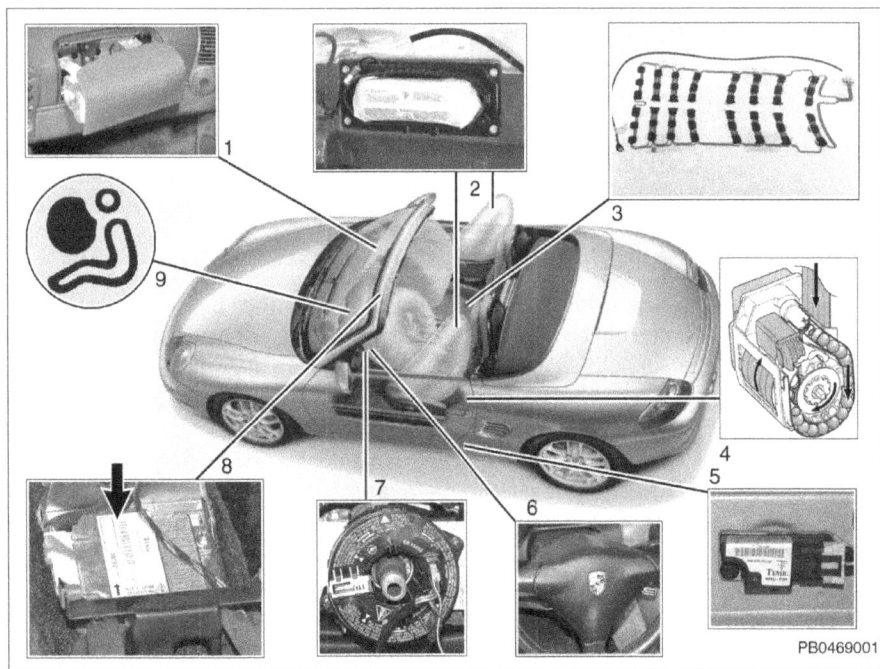

Airbag system components

1. **Passenger airbag**
2. **Side impact (door-mounted) airbag**
   - Optional from 1998
3. **Seat occupancy sensor**
   - Some 1997 models
   - All models from 1998
4. **Pyrotechnic seat belt reel**
   - 2002 and later models
5. **Side impact airbag crash sensor**
6. **Driver airbag**
7. **Driver airbag contact spring**
8. **Airbag control module and frontal crash sensor**
9. **Airbag warning light in instrument cluster**

Airbag deployment

Airbag deployment depends on the angle and severity of an accident. Front airbags deploy in case of frontal impact and side airbags deploy in case of lateral impact. Front airbags quickly deflate after deployment so that they do not obstruct the view of occupants. Side airbags are not designed to self-deflate.

Seat Belts, Airbags 69-3
General

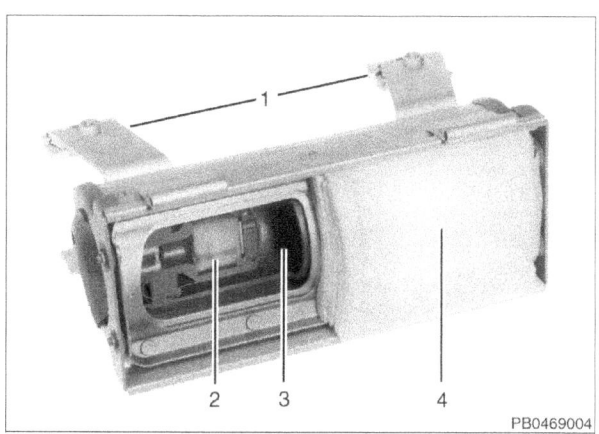

Passenger airbag

◀ The airbag unit in the passenger side of the dashboard inflates using argon and helium gas. The use of these inert gases instead of the more traditional sodium azide gas results in reduced hazard to passenger health and simpler production and recycling.

1. Passenger airbag mounting brackets
2. Igniter capsule
3. Gas chamber
4. Folded airbag

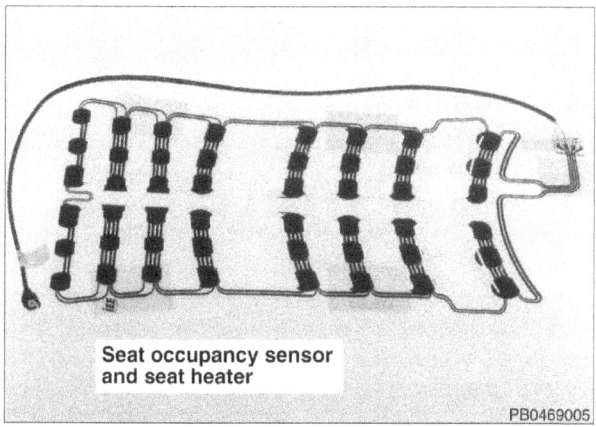

Seat occupancy sensor and seat heater

Smart airbags

Some 1997 models and all models from 1998 are equipped with smart airbag technology which automatically deactivates the passenger airbag if there is no seat occupant.

◀ The smart airbag system uses a sensor pad glued to the seat cushion foam under the seat heater. The sensor signals seat occupancy information to the SRS control module.

> **CAUTION—**
> - To avoid injury and prevent unwanted triggering of the passenger airbag, do not place a heavy object (such as a grocery bag) on the passenger seat.

Airbag control module

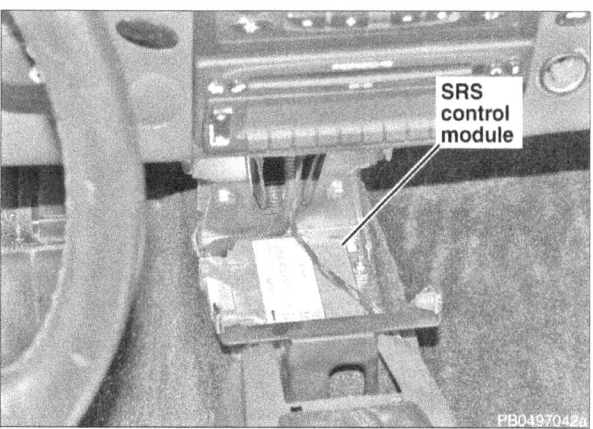

SRS control module

◀ The control module for the airbag system is located below the center dashboard, just ahead of the shifter.

The module combines the following functions:
- Electronic control of airbags and (if applicable) pyrotechnic seat belt reels.
- Frontal crash sensor.
- Shutting off fuel supply and unlocking of doors when a crash is detected.
- Airbag system fault detection.

When the ignition is switched ON, the airbag module requires approximately 10 seconds to check and store fault information.

◀ The airbag warning light in the instrument cluster illuminates for approx. 3 seconds when the ignition is switched ON and turns OFF if no airbag system faults are detected by the control module. The light illuminates when a fault is detected.

Use Porsche System Tester 2 (PST2) to download and diagnose faults. Clear fault memory once problems are corrected.

If a new control module is installed, use PST2 to code to vehicle.

> **NOTE—**
> - If the airbag system is deactivated, the airbag warning light stays ON continuously. Use PST2 to reset the airbag system.

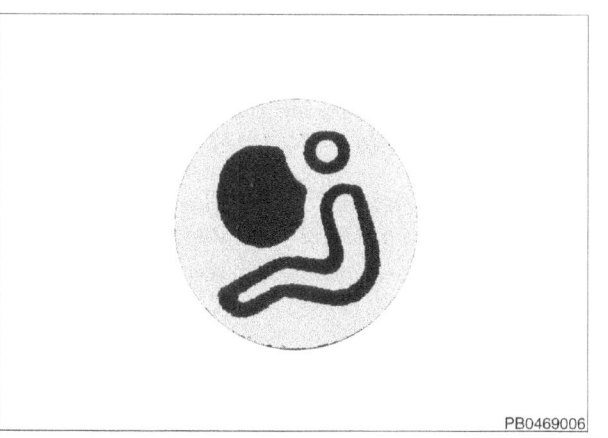

69-4 Seat Belts, Airbags

General

Airbag replacement

Porsche recommendations following an accident involving airbag deployment are in **Table a**.

Table a. Airbag component replacement (Porsche recommendations)	
Condition	**Replace**
Driver airbag triggered	Driver airbag contact spring Driver airbag
Passenger airbag triggered	Passenger airbag Airbag control module (after 3rd time airbag is triggered)
Side (door-mounted) airbag triggered	Side airbag Side airbag crash sensor (after 3rd time airbag is triggered)
Pyrotechnic seat belt reel triggered	Both seat belt reels

Warnings and Cautions

> *WARNING—*
> - *If the airbag warning light is ON, the airbags do not deploy in case of an accident. Be sure to have the airbag system inspected and repaired immediately.*
> - *Airbags and pyrotechnic seat belt reels are deployed by explosive devices. Handled improperly or without adequate safeguards, these components are very dangerous. Serious injury may result if airbag system service is attempted by persons unfamiliar with the Porsche airbag system and its approved service procedures.*
> - *If using a child seat, deactivate the passenger airbag. Airbag deactivation requires special programming available only through an authorized Porsche dealer.*
> - *Disconnect the battery and cover the negative (–) battery terminal with an insulator before starting diagnostic, troubleshooting or service work on electrical components, even if not associated with the airbags, and before doing any welding on the car.*
> - *When handling a control module or other electronic component, use approved methods of dissipating electrostatic charge.*
> - *After disconnecting the battery, wait 1 minute before beginning work on airbag components.*
> - *If an airbag or pyrotechnic seat belt reel is fired due to an accident, Porsche specifies that airbag system components be replaced. For more information on post-collision airbag service, see **Table a** and consult an authorized Porsche dealer.*
> - *Do not fire an airbag unit or pyrotechnic seat belt reel prior to disposal. It must be fired by a special disposal company or shipped back to Porsche in the packaging of the new components.*
> - *When removing a fired airbag unit, avoid contact with the skin; wear gloves. In case of skin contact, wash with water.*
> - *Do not allow airbag system components to come in contact with cleaning solutions or grease.*
> - *Do not subject airbag components to temperatures above 194°F (90°C).*

Seat Belts, Airbags 69-5

Seat Belt Repairs

> **WARNING—**
> - When reconnecting the battery, make sure no one is inside the vehicle.
> - Store an airbag unit with the padded side facing up. Do not leave an airbag unit unattended.
> - If the airbag unit or airbag control module is dropped from a height of ½ meter (1½ ft) or more, do not use the component.
> - Remove the airbag control module from the vehicle before performing any welding.
> - Do not attempt to repair airbag or pyrotechnic components.
> - Replace seat belt if subject to occupant loading in a collision.
> - Do not modify or repair seat belts. Do not modify seat belt mounting points.
> - Do not bleach or dye seat belt webbing. Webbing that is severely faded or redyed does not meet strength requirements.
> - Clean belts with a luke-warm soap solution only.
> - Periodically inspect seat belts for webbing defects such as cuts or pulled threads.
> - Immediately after replacing a damaged or worn seat belt, destroy the old belt to prevent it from being reused.

> **CAUTION—**
> - When working on electrical components, disconnect the negative (−) cable from the battery and insulate the cable end to prevent accidental reconnection.
> - Prior to disconnecting the battery, read the battery disconnection cautions given in **001 Warnings and Cautions**.
> - To prevent marring trim when working on interior components, work with a plastic prying tool or wrap a screwdriver tip with tape.

SEAT BELT REPAIRS

Seat belt reel, removing and installing

− 2002 - 2004 models: Disconnect negative (−) battery cable and cover battery terminal to keep cable from accidentally contacting terminal.

> **WARNING—**
> - 2002 - 2004 models: The seat belt reel contains the pyrotechnic (explosive charge) belt tensioner device and is a component in the Multiple Restraint System (MRS). Read **Warnings and Cautions** in this repair group before proceeding with seat belt removal.
> - After disconnecting the battery, wait 1 minute before beginning work on airbag components.

> **CAUTION—**
> - Prior to disconnecting the battery, read the battery disconnection cautions in **00 Warnings and Cautions**.

69-6 Seat Belts, Airbags

Seat Belt Repairs

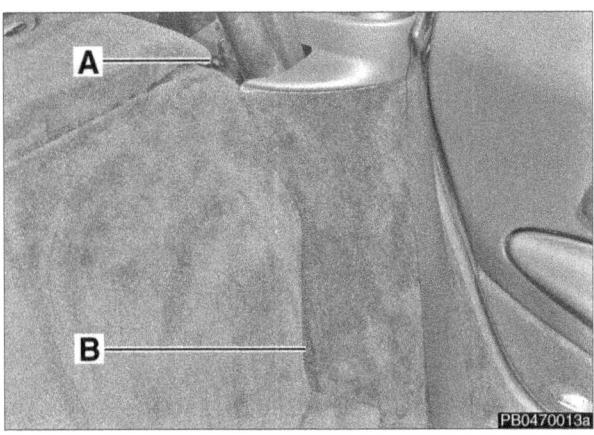

◄ Lean seat backrest forward. Working at B-pillar trim:
- Loosen upper trim retaining screw (**A**).
- Remove lower trim screw (**B**).
- Slide trim up and out of car.

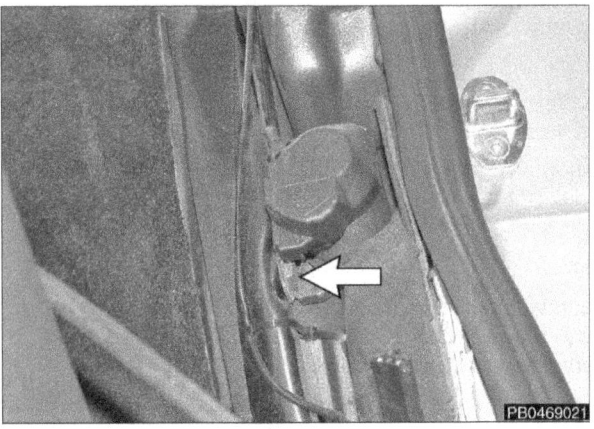

– 2002 - 2004 models with pyrotechnic seat belt reel: Working at seat belt reel, pry out fuse and detach reel electrical connector.

◄ Remove reel mounting bolt (**arrow**) at roll-bar assembly.

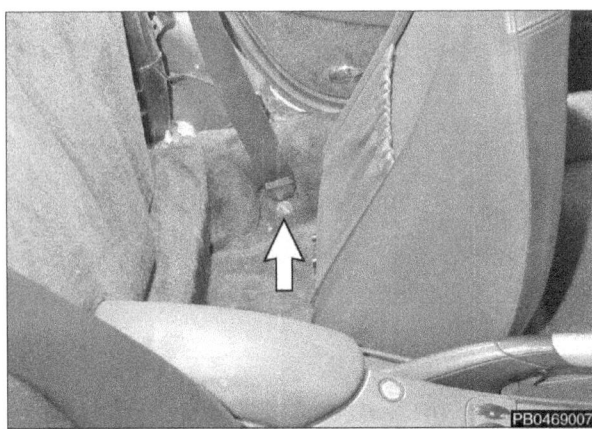

◄ Remove lower belt anchor bolt trim cover near floor. Remove belt anchor bolt (**arrow**).

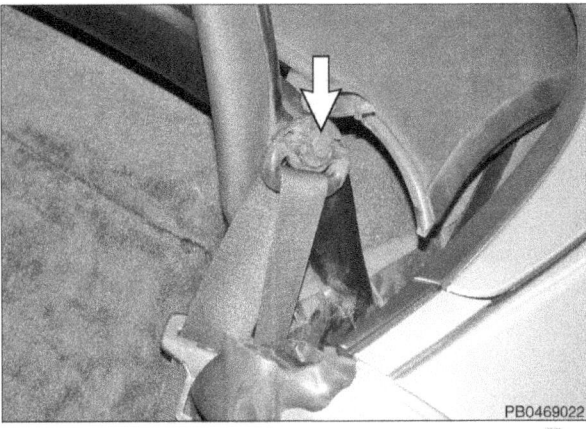

◄ Pry open plastic trim and remove upper belt anchor bolt (**arrow**) from roll-bar assembly. Remove seat belt reel.

Seat Belts, Airbags 69-7

Seat Belt Repairs

- Installation is reverse of removal. Bear in mind the following:
  - Place spacers and washers in correct order, as removed.
  - After installing seat belt reel, pull on belt to check winding ability of reel.
  - 2002 - 2004 models: Read out and reset fault memory using Porsche System Tester 2 (PST2).

Tightening torque	
Seat belt mounting bolts	50 Nm (37 ft-lb)

Seat belt buckle, removing and installing

The seat belt buckle (lock) is bolted to the inboard seat base.

- Remove seat. See **72 Seats**.

 Working at inboard seat base, remove belt buckle mounting bolt (**arrow**).

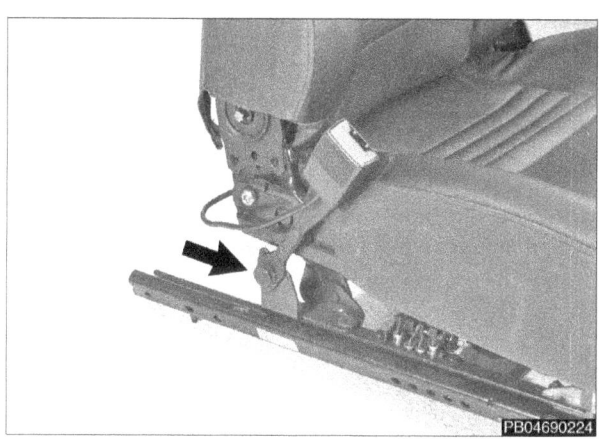

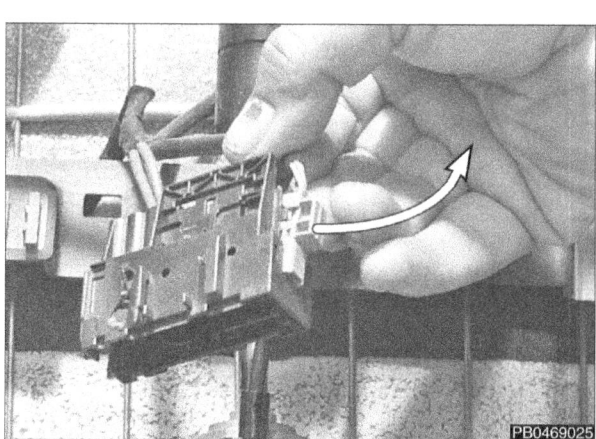

 Turn seat over. Detach seat belt buckle microswitch connectors from seat multi-connector plug.

- Installation is reverse of removal.

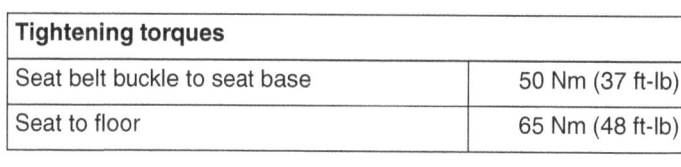

Tightening torques	
Seat belt buckle to seat base	50 Nm (37 ft-lb)
Seat to floor	65 Nm (48 ft-lb)

69-8 Seat Belts, Airbags

Airbag Repairs

AIRBAG REPAIRS

See **Airbag replacement** in this repair group for Porsche recommendations on airbag component replacement subsequent to an accident.

Airbag control module, removing and installing

The airbag system (SRS or MRS) control module and frontal crash sensor are combined in a unit located at the front of the center console, under the center dashboard.

- Disconnect negative (−) battery cable and cover battery terminal to keep cable from accidentally contacting terminal.

> **WARNING—**
> - After disconnecting the battery, wait 1 minute before beginning work on airbag components.

> **CAUTION—**
> - Prior to disconnecting the battery, read the battery disconnection cautions in **00 Warnings and Cautions**.

◄ Use plastic prying tool to gently pry out CD caddy, front storage bin and lower trim from lower center dashboard.

> **NOTE—**
> - 1998 Boxster center dashboard is illustrated. Other models are similar.

◄ Gently pry off center dashboard plastic and carpet covered trim (**arrows**).

> **NOTE—**
> - Pry off each panel at the front, then unclip the rear.

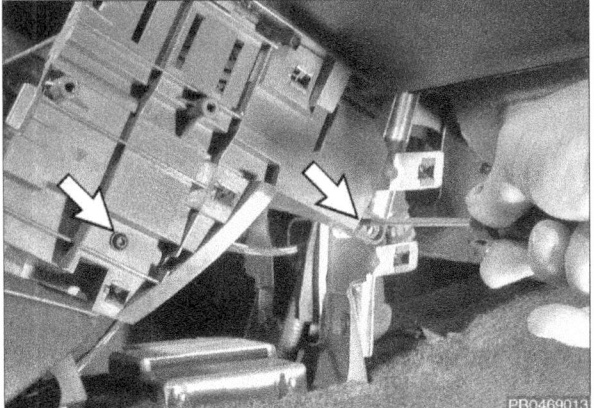

◄ Remove center console frame mounting Torx screws (**arrows**) from both sides. Lift off plastic frame.

Seat Belts, Airbags 69-9
Airbag Repairs

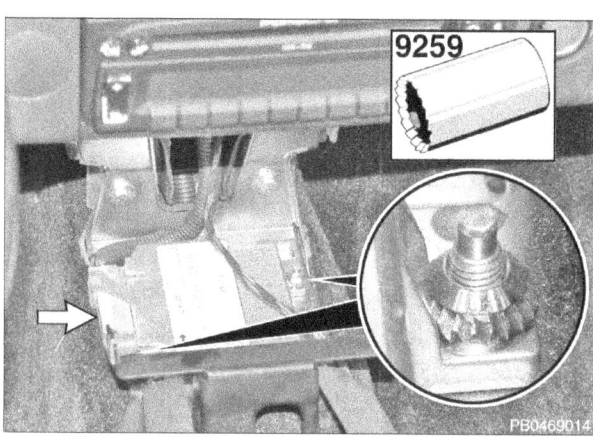

◀ Working at airbag control module:
- Unlock and detach harness connector (**arrow**).
- Remove 3 module mounting nuts and lift off module.

NOTE—
- *2 of the 3 mounting fasteners are shear nuts (**inset**). Use Porsche special tool 9259 or equivalent to remove.*

– When reinstalling control module:
- Make sure module and bracket attachments surfaces are bright and clean.
- Use ¼ Allen key to tighten shear nuts until they shear.

Tightening torque	
Airbag control module to mounting plate (M6 nut)	10 Nm (8 ft-lb)

– Read out fault memory using Porsche System Tester 2 (PST2). Then use PST2 to reset airbag system.

Side impact crash sensor, removing and installing

The side impact crash sensor is located in the door sill, underneath sill trim, carpet and plastic sill frame.

– Disconnect negative (–) battery cable and cover battery terminal to keep cable from accidentally contacting terminal.

WARNING—
- *After disconnecting the battery, wait 1 minute before beginning work on airbag components.*

CAUTION—
- *Prior to disconnecting the battery, read the battery disconnection cautions in **00 Warnings and Cautions**.*

– Remove seat. See **72 Seats**.

– Remove right or left sill trim panel. See **70 Interior Trim**.

◀ Remove seatbelt lower anchor bolt. Peel back door sill and floor carpeting and remove sill frame fasteners (**arrows**). Lift out plastic sill frame.

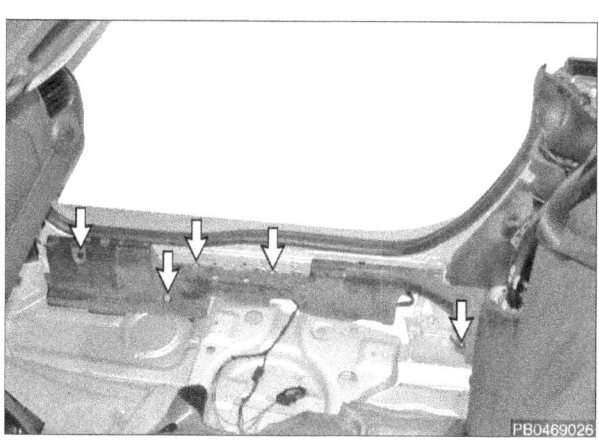

69-10 Seat Belts, Airbags

Airbag Repairs

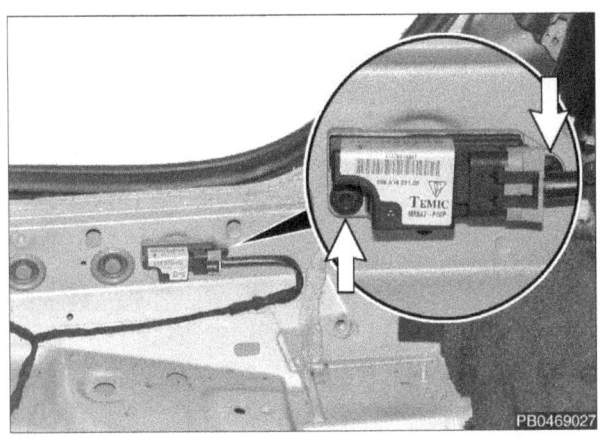

◄ Detach crash sensor electrical connector. Remove sensor mounting bolts (**arrows**) and lift off sensor.

– Installation is reverse of removal.

Tightening torques	
Seat to floor	65 Nm (48 ft-lb)
Side impact crash sensor to sill	10 Nm (8 ft-lb)

– Read out and reset fault memory using Porsche System Tester 2 (PST2).

Driver airbag, removing and installing

NOTE—
- *The photos illustrate repairs on a 1998 Boxster. Other models are similar.*

– Disconnect negative (–) battery cable and cover battery terminal to keep cable from accidentally contacting terminal.

WARNING—
- *After disconnecting the battery, wait 1 minute before beginning work on the airbag.*

CAUTION—
- *Prior to disconnecting the battery, read the battery disconnection cautions in* **00 Warnings and Cautions**.

◄ Extend steering column as far as possible. Use Torx T30 driver to remove airbag mounting screws from back of steering wheel.

◄ Lift airbag off center of steering wheel. Pry electrical connector from airbag.

WARNING—
- *Store airbag with the padded side facing up. Do not leave airbag unit unattended.*

Seat Belts, Airbags 69-11

Airbag Repairs

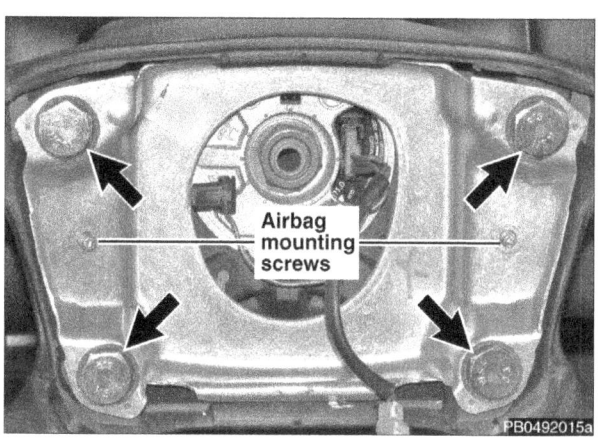

◁ Prior to remounting airbag, replace self-locking airbag mounting screws:

- Remove airbag mounting plate bolts (**arrows**).
- Lift off plate and replace airbag mounting screws from other side.
- Reinstall mounting plate.

Tightening torque	
Airbag mounting plate to steering wheel (M5 x 20 bolt)	5 Nm (4 ft-lb)

NOTE—

- *1998 4-spoke steering wheel is illustrated. 3-spoke wheel is similar.*

– Reinstall airbag after connecting electrical connector.

CAUTION—
- *Do not pinch airbag contact spring and horn harnesses.*

Tightening torque	
Airbag to mounting plate (M6 x 16 screw)	10 Nm (8 ft-lb)

Driver airbag contact spring, removing and installing

– Disconnect negative (–) battery cable and cover battery terminal to keep cable from accidentally contacting terminal.

WARNING—
- *After disconnecting the battery, wait 1 minute before beginning work on the airbag.*

CAUTION—
- *Prior to disconnecting the battery, read the battery disconnection cautions in* **00 Warnings and Cautions**.

– Remove driver airbag. See **Driver airbag, removing and installing** in this repair group.

◁ Working at steering wheel center:

- Detach electrical connectors (**A**).
- Remove steering wheel mounting nut (**B**).

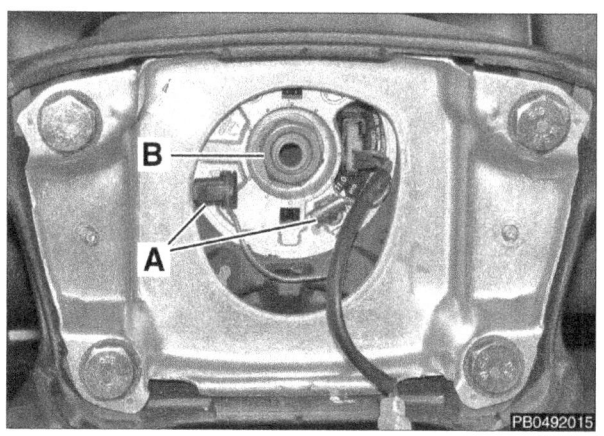

69-12 Seat Belts, Airbags

Airbag Repairs

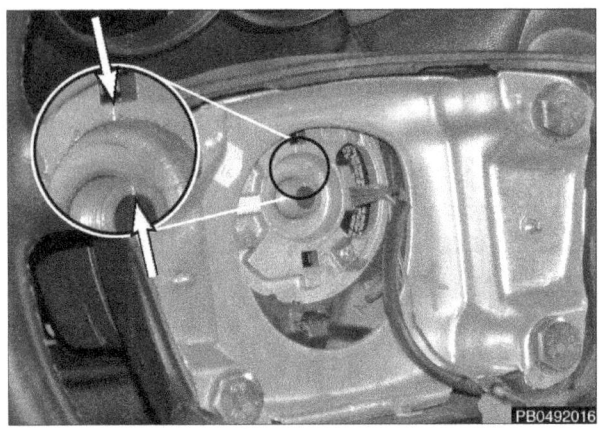

◄ Turn wheels to straight ahead position and mark position of steering wheel splines in relation to shaft splines (**arrows**). Remove steering wheel with front wheels in straight ahead position.

NOTE—
* *To prevent damage caused by unintended rotation, airbag contact spring locks when steering wheel is removed. Contact spring unlocks automatically when steering wheel is refitted.*

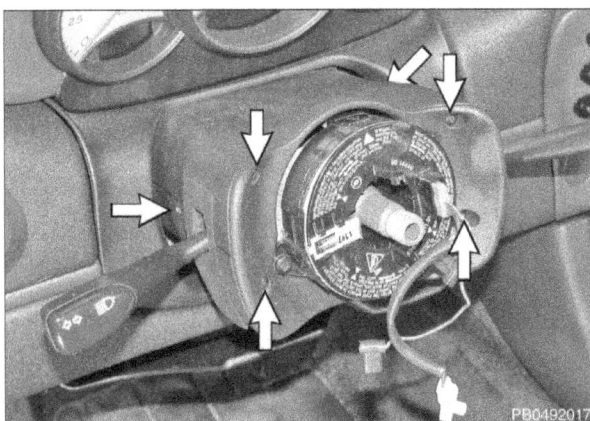

◄ Remove steering column cover fasteners (**arrows**). Remove covers.

◄ Working at airbag contact spring:
* Separate electrical connectors (**A**).
* Remove contact spring mounting screws (**B**). Slide contact spring off steering shaft.

◄ When reinstalling, prior to installing steering wheel, make sure airbag contact spring alignment **arrows** line up for steering wheel center position.

− Fit steering wheel using previously made spline alignment marks and make sure upper spokes of wheel are horizontal. Install and tighten steering wheel mounting nut.

CAUTION—
* *Do not pinch airbag contact spring and horn harnesses.*

Tightening torque	
Steering wheel to steering shaft	46 Nm (34 ft-lb)

Seat Belts, Airbags 69-13

Airbag Repairs

- Reattach airbag and horn harness connectors.

- Remount airbag using new mounting screws. See **Driver airbag, removing and installing** in this repair group.

> *CAUTION—*
> * Do not pinch airbag contact spring and horn harnesses.

Tightening torque	
Airbag to mounting plate (M6 x 16 screw)	10 Nm (8 ft-lb)

Passenger airbag, removing and installing

The passenger airbag is mounted to the dashboard support frame. Access to the mounting bolts is from underneath the dashboard.

- Disconnect negative (–) battery cable and cover battery terminal to keep cable from accidentally contacting terminal.

> *WARNING—*
> * After disconnecting the battery, wait 1 minute before beginning work on the airbag.

> *CAUTION—*
> * Prior to disconnecting the battery, read the battery disconnection cautions in **00 Warnings and Cautions**.

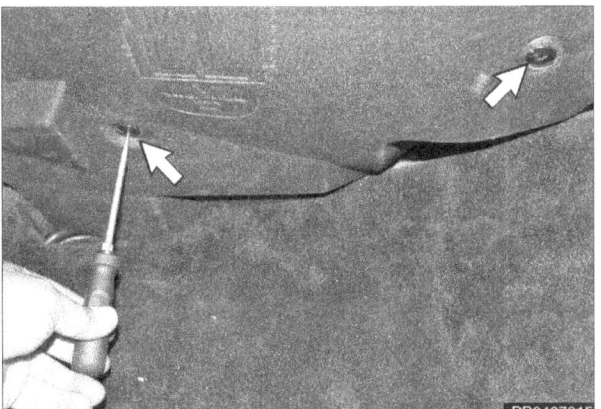

◄ Working in passenger side footwell, unscrew upper trim panel plastic fasteners (**arrows**). Pull down trim.

◄ Remove right air duct from under dashboard.

69-14 Seat Belts, Airbags

Airbag Repairs

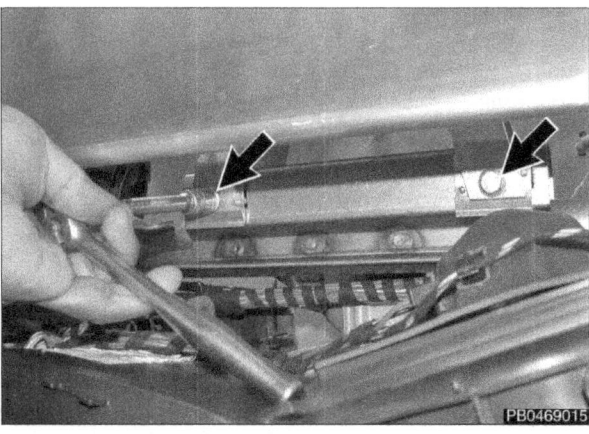

◄ Reaching up from underneath dashboard, remove airbag mounting bolts (**arrows**).

◄ Pull airbag out of dashboard. Carefully pry electrical connector from airbag (**arrow**).

> **WARNING—**
> - Store airbag with the padded side facing up. Do not leave airbag unit unattended.

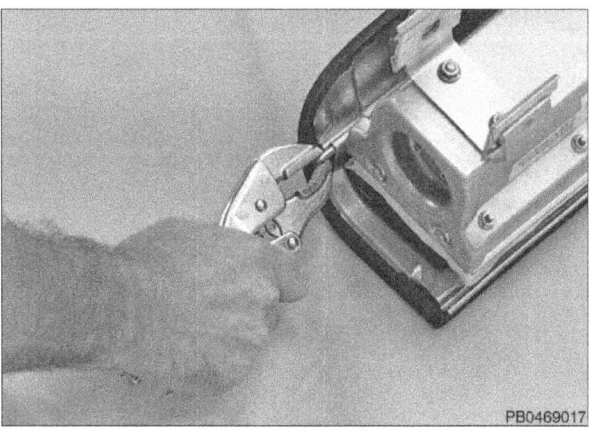

◄ If necessary, pull out pin to separate dashboard trim cover from airbag unit.

– Installation is reverse of removal.

Tightening torque	
Passenger airbag mount to dashboard frame (M8 x 95 mm bolt)	20 Nm (15 ft-lb)

Side impact (door-mounted) airbag, removing and installing

– Disconnect negative (–) battery cable and cover battery terminal to keep cable from accidentally contacting terminal.

> **WARNING—**
> - After disconnecting the battery, wait 1 minute before beginning work on the airbag.

> **CAUTION—**
> - Prior to disconnecting the battery, read the battery disconnection cautions in **00 Warnings and Cautions**.

Seat Belts, Airbags 69-15

Airbag Repairs

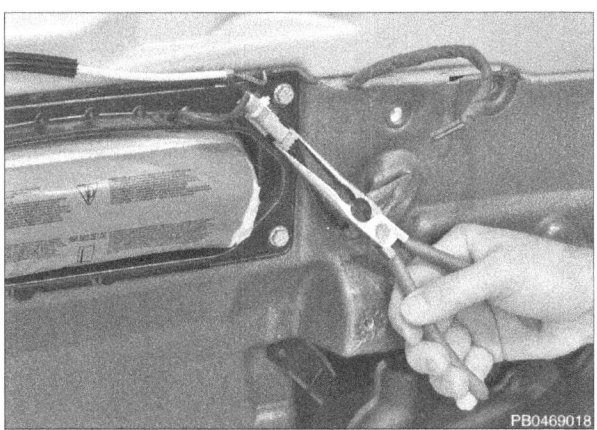

- Remove inside door trim panel. See **57 Door and Locks**.

◄ Separate airbag harness: Use needle nose pliers to gently squeeze tabs on connector.

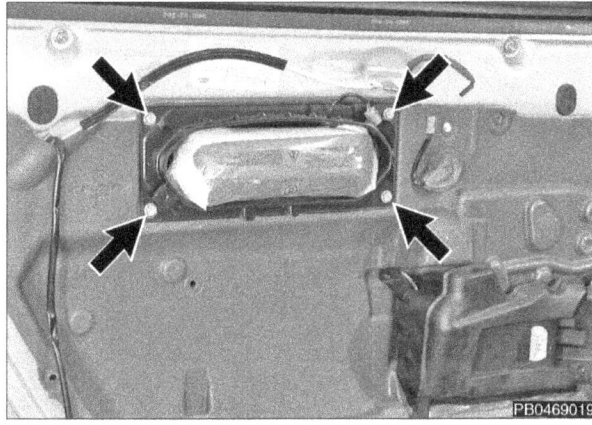

◄ Remove 4 airbag mounting bolts (**arrows**) and remove airbag.

> **WARNING—**
> - Store airbag with the padded side facing up. Do not leave airbag unit unattended.

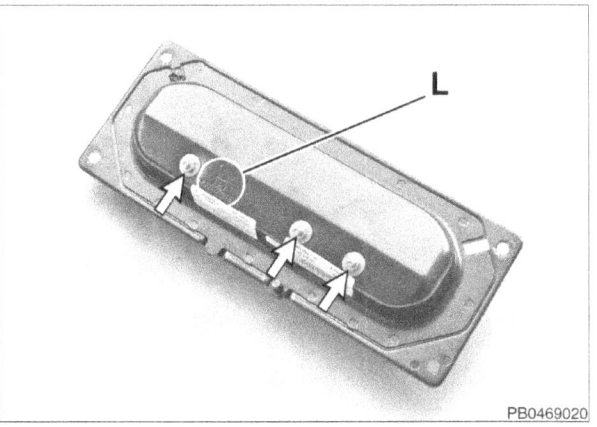

◄ If replacing airbag, remove 3 fasteners (**arrows**) to separate from airbag carrier.

> **NOTE—**
> - Left or right airbag carrier is identified by **L** or **R** stamped on back.

- Installation is reverse of removal. Use new self-locking bolts for mounting airbag to door.

Tightening torque	
Airbag to door (M6 x 16 mm bolt)	10 Nm (8 ft-lb)

70 Interior Trim

GENERAL 70-1
DASHBOARD 70-1
   Dashboard, removing and installing 70-1
CENTER CONSOLE 70-5
   Center console, removing and installing 70-5

OTHER TRIM 70-6
   B-pillar trim, removing 70-6
   Door sill trim, removing 70-6

GENERAL

This repair group covers interior trim removal and installation.

See the following repair groups for additional information:
- **57 Doors and Locks**
- **72 Seats**
- **69 Seat Belts, Airbags**
- **91 Radio and Communication**
- **48 Steering**
- **90 Instruments, Horns**

DASHBOARD

Dashboard, removing and installing

> *WARNING—*
> - *Cars covered by this manual are equipped with airbags mounted on the steering wheel and in the dashboard. When servicing these components, disconnect the negative battery terminal (-). See* **69 Seat Belts, Airbags** *for cautions and procedures relating to the airbag system.*

> *NOTE—*
> - *1998 Boxster dashboard is illustrated. Other models are similar.*

– Disconnect negative (–) battery cable and cover battery terminal to keep cable from accidentally contacting terminal.

> *CAUTION—*
> - *Prior to disconnecting the battery, acquire the radio anti-theft code.*
> - *Read the battery disconnection cautions in* **00 Warnings and Cautions**.

70-2 Interior Trim

Dashboard

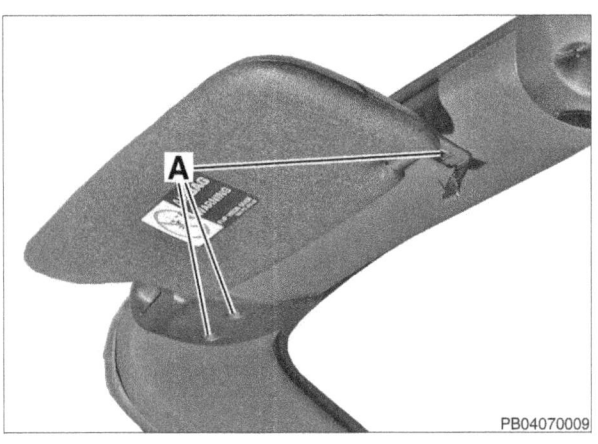

◀ Working at sun visors, pry open cover at inside hinge point. Remove mounting screws (**A**) and remove sun visor assemblies.

– Pry up edge of A-pillar trim and unsnap trim from pillar.

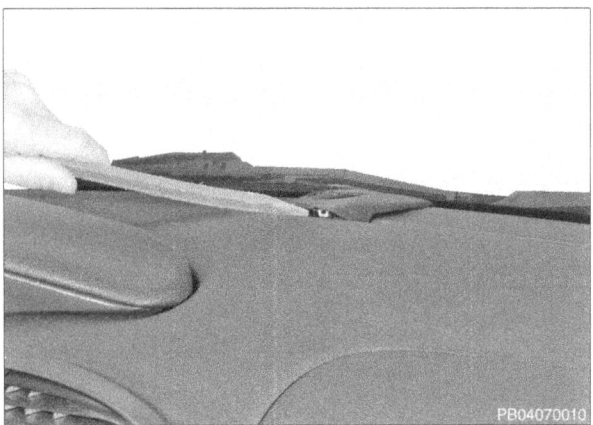

◀ Pry up alarm readiness indicator. Remove cover and unclip sensor from bracket.

– Using a plastic trim removal tool, pry up left and right side defrost vent panels from dashboard.

– Remove mounting screws and remove alarm readiness indicator bracket from dashboard.

– Remove dashboard mounted speakers and radio or PCM display unit. See **91 Radio and Communication**.

– Remove driver and passenger side airbags. See **69 Seat Belts, Airbags**.

– Remove steering wheel and steering column stalk switch. See **48 Steering**.

– Remove instrument cluster. See **90 Instruments, Horns**.

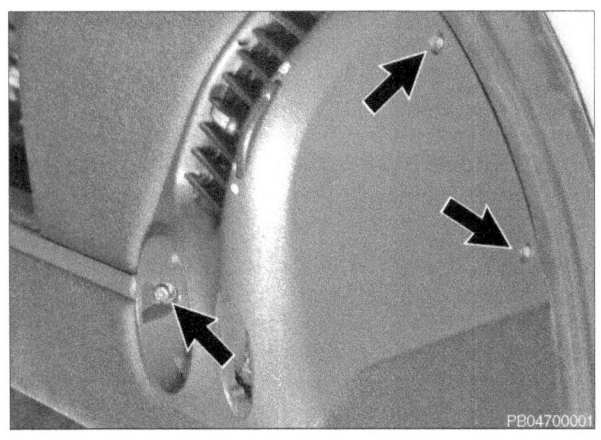

◀ Working at right side dashboard vent, remove cover at interior temperature sensor. With passenger door open, remove right side vent mounting screws (**arrows**). Remove vent from dash. Unclip interior temperature sensor through rear of vent.

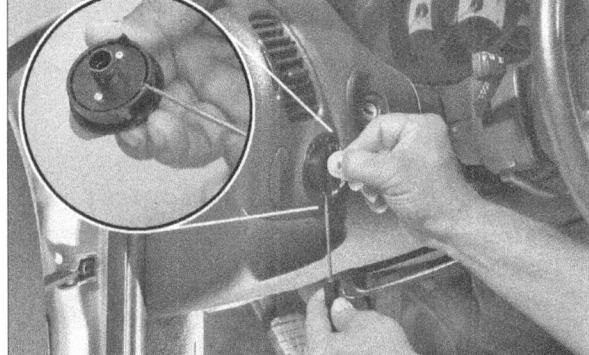

◀ Working at headlight switch, pull out light switch knob to stop. Use sharp thin tool to press in locking clip on underside of knob to disengage knob from switch.

Interior Trim 70-3
Dashboard

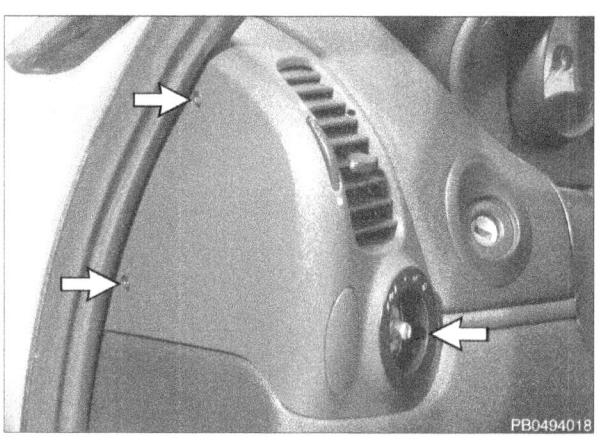

◀ Remove three left side air vent mounting screws (**arrows**).
- Pull vent and light switch together out of dashboard.
- Disconnect switch electrical connector.

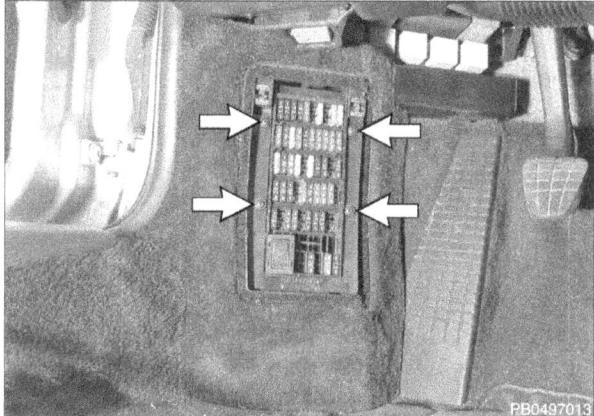

◀ Remove fuse panel cover and remove retaining screws (**arrows**). Remove left side kick panel.

◀ Remove left and right switch panels at center of dashboard. Disconnect electrical harness connectors from switches.

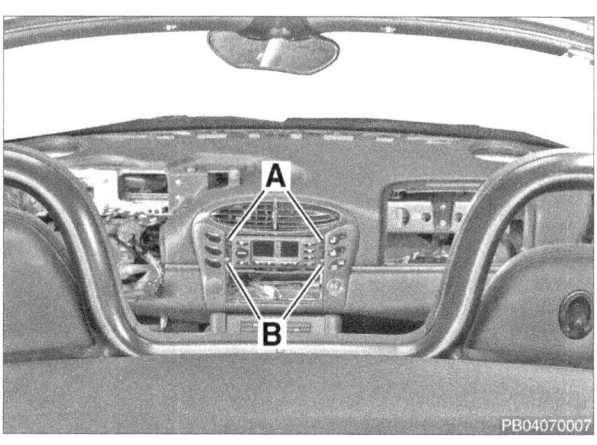

◀ Remove trim ring around heating and air-conditioning control panel. Remove screws (**A**) and slide control panel from dashboard. Remove screws (**B**) and remove control panel brackets.

NOTE—
- *Early model dashboard shown. Removal procedures for later model cars with heating and air-conditioning control panel mounted in lower console is similar.*

— Remove PCM display unit retaining bracket, if equipped. Raise retaining clip of bracket and guide bracket out through frame. Disconnect ground cable.

70-4 Interior Trim

Dashboard

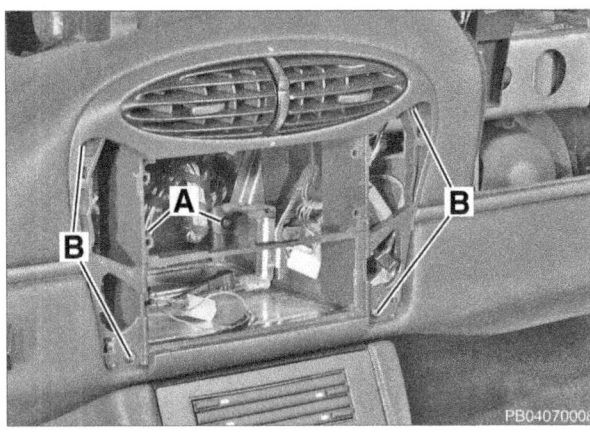

◄ Remove center vent and support frame. Remove lower rear fastening screws (**A**) and outer screws (**B**). Guide central vent out of dashboard.

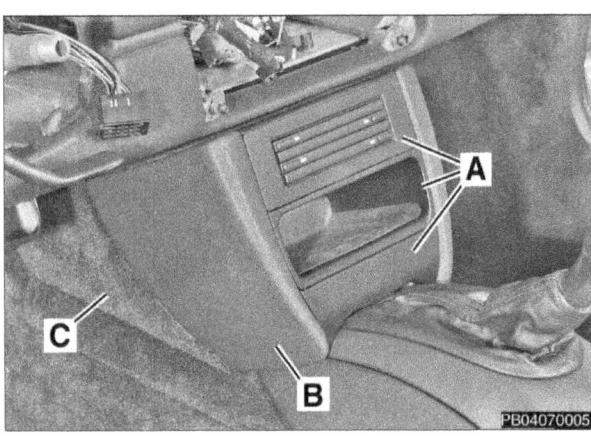

◄ Remove compact disc caddy, storage bin, and trim blank (**A**) by prying out with a plastic trim tool. Pry off console side panels (**B**) and carpeted trim panels (**C**).

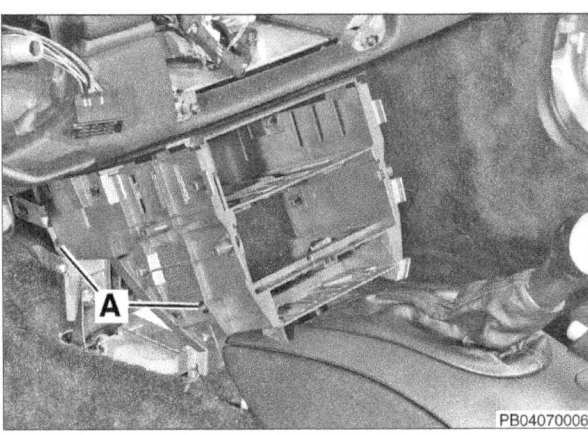

◄ Remove lower console frame mounting screws at left and right sides (**A**). Remove frame from console.

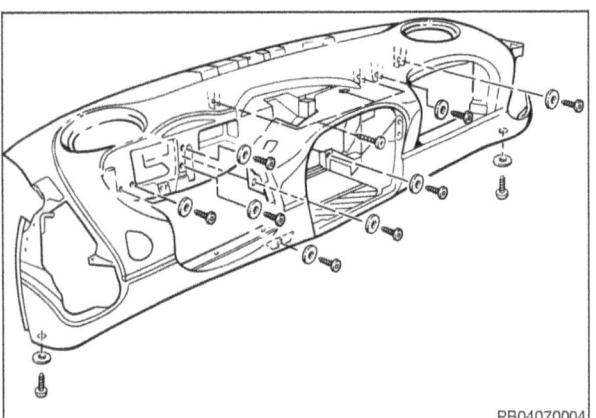

◄ Remove dashboard retaining screws and remove dashboard from support frame.

- Installation is reverse of removal, noting the following:
  - Check all electrical harness connections, including radio/PCM bracket ground connections, airbag connections, and multi-pin connectors at instrument cluster and steering column stalk switch.

Interior Trim 70-5

Center Console

CENTER CONSOLE

Center console, removing and installing

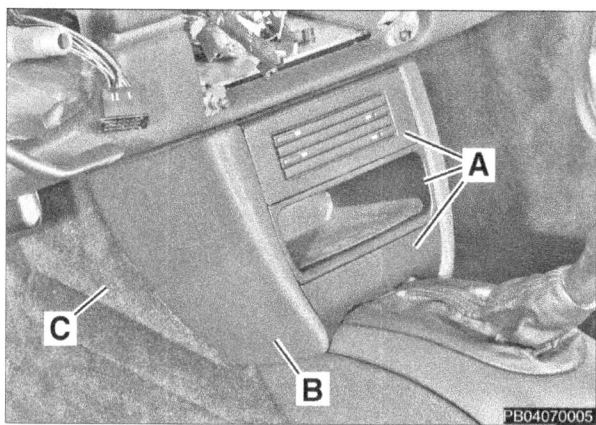

◄ Remove compact disc caddy, storage bin, and trim blank (**A**) by prying out with a plastic trim tool. Pry off console side panels (**B**) and carpeted trim panels (**C**).

– Remove ashtray from center console.

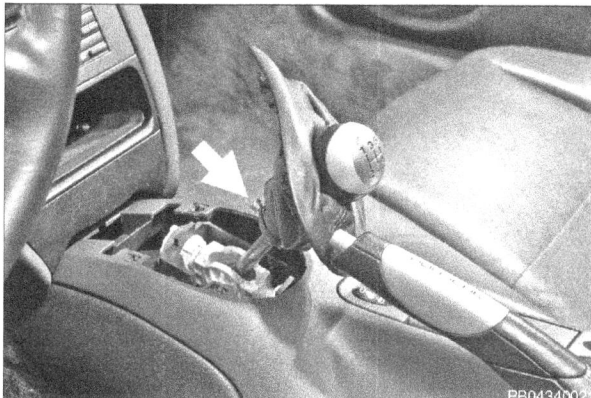

◄ Remove shift knob and boot:
  • Pry up shift lever boot at center console.
  • Pull shift boot over shift knob, turning boot inside out.
  • Loosen retaining screw (5 mm Allen) and slide knob off shift lever.

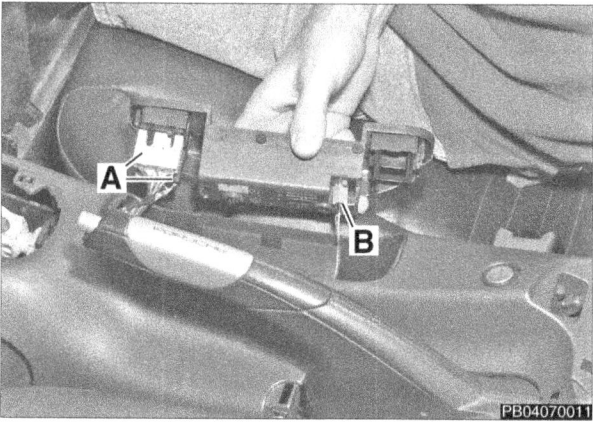

◄ Working in ashtray recess, remove power window switch panel retaining screws and lift panel from console. With panel free, unplug electrical harness connectors for power window switches (**A**) and ash tray light (**B**).

– Open console compartment cover behind parking brake lever and remove rubber insert. Remove retaining screw at bottom rear of compartment and remove bottom panel.

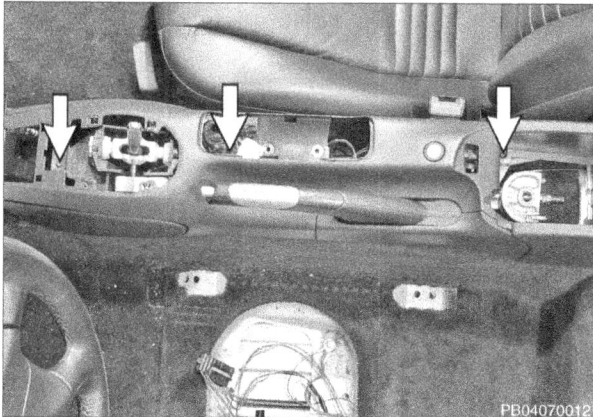

◄ Remove three remaining console retaining screws (**arrows**) and lift console from center tunnel.

NOTE—
  • *Driver's seat shown removed for clarity.*

– Installation of center console is reverse of removal.

70-6 Interior Trim

Other Trim

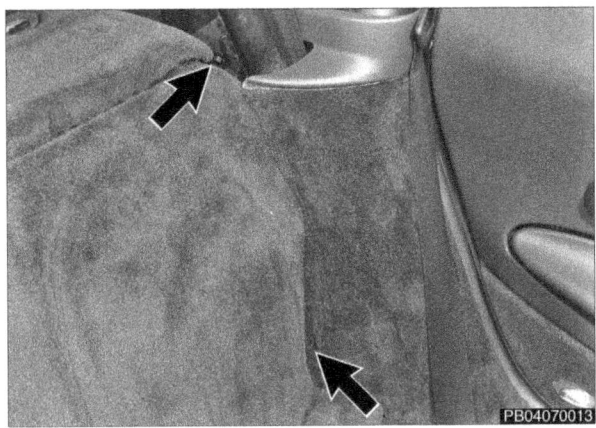

OTHER TRIM

B-pillar trim, removing

◀ Remove two retaining screws (**arrows**), and guide trim panel up from interior.

NOTE—
* *Interior shown with driver's seat removed for clarity. Seat does not need to be removed to remove trim panel.*

Door sill trim, removing

Driver side

— Remove driver seat. See **72 Seats**.

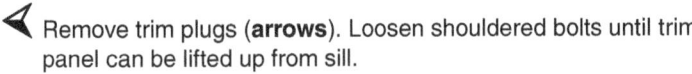

◀ Remove trim plugs (**arrows**). Loosen shouldered bolts until trim panel can be lifted up from sill.

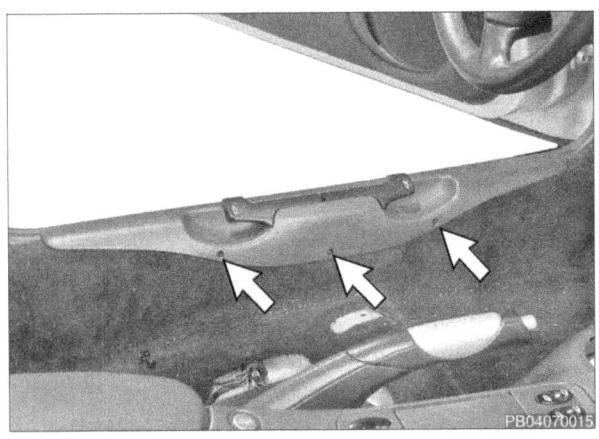

Passenger side

◀ Working at passenger side door sill, remove rubber compartment liner and remove retaining screws (**arrows**).

NOTE—
* *Interior shown with seat removed for clarity. Seat does not need to be removed to remove trim panel.*

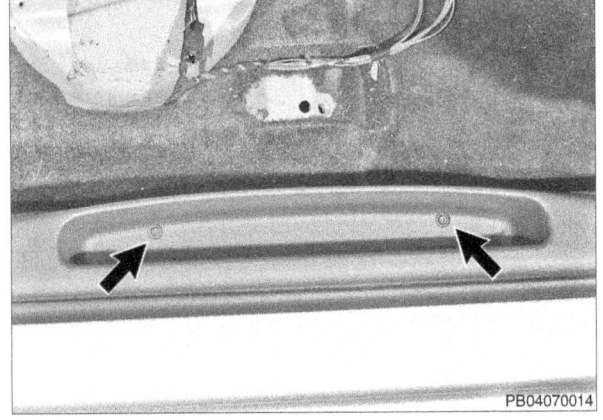

72 Seats

GENERAL 72-1
SEATS 72-1
   Seats, removing and installing 72-1

GENERAL

This repair group covers seat removal and installation.

SEATS

Seat, removing and installing

- Disconnect negative (−) battery cable and cover battery terminal to keep cable from accidentally contacting terminal.

> **CAUTION—**
> - *Prior to disconnecting the battery, acquire the radio anti-theft code.*
> - *Read the battery disconnection cautions in **00 Warnings and Cautions**.*

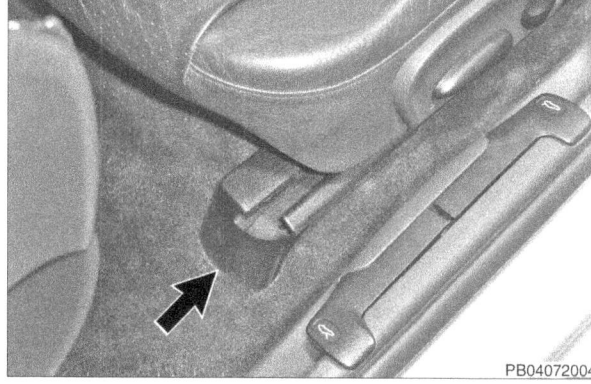

◀ With seat slid to far rear position, remove plastic trim caps at front end of both inside and outside (**arrow**) seat rails.

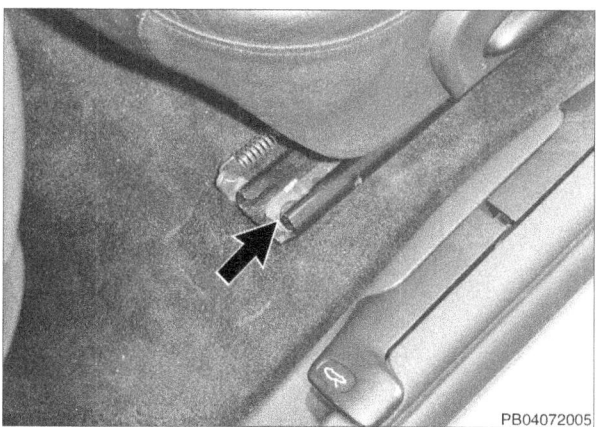

◀ Using an E12 Torx socket, remove front seat retaining bolts (**arrow**) for inside and outside rails.

72-2 Seats

Seats

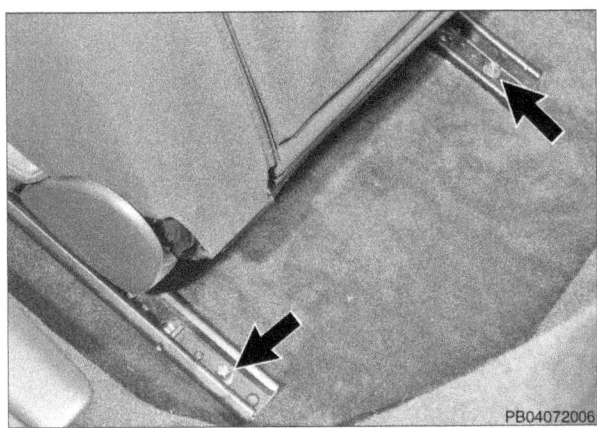

◄ Slide seat to far forward position. Remove rear seat mounting bolts (**arrows**).

◄ With seat unbolted, tilt seat sideways to reveal power seat electrical harness connector (**arrow**) under seat frame.

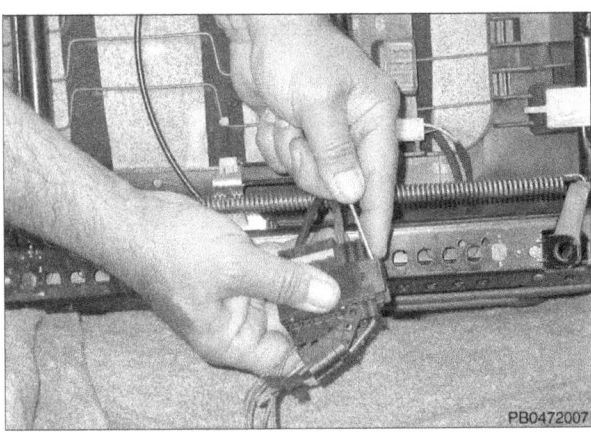

◄ Slide harness connector retaining lock outward to disconnect harness connector.

− With connector removed and convertible top down, carefully remove seat from passenger compartment.

CAUTION—
- *Use a blanket on door sill to protect vehicle finish when removing seat.*

− Installation of seat is reverse of removal. Be sure to locate power seat electrical harness out of the way of sliding seat rails.

Tightening torque	
Seat mounting bolt (M10 x 28)	65 Nm (48 ft-lb)

80 Heating

GENERAL 80-1
   Heating system 80-1
   Heating components 80-1
   Troubleshooting 80-2
   Warnings and Cautions 80-2

HEATER CORE 80-3
   Heater core, removing and installing 80-3

GENERAL

This repair group covers the heating system.

For additional information, see:
- **19 Cooling System**
- **87 Air-conditioning**

Heating system

The heating and air-conditioning system in Boxster models is adjusted by a control module integrated with the dashboard-mounted control panel.

Heating is regulated by airflow flaps in the A/C and heater housing. There is no heater valve. Engine coolant always flows through the heater core (heat exchanger) and, therefore, interior heat is quickly available.

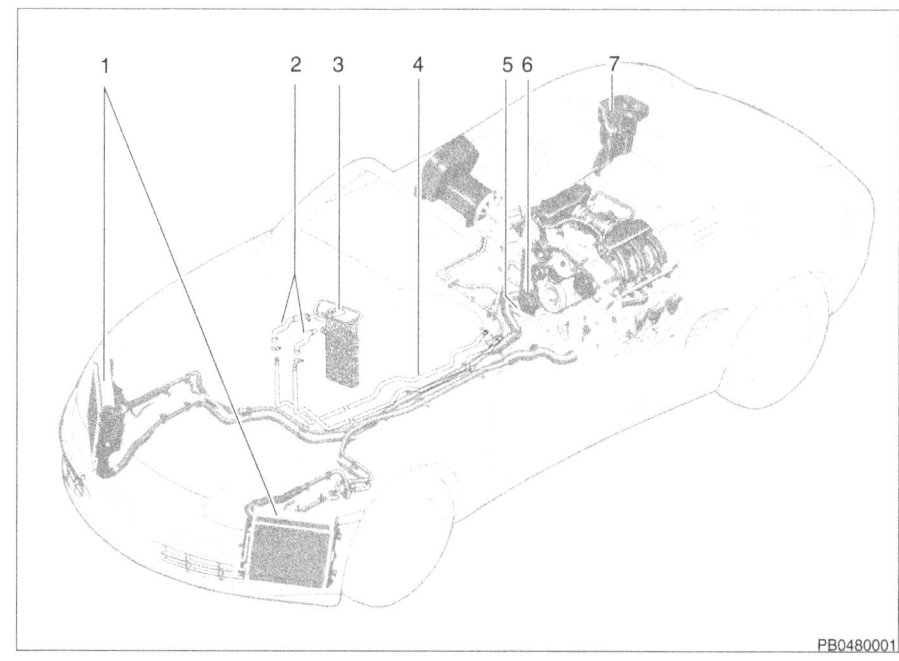

Heating components

1. Radiators
2. Heater core hoses
3. Heater core
4. Coolant lines to heater core
5. Coolant thermostat
6. Coolant pump
7. Coolant reservoir

80-2 Heating

General

Troubleshooting

Begin heating system troubleshooting with an analysis of the problem, then attempt to track the problem to its logical conclusion, and repair the fault.

Example: If the heater is not supplying hot air, the problem could be a faulty engine coolant thermostat.

Table a provides common symptoms and possible corrective actions to aid in diagnosing simple faults in the heating system. For air-conditioning system troubleshooting, see **87 Air-conditioning**.

Table a. Heating system troubleshooting		
Symptom	Probable cause	Corrective action
Blower does not work.	Blower fuse blown.	Inspect blower fuse (fuse D6, 30A, in main fuse panel, behind cover in left kick panel). Replace if necessary. See **97 Fuses, Relays, Component Locations**.
	Blower resistor (power stage) faulty.	Test blower resistor. Replace if necessary. See **87 Air-conditioning**.
	Blower motor wiring faulty.	Troubleshoot electrical harness. Repair as necessary.
	Blower motor faulty.	Test motor. Replace if necessary. See **87 Air-conditioning**.
Blower does not operate in lower speeds.	Blower resistor (power stage) faulty.	Test blower resistor. Replace if necessary. See **87 Air-conditioning**.
	Blower switch faulty.	Replace A/C and heating control panel. See **87 Air-conditioning**.
Heater does not produce hot air.	Engine coolant thermostat faulty.	Test engine coolant thermostat. Replace if necessary. See **19 Cooling System**.
	Temperature control flap malfunction.	Test temperature control flap motor. Replace if necessary.
Blower deposits oily film on inside of windshield; sweet odor detected from vents.	Heater core leaking coolant.	Pressure test cooling system. See **19 Cooling System**. Replace heater core if necessary.

Warnings and Cautions

Observe the following warnings and cautions when working with engine coolant.

> *WARNING—*
> - *At normal operating temperature the cooling system is pressurized. Allow the engine to cool thoroughly (a minimum of one hour), then loosen the cooling system pressure cap slowly to allow safe release of pressure.*
> - *Releasing cooling system pressure lowers the boiling point of coolant and it may boil suddenly. Use heavy gloves and wear eye and face protection to guard against scalding.*
> - *Use extreme care when draining and disposing of engine coolant. Coolant is poisonous and lethal to humans and pets. Pets are attracted to coolant because of its sweet smell and taste. Seek medical attention immediately if coolant is ingested.*

Heating 80-3

Heater Core

CAUTION—
- *Dispose of coolant in an environmentally safe manner.*
- *When replacing cooling or heating system components, replace spring type hose clamps with screw type clamps.*
- *Prior to disconnecting the battery, read the battery disconnection cautions given in* **00 Warnings and Cautions***.*

HEATER CORE

The heater core is located at the top front of the A/C and heating housing and easily accessible from under the cowl. There is no need to remove the housing to remove the heater core.

Heater core, removing and installing

WARNING—
- *Due to risk of personal injury, be sure the engine is cold before beginning work on the cooling system.*

◄ Working at engine service tray in rear trunk, loosen coolant expansion reservoir cap (**arrow**) slowly to allow coolant pressure to vent. Remove cap.

— Remove front cowl covers under front trunk lid.

— Disconnect negative (–) battery cable and cover battery terminal to keep cable from accidentally contacting terminal.

CAUTION—
- *Prior to disconnecting the battery, read the battery disconnection cautions in* **00 Warnings and Cautions***.*
- *Be sure to have the radio anti-theft code before disconnecting the battery.*

— Remove complete wiper assembly. See **92 Wipers and Washers**.

◄ Remove right front strut tower reinforcement brace fasteners (**arrows**) and remove brace.

80-4 Heating

Heater Core

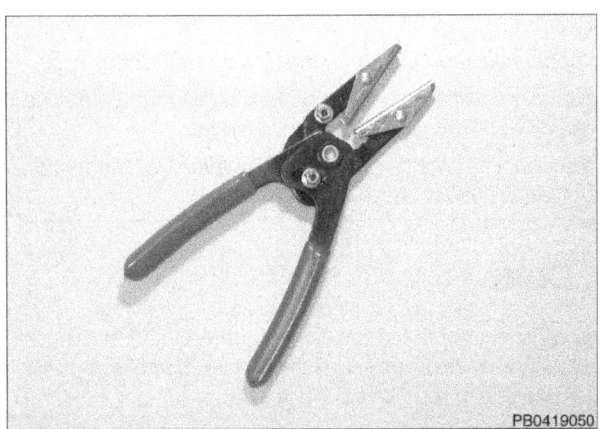

◄ Use hose pinch tools to clamp shut heater core hoses behind battery.

◄ Loosen heater hose clamps (**arrows**) at heater core and pull hoses off.

– Carefully remove cover over heater core. Release heater core and pull it out.

> **CAUTION—**
> • Soak up spilling coolant with shop towels.

– Installation is reverse of removal. Bear in mind the following:
  • Replace cover over heater core if damaged. Seal airtight with butyl adhesive.
  • Replace spring type hose clamps with screw type.
  • Replace lost coolant, then run engine and bleed system. See **19 Cooling System**.

87 Air-conditioning

GENERAL 87-1
 A/C and heater system 87-2
 A/C and heater control panel 87-3
 Refrigerant and refrigerant oil 87-3
 Fault detection 87-4
 Sensor default values 87-4
 Troubleshooting 87-4
 A/C system warnings and cautions 87-5

A/C SYSTEM COMPONENT REPLACEMENT .. 87-7
 A/C system components 87-7
 Blower, removing and installing 87-8
 Blower resistor (power stage),
  removing and installing 87-9
 Compressor, removing and installing .. 87-9

TABLES

a. Climate control sensor default values 87-4
b. Air-conditioning troubleshooting 87-4

GENERAL

This repair group covers interior ventilation and air-conditioning. For additional information, see:

- **03 Service and Maintenance** for A/C compressor drive belt and ventilation microfilter service.
- **19 Cooling System** for cooling system fan(s) replacement.
- **80 Heating** for heater core replacement.
- **EWD Electrical Wiring Diagrams**.

Some A/C or heater system repairs procedures require that the A/C refrigerant charge be evacuated. A/C system recharging procedures are beyond the scope of this manual.

A/C and heater control panel

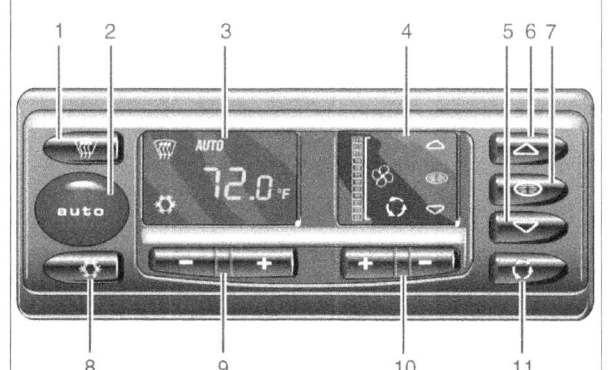

◄ The climate control system (A/C and heater) automatically regulates the interior temperature. Interior temperatures between 18°C (64°F) and 29°C (84°F) can be selected.

1. Windshield defroster control.
2. Automatic setting control: Returns system to automatic mode from any other mode.
3. Left display panel: Temperature setting and active left side buttons.
4. Right display panel: Blower setting and active right side buttons.
5. Footwell vent flap control.
6. Windshield defrost vent flap control.
7. Center and side vent flap control.
8. A/C button.
9. Temperature control.
10. Blower speed control.
11. Recirculation flap control.

87-2 Air-conditioning

General

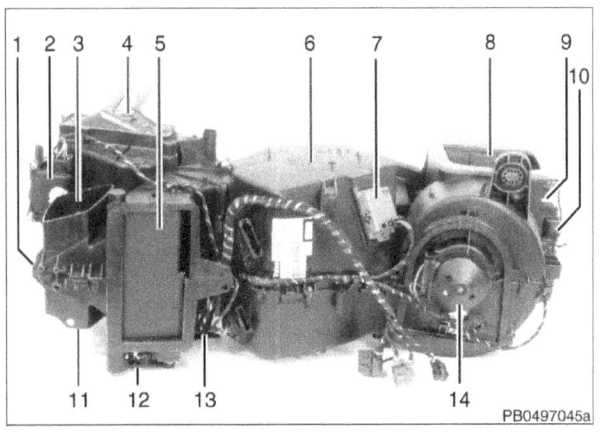

A/C and heater system

◀ The integrated climate control system combines heating and air-conditioning functions in one housing under the center and right side dashboard.

1. Footwell air temperature sensor
2. Footwell / defrost flap motor
3. Defrost duct
4. Heater core
5. Center and side duct
6. A/C evaporator housing
7. Blower resistor (power stage)
8. Fresh air intake
9. Fresh air temperature sensor
10. Fresh air / recirculation flap motor
11. Footwell duct
12. Center flap motor
13. Temperature control flap motor
14. Blower

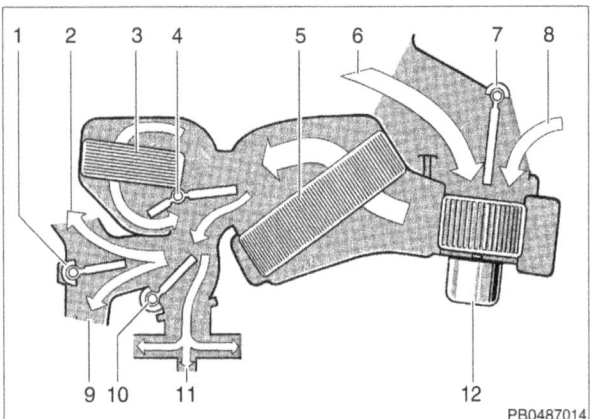

◀ Air flow through the A/C and heater housing is controlled by electric motor-controlled flaps.

1. Footwell / defrost flap
2. Defrost duct
3. Heater core
4. Temperature control flap
5. A/C evaporator
6. Fresh air intake
7. Fresh air / recirculation flap
8. Recirculated air intake
9. Footwell duct
10. Center and side ducts flap
11. Center and side ducts
12. Blower

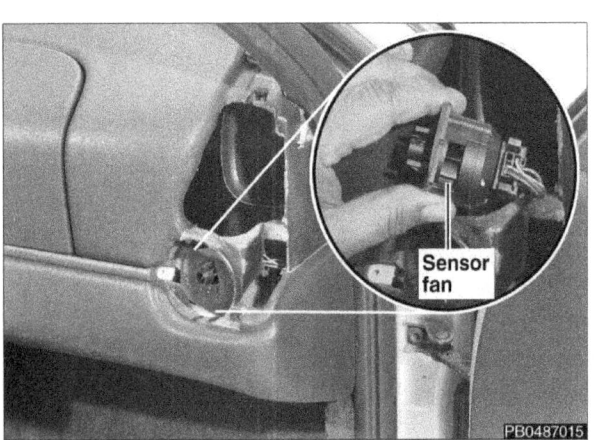

◀ The interior temperature sensor is behind the right dashboard vent. A small electric fan in the sensor draws cabin air past the temperature detector.

In addition to detecting inside and ambient temperature, the climate control system also responds to solar heating, measured by a solar sensor in the dashboard.

Air-conditioning 87-3

General

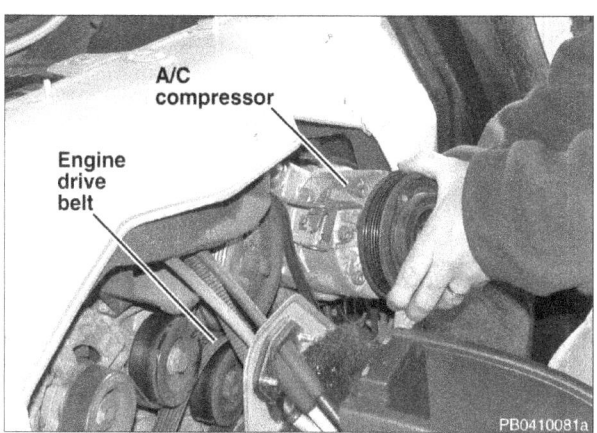

The cooling output of the air-conditioning system is generated by a flow-regulated compressor in the engine compartment. The compressor is driven by the engine drive belt.

Refrigerant and refrigerant oil

The air-conditioning refrigerant in the Boxster is R134a. A synthetic oil, ND 8, is used to lubricate the A/C compressor.

A new A/C compressor is filled at the factory with refrigerant oil. If installing a new compressor, recharge with refrigerant using A/C recharging equipment. Do not reuse oil from old compressor.

A/C fluid capacities	
Refrigerant (R134a)	850 g (30 oz)
Refrigerant oil (ND 8)	195 ± 15 ml (6.6 ± 0.5 oz)

Fault detection

The A/C and heater control module self-diagnosis function continuously monitors system components and stores any faults in non-volatile fault memory (EEPROM). The fault memory can be accessed via Porsche Service Tester 2 (PST2) at the OBD II plug under the left side of the dashboard.

When a fault occurs for the first time, the fault counter is set to 50. The fault counter decreases by 1 when the fault is no longer present and the diagnostic conditions are true. The fault is cleared from the fault memory when the counter reaches 0.

Sensor default values

If there is a short or open circuit in an A/C component or sensor, the climate control system software uses defaults shown in **Table a** to enable emergency operation.

Table a. Climate control sensor default values		
Parameter	Possible fault in	Default value
Outside temperature	Sensor or instrument cluster	10°C (50°F)
Fresh air temperature	Sensor	10°C (50°F)
Interior temperature	Sensor	22°C (72°F)
Footwell temperature	Sensor	50°C (122°F)
Engine temperature	Sensor or instrument cluster	80°C (176°F)
Solar intensity	Sensor	150 w / sq meter
Vehicle speed	Instrument cluster	70 kph (40 mph)
Engine speed	Instrument cluster	1000 rpm

87-4 Air-conditioning

General

Troubleshooting

Table b provides common symptoms and possible corrective actions to aid in diagnosing simple faults in the A/C system. For heating system troubleshooting, see **80 Heating**.

Table b. Air-conditioning troubleshooting		
Symptom	Probable cause	Corrective action
Blower does not work.	Blower fuse blown.	Inspect blower fuse (fuse D6, 30A, in main fuse panel, behind cover in left kick panel). Replace if necessary. See **97 Fuses, Relays, Component Locations**.
	Blower resistor (power stage) faulty.	Test blower resistor. Replace if necessary.
	Blower motor wiring faulty.	Troubleshoot electrical harness. Repair as necessary.
	Blower motor faulty.	Test motor. Replace if necessary.
Blower does not operate in lower speeds.	Blower resistor (power stage) faulty.	Test blower resistor. Replace if necessary.
	Blower switch faulty.	Replace A/C and heating control panel.
Air-conditioning system does not produce cold air.	A/C compressor not engaging.	Check A/C compressor clutch electrical connector at left of engine compartment. Repair or reconnect as necessary.
	Refrigerant charge too low or too high.	Test A/C system pressures. Repair as necessary.
	Temperature control flap malfunction.	Test temperature control flap and motor. Repair or replace as necessary.
Air-conditioned air smells musty.	Mold growing in A/C evaporator vanes.	Apply disinfectant using evaporator cleaning wand.

A/C system warnings and cautions

> *WARNING—*
> - *Wear hand and eye protection (gloves and goggles) when working around the A/C system. If refrigerant oil comes in contact with your skin or eyes:*
>   *-Remove contact lenses, if worn.*
>   *-Do not rub skin or eyes.*
>   *-Immediately flush skin or eyes with cool water for 15 minutes.*
>   *-Rush to a doctor or hospital. Do not induce vomiting.*
> - *Work in a well ventilated area. Switch on building exhaust / ventilation system when working on the A/C system.*
> - *If you inhale refrigerant oil, breathe fresh air immediately and then seek medical attention.*
> - *Do not expose any component of the A/C system to high temperatures (above 80°C / 176°F). Excessive heat could burst the system.*
> - *Keep refrigerant away from open flames. Poisonous gas is produced if it burns. Do not smoke near refrigerant gases for the same reason.*
> - *The A/C system is filled with refrigerant gas which is under pressure. Pressurized refrigerant in the presence of oxygen may form a combustible mixture. Do not introduce compressed air into any refrigerant container.*
> - *Electric welding near refrigerant hoses causes R134a to decompose. Discharge system before welding.*
> - *Dispose of refrigerant oil as hazardous waste.*

Air-conditioning 87-5

General

CAUTION—
- *US law requires that any person who services a motor vehicle air-conditioner is properly trained and certified, and uses approved refrigerant recycling equipment. Technicians must complete an EPA-approved recycling course to be certified.*
- *It is recommended that all A/C service be left to an authorized Porsche dealer or other qualified A/C service facility.*
- *State and local governments may have additional requirements regarding A/C servicing. Always comply with state and local laws.*
- *Leak test and repair any A/C system which is known to lose its charge.*
- *Do not top off a partially charged refrigerant system. Discharge system, evacuate and then recharge system.*
- *Do not use R12 refrigerant, refrigerant oils or system components in R134a system. Component damage and system contamination results.*
- *Refrigerant oil (PAG oil) is extremely hygroscopic (water-absorbent). Do not store in open container.*
- *The mixture of refrigerant oil and refrigerant (R134a) attacks some metals and alloys (for example, copper) and breaks down certain hose materials. Use only hoses and pipes that are identified with a green mark (stripe) or the lettering "R134a".*
- *Immediately plug open connections on A/C components and pipes to prevent dirt and moisture contamination.*
- *Do not steam clean A/C condensers or evaporators. Use only cold water or compressed air.*
- *Do not reuse refrigerant oil.*
- *Do not store refrigerant oil in vicinity of flames, heat sources or strongly oxidizing materials.*
- *In case of fire, use carbon dioxide (CO_2) extinguisher or sprayed water jet on A/C components.*
- *After refilling or testing refrigerant lines, replace protective caps. They serve as additional seals.*
- *In winter, run the A/C compressor at least 10 minutes each month to maintain a film of oil on its moving parts and to ensure seal integrity.*

87-6 Air-conditioning

A/C System Component Replacement

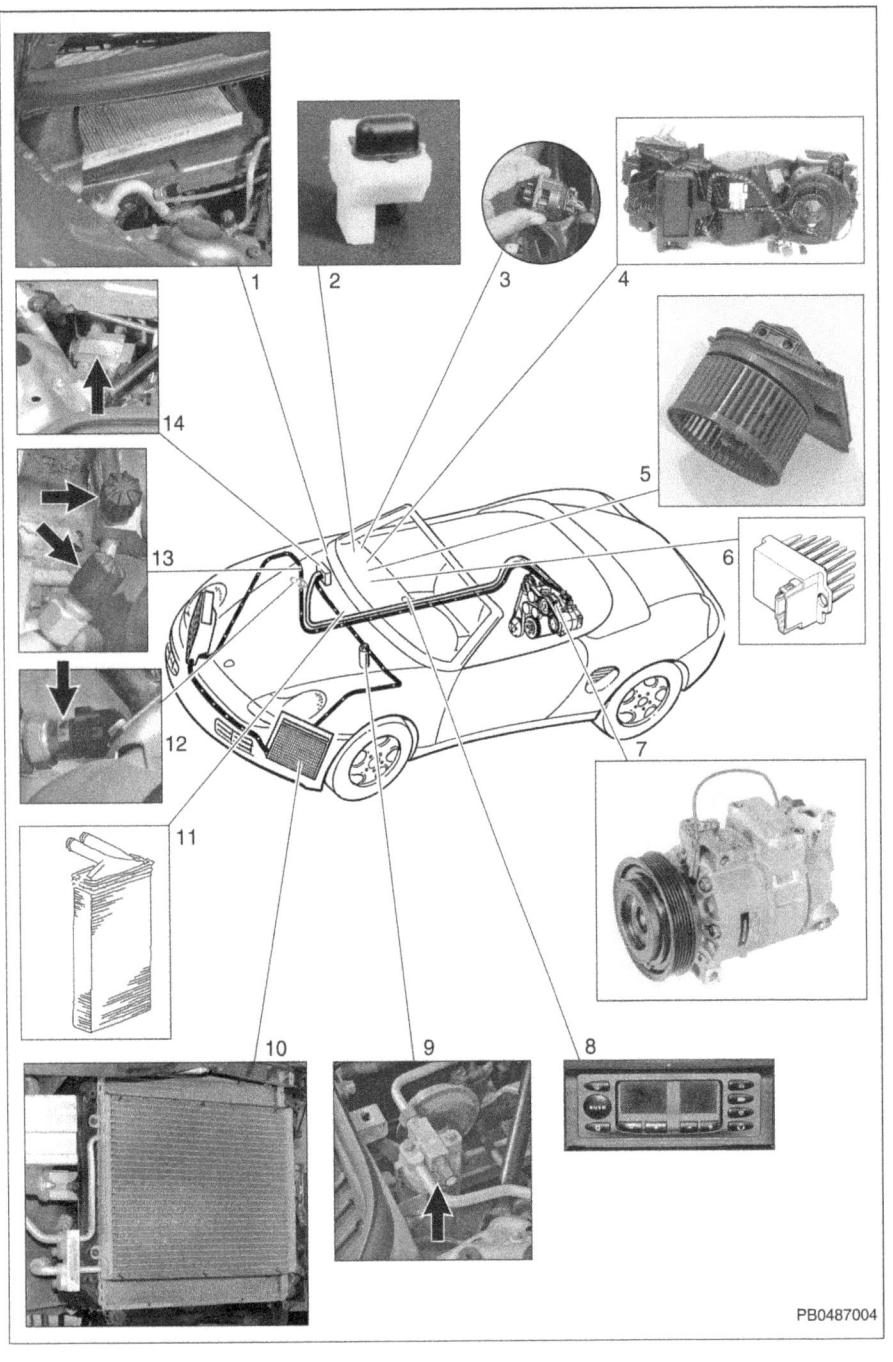

A/C SYSTEM COMPONENT REPLACEMENT

A/C system components

1. **Ventilation microfilter**
   - Under right cowl.
   - See **03 Service and Maintenance**

2. **Solar sensor**

3. **Interior temperature sensor**

4. **A/C and heater housing**
   - Incorporates A/C evaporator and air flow flaps.
   - Blower and blower resistor, flap motors, fresh air and footwell temperature sensor are attached.
   - Heater core inserted under cowl.
   - See **70 Trim—Interior** for removal procedure.

5. **Blower**
   - See **Blower, removing and installing** in this repair group.

6. **Blower resistor (power stage)**
   - See **Blower resistor (power stage), removing and installing** in this repair group.

7. **Compressor**
   - See **Compressor, removing and installing** in this repair group.

8. **A/C and heater control panel and module**
   - Center dashboard.

9. **Receiver-drier**
   - Under left cowl.

10. **Condenser**
    - Left and right behind front bumper cover.
    - See **63 Bumpers** for bumper cover removal.

11. **Heater core**
    - See **80 Heating.**

12. **Pressure switch**
    - Under right cowl.

13. **Refrigerant service connections**
    - Under right cowl.

14. **Expansion valve**
    - Under right cowl.

Air-conditioning **87-7**

A/C System Component Replacement

Blower, removing and installing

The blower is in the A/C and heater housing on the right side, under the dashboard.

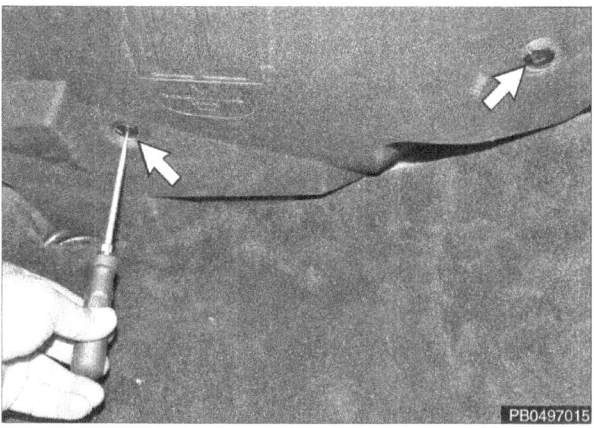

◄ Working in passenger compartment underneath right side of dashboard, unscrew upper footwell trim panel plastic fasteners (**arrows**). Pull down trim.

◄ Remove right air duct from under dashboard.

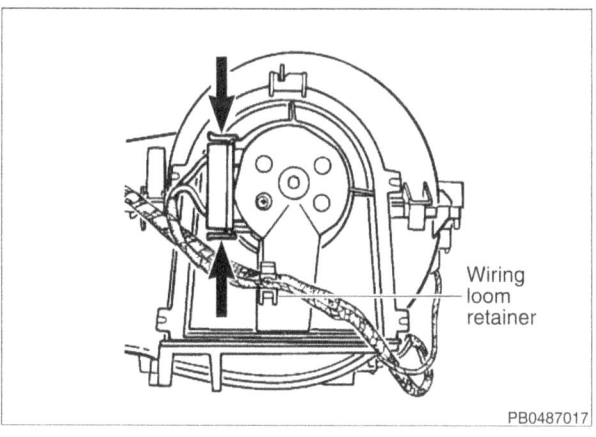

◄ Reach up into opening under right side dashboard and unclip (**arrows**) harness connector at blower. Detach wiring loom from plastic retainer at base of blower.

◄ Remove blower cover mounting screws (**arrows**) while supporting blower. Remove cover and lower blower out of housing.

– Installation is reverse of removal. Do not overtighten blower cover mounting screws into plastic housing.

87-8 Air-conditioning

A/C System Component Replacement

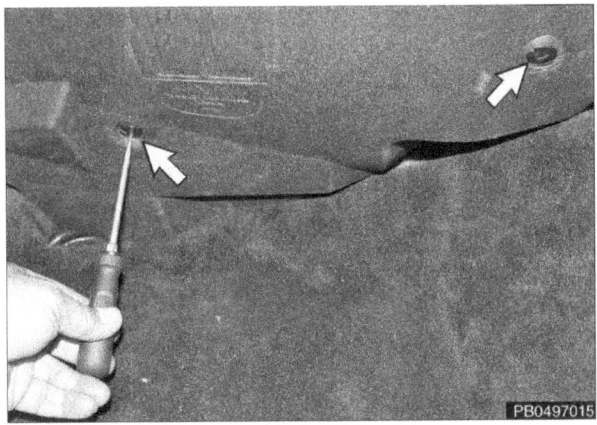

Blower resistor (power stage), removing and installing

The blower resistor is above and to the left of the blower in the A/C and heater housing.

◄ Working in passenger compartment underneath right side of dashboard, unscrew upper footwell trim panel plastic fasteners (**arrows**). Pull down trim.

◄ Remove right air duct from under dashboard.

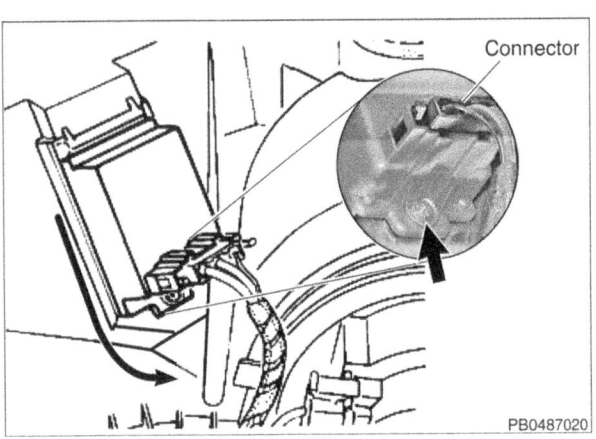

◄ Reach up into opening under right side of dashboard, to left of blower motor.
- Remove resistor mounting screw (**arrow**).
- Slide resistor down and tilt out of housing.
- Lower resistor and detach connector.

NOTE—
- *For improved access to resistor, remove passenger airbag in dashboard. See **69 Seat Belts, Airbags**.*

– Installation is reverse of removal. Do not overtighten blower cover mounting screws into plastic housing.

Air-conditioning 87-9

A/C System Component Replacement

Compressor, removing and installing

Removing

- Following manufacturer's instructions, connect an approved refrigerant recovery / recycling / recharging unit to A/C system and discharge system.

> *WARNING—*
> - *Do not attempt to discharge or charge the A/C system without proper equipment and training. Damage to the vehicle and personal injury may result.*

◄ Move convertible top and convertible-top storage lid to service position. See **03 Service and Maintenance**.

- Remove left seat. See **72 Seats**.

- Disconnect negative (–) battery cable and cover battery terminal to keep cable from accidentally contacting terminal.

> *CAUTION—*
> - *Prior to disconnecting the battery, acquire the radio anti-theft code.*
> - *Read the battery disconnection cautions in **00 Warnings and Cautions**.*

- Remove carpeted panels from top and front engine access covers.

◄ Twist engine top cover hold-downs (**arrows**) and open cover.

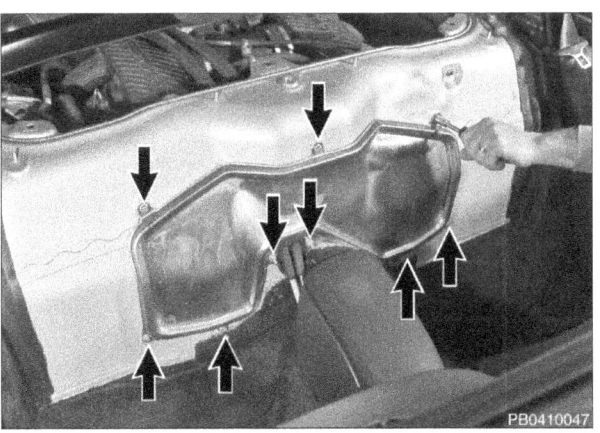

◄ Undo engine access cover mounting fasteners at rear of passenger compartment (**arrows**) and lift out cover.

> *CAUTION—*
> - *Access cover edges are sharp.*

87-10 Air-conditioning

A/C System Component Replacement

◀ Use 24 mm wrench to rotate engine drive belt tensioner clockwise. Slip belt off air-conditioning compressor (upper left pulley, upper right in photo).

NOTE—
* *If taking drive belt off, mark direction of rotation for reinstallation.*

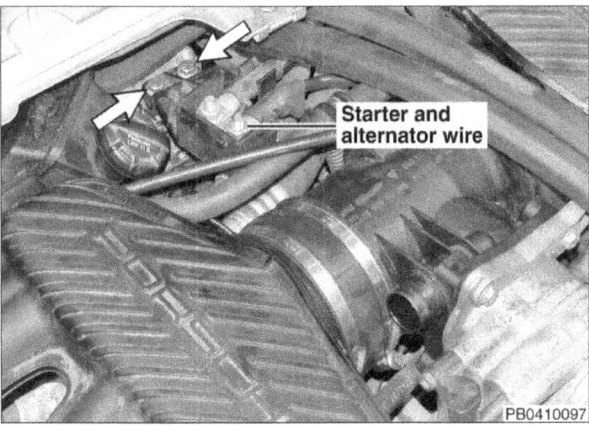

◀ Working at front center of engine:
* Flip open B+ junction box cover.
* Detach starter and alternator wire at B+ junction and set aside.
* Remove B+ junction mounting bolts (**arrows**) and move junction box aside.

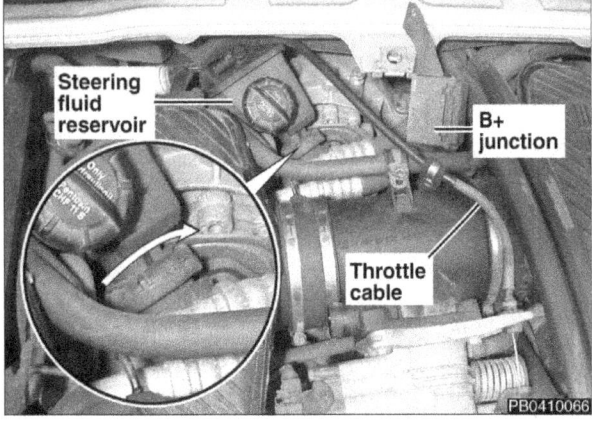

◀ Remove power steering reservoir:
* Loosen and move aside throttle cable, if applicable.
* Use clean syringe to siphon out Pentosin® fluid from power steering reservoir.
* Unscrew (**arrow**) bayonet lock at bottom of reservoir and lift off. Be prepared for dripping fluid.
* After removing reservoir, plug open steering fluid port.

CAUTION—
* *Do not spill Pentosin® fluid. If coolant hoses come in contact with Pentosin® fluid, wash off immediately.*

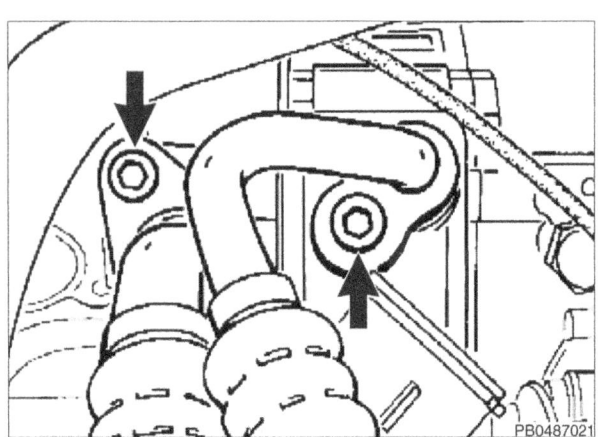

◀ Working from above and to right side of compressor, remove refrigerant line mounting bolts (**arrows**). Plug open refrigerant lines and ports.

Air-conditioning 87-11

A/C System Component Replacement

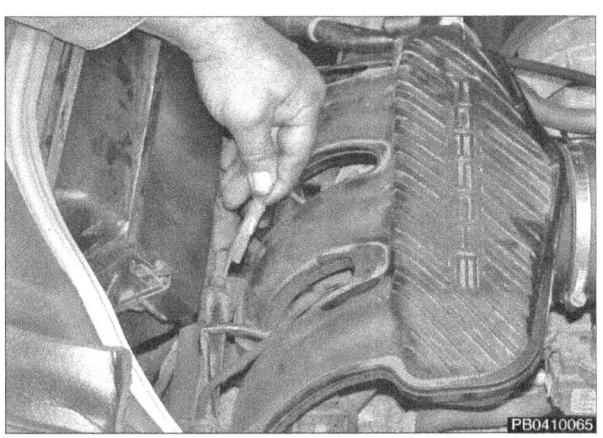

◄ Working at left of engine, detach air-conditioning compressor clutch electrical connection:
  • Cut wire tire holding connector to harness loom next to cylinder 4 and 5 intake runners.
  • Pull connector apart.

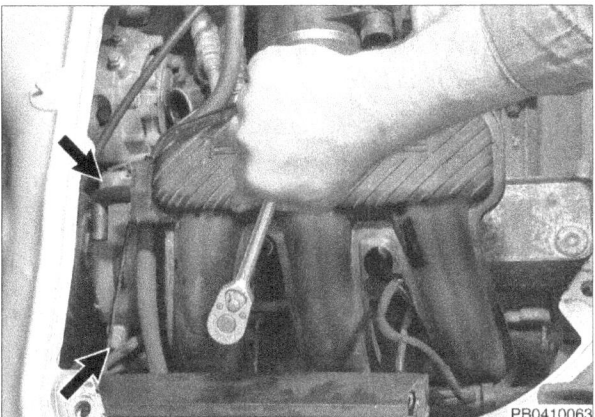

◄ Remove compressor mounting bolts (**arrows**) from above.

NOTE—
• *Rear compressor mounting bolt is accessible between cylinders 4 and 5 intake runners.*

◄ Slide compressor forward out of engine compartment to remove.

Installing

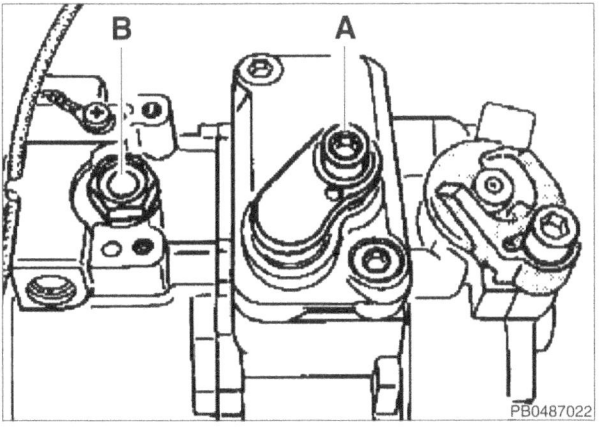

◄ New compressor is filled with the correct volume of oil for entire refrigerant circuit and pressurized. Prior to installation:
  • Open pressure cap (**A**) to relieve pressure. Use shop towel to prevent oil spray.
  • Open oil filler (**B**) and drain approx. 80 ml (2.7 oz) of oil into measuring glass.
  • Reinstall sealing plugs using new O-rings.

– Install compressor to engine.

Tightening torque	
Compressor to engine crankcase (M8)	28 Nm (21 ft-lb)

A/C System Component Replacement

- Reattach compressor clutch harness connector.

- When reattaching refrigerant lines to compressor, use new O-ring seals. Lubricate with refrigerant oil.

Tightening torque	
Refrigerant line to compressor (M8)	23 Nm (17 ft-lb)

- If installing a new compressor, or if A/C system has been open for more than 24 hours, replace receiver - drier unit.

- Install engine drive belt, noting previously made direction mark. See **03 Service and Maintenance**.

- Recharge A/C system following equipment manufacturer's instructions. Be sure to use correct type and volume of refrigerant oil. See **Refrigerant and refrigerant oil** in this repair group.

Tightening torque	
Filler plug to compressor (M10)	23 - 26 Nm (17 - 19 ft-lb)

90 Instruments, Horns

GENERAL 90-1
   Electrical system safety precautions 90-1
INSTRUMENT CLUSTER 90-2
   Instrument cluster versions 90-2
   Instrument cluster layout (1997 Boxster) 90-3
   Instrument cluster layout (2001 Boxster) 90-4
   Instrument cluster operation 90-5
   Warning lights 90-8

INSTRUMENT CLUSTER REPAIRS 90-8
   Instrument cluster, removing and installing 90-8
   Instrument cluster bulb
     removing and installing (1997 - 2000) 90-9
HORNS 90-10
   Horn, removing and installing 90-10

TABLE
a. Boxster instrument bulbs 90-9

GENERAL

This repair group covers the instrument cluster, warning lights and horns. See **91 Radio and Communication** for information about on-board computer and PCM (navigation system).

Electrical system safety precautions

Read the following and the additional warnings and cautions in **97 Fuses, Relays, Component Locations** before doing any work on your electrical system.

> *CAUTION—*
> - *Before disconnecting the battery, read the battery disconnection cautions in* **00 Warnings and Cautions**.
> - *Read out and record control module fault memory before disconnecting the battery.*
> - *Record the radio station presets and obtain the radio activation code before disconnecting the battery.*
> - *Switch the ignition OFF and remove the negative (-) battery cable before removing any electrical components.*
> - *Connect and disconnect electrical connectors and ignition test equipment leads only while the ignition is OFF.*
> - *Only use a digital multimeter for electrical tests.*

Instrument Cluster

INSTRUMENT CLUSTER

The Boxster instrument cluster consists of three overlapping circle gauges and a row of warning and information lights underneath.

Instrument cluster versions

The cluster is made in several versions:
- Canada, Great Britain, Japan, Saudi Arabia, USA
- With or without on-board computer
- With or without Tiptronic gear range indicator
- Black with white letters or silver with black letters

The analog speedometer is set either as kph or mph. The digital displays may be reset either way in any version.

In 2001 a redesigned cluster was provided with slightly different displays. In this version, important data is sent to the cluster over two CAN data buses:
- CAN-C interface for driveline data
- CAN-B interface for vehicle interior data

The redesigned instrument cluster provides self-diagnosis capability and fault memory, accessible via Porsche System Tester 2 (PST2) or equivalent.

Instruments, Horns 90-3

Instrument Cluster

Instrument cluster layout (1997 Boxster)

1. Analog speedometer
2. Cruise control light
3. Odometer, trip odometer
4. Left setting button
5. Tachometer
6. Turn signal indicator
7. On-board computer:
   - Ambient (outside) temperature
   - Average fuel consumption
   - Average speed
   - Calculated range on remaining fuel
   - Cruise control over-speed signal
   - Digital speedometer
8. High beam indicator light
9. Instrument illumination sensor
10. Right setting button
11. Clock
12. Tiptronic gear range indicator
13. Coolant temperature gauge and warning light
14. Fuel level gauge and warning light
15. Traction control OFF warning
16. Traction control intervention light
17. Spoiler extended indicator
18. Brake pad wear warning
19. Parking brake warning
20. Brake fluid level warning
21. ABS warning
22. Engine oil pressure warning
23. Battery and charging system warning
24. Washer fluid level warning
25. Seat belt warning
26. Airbag warning
27. Convertible top warning
28. Check engine warning
29. Engine oil level gauge

90-4 Instruments, Horns

Instrument Cluster

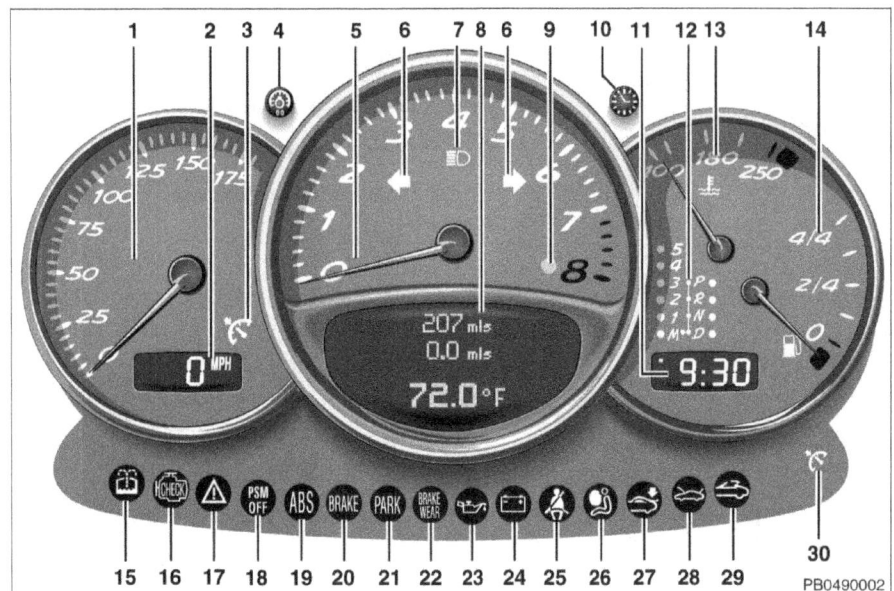

Instrument cluster layout (2001 Boxster)

1. Analog speedometer
2. Odometer, trip odometer
3. Cruise control light (Boxster)
4. Left setting button
5. Tachometer
6. Turn signal indicator
7. High beam indicator light
8. On-board computer:
   - Ambient (outside) temperature
   - Average fuel consumption
   - Average speed
   - Calculated range on remaining fuel
   - Cruise control over-speed signal
   - Digital speedometer
   - Driver settings
   - Engine oil level gauge
   - Stored warnings
9. Instrument illumination sensor
10. Right setting button
11. Clock
12. Tiptronic gear range indicator
13. Coolant temperature gauge and warning light
14. Fuel level gauge and warning light
15. Washer fluid level warning
16. Check engine warning
17. PSM intervention light
18. PSM OFF warning
19. ABS warning
20. Brake fluid level warning
    Brake distribution warning
21. Parking brake warning
22. Brake pad wear warning
23. Engine oil pressure warning
24. Battery and charging system warning
25. Seat belt warning
26. Airbag warning
27. Spoiler extended indicator
28. Trunk lid(s) open warning
29. Convertible top warning
30. Cruise control light (Boxster S)

Instruments, Horns 90-5
Instrument Cluster

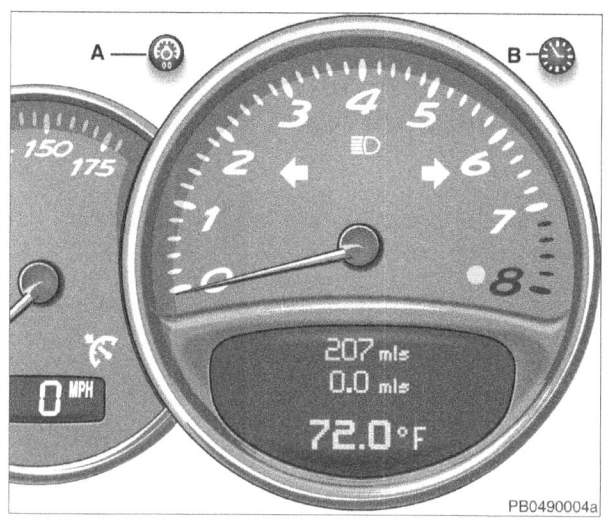

Instrument cluster operation

If the ignition is switched OFF but the ignition key left in the switch, the instrument cluster displays remain lit. Once the key is removed, the displays switch off after 4 minutes.

◀ Activate the cluster display without the ignition key as follows:
- Switch headlights ON.
- Press left setting button (**A**).
- Press right setting button (**B**).

If the key is not inserted, display is switched OFF after 4 minutes.

Gauges

Gauges in the cluster are powered by stepper motors featuring high torque, low hysteresis and high accuracy. The gauge needle has no return spring, so the needle is actively reset when ignition is switched OFF. If cluster power is interrupted, the gauge needles remain at their positions until voltage is restored.

Vehicle speed data

Vehicle speed, derived from ABS wheel speed sensor input, is displayed in both digital and analog display.

Speedometer reset

Digital speedometer and odometer display may be switched from mph and miles to kph and kilometers:

◀ Turn right setting button (**B**) counterclockwise and hold for 5 seconds.

NOTE—
- *This operation does not change the data stored in on-board computer.*

Reset trip odometer:

– Press and hold left setting button (**A**) for approx. 1 second.

Oil level indicator

◀ 1997 - 2001: Segmented display is at bottom of right gauge.

2001 - 2004: Level indicator is displayed by on-board computer in bottom of center gauge when ignition is switched ON.

For additional details, see **17 Lubrication System**.

90-6 Instruments, Horns

Instrument Cluster

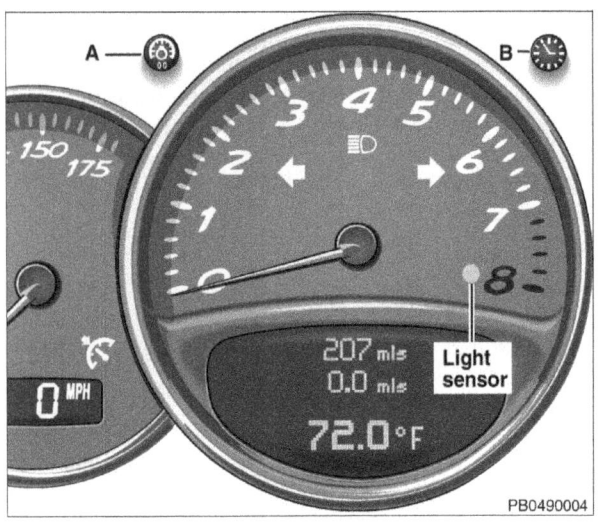

Tachometer

The tachometer signal is provided by the engine control module (ECM).

Instrument illumination

◄ When the lights are ON, adjust instrument and switch symbol illumination using left setting button (**A**).

In addition, instrument illumination is automatically adjusted by the ambient light sensor in the tachometer.

Coolant temperature

◄ If coolant temperature is too high (>120°C / 264°F), the coolant temperature warning light illuminates. If coolant level is too low, the warning light flashes.

If a fault occurs in the coolant temperature sensor circuit (short to ground) the maximum temperature is displayed.

If there are faults in the engine compartment blower system, the coolant warning light also flashes.

Fuel gauge

At 9 liter (2.4 US gal) the low fuel warning glows. If a fault occurs in the fuel level sensor circuit (a wire break or short to ground) the low fuel light flashes.

Clock

The clock, a digital display with either 24-hour or 12-hour mode, turns off 4 minutes after ignition is switched OFF or when the vehicle is locked.

When the ignition is switched ON, the clock temporarily changes to display waiting time for oil level measurement.

◄ Set hours:
- Press right setting button for approx. 1 second and release. Hours flash.
- Advance hour: Turn button clockwise.
- Retard hour: Turn button counterclockwise.
- Turn and hold button for rapid scrolling through hours.

— Set minutes:
- Press right setting button again and hold for approx. 1 second. Minutes flash.
- Advance or retard minutes by turning clockwise or counterclockwise.

Time setting mode automatically stops after 1 minute. press right setting button to resume.

Instruments, Horns 90-7

Instrument Cluster

To set seconds accurately, press setting button again at precise time.

On-board computer and navigation

The on-board computer display, when installed, is in the center gauge of the instrument cluster, below the tachometer.

Starting with 1998 models, an optional information and navigation system known as Porsche Communication Management (PCM) was available. This system display, when installed, also displays on-board computer information. See **91 Radio and Communication** for additional details.

Tiptronic gear indicator

 When the ignition is ON, Tiptronic gear selector position is displayed.

In D or M position, the currently selected gear is also displayed.

If the selector lever is located between selection positions or not properly engaged, the position light flashes.

In case of a fault in the Tiptronic system, the 4th gear display flashes.

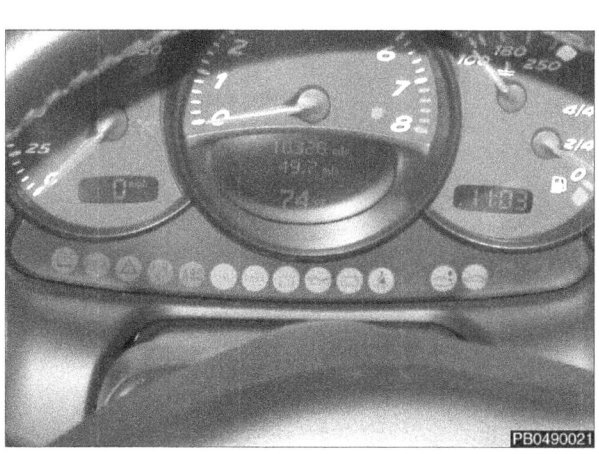

Warning lights

 Warning lights are activated for self-test when ignition is switched ON.

After 30 seconds, the following bulbs, activated by the instrument cluster wiring, are switched OFF. This protects the cluster wiring from overheating:

- Spoiler warning
- Brake pad wear warning
- Brake fluid level warning
- Oil pressure warning
- Washer fluid level warning
- Coolant temperature warning
- Low fuel level warning

Bulbs directly activated by other control modules or switches may remain lit for longer than 30 seconds.

90-8 Instruments, Horns

Instrument Cluster Repairs

INSTRUMENT CLUSTER REPAIRS

Instrument cluster, removing and installing

- Disconnect negative (–) battery cable and cover battery terminal to keep cable from accidentally contacting terminal.

> **CAUTION—**
> - Prior to disconnecting the battery, read the battery disconnection cautions in **00 Warnings and Cautions**.
> - Be sure to have the radio anti-theft code before disconnecting the battery.

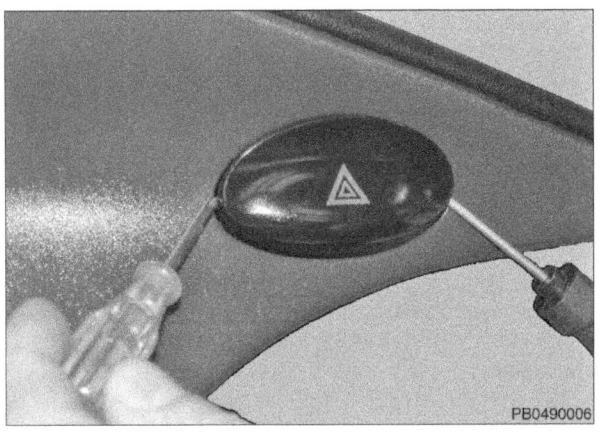

◀ Press in hazard warning switch so that button projects out. Insert two small screwdrivers into openings on sides of switch and gently pry out button.

◀ Use small screwdriver to press right side switch retaining clip. Pull switch out of instrument cluster with pliers to access right side cluster mounting screw.

- At left upper edge of cluster, remove trim plug or hand-free microphone to access left side cluster mounting screw. Detach microphone harness, if applicable.

◀ Remove cluster mounting Torx screws (**arrows**).

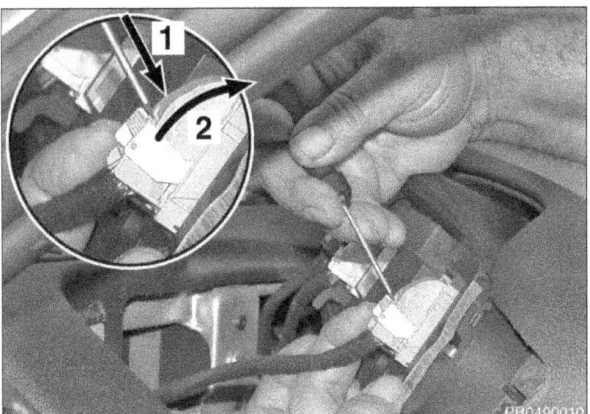

◀ Lift cluster and surround frame up and detach electrical connectors in back.
   - Use small screwdriver to release (**1**) connector lock lever.
   - Swing lever (**2**) to release connector.

- Release hazard warning switch plug from cluster and remove cluster.

- Installation is reverse of removal.

- Connect battery and use Porsche System Tester 2 (PST2) or equivalent scan tool to check cluster function.

Instruments, Horns 90-9

Instrument Cluster Repairs

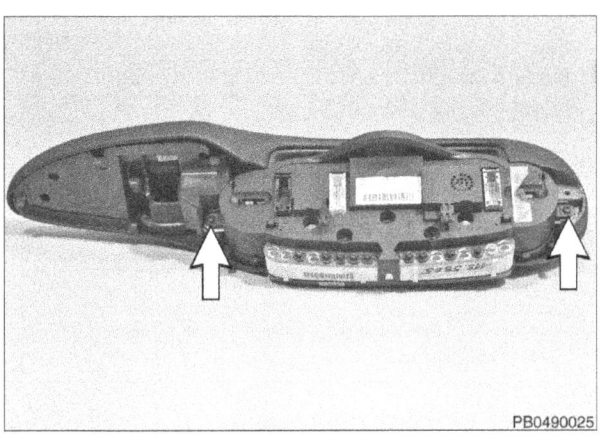

Instrument cluster bulb, removing and installing (1997 - 2000)

– Remove instrument cluster. See **Instrument cluster, removing and installing** in this repair group.

◄ Place cluster face down on clean work surface. Remove cluster mounting screws (**arrows**) and separate cluster from surround frame.

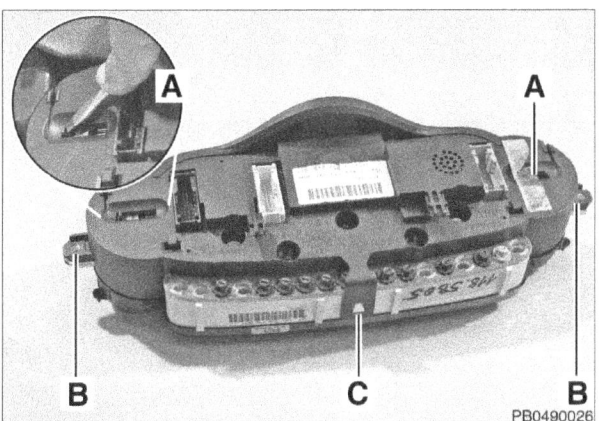

◄ If necessary, separate cluster back plate from cluster face:
- Pull left and right setting buttons straight off.
- Use screwdriver to press sliding tabs (**A**) outward.
- Pry off metal brackets (**B**).
- Release bottom clip (**C**).
- Lift cluster back gently and evenly off cluster face.

NOTE—
- *Newer version cluster: Remove indicator light housing locking clips to separate cluster components.*

CAUTION—
- *Do not touch dial gauges or solder joints.*

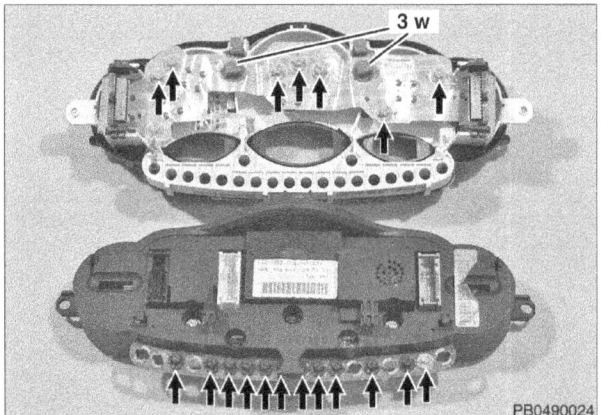

◄ **Arrows** indicate locations of instrument bulbs.

Table a. Boxster instrument bulbs	
Cluster illumination bulb	3 w
Instrument bulb: • Tan base • Black base	 EBS-P/4-A 1.5 w EBS-P/4-A 1.2 w

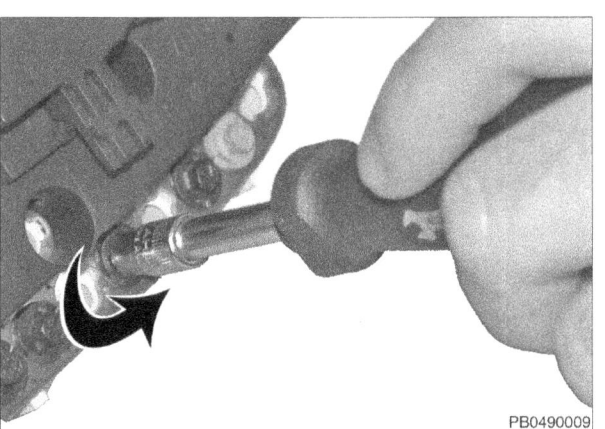

◄ Using 6 mm socket, twist instrument bulb holder to remove.

CAUTION—
- *Bulbs with excessive wattage could melt cluster housing. Replace defective bulbs with bulbs of the same wattage.*

– Cluster assembly and installation is reverse of removal.

Horns

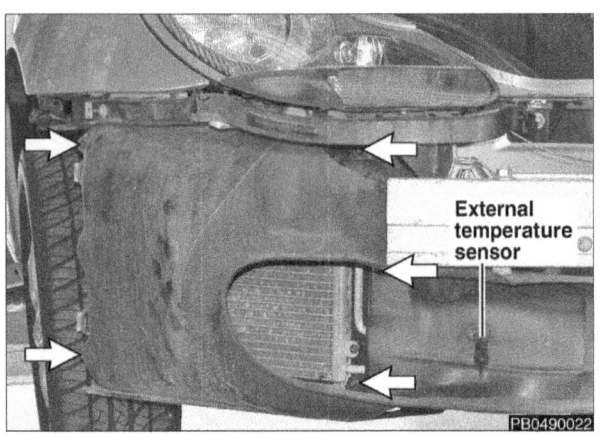

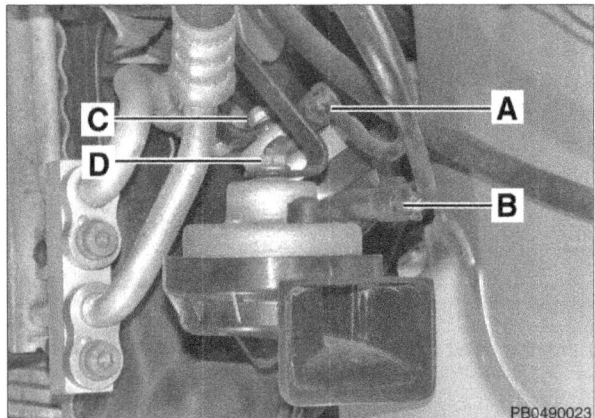

Horns

The high-pitched and low-pitched horns are installed in front of the vehicle next to the right side radiator.

Horn, removing and installing

− Raise front of vehicle and support safely.

> **CAUTION—**
> • Make sure the car is stable and well supported at all times. A floor jack is not adequate support.

− Remove front bumper cover. See **63 Bumpers**.

◄ Remove right radiator air duct:
  • Push external temperature sensor harness and grommet through air duct housing.
  • Remove duct mounting fasteners (**arrows**) and lift off air duct,

◄ Disconnect horn harness(es) and remove horn mounting fastener(s) as needed:
  • High pitched horn connector (**A**)
  • Low pitched horn connector (**B**)
  • High pitched horn fastener (**C**)
  • Low pitched horn fastener (**D**)

− Installation is reverse of removal. Bear in mind the following:
  • Fit serrated washer between horn and horn bracket.
  • Make sure horns do not touch one another or parts of body.

Tightening torque	
Horn to horn bracket (M6)	10 Nm (7 ft-lb)

91 Radio and Communication

GENERAL 91-1
  Communication components................. 91-1
  MOST-bus (media oriented systems transport) . 91-2
  Parking assistant........................ 91-2
SOUND SYSTEM 91-3
  Sound system components................. 91-4
  Radio, removing and installing 91-4

Amplifier, removing and installing 91-5
Speakers................................. 91-6
ON-BOARD COMPUTER 91-7
  On-board computer display, scrolling 91-7
NAVIGATION 91-8
  PCM display unit, removing and installing 91-8

GENERAL

This repair group covers radio and communication components installed in Boxster models. Most of these systems are optional. See **90 Instruments, Horns** for information on the instrument cluster.

Communication components

Boxster models are equipped with a variety of instrument cluster configurations which may include on-board computer and navigation equipment (Porsche Communication Management or PCM). See **Table a** for application information.

Table a. Information and communication equipment	
Year	Component or system
1997 • Standard • Optional	Instrument cluster with no on-board computer Instrument cluster with on-board computer
1998 - 2002 • Standard • Optional • Optional	Instrument cluster with no on-board computer Instrument cluster with on-board computer PCM incorporating on-board computer and navigation
2003 - 2004 • Standard • Optional • Optional	Instrument cluster with no on-board computer Instrument cluster with on-board computer PCM version 2 incorporating on-board computer and navigation

In addition, the following options are offered:
- **Telephone** in the center console
- **Parking assistant** with 4 ultrasonic sensors in the rear bumper cover and a control module under the left seat

91-2 Radio and Communication

General

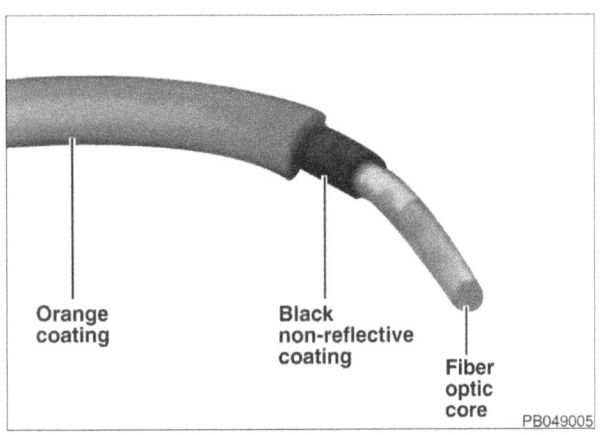

MOST-bus (media oriented systems transport)

◀ Beginning with 2003 models, sound system and navigation components are connected using a fiber optic network known as MOST-bus. MOST transmission is only used for multimedia data.

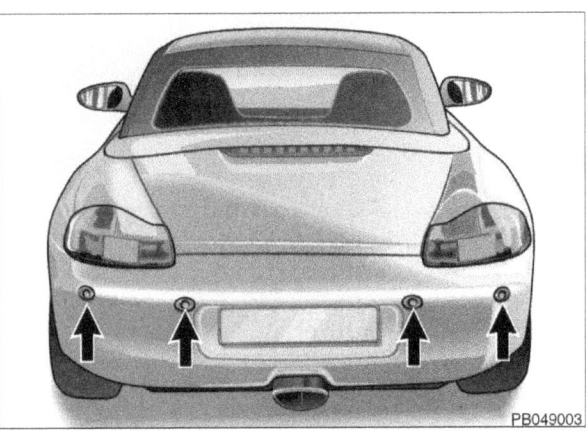

Parking assistant

◀ In a vehicle equipped with this optional system, four ultrasonic sensors (**arrows**) are fitted in the rear bumper.

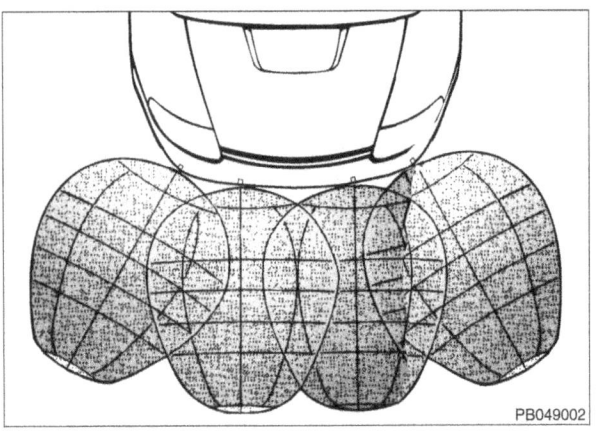

◀ When the vehicle is backed up, each sensor emits an ultrasonic beam with a horizontal beam width of approx. 120° and a vertical beam width of 60°. When an obstacle bounces any of the beams back to the sensor, an intermittent acoustic signal is sounded to warn the driver. The intervals between signals shortens until the distance to the obstacle reaches 0.3 meters (1 ft), at which time the signal becomes continuous.

The parking assistant control module continually performs self-diagnosis. Faults set in the module may be read out using Porsche System Tester 2 (PST2).

Radio and Communication 91-3

Sound System

SOUND SYSTEM

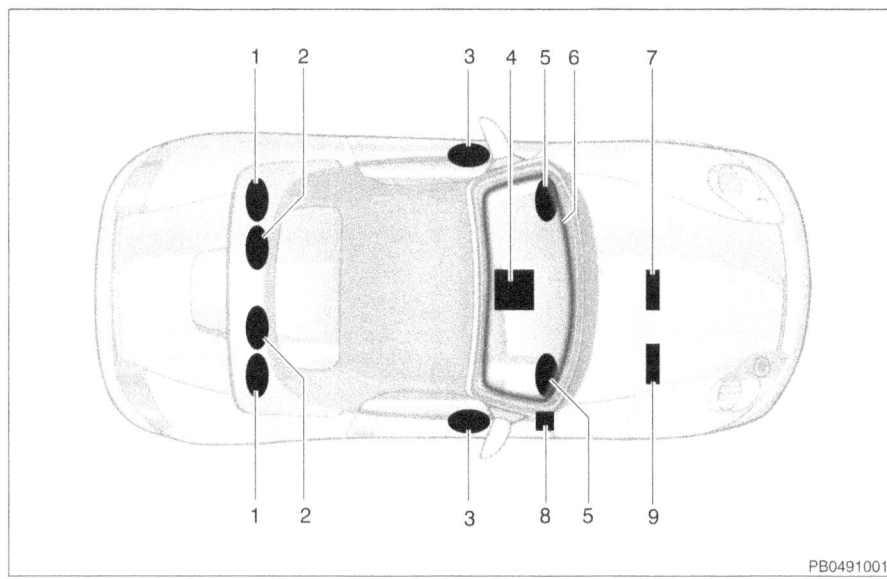

Sound system components

1. Midrange speaker (2003 - 2004)
2. Woofer (2003 - 2004)
3. Door speaker
4. Radio
5. Dashboard speaker assembly (midrange and tweeter)
6. Windshield antenna
7. Amplifier
8. Antenna amplifier
9. CD changer (2003 - 2004)

The base Boxster sound system consists of radio, cassette or CD player, power amplifier and 4 speakers. However, a number of different optional radio and sound system configurations were used, depending on year, model and options.

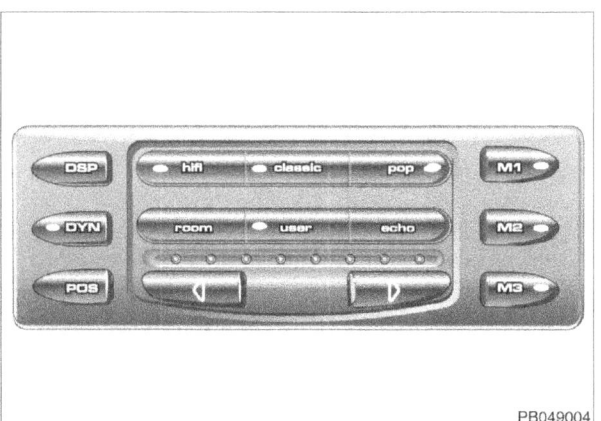

◄ **Digital sound processing (DSP)** provides equalization and other digital sound enhancements. The DSP controller is housed in front of the center console.

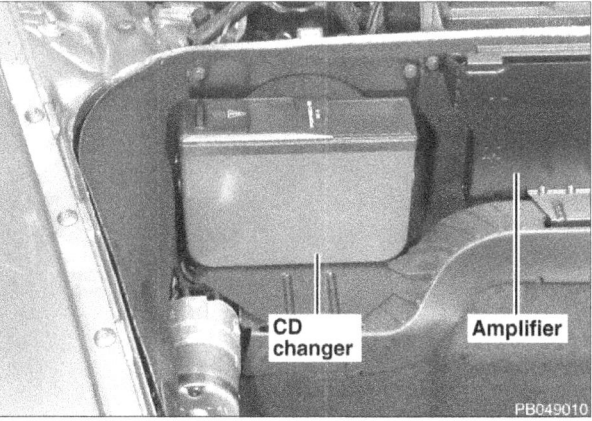

◄ **6-disc CD changer** with 6-second skip protection is housed in the front trunk.

6-channel digital power amplifier in the front trunk provides:
- 2 x 40 watts output for dashboard mounted speakers
- 2 x 70 watts output for door mounted speakers

The volume control in this system dynamically adapts to average driving speed, providing a constant volume impression and balanced sound image.

91-4 Radio and Communication

Sound System

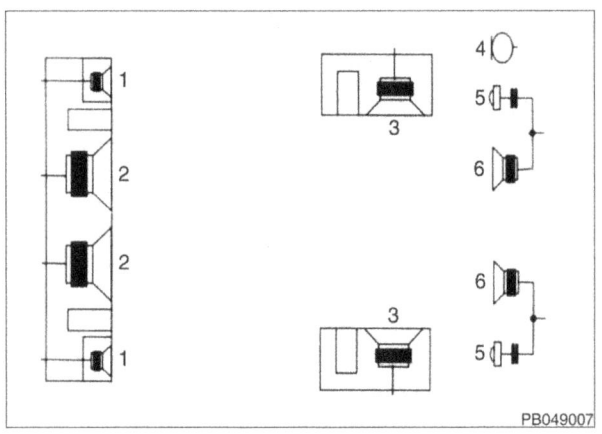

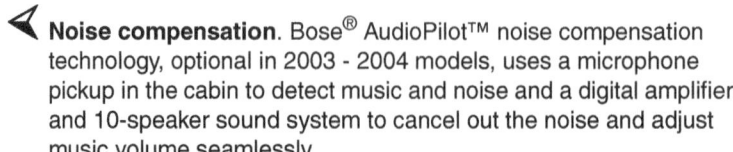

Noise compensation. Bose® AudioPilot™ noise compensation technology, optional in 2003 - 2004 models, uses a microphone pickup in the cabin to detect music and noise and a digital amplifier and 10-speaker sound system to cancel out the noise and adjust music volume seamlessly.

1. 6.4 cm midrange speaker in 0.2 liter housing
2. 13.3 cm woofer in 11 liter base reflex housing
3. 11.4 bass-midrange speaker in 5.5 liter door-mounted base-reflex housing
4. Noise compensation system microphone
5. 4.3 cm tweeter
6. 8.9 cm midrange speaker

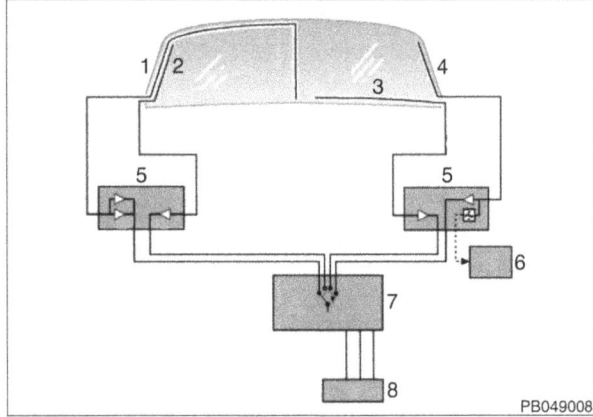

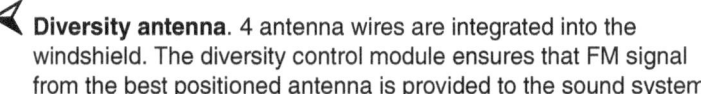

Diversity antenna. 4 antenna wires are integrated into the windshield. The diversity control module ensures that FM signal from the best positioned antenna is provided to the sound system.

1. Antenna 1 (AM and FM)
2. Antenna 2
3. Antenna 3
4. Antenna 4 (sound system and keyless entry)
5. Antenna multiplex
6. Central locking control module
7. Antenna amplifier and diversity control module
8. Radio interface

The diversity antenna is controlled by a special antenna amplifier and is integrated into the second generation PCM (2003 - 2004 models).

Radio, removing and installing

Radio activation code and radio preset may be lost when radio is removed or power to the radio is interrupted.

> **CAUTION—**
> * Record the radio station presets and obtain the radio activation code before disconnecting the radio.

– Turn off ignition and remove ignition key.

– Remove radio face plate (if applicable).

Use Porsche special tool set 9570 in slots (**inset**) at base of radio.
* Push tools into slots and lever away from each other to release radio mounting clips.
* Pull radio out of dashboard.

Radio and Communication 91-5

Sound System

◄ Detach antenna and harness connections (**arrows**) and remove.

– When installing, connect antenna and harness connections, then press radio into dashboard until it snaps firmly in place.

– Set radio activation code and presets recorded earlier (if applicable).

Amplifier, removing and installing

The power amplifier is in the front trunk.

– Turn off ignition and remove ignition key.

– Open front trunk. Remove spare tire.

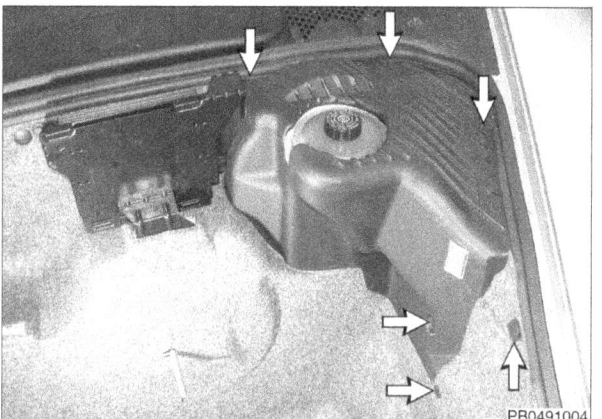

◄ Remove brake booster and master cylinder plastic cover fasteners (**arrows**). Lift out plastic cover.

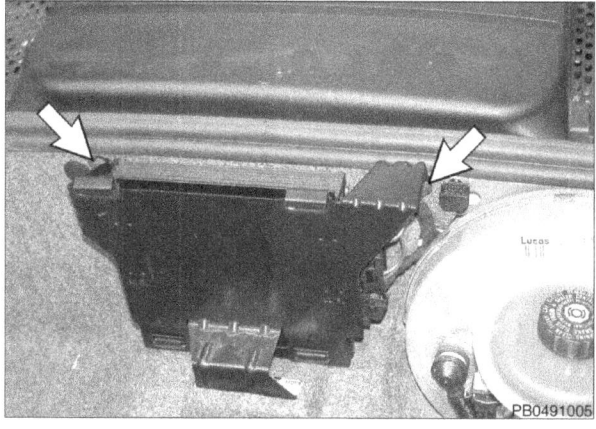

◄ Remove plastic cover over harness connectors from left side of power amplifier. Then remove 2 mounting bolts (**arrows**).

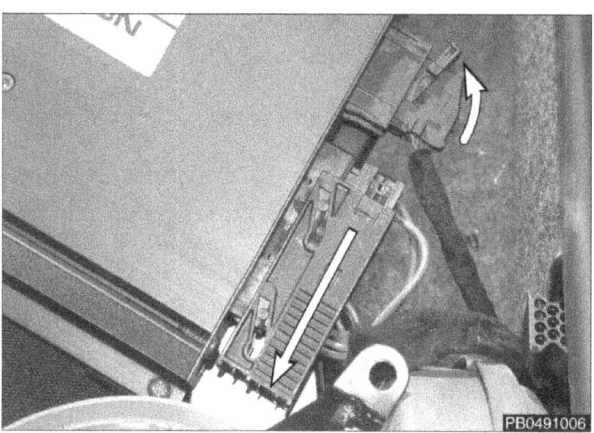

◄ Lift out amplifier and rotate to gain access to electrical connectors. Slide connector locks (**arrows**) to disconnect.

– Installation is reverse of removal.

91-6 Radio and Communication

Sound System

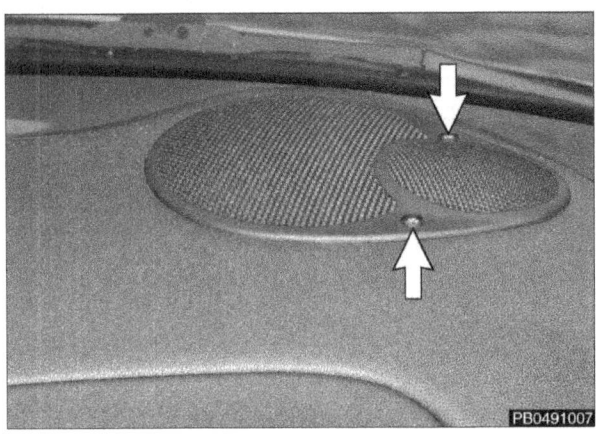

Speakers

Dashboard speaker assembly, removing and installing

◄ Working at left or right end of dashboard, remove speaker mounting Torx screws (**arrows**).

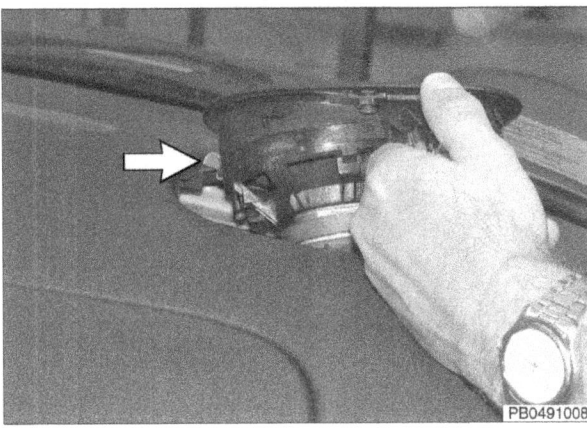

◄ Tilt outboard edge of speaker to clear securing hook (**arrow**) from under dashboard.

- Detach electrical connector and remove speaker.
- Installation is reverse of removal.

Door-mounted speaker, removing and installing

> **WARNING—**
> - Cars covered by this manual are equipped with side-impact airbags in the doors. When servicing door mounted components, disconnect the negative (-) battery terminal. See **69 Seat Belts, Airbags** for cautions and procedures relating to the airbag system.

- Disconnect negative (–) battery cable and cover battery terminal to keep cable from accidentally contacting terminal.

> **CAUTION—**
> - Prior to disconnecting the battery, acquire the radio anti-theft code.
> - Read the battery disconnection cautions in **00 Warnings and Cautions**.

- Remove door panel. See **57 Doors and Locks**.

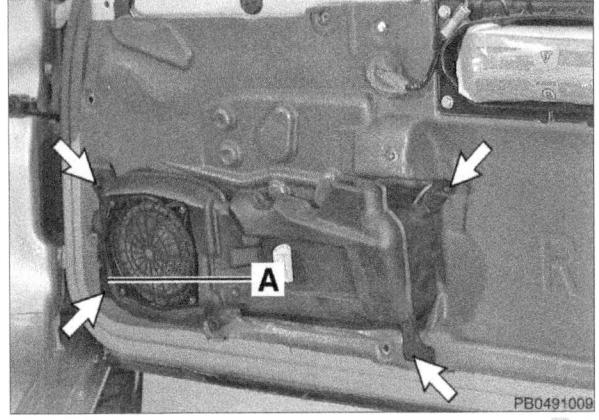

◄ Detach speaker harness connector (**A**). Remove speaker enclosure mounting screws (**arrows**). Lift enclosure and speaker out of door.

- Installation is reverse of removal.

Radio and Communication 91-7

On-board Computer

ON-BOARD COMPUTER

The on-board computer, when equipped, is in the center gauge of the instrument cluster, below the tachometer.

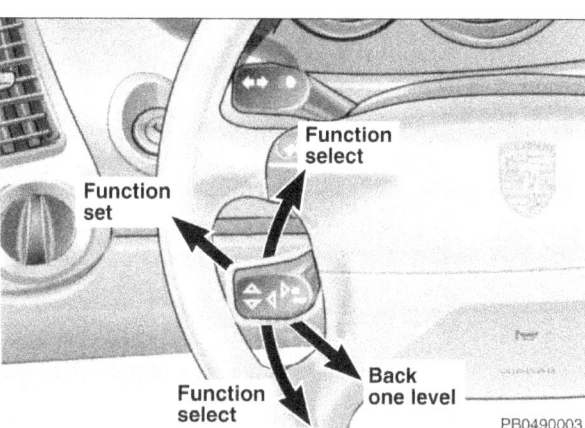

On-board computer display, scrolling

Turn ignition ON, then scroll on-board computer with lever to left of steering column:
- Select on-board computer function: Swing lever **up** or **down**
- Set function: Push lever **forward**.
- Move back one or more selection levels: Pull lever **backward**.

Available on-board computer functions are:
- Ambient (outside) temperature
- Average fuel consumption
- Average speed
- Calculated range on remaining fuel
- Cruise control speed exceeded (accompanied by warning chime)
- Digital speedometer

Additional functions added in 2001 models include:
- Driver settings
- Engine oil level gauge
- Stored warnings
- Additional information on illuminated warning lights:
  -ABS or PSM failure
  -Airbag failure
  -Brake fluid warning
  -Brake pad wear
  -Convertible top failure
  -Engine coolant level
  -Engine oil level
  -Engine oil pressure
  -Engine temperature
  -Spoiler failure
  -Tiptronic failure

91-8 Radio and Communication

Navigation

NAVIGATION

◀ Porsche Communication Management (PCM), is available as an option. The navigation system is based on Global Positioning System (GPS) satellite technology which can locate the vehicle position to within approx. 100 meters (300 ft). PCM interacts with CD based digitized street map to provide driving instructions with voice output.

Basic PCM functions are:
- Radio receiver with cassette player
- Navigation
- On-board computer
- Air-conditioning display

◀ The second generation PCM, installed in 2003 - 2004 models, features the following:
- CD-ROM drive
- 5.8 inch thin-film transistor (TFT) color monitor in 16 : 9 ratio
- 12-button quick function control panel
- Dot-matrix on-board computer in instrument cluster display
- Diversity antenna

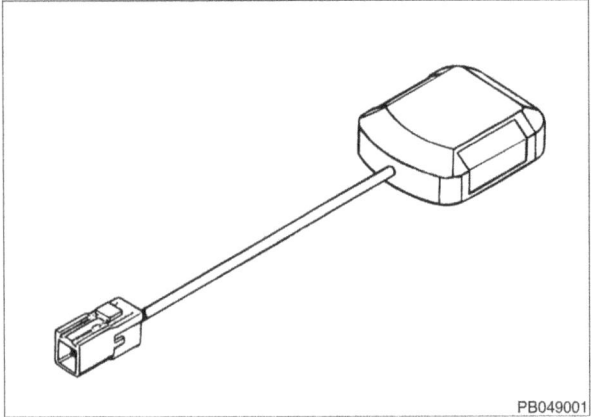

◀ The PCM antenna is integrated into the dashboard.

PCM display unit, removing and installing

– Turn ignition OFF and remove ignition key.

Version 1

◀ Use plastic prying tool at lower edge of switch bezels (**arrows**) to pry off bezels.

– Remove center vent mounting screws and remove vent.

– Carefully pull out display unit and disconnect harness connector.

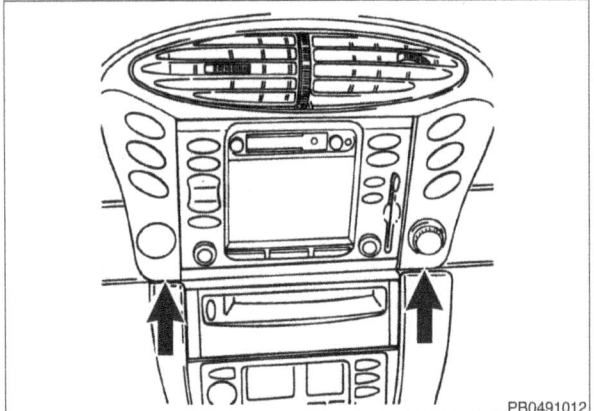

Radio and Communication 91-9
Navigation

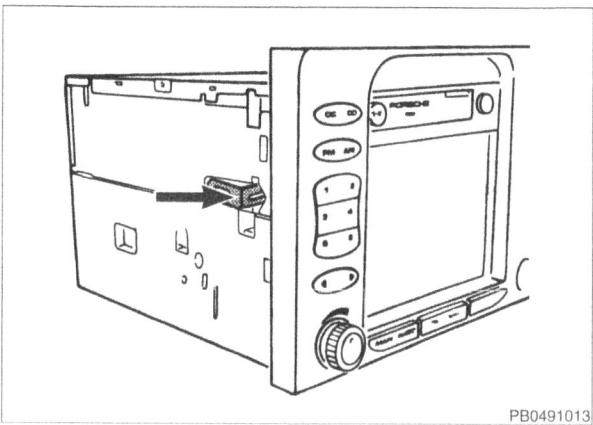

Version 2

- Unclip switch bezels from sides of PCM display unit and allow to dangle from harnesses.

◄ Reach through openings at sides of display unit and press in retaining clips (**arrow**).

- Carefully pull out display unit and disconnect harness connector.

- When installing, route electrical leads as before. Make sure they do not get pinched.

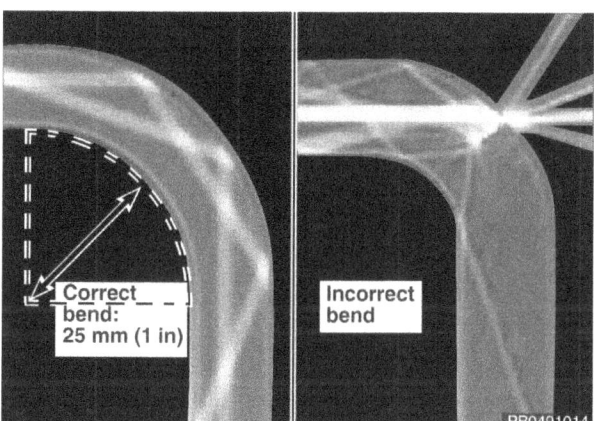

◄ Make sure any bend in MOST-bus fiber optic harness has a radius of 25 mm (1 in) or greater.

> **CAUTION—**
> - Signal transmission through an optical fibers that is bent excessively is impaired.

- After installing new display unit, activate PCM using Porsche System Tester 2 (PST2).

92 Wipers and Washers

GENERAL 92-1	WIPER REPAIRS 92-4
Rain sensor............................. 92-1	Wiper assembly components 92-4
Cautions 92-2	Wiper assembly, removing and installing 92-5
WIPER ARMS AND BLADES 92-2	Wiper motor, removing and installing 92-7
Wiper blade cleaning problems............... 92-2	WASHER REPAIRS 92-9
Wiper blade, replacing 92-2	Windshield washer nozzles.................. 92-9
Wiper arm, removing and installing........... 92-3	Windshield washer nozzle, adjusting 92-9
WIPER AND WASHER CONTROLS 92-3	Windshield washer nozzle, removing and installing 92-9
Wiper and washer system fuse............... 92-3	Washer fluid reservoir, removing and installing 92-10
Wiper and washer system relays 92-3	Washer pump, removing and installing........ 92-11
Wiper and washer switch 92-4	Washer fluid level sensor, removing and installing 92-12
Intermittent wipe interval control 92-4	

GENERAL

This repair group covers wiper blade service and wiper assembly repairs. Also covered are repairs to the windshield and headlight washer assemblies.

For additional information see the following:

- **48 Steering** for steering column stalk switch assembly replacement
- **97 Fuses, Relays, Component Locations**
- **EWD Electrical Wiring Diagrams**

Rain sensor

 An option in 2001 and later cars, the rain sensor system uses infrared to detect precipitation (rain or snow) on the windshield and automatically adjust wiper speed.

1. Infrared transmitter
2. Infrared detector
3. Signal generator
4. Amplifier
5. Output signal

The rain sensor is at the center of the windshield, at the top. The rain sensor control module is in the left cowl area, under the front trunk lid.

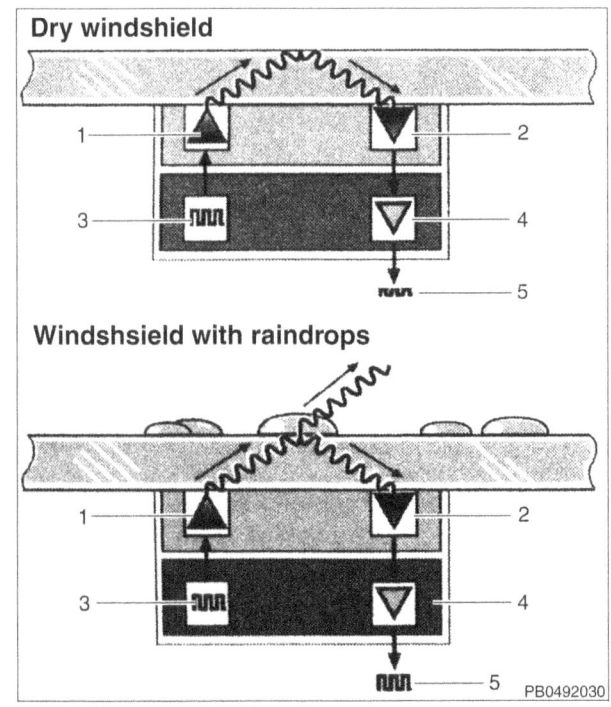

Wipers and Washers

Wiper Arms and Blades

Cautions

> **CAUTION—**
> - Switch the ignition OFF and disconnect the negative (-) battery cable before removing any electrical components.
> - Prior to disconnecting the battery, read the battery disconnection cautions in **00 Warnings and Cautions**.
> - Prior to disconnecting the battery:
>   -Obtain the radio operation code, if applicable.
>   -Record radio presets.
> - Only use a digital multimeter for electrical tests.
> - Do not operate the windshield wipers with the front trunk lid open. The wiper arm may scrape.

WIPER ARMS AND BLADES

Wiper blade cleaning problems

Common problems with the windshield wipers include streaking or sheeting, water drops after wiping, and blade chatter.

Streaking is usually caused when wiper blades are coated with road film or car wash wax.

– Clean blades using soapy water. If cleaning does not cure problem, replace blades.

Drops that remain behind after wiping are usually caused by oil, road film, or diesel exhaust residue on the glass.

– Use an alcohol or ammonia solution or other non-abrasive cleaner to clean windshield.

Chatter may be caused by dirty or worn blades, or by wiper arms that are out of alignment.

– Clean blades and windshield as described above.

– Bend wiper arm so that there is even pressure along blade, and so that blade is perpendicular to windshield at rest.

– If problems persist, replace blades and wiper arms.

Wiper blade, replacing

– Make sure ignition switch is OFF.

– Pivot wiper arm off glass.

◄ Position wiper blade approximately perpendicular to wiper arm.
  • Depress retaining tab (**A**).
  • Slide blade plastic retaining clip out of arm (**B**).

– Installation is reverse of removal. Install wiper blade clip to wiper arm until it clicks into position.

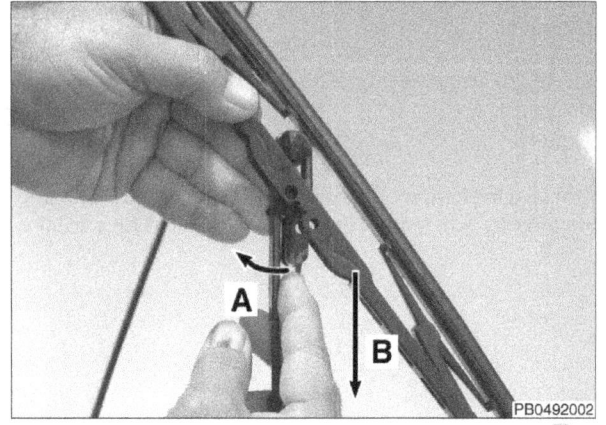

Wipers and Washers 92-3

Wiper and Washer Controls

Wiper arm, removing and installing

- Make sure wipers are in PARK position and ignition switch is OFF. Open engine hood.
- Remove plastic trim plug from base of wiper arm.

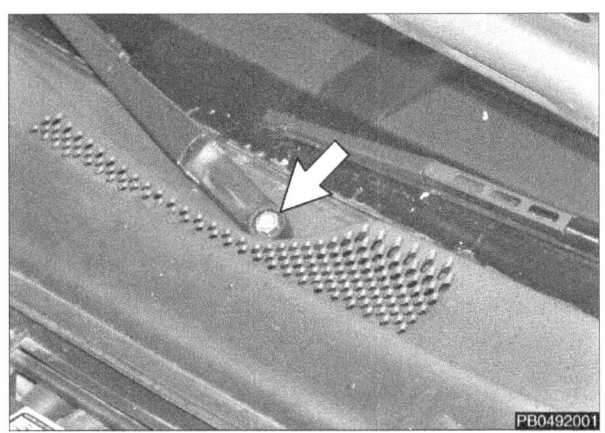

◄ Remove wiper arm mounting nut (**arrow**).

- Mark position of wiper arm on shaft to aid installation.
- Remove wiper arm.

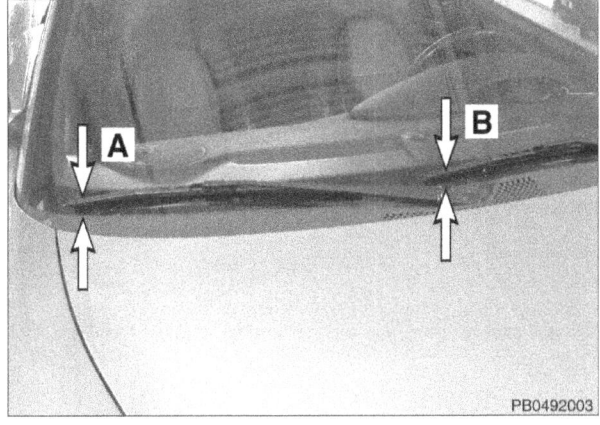

◄ Installation is reverse of removal. Tighten mounting nuts after adjusting position of arms.

NOTE—
- *The right wiper arm is longer.*

Wiper arm position	
Distance from wiper blade tip to lower edge of windshield:	
• Right wiper (**A**)	approx. 12 mm (½ in)
• Left wiper (**B**)	approx. 25 mm (1 in)

Tightening torque	
Wiper arm to shaft (M8)	17 Nm (13 ft-lb)

WIPER AND WASHER CONTROLS

Wiper and washer system fuse

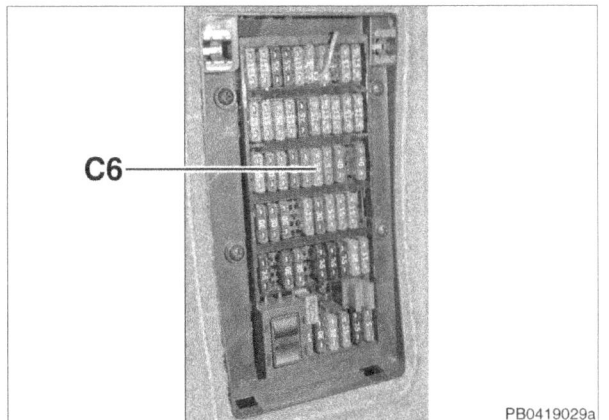

◄ In main fuse panel, behind left footwell cover.

Wiper and washer fuse	
Location	Main fuse panel
Number	C6
Rating	25 A

Wiper and washer system relays

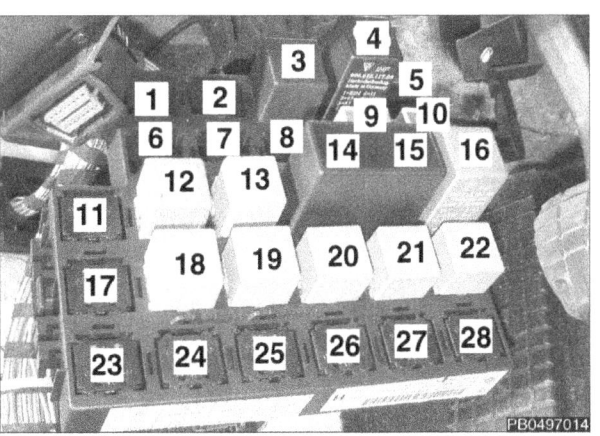

◄ In relay panel 1, above left kick panel.

NOTE—
- *To gain access to relays, see **97 Fuses, Relays, Component Locations**.*

Wiper and washer system relays	
Headlight washer (if equipped)	Relay 8
Intermittent wiper	Relay 16

92-4 Wipers and Washers

Wiper Repairs

Wiper and washer switch

Switch positions are as follows:
1. Wipers parked
2. Slow wipe
3. Fast wipe
4. Intermittent wipe
5. Windshield wash (short duration)
6. Windshield wash and wipe (long duration)
7. Headlight wash (if equipped)

The wiper and washer stalk switch is one component of an integrated steering column switch assembly which includes headlight dimmer and turn signal switches and, if equipped, on-board computer controller and cruise control switch. Steering column stalk switch replacement is covered in **48 Steering**.

Intermittent wipe interval control

Beginning with 2001 Boxster S models, the intermittent wipe interval can be adjusted with a potentiometer (**arrow**) in the center dashboard.

NOTE—
- *In a vehicle with rain sensor, the thumb wheel controls rain sensor system sensitivity.*

WIPER REPAIRS

Wiper assembly components

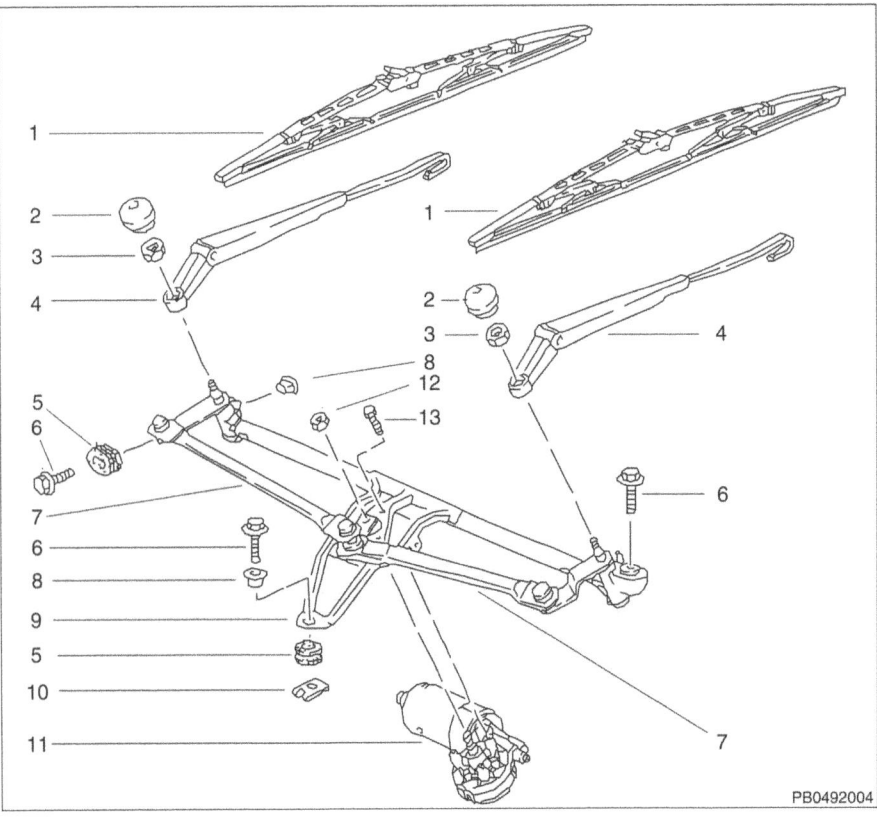

1. **Wiper blade**
2. **Plastic cap**
3. **M8 nut**
   - tighten to 17 Nm (13 ft-lb)
4. **Wiper arm**
5. **Grommet**
6. **M6 screw**
   - tighten to 10 Nm (7 ft-lb)
7. **Wiper assembly link**
8. **Sleeve**
9. **Wiper assembly frame**
10. **Speed nut**
11. **Wiper motor**
12. **M8 nut**
    - tighten to 17 Nm (13 ft-lb)
13. **M6 wiper assembly mounting screw**
    - tighten to 8 Nm (6 ft-lb)

Wipers and Washers 92-5

Wiper Repairs

Wiper assembly, removing and installing

– Disconnect negative (–) battery cable and cover battery terminal to keep cable from accidentally contacting terminal.

> **CAUTION—**
> - *Prior to disconnecting the battery, read the battery disconnection cautions in **00 Warnings and Cautions**.*
> - *Be sure to have the radio anti-theft code before disconnecting the battery.*

◀ Working inside front trunk, remove securing screws (**arrows**) for cowl covers to left and right of battery. Lift out covers.

– Remove wiper arms. See **Wiper arm, removing and installing** in this repair group.

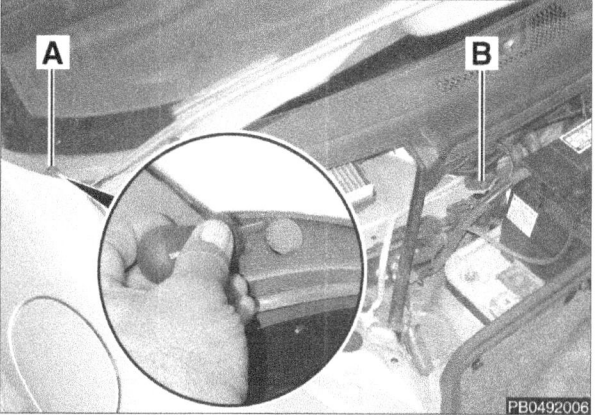

◀ Remove upper cowl cover fasteners:
- Use small screwdriver to pry off round trim covers at corners of windshield (**inset**).
- Remove plastic 10 mm nuts (**A**).
- Remove Torx fasteners (**B**).

> **NOTE—**
> - *Right side illustrated. Left side is similar.*

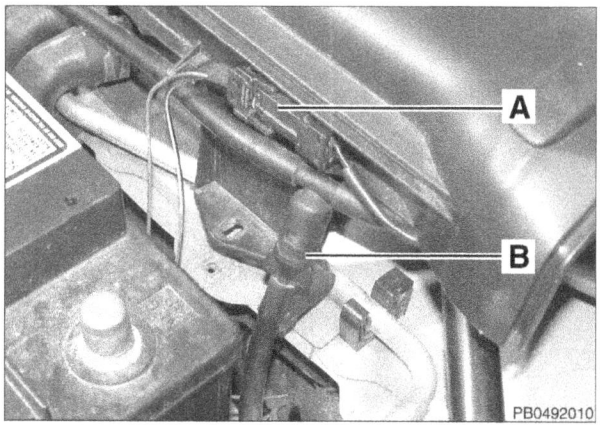

◀ Lift up cowl cover slightly:
- Disconnect left and right washer nozzle heater electrical connectors (**A**).
- Pull apart washer nozzle fluid connector (**B**).

– Slide cowl cover off toward rear of vehicle, under rear edge of front trunk lid.

◀ Remove left front strut tower reinforcement brace fasteners (**arrows**) and remove brace.

92-6 Wipers and Washers

Wiper Repairs

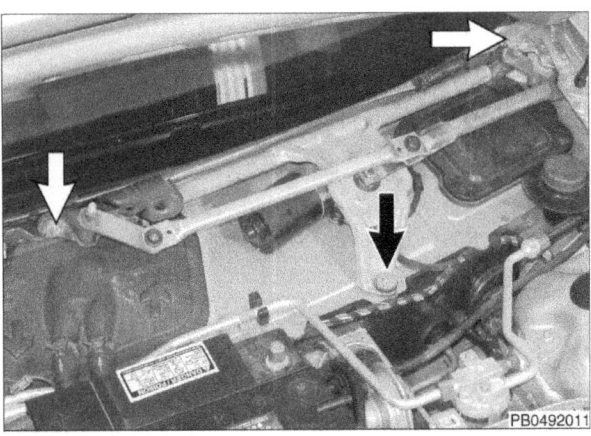

◀ Remove wiper assembly mounting bolts (**arrows**).

◀ Turn over assembly and disconnect electrical connector at wiper motor.

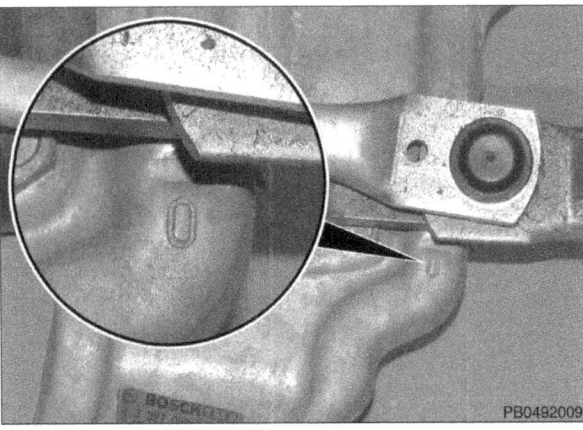

◀ Prior to reinstalling, make sure wiper motor is in PARK position:
- Line up wiper motor arm with **0** mark on wiper assembly frame.

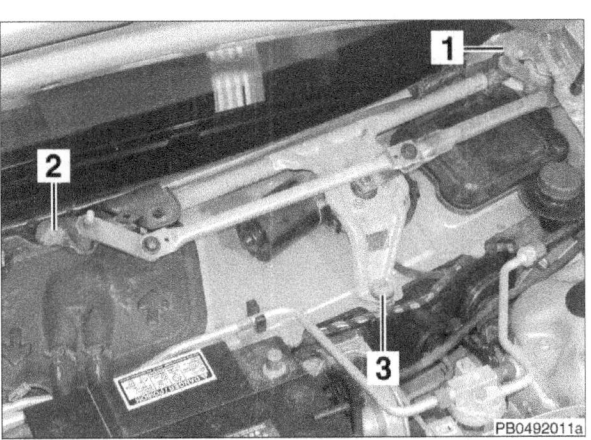

◀ Install wiper assembly and tighten mounting bolts in sequence **1 - 2 - 3**.

Tightening torque	
Wiper assembly to cowl (M6)	10 Nm (7 ft-lb)

Wipers and Washers 92-7

Wiper Repairs

- Remainder of installation is reverse of removal. Keep in mind the following:
  - Place vehicle on its wheels before tightening strut tower reinforcement fasteners.
  - Position wiper arms on wiper pivots. See **Wiper arm, removing and installing** in this repair group.
  - Reconnect battery and test wipers to make sure that wipers park correctly.

Tightening torque	
Wiper arm to shaft (M8)	17 Nm (13 ft-lb)

Wiper motor, removing and installing

- Disconnect negative (–) battery cable and cover battery terminal to keep cable from accidentally contacting terminal.

> **CAUTION—**
> - *Prior to disconnecting the battery, read the battery disconnection cautions in* **00 Warnings and Cautions**.
> - *Be sure to have the radio anti-theft code before disconnecting the battery.*

◂ Working inside front trunk, remove securing screws (**arrows**) for cowl covers to left and right of battery. Lift out covers.

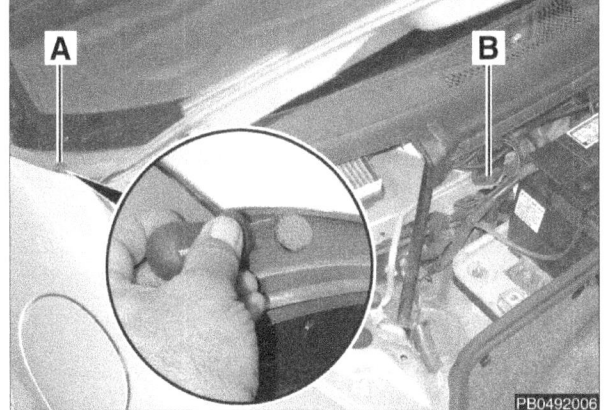

◂ Remove upper cowl cover fasteners:
- Use small screwdriver to pry off round trim covers at corners of windshield (**inset**).
- Remove plastic 10 mm nuts (**A**).
- Remove Torx fasteners (**B**).

> **NOTE—**
> - *Right side of cover illustrated. Left side is similar.*

◂ Lift up cowl cover slightly:
- Disconnect left and right washer nozzle heater electrical connectors (**A**).
- Pull apart washer nozzle fluid connector (**B**).

- Slide cowl cover off toward rear of vehicle, under rear edge of front trunk lid.

92-8 Wipers and Washers

Wiper Repairs

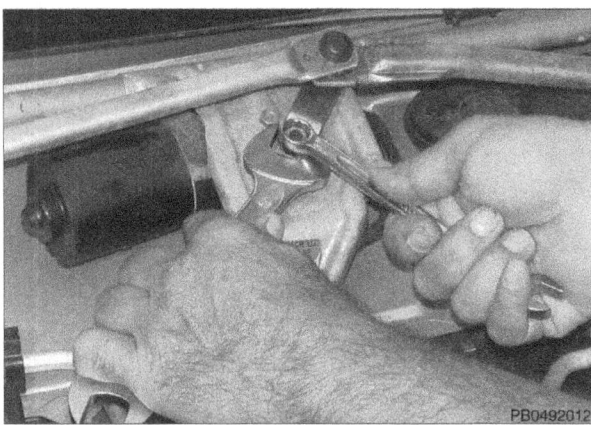

◀ Loosen and remove wiper motor arm mounting nut. Counterhold arm with 21 mm wrench. Detach arm from splined shaft of motor.

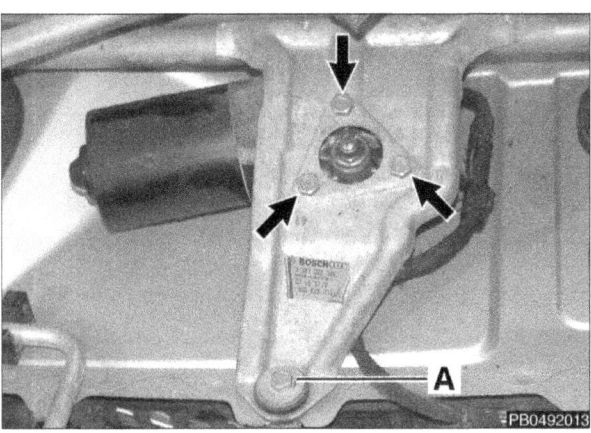

◀ Remove wiper motor:
- Unscrew mounting bolts (**arrows**).
- Remove lower wiper assembly mounting bolt (**A**).
- Carefully tilt up assembly and slide out wiper motor.
- Disconnect harness connector and remove motor.

– When reinstalling, reattach harness connector, then slide motor into position and install bolts.

Tightening torques	
Wiper assembly to cowl (M6)	10 Nm (7 ft-lb)
Wiper motor to wiper assembly (M6)	8 Nm (6 ft-lb)

– Reattach negative (–) battery cable and turn ignition ON. Make sure wiper switch is OFF.

– Once wiper motor arrives at PARK position, turn ignition OFF and disconnect negative (–) battery cable.

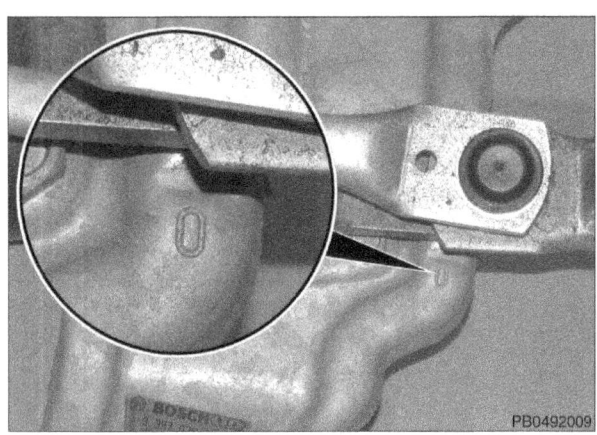

◀ Reattach wiper motor arm to motor splined shaft, lining up wiper motor arm with **0** mark on wiper assembly frame. Counterhold arm with 21 mm wrench.

Tightening torque	
Wiper motor arm to wiper motor splined shaft (M8)	17 Nm (13 ft-lb)

– Remainder of installation is reverse of removal. Reconnect battery and test wipers to make sure that wipers park correctly.

Wipers and Washers 92-9

Washer Repairs

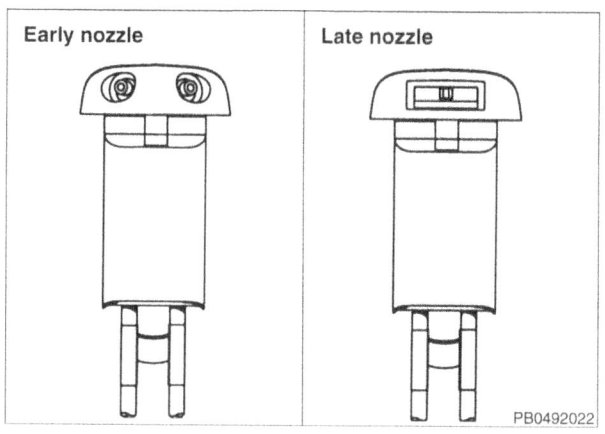

WASHER REPAIRS

Windshield washer nozzles

◂ The later style windshield washer nozzles may be installed as a replacement for the early style nozzles. Replace nozzles in pairs.

Windshield washer nozzle, adjusting

Only the early style washer nozzle can be adjusted.

◂ Use a pin tool, safety pin or needle to direct nozzle openings.

Windshield washer nozzle, removing and installing

◂ Working inside front trunk, remove securing screws (**arrows**) for cowl covers to left and right of battery. Lift out covers.

– Remove wiper arms. See **Wiper arm, removing and installing** in this repair group.

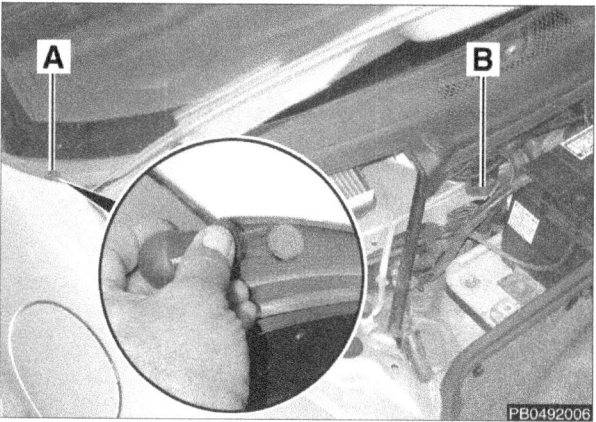

◂ Remove upper cowl cover fasteners:
- Use small screwdriver to pry off round trim covers at corners of windshield (**inset**).
- Remove plastic 10 mm nuts (**A**).
- Remove Torx fasteners (**B**).

NOTE—
- *Right side illustrated. Left side is similar.*

92-10 Wipers and Washers

Washer Repairs

◄ Lift up cowl cover slightly:
  - Disconnect left and right washer nozzle heater electrical connectors (**A**).
  - Pull apart washer nozzle fluid connector (**B**).

– Slide cowl cover off toward rear of vehicle, under rear edge of front trunk lid.

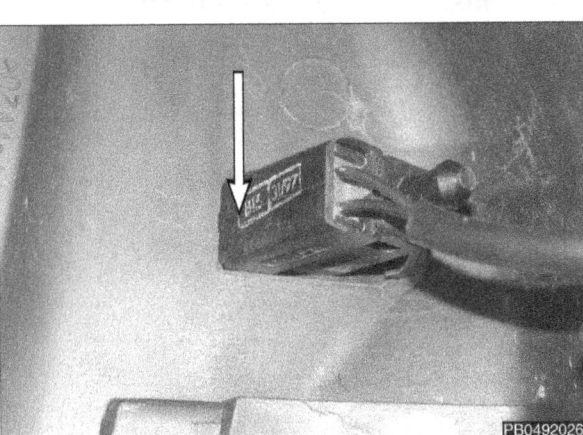

◄ For nozzle replacement, place cowl cover upside down on work bench.
  - Detach fluid hose from nozzle.
  - Press nozzle at edge (**arrow**) to disengage from cowl.

– Installation is reverse of removal.

Washer fluid reservoir, removing and installing

The washer fluid reservoir, under the left front fender, contains the windshield washer pump, the washer fluid level sensor and, if equipped, the headlight washer pump.

The washer reservoir is attached via a plastic tube and grommet to the washer fluid filler neck in the front trunk. Before removing, siphon as much fluid as possible from reservoir.

– To gain access to reservoir, raise front of car and support safely.

> **WARNING—**
> - Make sure the car is stable and well supported at all times. Use a professional automotive lift or jack stands designed for the purpose. A floor jack is not adequate support.

– Remove left front wheel.

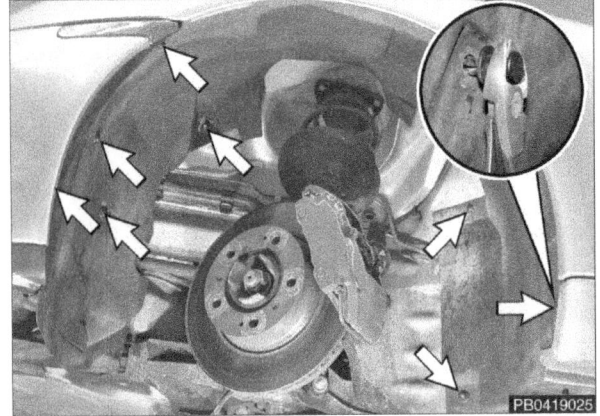

◄ Remove left front wheel well trim fasteners (**arrows**).
  - To disengage plastic rivet, pull out rivet lock (**inset**).
  - Remove plastic 10 mm nuts.

Wipers and Washers 92-11

Washer Repairs

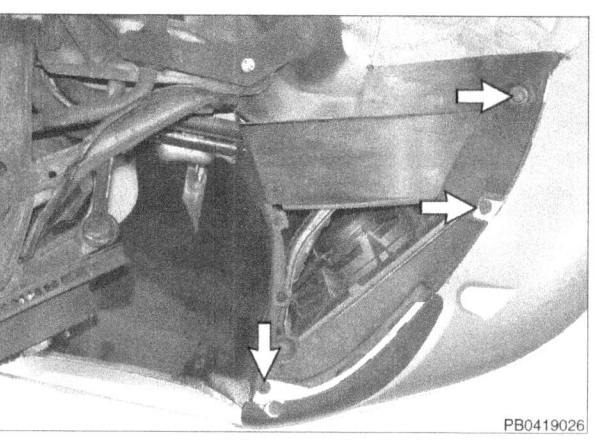

◀ Working underneath front fender, remove remainder of wheel well trim fasteners (**arrows**). Lift off trim.

◀ Working at top of fender housing, remove reservoir mounting bolt (**arrow**).

- Detach reservoir from filler neck and detach lower mounting pins.

CAUTION—
- *Make sure filler neck grommet remains on reservoir bore and does not fall inside reservoir.*

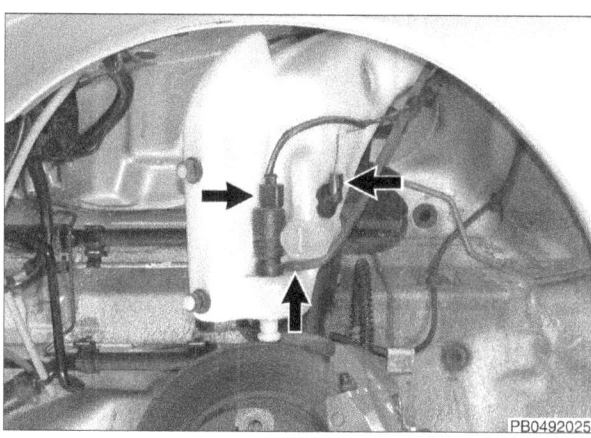

◀ Detach electrical connectors and washer hose(s) (**arrows**) and remove reservoir. Be prepared to catch dripping washer fluid.

– Installation is reverse of removal.

Washer pump, removing and installing

The windshield washer pump and, if equipped, the headlight washer pump are each pressed into a grommet in the washer fluid reservoir. Before removing, siphon as much fluid as possible from reservoir.

– Remove washer fluid reservoir. See **Washer fluid reservoir, removing and installing** in this repair group.

Washer Repairs

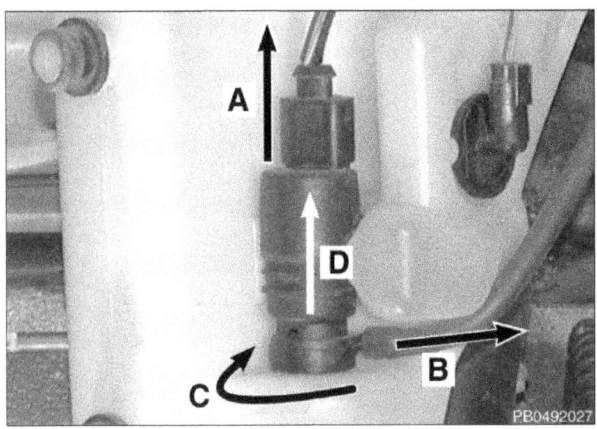

◄ Remove washer pump:
- Detach harness connector (**A**).
- Detach fluid hose (**B**).
- Twist pump clockwise (**C**).
- Slide pump out of reservoir grommet (**D**). Be prepared to catch dripping washer fluid.

– Check condition of sealing grommet and washer fluid hose. Replace as necessary.

– Installation is reverse of removal.

Washer fluid level sensor, removing and installing

The washer fluid level sensor is pressed into a grommet in the washer fluid reservoir. Before removing, siphon as much fluid as possible from reservoir.

– Remove washer fluid reservoir. See **Washer fluid reservoir, removing and installing** in this repair group.

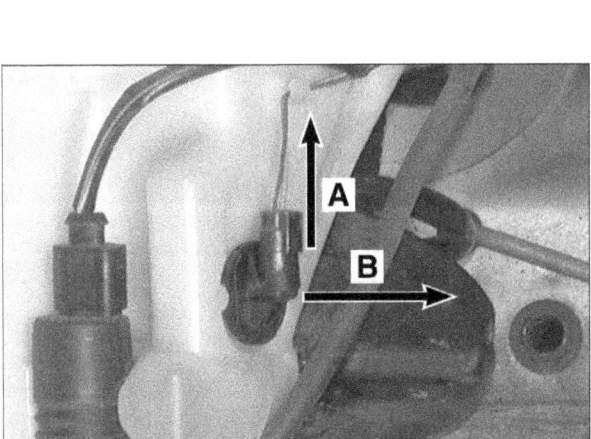

◄ Remove level sensor:
- Detach harness connector (**A**).
- Slide sensor out of reservoir grommet (**B**). Be prepared to catch dripping washer fluid.

– Check condition of sealing grommet. Replace as necessary.

– Installation is reverse of removal.

94 Exterior Lights

GENERAL 94-1
  Headlight design 94-1
  Litronic system 94-2
  Headlight troubleshooting 94-3
  Exterior bulb applications 94-4
  Warnings and Cautions..................... 94-4

LIGHTING CONTROLS...................... 94-5
  Light switch, removing and installing. 94-5
  Hazard warning switch, removing and installing . 94-6
  Brake light switch, removing and installing 94-6
  Lighting fuses 94-7

FRONT LIGHTING REPAIRS 94-7
  Headlight aim, adjusting 94-7

Headlight assembly,
  removing and installing 94-8
Headlight bulb, replacing (halogen) 94-9
Front foglight bulb, replacing 94-10
Front turn signal / parking light bulb,
  replacing 94-11
Side turn signal bulb
  removing and installing 94-11

REAR LIGHTING REPAIRS 94-12
  Taillights, servicing 94-12
  Center brake light, servicing 94-12
  License plate light bulb, removing and installing 94-13

TABLE

a. Boxster exterior bulb applications................ 94-4

GENERAL

This repair group covers exterior lighting repairs and bulb replacement for Boxster models.

For additional information, see:

- **34 Manual Transmission** for manual transmission back-up light switch
- **37 Automatic Transmission** for automatic transmission back-up light switch function (automatic transmission gear position switch)
- **97 Fuses, Relays, Component Locations**
- **EWD Electrical Wiring Diagrams**

Headlight design

The Boxster headlight modules integrate high and low beams, foglights, turn signal lights and (as an option) headlight washers.

The high and low beams use plastic reflectors. The ploycarbonate front lenses feature high scratch and impact resistance and low weight.

94-2 Exterior Lights

General

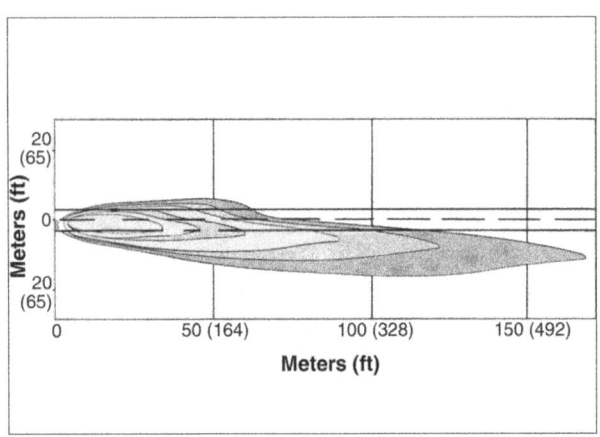

◄ Headlight reflector shape was computer designed for optimal road illumination and compact size.

Litronic system

Beginning with 1999 models, the Boxster was offered with an optional xenon low beam headlight system known as Litronic. The system offers the following features and benefits:

- Illumination approximately doubled with approx. 30% less power consumption.
- Improved and more homogeneous lateral illumination.
- Improved color vision due to higher xenon light color temperature.
- Improved high beam illumination through elevation of low beam.
- Raising and lowering of headlights when braking and accelerating.

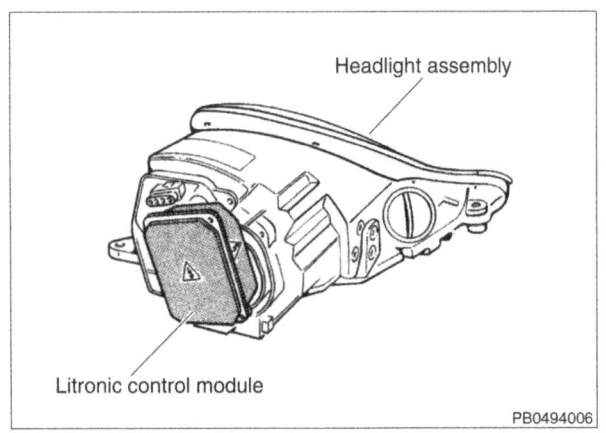

◄ Litronic control module is mounted to rear cover of headlight assembly.

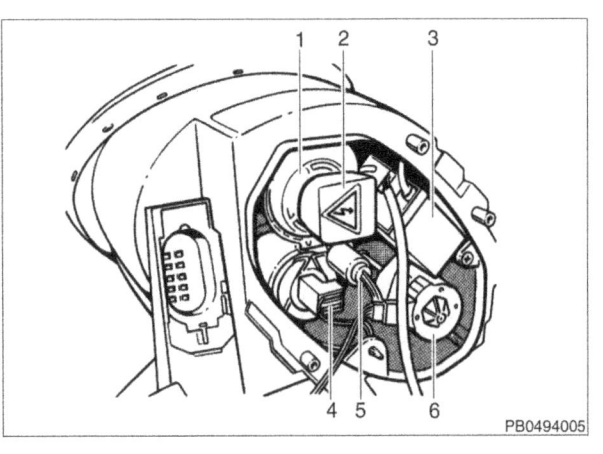

◄ Inside Litronic headlight assembly:
1. Low beam socket
2. Low beam plug
3. Xenon ignition unit
4. High beam bulb (H7 halogen)
5. Parking light bulb (H6 w halogen)
6. Headlight adjuster stepper motor

Exterior Lights 94-3

General

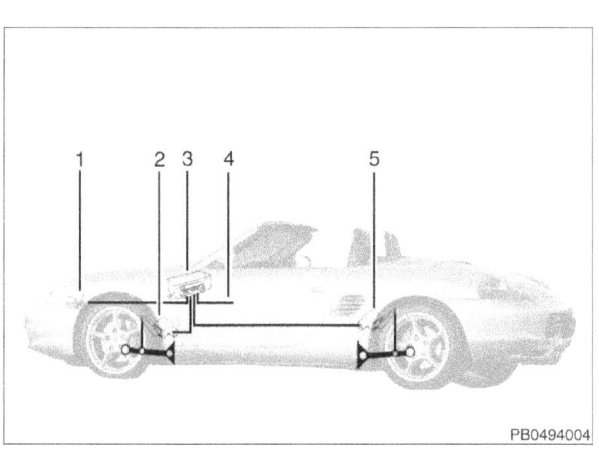

This system also includes dynamic headlight beam adjustment (AHBA) which corrects headlight beam angle to reduce blinding of oncoming traffic due to uneven vehicle load or the dynamic pitching motion of the car. AHBA components include:

1. Headlight adjuster stepper motor
2. Front suspension level sensor
3. AHBA control module
4. Signals: Wheel speed sensor input
5. Rear suspension level sensor

When high beams are switched ON, AHBA raises low beams by approx. 1.2° (2%) in 0.3 seconds. This substantially increases roadway illumination.

To help prevent dazzling of oncoming traffic due to dirty headlight lenses, the Litronic system is equipped with headlight washers as standard equipment.

Faults in the Litronic system are stored in the Litronic control module and are accessible via Porsche System Tester 2 (PST2).

NOTE—
* *A xenon headlight retrofit kit is available from Porsche with the same light output as the factory-installed Litronic system. However, it is not equipped with AHBA and cannot be diagnosed with (PST2).*

Headlight troubleshooting

Moisture inside headlights

Headlights are ventilated using moving air between the bumper cover and the lower part of the headlight assembly. There are also plastic vent lines between the headlight assemblies and the bottom of the front bumper cover.

- If excessive moisture builds up inside a headlight assembly, check the following:
  - Make sure gap between headlight and front bumper cover is clear.
  - Make sure front end cover (bra) is not impeding air flow over headlights.
  - Make sure plastic headlight vents under front bumper are intact and clear.

Xenon headlight problems

- In case of problems with one xenon headlight, swap over individual components (gas discharge bulb, ignition unit, control module) from the second headlight to isolate the defective component.

Dynamic headlight beam adjustment (AHBA) changes headlight aim automatically, adapting to vehicle load, acceleration and deceleration and road irregularity. A preliminary test of AHBA is as follows:

- Switch low beams ON. Switch ignition ON: Lights first dip all the way down, then adapt to vehicle load.

- Switch to high beam: Low beams rise perceptibly.

94-4 Exterior Lights

General

- If AHBA system fails any of these tests, have system checked and repaired at an authorized Porsche dealer or other qualified repair facility.

Exterior bulb applications

For convenience, exterior bulb applications for Boxster models are listed in **Table a**.

Table a. Boxster external bulb applications	
Back-up light	P 21 w
Brake light • Center • Corners	 W 3 w P 21 w
Foglight • Front • Rear	 H7 55 w P 21 w
Headlight • Halogen • Xenon (low beam)	 H7, 55 w Phillips D 2 S
License plate light	C5 w
Parking light • Front • Litronic	 W 5 w H6 w
Side marker light • Rear	W 5 w W 3 w
Taillight	R 5 w
Turn signal • Front • Rear	 MSCD 40 P 21 w, PY 21 w

Warnings and Cautions

> *WARNING—*
> - *Xenon bulbs operate at high voltages. When working on xenon headlight components (control module, ignition unit, gas discharge lamp, drive motor), be sure to switch lights OFF.*
> - *Operate xenon bulb only when installed in the headlight reflector.*
> - *An operating light, specially a headlight bulb, becomes very hot and will cause injury. Allow bulbs to cool before working on the lighting system.*

> *CAUTION—*
> - *Do not touch the glass of bulbs with bare skin. Dirt and skin oils may cause a bulb to fail prematurely. If necessary wipe bulb using a clean cloth dampened with rubbing alcohol.*
> - *Use only original equipment replacement bulbs. Non-original bulbs may cause false failure readings on the dashboard display.*
> - *Use only specified bulbs. A bulb with higher wattage may cause damage to bulb housing.*
> - *To avoid marring car paint or trim, work with a plastic prying tool or wrap screwdriver tip with tape.*
> - *Clean headlights with water and standard window cleaning agent.*

Exterior Lights 94-5

Lighting Controls

LIGHTING CONTROLS

Light switch, removing and installing

- Disconnect negative (–) battery cable and cover battery terminal to keep cable from accidentally contacting terminal.

> **CAUTION—**
> - Prior to disconnecting the battery, read the battery disconnection cautions in **00 Warnings and Cautions**.
> - Be sure to have the radio anti-theft code before disconnecting the battery.

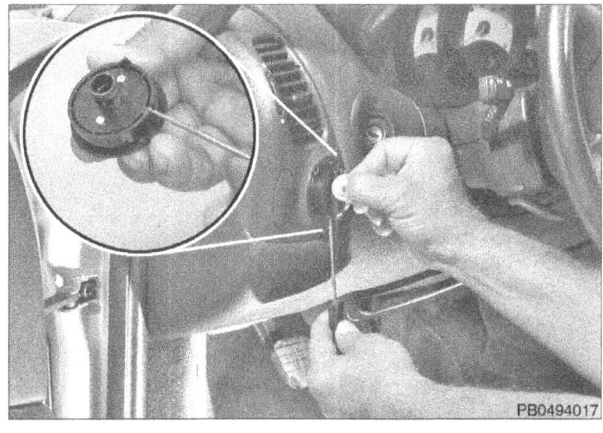

◄ Pull out light switch knob to stop. Use sharp thin tool to press in locking clip on underside of knob to disengage knob from switch.

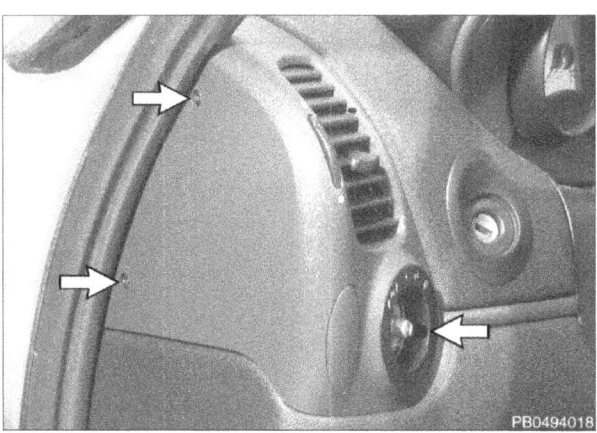

◄ Remove three left side air vent mounting screws (Torx T18) (**arrows**).
- Pull vent and light switch together out of dashboard.
- Disconnect switch electrical connector.

◄ Place vent on work bench and undo light switch mounting nut (M16 x 1) (**arrow**).

94-6 Exterior Lights

Lighting Controls

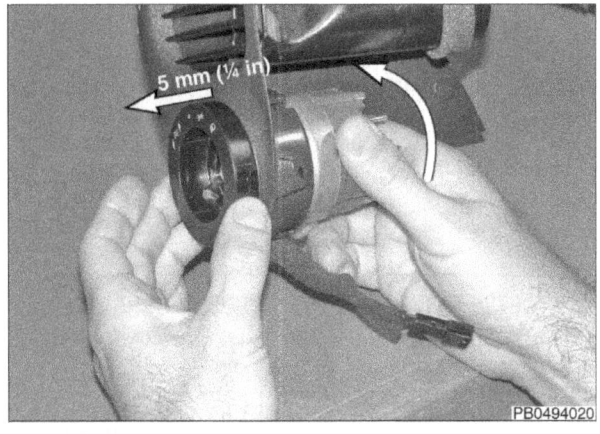

◄ Slide switch faceplate approx. 5 mm (¼ in) out of vent, then turn switch body slightly counterclockwise to remove.

– To install, engage light switch in side vent by turning it slightly in counterclockwise direction. Install mounting nut.

Tightening torque	
Light switch to side vent (M16 x 1 nut)	4 Nm (3 ft-lb)

– Remainder of installation is reverse of removal.

Hazard warning switch, removing and installing

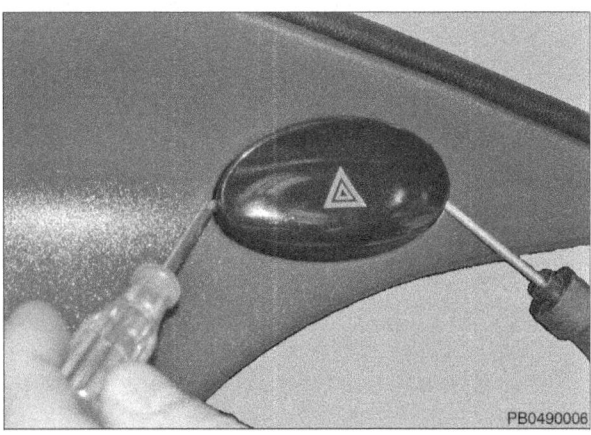

◄ Press in hazard warning switch so that button projects out. Insert two small screwdrivers into openings on sides of switch and gently pry out button.

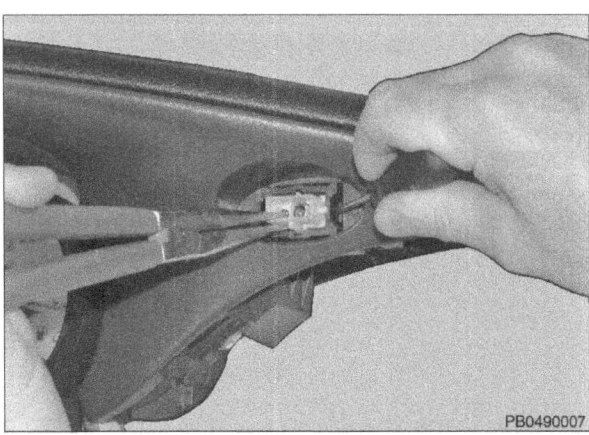

◄ Use small screwdriver to press right side switch retaining clip. Pull switch out of instrument cluster with pliers.

Brake light switch, removing and installing

◄ Working in passenger compartment under left side of dashboard:
  • Reach above brake pedal and twist brake light switch clockwise (**arrow**) to remove from bracket.
  • Disconnect harness connectors.

– Installation is reverse of removal.

Exterior Lights 94-7

Front Lighting Repairs

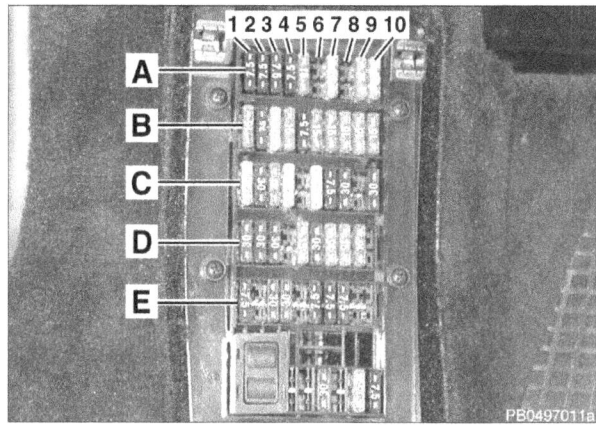

Lighting fuses

◀ Lighting system fuses in main fuse panel in left footwell.

Table b. Lighting system fuses

Protected circuit	Fuse	Rating in amps
Brake lights	B7	15
Daytime running lights	E1	7.5
Foglights	A7	25
Hazard warning lights	D7	15
Headlight vertical aim control	B10	15
Headlight washer	C9	25
Headlights: • High beam right, high beam control module • High beam left • Low beam right • Low beam left	A1 A2 A9 A10	15 15 15 15
Interior lights	C3	15
License plate lights • Canada • USA	 A8 A5	 7.5 15
Locating lights	A5	15
Side marker lights: • Right • Left	 A3 A4	 7.5 7.5
Turn signals	B6	15

FRONT LIGHTING REPAIRS

Headlight aim, adjusting

- Note conditions for adjusting headlight aim:
  - Vehicle on horizontal surface
  - Fuel tank full
  - Tire pressures at correct values
  - Approx. 75 kg (165 lb) on driver seat
  - Vehicle rolled a few feet to settle suspension

◀ Open front trunk and peel back trim on inner wheel housing next to headlight assembly. Pull back adjuster access hole plugs.

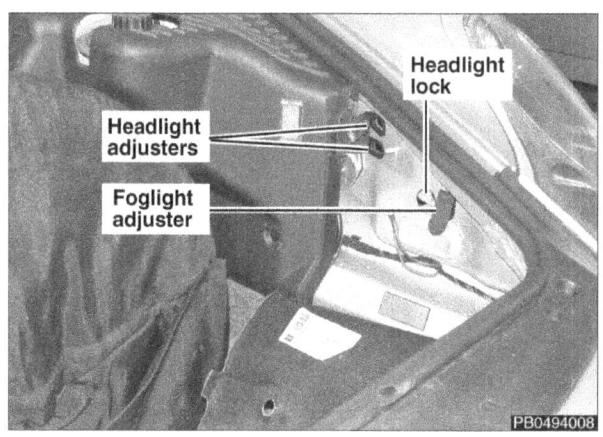

94-8 Exterior Lights

Front Lighting Repairs

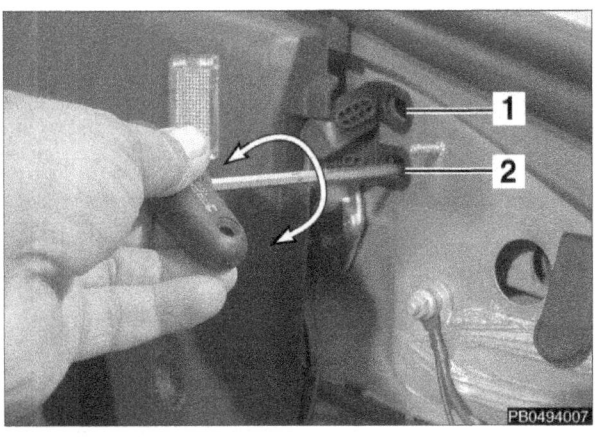

◄ Use 5 mm Allen wrench to adjust headlight aim:

1. Lateral adjustment: Turn upper adjuster
   - clockwise: Beam moves right.
   - counterclockwise: Beam moves left.
2. Height adjustment: Turn lower adjuster
   - clockwise: Beam moves down.
   - counterclockwise: Beam moves up.

– Litronic headlight: With headlights ON, cycle ignition ON and OFF then recheck headlight adjustment.

Headlight assembly, removing and installing

◄ Open front trunk and peel back trim on inner wheel housing next to headlight assembly.

◄ Use 5 mm socket (or special tool from vehicle tool kit) to detach headlight.

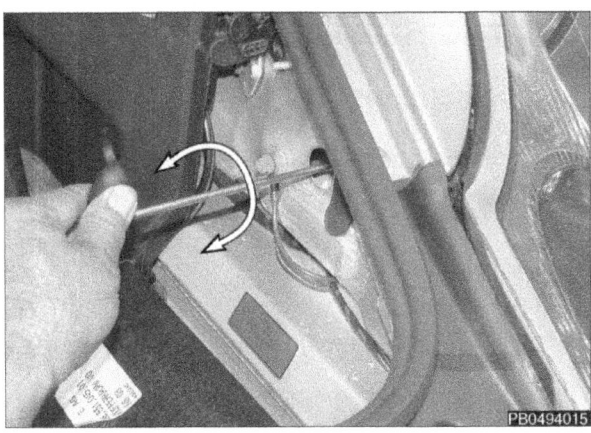

◄ Slide headlight forward out of fender. If necessary, twist headlight lock slightly to disengage.

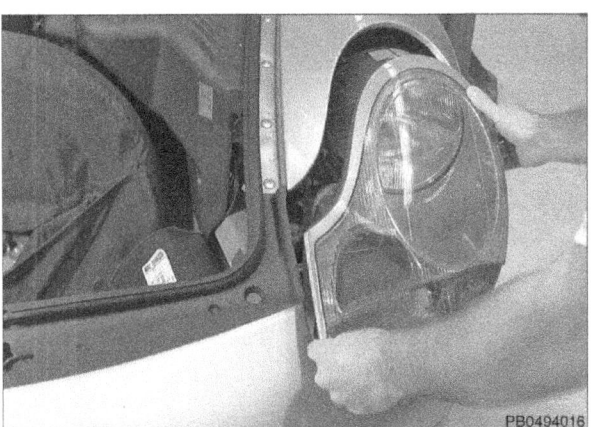

Exterior Lights 94-9

Front Lighting Repairs

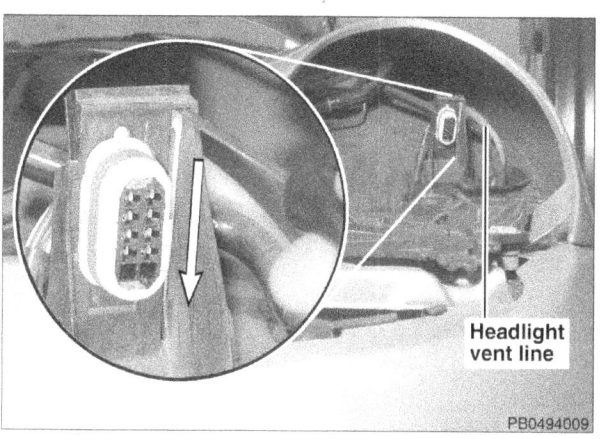

◀ If removing headlight carrier plate, unlock headlight connector (**inset**) with thin prying tool. Press plastic retaining clip downward (**arrow**) and release connector.

NOTE—
* *Prior to installing the headlight housing, make sure the headlight vent line is plugged into the back of the connector.*

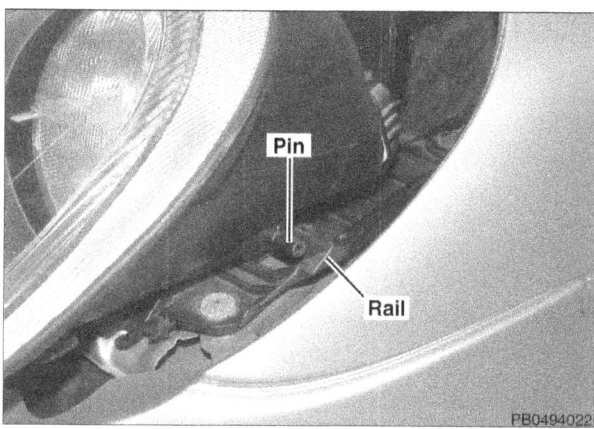

◀ When installing headlight, line up aligning pins on sides of headlight with headlight carrier plate rails. Press headlight firmly into fender, then turn headlight lock until it clicks.

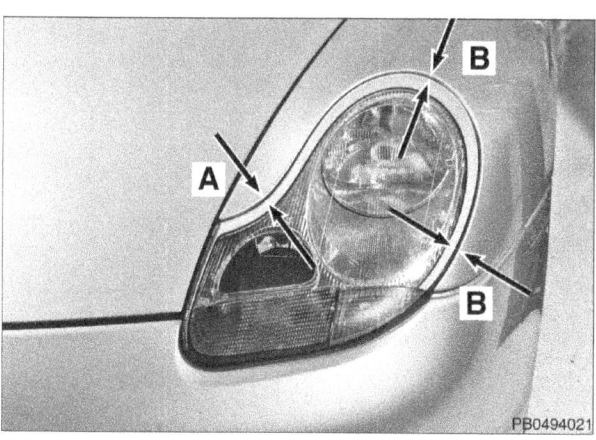

◀ Check level of headlight housing in relation to fender.

Headlight level	
Headlight inner edge to fender (**A**)	Same level
Headlight top or outer edge to fender (**B** or **C**)	2 mm (1/12 in) below level of fender

Headlight bulb, replacing (halogen)

WARNING—
* *An operating headlight bulb becomes very hot and will cause injury. Allow bulbs to cool before working on the lighting system.*

– Remove headlight assembly. See **Headlight assembly, removing and installing** in this repair group.

◀ Press clip to release rear cover.

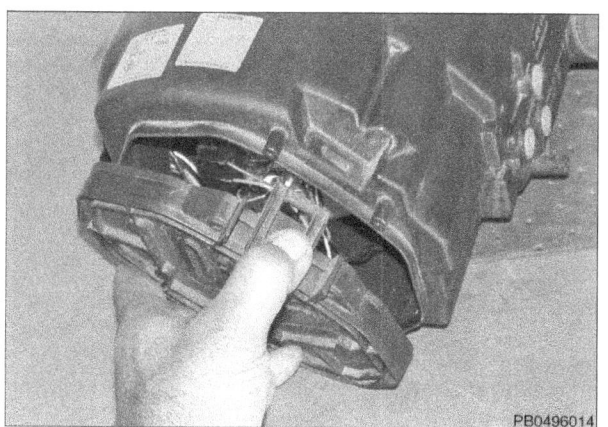

94-10 Exterior Lights

Front Lighting Repairs

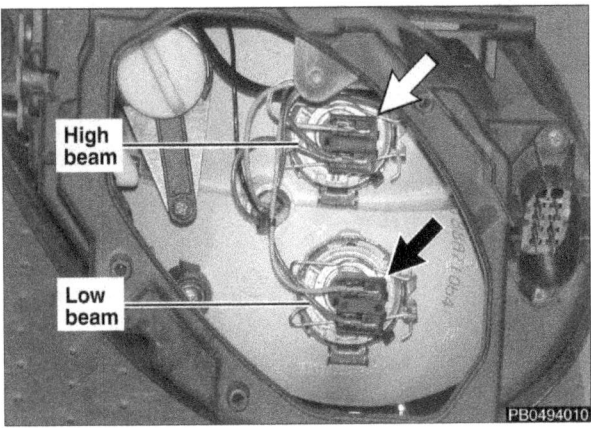

◄ Place headlight assembly face down on soft surface. Disconnect electrical connectors (**arrows**) from high or low beam bulbs.

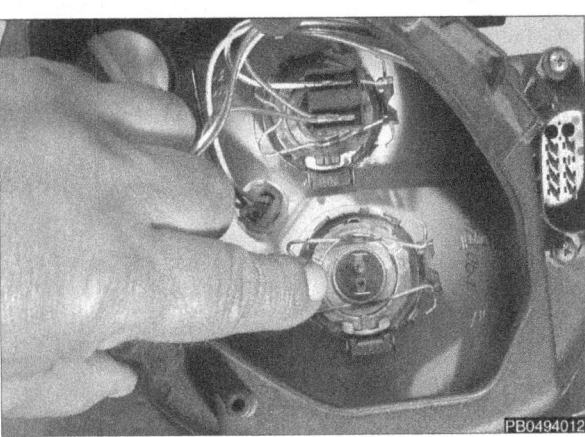

◄ Unclip bulb retainer spring clip.

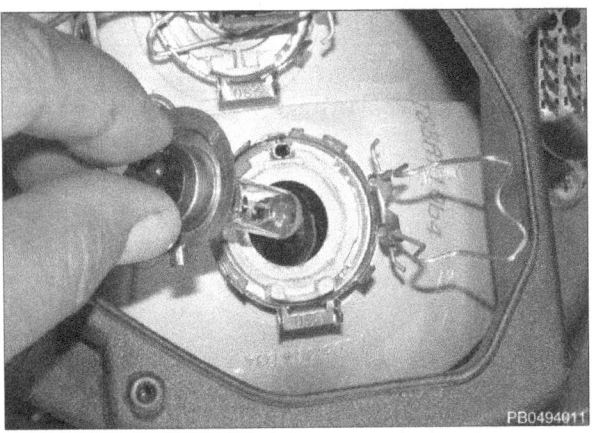

◄ Pull bulb out of socket and replace.

– Installation is reverse of removal.

Front foglight bulb, replacing

WARNING—
* *An operating bulb becomes very hot and will cause injury. Allow bulbs to cool before working on the lighting system.*

– Remove headlight assembly. See **Headlight assembly, removing and installing** in this repair group.

◄ Twist off foglight bulb cover. Pull off foglight bulb harness connector (**arrow**).

– Unclip bulb retainer spring clip.

Exterior Lights 94-11

Front Lighting Repairs

- Pull bulb out of socket and replace.
- Installation is reverse of removal.

Front turn signal / parking light bulb, replacing

> **WARNING—**
> - An operating bulb becomes very hot and will cause injury. Allow bulbs to cool before working on the lighting system.

- Remove headlight assembly. See **Headlight assembly, removing and installing** in this repair group.

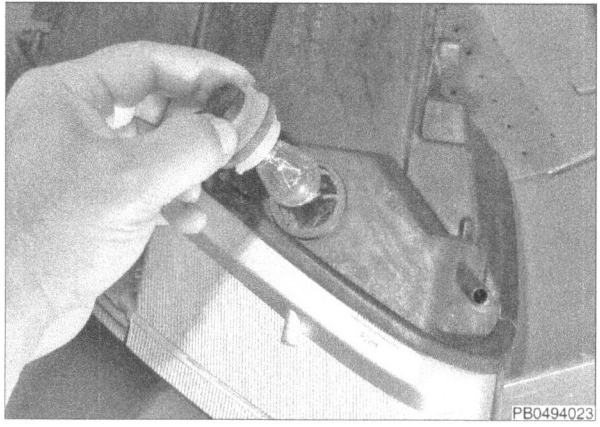

◀ Twist turn signal / parking light bulb socket and remove from headlight housing. Replace bulb.

- Installation is reverse of removal.

Side turn signal bulb, replacing

> **WARNING—**
> - An operating bulb becomes very hot and will cause injury. Allow bulbs to cool before working on the lighting system.

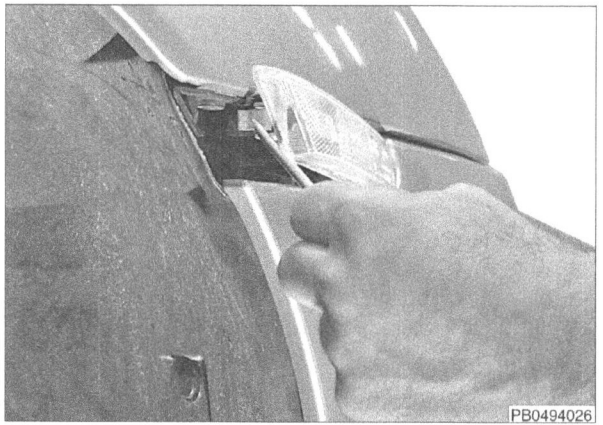

◀ Insert screwdriver in slot between front wheel housing liner and side turn signal socket. Press in retaining spring to disengage socket from fender.

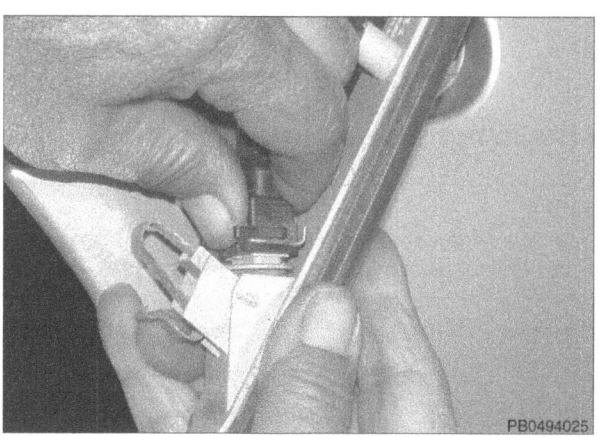

◀ Pull electrical harness off bulb socket. Twist out socket and replace bulb.

- Installation is reverse of removal.

94-12 Exterior Lights

Rear Lighting Repairs

REAR LIGHTING REPAIRS

Taillights, servicing

◄ The wrap-around taillights incorporate directional signals, brake lights, back-up lights, parking lights and rear side marker lights.

Taillight bulbs are accessible from inside rear trunk. Bulbs may be changed without tools.

> **WARNING—**
> - *An operating bulb becomes very hot and will cause injury. Allow bulbs to cool before working on the lighting system.*

- Open rear trunk lid and, if necessary, peel back trunk interior trim over taillight bulb holder.

◄ Press up on bulb holder retaining clip (**arrow**). Slide holder out of taillight assembly.

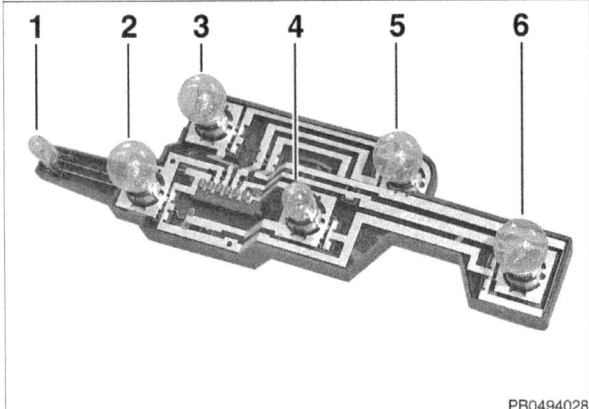

◄ Twist bulb to remove from socket (left side shown):
1. Rear side marker
2. Rear foglight
3. Turn signal
4. Taillight
5. Back-up light
6. Brake light

- Installation is reverse of removal.

Center brake light, servicing

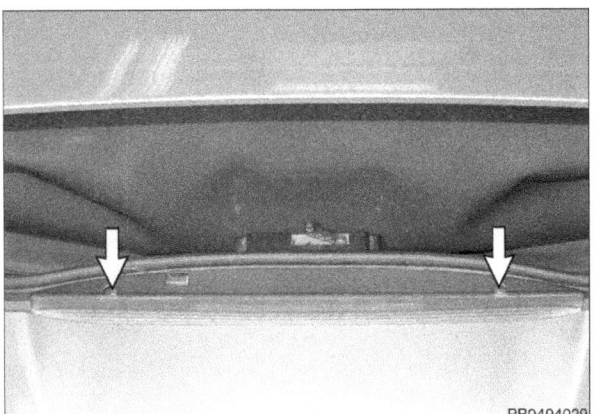

> **WARNING—**
> - *An operating bulb becomes very hot and will cause injury. Allow bulbs to cool before working on the lighting system.*

- Open convertible top until convertible top compartment lid is fully open. Remove ignition key.

◄ Remove center brake light strip mounting screws (**arrows**).

Exterior Lights 94-13

Rear Lighting Repairs

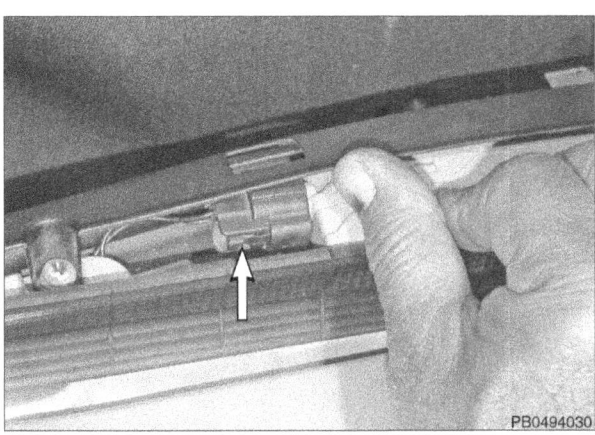

◀ Press harness connector clip (**arrow**) to detach harness from light strip.

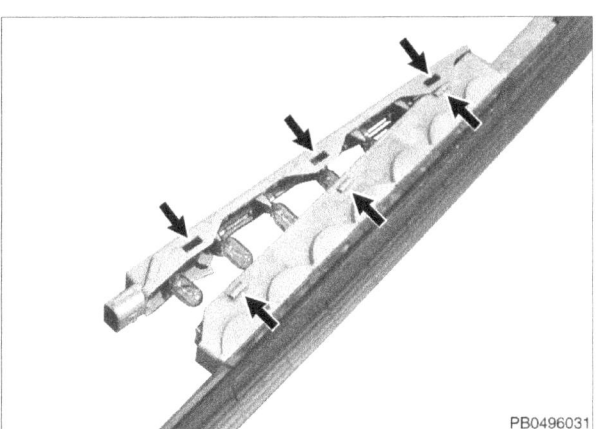

◀ Disengage bulb holder locking clips (**arrows**).

− Replace bulbs as necessary.

− Installation is reverse of removal.

License plate light bulb, removing and installing

> **WARNING—**
> • An operating bulb becomes very hot and will cause injury. Allow bulbs to cool before working on the lighting system.

◀ Remove license plate light mounting screws (**arrows**). Pull light assembly down and replace bulb.

− Installation is reverse of removal.

96 Interior Lights, Anti-theft

GENERAL 96-1

INTERIOR LIGHTING 96-1
   Door courtesy light bulb, replacing 96-2
   Dome light bulb, replacing. 96-2
   Trunk light bulb, replacing. 96-3

ANTI-THEFT SYSTEM 96-4
   Anti-theft system components. 96-4
   Vehicle security. 96-4
   Basic anti-theft system 96-5
   Anti-theft system with alarm 96-6
   Vehicle key 96-6
   Electronic immobilizer. 96-7

Electronic immobilizer codes. 96-8
Central locking 96-8
Central locking button 96-8
Alarm system 96-7
Alarm LED. 96-9
Passenger compartment monitor 96-10
Electronic immobilizer signal converter,
   removing and installing 96-10
Electronic immobilizer control module,
   removing and installing 96-12

TABLES
a.   Boxster interior bulb applications 96-1
b.   Alarm LED status 96-9

GENERAL

This repair group covers interior light bulb replacement as well as descriptions and repair information for the electronic immobilizer, central locking and alarm systems.

See the following repair groups for additional information:
- **57 Doors** for door locks
- **90 Instruments** for instrument cluster bulbs
- **94 Exterior Lights** for exterior bulbs

INTERIOR LIGHTING

The Boxster interior and front and rear trunks are illuminated by conventional lights. Courtesy lights are installed in the interior door trim.

Starting with 2001 models, light emitting diodes (LEDs) are used for orientation lighting. Installed in the top of the windshield and in the inside door panels, LEDs illuminate the interior almost imperceptibly, switching ON when the car is unlocked and OFF when it is locked or after 2 hours if the car is not used.

For convenience, interior bulb applications for Boxster models are listed in **Table a**.

Table a. Boxster interior bulb applications	
Dome light, door courtesy lights	W 5 w
Trunk light bulb (front or rear)	K 10 w

Interior Lighting

WARNING—
- *An operating light becomes very hot and will cause injury. Allow bulbs to cool before working on the lights.*

CAUTION—
- *Do not touch the glass of bulbs with bare skin. Dirt and skin oils may cause a bulb to fail prematurely. If necessary wipe bulb using a clean cloth dampened with rubbing alcohol.*
- *Use only original equipment replacement bulbs. Non-original bulbs may cause false failure readings on the dashboard display.*
- *Use only specified bulbs. A bulb with higher wattage may cause damage to bulb housing.*
- *To avoid marring car paint or trim, work with a plastic prying tool or wrap screwdriver tip with tape.*

Door courtesy light bulb, replacing

◀ Using plastic prying tool or screwdriver with tape wrapped around tip, carefully pry out door courtesy light socket.

– Detach electrical connector, remove light bulb from socket and replace.

– Installation is reverse of removal.

Dome light bulb, replacing

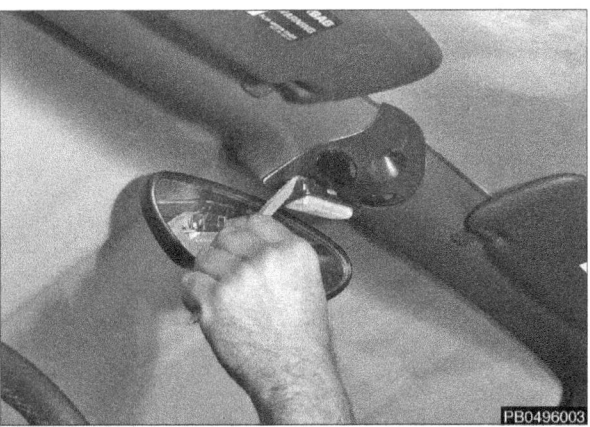

◀ Using plastic prying tool or screwdriver with tape wrapped around tip, carefully pry out dome light socket, left side first.

Interior Lights, Anti-theft   96-3

Interior Lighting

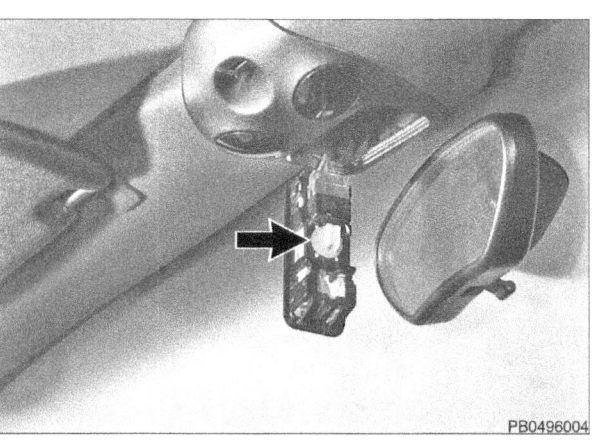

◄ Twist light bulb (**arrow**) counterclockwise to remove from socket and replace.

– Installation is reverse of removal.

Trunk light bulb, replacing

◄ Front trunk: Using plastic prying tool or screwdriver with tape wrapped around tip, carefully pry out trunk light socket.

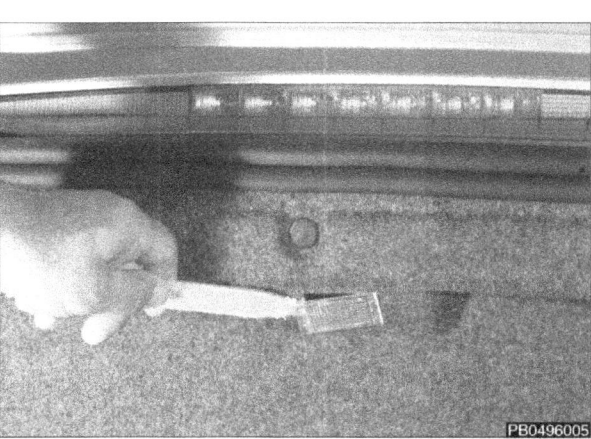

◄ Rear trunk: Using plastic prying tool or screwdriver with tape wrapped around tip, carefully pry out trunk light socket.

– Remove light bulb from socket and replace.

– Installation is reverse of removal.

Anti-theft System

ANTI-THEFT SYSTEM

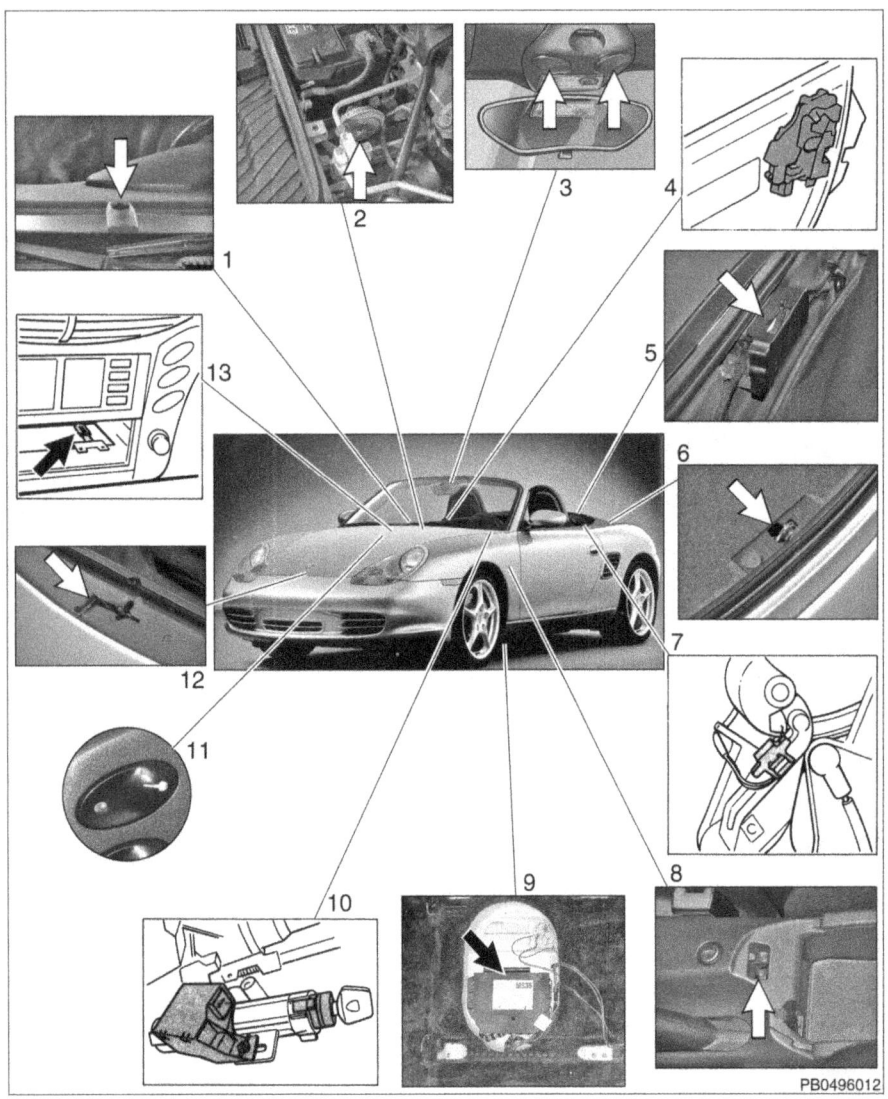

Anti-theft system components

1. **Alarm readiness LED**
   - center dashboard
2. **Alarm horn**
   - left of battery
3. **Infrared passenger compartment monitor**
   - above rear view mirror
4. **Door lock module**
   - inside door at door lock
5. **Convertible top compartment lid lock microswitch**
6. **Rear trunk lock microswitch**
7. **Convertible top position microswitch**
8. **Center console locking compartment lock microswitch**
9. **Control module (electronic immobilizer, alarm)**
   - under left seat
10. **Electronic immobilizer signal converter and induction coil**
    - at ignition switch
11. **Central locking switch**
12. **Front trunk lock microswitch**
13. **Radio alarm contact**

Vehicle security

The basic Boxster anti-theft system consists of electronic immobilizer (also referred to as *drive block*) and integrated central locking. The system performs the following functions:

- Electronic immobilization using key transponder and signal converter
- Central locking
- Interior lighting
- Power windows
- Consumer deactivation

There is an optional anti-theft system with the following additional security and convenience functions:

- Alarm system
- Infrared passenger compartment monitoring
- Radio anti-theft
- Rear trunk lid release

Interior Lights, Anti-theft 96-5

Anti-theft System

2001 models added the following modifications and options:
- Both front and rear trunk lids are equipped with power release instead of a mechanical cable.
- Seat and rear-view mirror memory module remote controlled by signal from door key.
- Passenger outside rear view mirror tilts down when reverse gear is engaged.

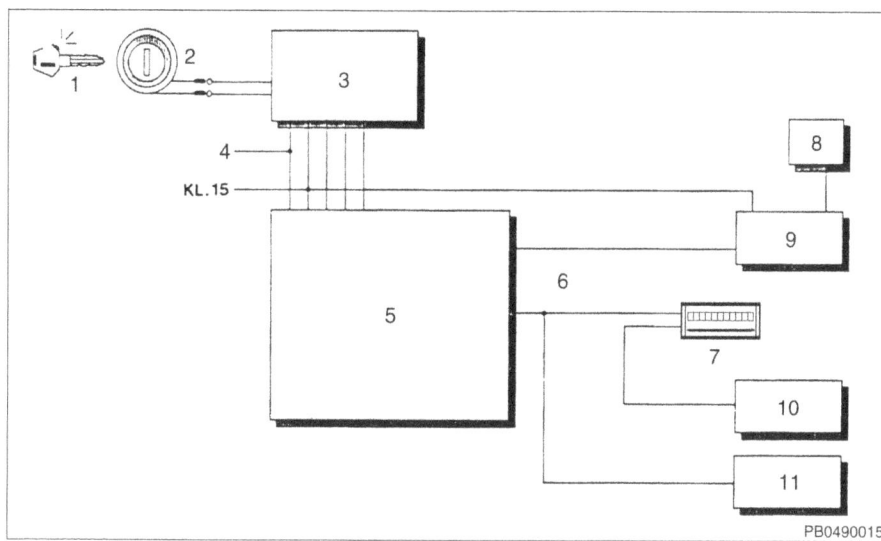

Basic anti-theft system

1. **Key with transponder**
2. **Induction coil**
   - around ignition lock
3. **Electronic immobilizer signal converter**
   - attached to steering lock cylinder
4. **Buzzer contact, terminal 86s**
5. **Electronic immobilizer and central locking control module**
   - under driver seat
6. **Data lines**
7. **OBD II (diagnosis) socket**
8. **Starter**
9. **Engine control module (ECM)**
10. **Other control modules**
11. **Tiptronic control module**

Interior Lights, Anti-theft

Anti-theft System

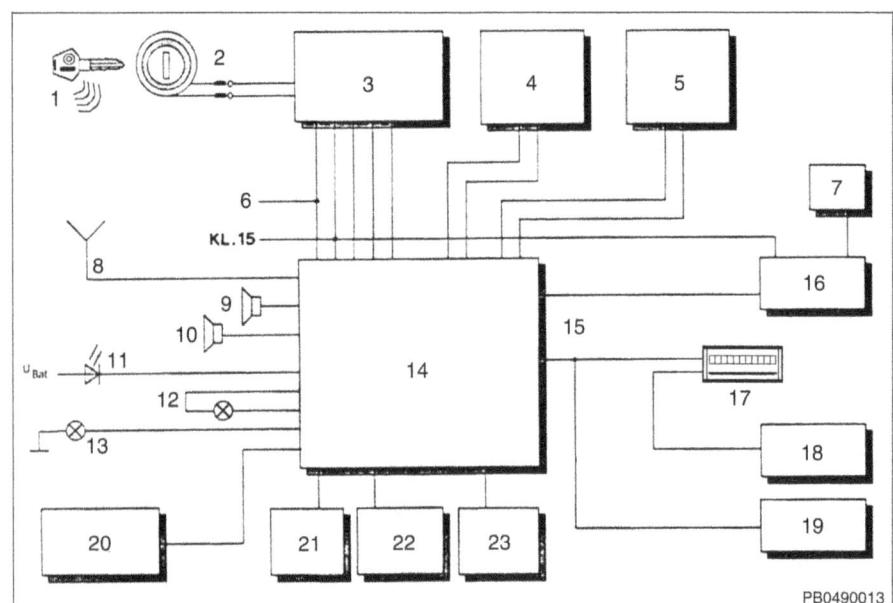

Anti-theft system with alarm

1. Key with transponder and transmitter
2. Induction coil
   - around ignition lock
3. Electronic immobilizer signal converter
   - attached to steering lock cylinder
4. Infrared passenger compartment monitor
5. Tilt sensor (optional)
6. Buzzer contact, terminal 86s
7. Starter
8. Key transmitter antenna
9. Alarm horn
   - left of battery
10. Alarm siren (optional)
11. Alarm readiness LED
    - center top dashboard
12. Interior lighting
13. Turn signal indicators
14. Electronic immobilizer, central locking and alarm control module
    - under driver seat
15. Data lines
16. Engine control module (ECM)
17. OBD II (diagnosis) socket
18. Other control modules
19. Tiptronic control module
20. Alarm switch contacts
21. Door lock modules
22. Rear trunk lid release
23. Fuel tank cap release

Vehicle key

 1997 - 2000: Keyless entry key is equipped with the following:

1. LED
2. Central locking button: Doors unlock and lock
3. Rear trunk unlock button (press for approx. 3 seconds)

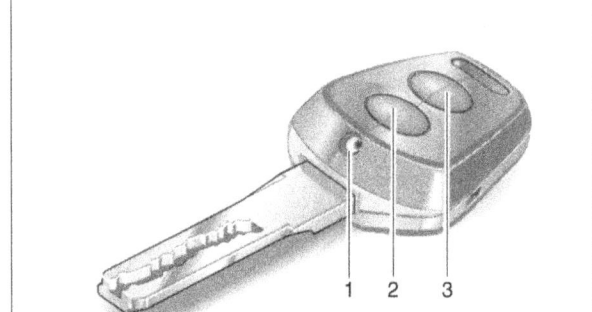

Interior Lights, Anti-theft 96-7

Anti-theft System

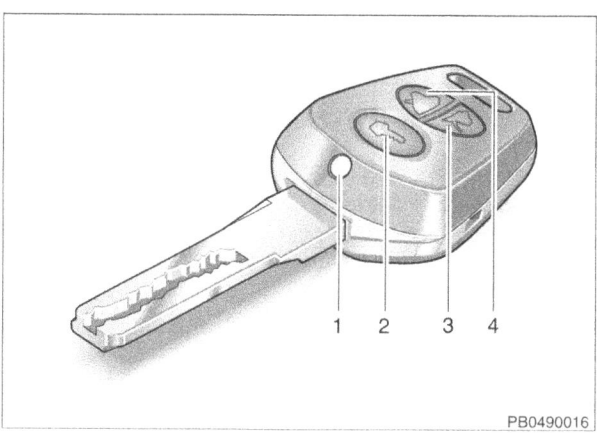

2001 - 2004: Keyless entry key is equipped with the following:
1. LED
2. Central locking button: Doors unlock and lock
3. Front trunk unlock button (press for approx. 3 seconds)
4. Rear trunk unlock button (press for approx. 3 seconds)

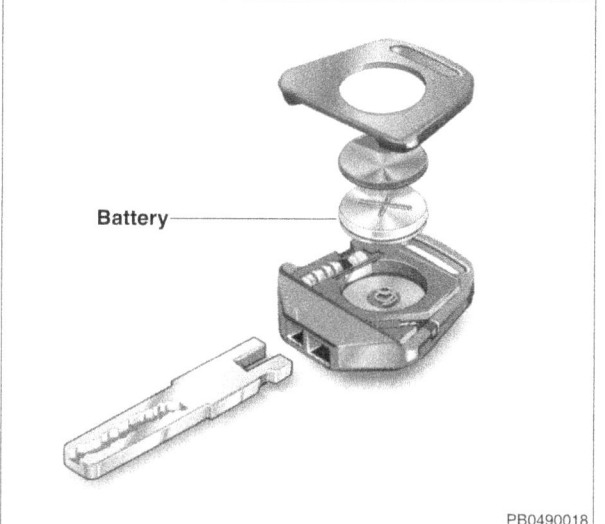

The key transmitter is powered by a battery in the key fob. Pry off key fob cover gently using small screwdriver or fingernail.

Keyless entry transmitter battery	
1997 - 2000 model	625U, 1.5 v
2001 - 2004 model	Lithium CR 2032 3 v

If the car is not started or unlocked with the remote control key for 5 days, the keyless entry function switches off to prevent draining the car battery. In this case:

— 1997 - 2000:
- Unlock car with key in door lock.
- Switch ignition ON to reactivate keyless entry.

— 2001 - 2004:
- Unlock car with key in door lock. Leave door closed to avoid triggering alarm.
- Press central locking button on key to reactivate keyless entry.

Electronic immobilizer

NOTE—
- *In Porsche technical literature the electronic immobilizer is referred to as drive block.*
- *There is one control module for the immobilizer, central locking and alarm (if equipped).*

When the key is used to switch ignition ON, the transponder in the key exchanges a unique, preassigned code with the induction coil surrounding the ignition switch. The code is transmitted to the signal converter attached to the steering lock housing.

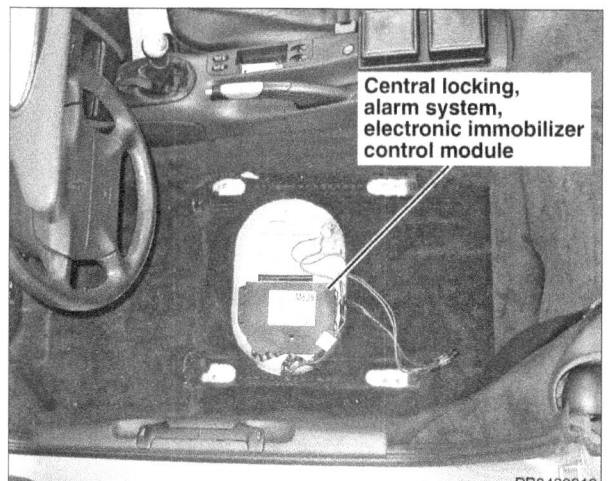

The code is converted to a digital signal and transmitted to the electronic immobilizer control module under the driver seat.
- If the immobilizer module recognizes the code as valid, it signals the engine control module (ECM) to proceed with engine starting.
- If no valid code is detected, starting is prevented by ECM: Starter, fuel injection and ignition functions are disabled.

Key coding can only performed using Porsche System Tester 2 (PST2). Up to 4 different keys may be coded.

Interior Lights, Anti-theft

Anti-theft System

Electronic immobilizer codes

The electronic immobilizer codes and serial numbers are written to the engine control module (ECM) and immobilizer (alarm) module. at the factory. The information is recorded in the vehicle master file. It is available to the legitimate vehicle owner or agent using the integrated Porsche trader processing system (IPAS).

If the control modules are changed, reenter the necessary code data using PST2. Note that:
- The immobilizer (alarm) module can only be programmed once.
- The ECM can be reprogrammed repeatedly.

Central locking

– Actuate central locking in these ways:
- Use key in driver door lock
- Use central locking button on dashboard
- Vehicle with alarm system: Use key transmitter.

Central locking has three possible states:
- **Unlocked**
- **Locked**: Doors can be opened from inside. Actuate door lock *twice* within one second via key or keyless entry transmitter. Only alarm system is active, not passenger compartment monitoring.
- **Door-safe**: Doors cannot be opened from inside. Actuate door lock *once* via key or keyless entry transmitter. Alarm system and passenger compartment monitoring (if equipped) are active.

If the driver door is not closed properly, the vehicle cannot be locked.

If the vehicle is unlocked using the keyless entry transmitter but none of the doors are opened within one minute, doors are relocked automatically, but the alarm is not reactivated.

In an accident, the doors are unlocked by a signal from airbag control module.

Central locking button

 Both doors can be lock and unlocked using the central locking button on the center dashboard. If the lock elements of the two doors are in different positions, they are synchronized before central locking triggers.

The central locking button LED illuminates as follows:
- Doors locked: Lights up for 10 seconds.
- Door-safe: Flashes twice per second for 10 seconds.

If doors are locked with central locking button, they can be opened using inside door handle:
- Pull inside door handle once: Door unlocks.
- Pull inside door handle again: Door opens.

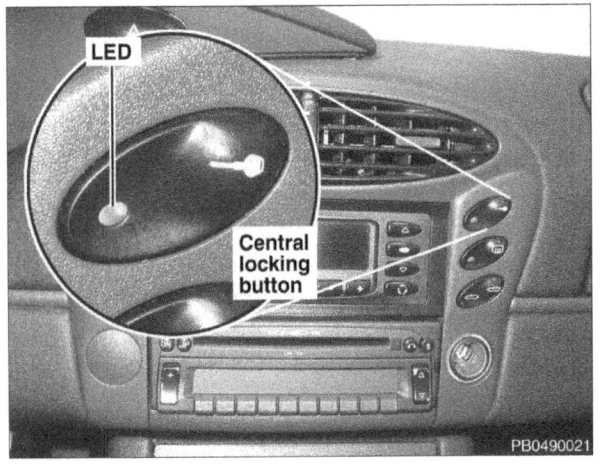

Interior Lights, Anti-theft 96-9

Anti-theft System

Alarm system

When the doors are locked in door-safe mode, the optional alarm system and passenger compartment monitor are activated.

NOTE—
- *Door-safe is the locked mode when the door lock or keyless entry transmitter is activated once. See* **Central locking** *in this repair group.*

When the doors are locked by activating the key or remote transmitter twice, only the alarm is activated. Select this mode if people or animals are in the vehicle. The doors are locked but can be opened from inside using the central locking button.

The alarm monitors the following components:
- Both doors
- Both trunk lids
- Convertible top compartment lid
- Radio
- Center console storage compartment
- Passenger compartment

If the alarm is activated, the following cause it to trigger:
- Activation of ignition switch
- Key with invalid transponder code in ignition
- Radio removal
- Disassembly of alarm horn
- Disassembly of passenger compartment monitor
- Interruption of voltage to alarm control module

The alarm can be triggered 10 seconds after it is activated. The alarm horn, hazard lights and interior lighting sound and flash when the alarm is triggered.

If the passenger door is not closed properly when the doors are locked, the alarm horn sounds briefly. If the driver door is not closed properly, the vehicle cannot be locked.

Alarm LED

 The alarm LED assembly, located in the center top dashboard, also contains the keyless entry system antenna.

The status of the alarm system is indicated by the alarm LED. See **Table b**.

Table b. Alarm LED status	
Alarm status	**LED signal**
Radio signal reception	Flashes 5 times per second.
Door-safe mode (alarm ON, passenger compartment monitor ON)	Flashes 2 times per second for 10 seconds; then flashes once every 2 seconds.
Locked (alarm ON)	Lights for 10 seconds; then flashes once every 2 seconds.

96-10 Interior Lights, Anti-theft

Anti-theft System

Table b. Alarm LED status	
Alarm status	**LED signal**
Emergency (locked but faults in alarm or locking system)	Lights up for 10 seconds; then flashes twice every 2 seconds.
Alarm triggered	Unlit for 10 seconds; then flashes twice every 2 seconds. Brief alarm horn signal.

Passenger compartment monitor

The passenger compartment monitor is an infrared emitter. When the alarm system is activated, the monitor is activated, performs a self-test and calibrates the inside of the passenger compartment within 10 seconds. The sensor then is ready to detect movement inside the vehicle.

If the convertible top is open, the monitor is disabled.

A double horn signal during locking indicates a fault in the passenger compartment monitor. Troubleshoot using PST2.

Electronic immobilizer signal converter, removing and installing

- Disconnect negative (–) battery cable and cover battery terminal to keep cable from accidentally contacting terminal.

> *CAUTION—*
> - *Prior to disconnecting the battery, read the battery disconnection cautions in 00 Warnings and Cautions.*
> - *Be sure to have the radio anti-theft code before disconnecting the battery.*

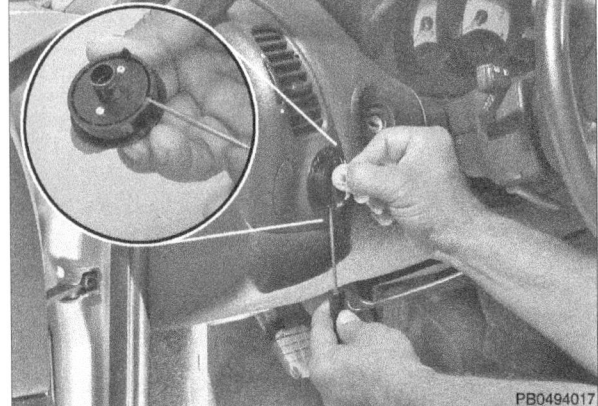

◀ Pull out light switch knob to stop. Use sharp thin tool to press in locking clip on underside of knob to disengage knob from switch.

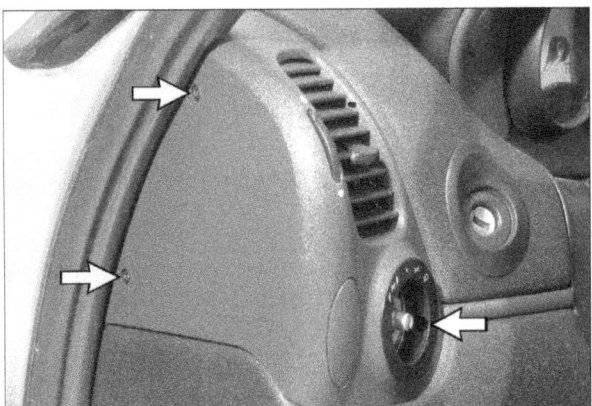

◀ Remove three left side air vent mounting screws (Torx T18) (**arrows**).
- Pull vent and light switch together out of dashboard.
- Disconnect switch electrical connector and set vent aside.

Interior Lights, Anti-theft   96-11

Anti-theft System

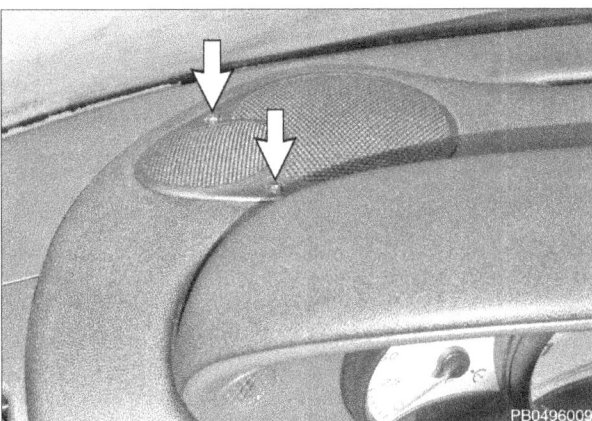

◀ Remove left dashboard speaker mounting screws (**arrows**).

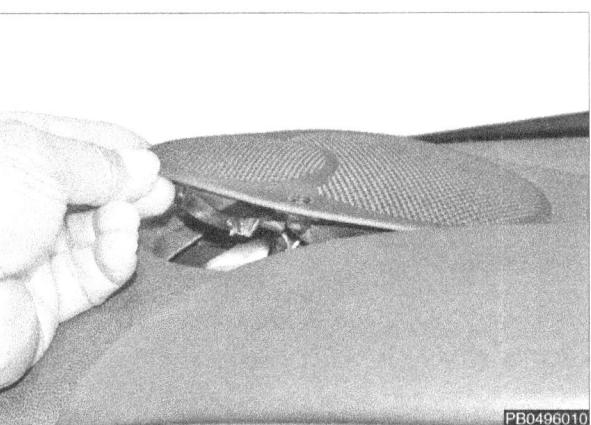

◀ Tilt speaker up from left side to clear securing arrow from under dashboard.

− Detach electrical connector and remove speaker.

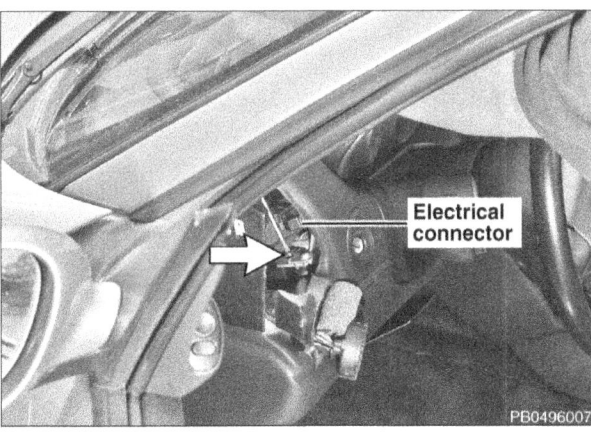

◀ Working through left speaker opening:
- Insert Phillips screwdriver and rotate locking clip (**arrow**) ¼ turn counterclockwise to release signal converter.
- Slide signal converter rearward out of bracket.
- Detach electrical connector and remove converter housing through vent opening.

− Installation is reverse of removal. Make sure electrical harnesses are routed as before.

96-12 Interior Lights, Anti-theft

Anti-theft System

Electronic immobilizer control module, removing and installing

NOTE—
- *There is one combined control module for the electronic immobilizer, central locking and alarm system (if equipped).*
- *The control module is coded to the original vehicle. It cannot be recoded to a new vehicle.*
- *Coding new vehicle keys or new control module to the each other and the vehicle requires the use of Porsche System Tester 2 (PST2).*

– Disconnect negative (–) battery cable and cover battery terminal to keep cable from accidentally contacting terminal.

> **CAUTION—**
> - *Prior to disconnecting the battery, read the battery disconnection cautions in* **00 Warnings and Cautions**.
> - *Be sure to have the radio anti-theft code before disconnecting the battery.*

– Remove left seat. See **72 Seats**.

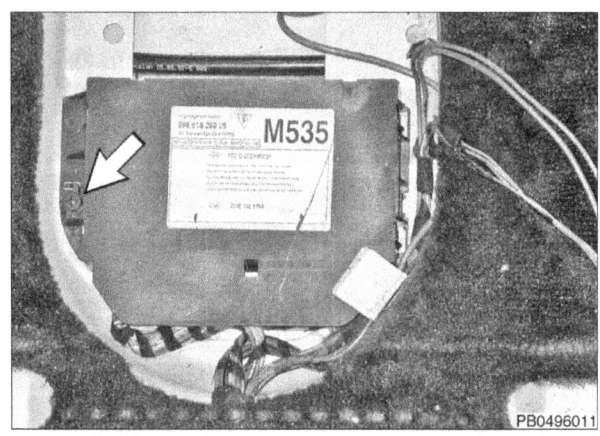

◀ Rotate control module locking clip (**arrow**) ¼ turn counterclockwise to release. Slide control module securing clip off floor anchor and turn module over.

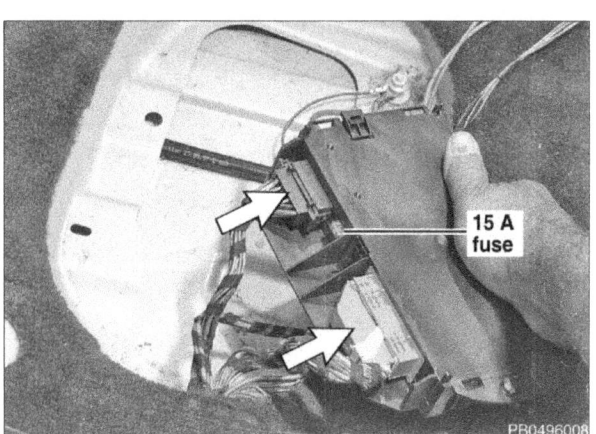

◀ Detach electrical connectors and remove module.

NOTE—
- *If experiencing immobilizer or alarm problems, be sure to inspect the 15 A fuse in the module.*

– Installation is reverse of removal.

97 Fuses, Relays, Component Locations

GENERAL	97-1
Boxster electrical system	97-1
Battery notes	97-2
Electrical system safety precautions	97-2
ELECTRICAL COMPONENT LOCATIONS	**97-4**
Fuses	97-10
Relay panels	97-11
Ground locations	97-12
Engine compartment components	97-14
Front trunk components	97-16
Rear trunk components	97-17
Interior components	97-17
Components under vehicle	97-18
FUSE LOCATIONS AND RATINGS	**97-20**
High amperage fuses, accessing	97-20
High amperage fuse ratings	97-22
Main fuses	97-23
RELAY PANELS	**97-29**
Relay panel 1, accessing	97-29
Relay panel 2, accessing	97-29
GROUNDS	**97-30**

ELECTRICAL TROUBLESHOOTING	97-38
Voltage and voltage drops	97-38
Continuity, checking	97-40
Short circuits	97-40
Short circuit, testing with ohmmeter	97-41
Short circuit, testing with voltmeter	97-41

TABLES

a.	Electrical component locations	97-4
b.	Relay panel 1	97-11
c.	Relay panel 2	97-12
d.	1997 - 2000 ground points	97-12
e.	2001 ground points	97-13
f.	1997 - 1999 high amperage fuses	97-22
g.	2000 high amperage fuses	97-22
h.	2001 high amperage fuses	97-23
i.	1997 main fuse panel	97-24
j.	1998 main fuse panel	97-25
k.	1999 main fuse panel	97-26
l.	2000 main fuse panel	97-27
m.	2001 main fuse panel	97-28
n.	1997 - 2000 ground applications	97-30
o.	2001 ground applications	97-34

GENERAL

This repair group provides a brief description of the electrical system and covers fuses, relays, ground locations and electrical component locations, primarily via photos or illustrations. Also covered here are basic electrical system troubleshooting tips.

Boxster electrical system

Boxster models are equipped with a 12-volt, direct current (dc), negative-ground electrical system. The voltage regulator maintains the voltage in the system at approximately the 12 vdc rating of the battery, and all circuits are grounded by direct or indirect connection to the negative (-) terminal of the battery.

The complex nature of Boxster electrical systems requires a large number of fuses and electrical components. Locating the correct component in this array of equipment is an important first step in electrical diagnosis. Electrical equipment and accessories installed vary depending on model and model year. Always confirm that the proper electrical component has been properly identified.

Fuses, Relays, Component Locations

General

Almost all electrical circuits are protected by fuses. Fuses for individual components are primarily housed in the main fuse panel in the left footwell. For a full listing of fuses and fuse locations, see **Fuse Locations and Ratings** in this repair group.

Investigating and correcting ground problems often clears mysterious and difficult to trace electrical problem. For ground information, see **Ground locations** and **Grounds** in this repair group.

Battery notes

The battery is located under a cover in the front cowl area.

If the battery is disconnected and reconnected, the following systems or components may need to be reset or reinitialized:
- Alarm siren
- Control module fault memories
- ECM adaptation
- Automatic transmission control module adaptation
- Power window motor limit position (standardization)
- Trip odometer and clock
- On-board computer
- Radio
- PCM

See **Battery disconnection notes** in **27 Battery, Alternator, Starter** for procedures necessary to reinitialize components.

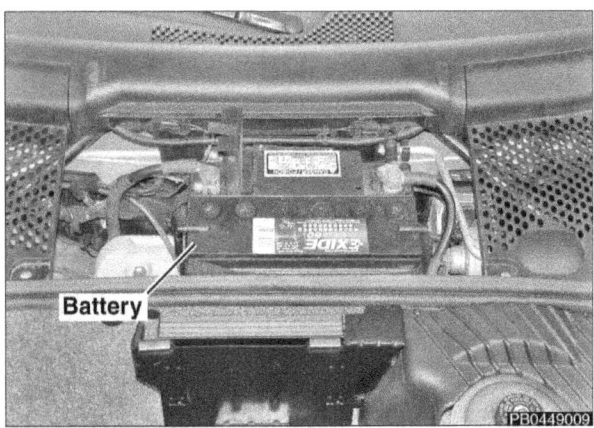

Electrical system safety precautions

Please read the following before doing any work on your electrical system.

> **WARNING—**
> - Airbags utilize explosive devices. Handle with extreme care. Refer to the warnings and cautions in **69 Seat Belts, Airbags**.
> - The ignition system of the car operates at lethal voltages. If you have a weak heart or wear a pacemaker, do not expose yourself to the ignition system electric currents. Take extra precautions when working on the ignition system or when servicing the engine while it is running or the key is ON. See **28 Ignition System** for additional ignition system warnings and cautions.

Fuses, Relays, Component Locations 97-3

General

CAUTION—
- *Before disconnecting the battery, read the battery disconnection cautions in* **00 Warnings and Cautions**.
- *Read out and record control module fault memories before disconnecting the battery.*
- *Record the radio station presets and obtain the radio activation code before disconnecting the battery.*
- *Do not disconnect the battery with the engine running.*
- *Switch the ignition OFF and remove the negative (-) battery cable before removing any electrical components. Connect and disconnect electrical connectors and ignition test equipment leads only while the ignition is switched OFF.*
- *Relay and fuse positions are subject to change and may vary from car to car. If questions arise, an authorized Porsche dealer is the best source for the most accurate and up-to-date information.*
- *Use a digital multimeter for electrical tests. Switch the multimeter to the appropriate function and range before making test connections.*
- *Many control modules are static sensitive. Static discharge damages them permanently. Handle the modules using proper static prevention equipment and techniques.*
- *To avoid damaging harness connectors or relay panel sockets, use jumper wires with flat-blade connectors that are the same size as the connector or relay terminals.*
- *Do not try to start the engine of a car which has been heated above 176°F (80°C) (for example, in a paint drying booth). Allow it to cool to normal temperature.*
- *Disconnect the battery before doing any electric welding on the car.*
- *Do not wash the engine while it is running, or any time the ignition is ON.*
- *Choose test equipment carefully. Use a digital multimeter with at least 10 MΩ input impedance, or an LED test light. An analog meter (swing-needle) or a test light with a normal incandescent bulb may draw enough current to damage sensitive electronic components.*
- *Do not use an ohmmeter to measure resistance on solid state components such as control modules.*
- *Disconnect the battery before making resistance (ohm) measurements on a circuit.*

Electrical Component Locations

ELECTRICAL COMPONENT LOCATIONS

Table a is a cross-referenced listing of Boxster electrical components. Figure numbers in the 3rd column in the table refer to photographs and illustrations immediately following the table.

NOTE—
- *Every component is not installed in every car.*
- *Due to changes in production, component locations may vary from what is illustrated. Consult your Porsche dealer for the latest information.*
- *Component location photos are sourced from a 1998 Boxster with standard transmission, unless noted otherwise.*
- *Porsche refers to the automatic transmission gear position switch as the Tiptronic multifunction switch.*

Table a. Electrical component locations		
Component	Location	See
A/C	see also Heater	
A/C and heater blower	A/C and heater housing under right side of dashboard	Fig. 27
A/C and heater blower relay	Relay panel 1, above left kick panel	Fig. 5
A/C and heater blower resistor (power stage)	Above and left of blower in housing	Fig. 27 Fig. 28
A/C and heater cabin temperature sensor	Right side dashboard	Fig. 27
A/C and heater control module	A/C and heater control panel in dashboard	
A/C and heater flap motors	A/C and heater housing under dashboard	Fig. 28
A/C and heater footwell temperature sensor	Left side of A/C and heater housing	Fig. 28
A/C and heater fresh air temperature sensor	Right side of A/C and heater housing	Fig. 28
A/C compressor relay	Relay panel 2, rear trunk	Fig. 6
A/C pressure switch	In front of air cleaner	Fig. 19
AAV	see Carbon canister shutoff valve	
ABS	see also PSM	
ABS / TC control module	Front trunk, left	Fig. 21
ABS / TC hydraulic unit	Front trunk, left	Fig. 21
ABS wheel speed sensor	Wheel bearing carrier	
Accelerator pedal module	Below knee protection bar near driver feet	
Air bag connector, passenger side	Left side of airbag behind right dashboard	Fig. 27
Airbag connector, driver side	Behind airbag in steering wheel	
Airbag contact spring, driver side	In steering wheel	
Airbag control module	Under center console	Fig. 26
Airbag sensor, side	Door sill, next to seat	
Airbag, driver side	In steering wheel	
Airbag, passenger side	In dashboard	
Airbag, side	In door trim panel	
Alarm system	see also Anti-theft see also Electronic immobilizer	
Alarm system horn	Under front cowl, left of battery	Fig. 19
Alarm system passenger compartment monitor	Top center of windshield	
Alarm system siren	Right side of cowl	
Alternator (generator)	Top right of engine	Fig. 13
Amplifier	Rear of front trunk	Fig. 19 Fig. 21
Antenna amplifier	Right A-pillar, next to outside rear view mirror	

Fuses, Relays, Component Locations 97-5

Electrical Component Locations

Table a. Electrical component locations		
Component	**Location**	**See**
Antenna amplifier, diversity	Navigation computer, center dashboard	
Antenna, GPS	Dashboard center	
Antenna, sound system	Windshield	
Antenna, telephone	Below cowl	
Anti-theft control module	Under left seat	Fig. 10
Ashtray light	In ashtray	
ATF coolant valve	Top center of engine, next to engine lifting eye	
Automatic transmission	*see also* Tiptronic	
Automatic transmission gear position switch	Left side of transmission	
Automatic transmission multifunction switch	*see* Automatic transmission gear position switch	
Automatic transmission selector lever switch	At transmission gear selector lever	
AY sensor	*see* PSM sensors, Rotation rate sensor	
Back-up light switch, automatic transmission	*see* Automatic transmission gear position switch	
Back-up light switch, manual transmission, Boxster	Top of transmission	Fig. 32
Back-up light switch, manual transmission, Boxster S	Rear of transmission	Fig. 33
Blower	*see* A/C and heater blower *see also* Engine compartment blower	
Blower power stage	*see* A/C and heater blower resistor (power stage)	
Brake fluid level sensor	Lower right of brake fluid reservoir	Fig. 21
Brake light switch	Below dashboard on pedal cluster	Fig. 25
Bridge plug	Relay panel 1, above left kick panel	Fig. 5
Camshaft sensor, cylinder 1 - 3 (right)	Rear of right cylinder head	
Camshaft sensor, cylinder 4 - 6 (left)	Front of left cylinder head	Fig. 17
Carbon canister shutoff valve	Top of carbon canister, right front wheel housing under liner	Fig. 31
CD changer	Front trunk right	Fig. 22
Central locking button	Center dashboard	
Central locking control module	Under left seat	Fig. 10
Child seat recognition sensor	Child seat anchor	
Clutch interlock switch (starter immobilizer)	Above clutch pedal	Fig. 25
Clutch switch (cruise control)	Above clutch pedal	
Convertible top closed microswitch	1997 - 1999: Left B-pillar 2000 - 2004: Right convertible top transmission	
Convertible top control module	Relay panel 1, above left kick panel	Fig. 5
Convertible top drive motor	Center rear of convertible top compartment	
Convertible top lock microswitch	Top center of windshield frame	
Convertible top storage lid microswitch	At convertible top motor, center rear of convertible top compartment	
Convertible top switch	Center dashboard	
Cooling fan	*see* Engine cooling fan	
Crankshaft sensor	Right side of engine block at bellhousing	Fig. 16
Cruise control indicator	Instrument cluster	
Cruise control module	Above left footwell air duct	
Cruise control switch	Steering column stalk switch, right lower	
Current distributor	On bulkhead in front of right seat	Fig. 1
Daytime running lights double relay	Relay panel 1, above left kick panel	Fig. 5
Defrost flap motor	A/C and heater housing under dashboard	Fig. 28
Diagnosis plug	*see* OBD II plug	
Differential pressure sensor	Above fuel tank at tank locking collar	Fig. 20
DME control module	Rear trunk, left front	Fig. 23

97-6 Fuses, Relays, Component Locations

Electrical Component Locations

Table a. Electrical component locations		
Component	Location	See
DME main relay	Relay panel 2, rear trunk	Fig. 6
Door courtesy light	In door trim panel	
Door handle light, inner	LEDs in door panel	
Door lock	In door	
Door lock microswitch, inside	Inside door panel at inner door lock release	
Door lock microswitch, outside	In door on outside door handle	
Drive block	see Electronic immobilizer	
DSP amplifier	see Amplifier	
E-gas throttle housing	Rear of intake manifold	Fig. 14
Electronic immobilizer control module	Under left seat	Fig. 10
Electronic immobilizer signal transformer	On ignition switch	
Electronic immobilizer transponder	Around ignition lock cylinder	
Emergency trunk lid release jumper	Main fuse panel, behind cover in left footwell kick panel	
Engine compartment blower	Right side engine compartment	Fig. 14
Engine compartment blower relay	Relay panel 2, rear trunk	Fig. 6
Engine compartment temperature sensor	Top right of engine, between intake runners 1 and 2	Fig. 13
Engine coolant level sensor	Bottom of coolant reservoir	Fig. 23
Engine coolant temperature sensor	Left front of engine	
Engine cooling fan relays	Relay panel 1, above left kick panel	Fig. 5
Engine cooling fan resistor	On cooling fan harness behind and below fan	Fig. 30
Engine cooling fans	At radiators in front wheel housings	Fig. 30
Engine speed sensor	see Crankshaft sensor	
Engine temperature sensor	see Engine compartment temperature sensor	
FIO (fuel injector / ignition coil / oxygen sensor heater) relay	Relay panel 2, rear trunk	Fig. 6
Flasher	see Hazard warning flasher	
Foglight relay	Relay panel 1, above left kick panel	Fig. 5
Footwell flap motor	A/C and heater housing under dashboard	Fig. 28
Fresh air flap motor	A/C and heater housing under dashboard	Fig. 28
Fuel injector relay	see FIO (fuel injector / ignition coil / oxygen sensor heater) relay	
Fuel injector relay	Relay panel 2, rear trunk	Fig. 6
Fuel injectors cyl. 1 - 3 (right)	Right cylinder head	
Fuel injectors cyl. 4 - 6 (left)	Left cylinder head	
Fuel level sender	In fuel tank, under battery	
Fuel pump	In fuel tank, under battery	
Fuel pump relay	Relay panel 1, above left kick panel	Fig. 5
Fuel tank evaporative vent valve (EVAP purge valve)	Top of engine under left intake manifold	Fig. 15
Fuel tank filler door lock actuator	Underneath rear of right fender, behind right front wheel housing liner	
Fuel tank filler neck ground connection	Top of filler neck, behind right front wheel housing liner	Fig. 31
Fuel tank filling detection sensor	see ORVR reed valve	
Fuse panel, additional	Relay panel 2, rear trunk	Fig. 6
Fuse panel, high amperage fuses	see Current distributor	
Fuse panel, main	Behind cover in left footwell kick panel	Fig. 2
Fuses	see **Fuse Locations and Ratings** in this repair group	Fig. 1 Fig. 2 Fig. 3 Fig. 4
Generator	see Alternator (generator)	
GPS	see Navigation computer (PCM)	

Fuses, Relays, Component Locations

Electrical Component Locations

Table a. Electrical component locations

Component	Location	See
Grounds	see **Grounds** in this repair group	Fig. 7 Fig. 8
Hazard warning flasher	Relay panel 1, above left kick panel	Fig. 5
Headlight adjuster control module	Under dashboard, right	
Headlight adjuster control sensor	Left front and left rear control arm	
Headlight adjuster stepper motor	At headlight	
Headlight washer pump	Rear of washer fluid reservoir	
Headlight washer relay	Relay panel 1, above left kick panel	Fig. 5
Heater	see also A/C	
Heater blower	Under dashboard right	Fig. 27
Horn button	In driver airbag	
Horn relay	Relay panel 1, above left kick panel	Fig. 5
Horns	Right front, behind front bumper	
Idle control valve (1997 - 1999)	Left of center intake manifold	Fig. 13 Fig. 15
Ignition coil relay	see FIO (fuel injector / ignition coil / oxygen sensor heater) relay	
Ignition coils cyl. 1 - 3 (right)	Right cylinder head cover	
Ignition coils cyl. 4 - 6 (left)	Left cylinder head cover	
Infosystem	see Navigation system (PCM)	
Infrared sensor	see Alarm system passenger compartment monitor	
Intake air temperature (IAT) sensor	In mass air flow sensor	
Intake runner change-over valve	see Intake manifold resonance valve	
Interior motion detector	see Alarm system passenger compartment monitor	
Interior temperature sensor	see A/C and heater cabin temperature sensor	
Kick-down switch (automatic transmission)	Under accelerator pedal	
Klimatronic	see A/C and heater	
Knock sensors	On top of engine block	Fig. 17 Fig. 18
Lateral acceleration sensor	see PSM sensors, Rotation rate sensor	
Lifting switch	see Automatic transmission gear position switch	
Light switch assembly	left of steering wheel on dashboard	
Lighter	Below center dashboard	
Lights ON warning buzzer	Instrument cluster	
Litronic actuator	see Headlight adjuster stepper motor	
Litronic control module	Under left dashboard	
Litronic ignition module	Behind each headlight	
Luggage compartment	see Trunk	
Mass air flow sensor	Left of engine compartment, in air cleaner duct	
Maxi fuse	Relay panel 2, rear trunk	Fig. 6
MFI-DI	see DME	
Microphone, hands-free	Left of instrument cluster	
Microphone, noise compensation	Left A-pillar, next to outside rear view mirror	
Navigation computer (PCM)	Center dashboard	
Neutral safety switch	see Automatic transmission gear position switch	
OBD II plug	Under left of dashboard	
Oil level sensor	Connector at top of crankcase	Fig. 17
Oil pressure sensor	Top of right cylinder head	
Oil temperature sensor	Connector at top of crankcase	Fig. 17
On-board computer control	Steering column stalk switch, left lower	
ORVR reed valve	Fuel tank filler neck, behind right front wheel housing liner	

Electrical Component Locations

Table a. Electrical component locations

Component	Location	See
Outside rear view mirror	Left or right door	
Outside rear view mirror heater	In mirror	
Outside rear view mirror heater relay	Relay panel 1, above left kick panel	Fig. 5
Outside rear view mirror memory module	see Seat position and mirror memory module	
Outside temperature sensor	Right of front bumper	
Oxygen sensor heater relay	see FIO (fuel injector / ignition coil / oxygen sensor heater) relay	
Oxygen sensor test harness	Relay panel 2, rear trunk	Fig. 6
Oxygen sensor, post-catalyst (1997 - 1999)	Back of catalyst	
Oxygen sensor, post-catalyst (2000 - 2004)	In exhaust pipe after front catalyst	
Oxygen sensor, precatalyst (1997 -1999)	In exhaust pipe before catalyst	
Oxygen sensor, precatalyst (2000 - 2004)	Front of front catalyst	
Parking assistant control module	Under left seat	
Parking assistant sensors	In rear bumper	
PCM	see Navigation computer (PCM)	
Pressure switch	see A/C	
PSM	see also ABS	
PSM brake fluid pressure sensor	On PSM control module and hydraulic unit	
PSM control module and hydraulic unit	Left side of front trunk	Fig. 21
PSM precharge pump	Right side of front trunk	Fig. 22
PSM sensors • Rotation rate sensor • Steering angle sensor	 Under center console Steering column	
PSM switch	Dashboard center	
Rain sensor	Top center of windshield	
Rain sensor control module	Left cowl, under front trunk lid	
Rear window defogger relay	Relay panel 1, above left kick panel	Fig. 5
Recirculation flap motor	A/C and heater housing under dashboard	Fig. 28
Relay panel 1	Above left kick panel	Fig. 5
Relay panel 2	Rear trunk, left side	Fig. 6
Rotation rate sensor	see PSM	
Seat belt lock microswitch	Seat belt lock	
Seat position and mirror memory module	Under driver seat	
Seat position memory switch	Left door sill	
Seat position motor	In seat	
Secondary air pump	Top right engine compartment	Fig. 12
Secondary air pump fuse	Relay panel 2, rear trunk	Fig. 3 Fig. 6
Secondary air pump relay	Relay panel 2, rear trunk	Fig. 6
Secondary air valve	Right intake manifold, between cyl. 1 and cyl. 2 runners	Fig. 12
Signal transformer	see Electronic immobilizer signal transformer	
Siren	see Alarm system siren	
Speakers, bass reflex	In doors	
Speakers, bass reflex (2003)	Behind seats	
Speakers, midrange	Left and right of dashboard	
Speakers, midrange (2003)	Behind seats	
Speakers, midrange (2003)	In doors	
Speakers, tweeter	Left and right of dashboard	
Spiral spring	see Airbag contact spring, driver	
Spoiler motor	Behind cover under rear trunk lock	
Spoiler relays	Relay panel 2, rear trunk	Fig. 6

Fuses, Relays, Component Locations 97-9

Electrical Component Locations

Table a. Electrical component locations

Component	Location	See
Spoiler switch	Main fuse panel, left footwell kick panel	Fig. 2
Starter	Above engine at bellhousing	
Starter relay	Relay panel 2, rear trunk	Fig. 6
Steering angle sensor	see PSM sensors	
Stop light switch	see Brake light switch	
Sun sensor	Center dashboard	
TC	see ABS / TC	
Telephone speaker changeover relay	Relay panel 1, above left kick panel	Fig. 5
Temperature control flap motor	see A/C and heater flap motors	
Terminal X relay	Relay panel 1, above left kick panel	Fig. 5
Throttle housing	Rear of intake manifold	Fig. 14
Throttle position sensor	In throttle valve housing, rear of intake manifold	Fig. 15
Tilt sensor	see Alarm system tilt sensor	
Tiptronic control module	Front of rear trunk	
Tiptronic gear selector	see Automatic transmission gear position switch	
Tiptronic paddle switch	On steering wheel	
Traction control	see ABS / TC	
Transponder coil	see Electronic immobilizer transponder	
Trunk light, front	Left side, front trunk	
Trunk light, rear	Front trim, rear trunk	
Trunk lock microswitch, front	Front trunk lock	
Trunk lock microswitch, rear	Rear trunk lock	Fig. 24
Turn signal / high beam switch	Steering column stalk switch, left upper	
Ultrasonic sensor	see Parking assistant sensor	
Vehicle ride height sensor	see Headlight adjuster control sensor	
Washer fluid level switch	Washer fluid reservoir, behind left front wheel housing liner	Fig. 29
Wheel speed sensor	see ABS wheel speed sensor	
Window motor	In door	
Window motor switch	Center console	
Windshield washer fluid pump	Rear of washer fluid reservoir, behind left front wheel housing liner	Fig. 19
Windshield washer nozzle, heated	Under cowl	Fig. 19
Windshield wiper motor	Under left cowl	Fig. 19
Wiper / washer switch	Steering column stalk switch, right upper	
Wiper intermittent relay	Relay panel 1, above left kick panel	Fig. 5
Xenon headlights	see Litronic	
Yaw sensor	see PSM sensors, Rotation rate sensor	

97-10 Fuses, Relays, Component Locations

Electrical Component Locations

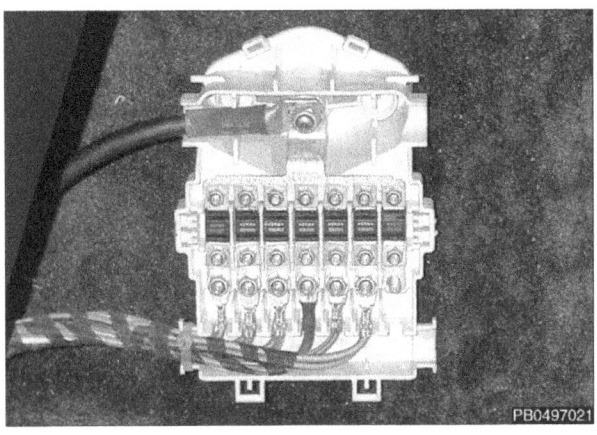

Fuses

Fig. 1 Current distributor

High on bulkhead in front of right seat.
- For access information, see **High amperage fuses, accessing** in this repair group.
- For high amperage fuse (fusible link) ratings, see **High amperage fuse ratings** in this repair group.

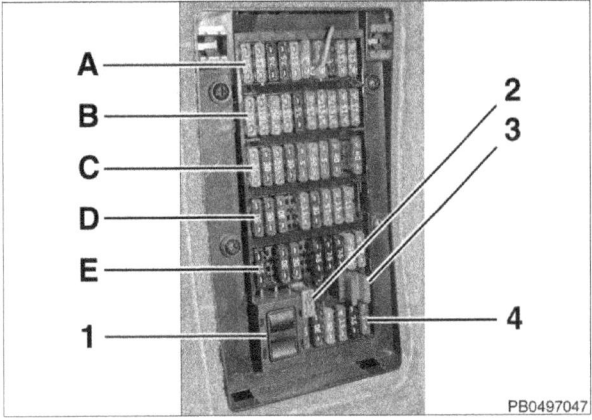

Fig. 2 Main fuse panel

Behind cover in left footwell kick panel. For fuse ratings in rows **A**, **B**, **C**, **D** and **E**, see **Main fuses** in this repair group.

1. Spoiler switch
2. Emergency trunk lid release
3. Fuse puller
4. Spare fuses

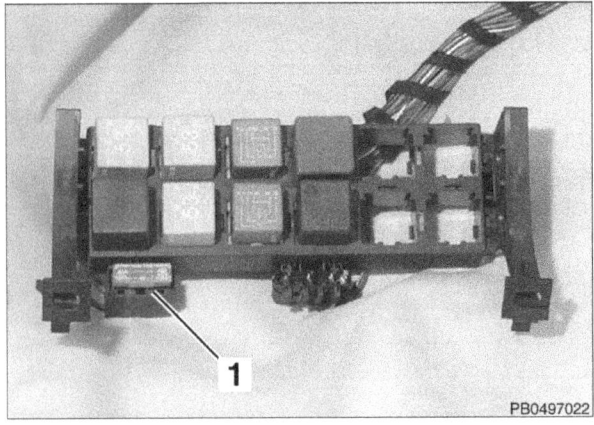

Fig. 3 Secondary air pump fuse

Relay panel 2, rear trunk, behind left trim. To gain access, see **Relay panel 2, accessing** in this repair group.

1. Secondary air pump fuse (maxi fuse), 40 A rating

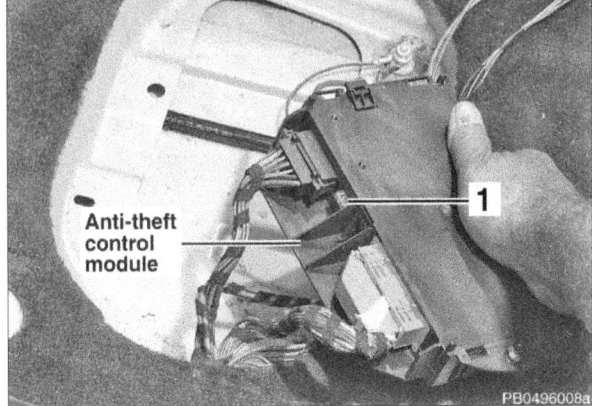

Fig. 4 Anti-theft control module fuse

Under left seat.

1. Anti-theft system fuse, 15 A rating

Fuses, Relays, Component Locations

Electrical Component Locations

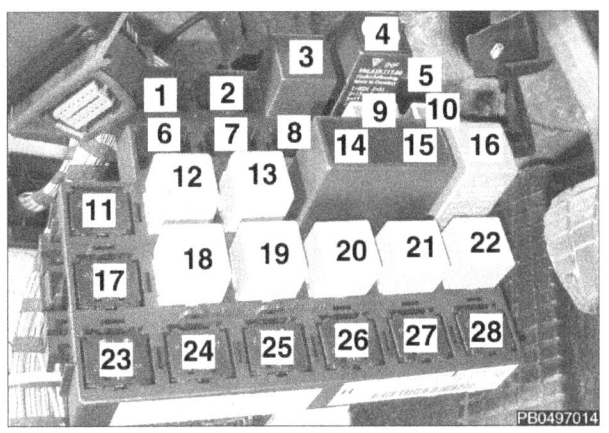

Relay panels

CAUTION—
* *Relay positions are subject to change and may vary from car to car. If questions arise, an authorized Porsche dealer is the best source for the most accurate and up-to-date information.*

Fig. 5   Relay panel 1

 Main relay and bridge plugs above left footwell kick panel. To gain access to panel, see **Relay panel 1, accessing** in this repair group.

Table b. Relay panel 1

Position	Function
1	Not used
2	Not used
3	Hazard warning flasher
4	Outside rear view mirror heater Rear window defogger (hardtop)
5	1997 - 1998: Telephone speaker changeover relay 1999 - 2001: Not used
6	Daytime running lights
7	Daytime running lights
8	Headlight washer
9	Terminal XE
10	Horns

11	Bridge plugs:
	   Terminal 15 or 31
	   Terminal 58d

12	Foglights (USA, Japan)
13	Fuel pump
14	Convertible top control module
15	
16	Wiper intermittent control

17	Bridge plugs:
	   Terminal 86S
	   Terminal X, wipers
	   Terminal 15 brake light fuse input

18	A/C and heater system

Position	Function
19	Engine cooling fan, left, low speed
20	Engine cooling fan, left, high speed
21	Engine cooling fan, left, high speed
22	Engine cooling fan, right, high speed

23	Bridge plugs:
	   Speedometer terminal A
	   Parking brake signal
	   Terminal 31 electronics ground
	   Terminal 15

24	Bridge plugs:
	   Foglight
	   Terminal 56 right
	   Telephone, cell phone mute
	   Terminal TN

25	Bridge plugs:
	   Terminal 56a
	   Terminal 54 brake lights
	   Parking brake warning
	   Terminal 31d

26	Bridge plugs:
	   Consumer cut-off
	   K2 lead

27	Bridge plugs
	   Terminal 30
	   Terminal 58

28	Bridge plug: Terminal 31

97-12 Fuses, Relays, Component Locations

Electrical Component Locations

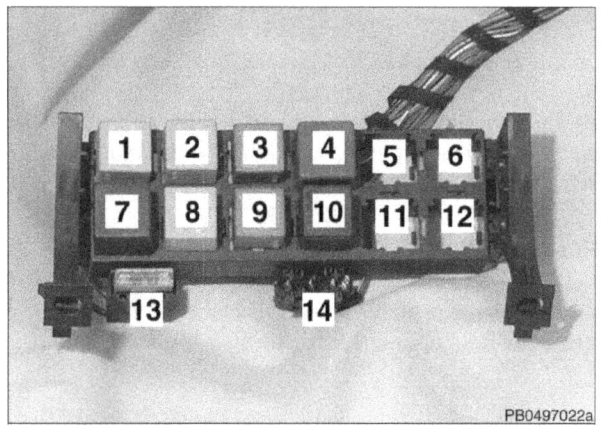

Fig. 6 Relay panel 2

◀ In rear trunk. To gain access to panel, see **Relay panel 2, accessing** in this repair group.

Table c. Relay panel 2	
Position	Function
1	DME main relay
2	Fuel injectors Ignition coils Oxygen sensor heaters
3	Spoiler extension
4	A/C compressor
5	Not used
6	1997, 1999: not used 1998, 2001: Back-up lights jumpers
7	Starter
8	Engine compartment blower
9	Spoiler retraction
10	1997: Spoiler retraction 1998 - 2001: Secondary air pump
11	1997: Secondary air pump 1998 - 2001: Not used
12	Not used
13	Secondary air pump fuse (40 A) (maxi fuse)
14	Oxygen sensor test harness

Ground locations

Fig. 7 1997 - 2000 grounds

◀ Grounds are found throughout the body. Ground applications by circuit are in **Table n**.

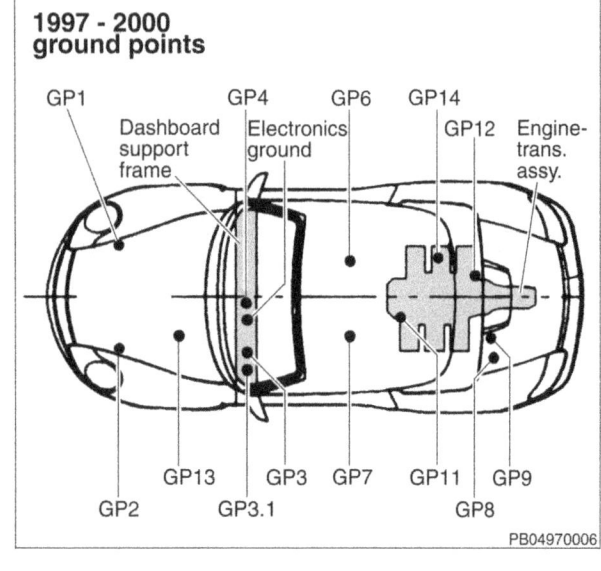

Table d. 1997 - 2000 ground points		
No.	Location	See
GP1	Right front inner fender (front trunk)	Fig. 9
GP2	Left front inner fender (front trunk)	Fig. 9
GP3	Above steering column on dashboard support frame	
GP3.1	Left dashboard on support frame	
GP4	Center dashboard on support frame	
GP6	Under right seat	
GP7	Under left seat	Fig. 10
GP8	Left front of rear trunk	Fig. 11
GP9	Left front of rear trunk	Fig. 11
GP11	Front of left cylinder head	
GP12	Rear of right cylinder head	Fig. 16
GP13	Under cowl, left of battery	
GP14	Right side engine compartment	Fig. 12
Electronics	On dashboard support frame	

Fuses, Relays, Component Locations 97-13
Electrical Component Locations

Fig. 8 2001 grounds

◀ Grounds are found throughout the body. Ground applications by circuit are in **Table o**.

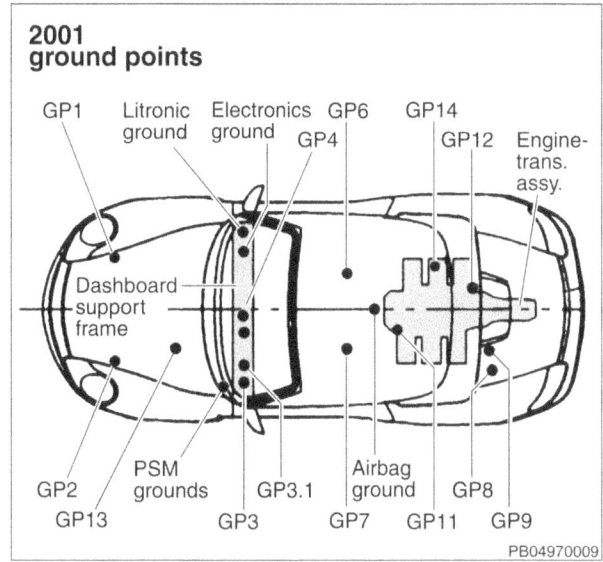

Table e. 2001 ground points		
No.	Location	See
GP1	Right front inner fender (front trunk)	Fig. 9
GP2	Left front inner fender (front trunk)	Fig. 9
GP3	Above steering column on dashboard support frame	
GP3.1	Left dashboard on support frame	
GP4	Center dashboard on support frame	
GP6	Under right seat	
GP7	Under left seat	Fig. 10
GP8	Left front of rear trunk	Fig. 11
GP9	Left front of rear trunk	Fig. 11
GP11	Front of left cylinder head	
GP12	Rear of right cylinder head	Fig. 16
GP13	Under cowl, left of battery	
GP14	Right side engine compartment	Fig. 12
Airbag	Between seats under center console	
Electronics	Right on dashboard support frame	
Litronic	Right on dashboard support frame	
PSM	Left of front trunk	Fig. 21

Fig. 9 Front trunk

◀ Behind left and right front trunk trim
1. GP1
2. GP2

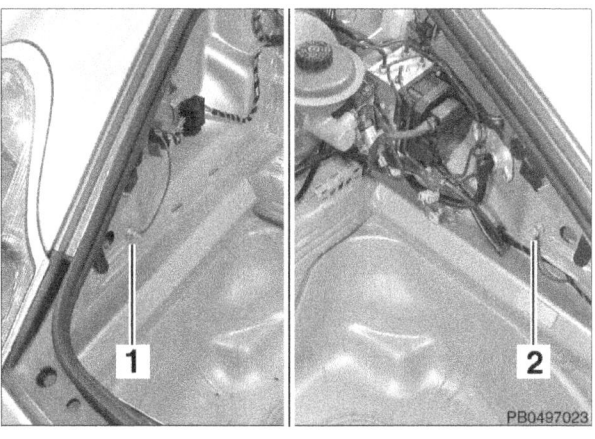

Fig. 10 Under left seat

◀ Take out left seat.
1. GP7
2. Anti-theft control module (electronic immobilizer, central locking, alarm system)

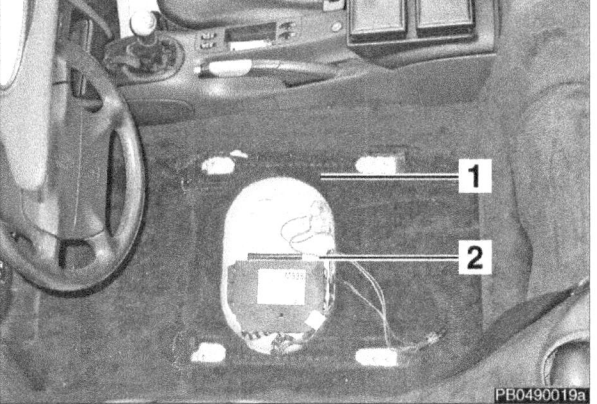

97-14 Fuses, Relays, Component Locations

Electrical Component Locations

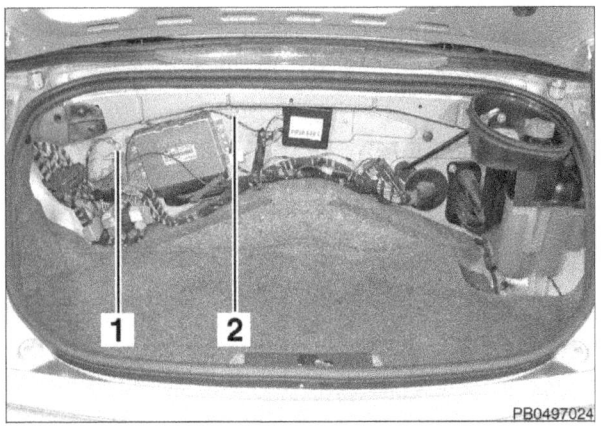

Fig. 11 Rear trunk

◄ Behind front trim panel on bulkhead
1. GP8
2. GP9

Fig. 12 Right side engine compartment

◄ At secondary air pump mounting
1. GP14
2. Secondary air pump valve
3. Secondary air pump

Engine compartment components

Fig. 13 Engine compartment, 1997 Boxster

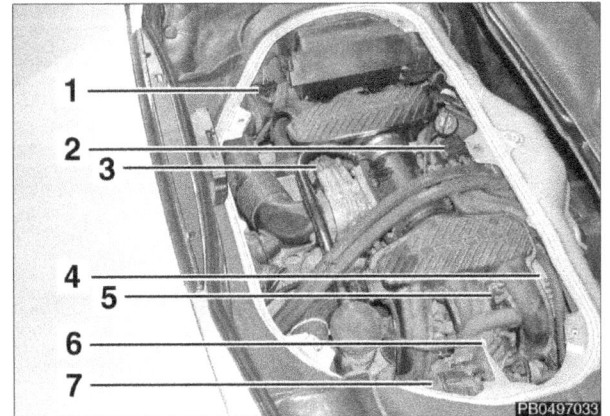

◄ Overview, top of engine
1. Mass air flow sensor
2. B+ junction box
3. Idle control valve
4. Alternator
5. Engine compartment temperature sensor
6. Engine to body ground strap
7. Secondary air pump

Fig. 14 Engine compartment, 2001 Boxster S

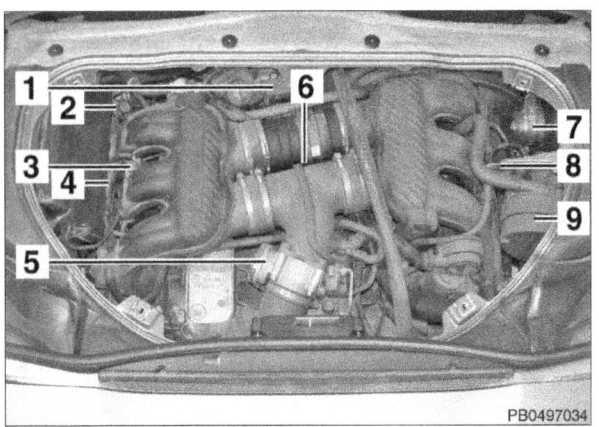

◄ Overview, top of engine
1. B+ junction box
2. Fuel pressure regulator
3. Fuel tank evaporative valve
4. A/C compressor clutch connector
5. E-gas throttle housing
6. Intake manifold resonance valve
7. Engine compartment blower
8. Engine to body ground strap
9. Secondary air pump

Fuses, Relays, Component Locations   97-15

Electrical Component Locations

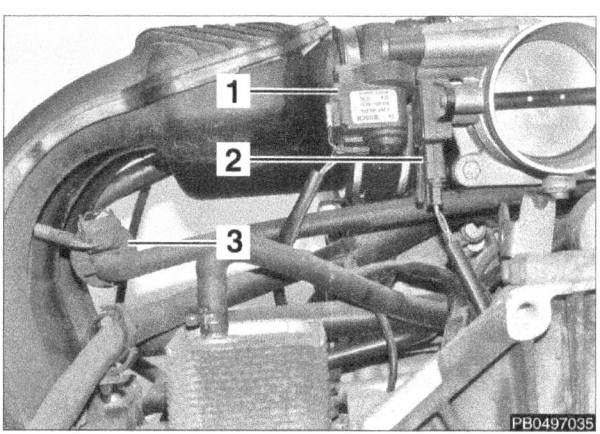

Fig. 15   Engine left side

◀ Under left intake manifold
1. Idle control valve
2. Throttle position sensor
3. Fuel tank evaporative vent valve (EVAP purge valve)

Fig. 16   Engine right side

◀ Behind right cylinder head
1. Crankshaft sensor
2. Right VarioCam harness connector (cylinders 1 -3)
3. Crankshaft sensor connector
4. GP12

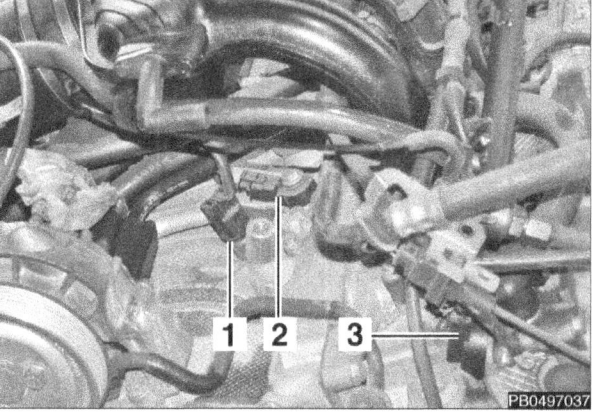

Fig. 17   Engine crankcase top left

◀ Under left intake manifold, behind A/C compressor
1. Engine oil level and oil temperature sensor
2. Left knock sensor (cylinders 4 - 6)
3. Left camshaft sensor (cylinder 4 - 6)

Fig. 18   Engine crankcase top right

◀ Under right intake manifold, behind alternator
1. Secondary air valve
2. Right knock sensor (cylinders 1 - 3)

97-16 Fuses, Relays, Component Locations

Electrical Component Locations

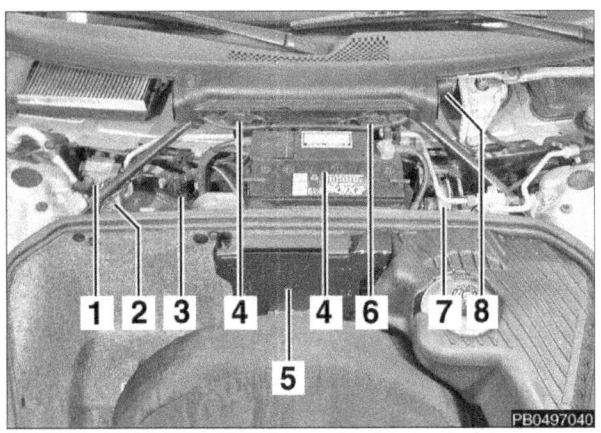

Front trunk components

Fig. 19 Under cowl

◄ Battery cover and left and right cowl covers removed
1. A/C evaporator valve
2. A/C pressure switch
3. Current distributor battery connection
4. Windshield washer nozzle heater harness connector
5. Sound system amplifier
6. Battery
7. Alarm horn
8. Windshield wiper motor

Fig. 20 Under battery tray

◄ Battery and battery tray removed
1. Fuel tank differential pressure sensor (EVAP system)
2. Fuel pump and fuel level sensor connector

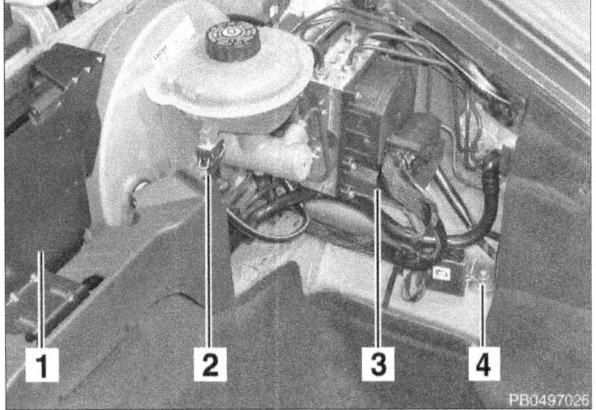

Fig. 21 Front trunk left side, 2001 Boxster S

◄ Behind plastic trim
1. Sound system amplifier
2. Brake fluid level sensor
3. PSM control module and hydraulic unit
4. PSM ground

Fig. 22 Front trunk right side, 2001 Boxster S

◄ Optional equipment
1. PSM precharge pump
2. 6-cd changer

Fuses, Relays, Component Locations 97-17

Electrical Component Locations

Rear trunk components

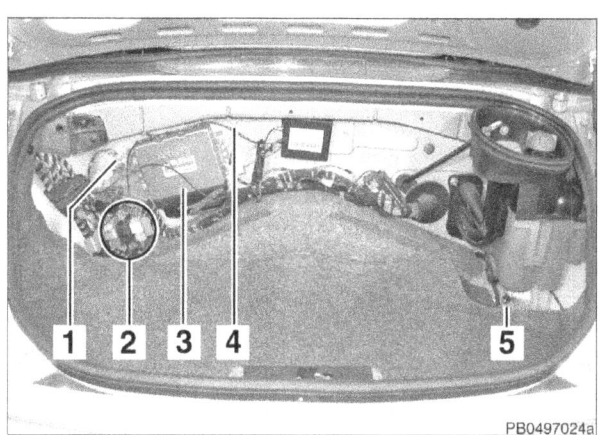

Fig. 23 Rear trunk front bulkhead

◄ Behind trim panel
1. GP8
2. Engine harness connectors
3. Engine control module (ECM)
4. GP9
5. Coolant level switch

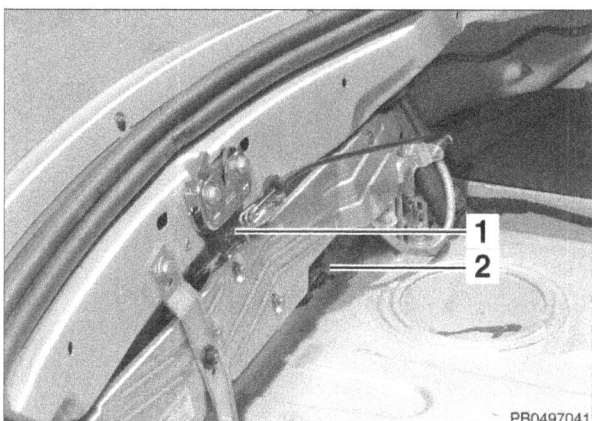

Fig. 24 Rear trunk sill

◄ Behind trim panels
1. Rear trunk lock microswitch
2. Spoiler motor

Interior components

Fig. 25 Under dashboard, left side

◄ Above pedal cluster
1. Clutch interlock switch (starter immobilizer)
2. Brake light switch

Fig. 26 Under front of center console

◄ Remove CD holder, storage bin and surrounding trim.
1. Airbag (SRS) crash sensor and control module

97-18 Fuses, Relays, Component Locations

Electrical Component Locations

Fig. 27 Dashboard, right side

◄ Some components behind trim
1. Passenger airbag connector
2. A/C and heater blower resistor (power stage)
3. A/C and heater blower
4. Right dashboard speaker
5. A/C and heater cabin interior temperature sensor

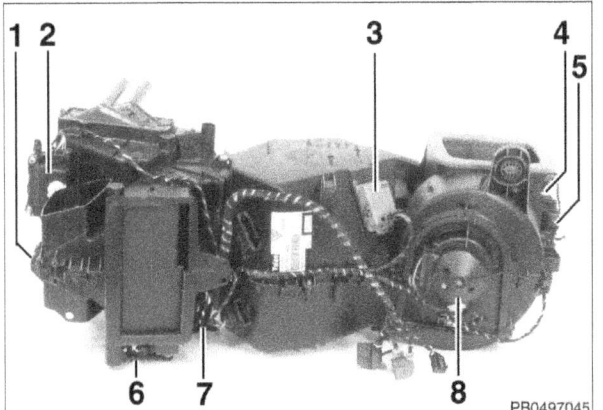

Fig. 28 A/C and heater housing

◄ Under dashboard
1. Footwell air temperature sensor
2. Footwell / defrost flap motor
3. A/C and heater blower resistor (power stage)
4. Fresh air temperature sensor
5. Fresh air / recirculation flap motor
6. Center flap motor (middle and side air vents)
7. Temperature control flap motor
8. Blower

Components under vehicle

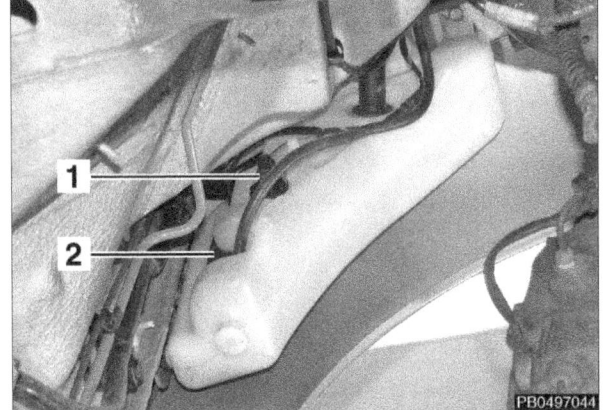

Fig. 29 Under left front wheel housing liner, rear

◄ Washer fluid reservoir
1. Washer fluid level sensor
2. Windshield washer pump

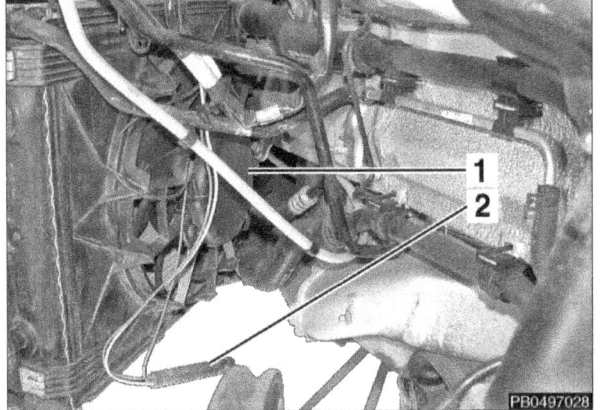

Fig. 30 Under left front wheel housing liner, front

◄ Radiator cooling fan assembly, left
1. Cooling fan
2. Cooling fan resistor

Fuses, Relays, Component Locations 97-19
Electrical Component Locations

Fig. 31 Under right front wheel housing liner, rear

◀ At fuel tank filler neck
1. Carbon canister shutoff valve
2. Fuel tank filler neck ground wire
3. Tank purge valve

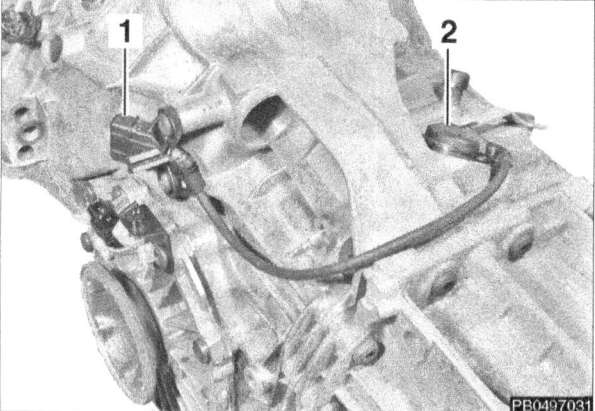

Fig. 32 Boxster, 5-speed manual transmission top

◀ Top of transmission
1. Back-up light switch connector
2. Back-up light switch

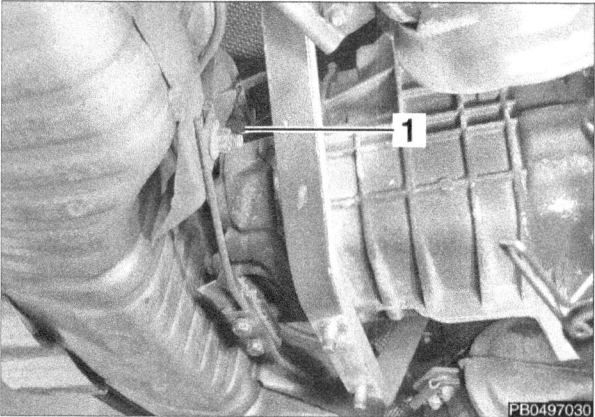

Fig. 33 Boxster S, 6-speed manual transmission top

◀ Rear top of transmission
1. Back-up light switch

97-20 Fuses, Relays, Component Locations

Fuse Locations and Ratings

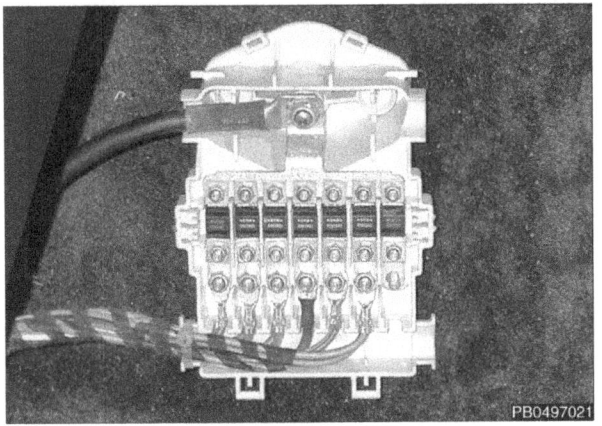

FUSE LOCATIONS AND RATINGS

High amperage fuses, accessing

◄ High amperage fusible links in the current distributor, located high on bulkhead in front of right seat, divide on-board networks and connect each directly to the battery. The benefits of this system are:
- Reduction of alternator current pulses
- Protection against two-way interference
- Short-circuit protection in main power circuits
- Isolation of lighting circuits from each other

> **CAUTION—**
> - *Porsche specifies that no work on the current distributor fuses may be performed except by an authorized Porsche dealer repair department.*
> - *A sealing label is attached to the current distributor cover by the manufacturer. If the cover is removed, attach a special service label to the cover upon completion of the work.*

Fusible link applications are in **Table f**, **Table g** and **Table h**.

To gain access to current distributor fusible links, proceed as follows:

– Disconnect negative (–) battery cable and cover battery terminal to keep cable from accidentally contacting terminal.

> **CAUTION—**
> - *Prior to disconnecting the battery, read the battery disconnection cautions in* **00 Warnings and Cautions**.
> - *Be sure to have the radio anti-theft code before disconnecting the battery.*

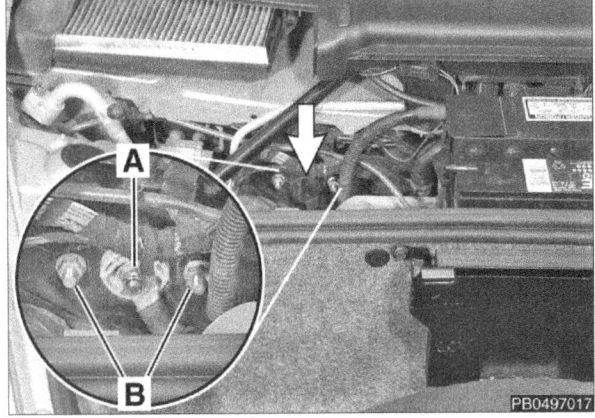

◄ Working in cowl area to right of battery:
- Remove positive cable connection plastic cover (**arrow**) at bulkhead.
- Detach positive cable (**A**) from current distributor.
- Remove current distributor mounting nuts (**B**).

Fuses, Relays, Component Locations 97-21

Fuse Locations and Ratings

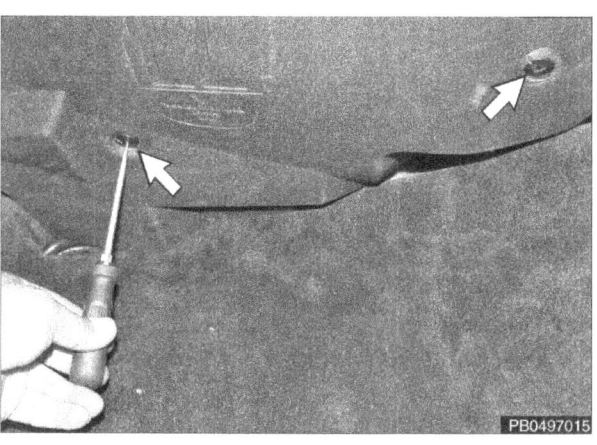

◄ Working in passenger compartment underneath right side of dashboard, unscrew upper footwell trim panel plastic fasteners (**arrows**). Pull down trim.

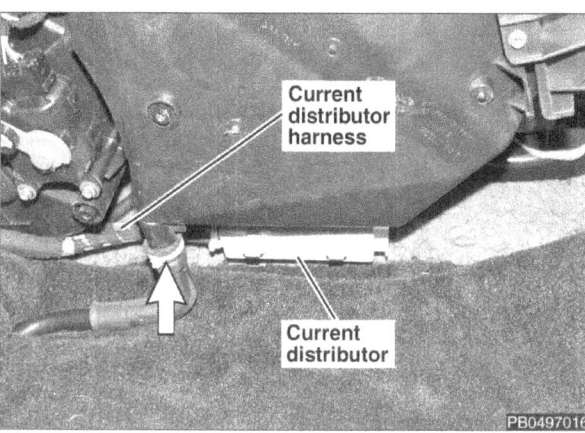

◄ Loosen or cut clamp (**arrow**) at A/C evaporator drain hose and detach hose. This allows current distributor harness to clear A/C and heater housing.

◄ Lower current distributor. Unclip cover at corners (**arrows**) and swing open.

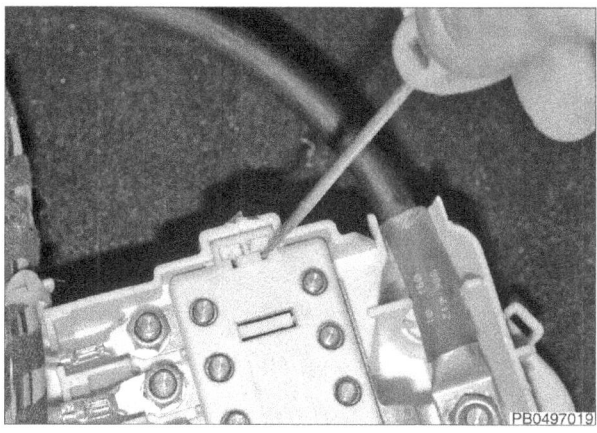

◄ Remove fusible link inner cover: Insert sharp tool in edge slots and unclip locking tabs.

97-22 Fuses, Relays, Component Locations

Fuse Locations and Ratings

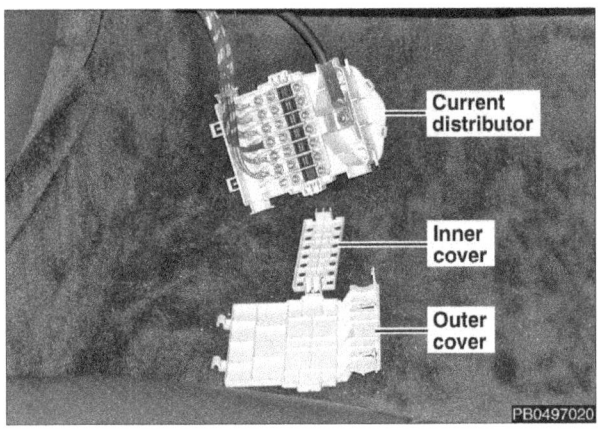

◀ Replace fusible links as necessary. New fusible links from Porsche include new M5 mounting nuts.

Tightening torque	
Fusible link to current distributor (M5)	4 Nm (3 ft-lb)

- When reassembling, replace inner cover with new cover from Porsche. Install outer cover and affix two service labels from Porsche.

- Check bulkhead seal and replace if necessary. Install current distributor and reattach battery positive cable using new self-locking nuts.

Tightening torques	
Current distributor to bulkhead (M8) (use new self-locking nuts)	15 Nm (11 ft-lb)
Positive cable to current distributor (M8) (use new self-locking nut)	15 Nm (11 ft-lb)

High amperage fuse ratings

Fusible links shown in **Table f**, **Table g** and **Table h** cover model years 1997 to 2001.

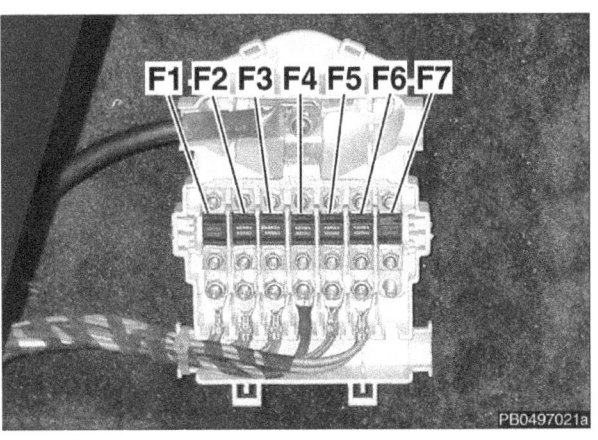

Table f. 1997 - 1999 high amperage fuses		
Fuse	Rating in amps	Protected circuit
F1	50	ABS
F2	80	On-board computer network 1
F3	80	Audio option pack Power windows Rear window defogger (hardtop)
F4	80	Ignition lock
F5	80	Engine electronics
F6	80	On-board computer network 2
F7	not used	

Table g. 2000 high amperage fuses		
Fuse	Rating in amps	Protected circuit
F1	50	ABS
F2	80	On-board computer network 1
F3	80	On-board computer network 2
F4	80	Ignition lock
F5	80	Engine electronics
F6	80	On-board computer network 3
F7	not used	

Fuses, Relays, Component Locations 97-23

Fuse Locations and Ratings

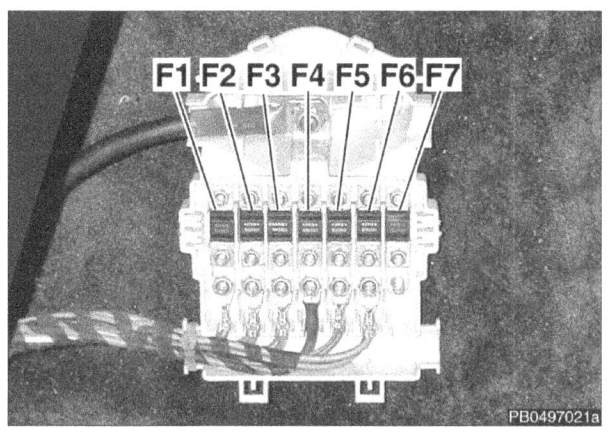

Table h. 2001 high amperage fuses		
Fuse	Rating in amps	Protected circuit
F1	50	PSM
F2	80	On-board computer network 1
F3	80	On-board computer network 2
F4	80	Ignition lock
F5	80	Engine electronics
F6	80	On-board computer network 3
F7	50	PSM

Main fuses

 Working in left footwell, pull open main fuse panel cover.

 Fuses are organized in five rows of 10. Be sure to replace a bad fuse with one of equivalent rating.

NOTE—

- *The main fuse tables on the following pages cover model years 1997 to 2001. Fuse locations in 2002 - 2004 Boxster models are similar to those in 2001 models.*

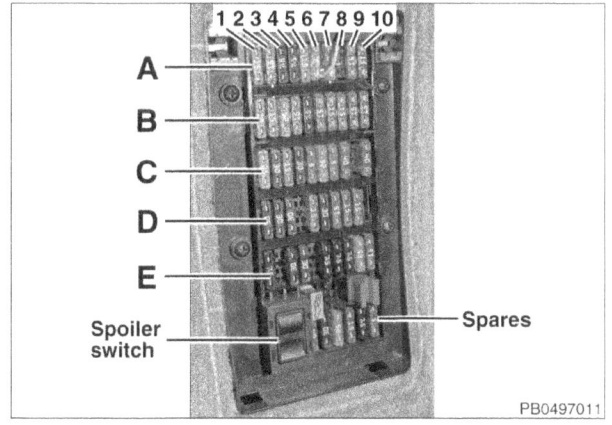

Fuses, Relays, Component Locations

Fuse Locations and Ratings

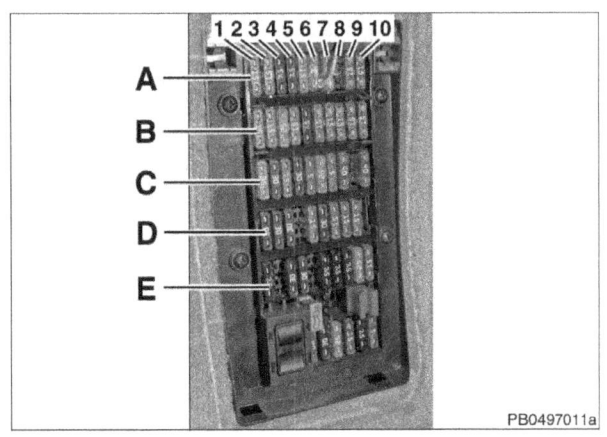

Table i. 1997 main fuse panel

Fuse	Rating in amps	Protected circuit
Row A		
A1	7.5	High beam right
A2	7.5	High beam left
A3	7.5	Side marker right
A4	7.5	Side marker left
A5	7.5	License plate lights
A6	25	Seat heater
A7	25	Foglights
A8	7.5	License plate lights (Canada)
A9	7.5	Low beam right
A10	7.5	Low beam left
Row B		
B1	15	Instruments OBD II (diagnosis) plug Tiptronic Traction control switch
B2	7.5	Radio
B3	25	Horns
B4	15	Engine compartment blower relay
B5	7.5	Backup lights
B6	15	Convertible top control module Turn signals
B7	15	Brake lights Cruise control
B8	15	Alarm control module DME control module
B9	15	ABS control module Traction control
B10	15	Instrument cluster diagnosis Mirror adjustment
Row C		
C1	25	DME main relay
C2	30	Ignition Oxygen sensors
C3	15	Alarm control module
C4	25	Fuel pump relay

Table i. 1997 main fuse panel

Fuse	Rating in amps	Protected circuit
C5	-	not used
C6	25	Wipers and washers
C7	7.5	Terminal X control
C8	30	Radiator fan right
C9	25	Headlight cleaning system
C10	30	Radiator fan left
Row D		
D1	30	Power windows
D2	30	Rear window, outside rear view mirror defogger
D3	30	Convertible top motor
D4	-	not used
D5	15	Lighter
D6	30	A/C and heater system relay
D7	15	DME control module Hazard warning lights
D8	15	Rear spoiler motor
D9	15	Sound system (option pack)
D10	7.5	Mounting point for retrofit
Row E		
E1	7.5	Radio Terminal 86s alarm control module
E2	-	not used
E3	30	Seat back control, left
E4	30	Seat back control, right
E5	-	not used
E6	7.5	Terminal 30 telephone
E7	7.5	A/C and heater system
E8	-	not used
E9	7.5	Terminal 15 telephone
E10	15	Tiptronic control module

Fuse Locations and Ratings

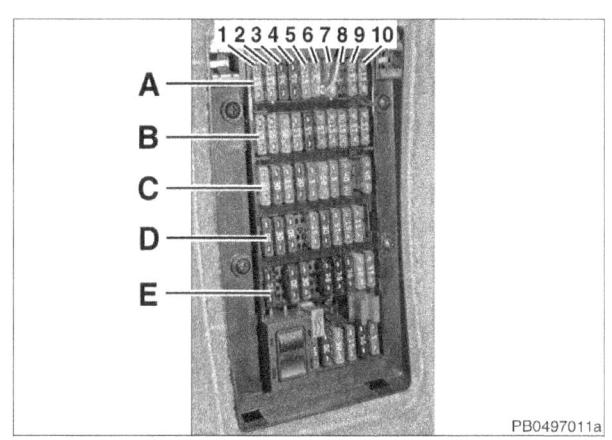

Table j. 1998 main fuse panel		
Fuse	Rating in amps	Protected circuit
Row A		
A1	7.5	High beam right
A2	7.5	High beam left
A3	7.5	Side marker right
A4	7.5	Side marker left
A5	15	Instrument lights License plate lights
A5	15	License plate lights Locating lights
A6	25	Seat heater
A7	25	Foglights
A8	7.5	License plate lights (Canada)
A9	7.5	Low beam right
A10	7.5	Low beam left
Row B		
B1	15	Instruments OBD II (diagnosis) plug Tiptronic Traction control switch
B2	7.5	Radio
B3	25	Horns
B4	15	Engine compartment blower relay
B5	7.5	Backup lights Seat and mirror memory control module Outside rear view mirror adjustment
B6	15	Convertible top control module Power windows Turn signals
B7	15	Brake lights Cruise control
B8	15	Alarm control module DME control module Tiptronic control module
B9	15	ABS control module Traction control
B10	15	Instrument cluster diagnosis

Table j. 1998 main fuse panel		
Fuse	Rating in amps	Protected circuit
Row C		
C1	25	DME main relay
C2	30	Ignition Oxygen sensors
C3	15	Alarm control module Power windows
C4	25	Fuel pump relay
C5	-	not used
C6	25	Wipers and washers
C7	7.5	Terminal X control
C8	30	Radiator fan right
C9	25	Headlight cleaning system
C10	30	Radiator fan left
Row D		
D1	30	Power windows
D2	30	Rear window, outside rear view mirror defogger
D3	30	Convertible top motor
D4	-	not used
D5	15	Lighter
D6	30	A/C and heater system relay
D7	15	DME control module Hazard warning lights
D8	15	Rear spoiler motor
D9	15	Sound system (option pack)
D10	7.5	Mounting point for retrofit
Row E		
E1	7.5	Radio Terminal 86s alarm control module
E2	7.5	Seat memory control module
E3	30	Seat back control, left Seat and mirror memory control module
E4	30	Seat back control, right Seat and mirror memory control module
E5	7.5	Navigation
E6	7.5	Navigation control module Terminal 30 telephone
E7	7.5	A/C and heater system
E8	7.5	Terminal 15 telephone
E9	-	not used
E10	-	not used

Fuse Locations and Ratings

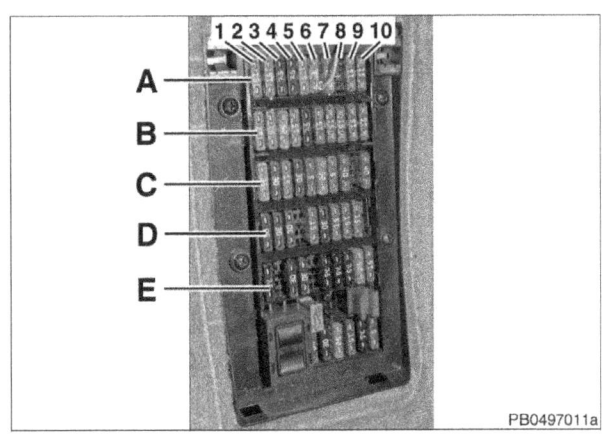

Table k. 1999 main fuse panel		
Fuse	Rating in amps	Protected circuit
Row A		
A1	15	High beam right High beam control module
A2	15	High beam left
A3	7.5	Side marker right
A4	7.5	Side marker left
A5	15	Instrument lights License plate lights Locating lights
A6	25	Seat heater
A7	25	Foglights
A8	7.5	License plate lights (Canada)
A9	15	Low beam right
A10	15	Low beam left
Row B		
B1	15	Instruments OBD II (diagnosis) plug Tiptronic Traction control switch
B2	7.5	Radio
B3	25	Horns
B4	15	Engine compartment blower relay
B5	7.5	Seat and mirror memory control module Outside rear view mirror adjustment
B6	15	Convertible top control module Power windows Turn signals
B7	15	Brake lights Cruise control
B8	15	Alarm control module DME control module Tiptronic control module
B9	15	ABS / TC control module
B10	15	Headlight vertical aim control Instrument cluster diagnosis

Table k. 1999 main fuse panel		
Fuse	Rating in amps	Protected circuit
Row C		
C1	25	DME main relay
C2	30	Fuel injection Ignition Oxygen sensor heaters
C3	15	Alarm control module Interior lights Power windows
C4	25	Fuel pump relay
C5	-	not used
C6	25	Wipers and washers
C7	7.5	Terminal X control
C8	30	Radiator fan right
C9	25	Headlight cleaning system
C10	30	Radiator fan left
Row D		
D1	30	Power windows
D2	30	Rear window, outside rear view mirror defogger
D3	30	Convertible top motor
D4	-	not used
D5	15	Lighter
D6	30	A/C and heater system relay
D7	15	Hazard warning lights
D8	15	Rear spoiler motor
D9	15	Sound system (option pack)
D10	7.5	Mounting point for retrofit
Row E		
E1	7.5	Cluster Daytime running lights Radio Terminal 86s alarm control module
E2	7.5	Seat and mirror memory control module
E3	30	Seat back control, left Seat and mirror memory control module
E4	30	Seat back control, right Seat and mirror memory control module
E5	7.5	Navigation
E6	7.5	ORVR Terminal 30 telephone
E7	7.5	A/C and heater system
E8	7.5	Navigation Terminal 15 telephone
E9	-	not used
E10	-	not used

Fuses, Relays, Component Locations 97-27
Fuse Locations and Ratings

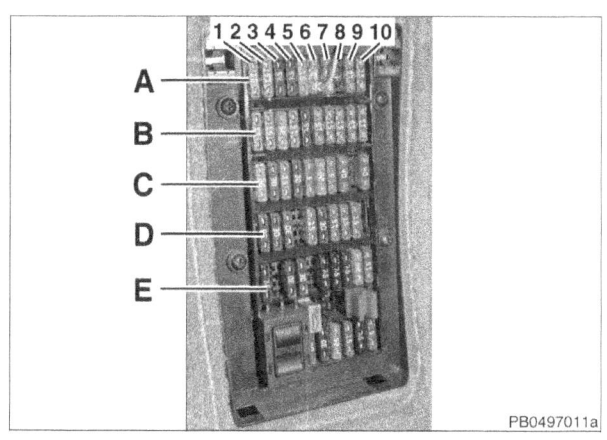

Table I. 2000 main fuse panel		
Fuse	Rating in amps	Protected circuit
Row A		
A1	15	High beam right High beam control module
A2	15	High beam left
A3	7.5	Side marker right
A4	7.5	Side marker left
A5	15	Instrument lights License plate lights Locating lights
A6	25	Seat heater
A7	25	Foglights
A8	7.5	License plate lights (Canada)
A9	15	Low beam right
A10	15	Low beam left
Row B		
B1	15	Instruments OBD II (diagnosis) plug Tiptronic Traction control switch
B2	15	Hazard warning system Turn signals
B3	25	Horns
B4	15	Engine compartment blower relay
B5	7.5	Back-up lights Seat and mirror memory control module Outside rear view mirror adjustment
B6	15	Convertible top control module Power windows Turn signals
B7	15	Brake lights Cruise control
B8	15	Alarm control module DME control module Tiptronic control module
B9	15	ABS / TC control module
B10	15	Headlight vertical aim control Instrument cluster diagnosis

Table I. 2000 main fuse panel		
Fuse	Rating in amps	Protected circuit
Row C		
C1	25	DME main relay
C2	30	Fuel injection Ignition Oxygen sensor heaters
C3	15	Alarm control module Interior lights Power windows
C4	25	Fuel pump relay
C5	5	Engine compartment blower stage 1
C6	25	Wipers and washers
C7	5	Terminal X control
C8	40	Radiator fan right
C9	25	Headlight cleaning system
C10	40	Radiator fan left
Row D		
D1	30	Power windows
D2	30	Outside rear view mirror defogger Rear window defogger (hardtop)
D3	30	Convertible top motor
D4	-	not used
D5	15	Lighter
D6	30	A/C and heater system relay
D7	15	Spoiler
D8	15	Radio
D9	15	Sound system (option pack)
D10	7.5	Mounting point for retrofit
Row E		
E1	7.5	Cluster Daytime running lights Radio Terminal 86s alarm control module
E2	7.5	Seat and mirror memory control module
E3	30	Seat back control, left Seat and mirror memory control module
E4	30	Seat back control, right Seat and mirror memory control module
E5	7.5	Navigation
E6	7.5	Navigation ORVR Terminal 30 telephone
E7	7.5	A/C and heater system
E8	7.5	Navigation Terminal 15 telephone
E9	-	not used
E10	-	not used

97-28 Fuses, Relays, Component Locations

Fuse Locations and Ratings

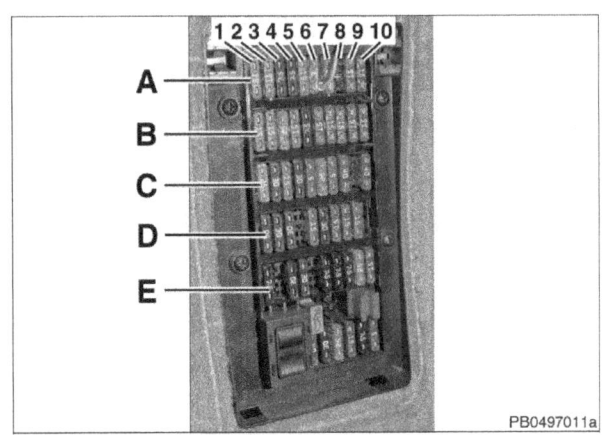

Table m. 2001 main fuse panel

Fuse	Rating in amps	Protected circuit
Row A		
A1	15	High beam right High beam control module
A2	15	High beam left
A3	7.5	Side marker right
A4	7.5	Side marker left
A5	15	Instrument lights License plate lights Locating lights
A6	25	Seat heater
A7	25	Foglights
A8	7.5	License plate lights (Canada)
A9	15	Low beam right
A10	15	Low beam left
Row B		
B1	15	Instruments OBD II (diagnosis) plug Tiptronic PSM
B2	15	Hazard warning system Turn signals
B3	25	Horns
B4	15	Engine compartment blower relay
B5	7.5	Back-up lights Seat and mirror memory control module Outside rear view mirror adjustment
B6	15	Convertible top control module Power windows Turn signals
B7	15	Brake lights Cruise control
B8	15	Alarm control module DME control module Tiptronic control module
B9	15	ABS / PSM control module

Table m. 2001 main fuse panel

Fuse	Rating in amps	Protected circuit
B10	15	Headlight vertical aim control Instrument cluster diagnosis Parking assistant
Row C		
C1	25	DME main relay
C2	30	Fuel injection Ignition Oxygen sensor heaters
C3	15	Alarm control module Interior lights Power windows
C4	30	Fuel pump relay
C5	5	Engine compartment blower stage 1
C6	25	Wipers and washers
C7	7.5	Terminal X control
C8	40	Radiator fan right
C9	25	Headlight cleaning system
C10	40	Radiator fan left
Row D		
D1	30	Power windows
D2	30	Outside rear view mirror defogger Rear window defogger (hardtop)
D3	30	Convertible top motor
D4	-	not used
D5	15	Lighter
D6	30	A/C and heater system relay
D7	15	Spoiler
D8	7.5	Radio and sound system (option pack)
D9	15	Amplifier
D10	5	Mounting point for retrofit
Row E		
E1	7.5	Cluster Daytime running lights Navigation Radio Terminal 86s alarm control module
E2	7.5	Seat and mirror memory control module
E3	30	Seat back control, left Seat and mirror memory control module
E4	30	Seat back control, right Seat and mirror memory control module
E5	7.5	Navigation
E6	7.5	Navigation ORVR Terminal 30 telephone
E7	7.5	A/C and heater system
E8	7.5	Navigation Terminal 15 telephone
E9	25	PSM control module
E10	15	PSM switch

Fuses, Relays, Component Locations 97-29

Relay Panels

RELAY PANELS

Most Boxster relays are in one of two locations:
- **Relay panel 1** above left kick panel
- **Relay panel 2** in rear trunk

Relay panel 1, accessing

◄ Working in left footwell, pull open main fuse panel cover.

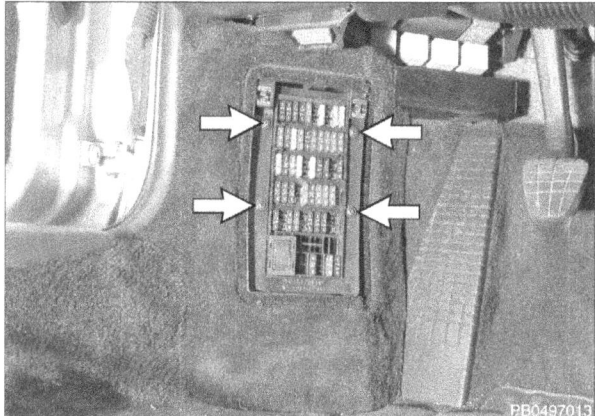

◄ Remove left kick panel mounting screws (**arrows**) and pull off kick panel.

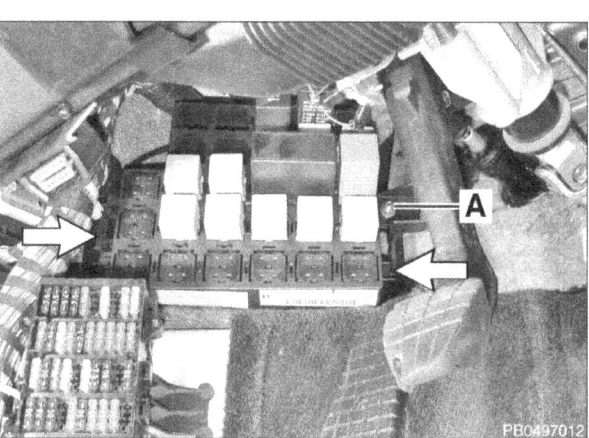

◄ Working to left of pedal cluster:
  - Remove relay panel mounting fastener (**A**).
  - Unclip relay panel from side clips (**arrows**) and lower.

NOTE—
- *Relay positions in relay panel 1 are shown in* **Fig. 5**.

Relay panel 2, accessing

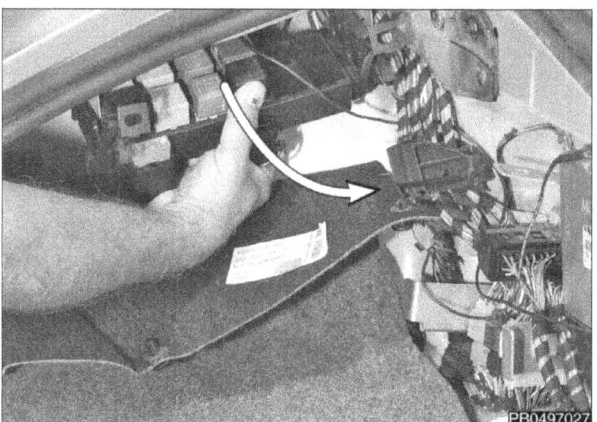

– Working in rear trunk, detach and fold back left side trim panel.

◄ Slide relay panel to right (**arrow**) to remove. Lower panel to trunk floor.

NOTE—
- *Relay positions in relay panel 2 are shown in* **Fig. 6**.

97-30 Fuses, Relays, Component Locations

Grounds

GROUNDS

Grounds are distributed throughout the vehicle body. Many are found under the interior carpets or behind trim panels. Several components grounds are often ganged. Ground positions vary among models. Lugs and connectors attached to ground are susceptible to damage and corrosion. Clean or renew as necessary.

Ground applications for individual electrical circuits are in **Table n** and **Table o**.

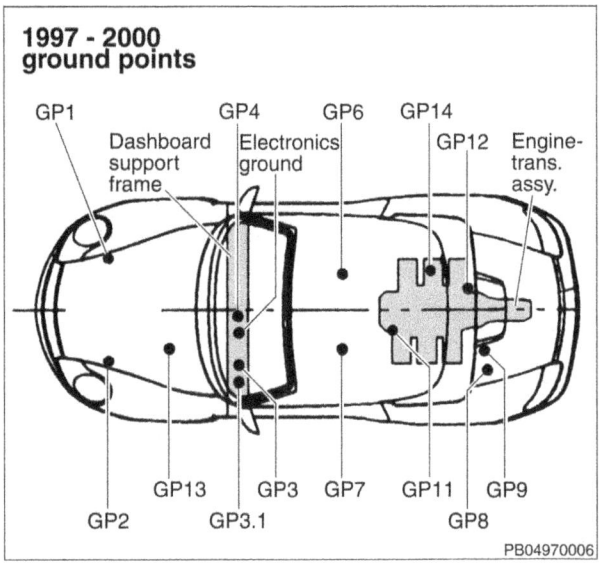

Table n. 1997 - 2000 ground applications

Circuit	GP1	GP2	GP3	GP3.1	GP4	GP6	GP7	GP8	GP9	GP11	GP12	GP13	GP14	Electronics
A/C and heater blower					✳									
A/C and heater control module					✳									
A/C compressor relay								✳						
A/C pressure switch					✳									
ABS control module		✳												✳
ABS / TC control module		✳												✳
Air circulation controls					✳									
Airbag control module							✳	✳						
Alarm siren					✳									
Alarm system control module							✳							
Amplifier					✳									
Ashtray light					✳									
Battery												✳		
Body to engine													✳	
Brake fluid level sensor			✳											
Brake pad wear indicator			✳											
Cell phone							✳							
Center brake light									✳					
Changeover relay				✳										
Cigarette lighter					✳									

Fuses, Relays, Component Locations — Grounds

Table n. 1997 - 2000 ground applications

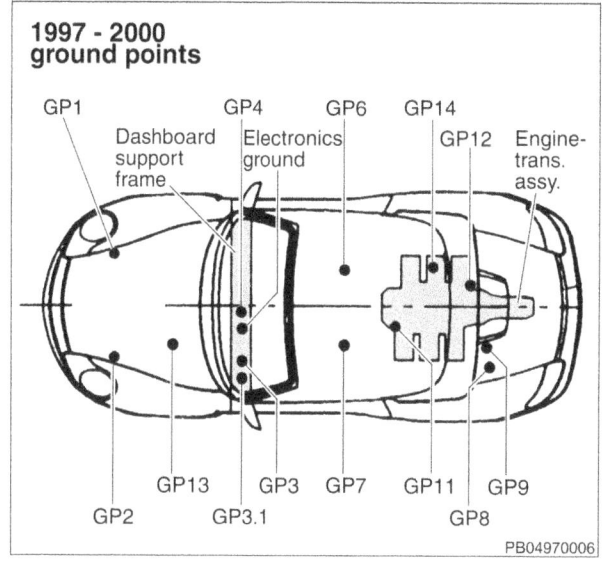

1997 - 2000 ground points

GP1 — Dashboard support frame
GP4 — Electronics ground
GP14, GP12 — Engine-trans. assy.
GP2, GP13, GP3.1, GP3, GP7, GP11, GP8, GP9, GP6

Circuit	GP1	GP2	GP3	GP3.1	GP4	GP6	GP7	GP8	GP9	GP11	GP12	GP13	GP14	Electronics
Convertible top microswitches			✻					✻						
Convertible top switch					✻									
Coolant level sensor									✻					
Cruise control module					✻									
Daylight running lights (Canada)				✻										
Engine control module (ECM)			✻						✻					
Door lock control module, left				✻										
Door lock control module, right					✻									
Electronic immobilizer			✻											
Engine compartment blower											✻			
Engine cooling fan, left		✻												
Engine cooling fan, right	✻													
Front and side lights, left		✻												
Front and side lights, right	✻													
Front trunk lock microswitch		✻												
Fuel filter					✻									
Fuel injector, ignition and oxygen sensor heater relay									✻					
Fuel pump			✻											
Fuel tank filler neck					✻									
Hands-free microphone				✻										
Hazard warning flasher				✻										
Hazard warning switch					✻									
Headlight washers				✻										
Heated windshield washer nozzles					✻									
Horn switch					✻									
Horns		✻												
Ignition coils, left (cyl. 4 - 6)											✻			

Fuses, Relays, Component Locations

Grounds

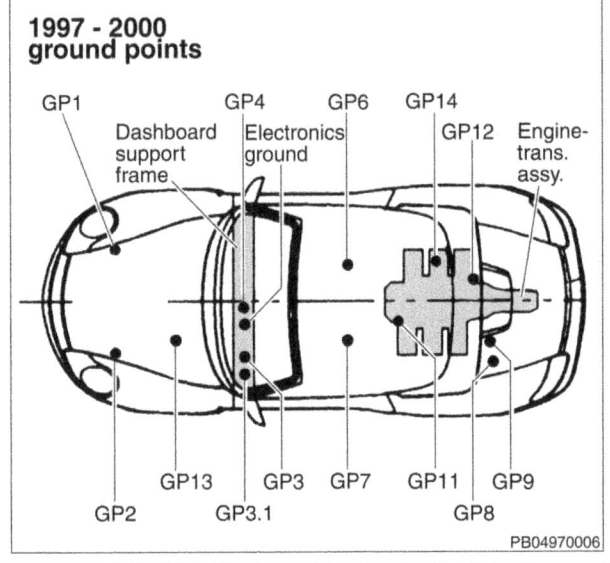

1997 - 2000 ground points

- GP1 Dashboard support frame
- GP4 Electronics ground
- GP6
- GP14
- GP12 Engine-trans. assy.
- GP2
- GP13
- GP3.1
- GP3
- GP7
- GP11
- GP8
- GP9

Table n. 1997 - 2000 ground applications

Circuit	GP1	GP2	GP3	GP3.1	GP4	GP6	GP7	GP8	GP9	GP11	GP12	GP13	GP14	Electronics
Ignition coils, right (cyl. 1 - 3)											✶			✶
Instrument cluster														✶
Interior lights			✶											
Interior monitor (infrared sensor)				✶										
Interior temperature sensor					✶									
License plate lights								✶						
Light switch assembly			✶											
Navigation					✶									
OBD II (diagnosis) plug				✶										✶
On-board computer							✶							✶
Outside rear view mirror, left					✶									
Outside rear-view mirror, right					✶									
Parking assistant control module							✶							
Parking brake warning switch					✶									
Power window switches					✶									
Radio					✶									
Rear window and outside rear-view mirror heater relay				✶										
Rear window defogger (hardtop)								✶						
Relay terminal XE				✶										
Seat and outside rear-view mirror memory control module				✶										
Seat, left								✶						
Seat, right							✶							
Spoiler									✶					
Spoiler switch				✶										
Switch package tray					✶									
Taillights												✶		
Telephone				✶										

Fuses, Relays, Component Locations

Grounds

Table n. 1997 - 2000 ground applications

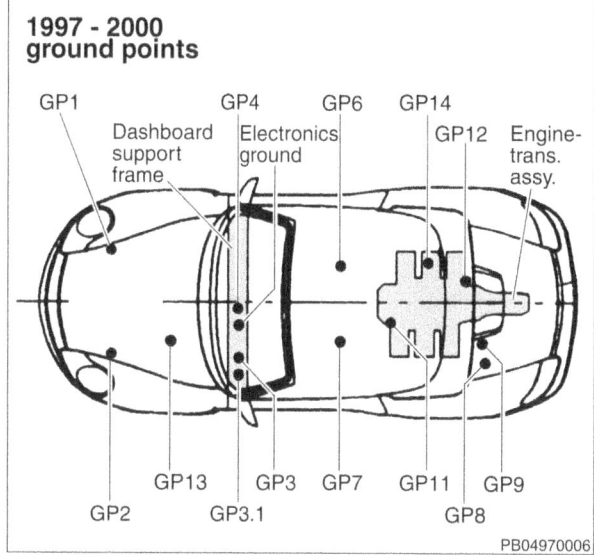

1997 - 2000 ground points

Circuit	GP1	GP2	GP3	GP3.1	GP4	GP6	GP7	GP8	GP9	GP11	GP12	GP13	GP14	Electronics
Terminal 31					✶									
Tiptronic control module					✶				✶					
Tiptronic selector switch									✶					
Vanity mirror lights				✶										
Washer fluid level indicator			✶											
Washer fluid motor			✶											
Windshield wiper motor			✶											
Wiper intermittent control					✶									

Fuses, Relays, Component Locations

Grounds

Table o. 2001 ground applications

2001 ground points

GP1, Litronic ground, Electronics ground, GP6, GP14, GP12, Engine-trans. assy., Dashboard support frame, GP2, GP13, PSM grounds, GP3.1, GP3, Airbag ground, GP7, GP8, GP11, GP9

Circuit	GP1	GP2	GP3	GP3.1	GP4	GP6	GP7	GP8	GP9	GP11	GP12	GP13	GP14	Airbag	Electronics	Litronic	PSM
A/C and heater blower					✻												
A/C and heater control module					✻												
A/C compressor relay								✻									
A/C pressure switch					✻												
ABS control module																	✻
Air circulation controls					✻												
Airbag control module							✻	✻						✻			
Alarm siren			✻	✻													
Alarm system control module								✻									
Amplifier					✻												
Anti-glare rear view mirror					✻												
Ashtray light					✻												
Battery												✻					
Body to engine													✻				
Brake fluid level sensor			✻														
Brake pad wear indicator			✻														
Cell phone							✻										
Center brake light									✻								
Cigar lighter					✻												
Clutch switch			✻														
Convertible top			✻		✻			✻									
Coolant level sensor									✻								
Cruise control module					✻												
Daylight running lights (Canada)					✻												
Engine control module (ECM)										✻							
Door lock								✻									
Door lock control module, left				✻													
Door lock control module, right					✻												

Fuses, Relays, Component Locations 97-35
Grounds

Table o. 2001 ground applications

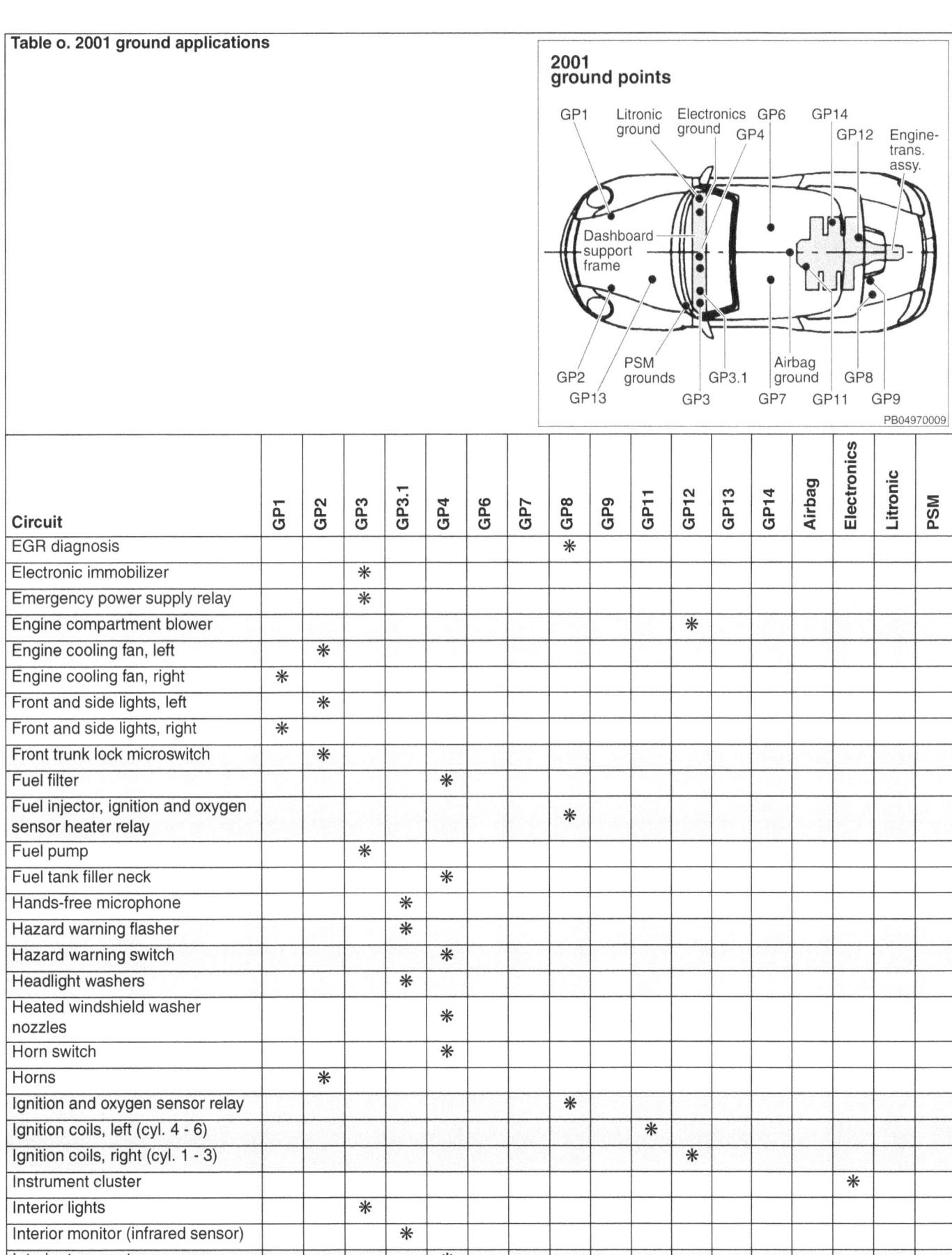

2001 ground points

Circuit	GP1	GP2	GP3	GP3.1	GP4	GP6	GP7	GP8	GP9	GP11	GP12	GP13	GP14	Airbag	Electronics	Litronic	PSM
EGR diagnosis								*									
Electronic immobilizer			*														
Emergency power supply relay			*														
Engine compartment blower											*						
Engine cooling fan, left		*															
Engine cooling fan, right	*																
Front and side lights, left		*															
Front and side lights, right	*																
Front trunk lock microswitch			*														
Fuel filter					*												
Fuel injector, ignition and oxygen sensor heater relay								*									
Fuel pump			*														
Fuel tank filler neck					*												
Hands-free microphone				*													
Hazard warning flasher				*													
Hazard warning switch					*												
Headlight washers				*													
Heated windshield washer nozzles					*												
Horn switch					*												
Horns		*															
Ignition and oxygen sensor relay								*									
Ignition coils, left (cyl. 4 - 6)											*						
Ignition coils, right (cyl. 1 - 3)												*					
Instrument cluster																*	
Interior lights			*														
Interior monitor (infrared sensor)				*													
Interior temperature sensor					*												

Fuses, Relays, Component Locations

Grounds

Table o. 2001 ground applications

2001 ground points: GP1 Litronic ground, GP6, GP14, GP12, Engine-trans. assy., Electronics ground, GP4, Dashboard support frame, GP2, PSM grounds, GP3.1, Airbag ground, GP8, GP13, GP3, GP7, GP11, GP9

Circuit	GP1	GP2	GP3	GP3.1	GP4	GP6	GP7	GP8	GP9	GP11	GP12	GP13	GP14	Airbag	Electronics	Litronic	PSM
Kick-down switch					✳												
License plate lights								✳									
Light switch assembly			✳														
Litronic control module																✳	
Navigation					✳												
OBD II (diagnosis) plug					✳										✳		
On-board computer					✳	✳									✳		
ORVR					✳												
Outside rear view mirror, left					✳												
Outside rear view mirror, right					✳												
Parking assistant control module							✳										
Parking brake warning switch					✳												
Power window switches					✳												
PSM (traction control)																	✳
Radio					✳												
Rain sensor system			✳														
Rear window and outside rear view mirror heater relay					✳												
Rear window defogger (hardtop)								✳									
Relay terminal XE					✳												
Seat and outside rear view mirror memory control module					✳			✳									
Seat heater, left								✳									
Seat heater, right						✳											
Seat belt lock microswitch								✳							✳		
Spoiler								✳									
Spoiler switch			✳														
Switch package tray					✳												
Taillights								✳									

Fuses, Relays, Component Locations 97-37

Grounds

Table o. 2001 ground applications

2001 ground points

GP1, Litronic ground, Electronics ground, GP6, GP14, GP4, GP12, Engine-trans. assy., Dashboard support frame, GP2, GP13, PSM grounds, GP3.1, GP3, Airbag ground, GP7, GP11, GP8, GP9

Circuit	GP1	GP2	GP3	GP3.1	GP4	GP6	GP7	GP8	GP9	GP11	GP12	GP13	GP14	Airbag	Electronics	Litronic	PSM
Telephone				*			*										
Secondary air pump											*						
Tiptronic control module					*				*								
Tiptronic selector switch					*				*								
Vanity mirror lights					*												
Washer fluid level indicator			*														
Washer fluid pump			*														
Wheel speed sensor, front			*														
Wheel speed sensor, rear								*									
Windshield wiper motor			*														
Wiper intermittent control module			*		*												
Wiper intermittent switch			*		*												

Electrical Troubleshooting

ELECTRICAL TROUBLESHOOTING

Four things are required for current to flow in any electrical circuit: a voltage source, wires or connections to transport the voltage, a load or device that uses the electricity, and a connection to ground.

Most problems can be found using a digital multimeter (volt/ohm/ammeter) to check the following:
- Voltage supply
- Breaks in the wiring (infinite resistance/no continuity)
- A path to ground that completes the circuit

Electric current is logical in its flow, always moving from the voltage source toward ground. Electrical faults can usually be located through a process of elimination. When troubleshooting a complex circuit, separate the circuit into smaller parts. General tests outlined below may be helpful in finding electrical problems. The information is most helpful when used with wiring diagrams.

Be sure to analyze the problem. Use wiring diagrams to determine the most likely cause. Get an understanding of how the circuit works by following the circuit from ground back to the power source.

When making test connections at connectors and components, use care to avoid spreading or damaging the connectors or terminals. Some tests may require jumper wires to bypass components or connections in the wiring harness. When connecting jumper wires, use blade connectors at the wire ends that match the size of the terminal being tested. The small internal contacts are easily spread apart, and this can cause intermittent or faulty connections that can lead to more problems.

Voltage and voltage drops

Wires, connectors, and switches that carry current are designed with very low resistance so that current flows with a minimum loss of voltage. A voltage drop is caused by higher than normal resistance in a circuit. This additional resistance actually decreases or stops the flow of current. A voltage drop can be noticed by problems ranging from dim headlights to sluggish wipers. Some common sources of voltage drops are corroded or dirty switches, dirty or corroded connections or contacts, and loose or corroded ground wires and ground connections.

A voltage drop test is a good test to perform if current is flowing through the circuit but the circuit is not operating correctly. A voltage drop test will help pinpoint a corroded ground strap or a faulty switch. Normally, there should be less than 1 vdc drop across most wires or closed switches. A voltage drop across a connector or short cable should not exceed 0.5 vdc.

Fuses, Relays, Component Locations 97-39

Electrical Troubleshooting

NOTE—
- *A voltage drop test is generally more accurate than a simple resistance check because the resistances involved are often too small to measure with most ohmmeters. For example, a resistance as small as 0.02 Ω would results in a 3 vdc drop in a typical 150 amp starter circuit. (150 amps x 0.02 Ω =3 vdc).*
- *Keep in mind that voltage with ignition key ON and voltage with engine running are not the same. With ignition ON and engine OFF (battery voltage), voltage should be approximately 12.6 vdc. With engine running (charging voltage), voltage should be approximately 14.0 vdc. Measure voltage at battery with ignition ON and then with engine running to get exact measurements.*

Voltage, measuring

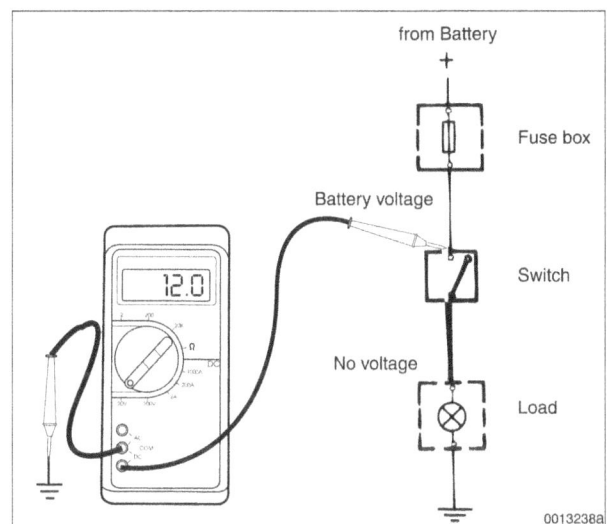

◀ Connect digital multimeter negative lead to a reliable ground point on car.

NOTE—
- *The negative (-) battery terminal is always a good ground point.*

– Connect digital multimeter positive lead to point in circuit you wish to measure.

– Voltage reading should not deviate more than 1 vdc from voltage at battery. If voltage drop is more than this, check for a corroded connector or loose ground wire.

Voltage drop, testing

Check voltage drop only when there is a load on the circuit, such as when operating the starter motor or turning on the headlights. Use a digital multimeter to ensure accurate readings.

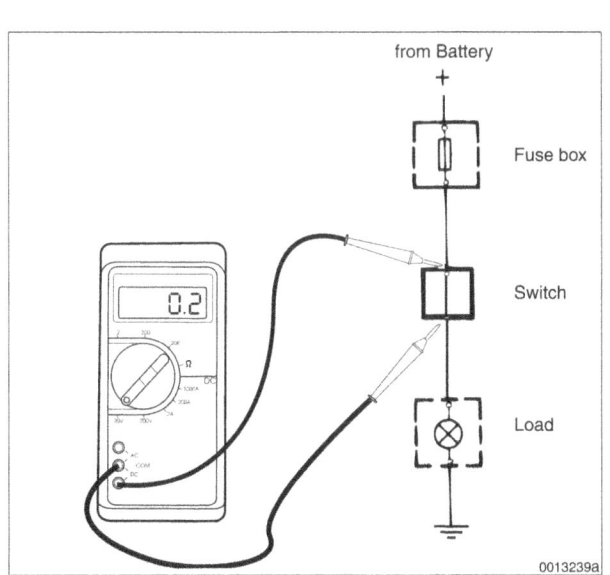

◀ Connect digital multimeter positive lead to positive (+) battery terminal or a positive power supply close to battery source.

– Connect digital multimeter negative lead to other end of cable or switch being tested.

– With power on and circuit working, meter shows voltage drop (difference between two points). This value should not exceed 1 vdc.

– Maximum voltage drop in an automotive circuit, as recommended by the Society of Automotive Engineers (SAE), is as follows:
- 0 vdc for small wire connections
- 0.1 vdc for high current connections
- 0.2 vdc for high current cables
- 0.3 vdc for switch or solenoid contacts

– On longer wires or cables, the drop may be slightly higher. In any case, a voltage drop of more than 1.0 vdc usually indicates a problem.

Electrical Troubleshooting

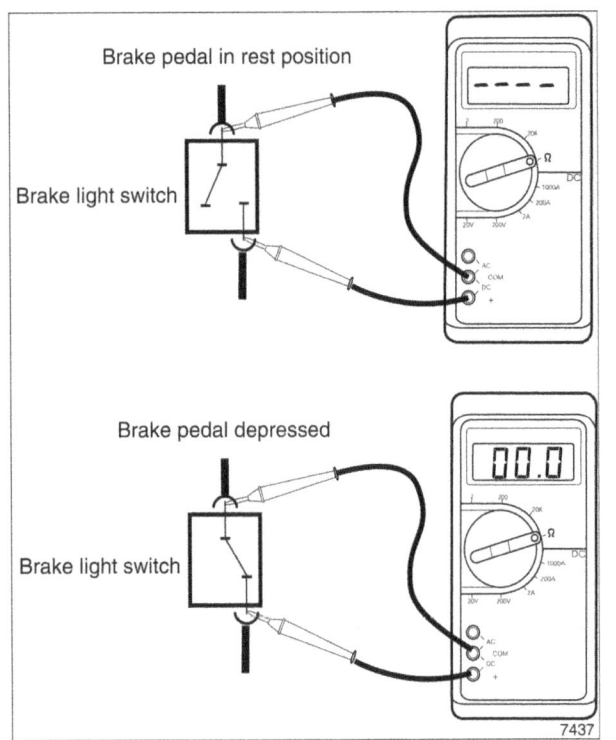

Continuity, checking

Use continuity test to check a circuit or switch. Because most automotive circuits are designed to have little or no resistance, a circuit or part of a circuit can be easily checked for faults using an ohmmeter. An open circuit or a circuit with high resistance does not allow current to flow. A circuit with little or no resistance allows current to flow easily.

When checking continuity, turn ignition OFF. On circuits that are powered at all times, disconnect battery. Using the appropriate wiring diagram, test circuit for faulty connections, wires, switches, relays and engine sensors by checking for continuity.

 Example: Test brake light switch for continuity:

- With brake pedal in rest position (switch open) there is no continuity (infinite Ω).
- With pedal depressed (switch closed) there is continuity (0 Ω).

Short circuits

Short circuits are exactly what the name implies. The circuit takes a shorter path than it was designed to take. The most common short that causes problems is a short to ground where the insulation on a positive (+) wire wears away and the metal wire is exposed. When the wire rubs against a metal part of the car or other ground source, the circuit is shorted to ground. If the exposed wire is live (positive battery voltage), a fuse blows and the circuit may be damaged.

Short circuits are often difficult to locate and may vary in nature. Short circuits can be found using a logical approach based on knowledge of the current path.

Use a digital multimeter to locate short circuits.

> **CAUTION—**
> - *In circuits protected with high rating fuses (25 amp and greater), wires or circuit components may be damaged before the fuse blows. Check for wiring damage before replacing fuses of this rating.*
> - *When replacing blown fuses, confirm correct fuse rating. See* **Fuse Locations and Ratings** *in this repair group.*

Fuses, Relays, Component Locations 97-41

Electrical Troubleshooting

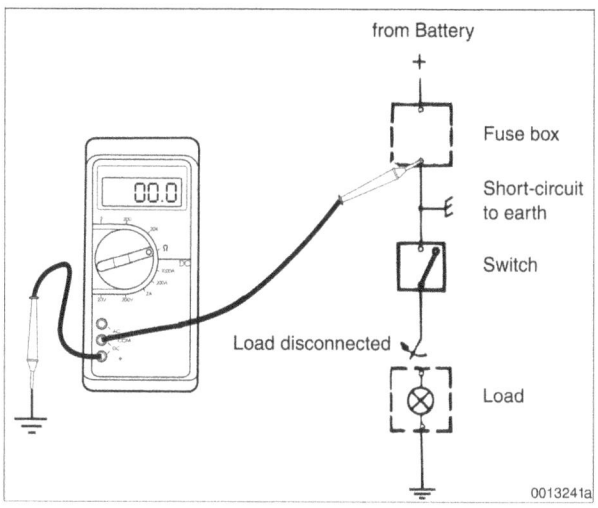

Short circuit, testing with ohmmeter

– Remove blown fuse from circuit and disconnect cables from battery. Disconnect harness connector from circuit load or consumer.

◄ Using an ohmmeter, connect one test lead to load side of fuse terminal (terminal leading to circuit) and other test lead to ground.

– If there is continuity to ground, there is a short to ground.

– If there is no continuity, work from wire harness nearest to fuse/relay panel and move or wiggle wires while observing meter. Continue to move down harness until meter displays a reading. This is location of short to ground.

– Visually inspect wire harness at this point for any faults. If no faults are visible, carefully slice open harness cover or wire insulation for further inspection. Repair any faults found.

Short circuit, testing with voltmeter

– Remove blown fuse from circuit. Disconnect harness connector from circuit load or consumer.

NOTE—

• *Most fuses power more than one consumer. Be sure all consumers are disconnected when checking for a short circuit.*

◄ Using a digital multimeter, connect test leads across fuse terminals. Make sure power is present in circuit. If necessary turn key on.

– If voltage is present at voltmeter, there is a short to ground.

– If voltage is not present, work from wire harness nearest to fuse/relay panel and move or wiggle wires while observing meter. Continue to move down harness until meter displays a reading. This is location of short to ground.

– Visually inspect wire harness at this point for any faults. If no faults are visible, carefully slice open harness cover or wire insulation for further inspection. Repair any faults found.

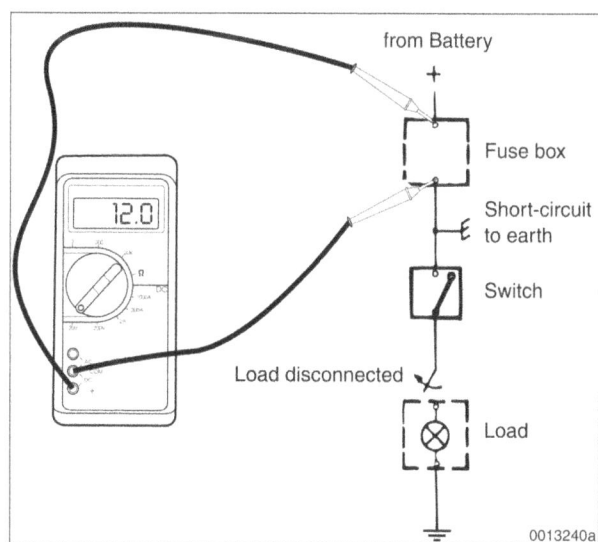

EWD Electrical Wiring Diagrams

GENERAL EWD-2	
WIRING CONVENTIONS EWD-2	
Using wiring diagrams. EWD-2	
Wiring color codes. EWD-2	
Circuit and terminal descriptions. EWD-4	
Porsche abbreviations and acronyms. EWD-5	
Bridge point (BS) assignments EWD-6	
Option codes. EWD-7	
X-connectors. EWD-8

ELECTRICAL WIRING DIAGRAMS (MODEL 1998)

- ABS / ASR. EWD-33
- Alarm system EWD-30
- Airbags EWD-41
- Central locking EWD-30
- Convertible top EWD-26
- Cruise control EWD-41
- Engine cooling, heating and air-conditioning EWD-20
- Engine management (Bosch M 5.2.2). EWD-38
- Heating and air-conditioning EWD-20
- Horns. EWD-18
- Instrument cluster EWD-15
- Lighting RoW. EWD-9
- Lighting USA / Canada EWD-12
- Parking assistant. EWD-27
- Power supply, fuses EWD-53
- Radio, telephone. EWD-44
- Seat and mirror control EWD-50
- Seat and mirror heating, rear window defogger EWD-23
- Spoiler. EWD-27
- Tiptronic EWD-35
- Navigation, infosytem EWD-47
- Windows EWD-19
- Windshield wipers, washers EWD-18

ELECTRICAL WIRING DIAGRAMS (MODEL 2001)

- ABS. EWD-103
- Airbag, MRS (LHD) EWD-99
- Airbag, MRS (RHD). EWD-101
- Air-conditioning EWD-74
- Alarm system (LHD) EWD-91
- Alarm system (RHD) EWD-95
- Central locking (LHD) EWD-91
- Central locking (RHD) EWD-95
- Convertible top EWD-89
- Engine management (Bosch ME 7.2), vehicle EWD-109
- Engine management (Bosch ME 7.2), engine EWD-114
- Heating system EWD-73
- Horns. EWD-72
- Instrument cluster EWD-64
- Lighting (RoW) EWD-56
- Lighting USA / Canada EWD-60
- Mirror adjustment (LHD) EWD-76
- Mirror adjustment (RHD) EWD-81
- Navigation, infosytem EWD-124
- Parking assistant. EWD-122
- Porsche Stability Management (PSM) EWD-104
- Power supply, fuses EWD-128
- Radio. EWD-118
- Radio, sound package, DSP. EWD-119
- Seat and mirror memory (LHD). EWD-78
- Seat and mirror memory (RHD) EWD-83
- Seats. EWD-86
- Spoiler. EWD-67
- Telephone, mobile phone EWD-127
- Tiptronic EWD-107
- Windows EWD-90
- Windshield wipers and washers, headlight cleaning . . . . EWD- 68
- Windshield wipers (with rain sensor), headlight cleaning, ... EWD- 70

TABLES

a. Electrical terminal designations (according to DIN 72 552 standard) EWD-4
b. Porsche abbreviations and acronyms EWD-5
c. Option codes EWD-7
d. X-connectors EWD-8

Electrical Wiring Diagrams

General

The wiring schematics given in this section represent detailed electrical circuit information for 1998 and 2001 Boxster and Boxster S models. Diagrams for additional model years are not included.

Wiring diagrams are necessary to troubleshoot and repair electrical or electronic circuits. Remember that electrical and electronic circuits often have more than one source of power and/or ground. In many cases the ground may also be a switched ground. Take time to study the schematics of the entire system to understand the circuit logic prior to circuit troubleshooting.

When working on electrical or electronic circuits, observe the following precautions.

> *CAUTION—*
> - *Airbags and pyrotechnic seat belt reels are deployed by explosive devices. Handled improperly or without adequate safeguards, these components are very dangerous. Before starting any work, refer to the warnings and cautions in **69 Seat Belts, Airbags**.*
> - *Prior to disconnecting the battery, read the battery disconnection cautions in **00 Warnings and Cautions**.*
> - *Connect and disconnect ignition system wires, multiple connectors and ignition test equipment leads only while the ignition is switched OFF. Keep clothing, hands, and feet dry if possible.*
> - *Switch a test meter to the appropriate function and range before making test connections.*
> - *Switch the ignition off and disconnect the negative (–) battery cable before removing electrical components.*

Wiring Conventions

Each diagram shows the wiring, connectors, terminals, and electrical or electronic components of the circuit. It also identifies the wires by color or terminal coding. The information contained under this heading is organized to help navigate and better understand the wiring diagrams contained in this repair group.

Using wiring diagrams

Using the wiring diagrams for the first time may require some interpretation. Abbreviations, acronyms and not-so-obvious conventions are used throughout the diagrams. These conventions are detailed in the sample wiring diagram below.

Wiring color codes

 Wire insulation colors in this section are abbreviated.

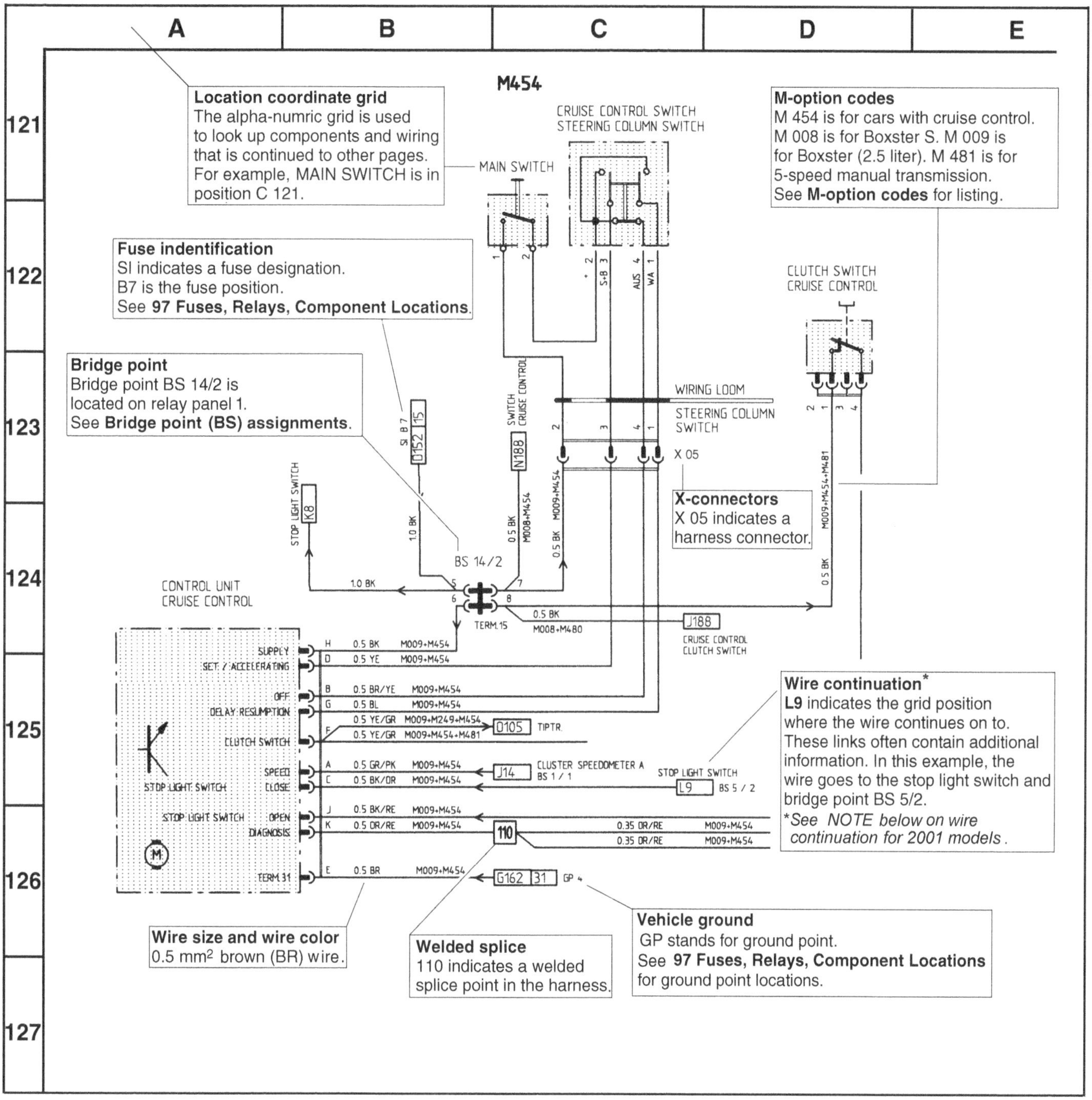

A
B
C
D
E
121
122
123
124
125
126
127
M454
Location coordinate grid
The alpha-numric grid is used to look up components and wiring that is continued to other pages. For example, MAIN SWITCH is in position C 121.
CRUISE CONTROL SWITCH
STEERING COLUMN SWITCH
MAIN SWITCH
M-option codes
M 454 is for cars with cruise control. M 008 is for Boxster S. M 009 is for Boxster (2.5 liter). M 481 is for 5-speed manual transmission.
See M-option codes for listing.
Fuse indentification
SI indicates a fuse designation.
B7 is the fuse position.
See 97 Fuses, Relays, Component Locations.
CLUTCH SWITCH
CRUISE CONTROL
Bridge point
Bridge point BS 14/2 is located on relay panel 1.
See Bridge point (BS) assignments.
WIRING LOOM
STEERING COLUMN SWITCH
X 05
X-connectors
X 05 indicates a harness connector.
STOP LIGHT SWITCH
K8
SI B7
D152 15
SWITCH CRUISE CONTROL
N188
BS 14/2
CONTROL UNIT
CRUISE CONTROL
TERM.15
J188
CRUISE CONTROL CLUTCH SWITCH
SUPPLY
SET / ACCELERATING
OFF
DELAY RESUMPTION
CLUTCH SWITCH
SPEED
STOP LIGHT SWITCH
CLOSE
OPEN
DIAGNOSIS
TERM 31
D105 TIPTR
J14 CLUSTER SPEEDOMETER A BS 1 / 1
STOP LIGHT SWITCH
L9 BS 5 / 2
110
G162 31 GP 4
Wire continuation*
L9 indicates the grid position where the wire continues on to. These links often contain additional information. In this example, the wire goes to the stop light switch and bridge point BS 5/2.
*See NOTE below on wire continuation for 2001 models.
Wire size and wire color
0.5 mm² brown (BR) wire.
Welded splice
110 indicates a welded splice point in the harness.
Vehicle ground
GP stands for ground point.
See 97 Fuses, Relays, Component Locations for ground point locations.

0.5 BN
VS 2 GP 3.1 17/D6
0.5 BK/RD
0.5 BK/RD
0.5 BK/RD
1.0 BK/OG
1.0 BK
CU PSM 10A/E8
CU ABS 10/E2
CU ME 12/E8
BS 6/2 10A/E12
BS 14/2 12/E16
0.75 BK/GN
CU ALARM-CL 8/A6
1.5 BN
CP 1 GP 8 17/E13

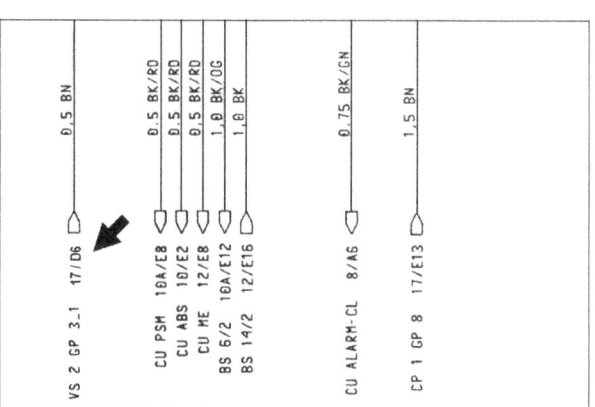

NOTE—

- *2001 wiring diagrams: The wire continuation link references a diagram number followed by a grid coordinate, as shown. For example, 17/D6 (**arrow**) indicate diagram 17, grid position D-6. The diagram numbers are identified on the top of each 2001 wiring page.*

EWD-4 Electrical Wiring Diagrams

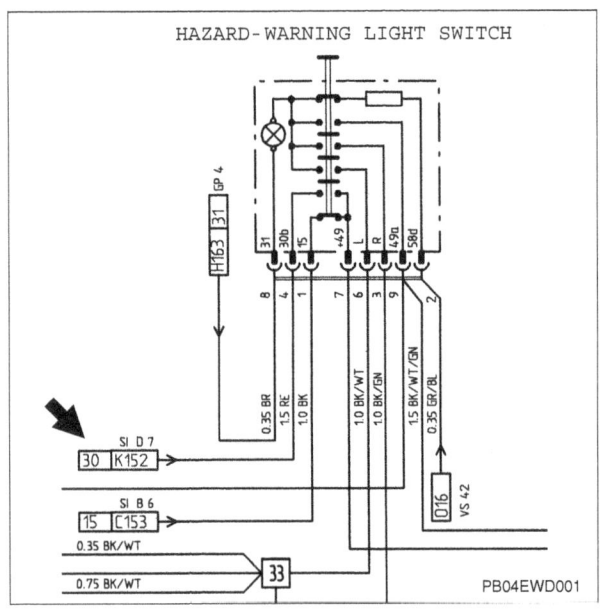

Circuit and terminal descriptions

◄ Porsche designates electrical circuits, including junctions and grounds, with unique designations, most of which follow the German DIN standard. See **Table a**.

For example, if a terminal is labeled 30 (**arrow**), it tells you that positive (+) voltage is supplied to that terminal at all times directly from the battery. The sample schematic also shows a few other commonly used terminal designations, such as terminal15 and terminal 31.

Table a. Electrical terminal designations (according to DIN 72 552 standard)

Number	Circuit description
1	Low voltage switched terminal of coil
4	High voltage center terminal of coil
X	Originates at ignition switch. Supplies power when ignition switch is in PARK, RUN or START position.
XE	Originates at ignition switch. Disrupts power when ignition switch is in START position.
15	Originates at ignition switch. Supplies power when ignition switch is in RUN or START position.
30	Battery positive (+) voltage. Supplies power whenever battery is connected. (Not dependent on ignition switch position, unfused)
31	Ground, battery negative (-) terminal
50	Starter control, supplies power from battery to starter solenoid when ignition switch is in START position only.
54	Stop light terminals
56	Headlight terminals
56a / 56b	High beam / low beam terminals
56d	Headlight flasher contact terminal
57	Parking lights terminal
58	Side marker lights, taillights, license plate lights, instrument panel lights
58d	Instrument panel lights
61	Alternator charge indicator lamp (generator D+)
85	Ground side (-) or relay coil
86	Power-in side (+) of relay coil
87	Power when relay coil is energized (usually terminal 87 connects to terminal 30 power when relay is energized)
D+	Alternator warning light and field energizing circuit

Electrical Wiring Diagrams EWD-5

Porsche abbreviations and acronyms

Acronyms and abbreviations used in the wiring diagrams and in this manual are summarized in **Table b**.

Table b. Porsche abbreviations and acronyms

Acronym	Component or system	Acronym	Component or system
A/C	air-conditioning	HA	rear axle
AAV	carbon canister shutoff valve	KL	terminal
ABD	automatic brake differential	KLS	combined steering column switch
ABS	antilock brakes	KVA	fuel consumption indicator
AHBA	headlight vertical aim control	LWR	headlight vertical aim control
AHVAC	Litronic headlights	MOST	media oriented systems transport (fiber optic bus)
ALWR	headlight vertical aim control	MRS	multiple restraint system
ASR	anti-skid regulation (see TC)	NTC	negative temperature coefficient
ATF	Outside temperature sensor	ORVR	fuel vapor extraction
BL	backup light	PCM	Porsche Communication Management (navigation or infosystem)
BKV	brake booster	POTI	potentiometer
BS	bridge point	PSM	Porsche stability management
CAN	controller area network (bus)	PST2	Porsche Service Tester 2
CLS	central locking system	RA	rear axle
CP	connecting point	RDK	tire pressure control
CU	control unit	RDS	radio station identification system
DDC	drive dynamics control system	RHD	right hand drive
DF	speed sensor	RoW	rest of world
DFA	engine speed sensor output	TC	traction control (see ASR)
DG	engine speed sensor	TI	injection time
DP	disconnecting point	TN	engine speed signal
DK	throttle valve	SI	fuse
DKG	throttle valve sensor	SRA	headlight cleaning system
DME	digital motor electronics	SRS	supplemental restraint system. *See also* MRS
DSP	digital sound processing	SS	speed sensor
EDB	electronic brake force distribution	VA	front axle
EDS	electronic differential lock	VDKD	convertible storage box lid
EKP	electronic fuel pump	VL	front left
ESO	engine speed sensor output	VR	front right
EV	fuel injectors	VS	connecting point (wire splice)
EVAP	evaporative control	WL	wiring loom (wiring harness)
FH	power window	WW	worldwide
FL	front left	ZK	ignition circuit
FOG	fog light	ZV	central locking system
FR	front right	HCS	headlight cleaning system
GP	ground point	HFM	hot-film mass air flow sensor
GND	ground	HA	rear axle
LED	light emitting diode	IC	ignition circuit
		IPAS	integrated Porsche processing system

EWD-6   Electrical Wiring Diagrams

Bridge point (BS) assignments

There are eight socket positions on relay panel 1 reserved for bridge points (BS). To see an example of a bridge point in a wiring diagrams, look in grid coordinate F5 (1998 diagrams). There you will see BS 5/1. Relay panel 1 is behind left side of dashboard. See **97 Fuses, Relays, Component Locations** for directions on accessing the panel.

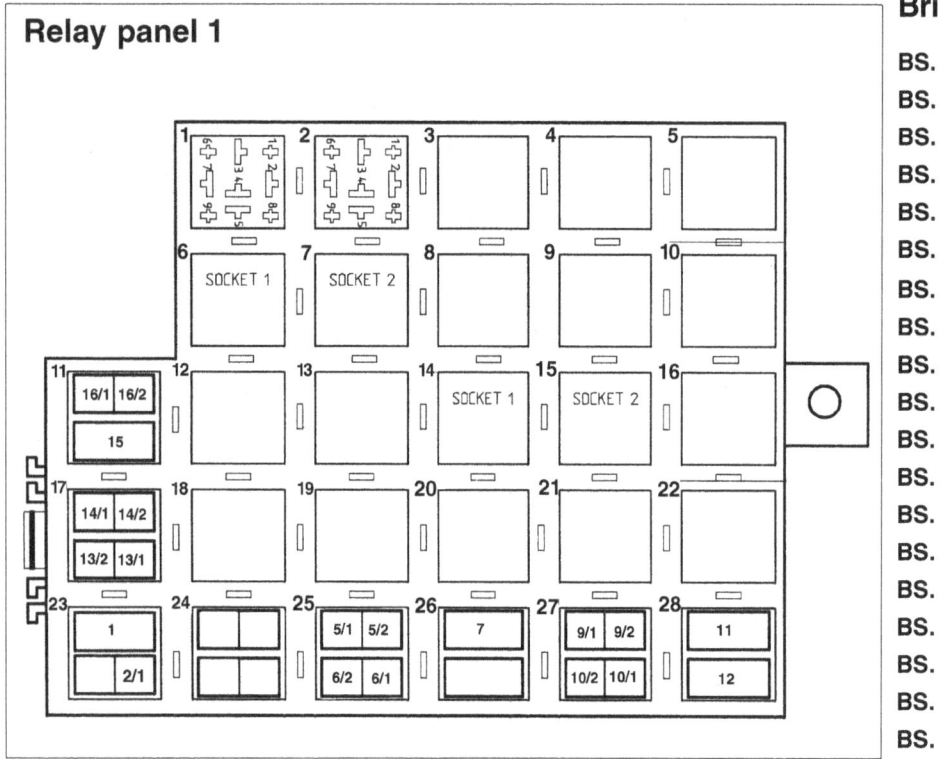

Bridge points

BS. 1 (relay socket 23)
BS. 2/1 (relay socket 23)
BS. 5/1 (relay socket 25)
BS. 5/2 (relay socket 25)
BS. 6/1 (relay socket 25)
BS. 6/2 (relay socket 25)
BS. 7 (relay socket 26)
BS. 9/1 (relay socket 27)
BS. 9/2 (relay socket 27)
BS. 10/1 (relay socket 27)
BS. 10 /2 (relay socket 27)
BS. 11 (relay socket 28)
BS. 12 (relay socket 28)
BS. 13/1 (relay socket 17)
BS. 13/2 (relay socket 17)
BS. 14/1 (relay socket 17)
BS. 14/2 (relay socket 17)
BS. 15 (relay socket 11)
BS. 16/1 (relay socket 11)
BS. 16/2 (relay socket 11)

Electrical Wiring Diagrams   EWD-7

Option codes

Table c lists option codes. These codes are referred to throughout the wiring diagrams.

NOTE—

- *For a more exhaustive listing of option codes, see* **02 Boxster Familiarization**.

Table c. Option codes

Code	Definition	Code	Definition
M 008	Boxster S, 3.2 liter	M 531	Central locking system, code lock
M 009	Boxster, 2.5 liter	M 534	Anti-theft system (ex. USA)
M 113	Daytime running lights (Canada)	M 535	Anti-theft system (USA)
M 139	Adjustable seat heating, left side	M 537	Lumbar support, left seat
M 193	Japan version	M 538	Lumbar support, right seat
M 222	Anti-skid regulator (ASR)	M 550	Hardtop
M 249	Automatic transmission (Tiptronic)	M 553	USA / Canada version
M 265	Rain sensor	M 562	Airbags, driver and passenger
M 273	Outside mirror (power adjustable)	M 563	Airbag, passenger
M 288	Headlight cleaning system	M 573	Air-conditioning system
M 340	Adjustable seat heating, right side	M 601	Litronic headlights
M 437	Comfort seat, left	M 614	Telephone preparation
M 438	Comfort seat, right	M 618	Telephone
M 441	Radio preparation	M 635	Parking assistant
M 454	Cruise control	M 651	Power windows
M 465	Rear foglight, left side	M 659	On-board computer
M 466	Rear foglight, right side	M 660	On-Board Diagnostics II (USA)
M 476	Brake pads, higher friction value	M 661	Stricter emissions concept
M 479	Australia version	M 662	Infosystem / navigation
M 480	6-speed manual transmission	M 664	On-board fuel vapor recovery system (ORVR)
M 481	5-speed manual transmission	M 680	Digital sound package
M 484	USA version, inscription in German	M 689	CD-changer
M 490	Audio option package	M 692	CD-changer
M 513	Lumbar support, right seat	MXE3/MXEH	Auto-dimming interior mirror

X-connectors

Porsche identifies harness connectors in the wiring diagrams using the letter X followed by a number(s). To see an example of an X-connector in a wiring diagrams, look in grid coordinate D4 (1998 diagrams). There you will see X 1/2, which is a 21-pin, black connector. **Table d** lists X-connectors and their locations.

Table d. X-connectors

Connector	Location, additional information	Connector	Location, additional information
X 1/1	In front trunk, front left, 21 pins, white	X 20	Center of dashboard, center console, 4 pins, black
X 1/2	In front trunk, front left, 21 pins, black	X 20/1	Below center of dashboard, 12 pins, black
X 3/1	In rear trunk, left side, 30 pins, black	X 20/1	Below center of dashboard, 10 pins, black
X 3/2	In rear trunk, left side, 21 pins, blue	X 24	2 pins, black
X 10	4 pins, black	X 59	In rear trunk, left side, 21 pins, white
X 11	A-pillar, driver's door connector, 46 pins, black	X 65	In driver's door, 15 pins, white
X 12	A-pillar, passenger's door connector, 46 pins, black	X 66	In driver's door, 15 pins, white
X 15	In body left side, ahead of rear wheel, 3 pins, black	X 2/1	In rear trunk, left side, 21 pins, black
X 16	Under driver's seat, 6,8 pins, white	X2/2	In rear trunk, left side, 21 pins, yellow
X 17	Under passenger seat, 6, 8 pins, black	X 2/3	In rear trunk, left side, 6 pins, black

Electrical Wiring Diagrams EWD-9

Lighting RoW
(page 1 of 3)

1998

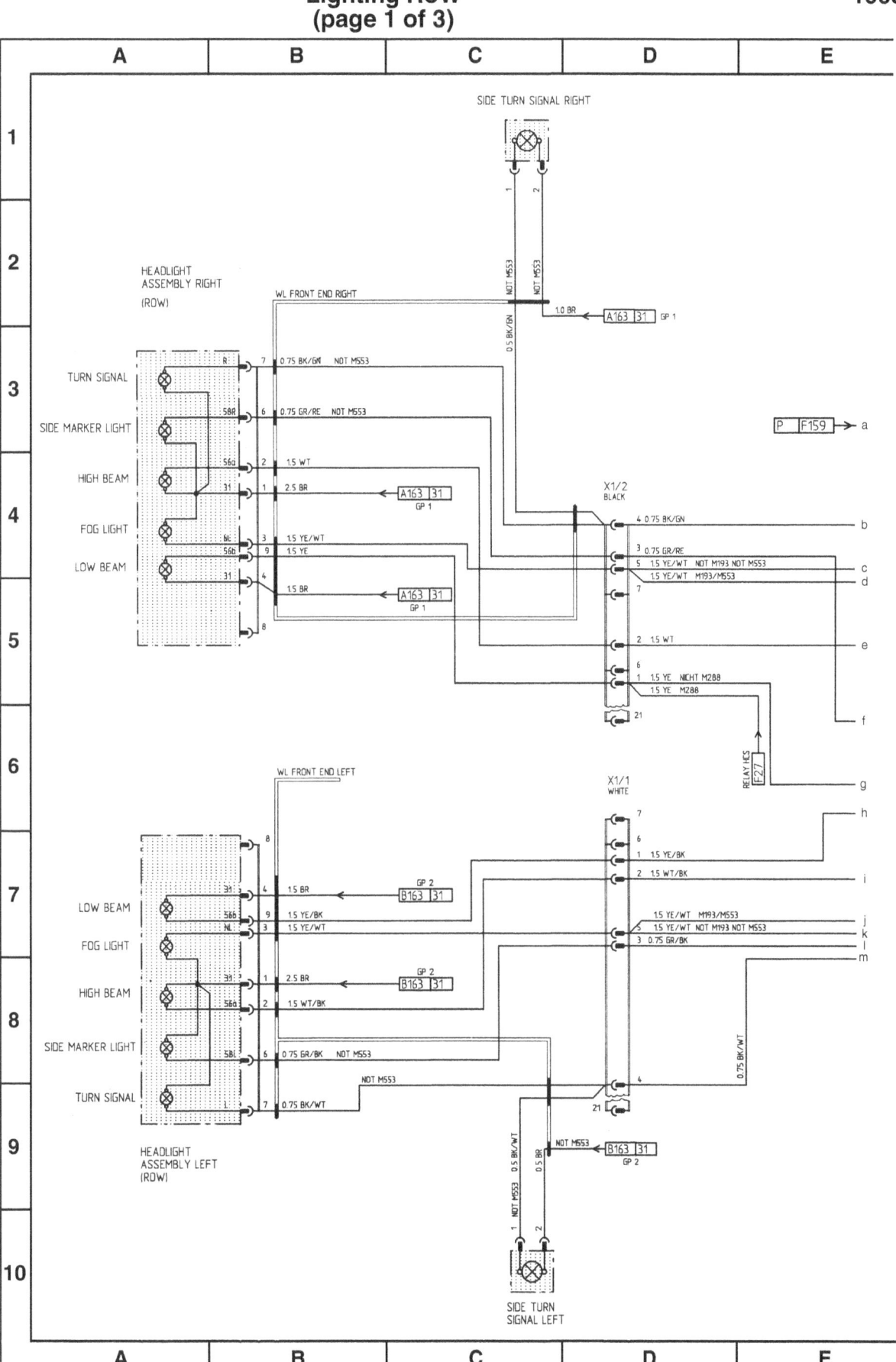

1998 Lighting RoW (page 2 of 3)

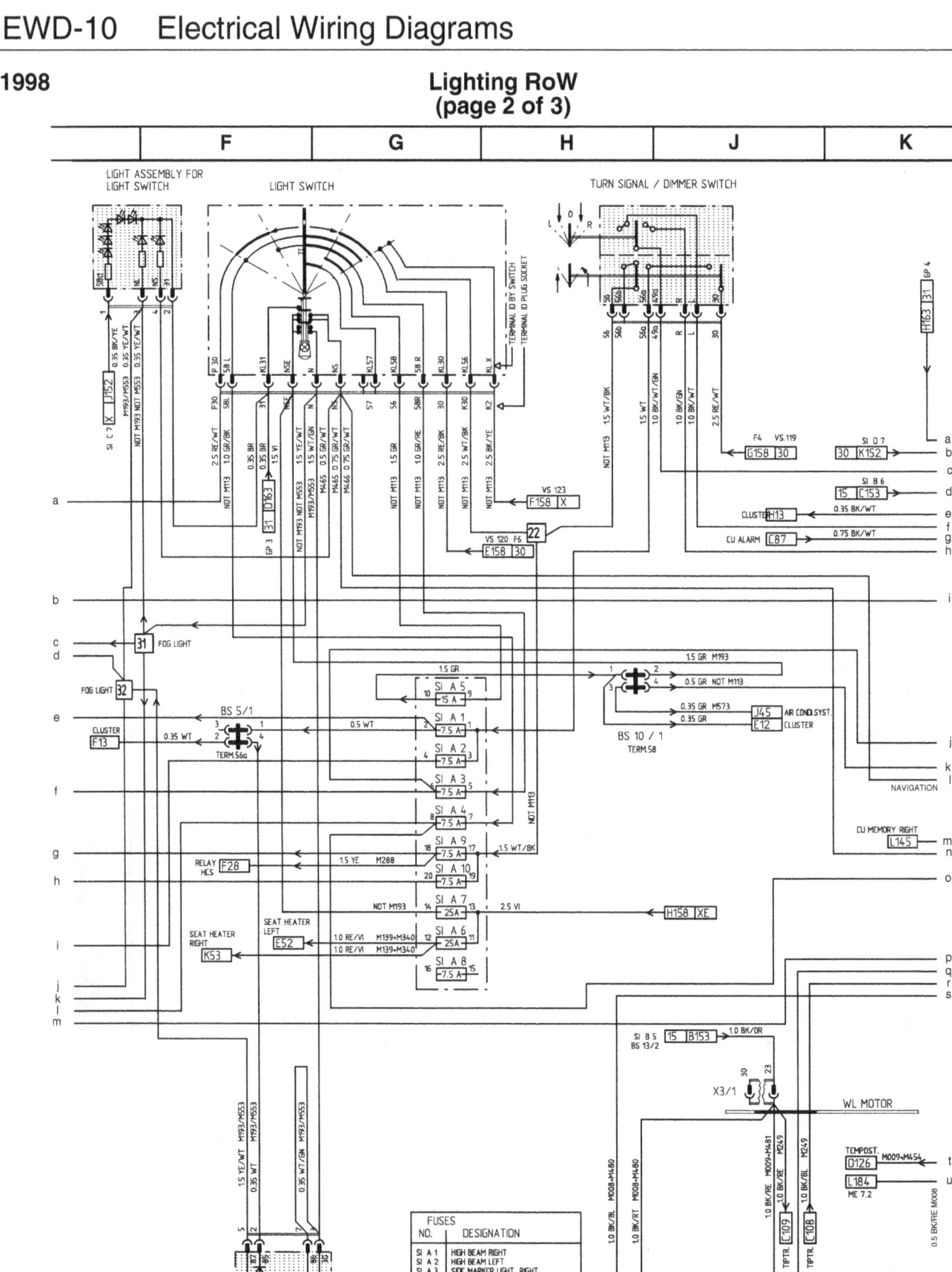

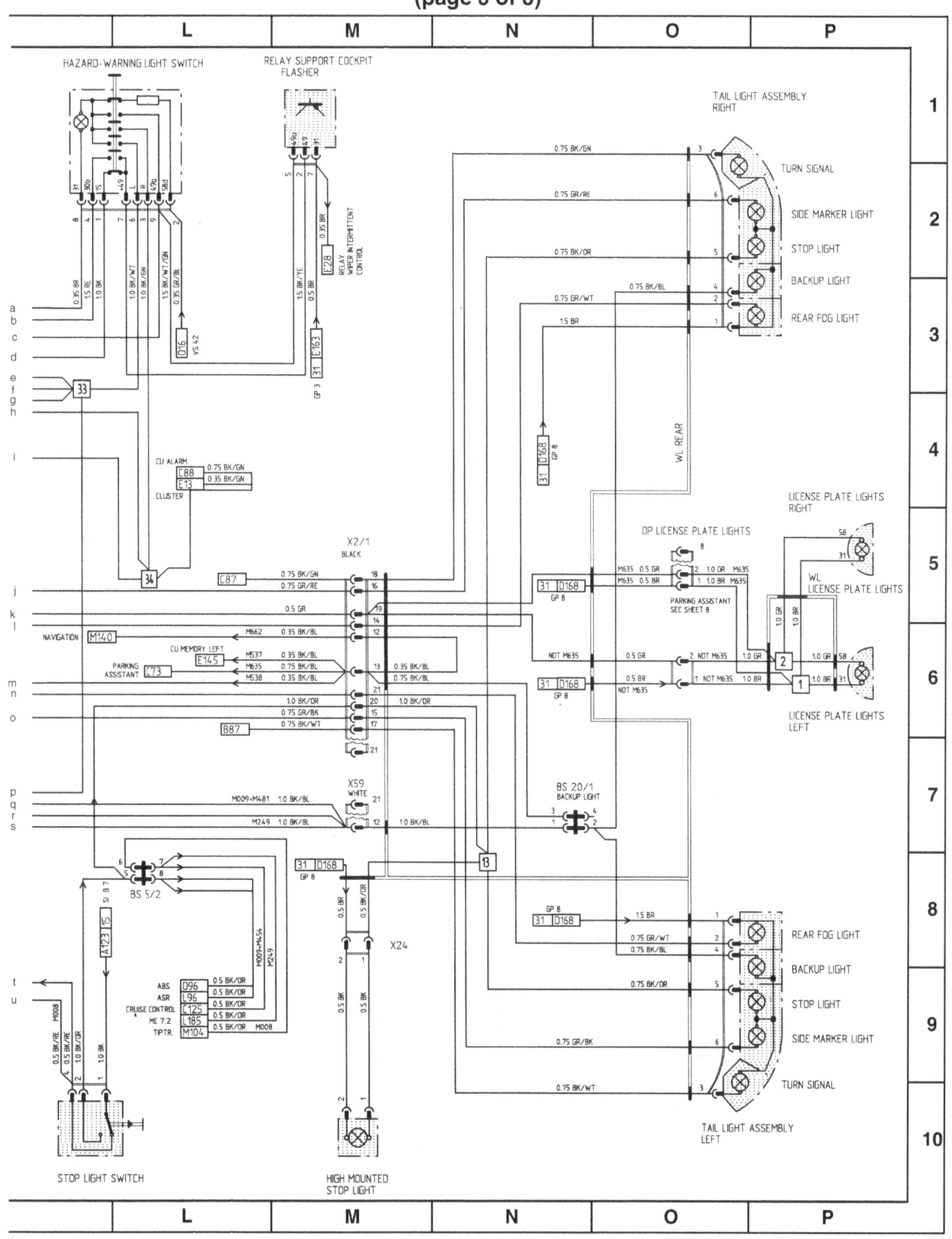

EWD-12 Electrical Wiring Diagrams

1998 — Lighting USA / Canada (page 1 of 3)

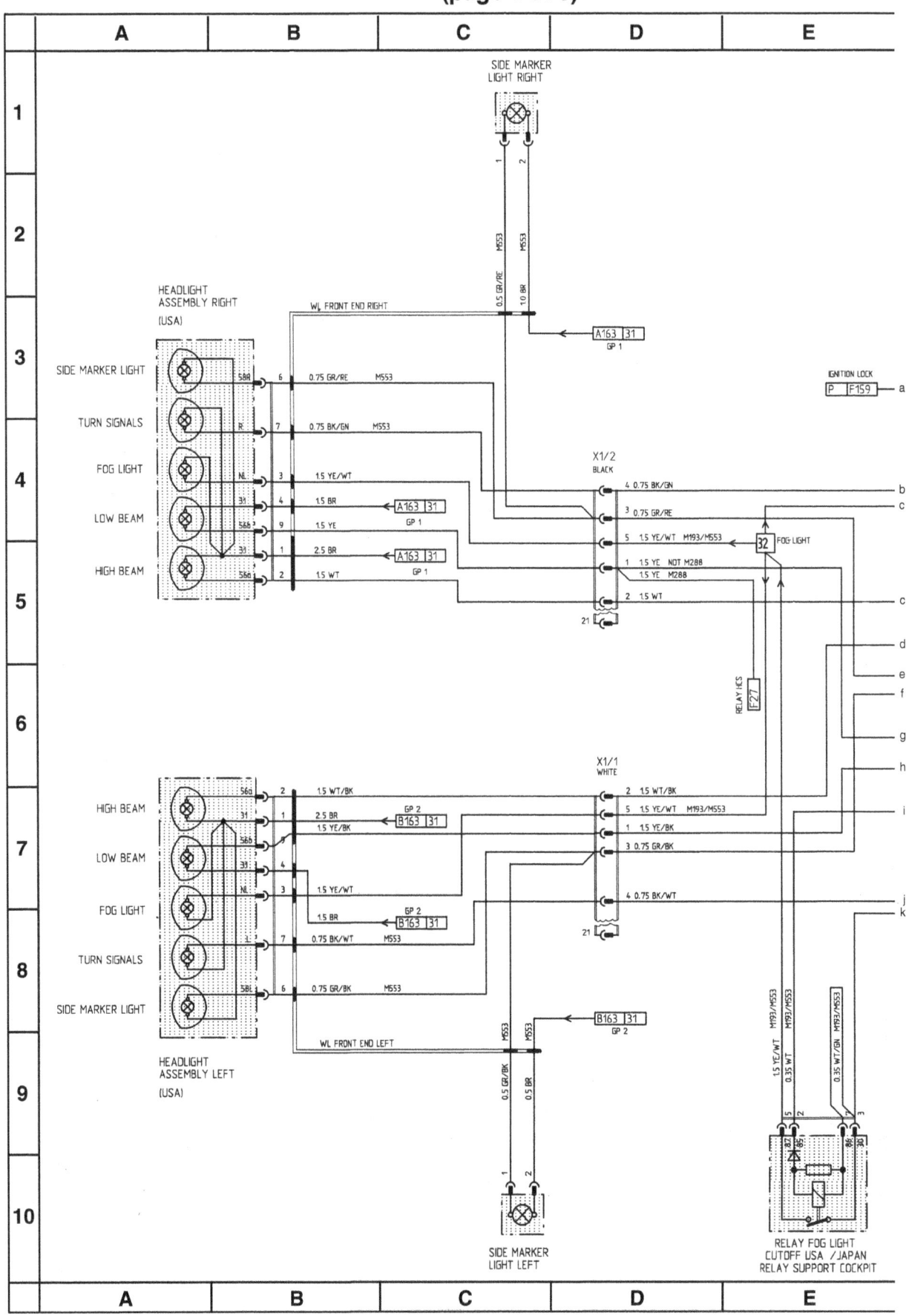

Lighting USA / Canada
(page 2 of 3)

1998

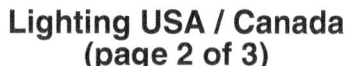

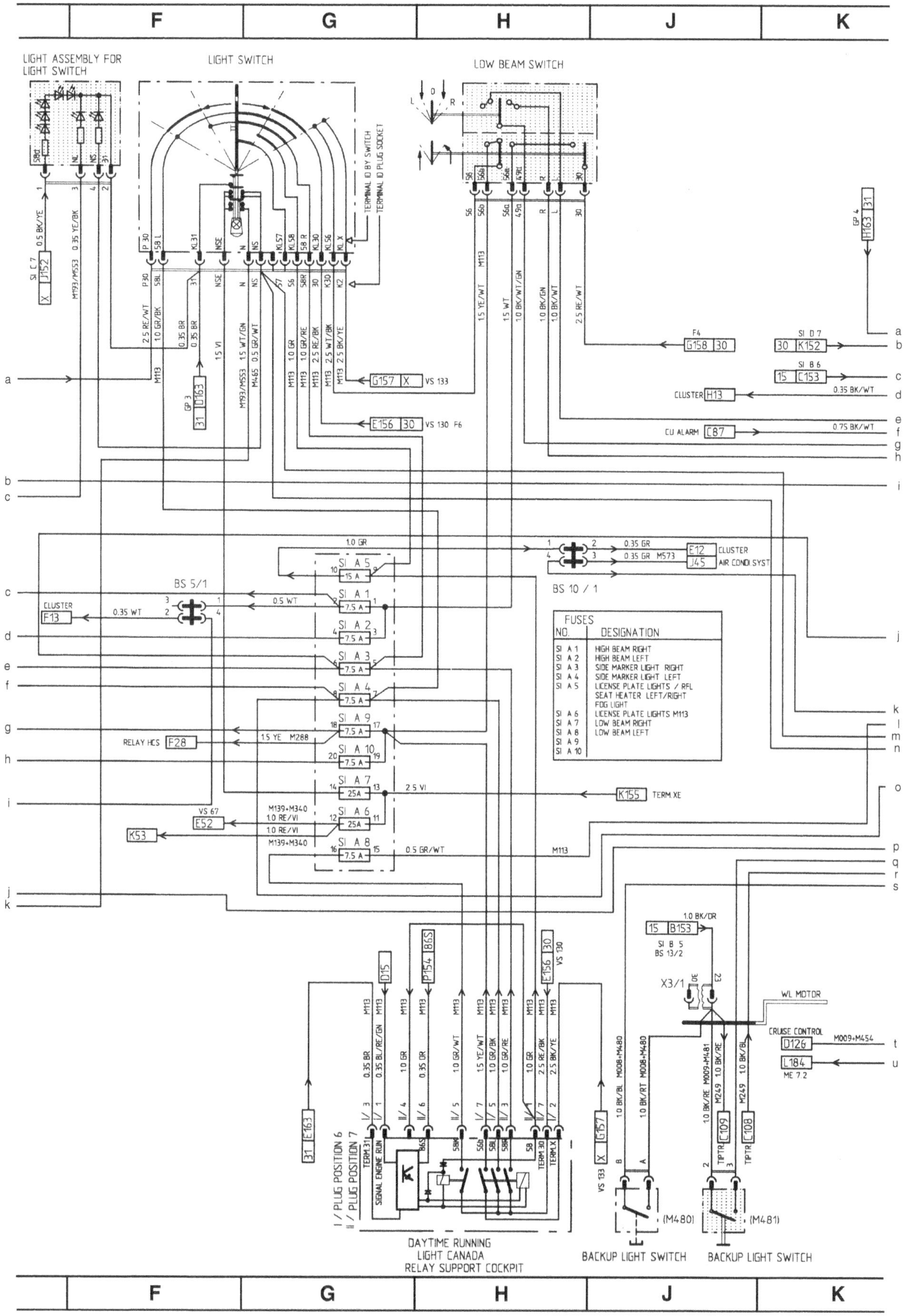

FUSES NO.	DESIGNATION
SI A 1	HIGH BEAM RIGHT
SI A 2	HIGH BEAM LEFT
SI A 3	SIDE MARKER LIGHT RIGHT
SI A 4	SIDE MARKER LIGHT LEFT
SI A 5	LICENSE PLATE LIGHTS / RFL SEAT HEATER LEFT/RIGHT FOG LIGHT
SI A 6	LICENSE PLATE LIGHTS M113
SI A 7	LOW BEAM RIGHT
SI A 8	LOW BEAM LEFT
SI A 9	
SI A 10	

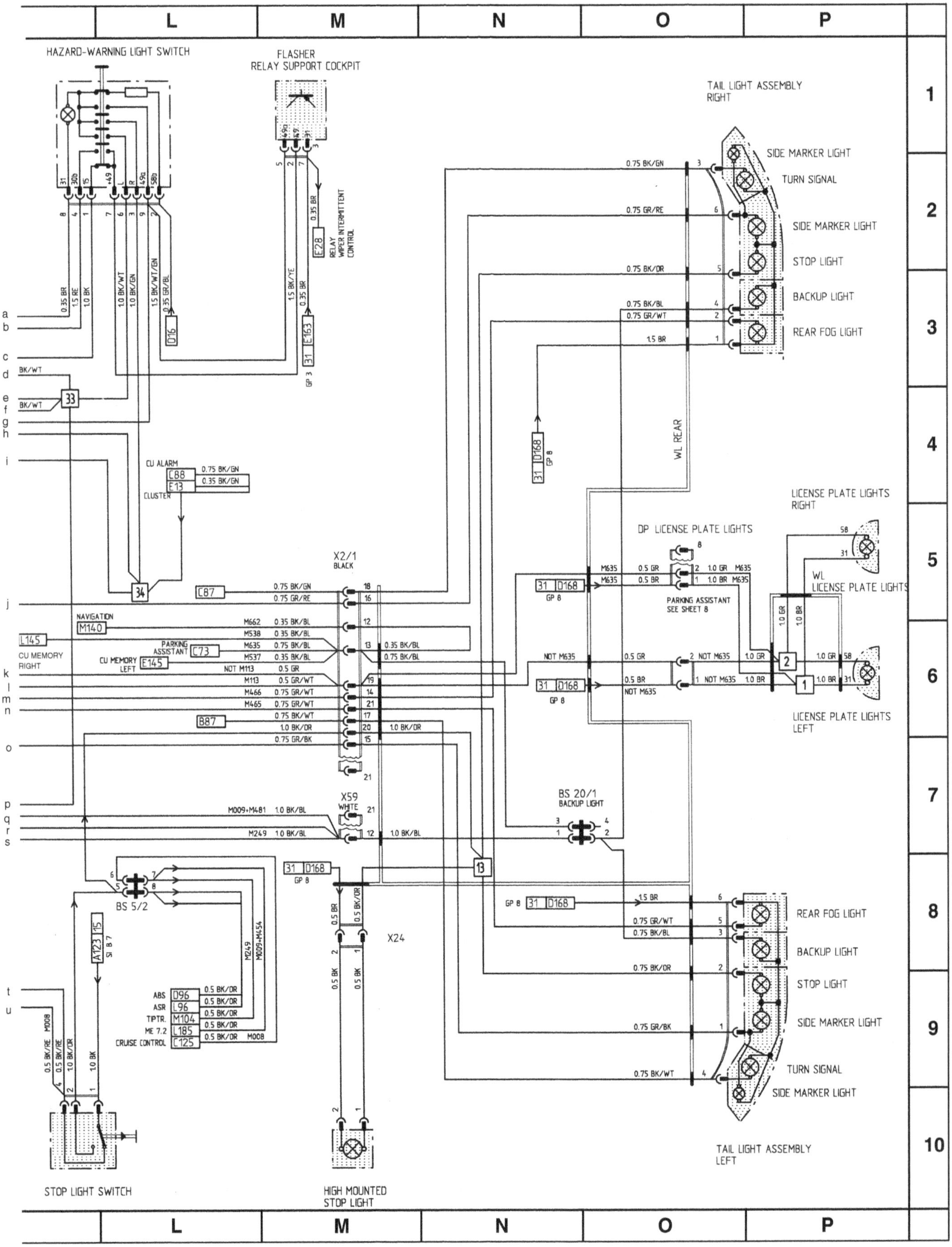

Electrical Wiring Diagrams EWD-15

Instrument cluster
(page 1 of 3)

1998

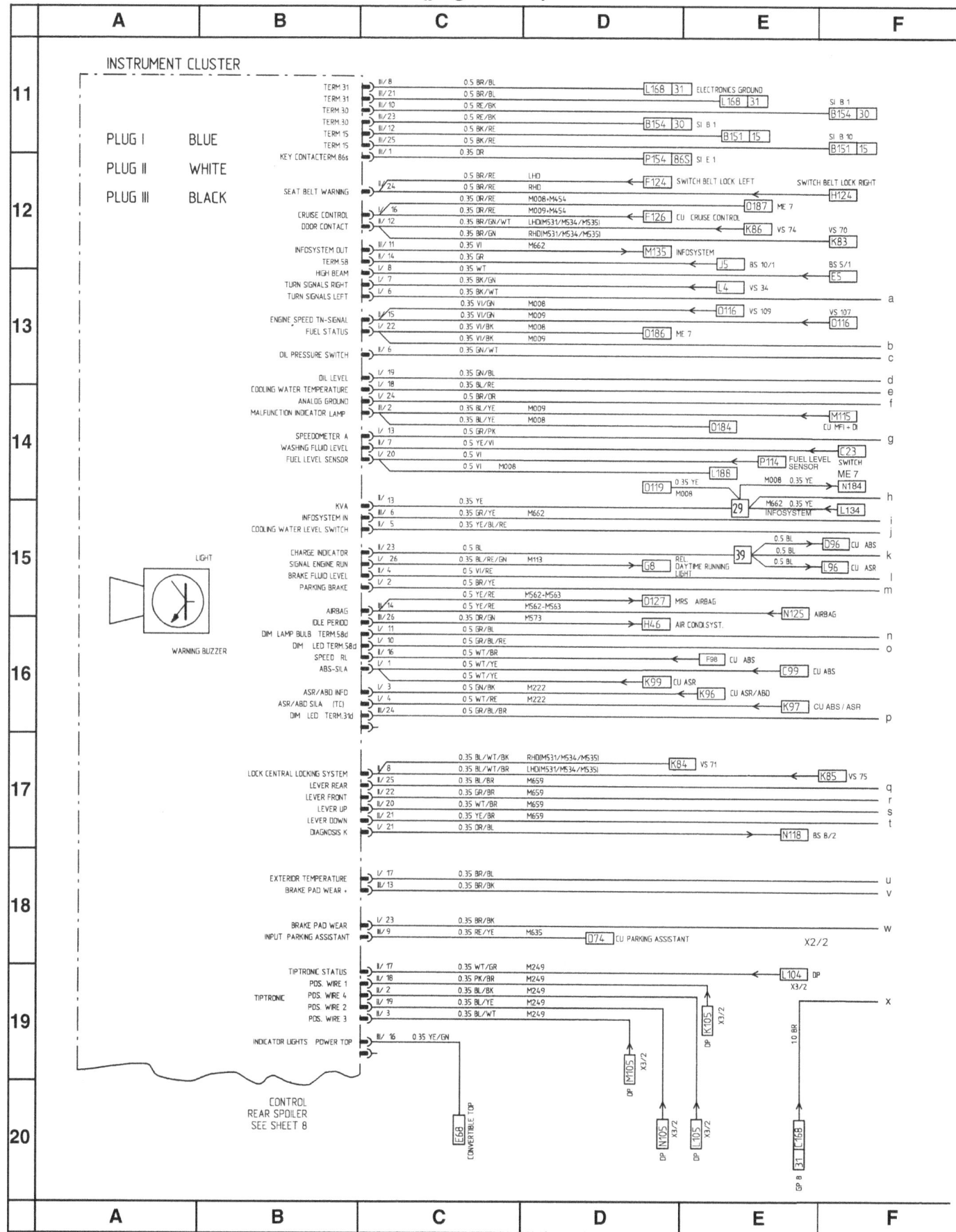

Instrument cluster
(page 2 of 3)

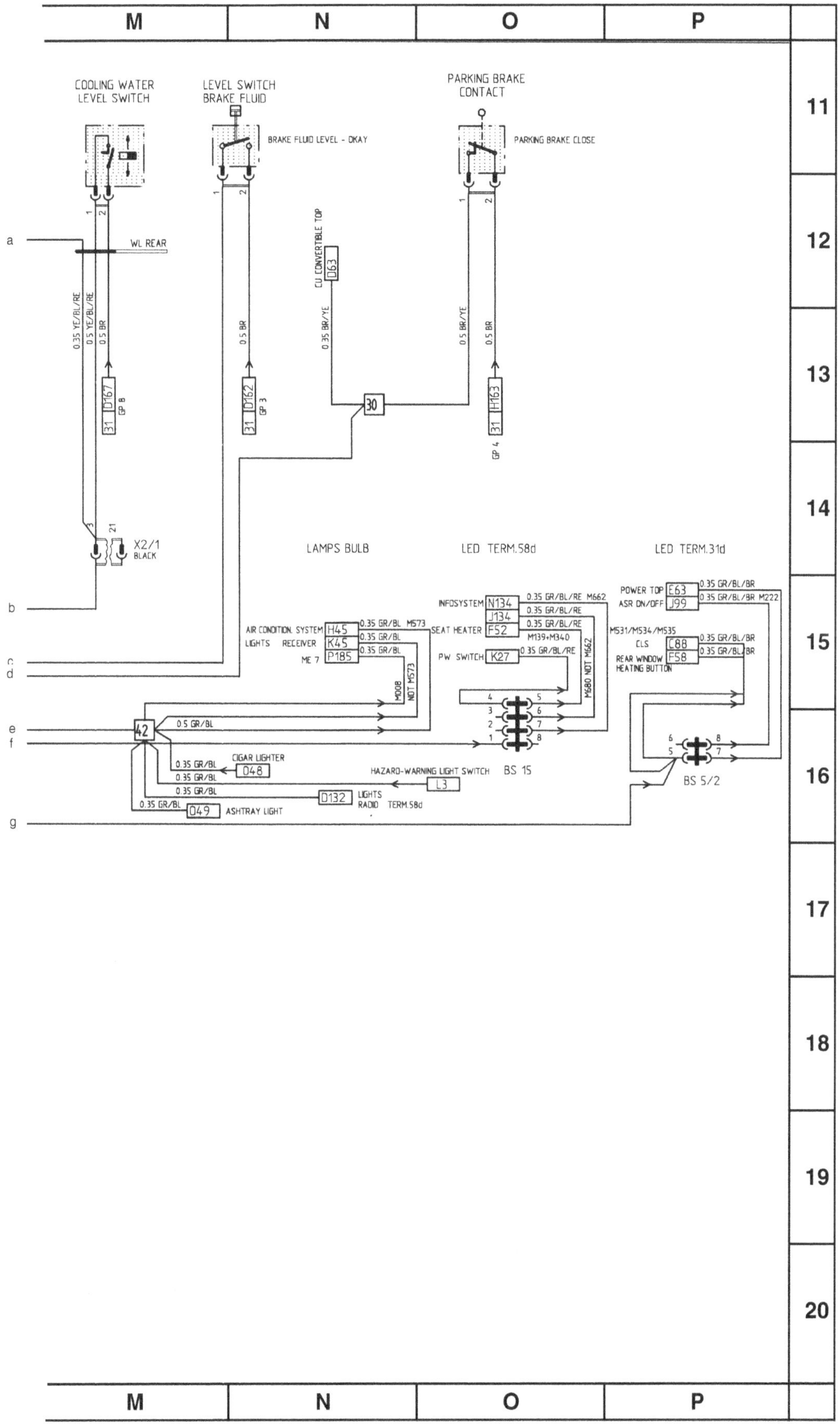

bentleypublishers.com

Instrument cluster
(page 3 of 3)

1998

EWD-17

EWD-18 Electrical Wiring Diagrams

1998   Windshield wipers, washers, horns

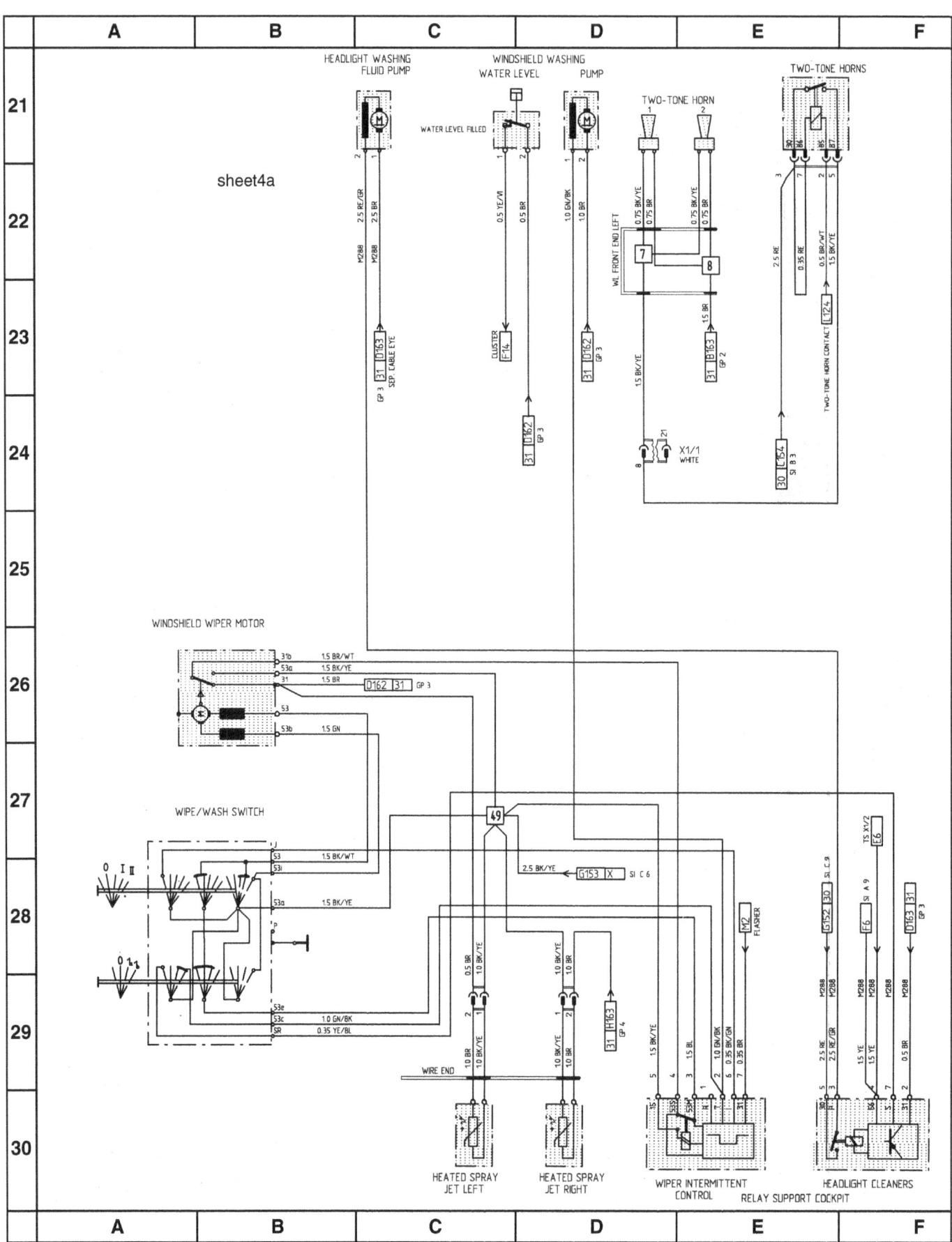

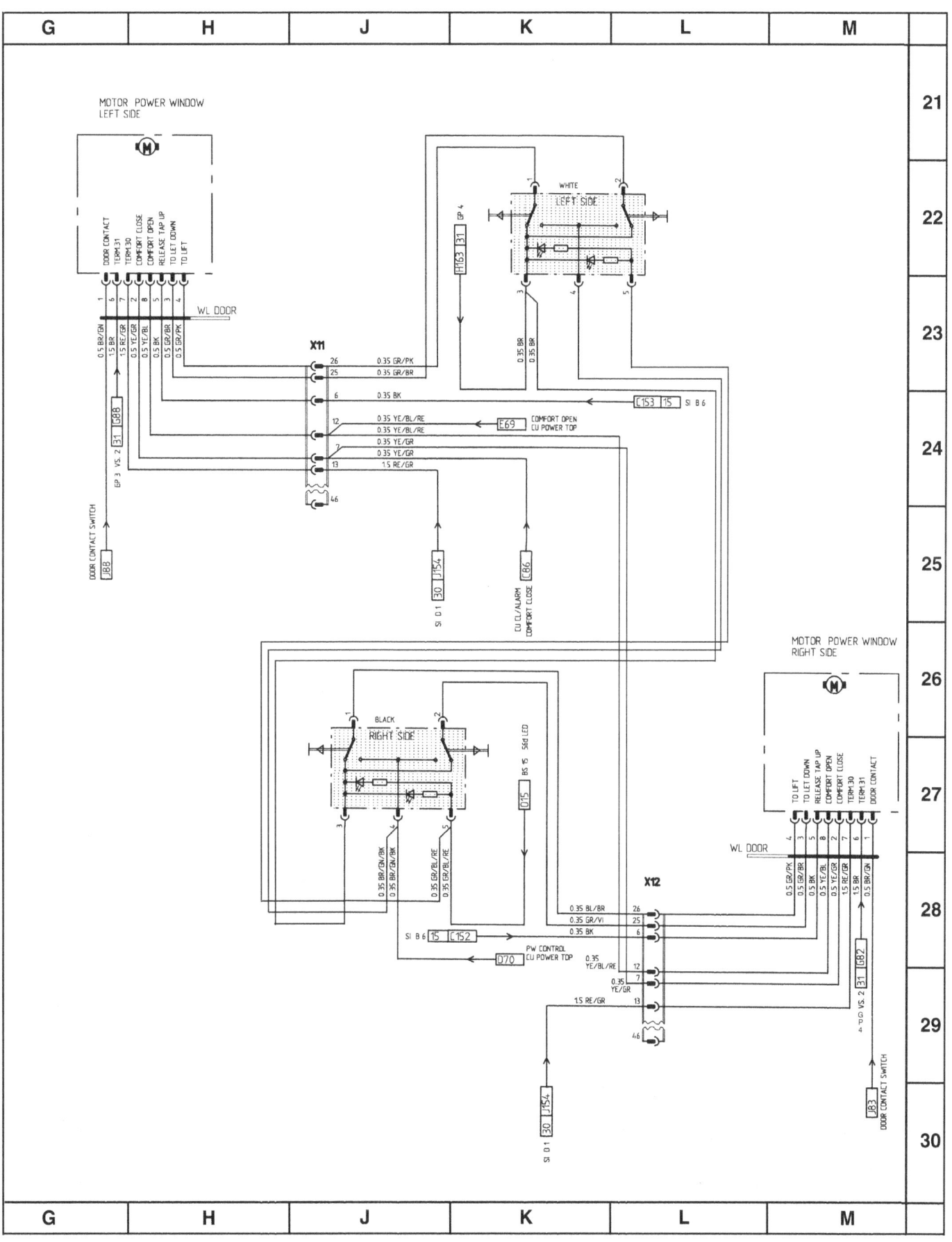

Electrical Wiring Diagrams EWD-19
Windows
1998

Electrical Wiring Diagrams

Engine cooling, heating, air-conditioning (page 1 of 3)

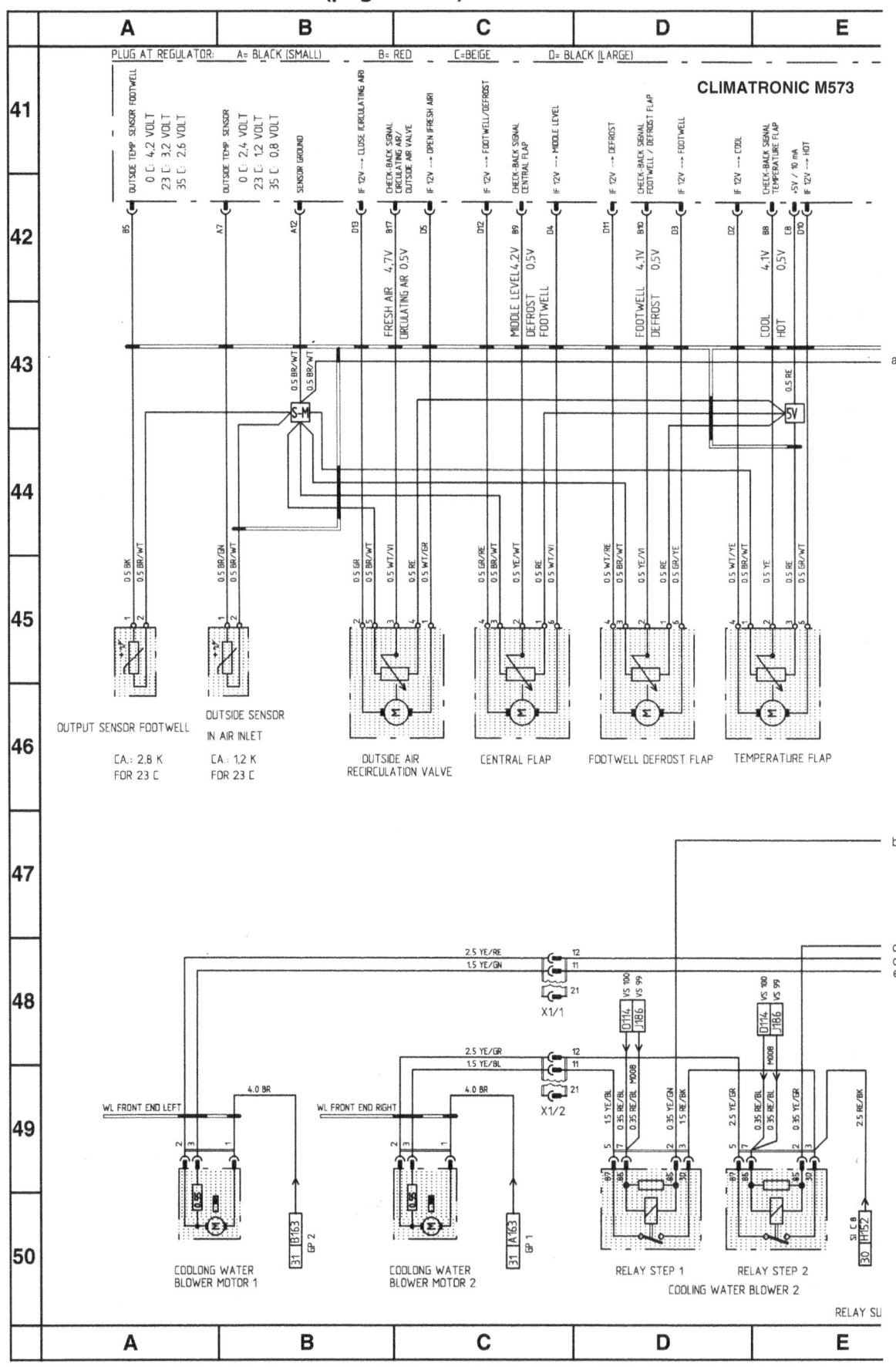

Electrical Wiring Diagrams EWD-21

Engine cooling, heating, air-conditioning (page 2 of 3)

1998

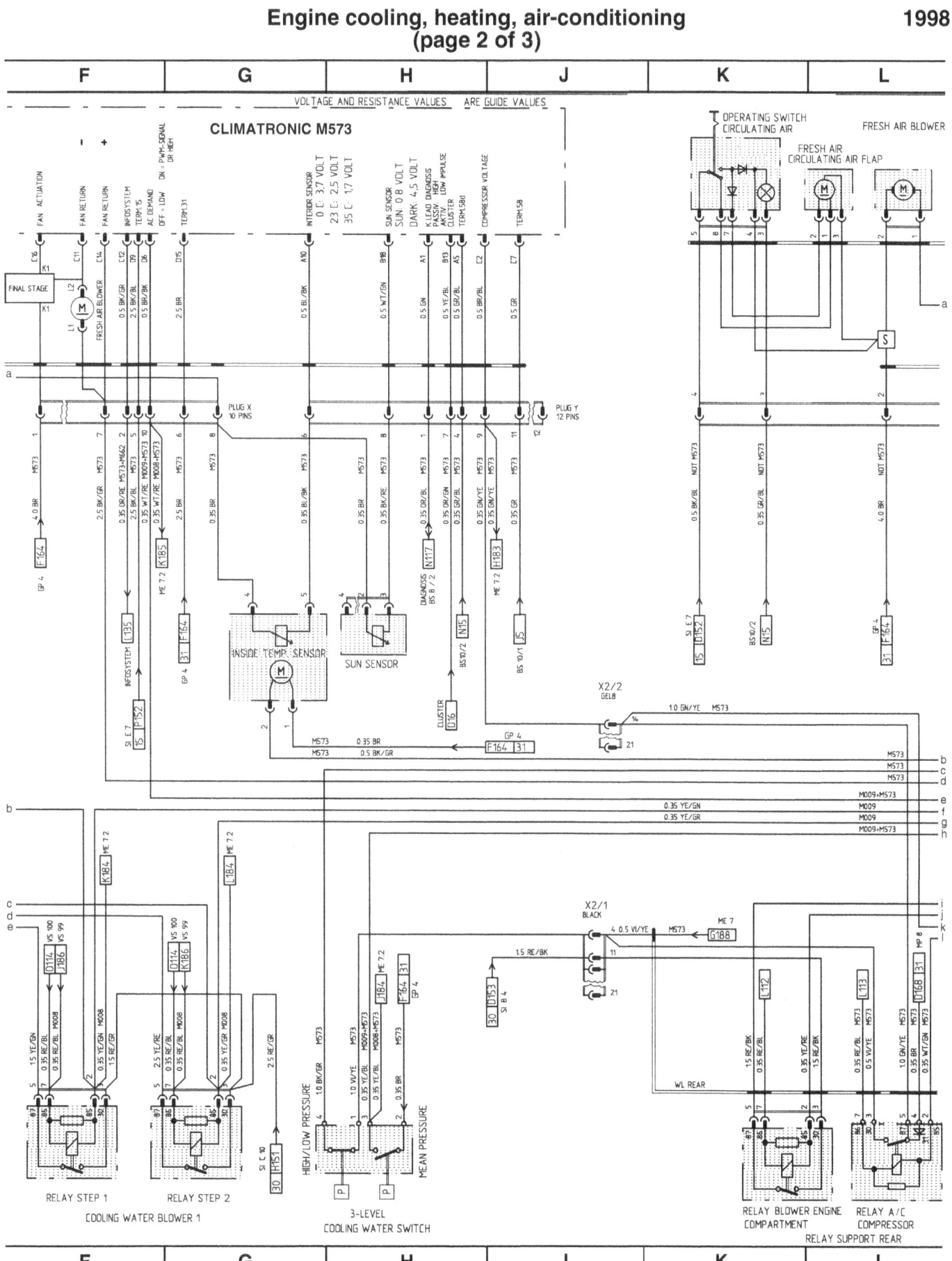

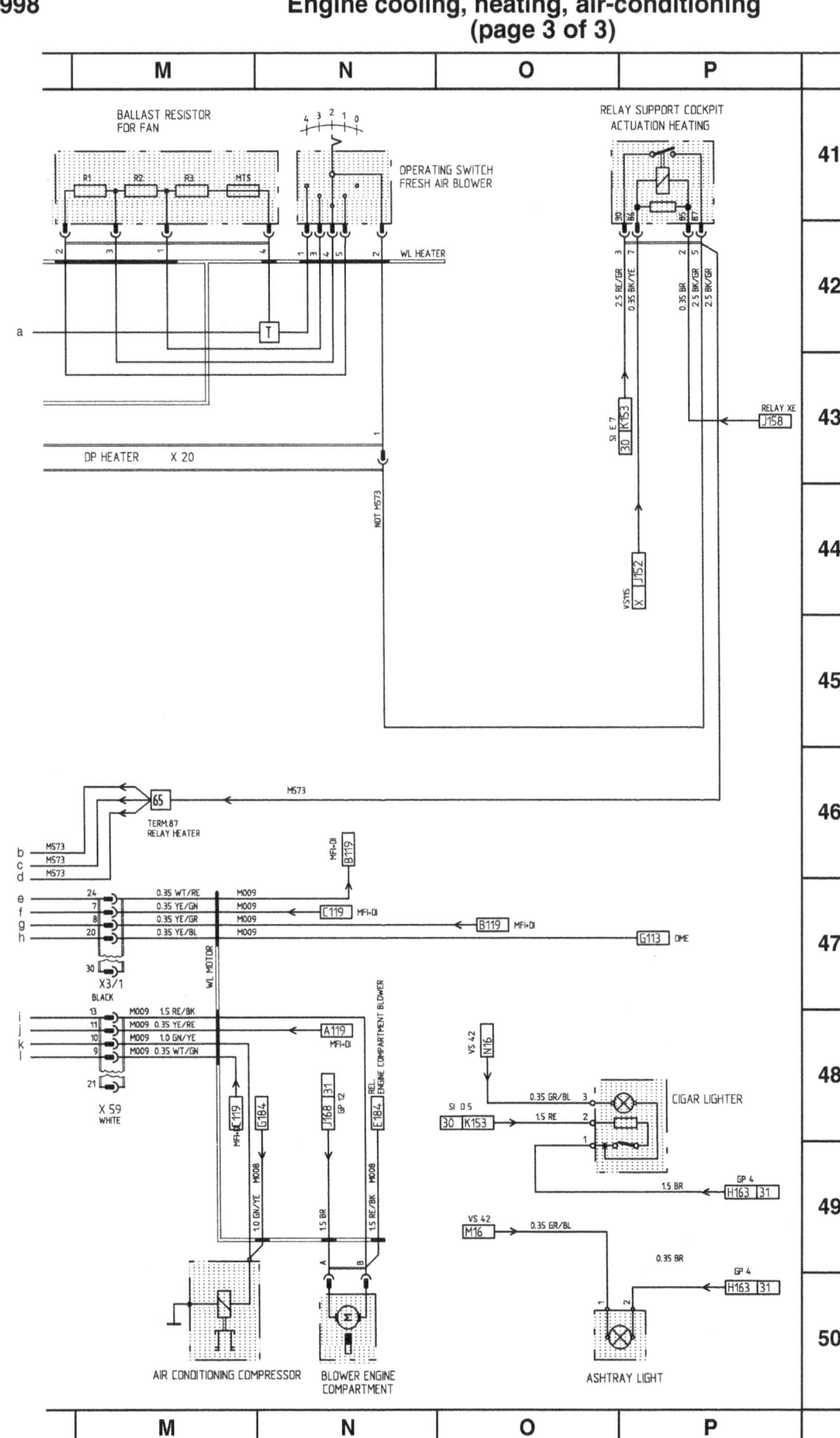

Electrical Wiring Diagrams EWD-23

Seat and mirror heating, rear window defogger (page 1 of 3)

1998

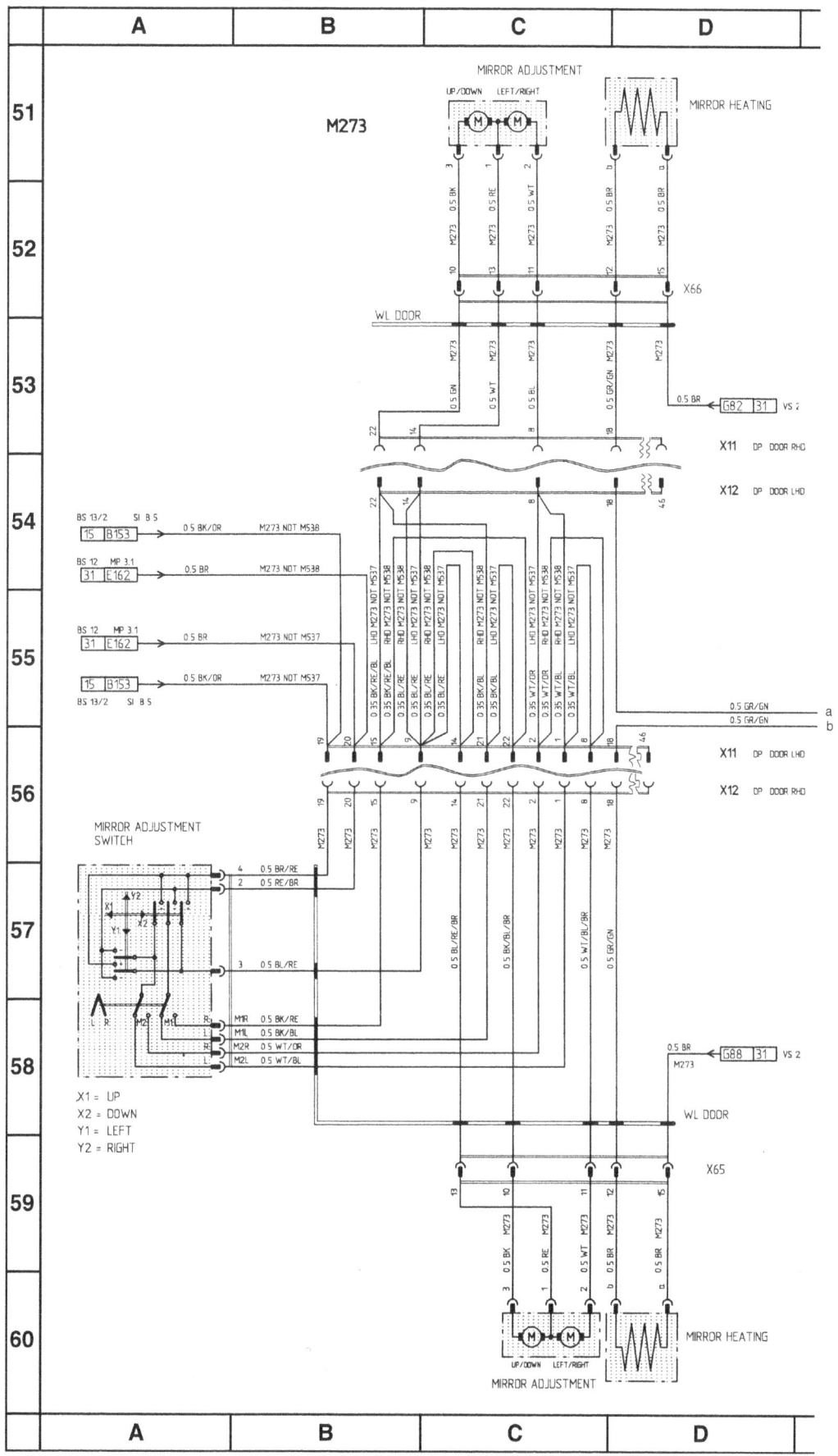

Electrical Wiring Diagrams

1998

Seat and mirror heating, rear window defogger
(page 2 of 3)

Electrical Wiring Diagrams EWD-25

Seat and mirror heating, rear window defogger (page 3 of 3)

1998

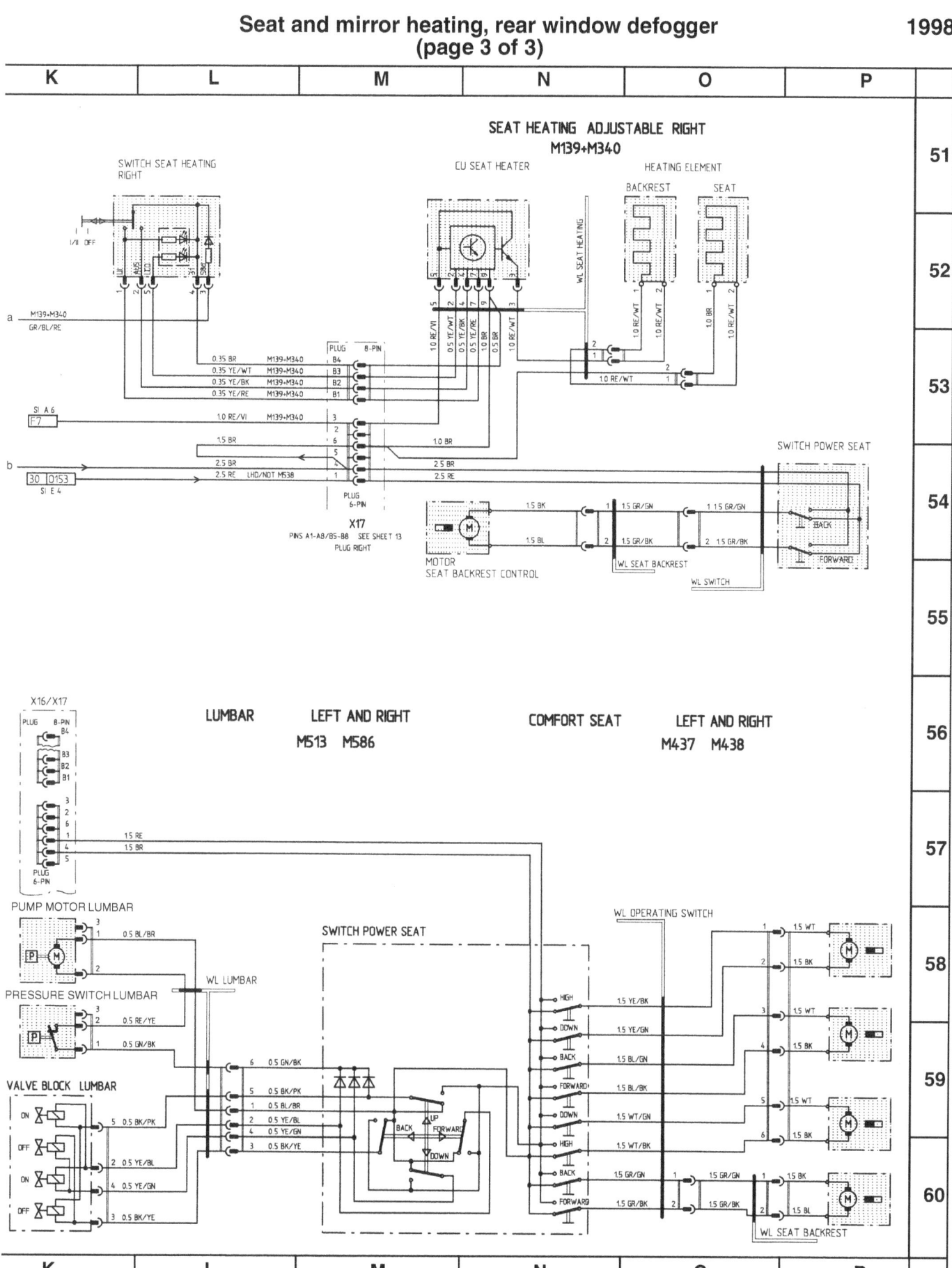

EWD-26 Electrical Wiring Diagrams

1998 Convertible top

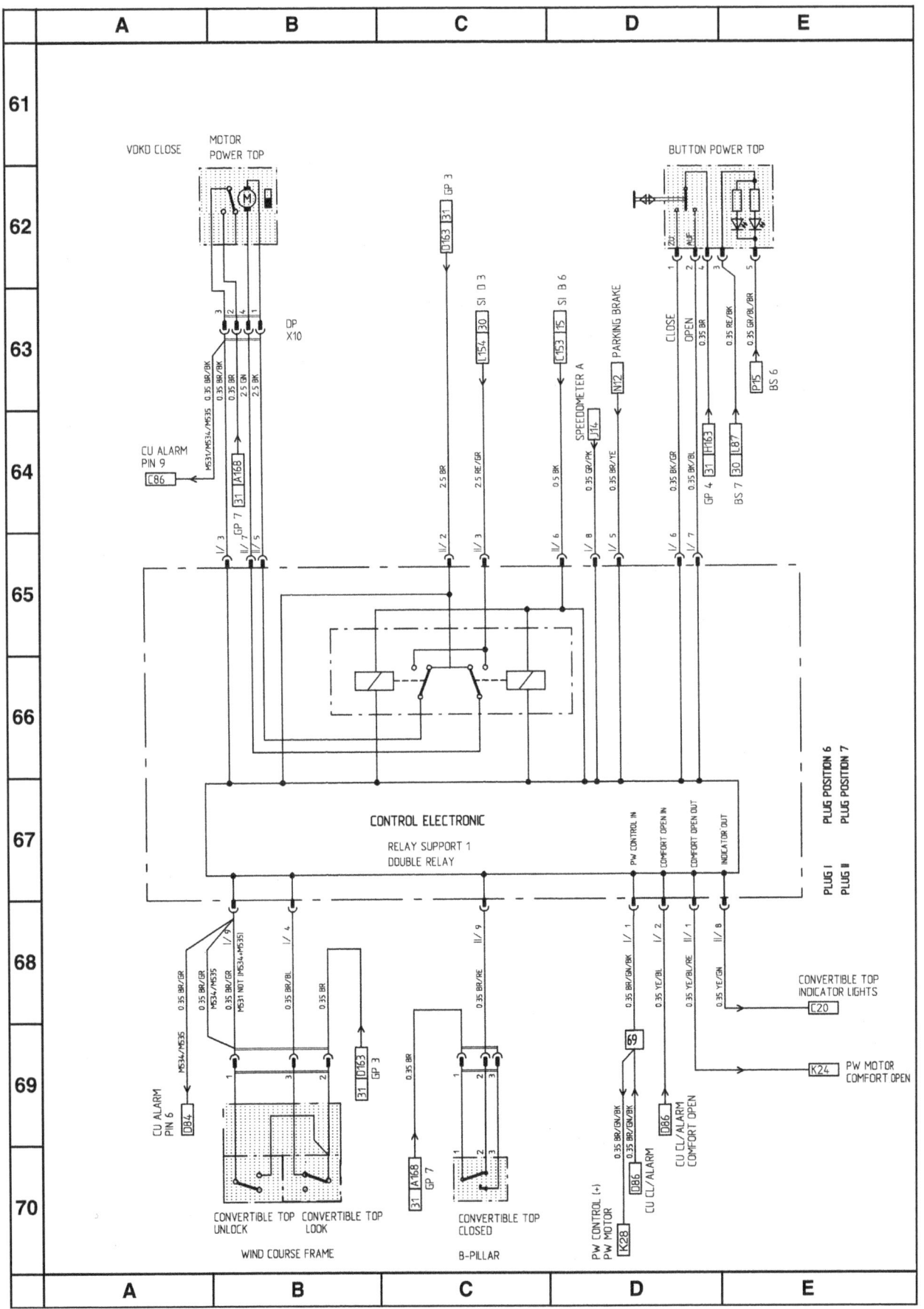

Parking assistant, rear spoiler (page 1 of 3)

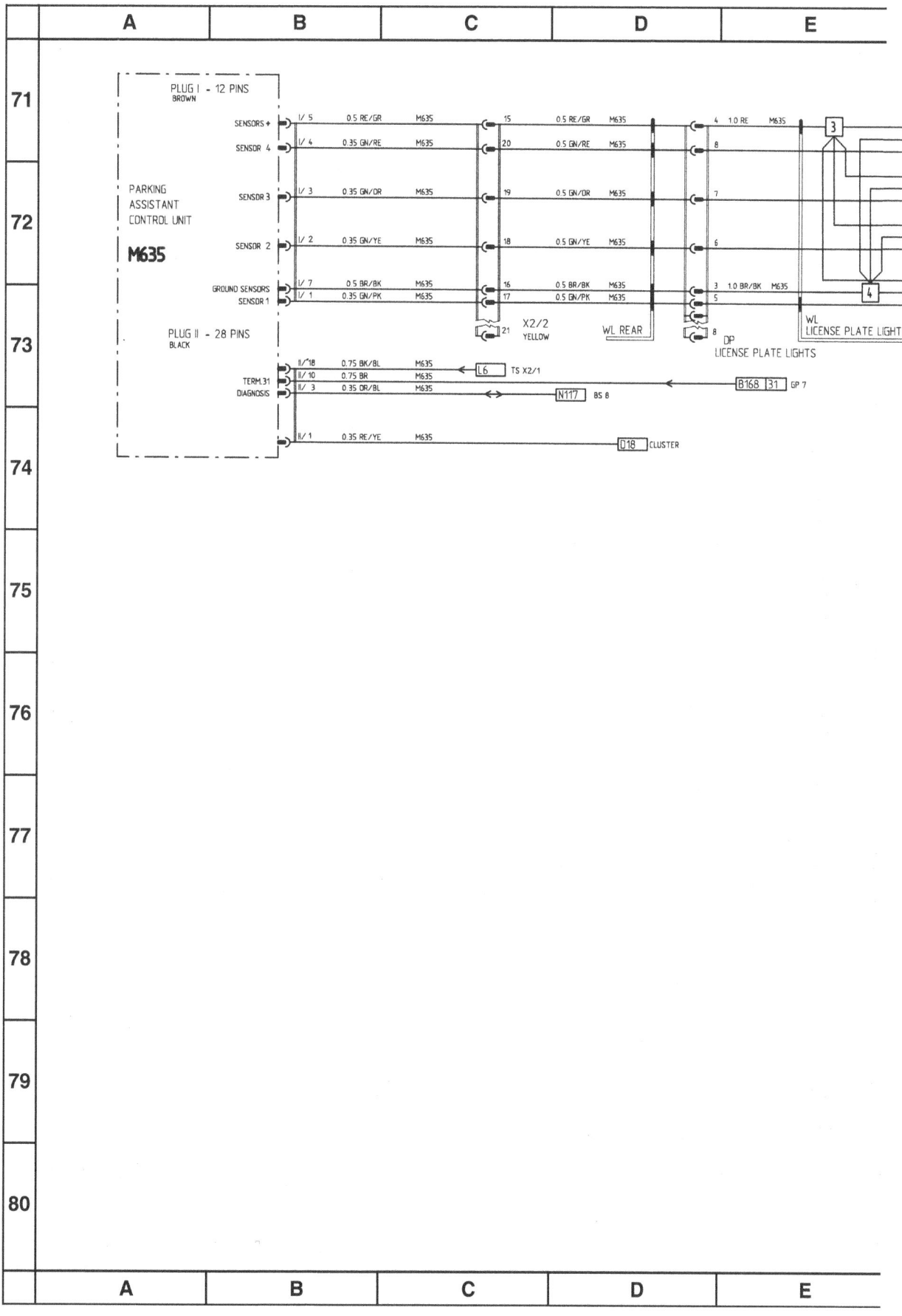

Parking assistant, rear spoiler
(page 2 of 3)

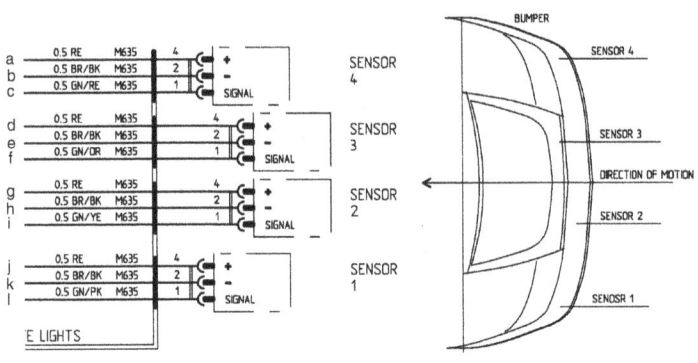

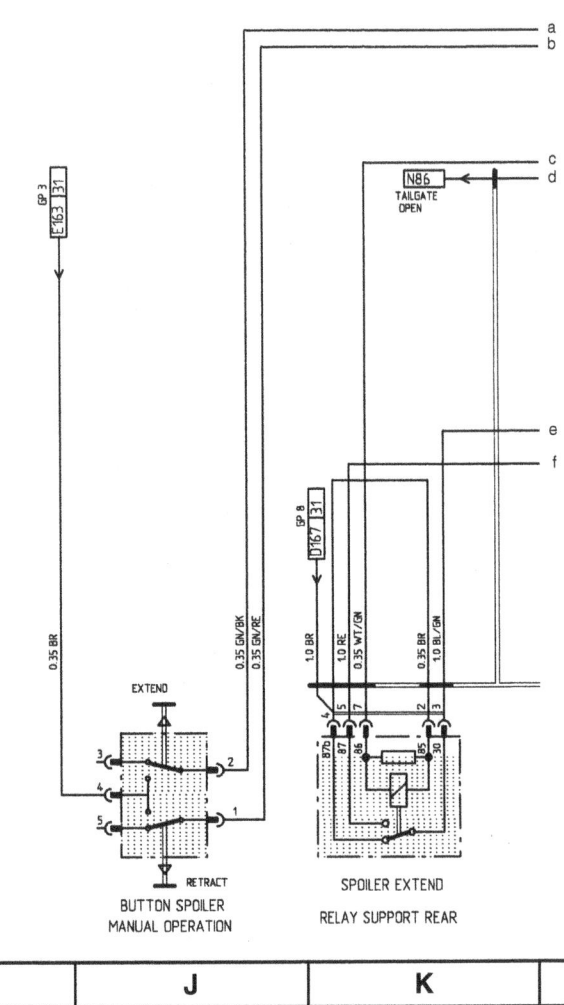

Parking assistant, rear spoiler
(page 3 of 3)

Electrical Wiring Diagrams EWD-29
1998

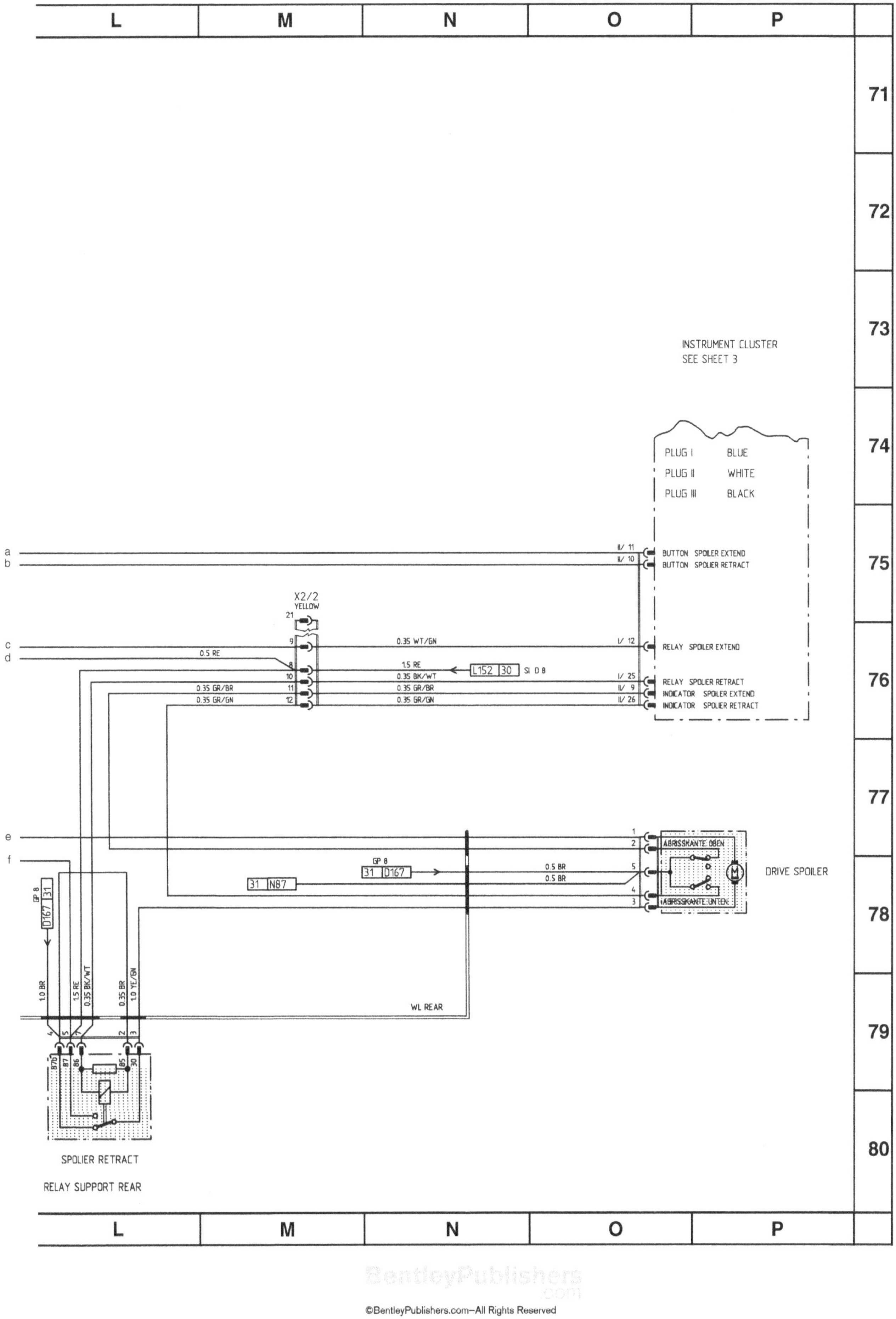

EWD-30 Electrical Wiring Diagrams

1998 — Central locking, alarm system (page 1 of 3)

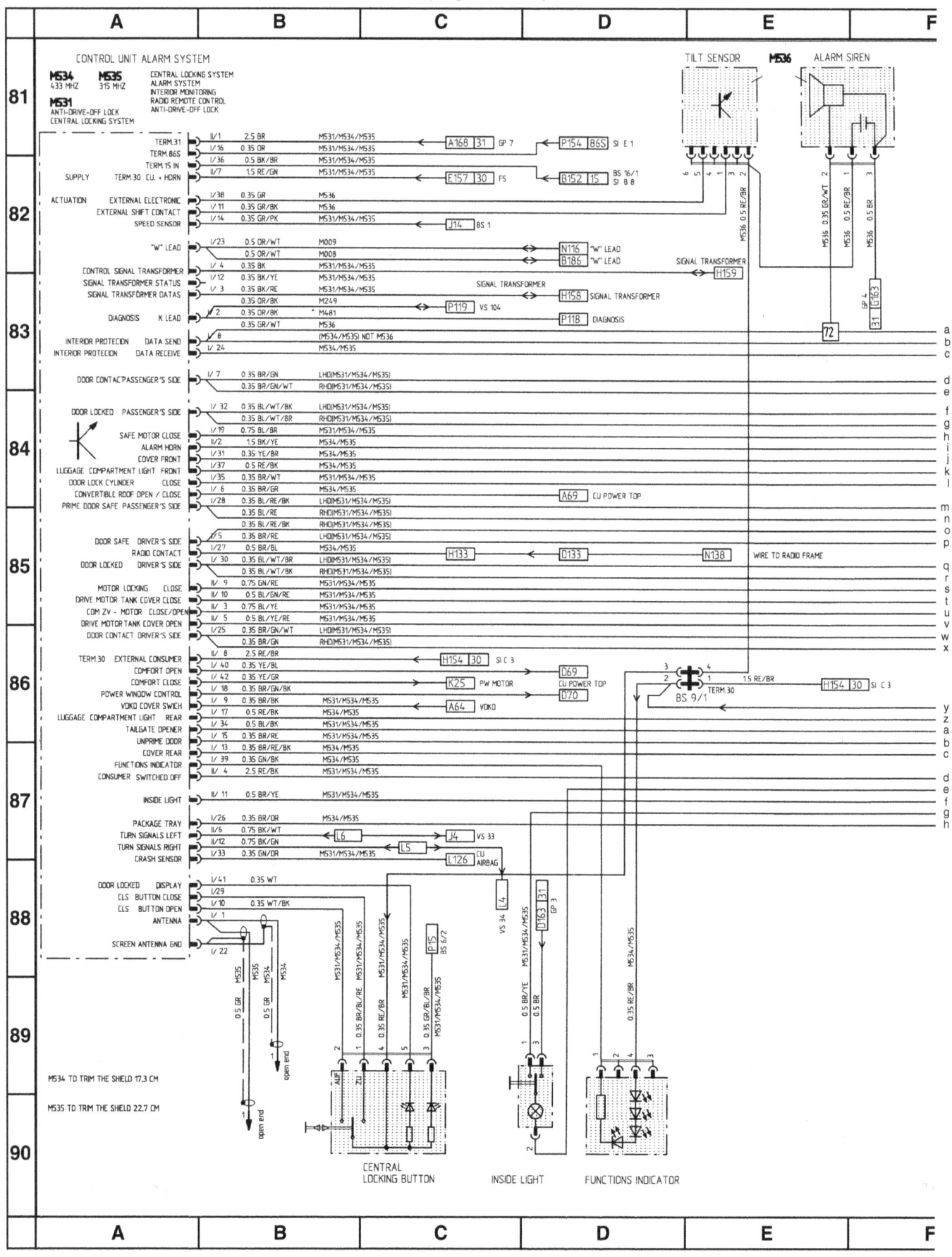

Electrical Wiring Diagrams EWD-31

Central locking, alarm system
(page 2 of 3)

1998

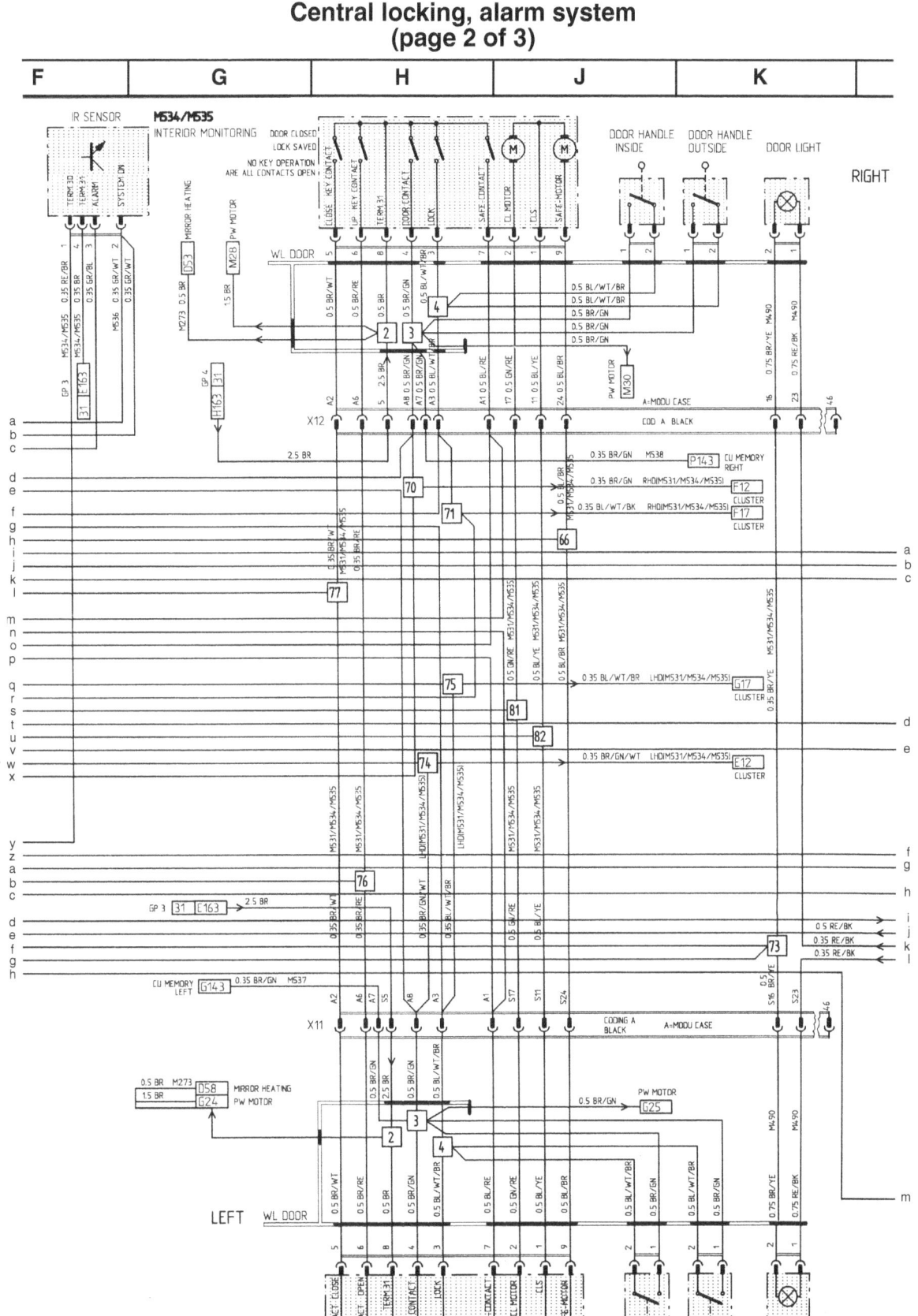

EWD-32 Electrical Wiring Diagrams

1998 Central locking, alarm system (page 3 of 3)

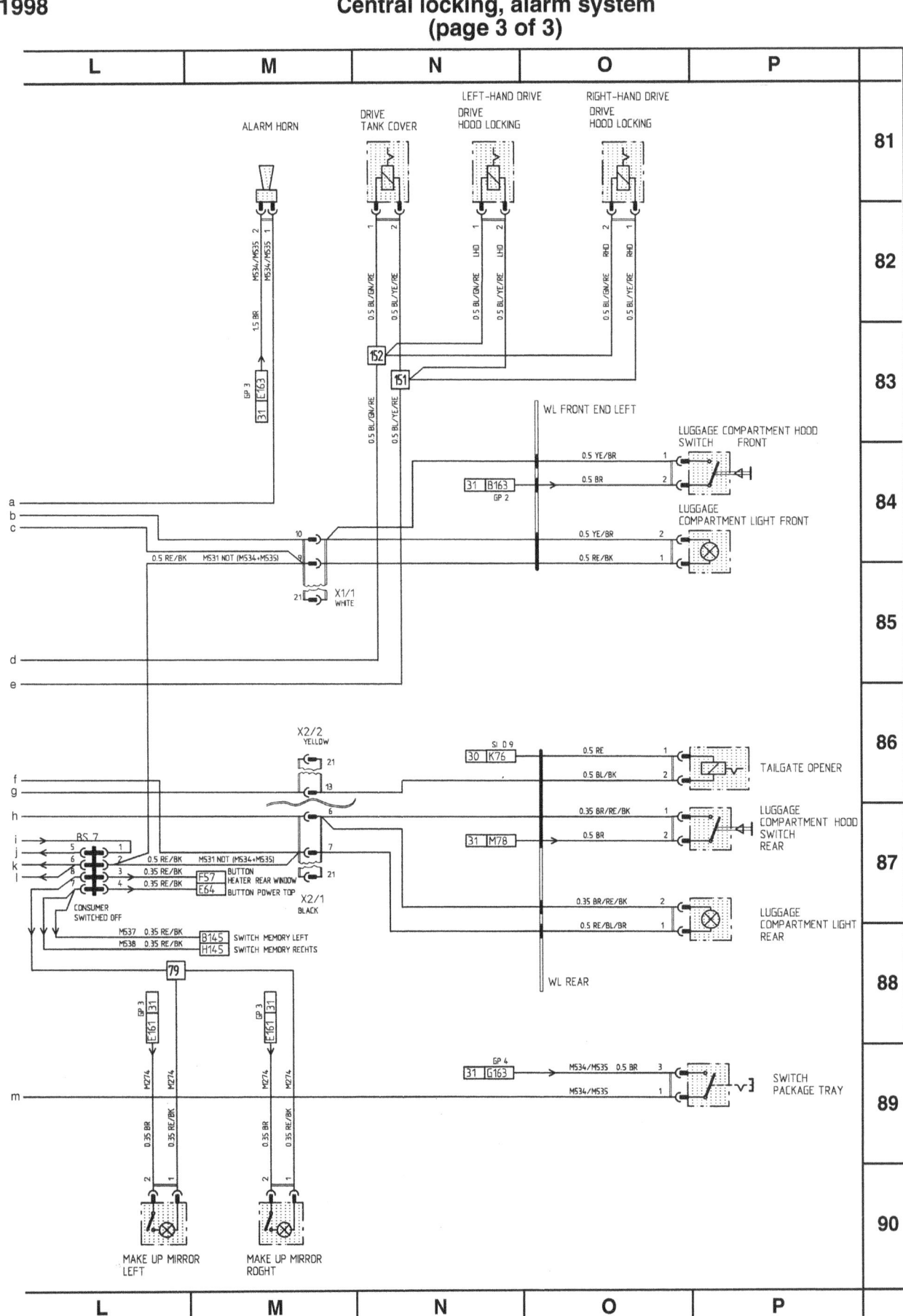

Electrical Wiring Diagrams   EWD-33

ABS / ASR
(page 1 of 2)

1998

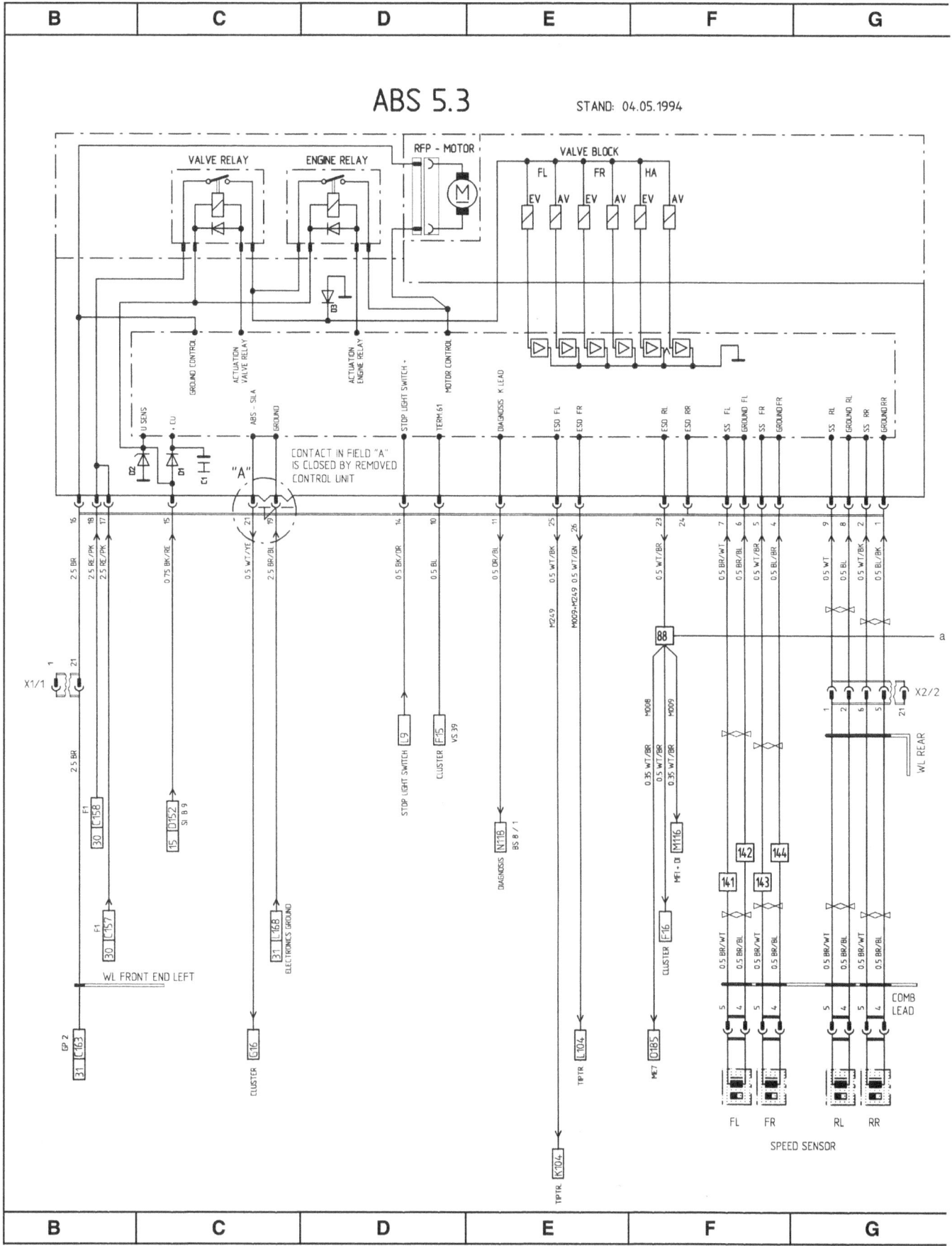

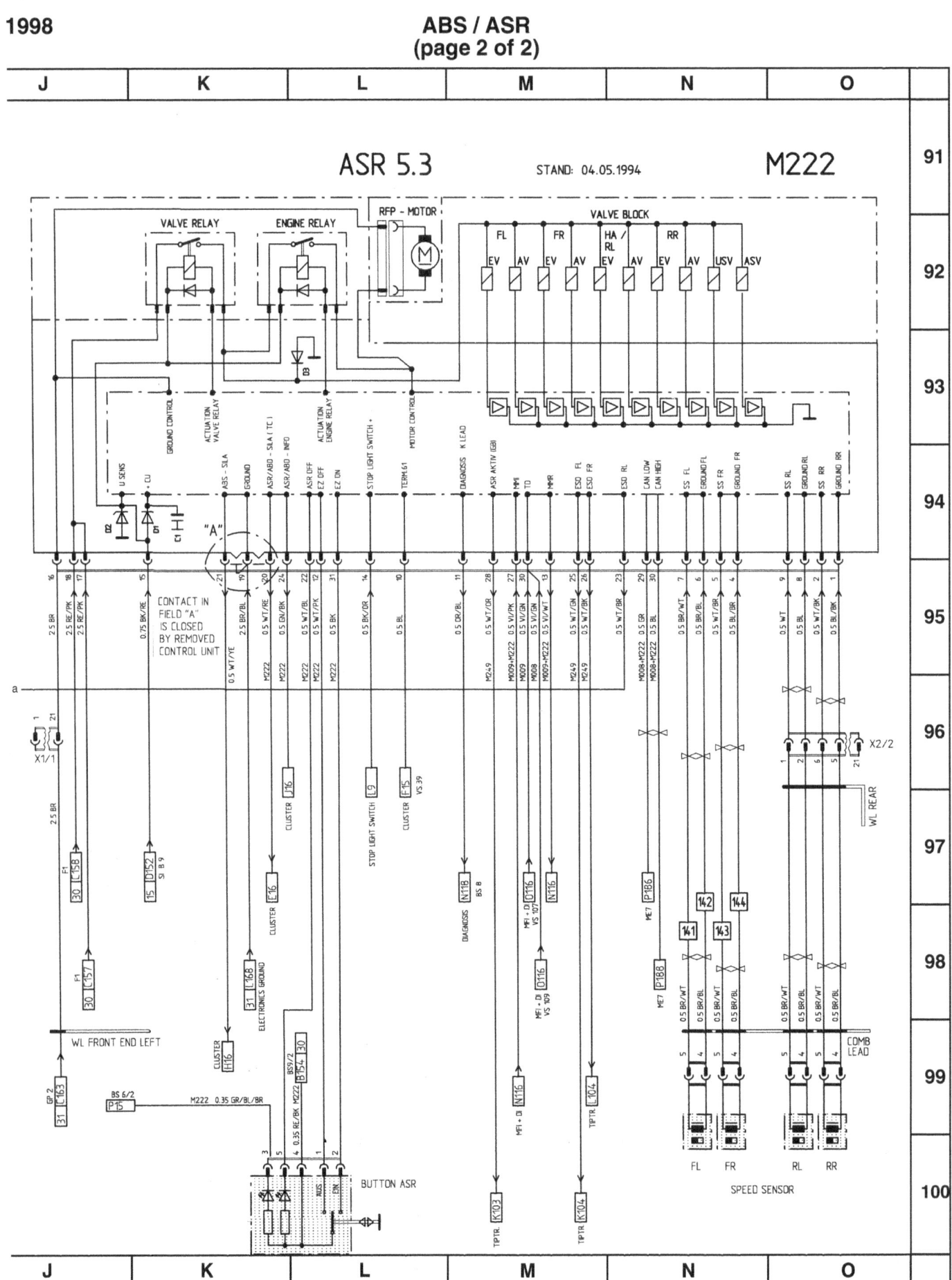

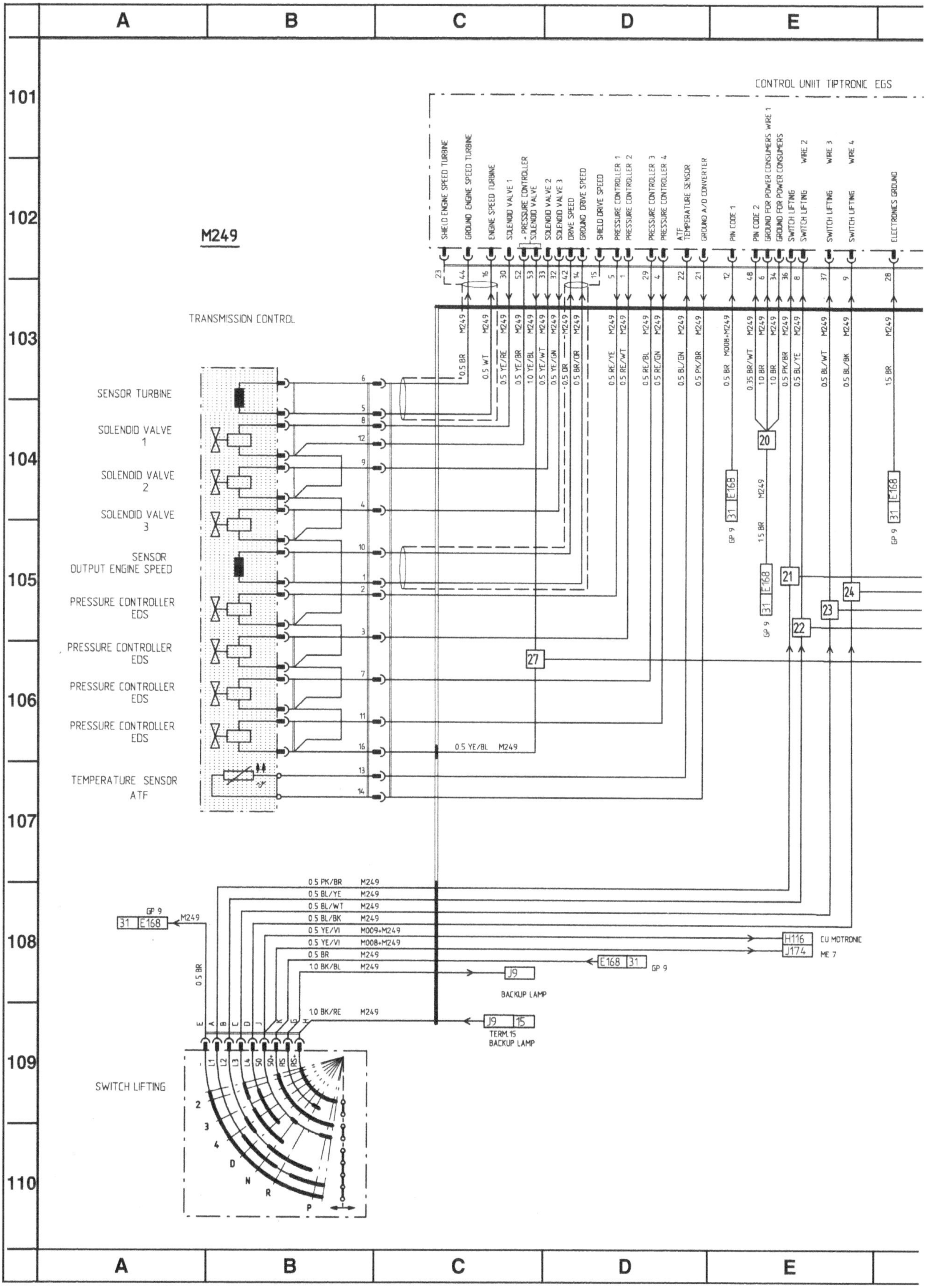

1998 Tiptronic (page 2 of 3)

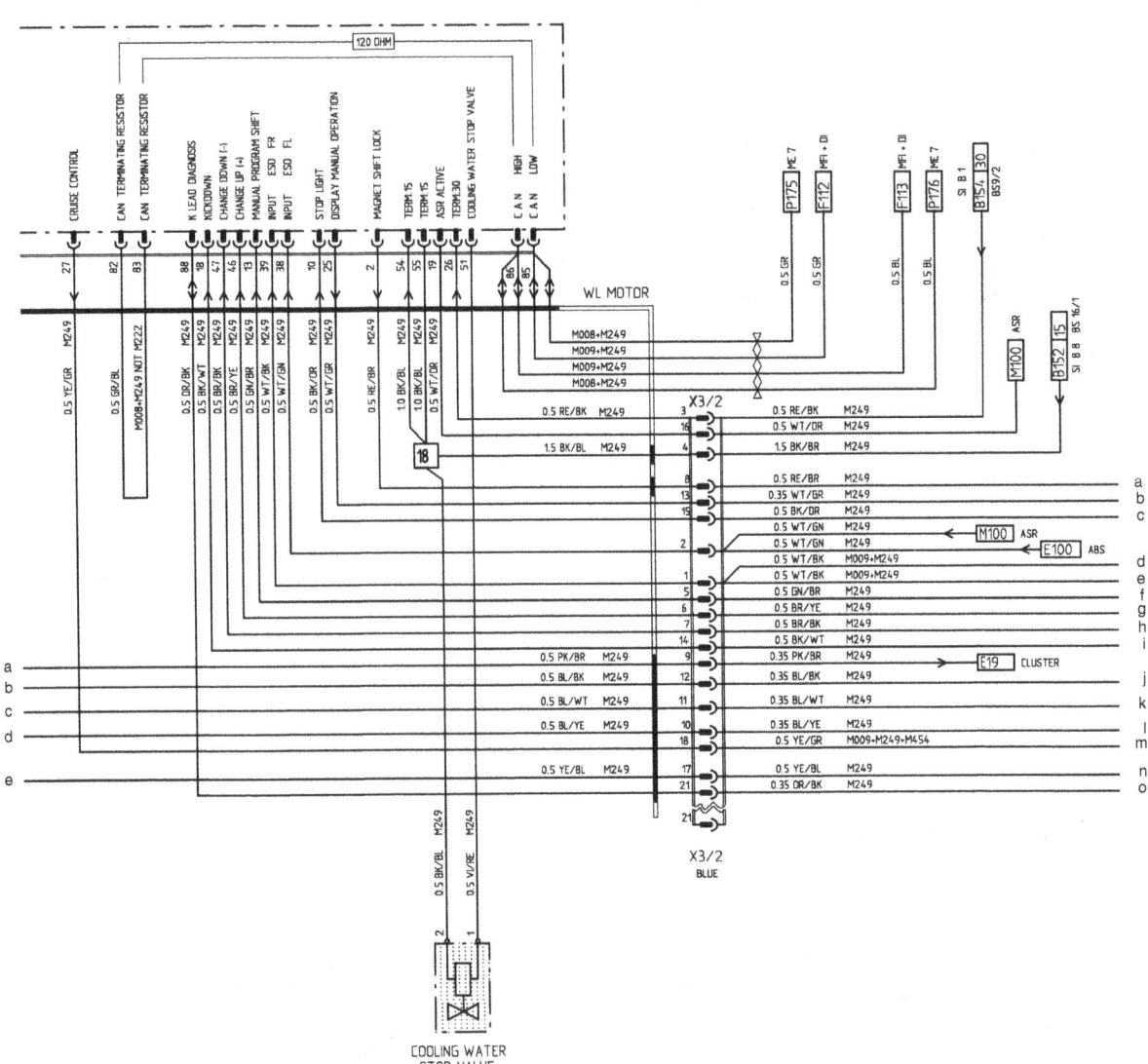

Electrical Wiring Diagrams EWD-37

Tiptronic
(page 3 of 3)

1998

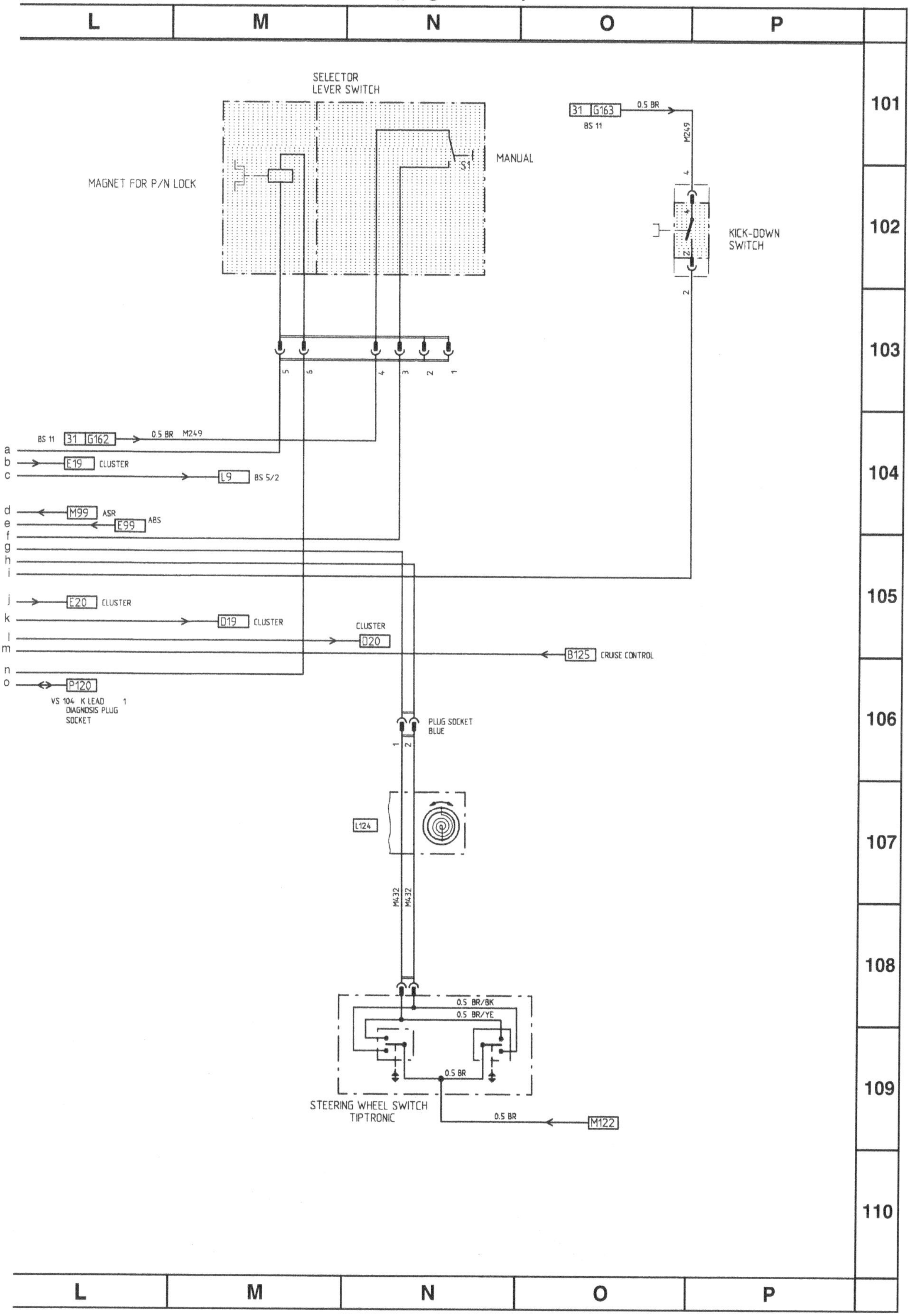

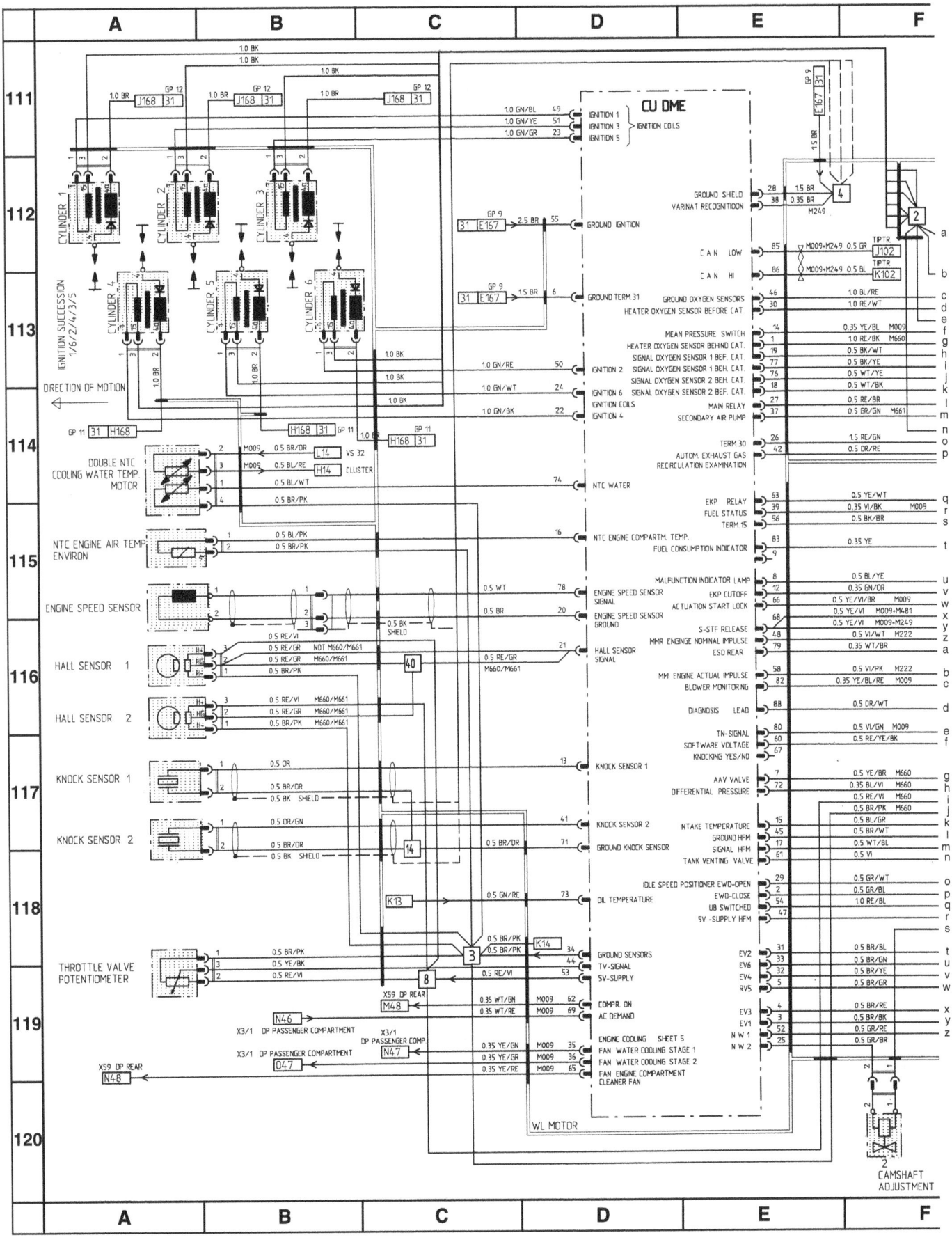

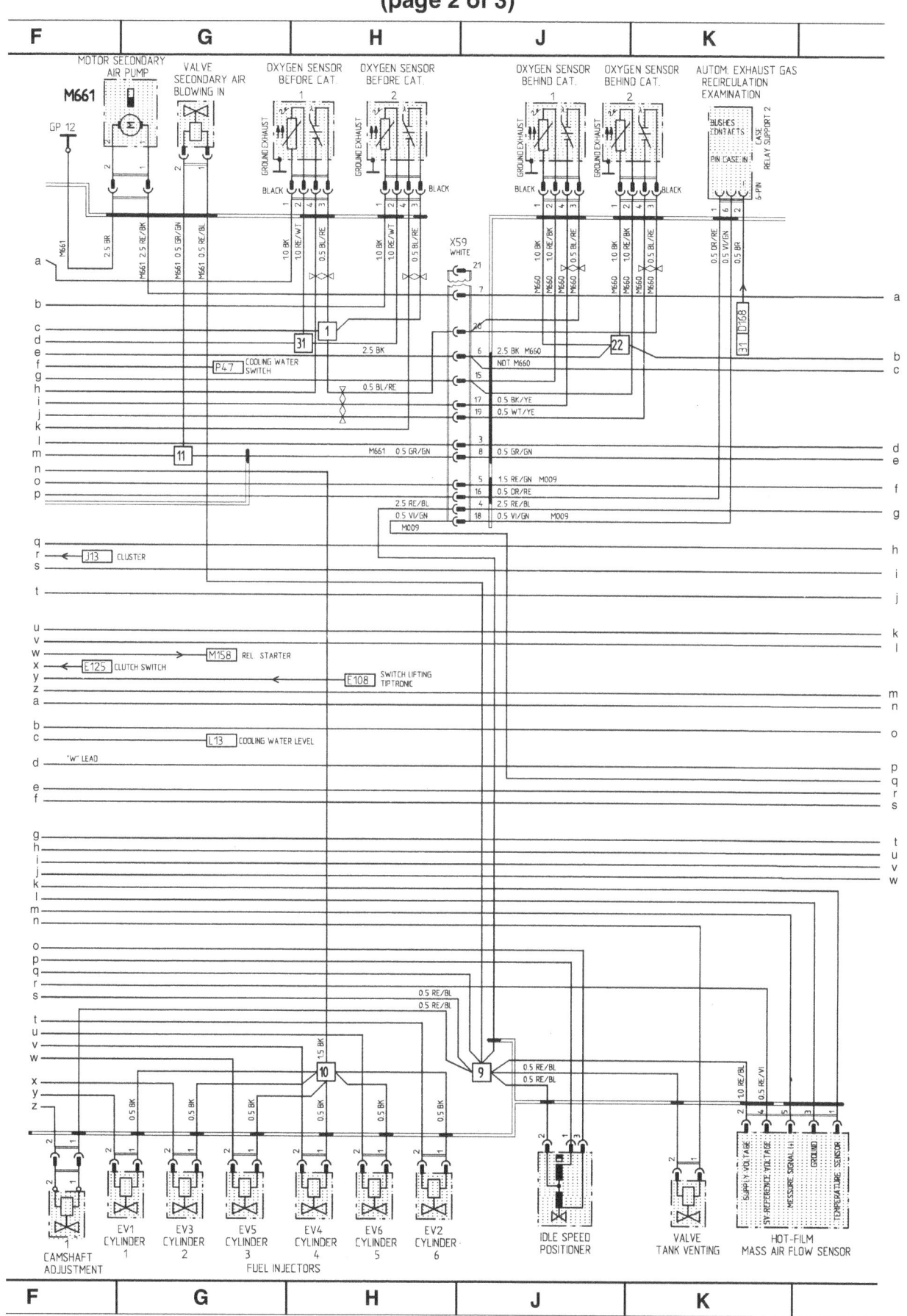

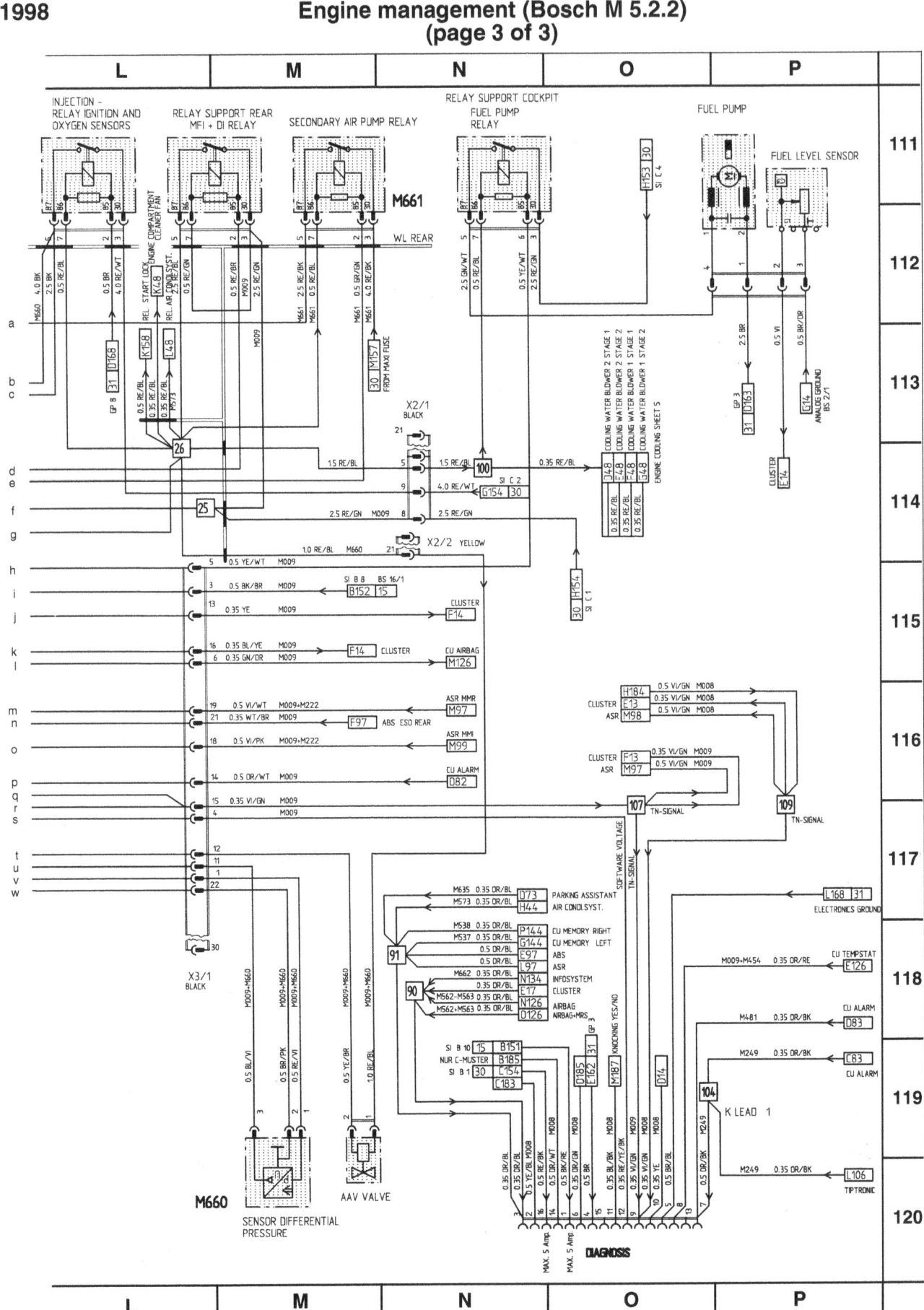

Electrical Wiring Diagrams EWD-41

Cruise control, airbags (page 1 of 3)

1998

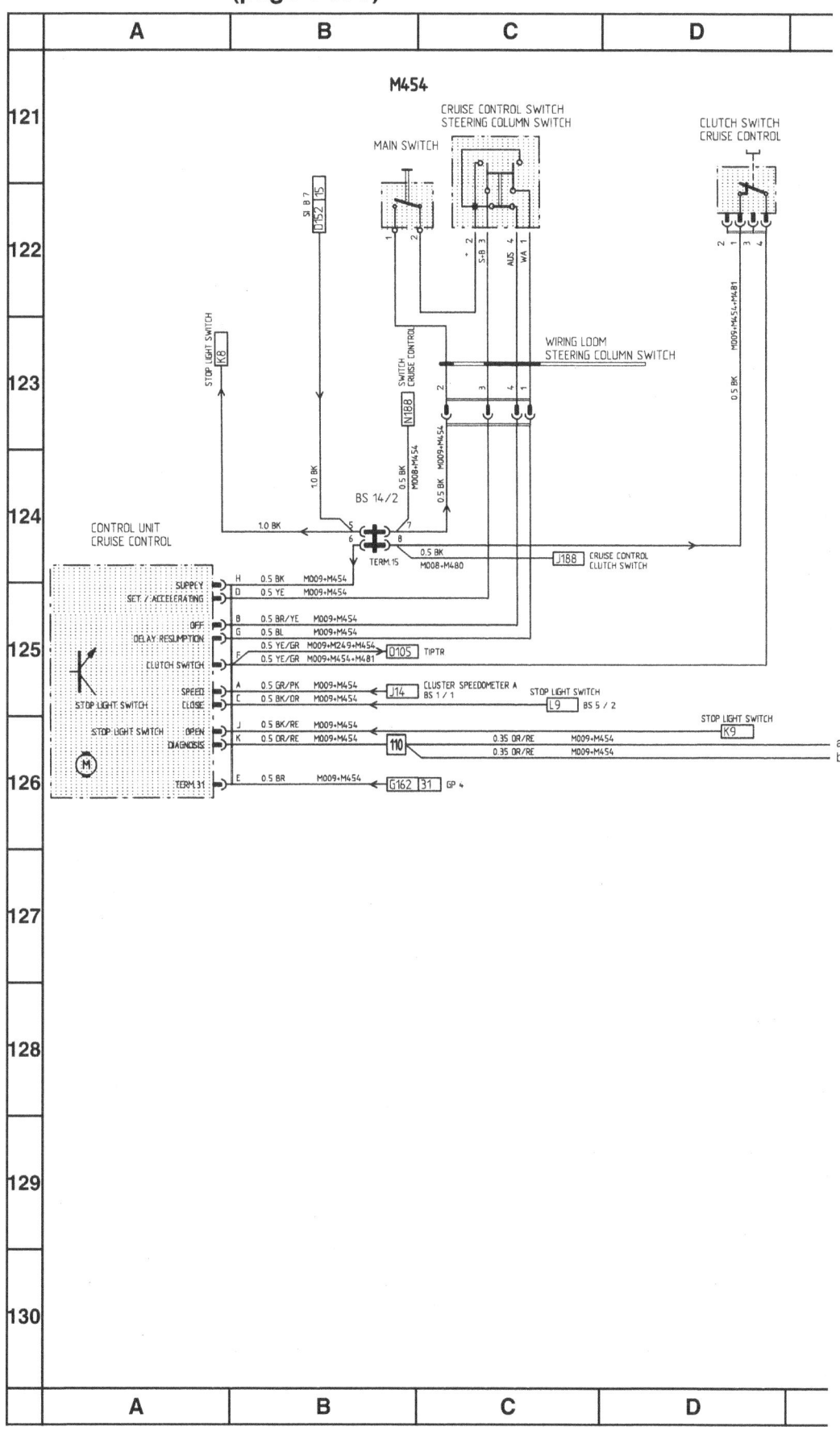

1998 Cruise control, airbags (page 2 of 3)

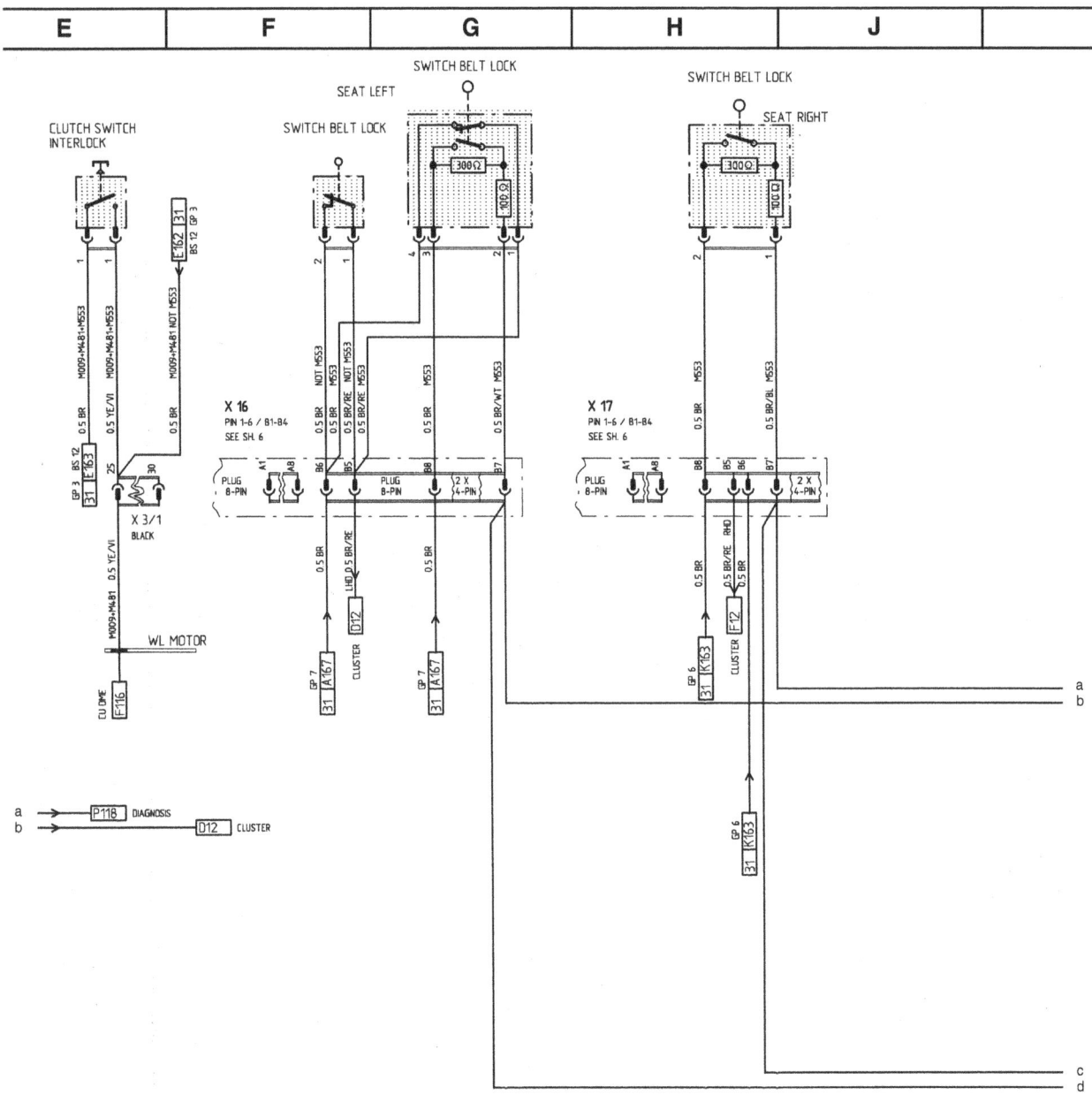

Electrical Wiring Diagrams EWD-43

Cruise control, airbags
(page 3 of 3)

1998

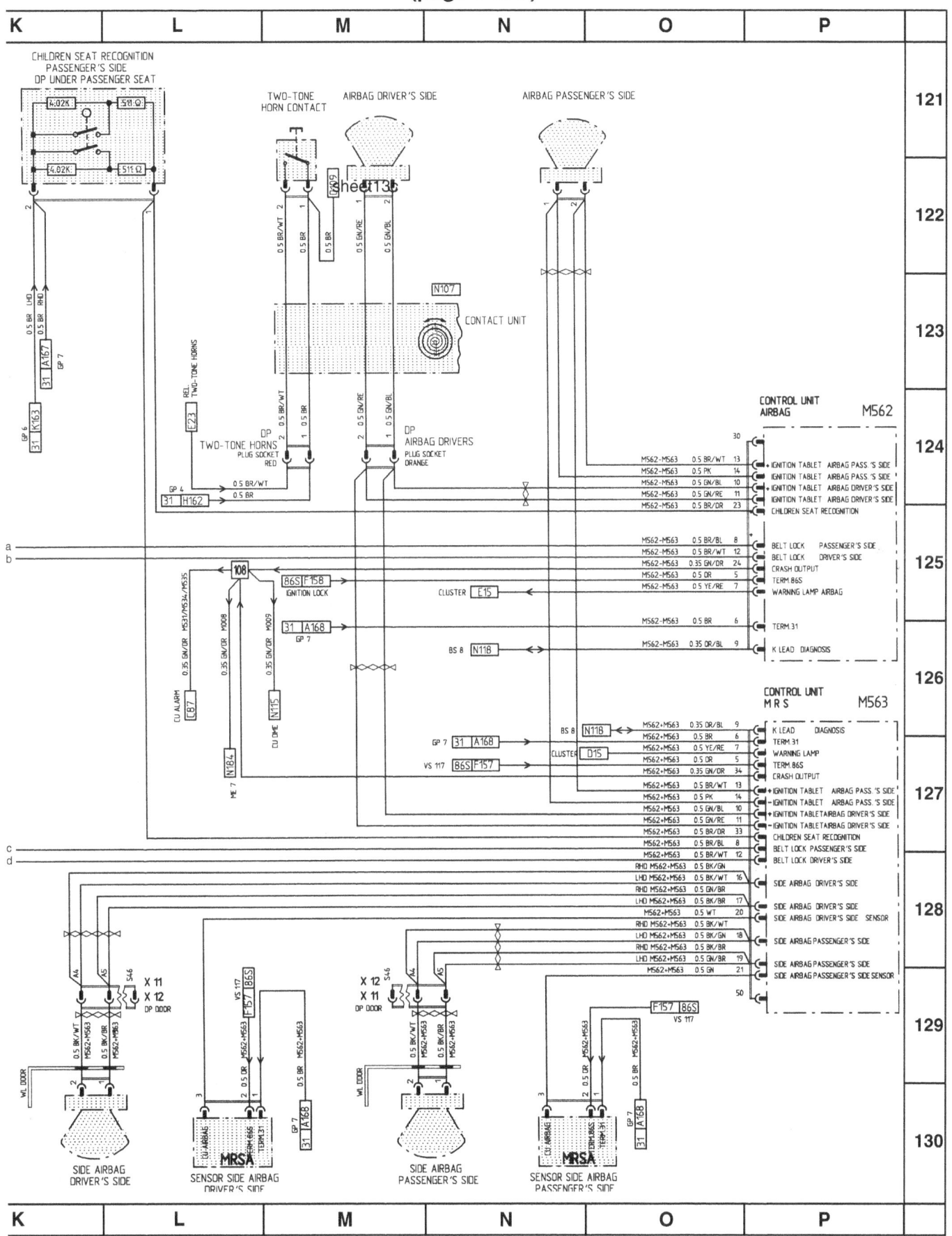

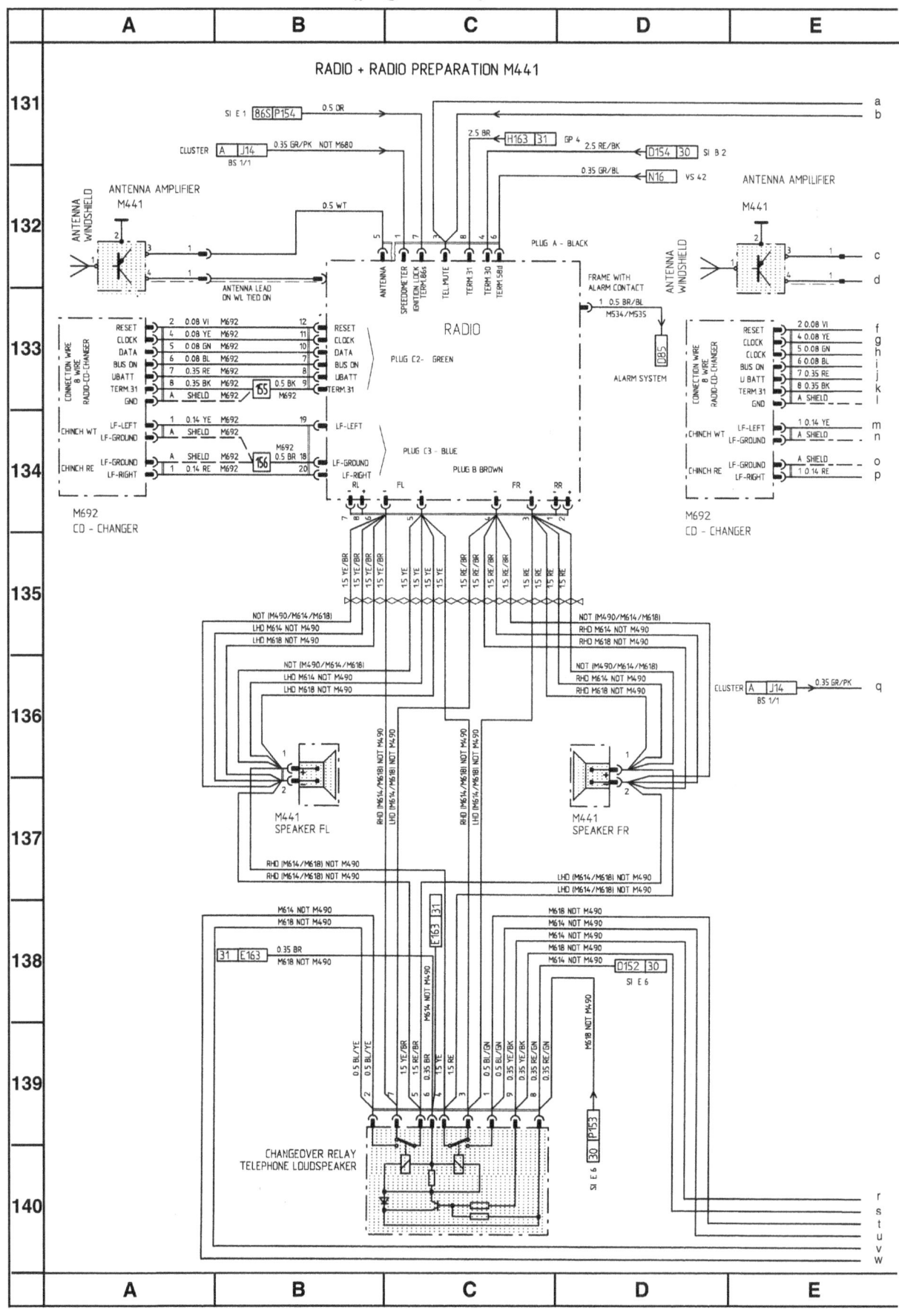

Radio, telephone
(page 1 of 3)

Electrical Wiring Diagrams EWD-45

Radio, telephone
(page 2 of 3)

1998

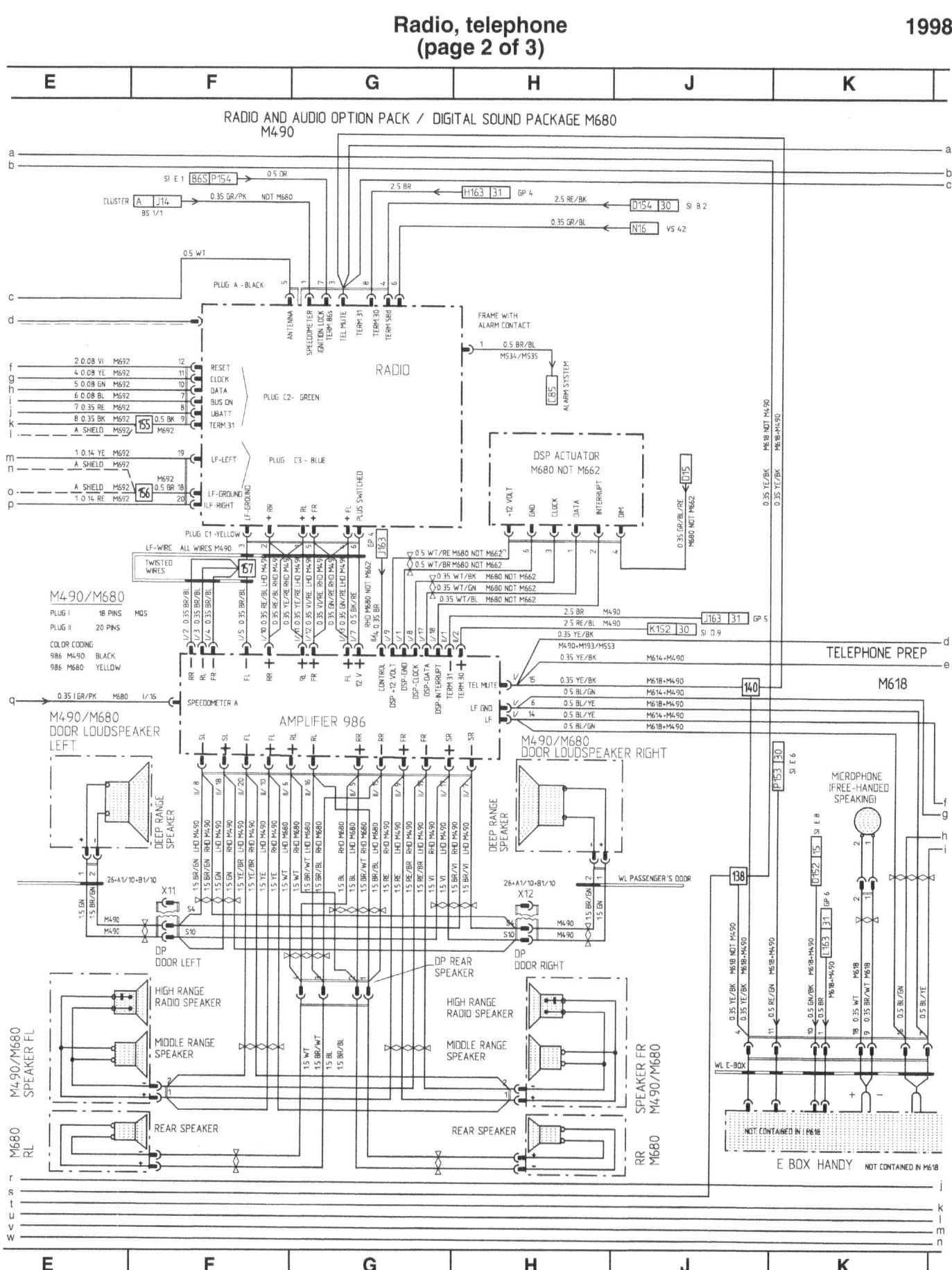

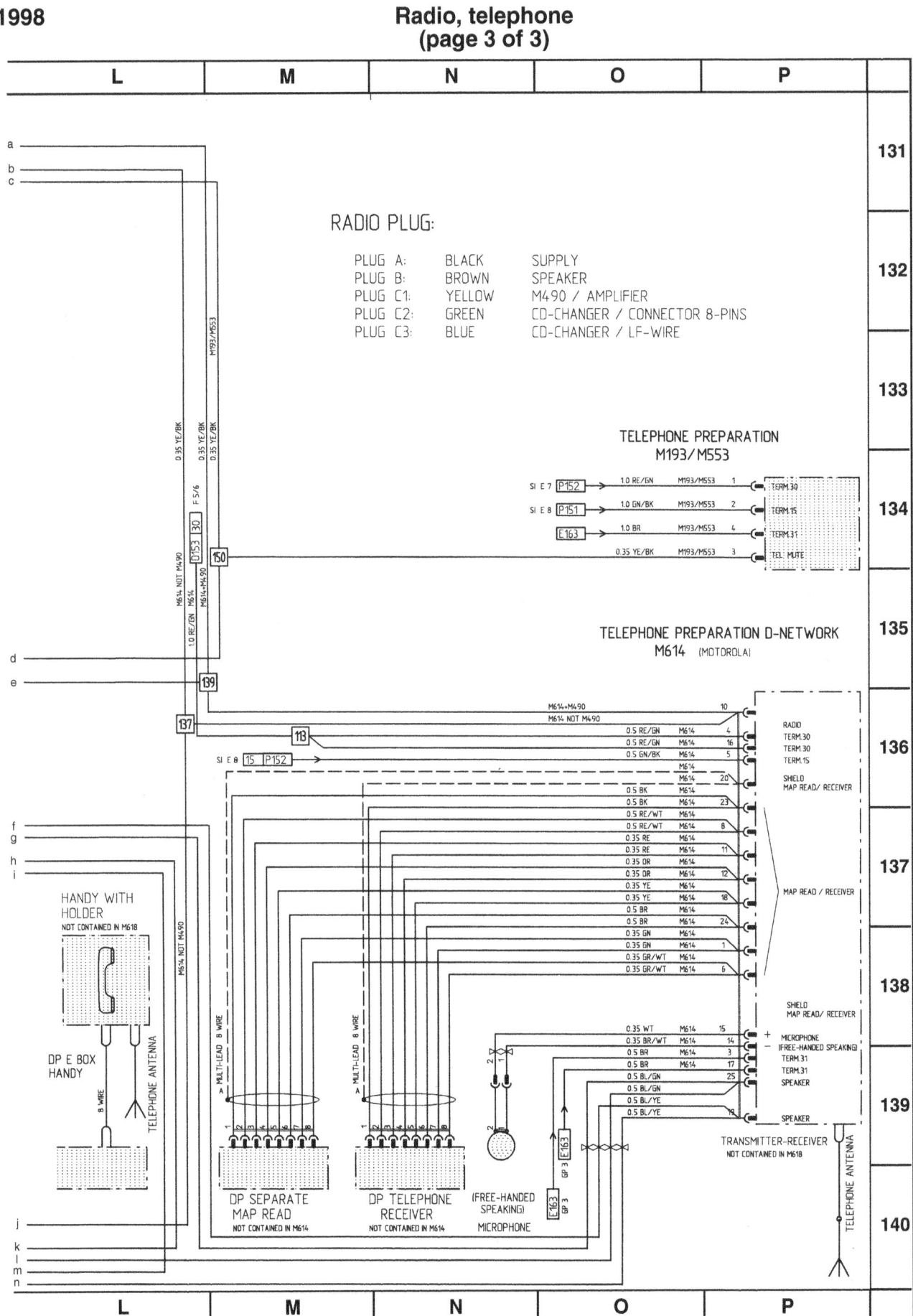

Navigation, infosystem
(page 1 of 3)

1998

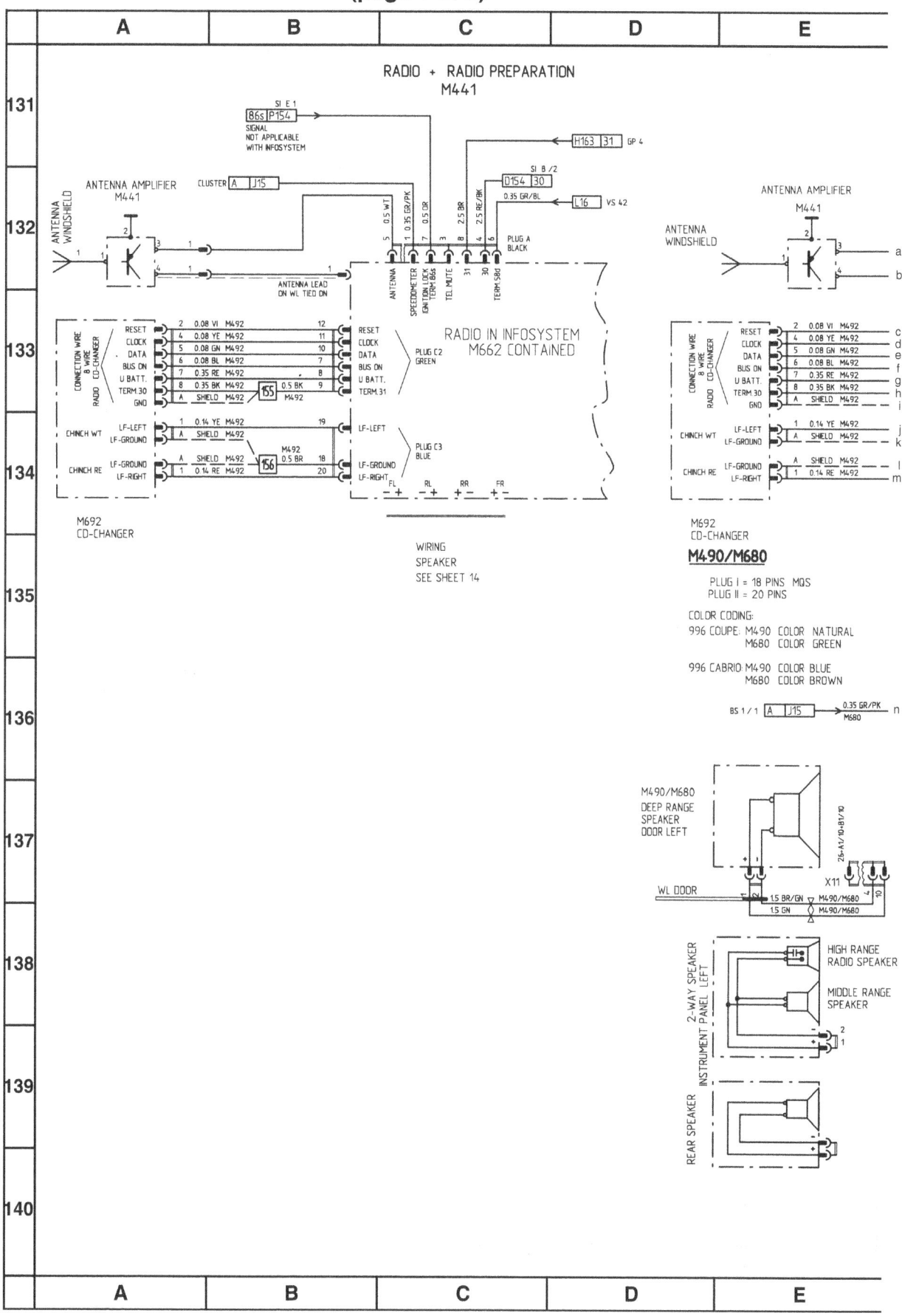

1998

Navigation, infosystem
(page 2 of 3)

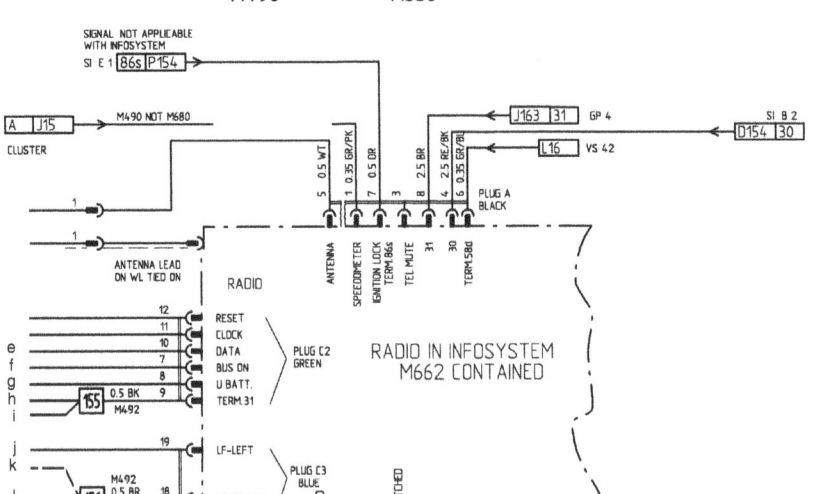

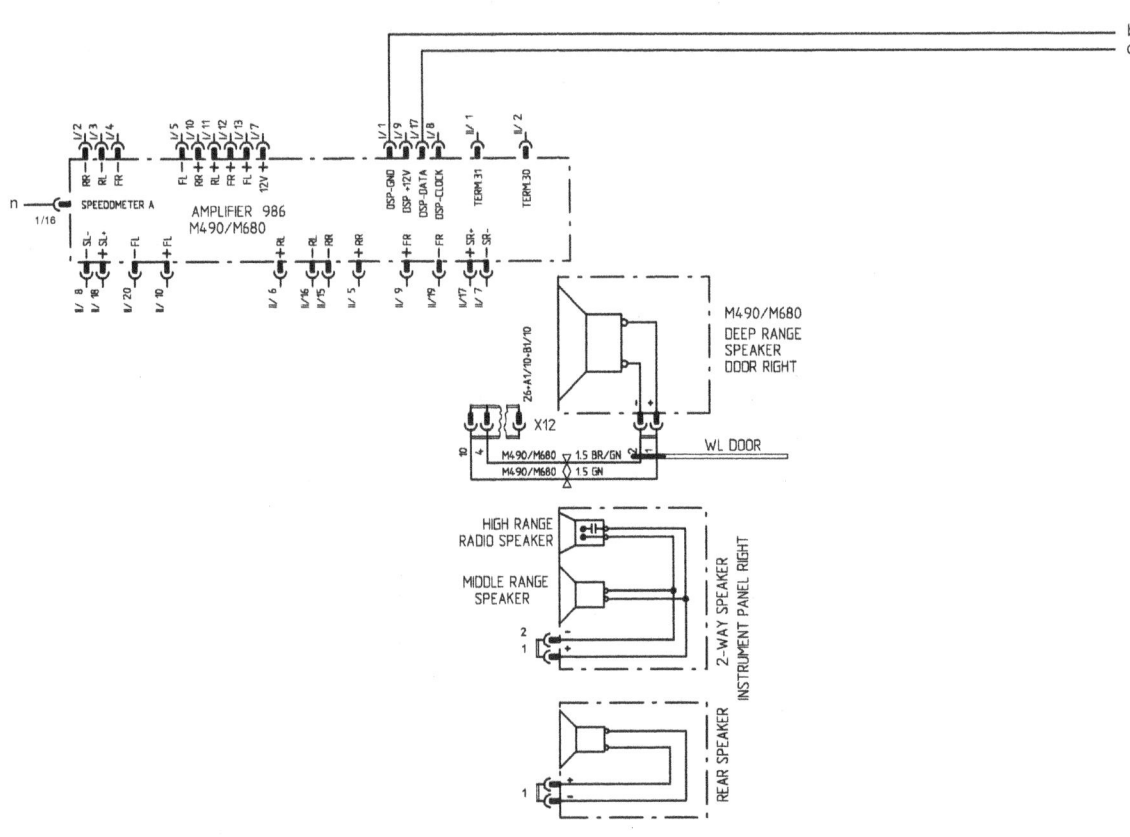

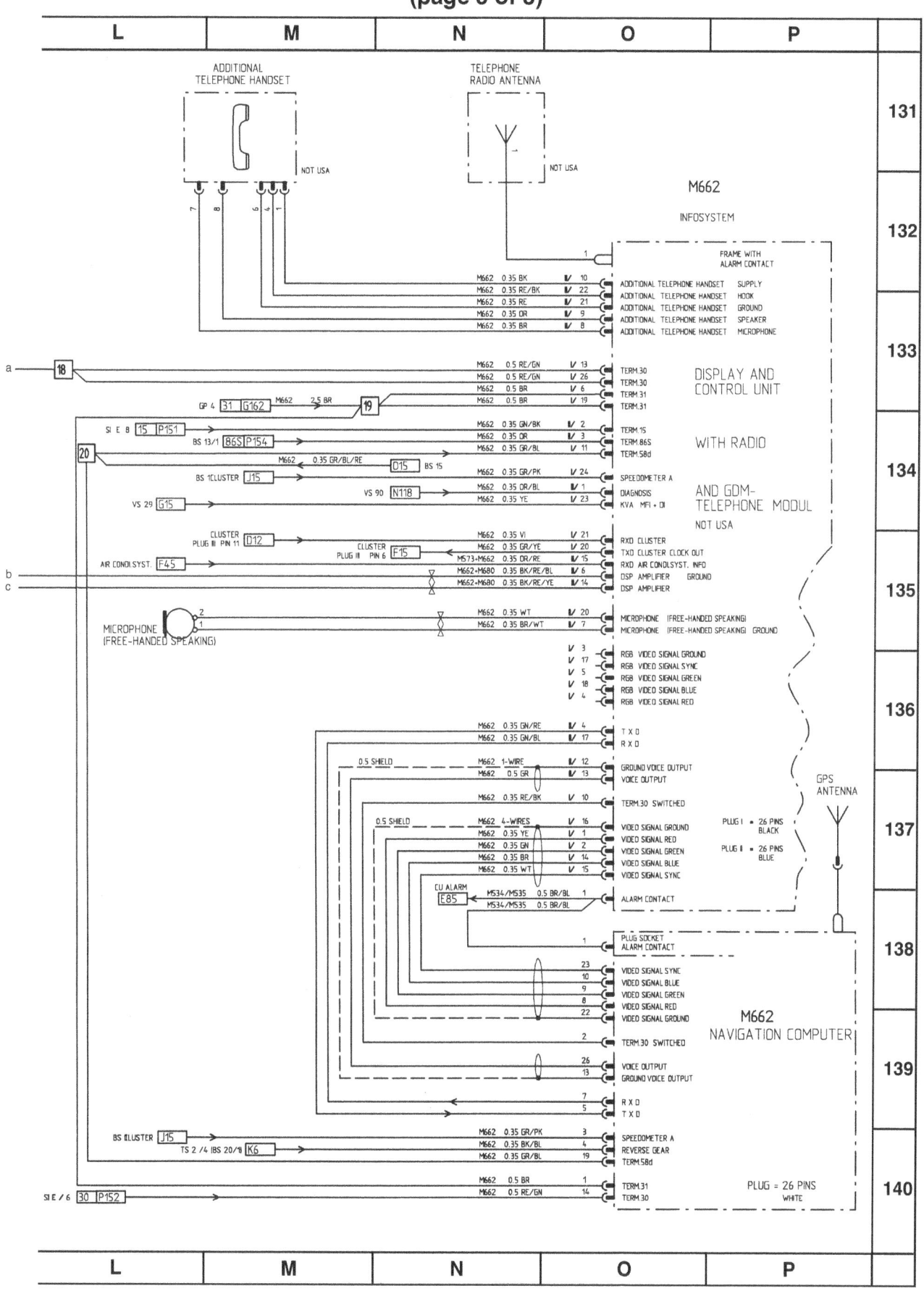

EWD-50 Electrical Wiring Diagrams

1998

Seat and mirror control
(page 1 of 3)

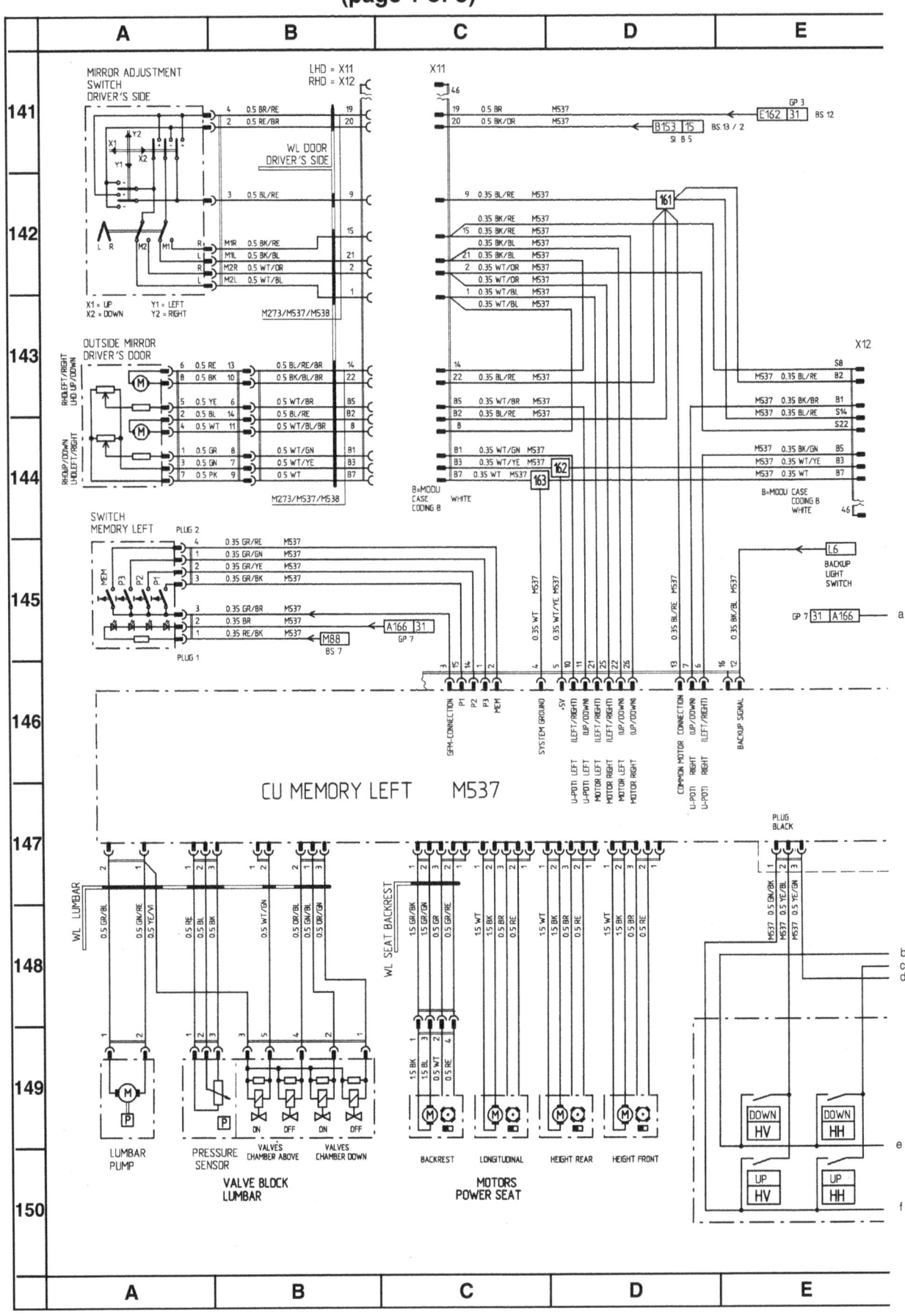

Electrical Wiring Diagrams EWD-51

Seat and mirror control (page 2 of 3)

1998

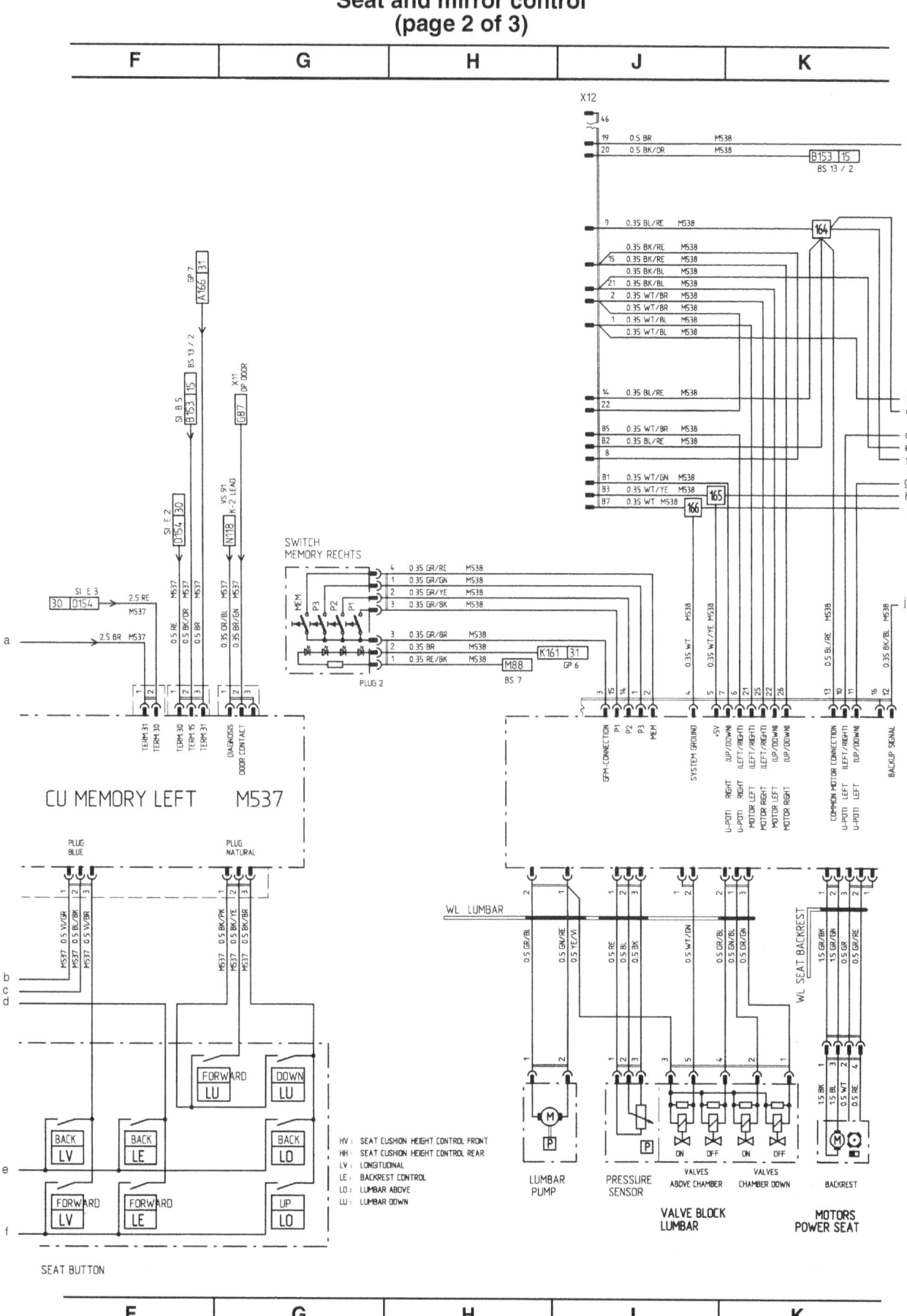

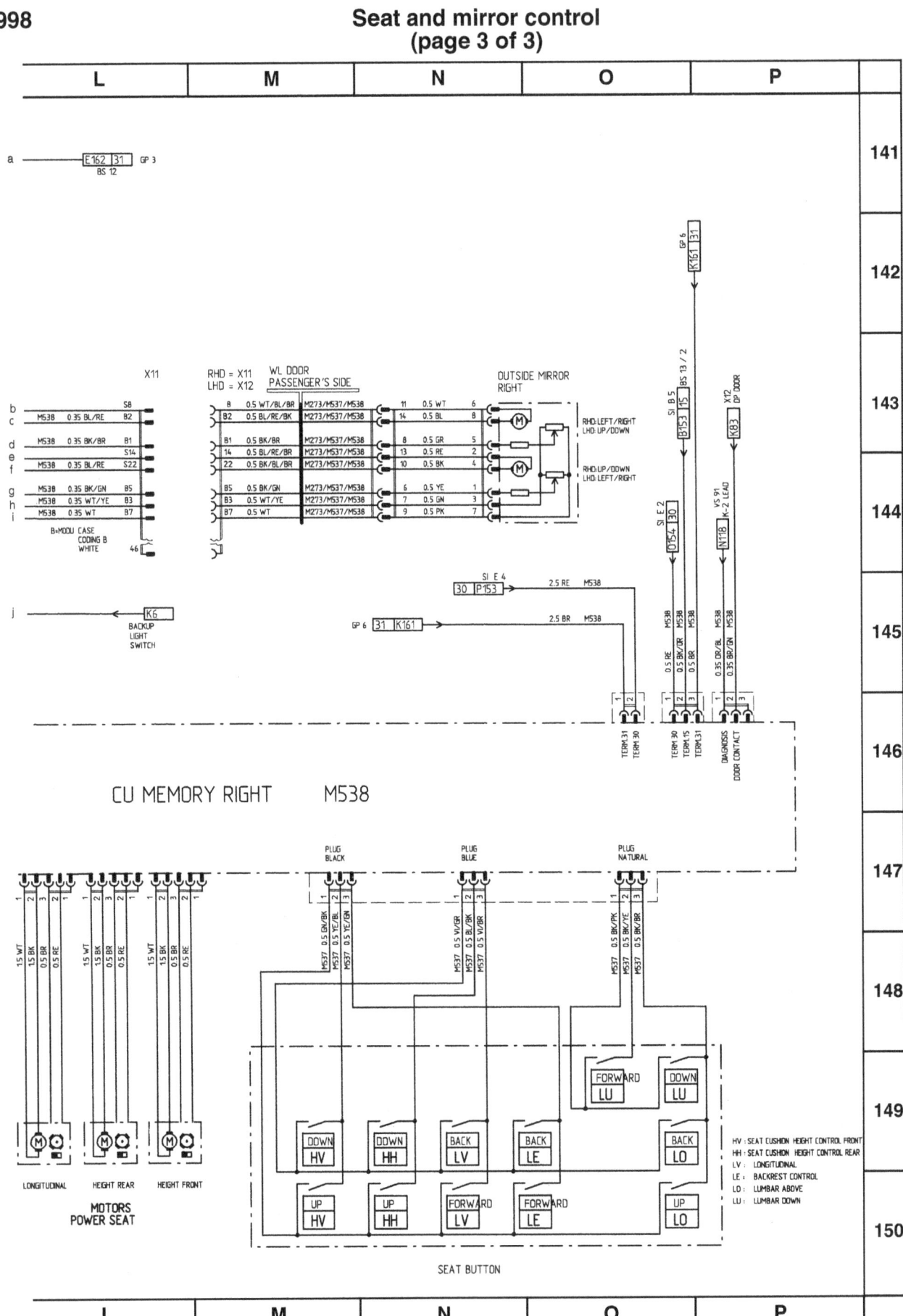

1998 Seat and mirror control (page 3 of 3)

Electrical Wiring Diagrams EWD-53

Power supply, fuses
(page 1 of 3)

1998

1998 Power supply, fuses (page 2 of 3)

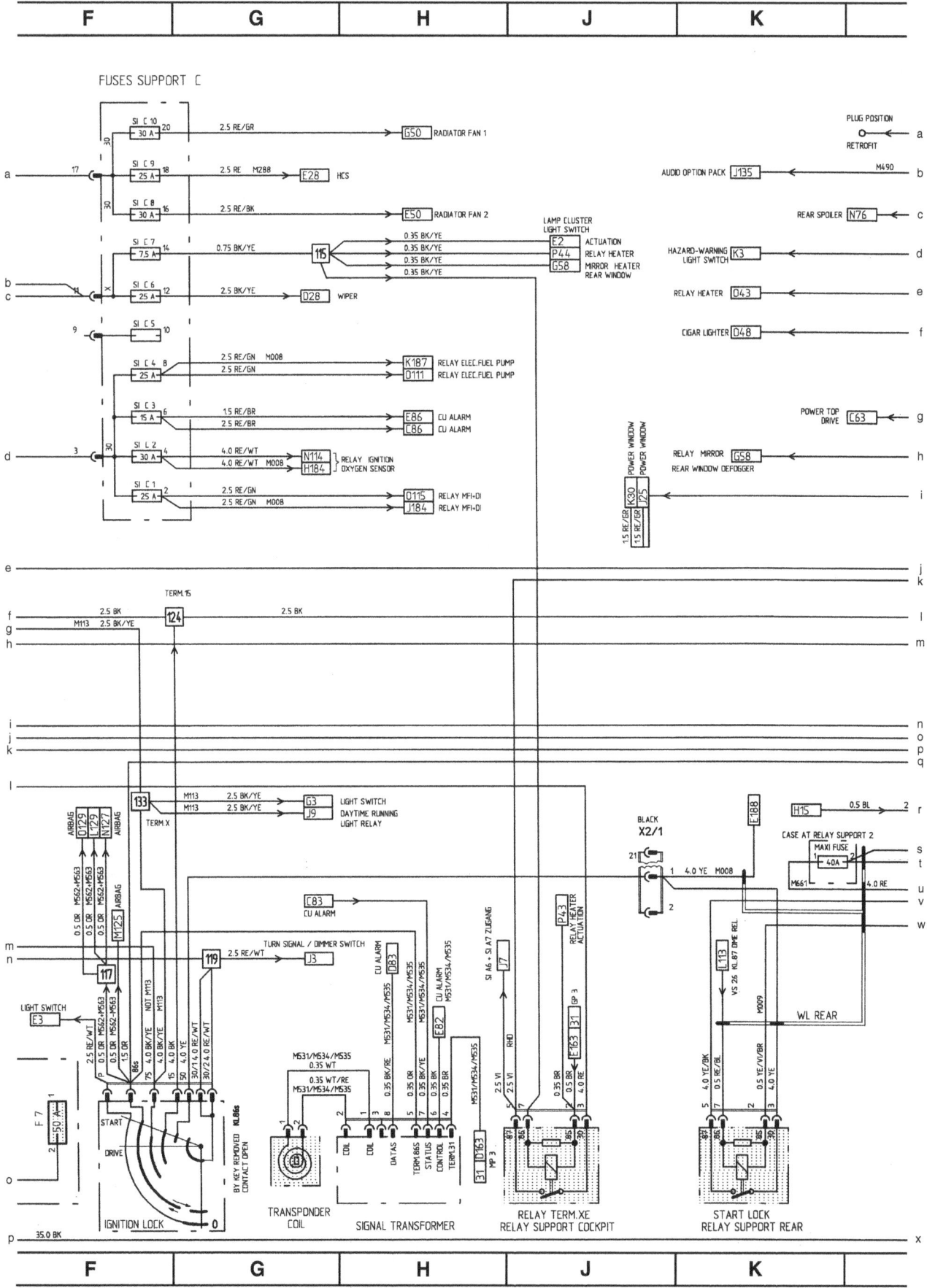

Power supply, fuses
(page 3 of 3)

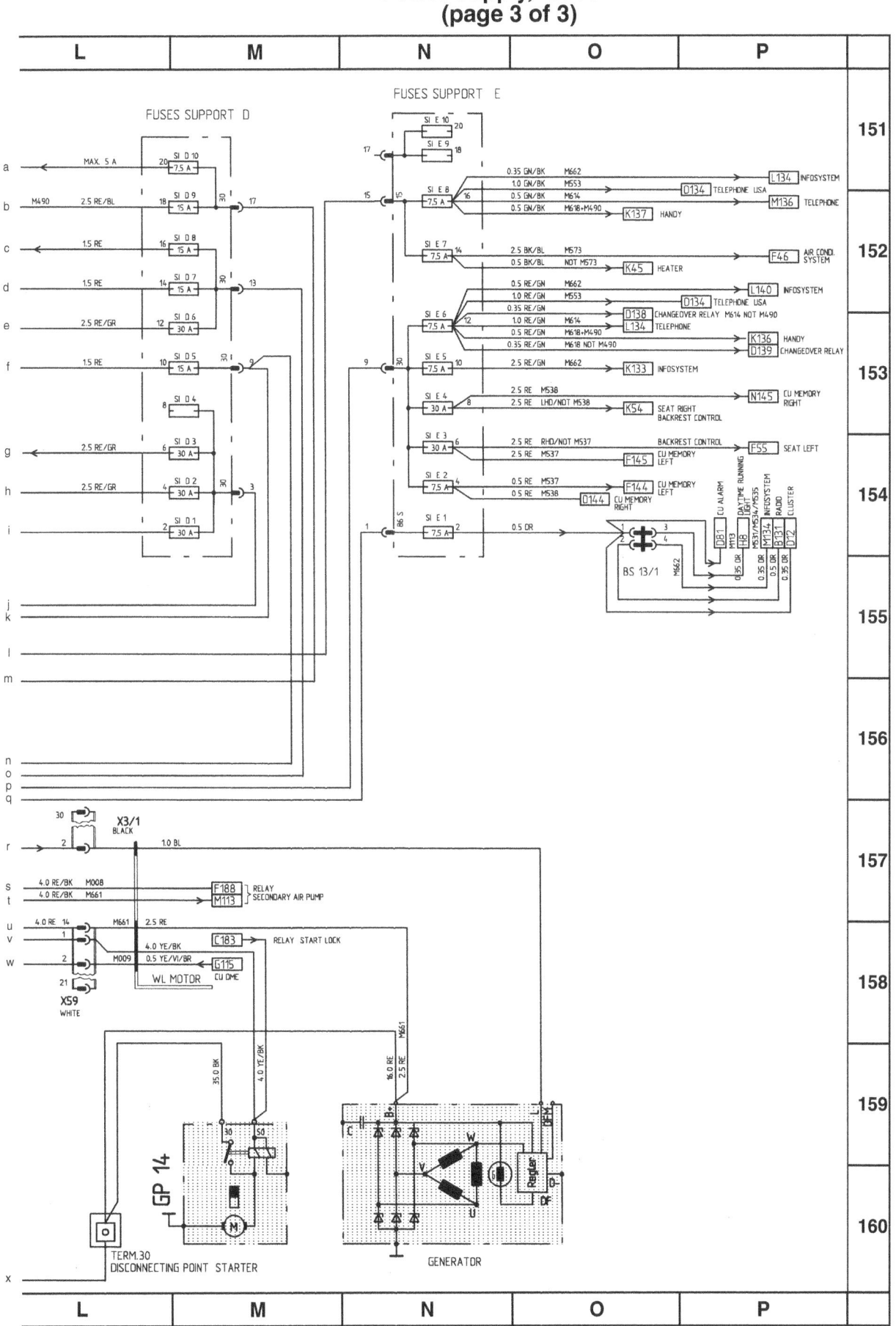

EWD-56   Electrical Wiring Diagrams

Diagram 1   Lighting (RoW)   2001
                                   (page 1 of 4)

	1			2	3	4

BLACK	BK	BK
BROWN	BR	BN
RED	RE	RD
ORANGE	OR	OG
YELLOW	YE	YE
GREEN	GN	GN
BLUE	BL	BU
VIOLET	VI	VT
GREY	GR	GY
WHITE	WT	WH
PINK	RS	PK
GOLD	-	GD
TURQUOISE	TK	TQ
SILVER	-	SR

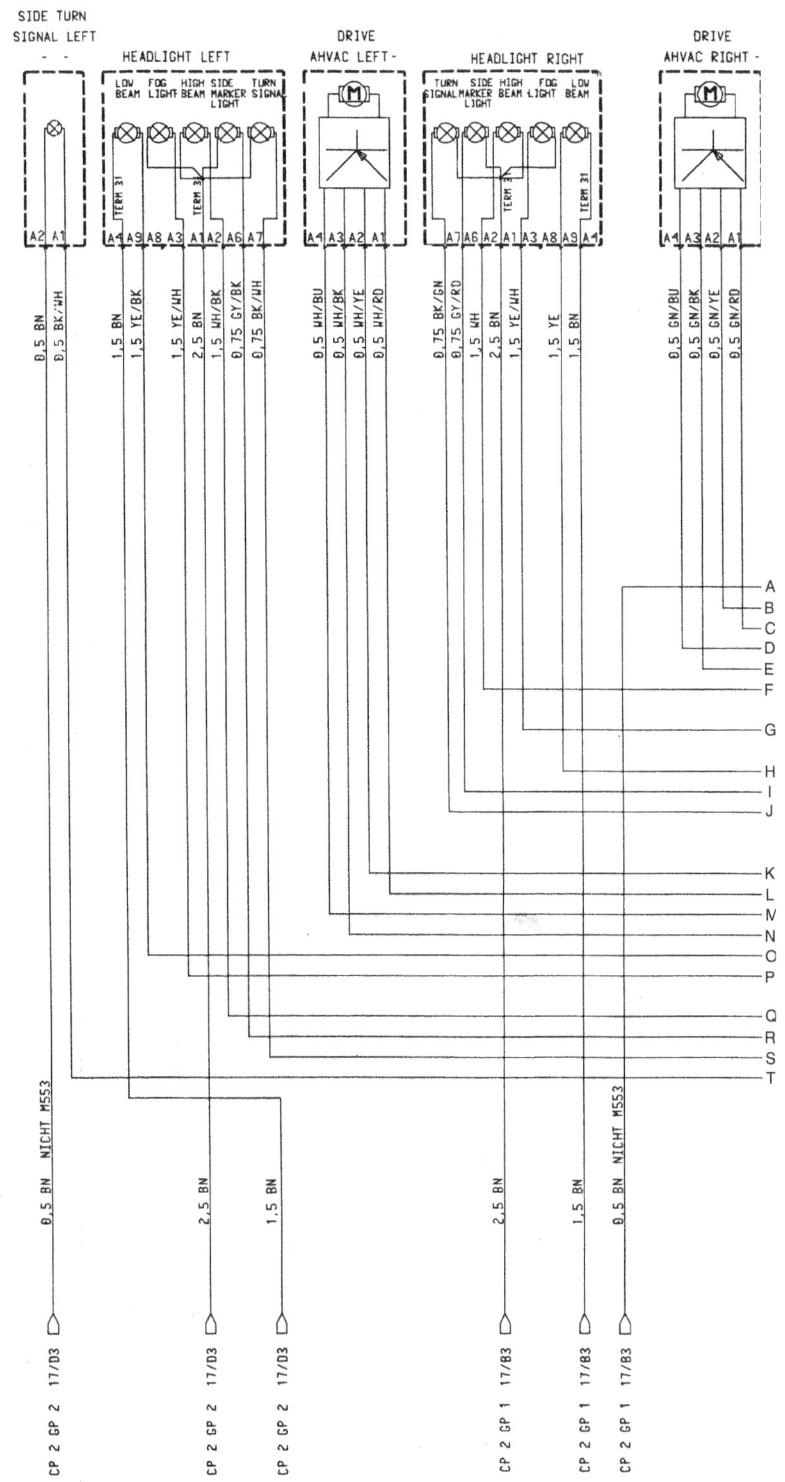

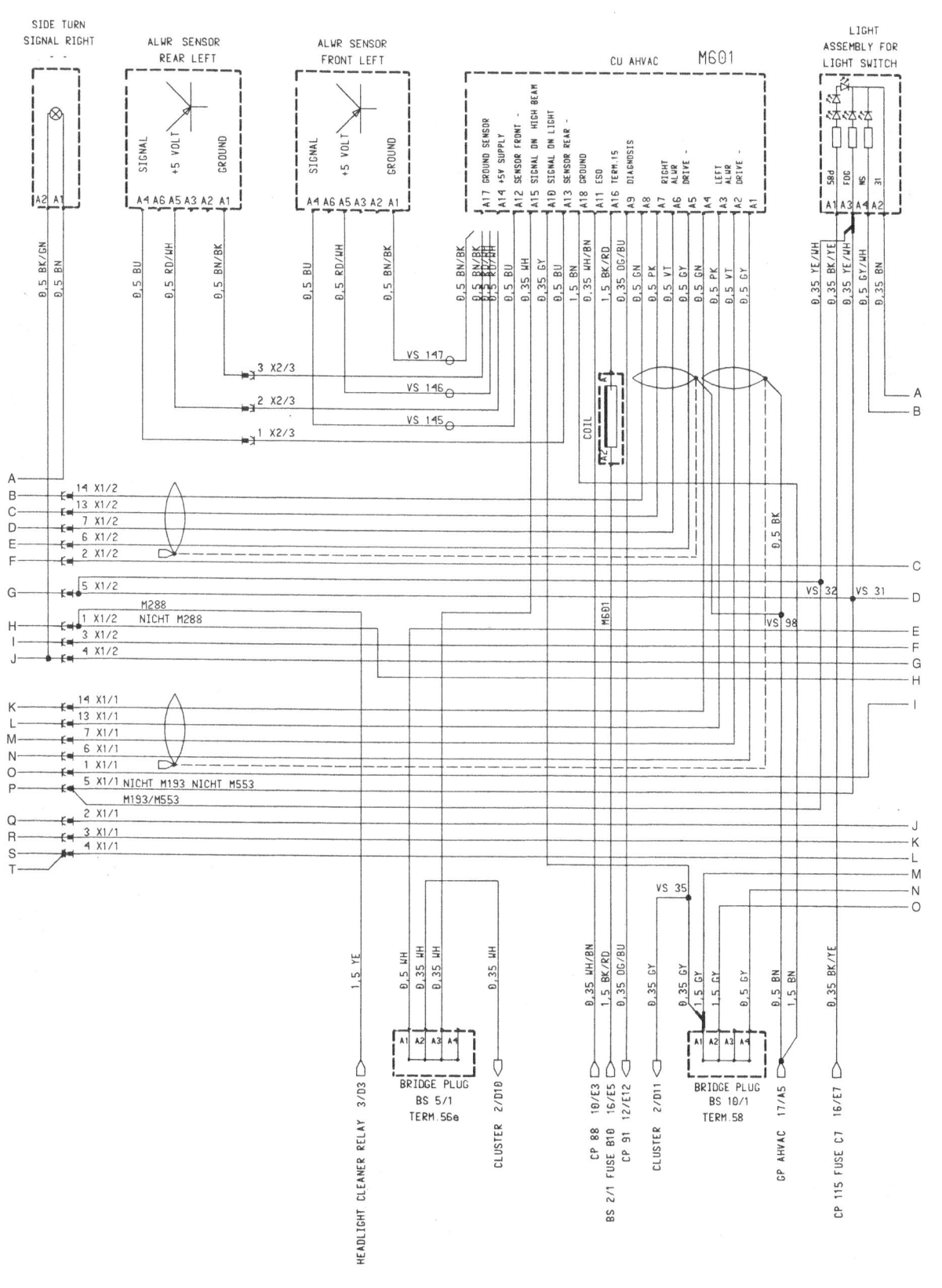

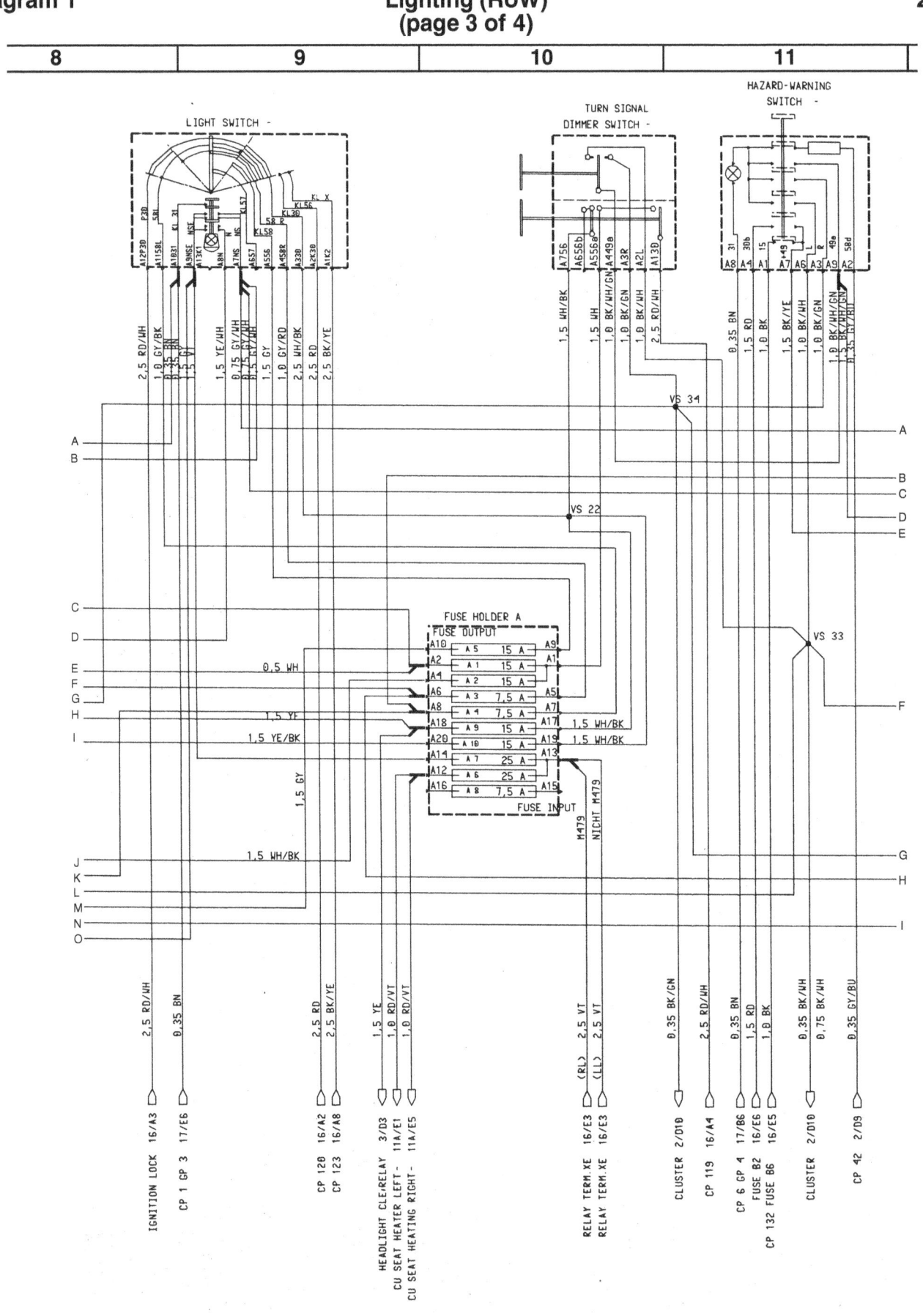

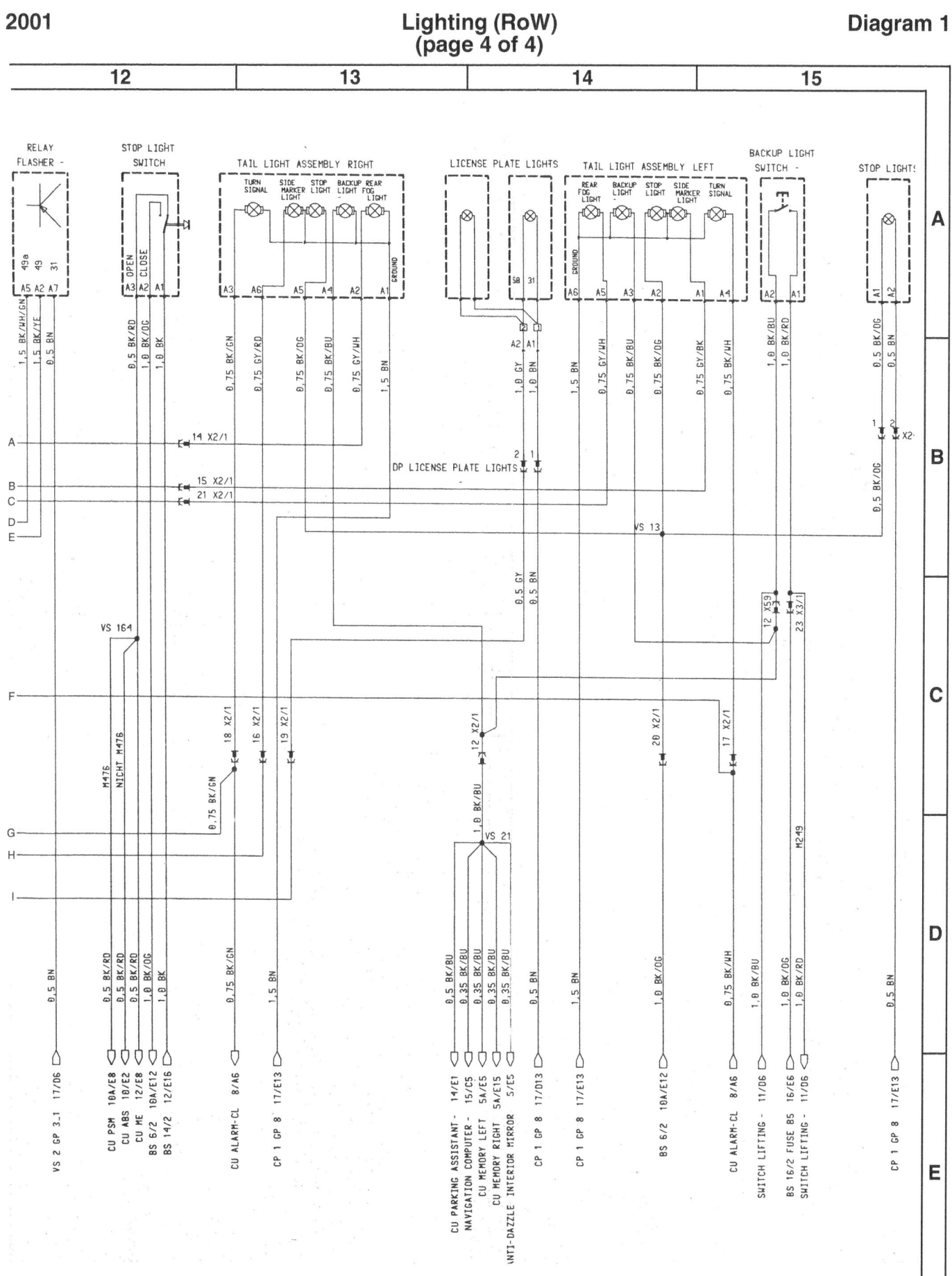

EWD-60 Electrical Wiring Diagrams

Diagram 1A Lighting USA / Canada (page 1 of 4) 2001

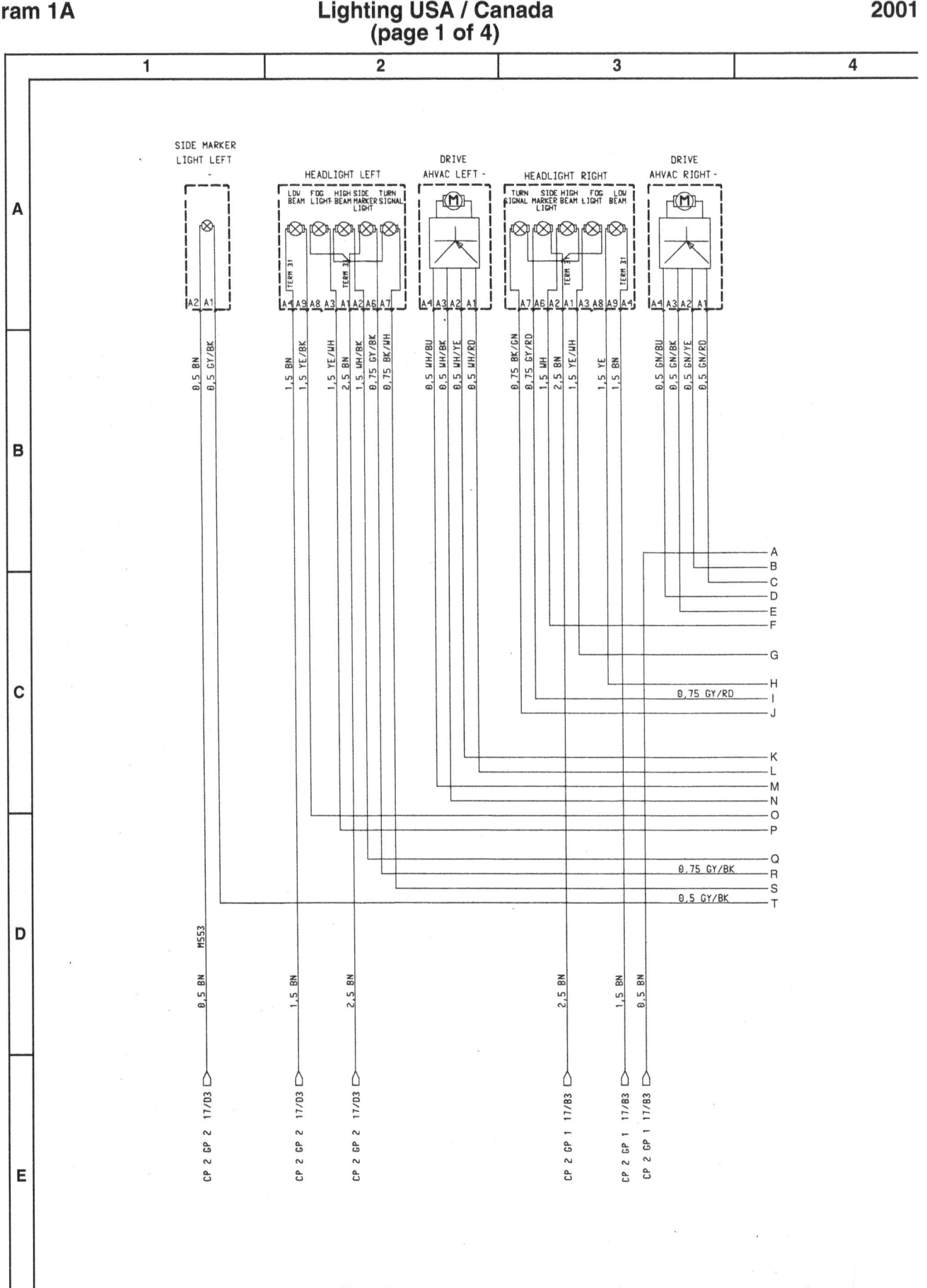

Electrical Wiring Diagrams EWD-61

2001 — Lighting USA / Canada (page 2 of 4) — **Diagram 1A**

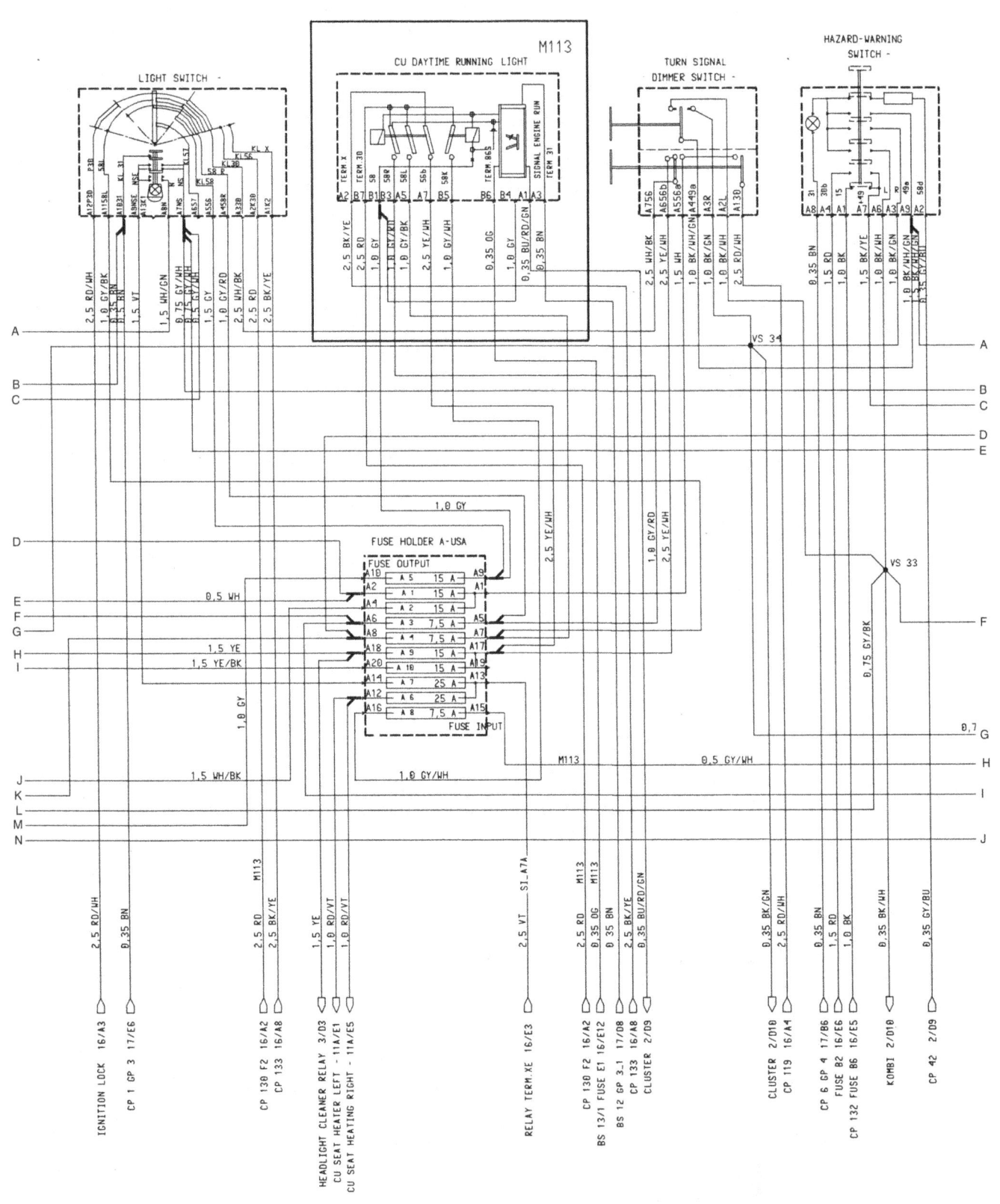

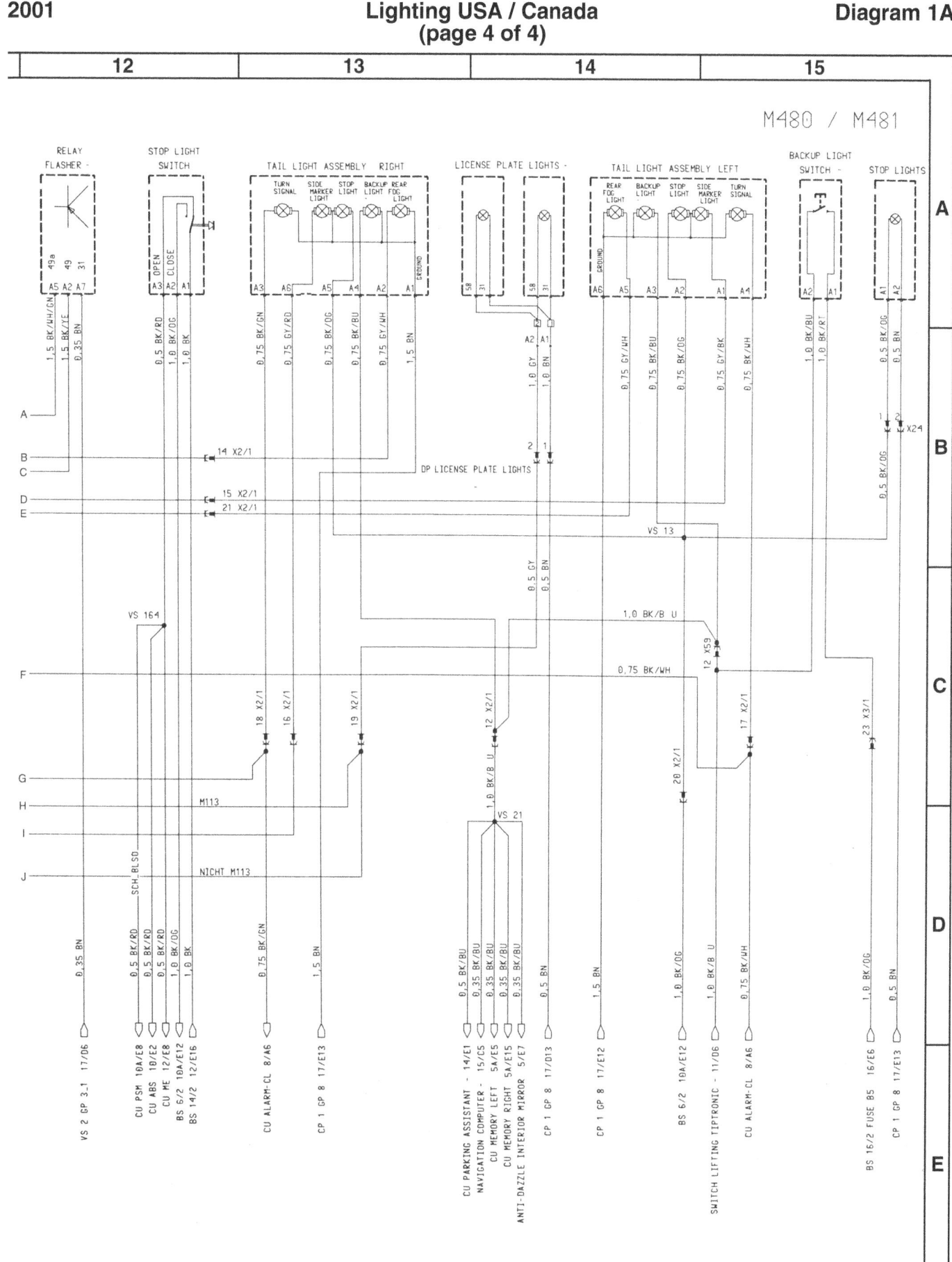

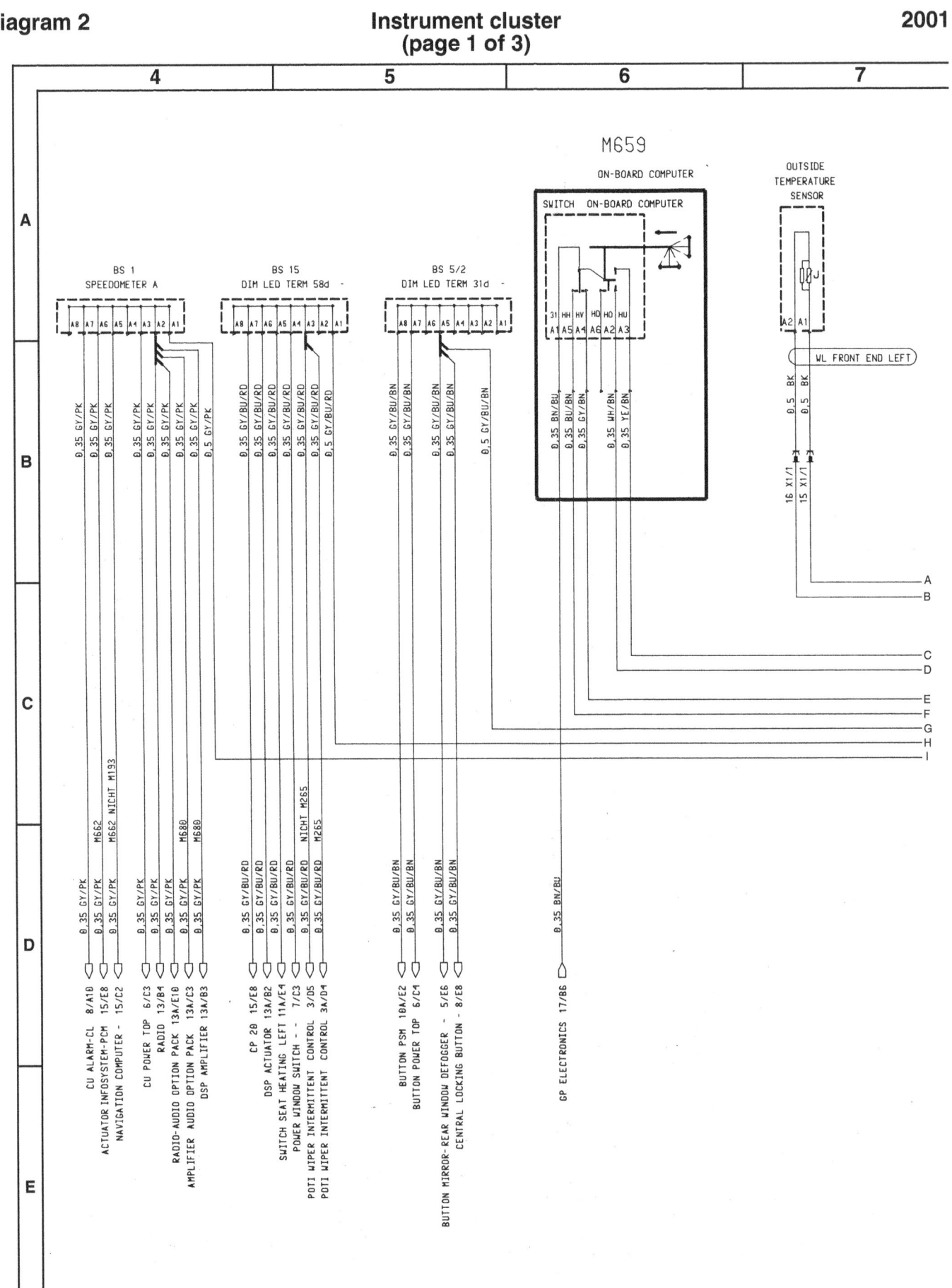

EWD-64　Electrical Wiring Diagrams

Diagram 2 — Instrument cluster (page 1 of 3) — 2001

Electrical Wiring Diagrams EWD-65

2001 Instrument cluster (page 2 of 3) Diagram 2

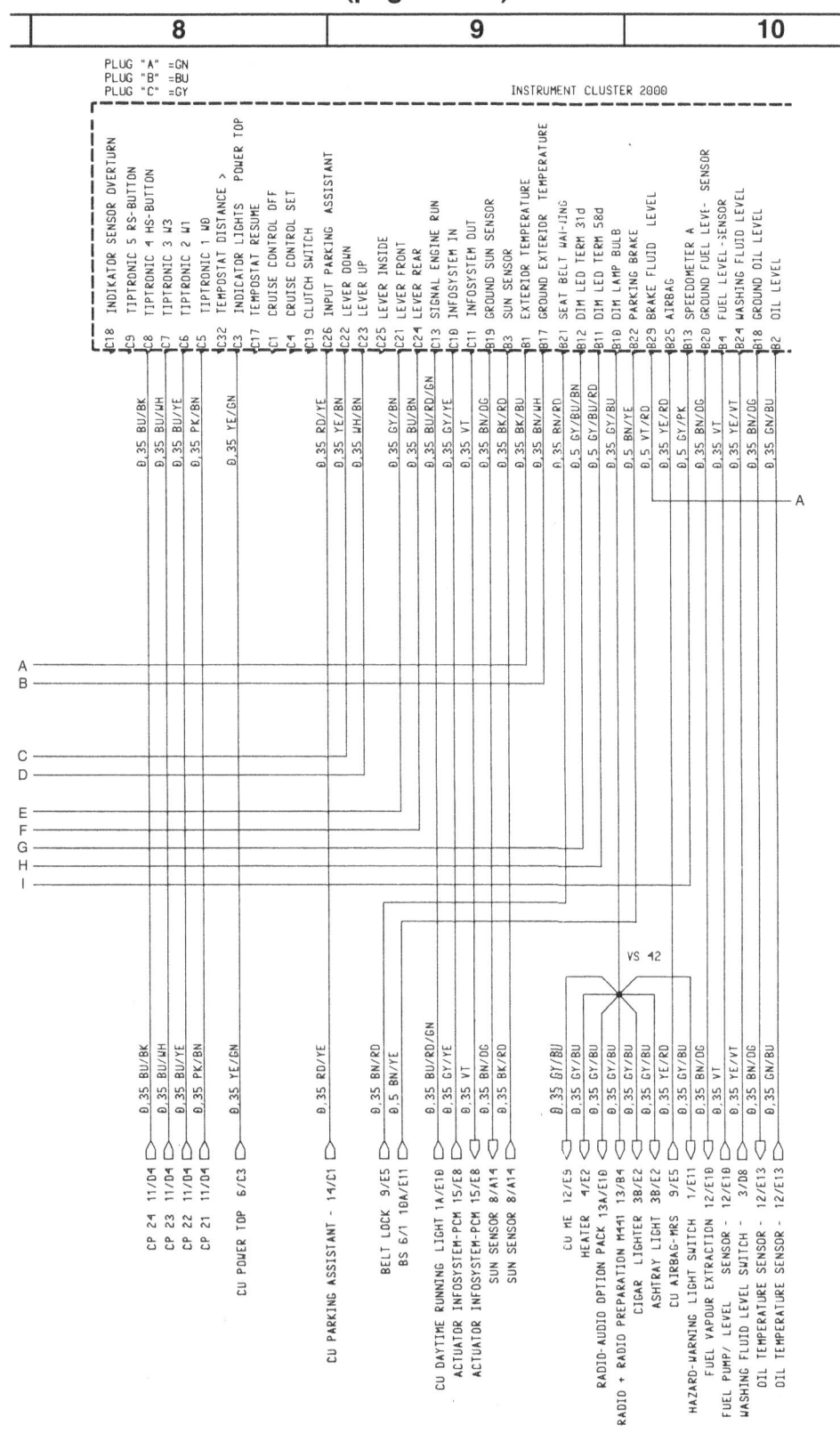

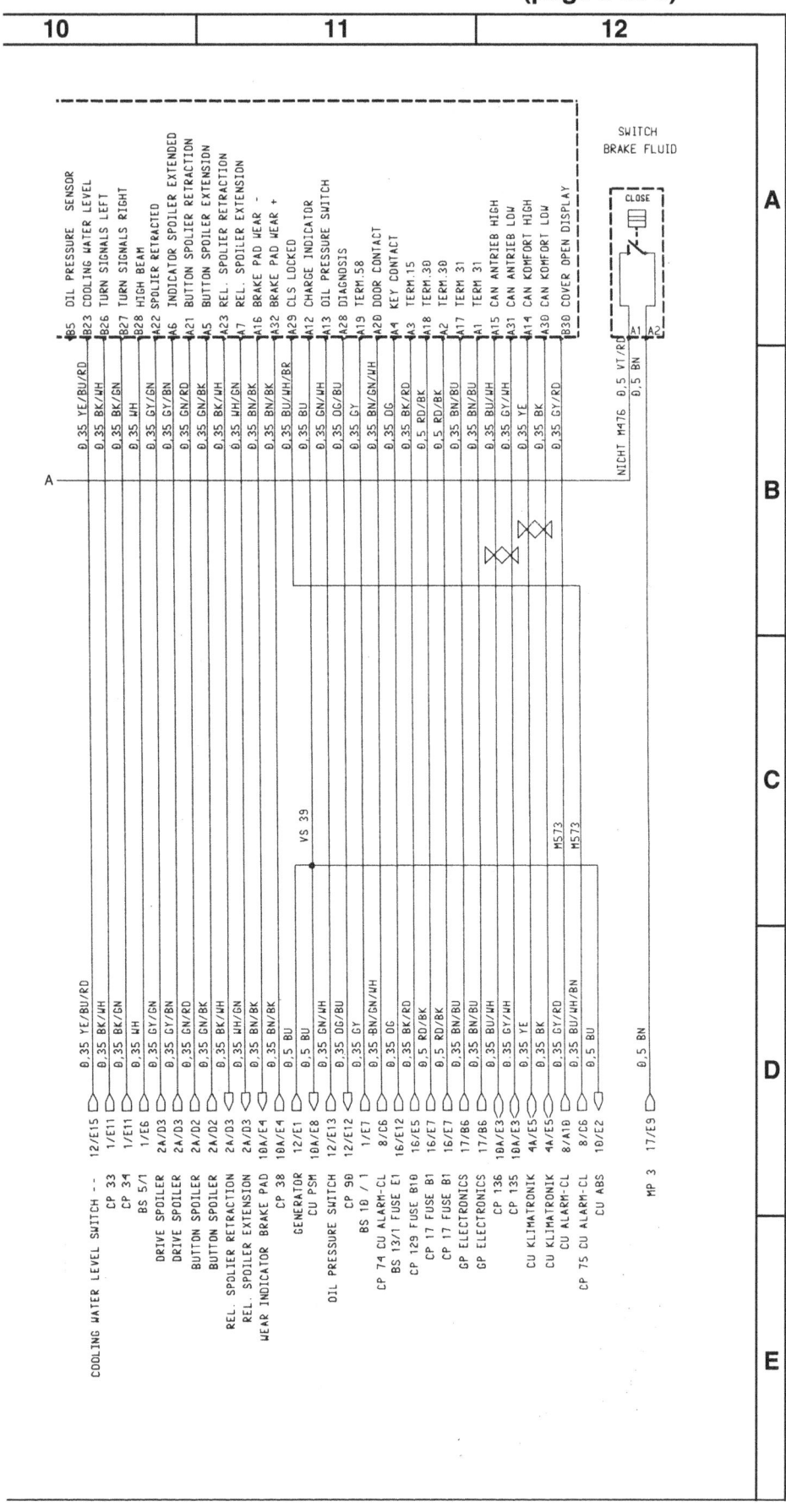

Diagram 2 — Instrument cluster (page 3 of 3) — 2001

Electrical Wiring Diagrams EWD-67

2001   **Spoiler**   **Diagram 2A**

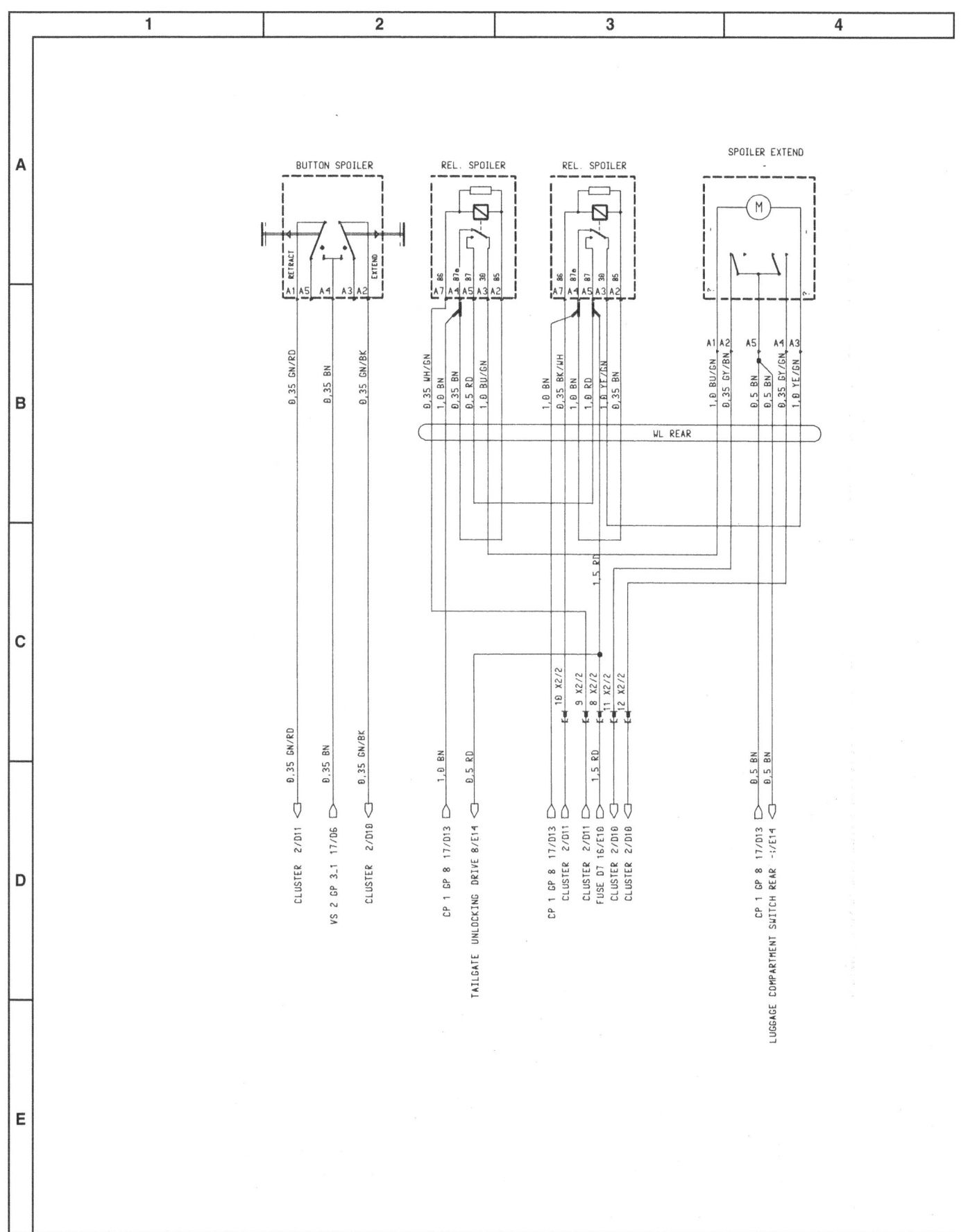

EWD-68 Electrical Wiring Diagrams

Diagram 3 — Windshield wipers and washers, headlight cleaning (page 1 of 2) — 2001

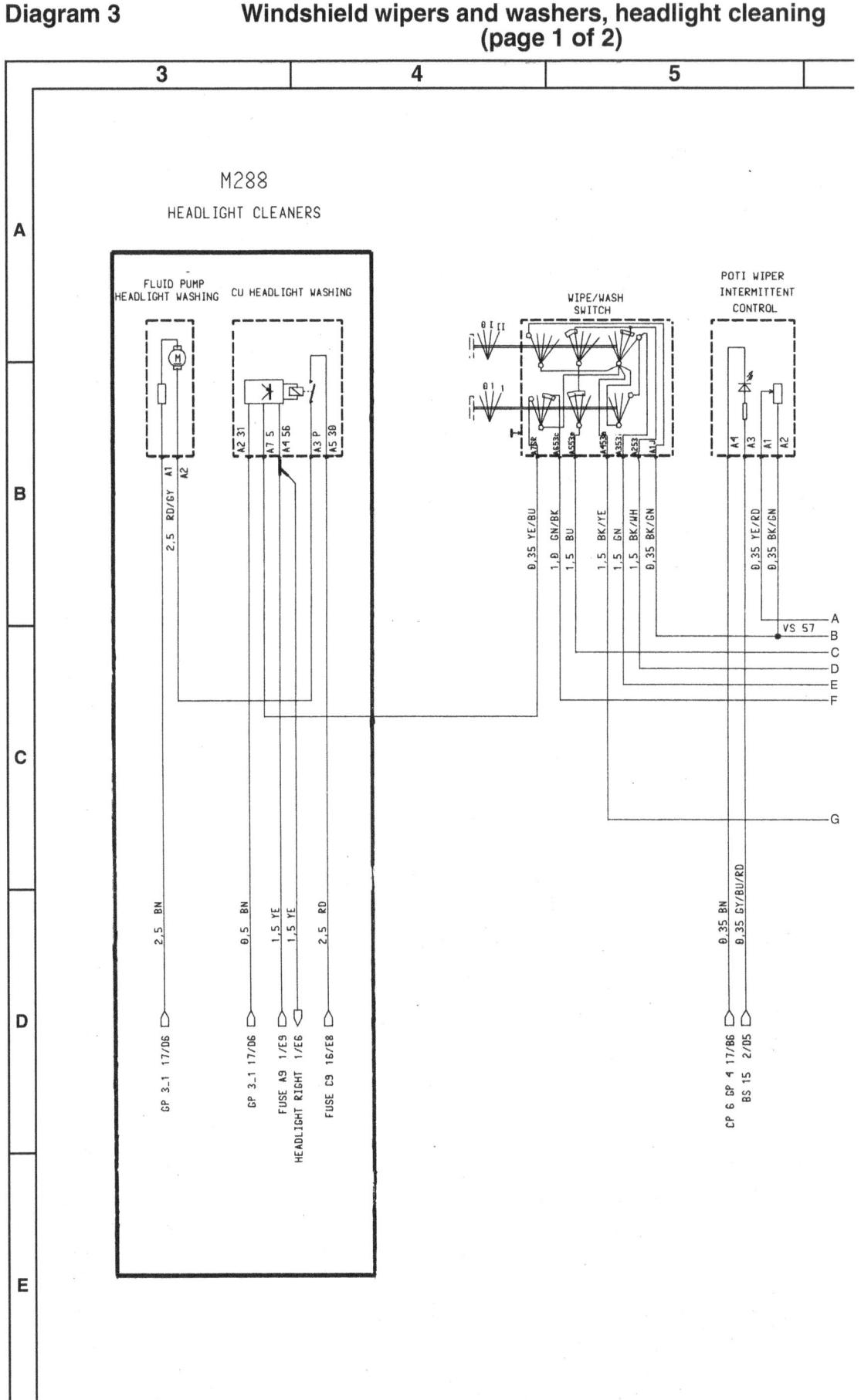

Electrical Wiring Diagrams EWD-69

2001 — Windshield wipers and washers, headlight cleaning (page 2 of 2) — **Diagram 3**

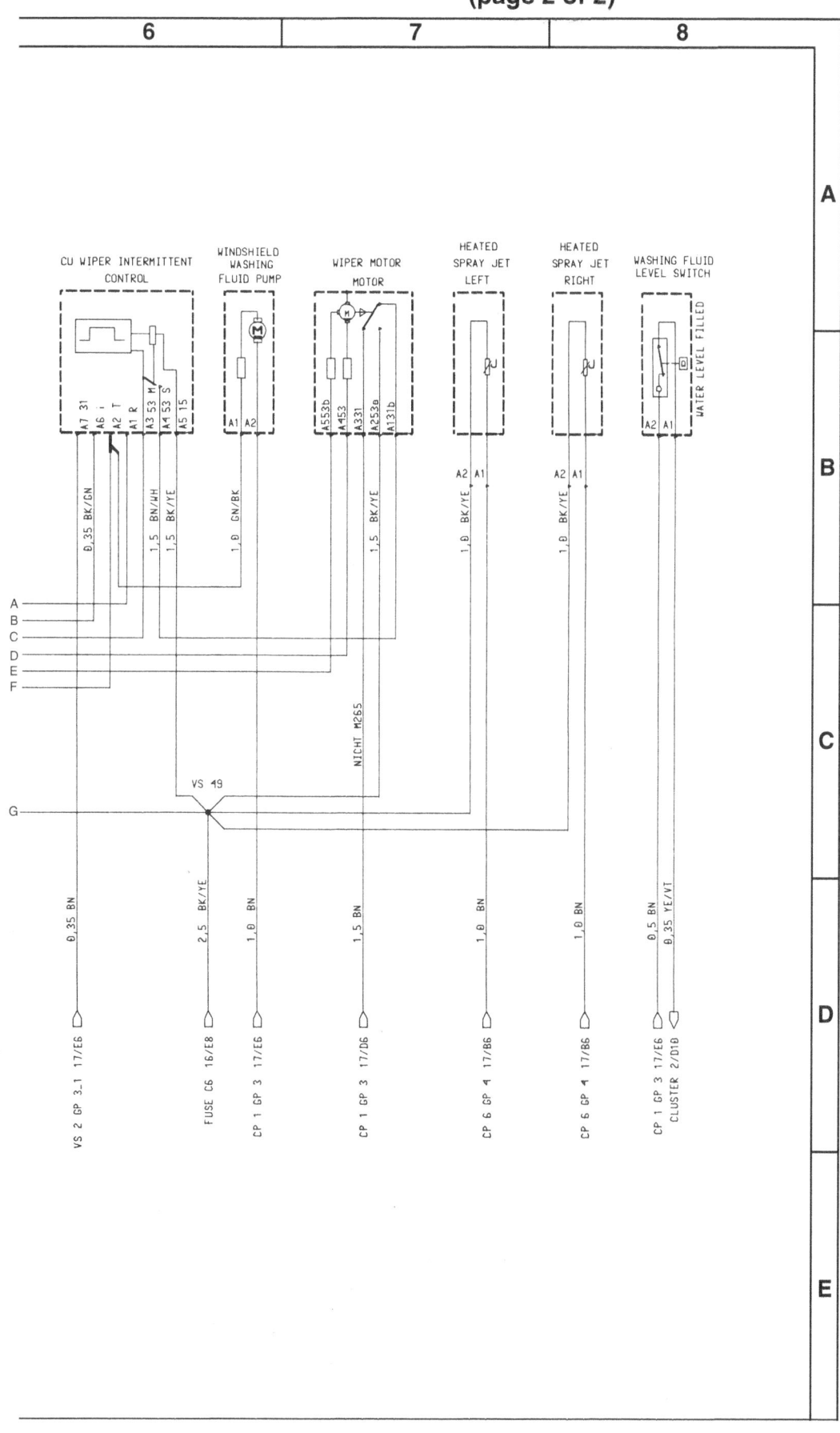

EWD-70 Electrical Wiring Diagrams

Diagram 3A — Windshield wipers (with rain sensor), headlight cleaning (page 1 of 2) — 2001

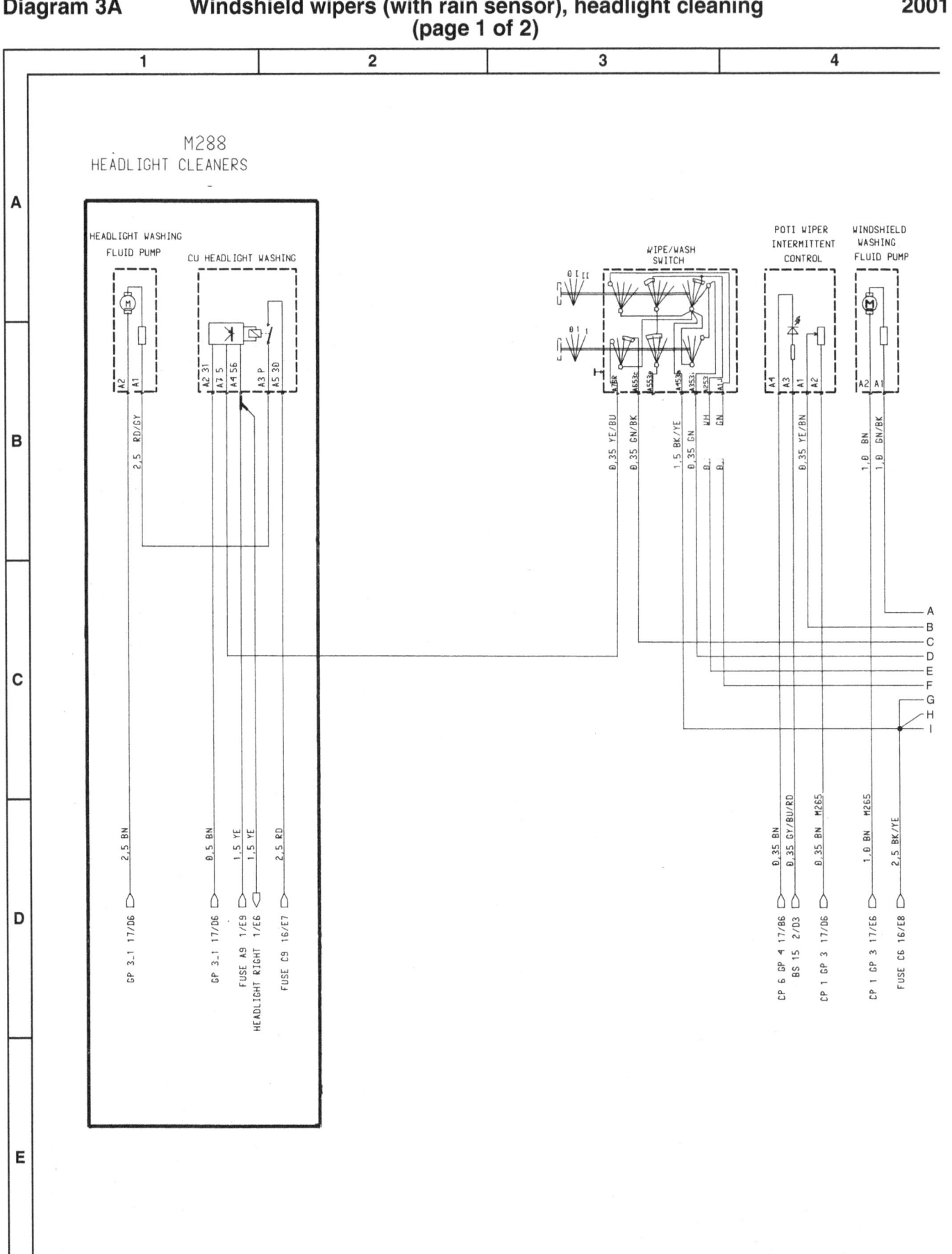

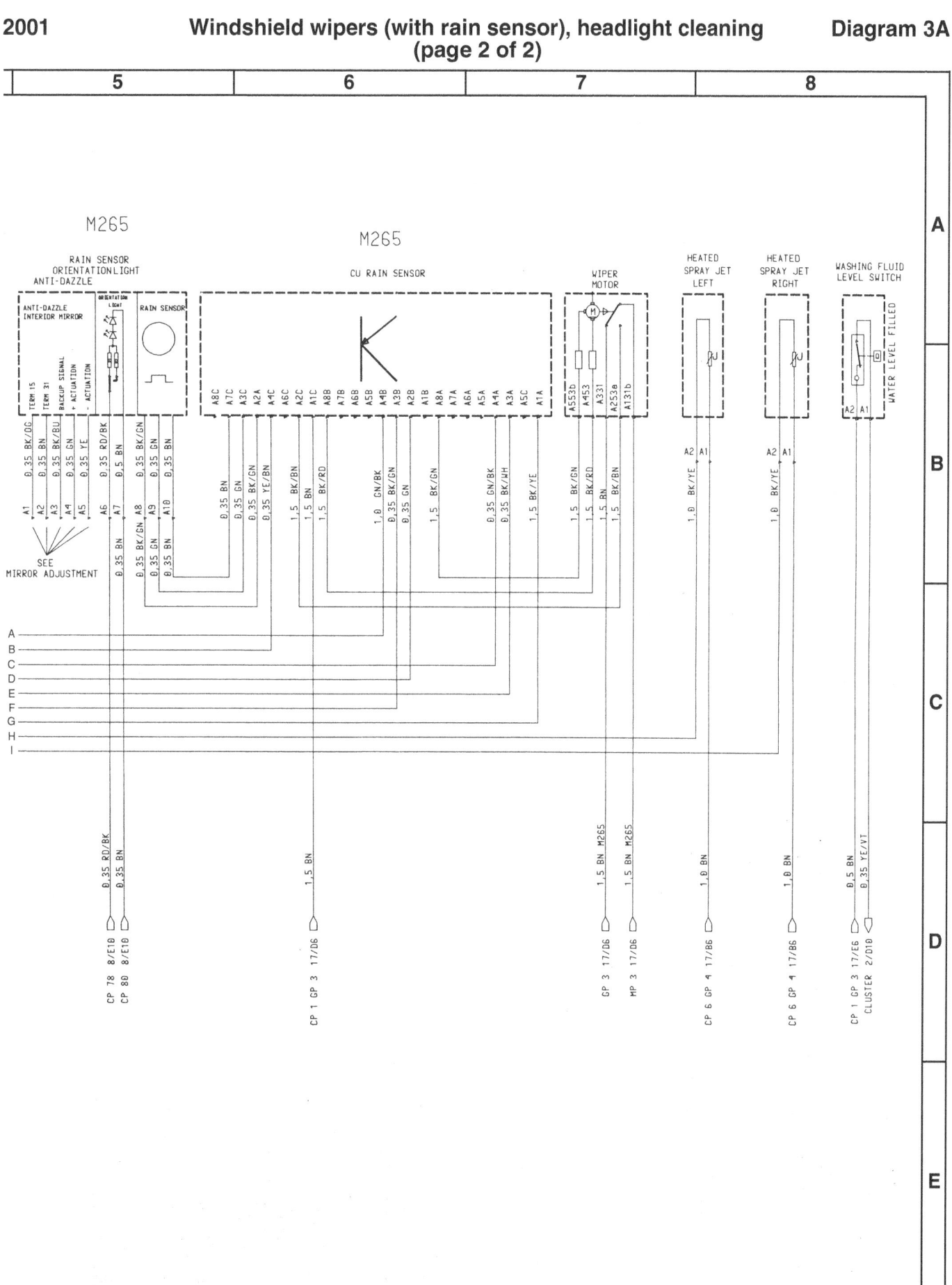

EWD-72 Electrical Wiring Diagrams

Diagram 3B Horns 2001

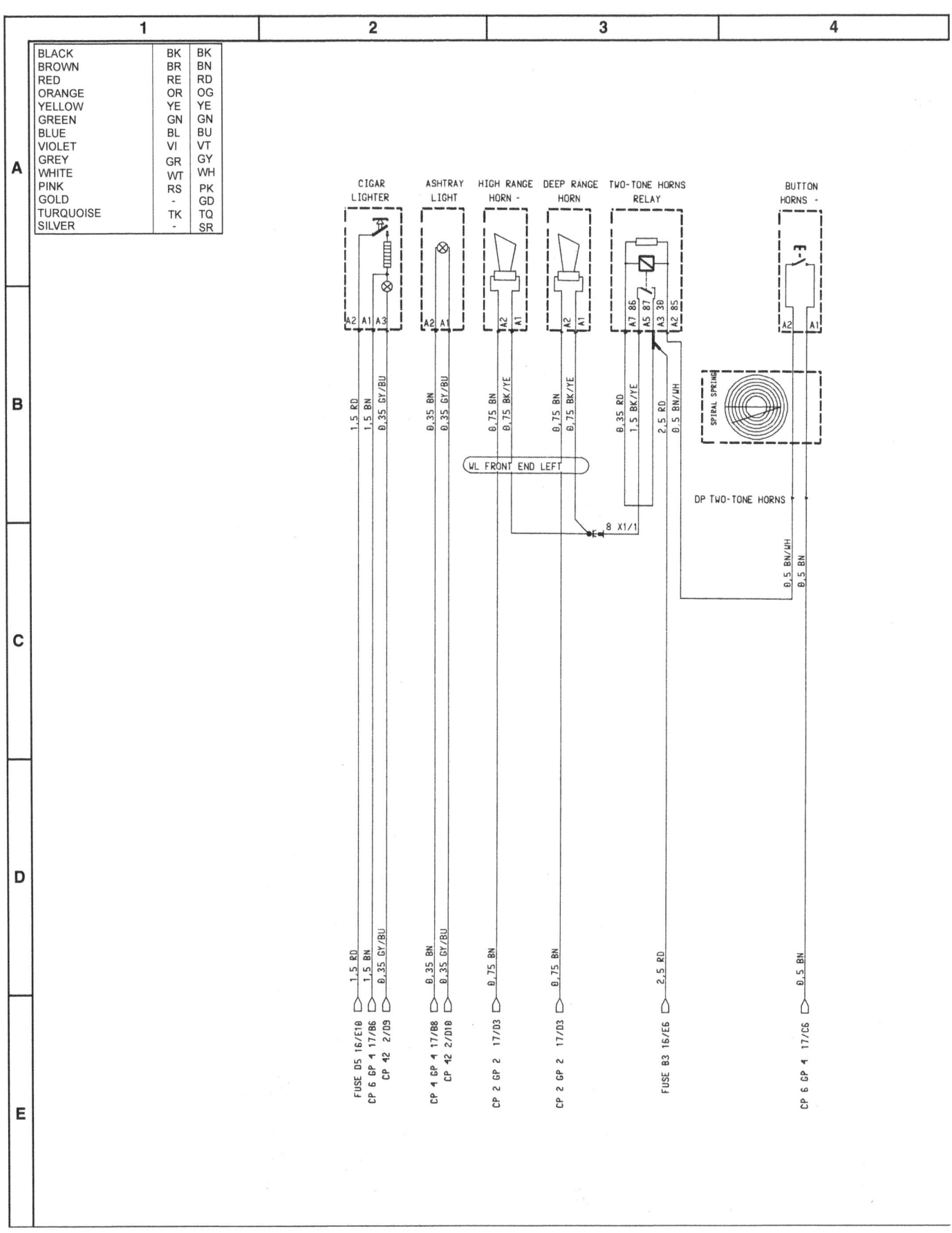

Electrical Wiring Diagrams EWD-73

2001 Heating system Diagram 4

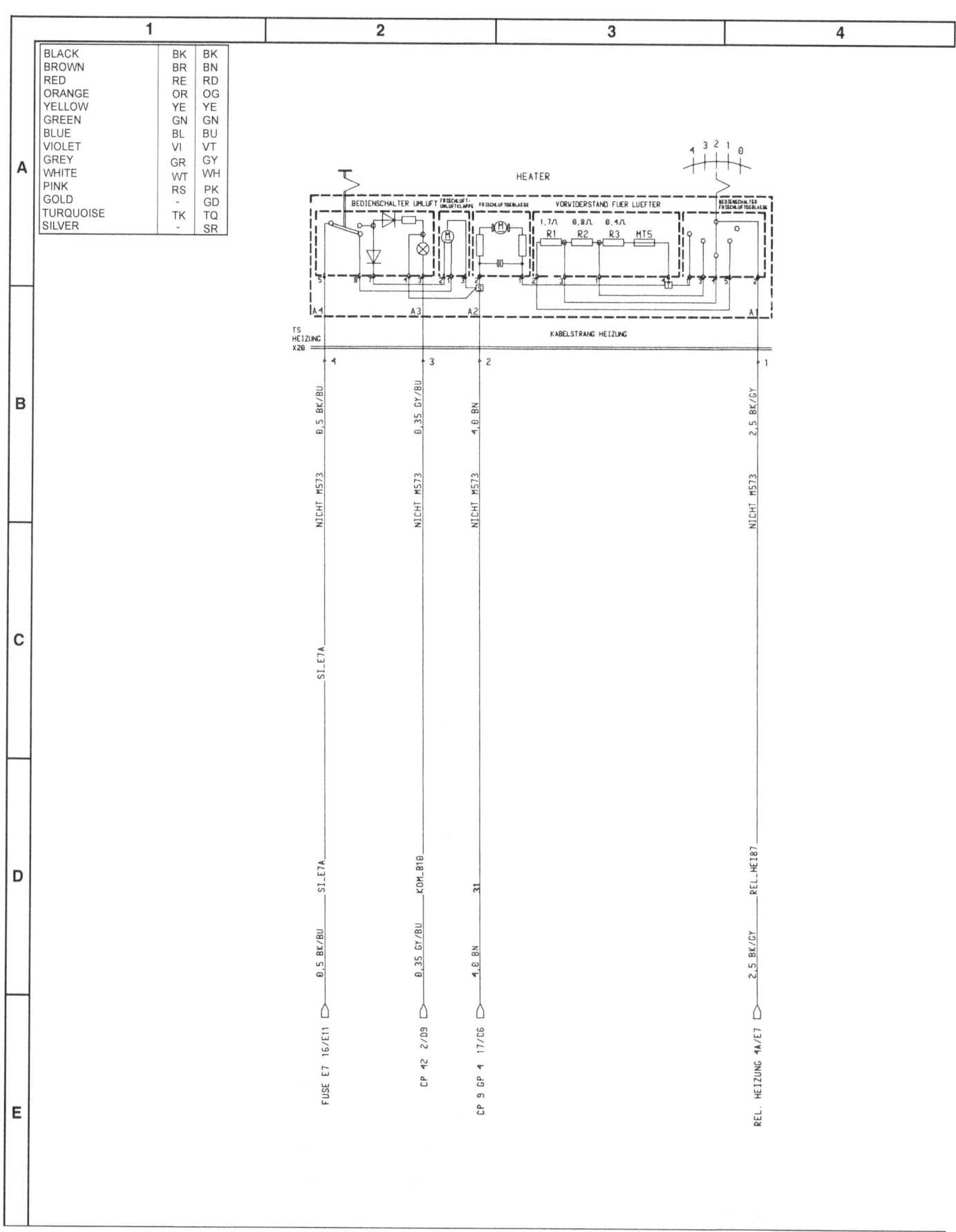

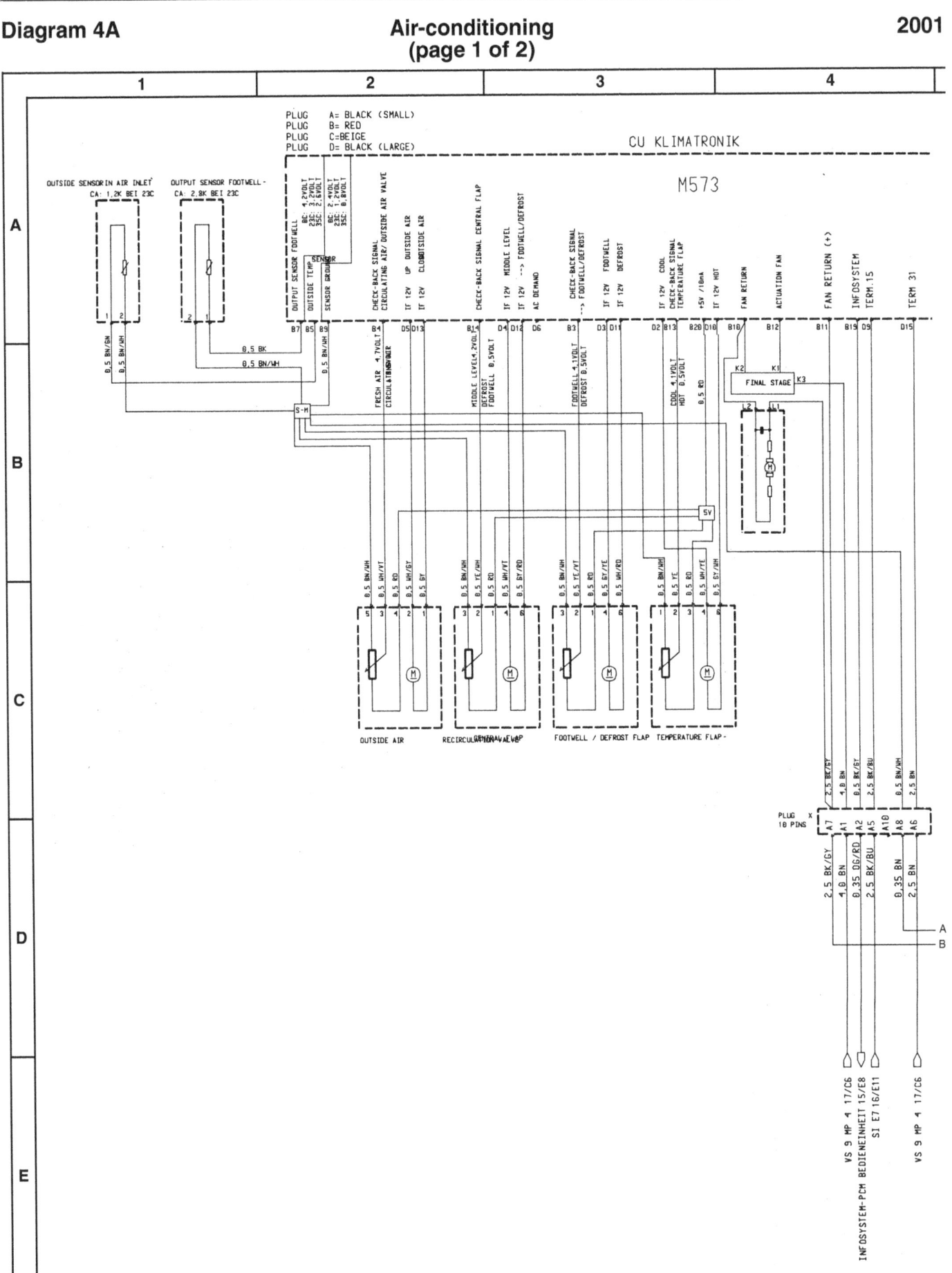

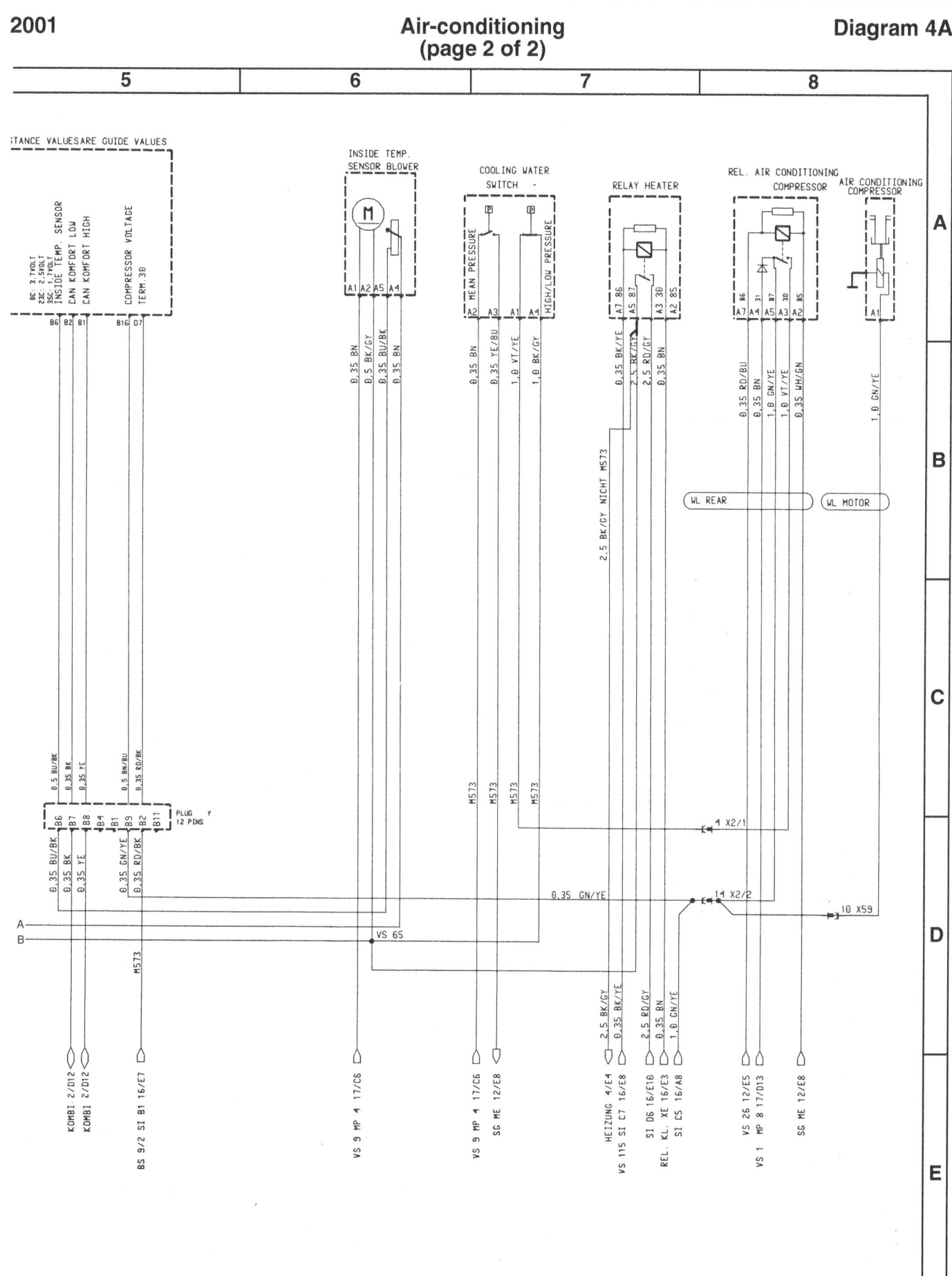

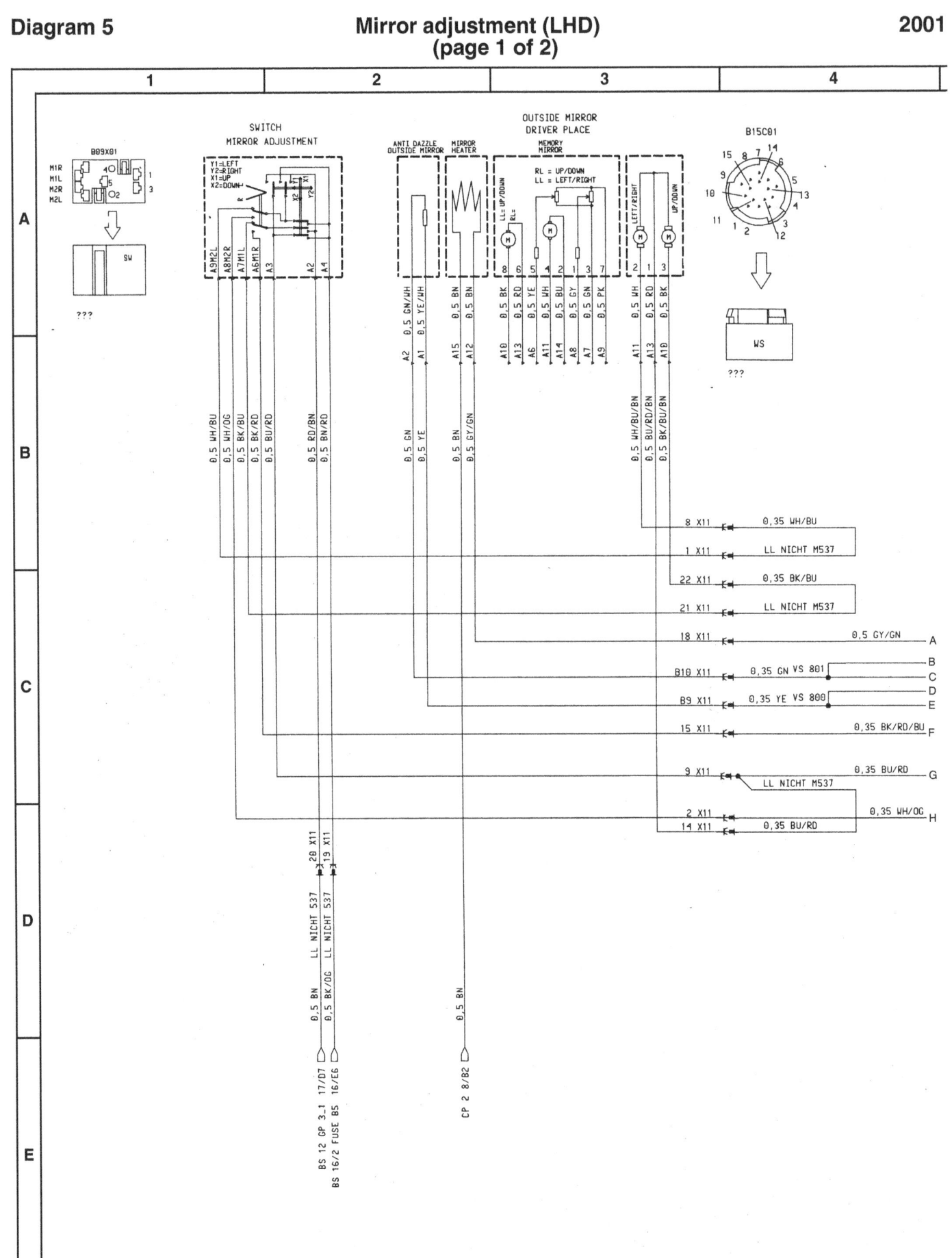

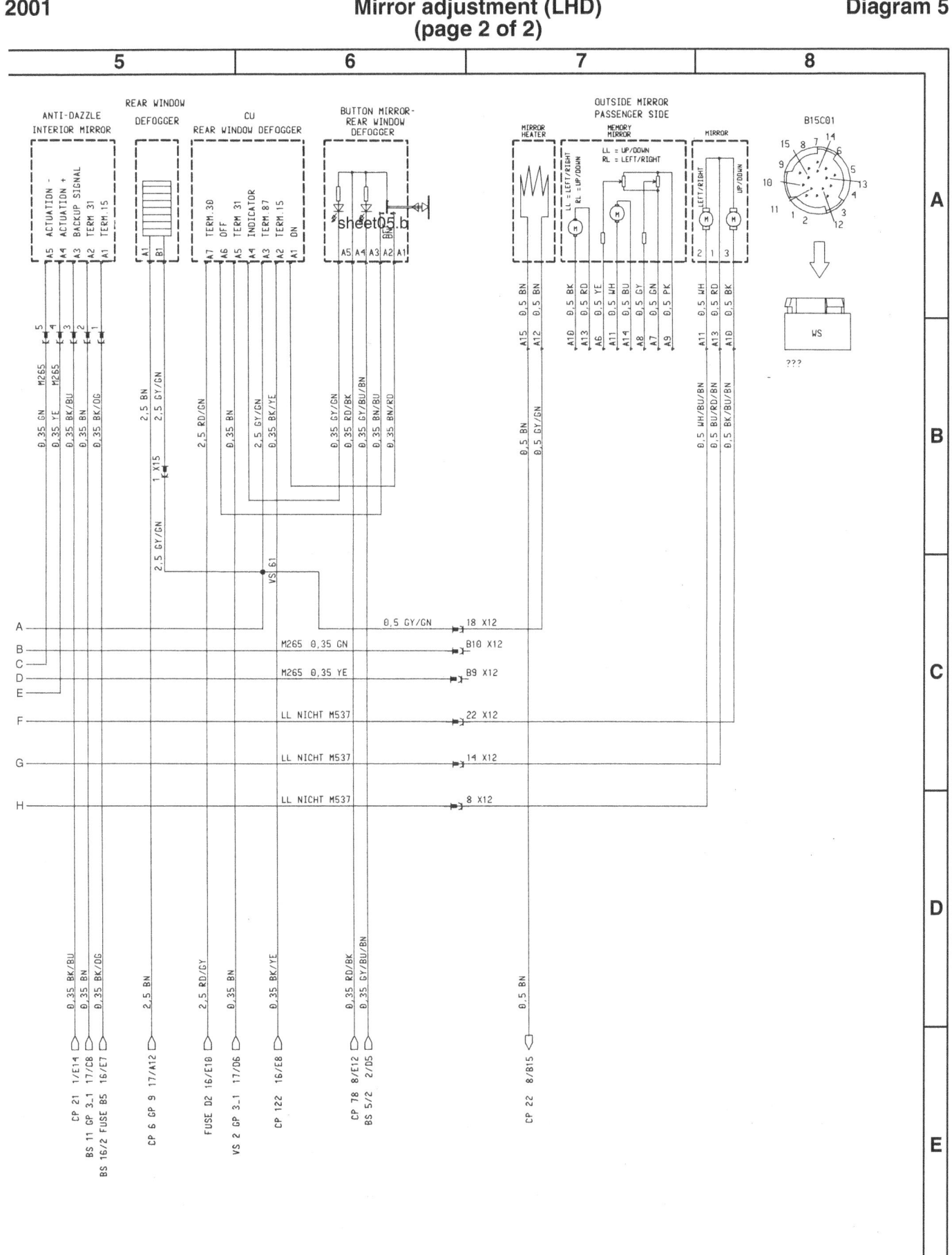

EWD-78 Electrical Wiring Diagrams

Diagram 5A — Seat and mirror memory (LHD) (page 1 of 3) — 2001

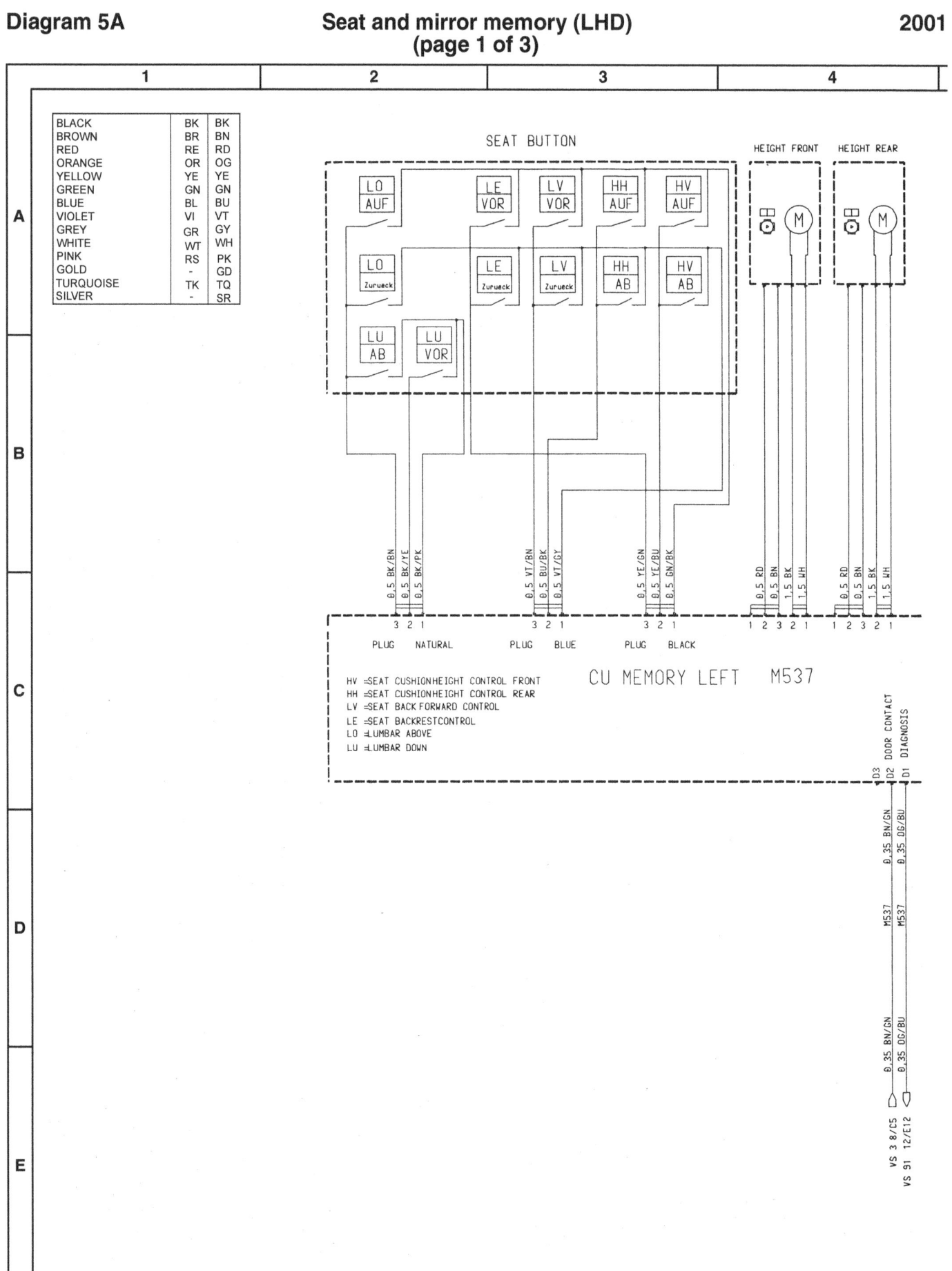

Electrical Wiring Diagrams EWD-79

2001 — Seat and mirror memory (LHD) (page 2 of 3) — Diagram 5A

EWD-80 Electrical Wiring Diagrams

Diagram 5A — Seat and mirror memory (LHD) (page 3 of 3) — 2001

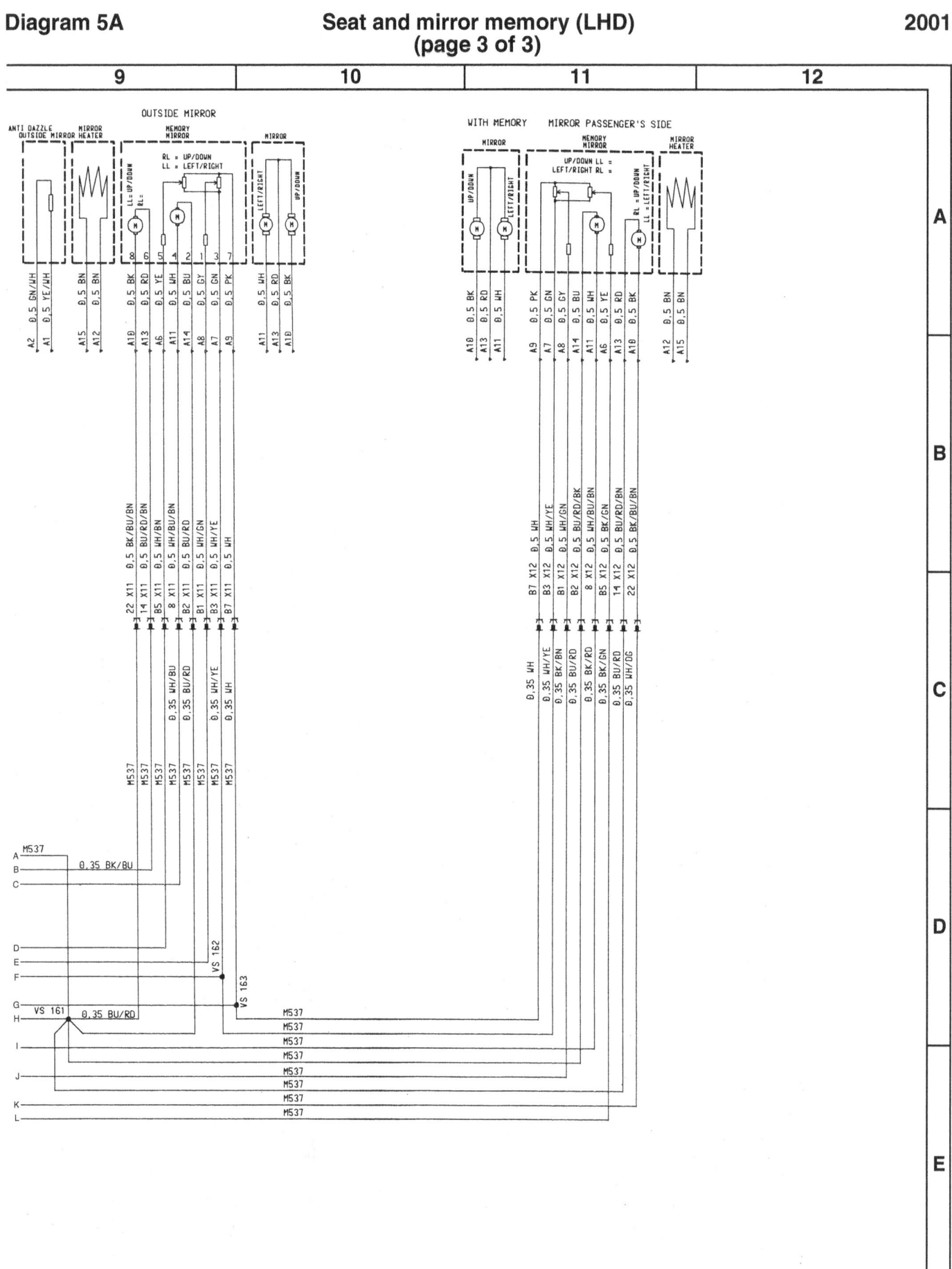

Electrical Wiring Diagrams EWD-81

2001 — Mirror adjustment (RHD) (page 1 of 2) — Diagram 5B

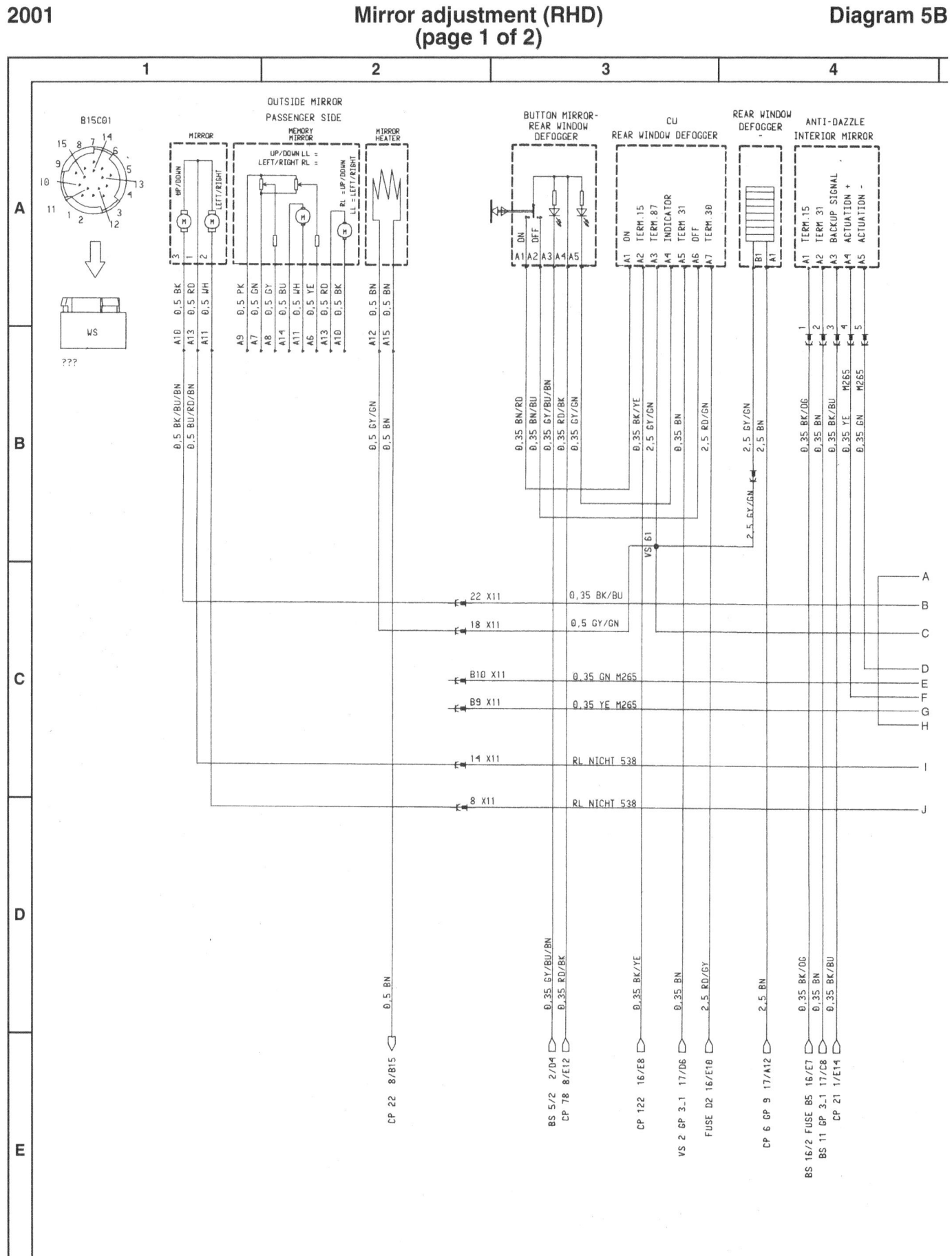

EWD-82 Electrical Wiring Diagrams

Diagram 5B — Mirror adjustment (RHD) (page 2 of 2) — 2001

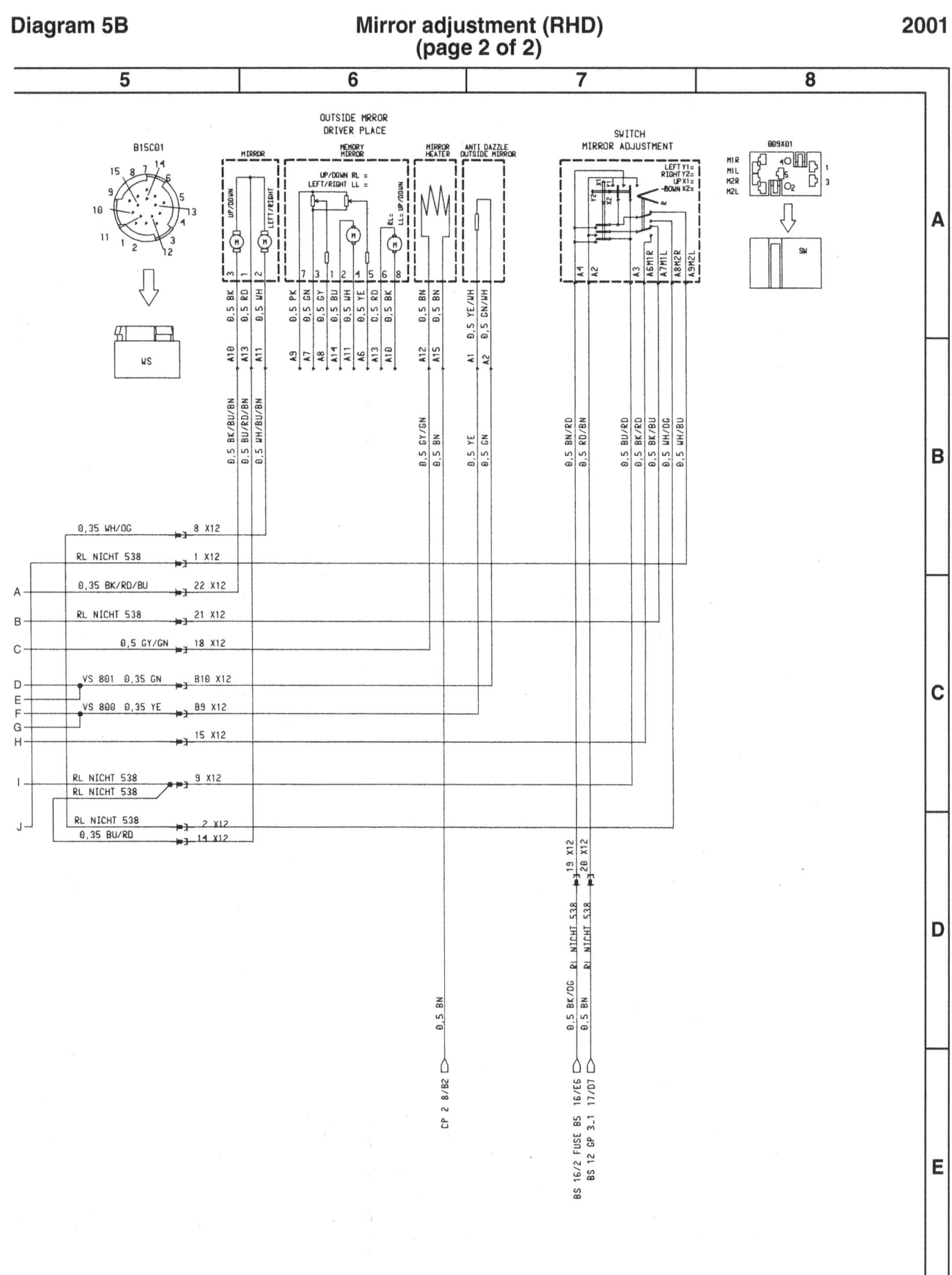

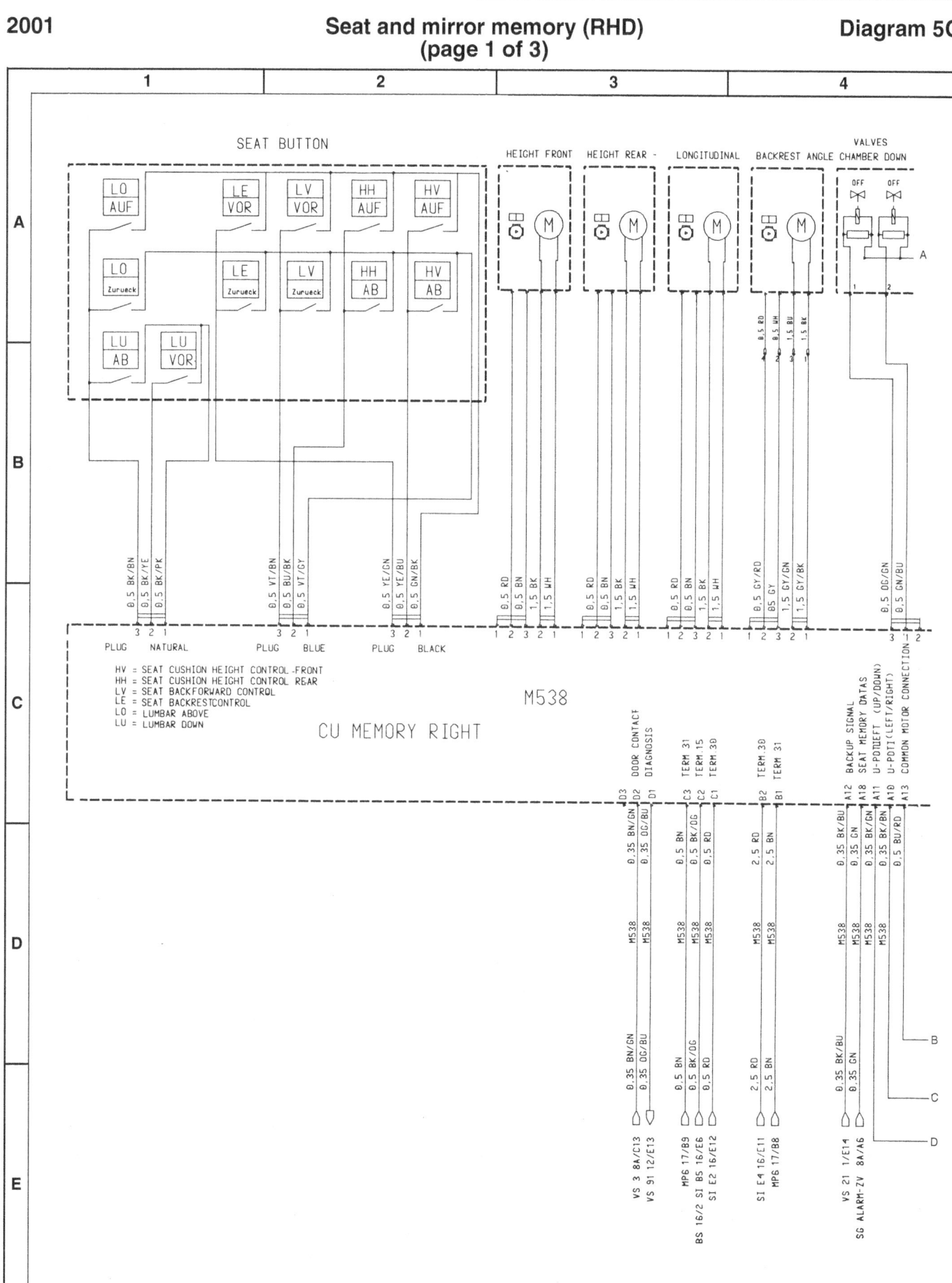

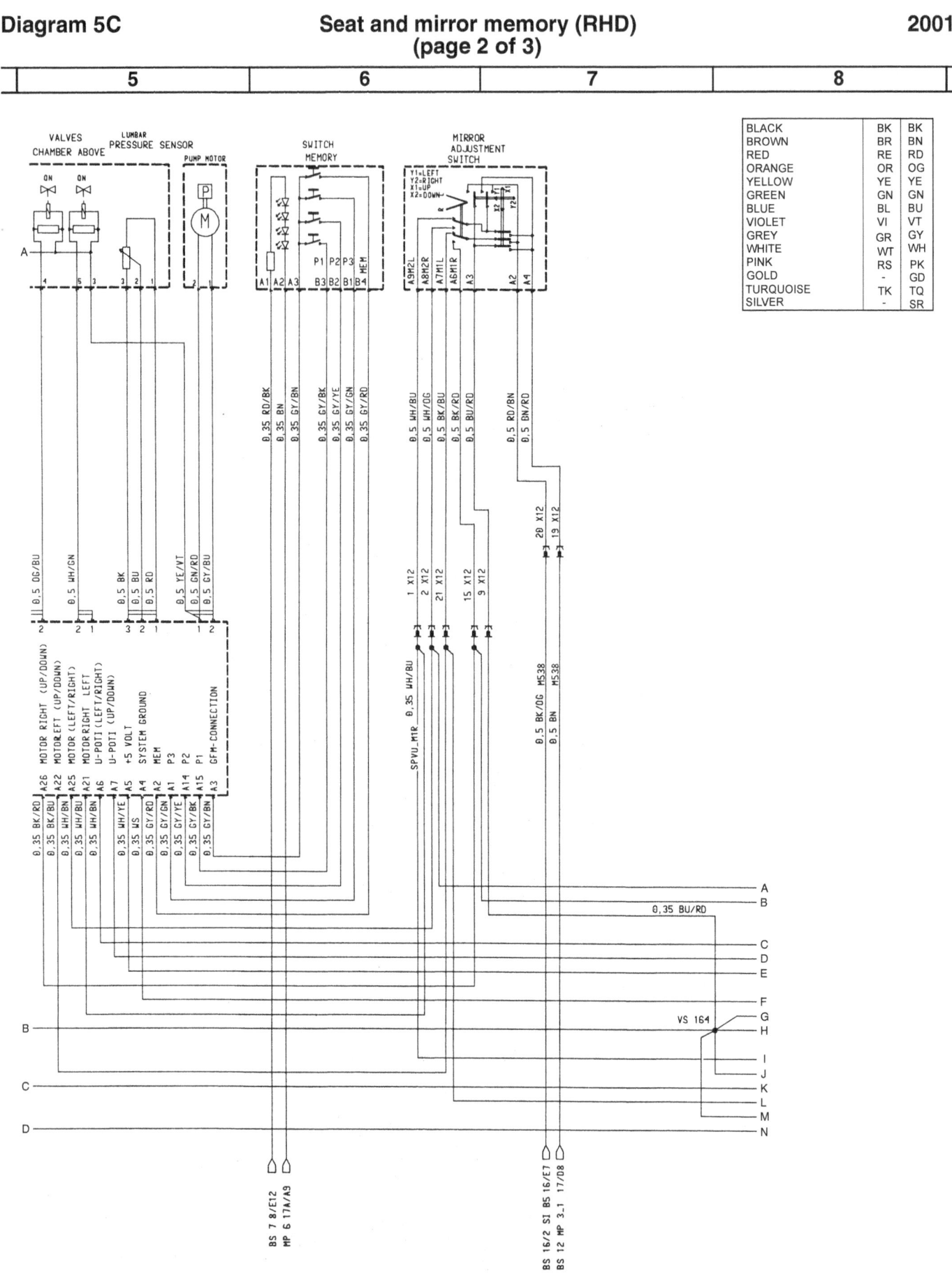

Electrical Wiring Diagrams EWD-85

2001 — Seat and mirror memory (RHD) (page 3 of 3) — **Diagram 5C**

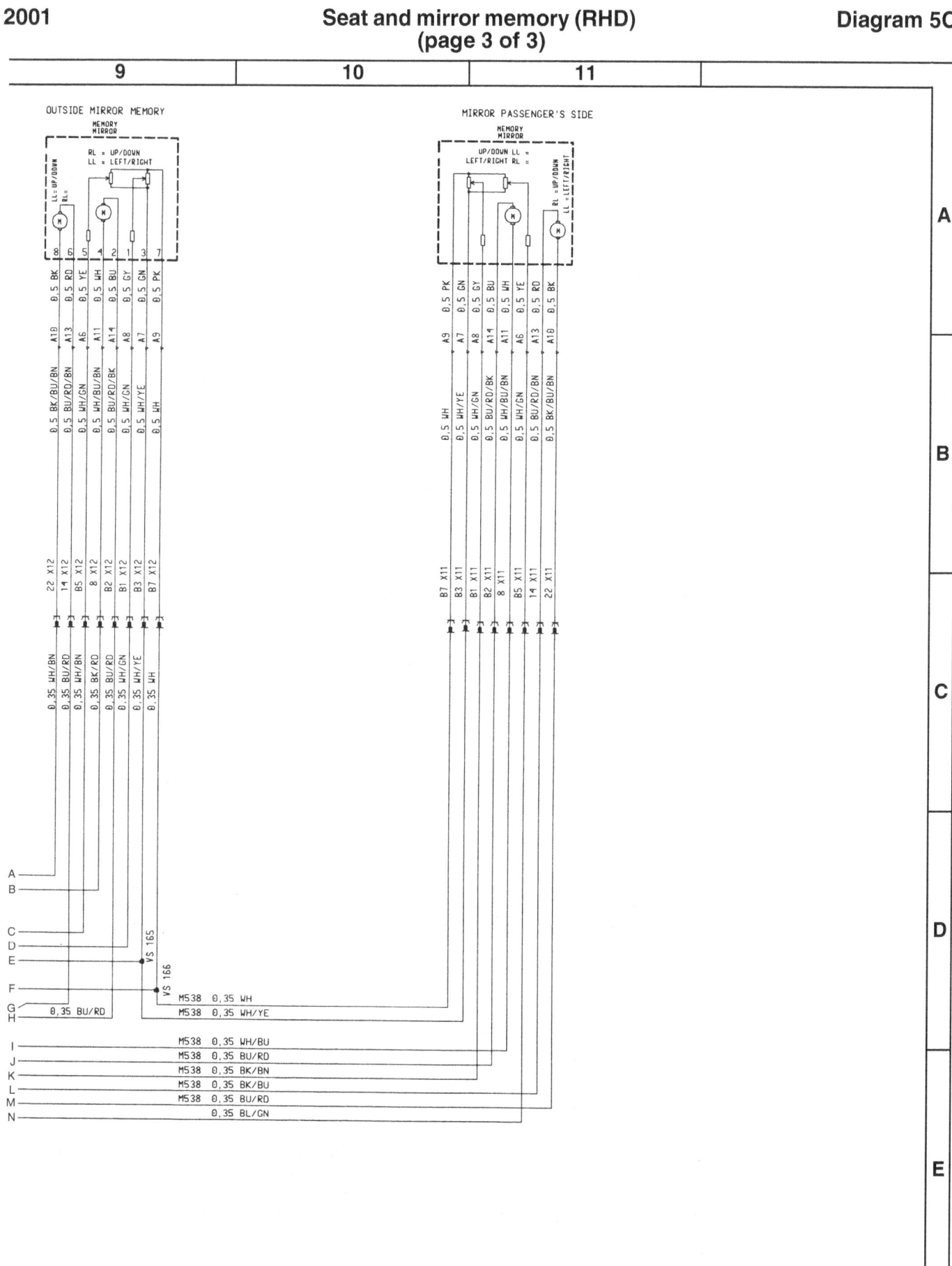

EWD-86 Electrical Wiring Diagrams

Diagram 5D Seats (page 1 of 3) 2001

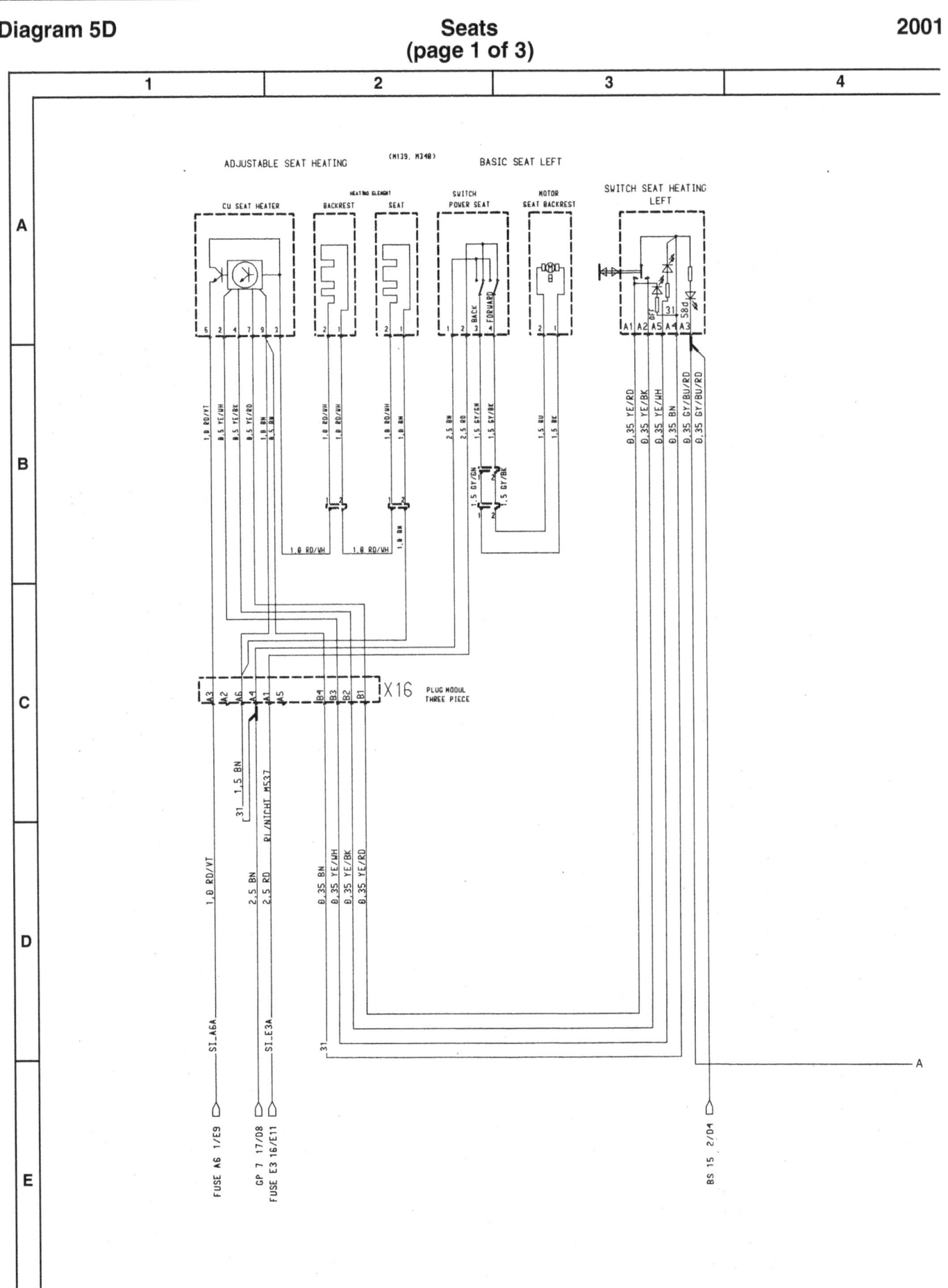

Electrical Wiring Diagrams EWD-87

2001 — Seats (page 2 of 3) — **Diagram 5D**

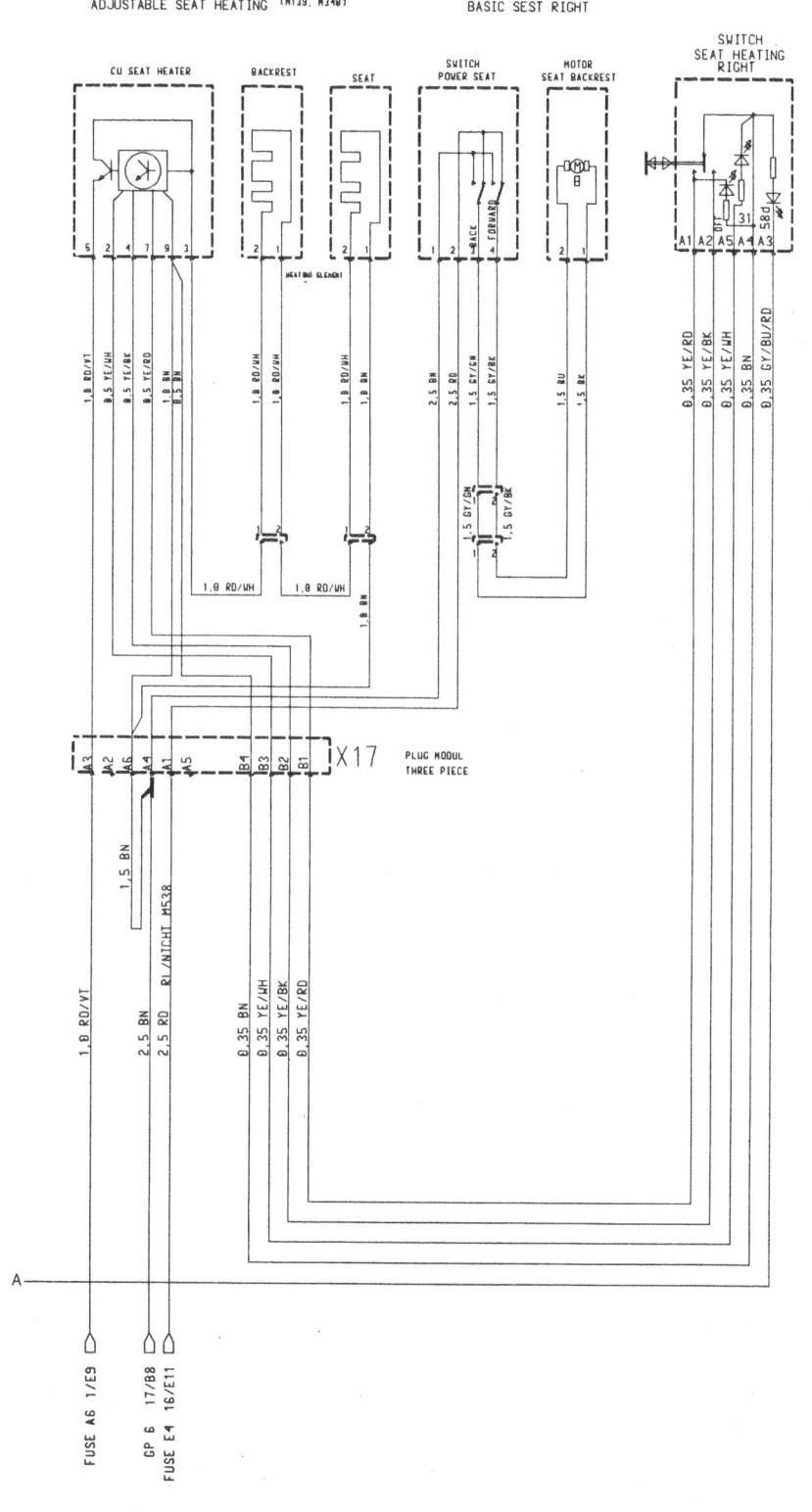

Diagram 5B — Seats (page 3 of 3)

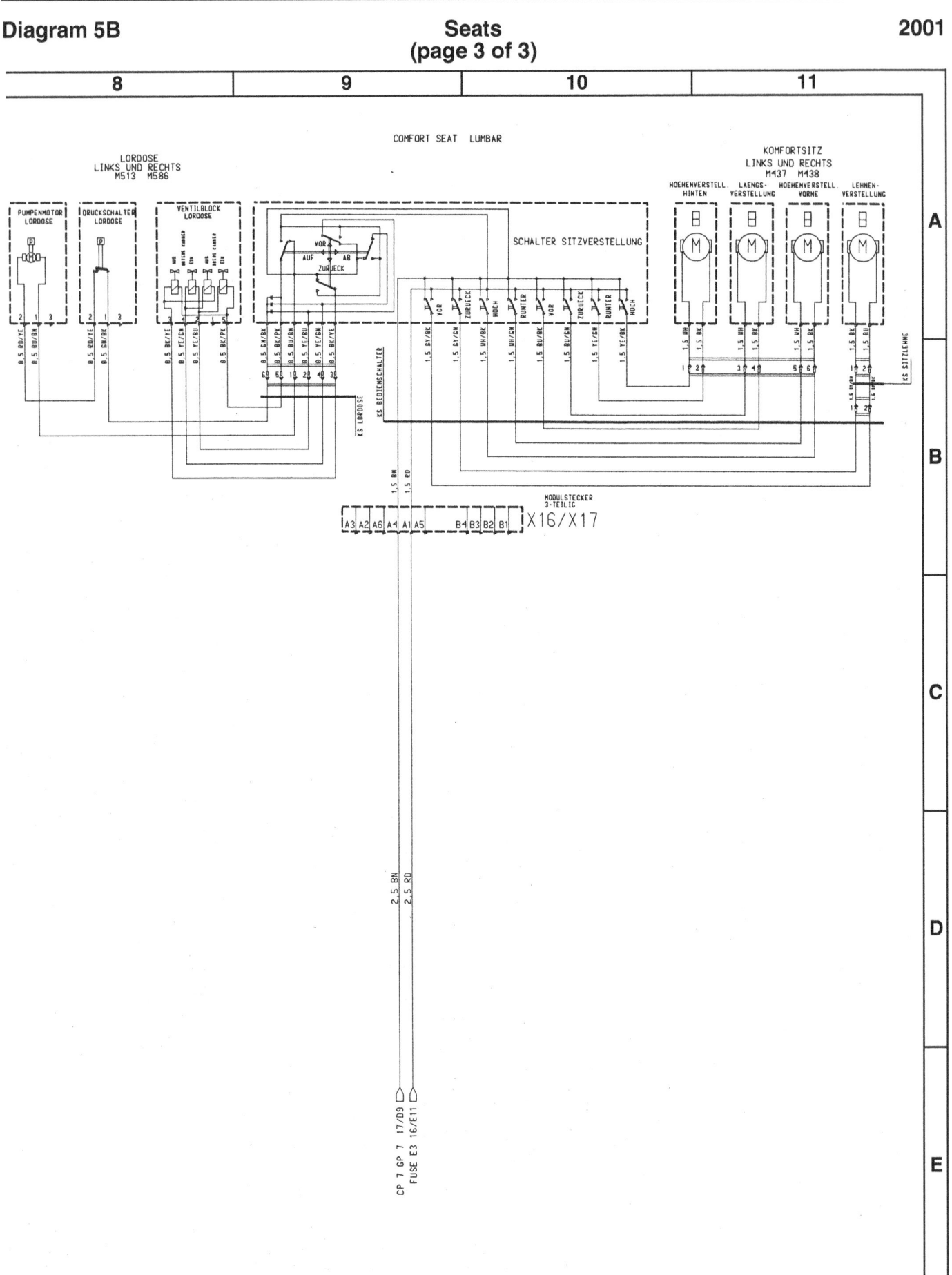

Electrical Wiring Diagrams EWD-89

2001 — **Convertible top** — Diagram 6

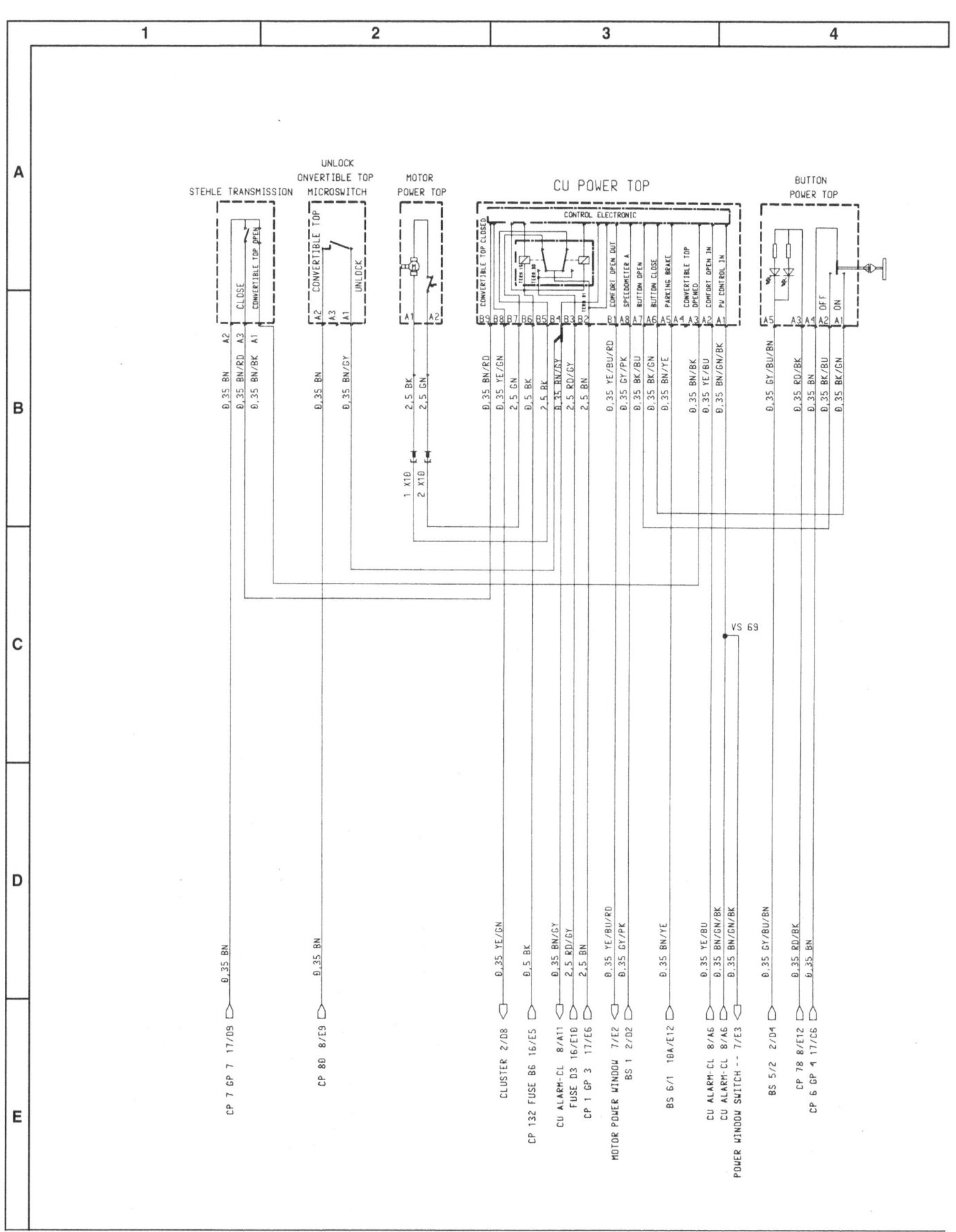

EWD-90   Electrical Wiring Diagrams

Diagram 7   Windows   2001

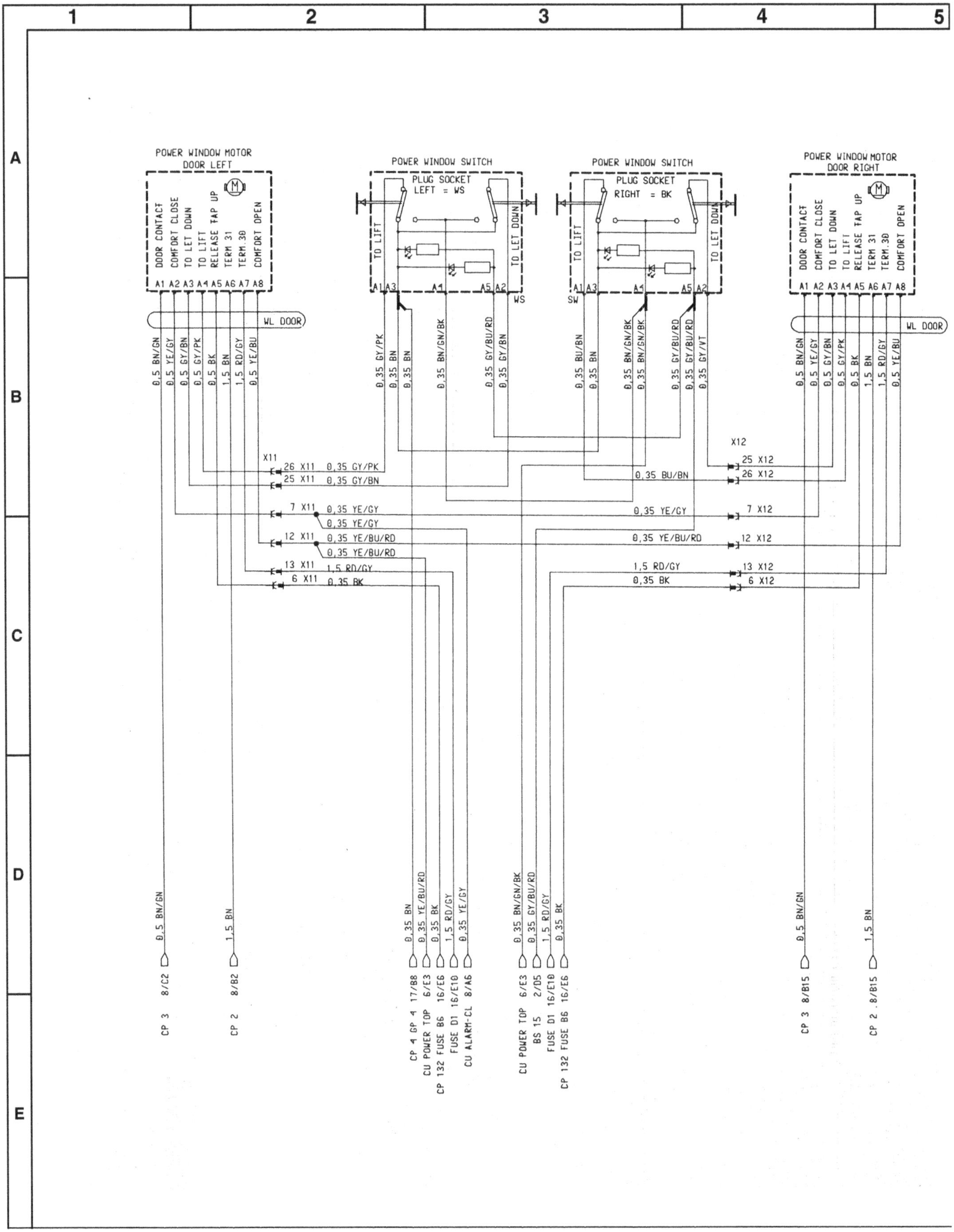

Electrical Wiring Diagrams EWD-91

2001 — Central locking, alarm system (LHD) (page 1 of 4) — Diagram 8

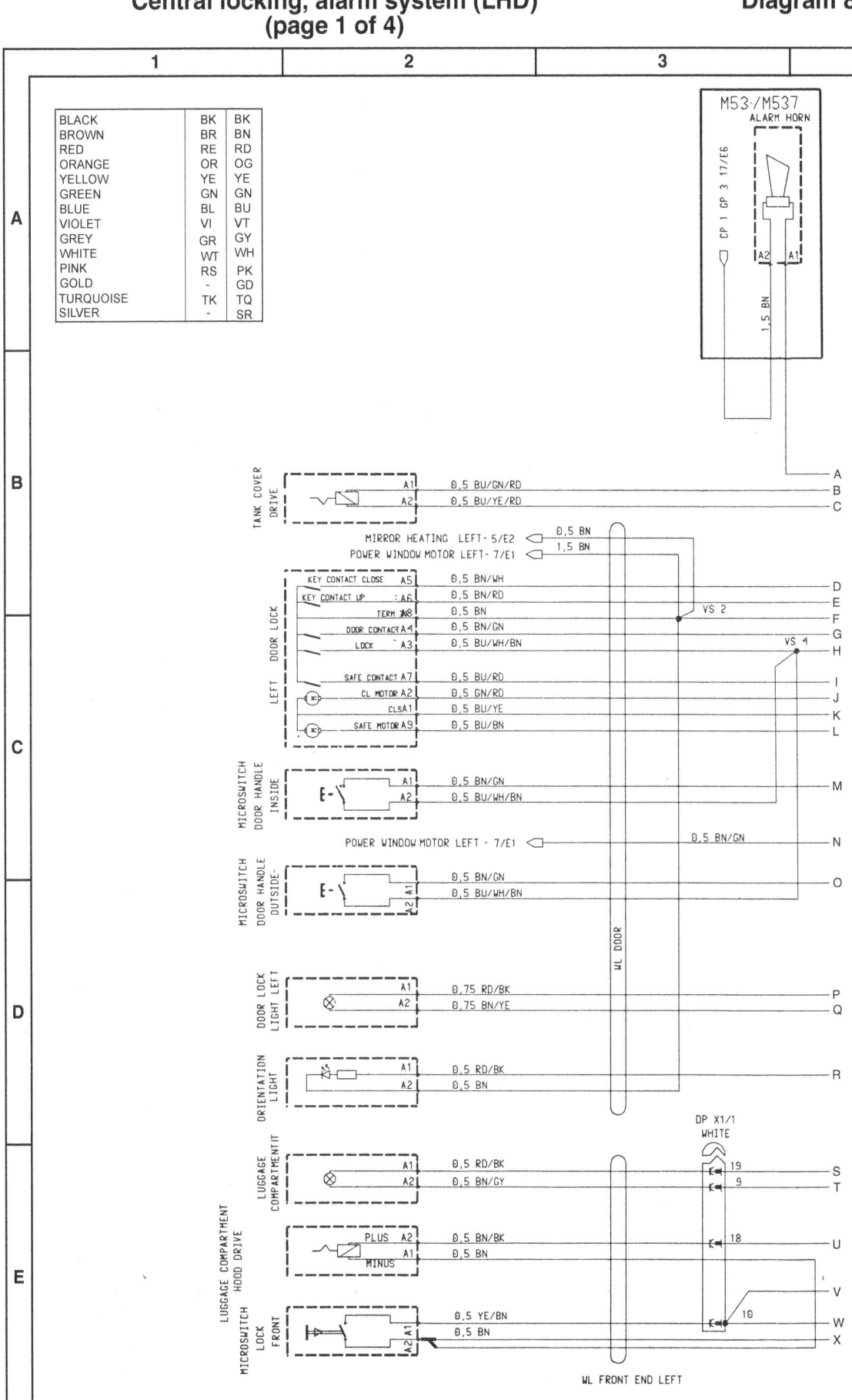

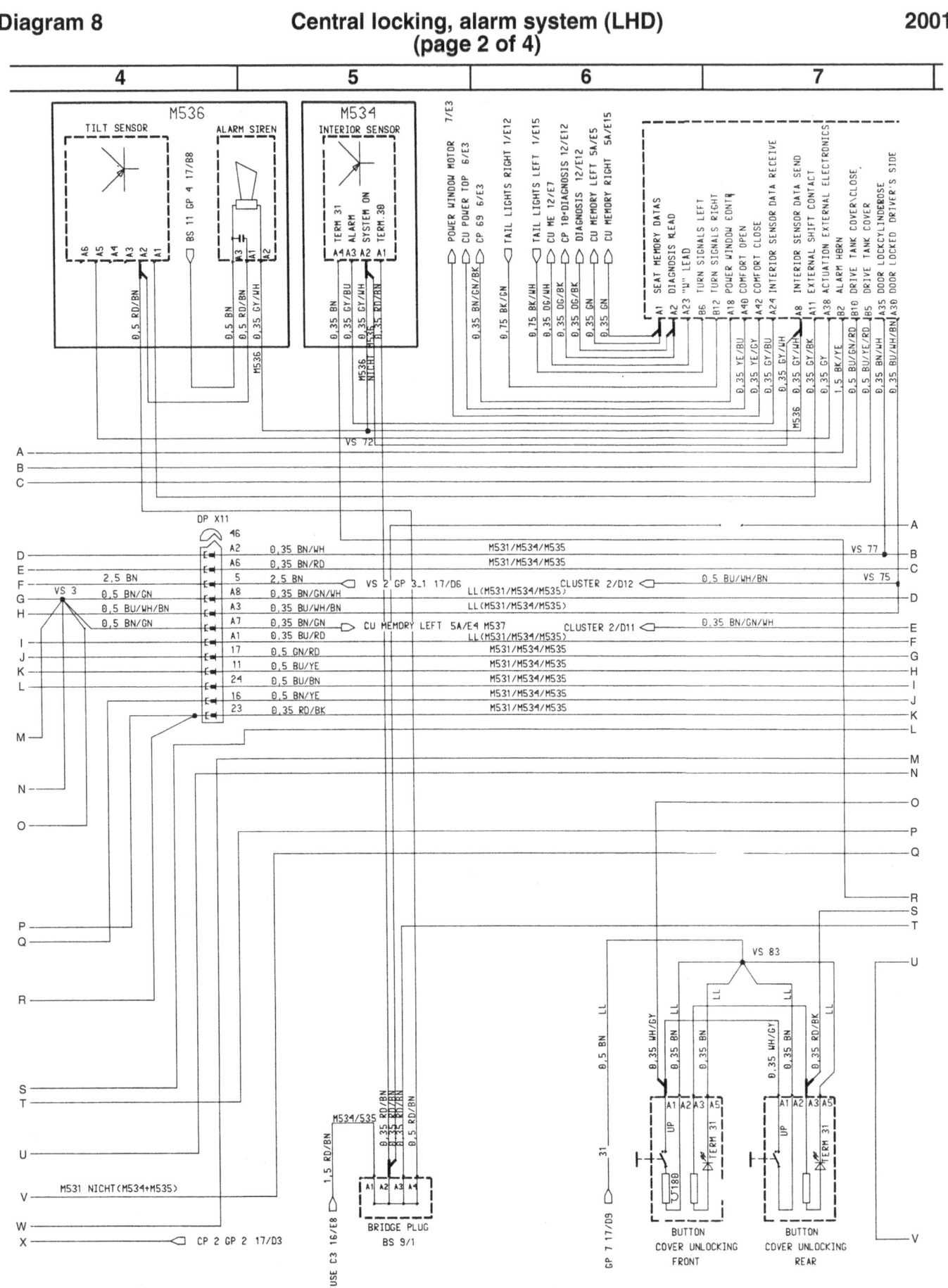

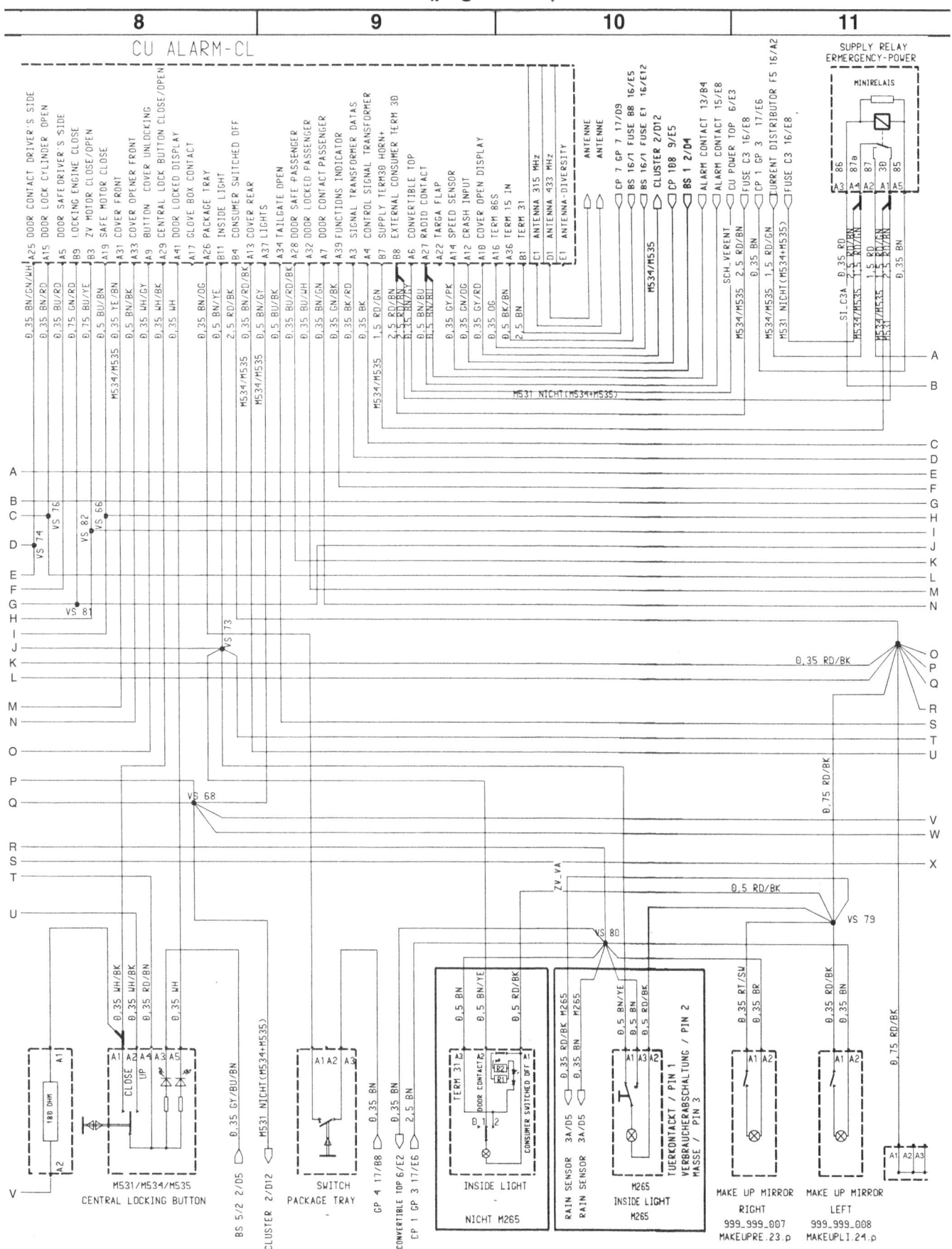

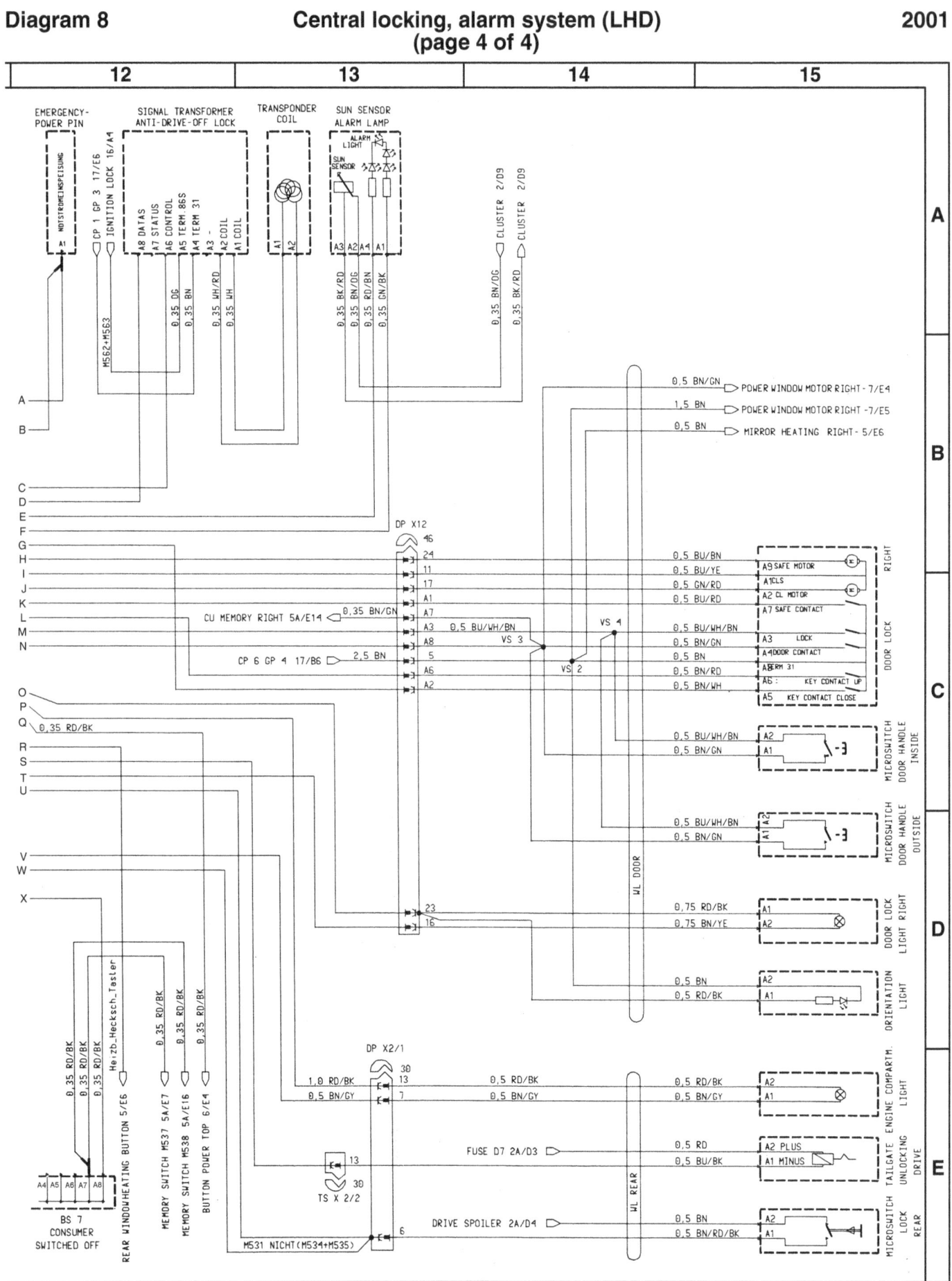

2001

Central locking, alarm system (RHD) (page 1 of 4)

Diagram 8A

BLACK	BK	BK
BROWN	BR	BN
RED	RE	RD
ORANGE	OR	OG
YELLOW	YE	YE
GREEN	GN	GN
BLUE	BL	BU
VIOLET	VI	VT
GREY	GR	GY
WHITE	WT	WH
PINK	RS	PK
GOLD	-	GD
TURQUOISE	TK	TQ
SILVER	-	SR

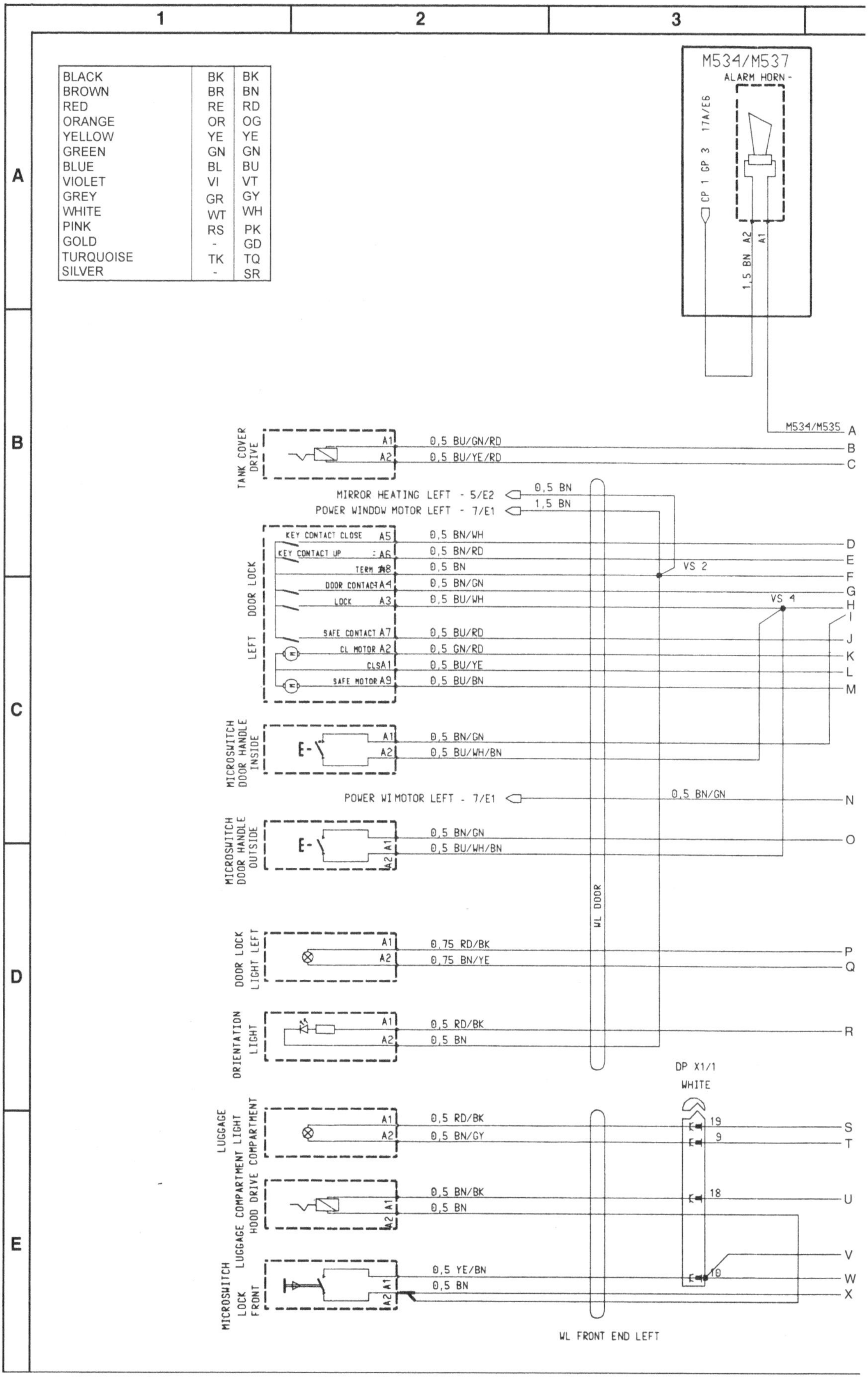

Diagram 8A — Central locking, alarm system (RHD) (page 2 of 4)

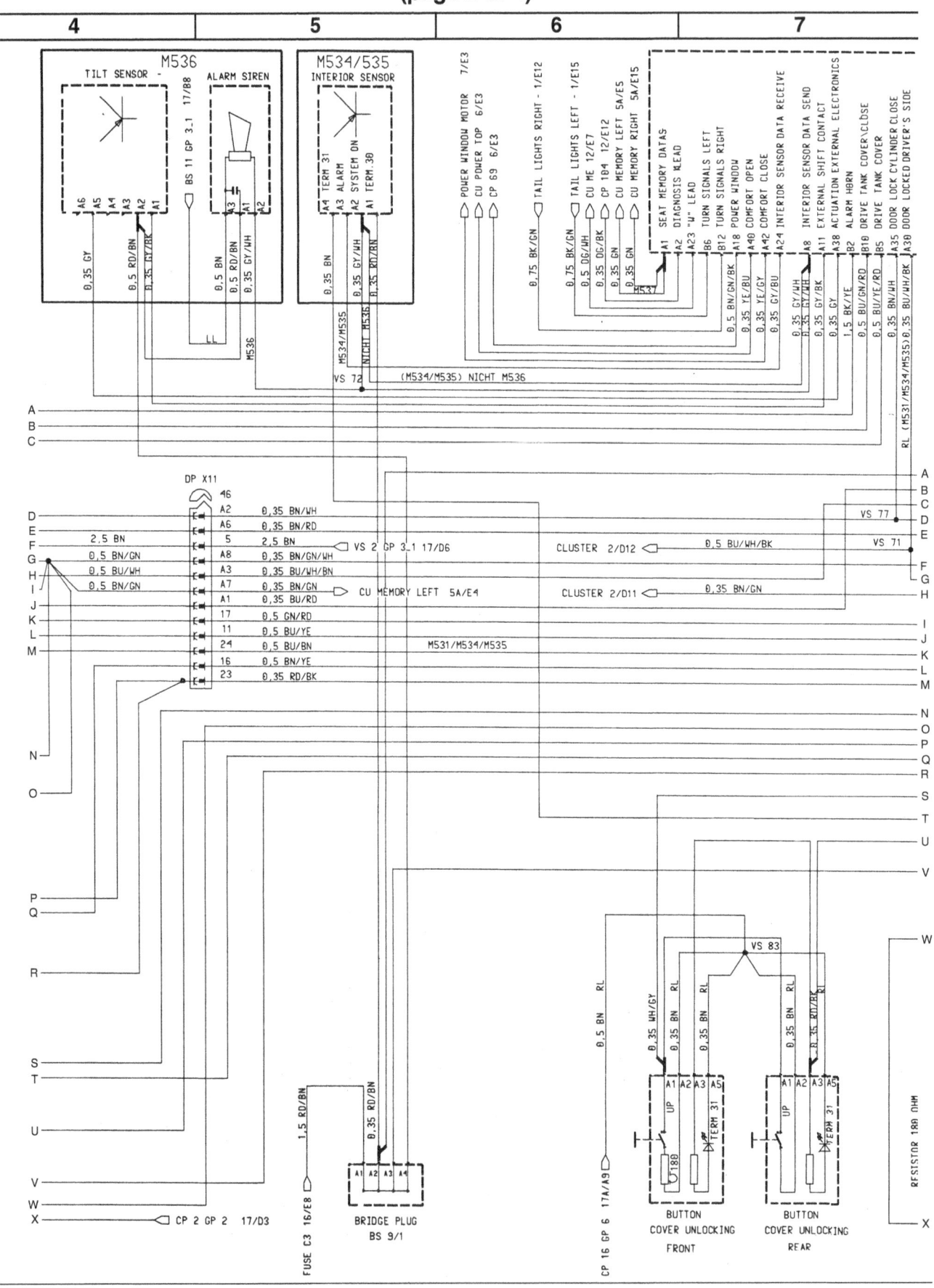

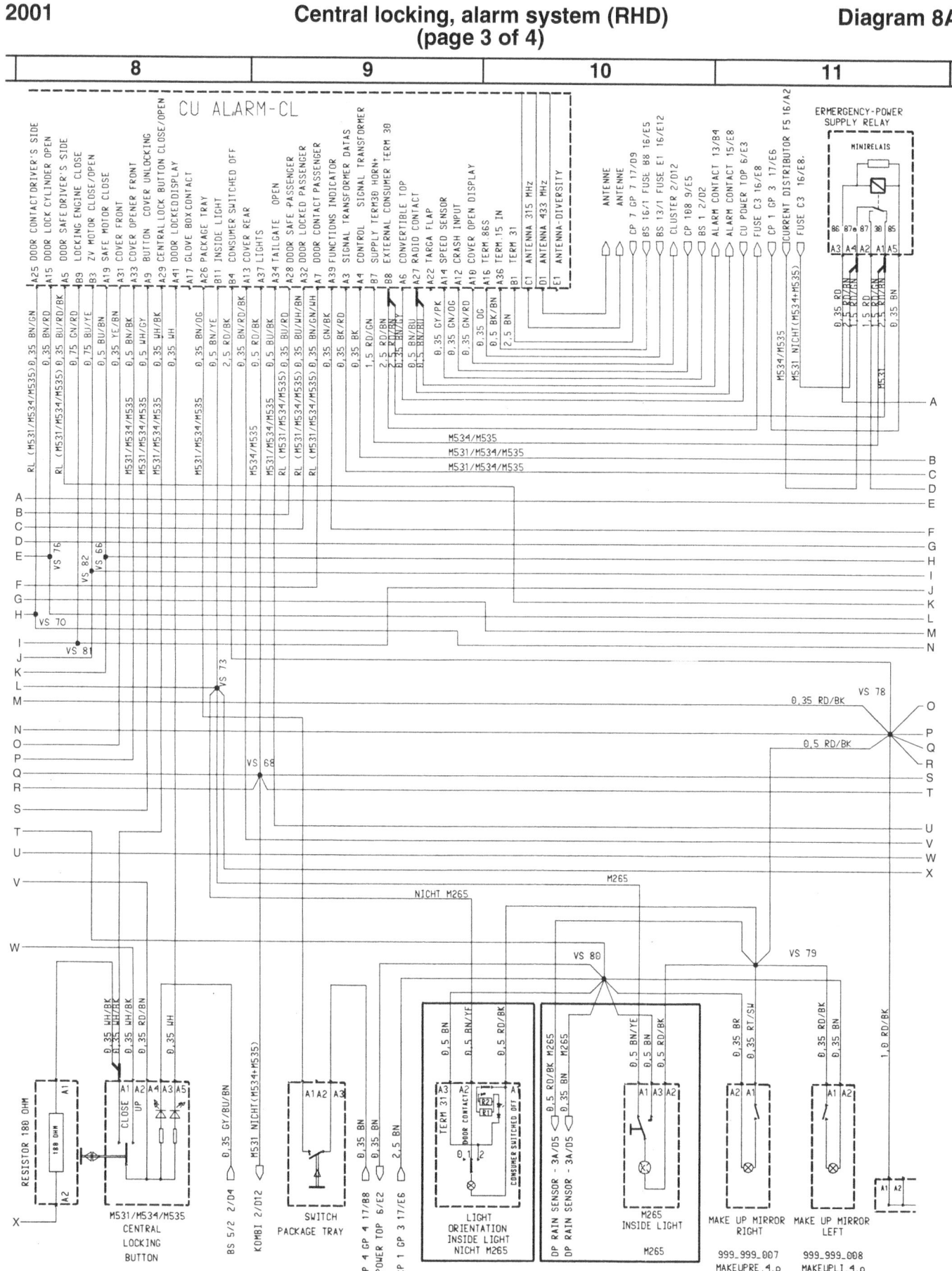

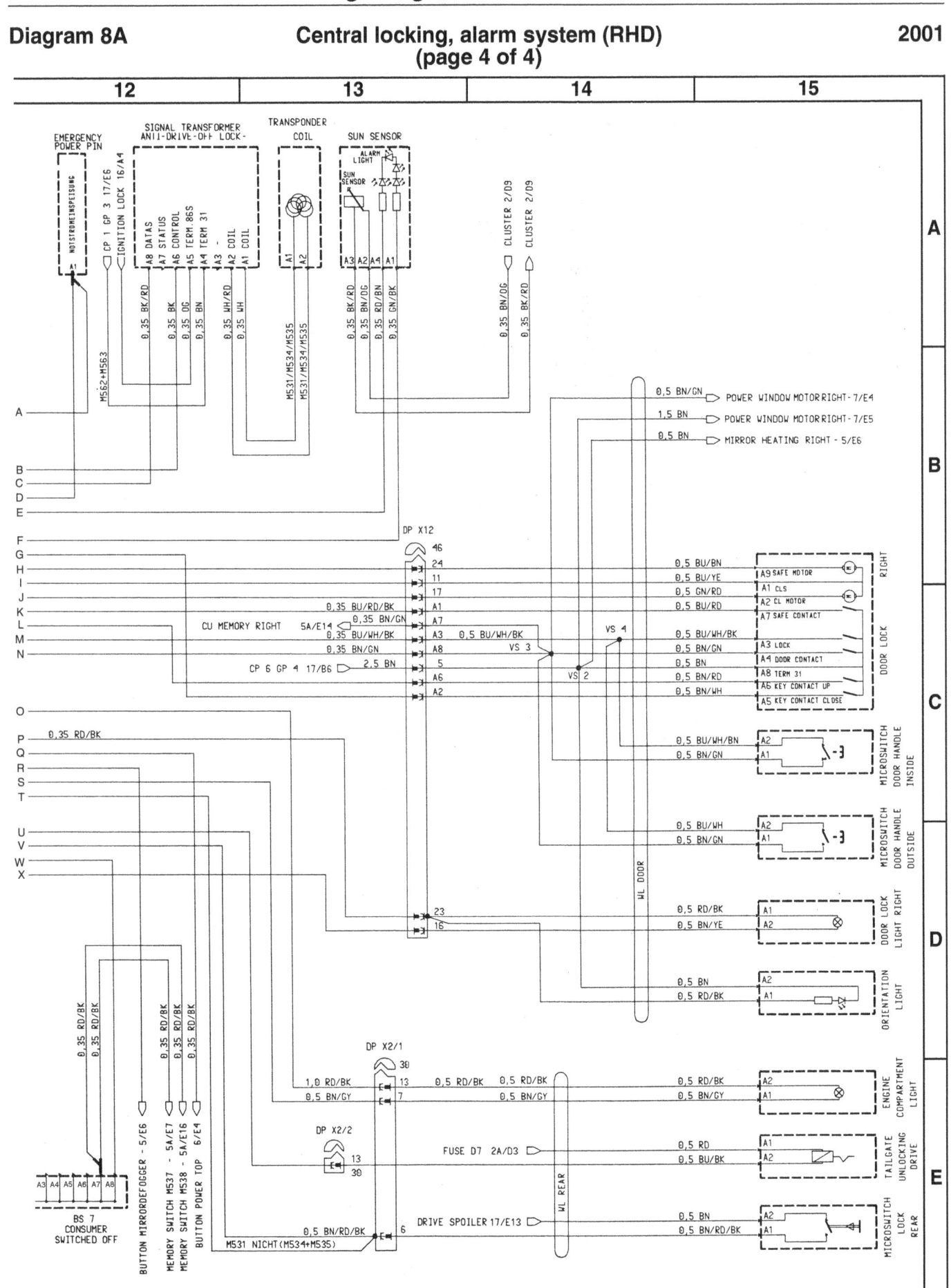

Electrical Wiring Diagrams EWD-99

2001 — Airbag, MRS (LHD) (page 1 of 2) — Diagram 9

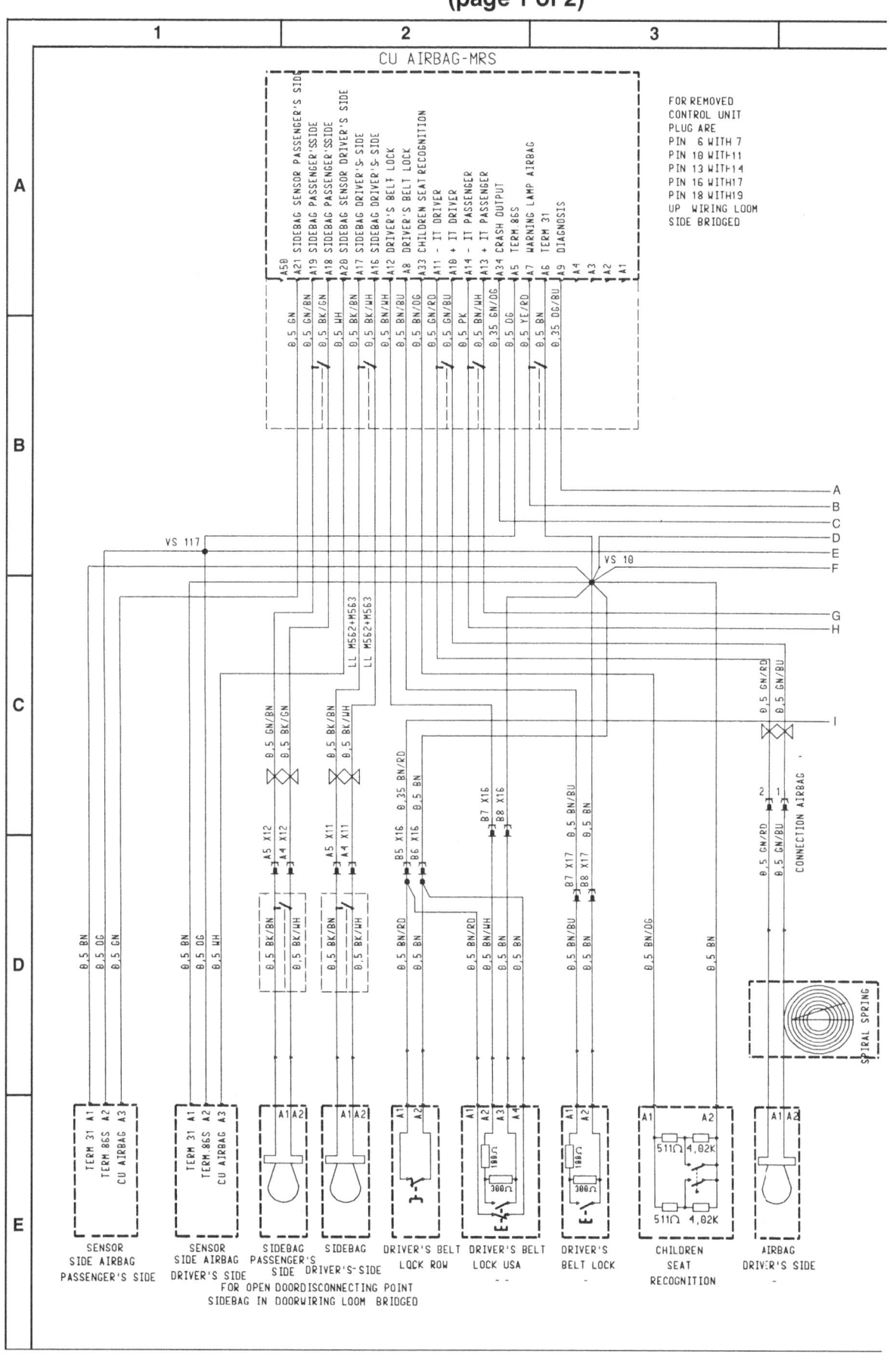

EWD-100 Electrical Wiring Diagrams

Diagram 9 — Airbag, MRS (LHD) (page 2 of 2) — 2001

BLACK	BK	BK
BROWN	BR	BN
RED	RE	RD
ORANGE	OR	OG
YELLOW	YE	YE
GREEN	GN	GN
BLUE	BL	BU
VIOLET	VI	VT
GREY	GR	GY
WHITE	WT	WH
PINK	RS	PK
GOLD	-	GD
TURQUOISE	TK	TQ
SILVER	-	SR

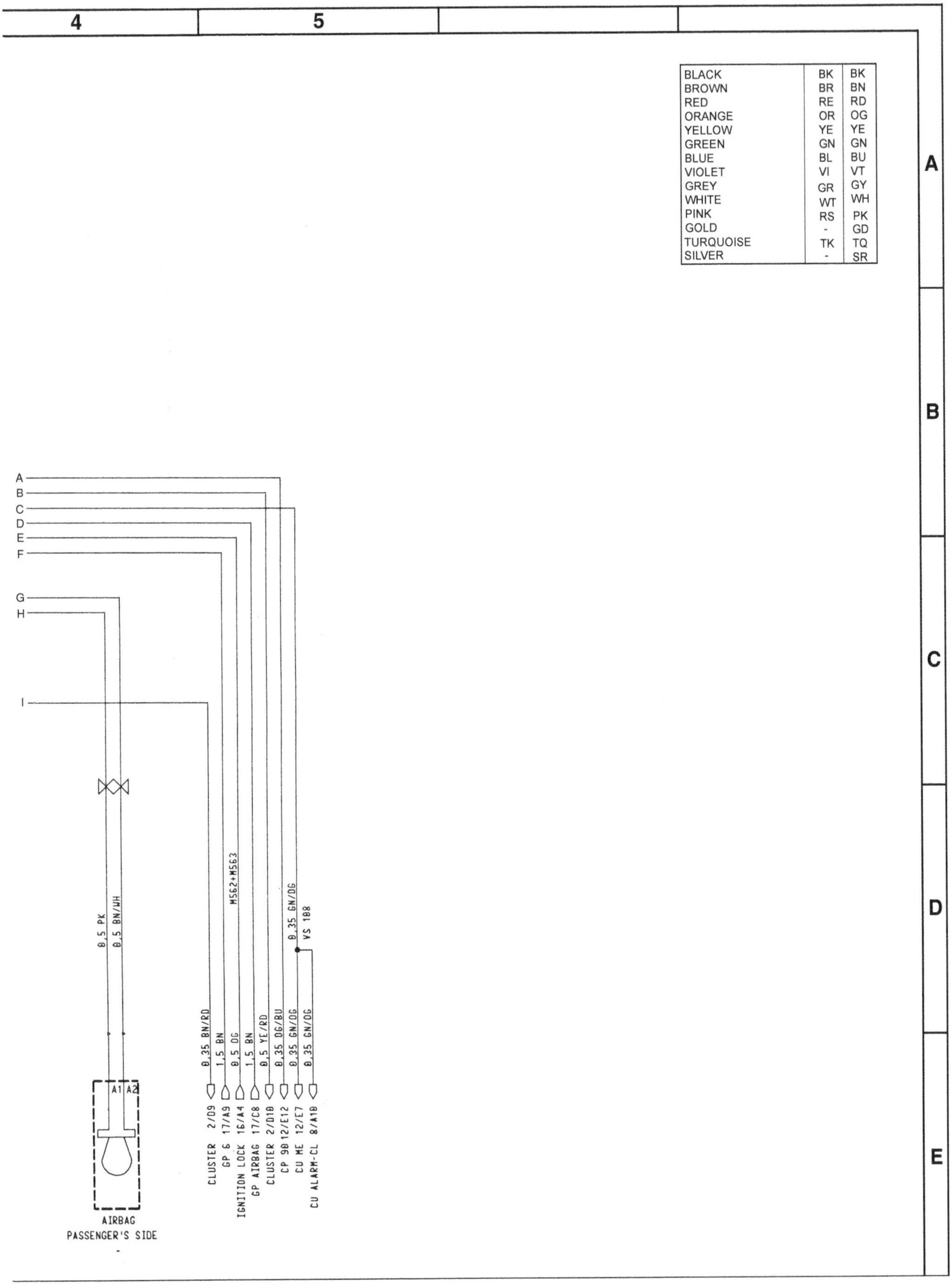

Wires (left to right at passenger-side airbag):
- 0.5 PK
- 0.5 BN/WH
- 0.35 BN/RD — CLUSTER 2/D9
- 1.5 BN — GP 6 17/A9
- 0.5 OG — IGNITION LOCK 16/A4
- 1.5 BN — GP AIRBAG 17/C8
- 0.5 YE/RD — CLUSTER 2/D18
- 0.35 OG/BU — CP 90 12/E12
- 0.35 GN/OG — CU ME 12/E7
- VS 188
- 0.35 GN/OG — CU ALARM-CL 8/A18

AIRBAG PASSENGER'S SIDE (A1 A2)

Electrical Wiring Diagrams EWD-101

2001 — Airbag, MRS (RHD) (page 1 of 2) — **Diagram 9A**

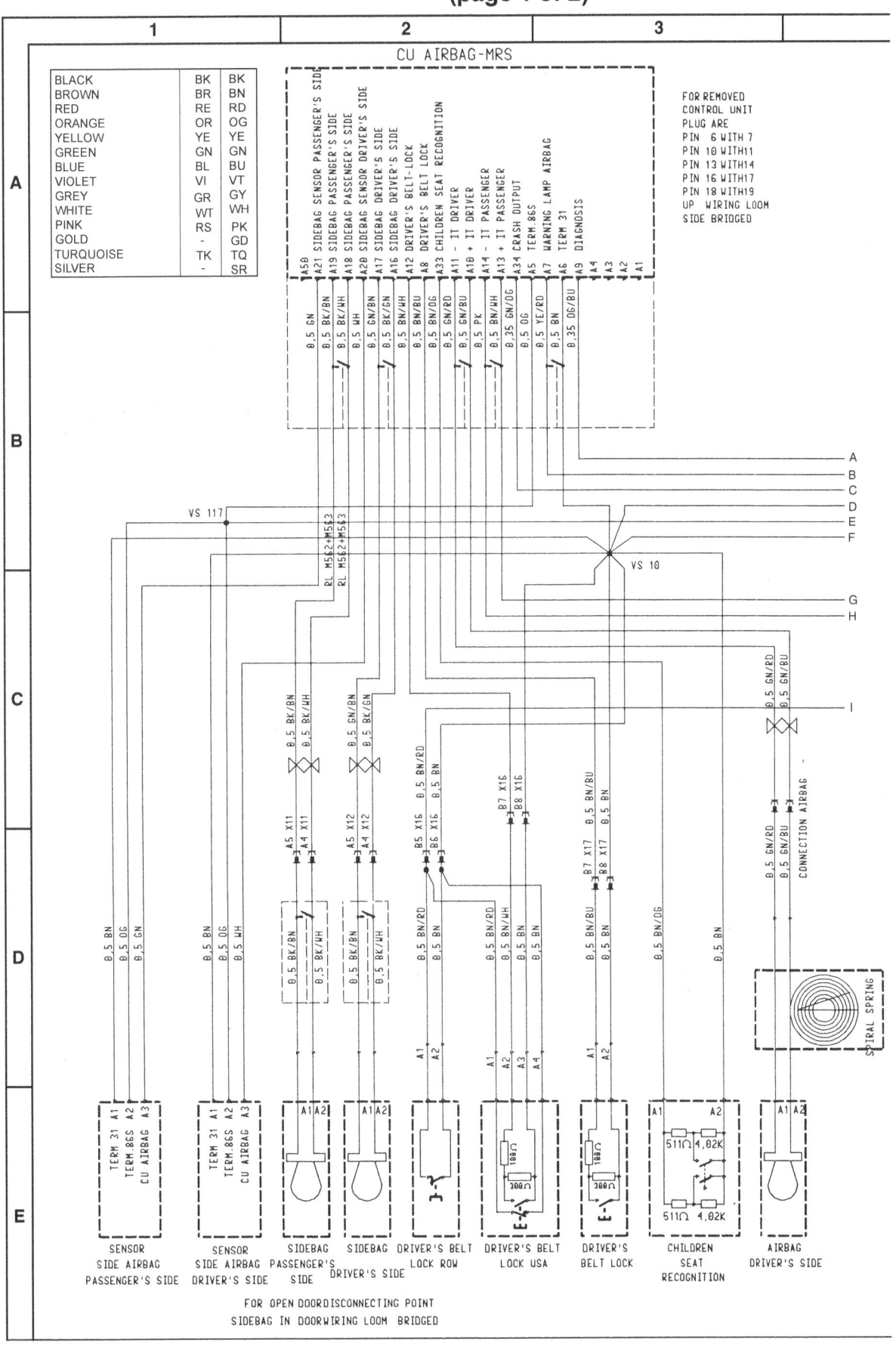

EWD-102 Electrical Wiring Diagrams

Diagram 9A Airbag, MRS (RHD) (page 2 of 2) 2001

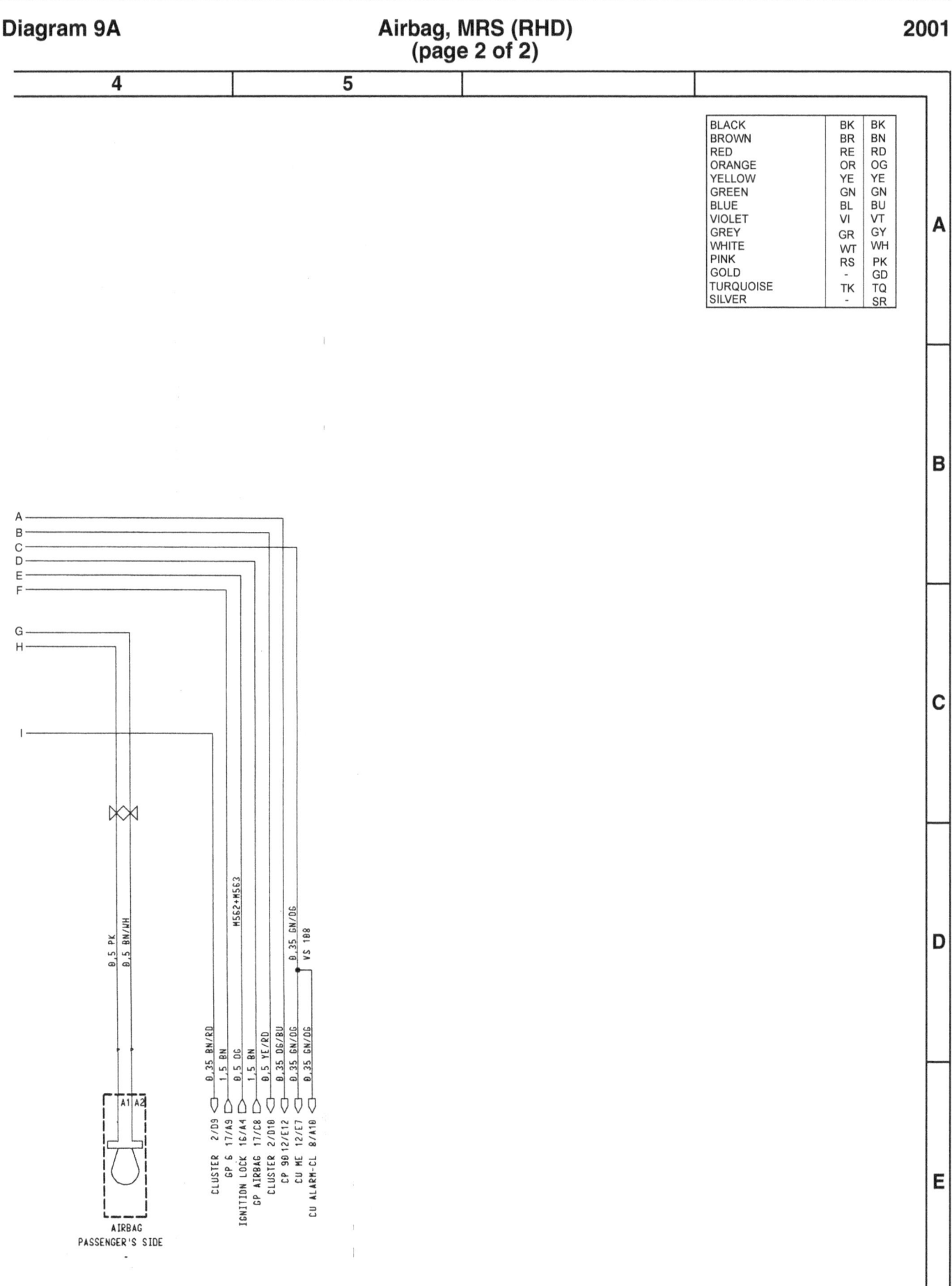

Electrical Wiring Diagrams EWD-103

2001 — ABS — Diagram 10

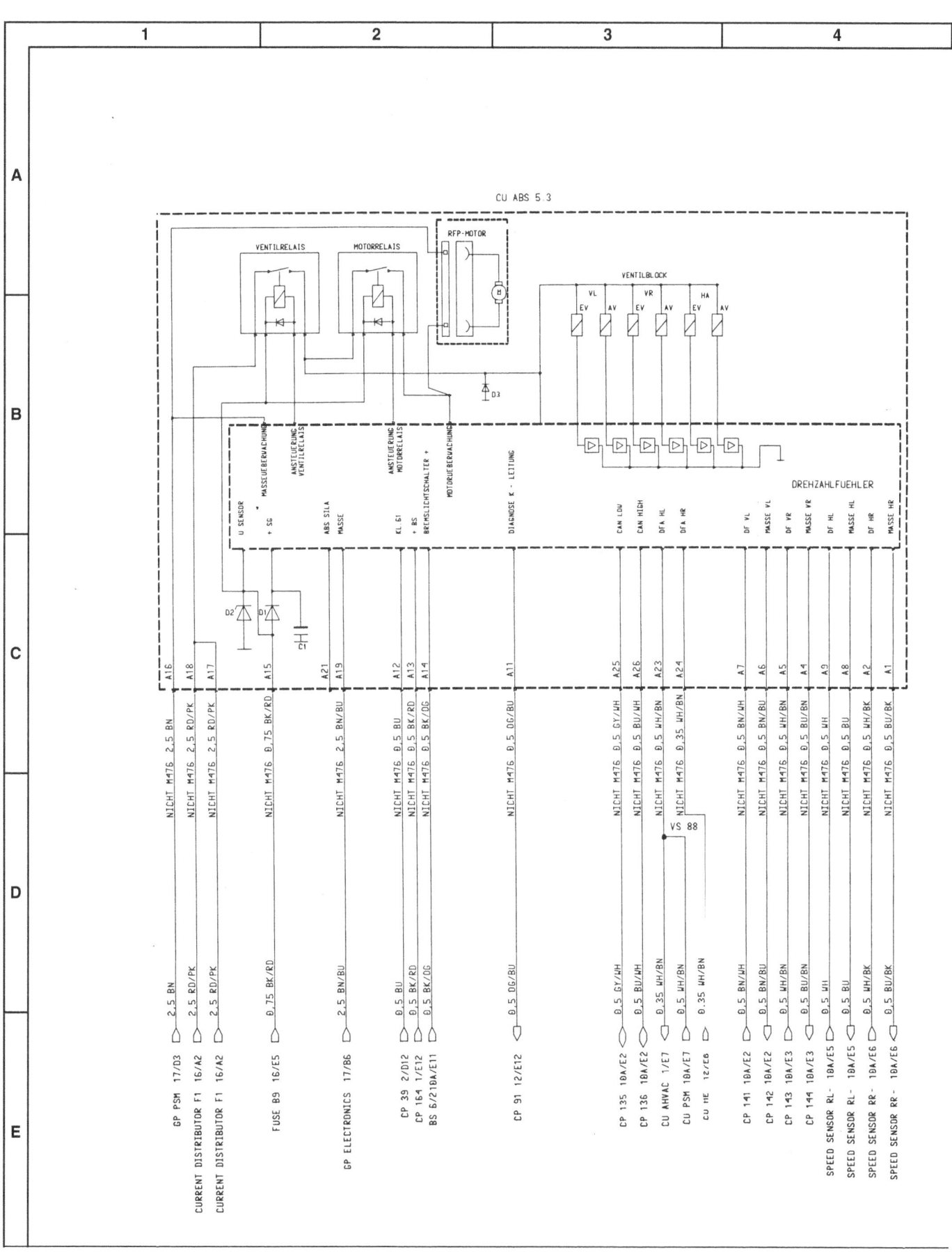

EWD-104 Electrical Wiring Diagrams

Diagram 10A — Porsche Stability Management (PSM) (page 1 of 3) — 2001

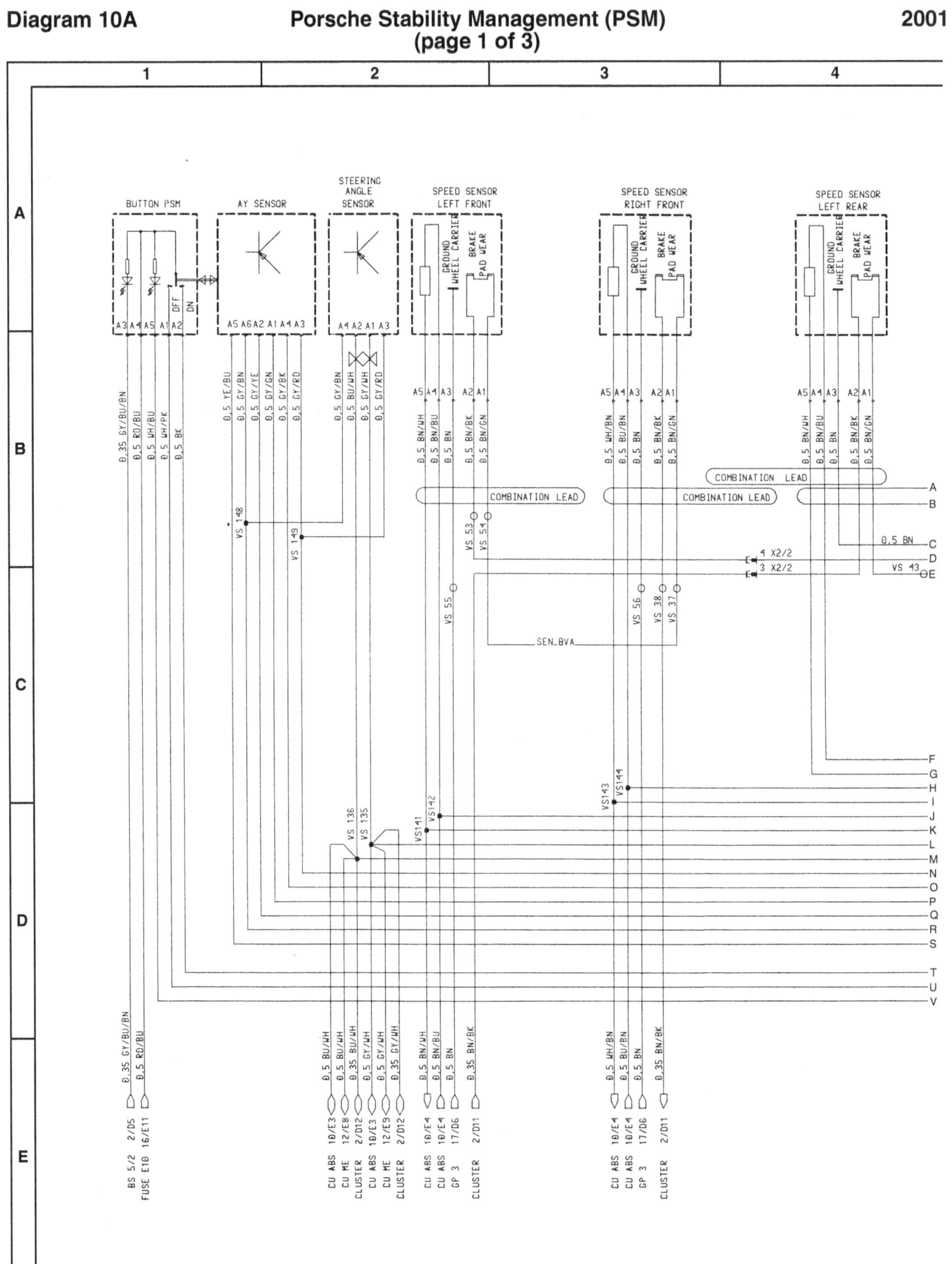

Electrical Wiring Diagrams EWD-105

2001 — Porsche Stability Management (PSM) (page 2 of 3) — Diagram 10A

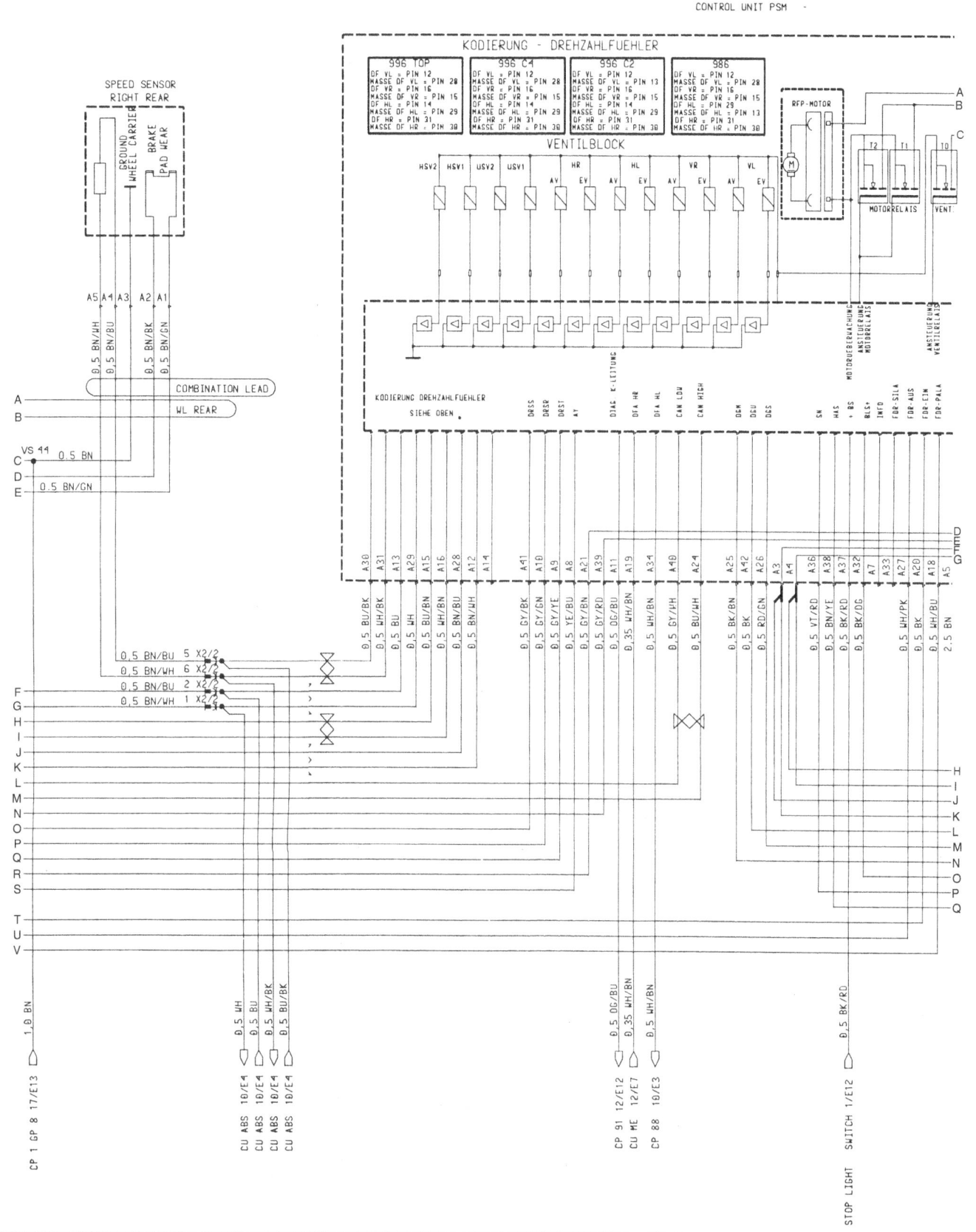

Diagram 10A — Porsche Stability Management (PSM) (page 3 of 3) — 2001

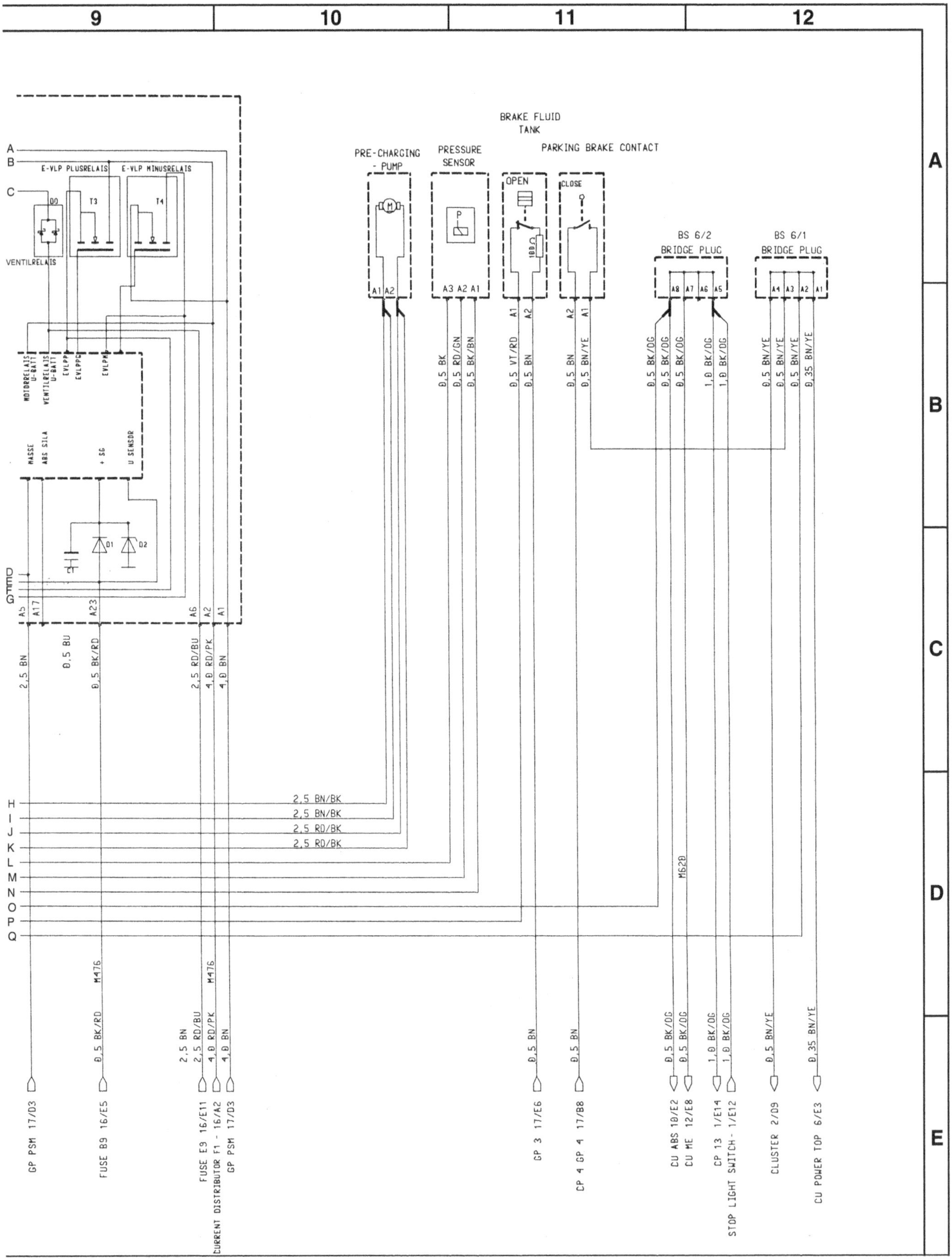

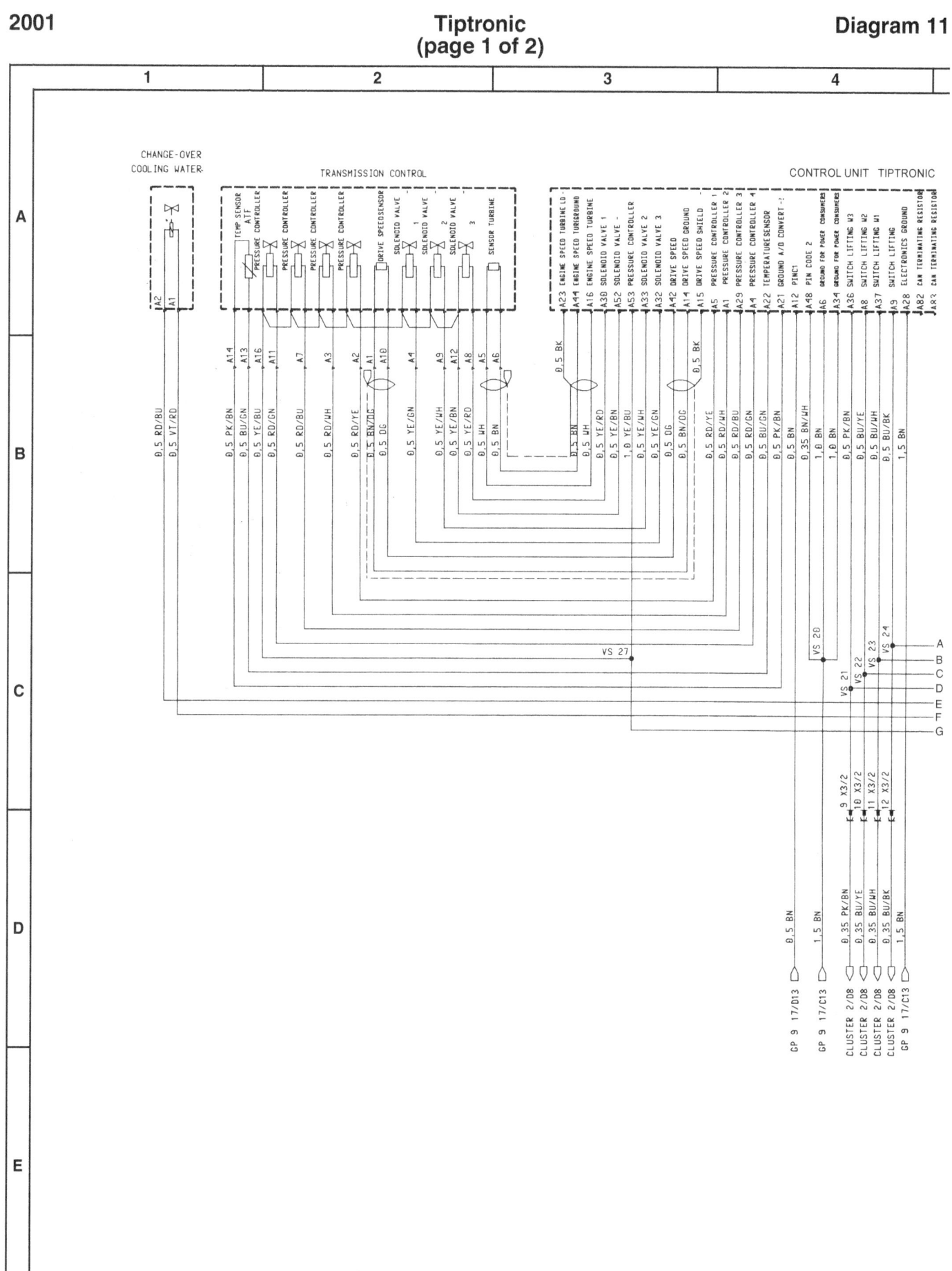

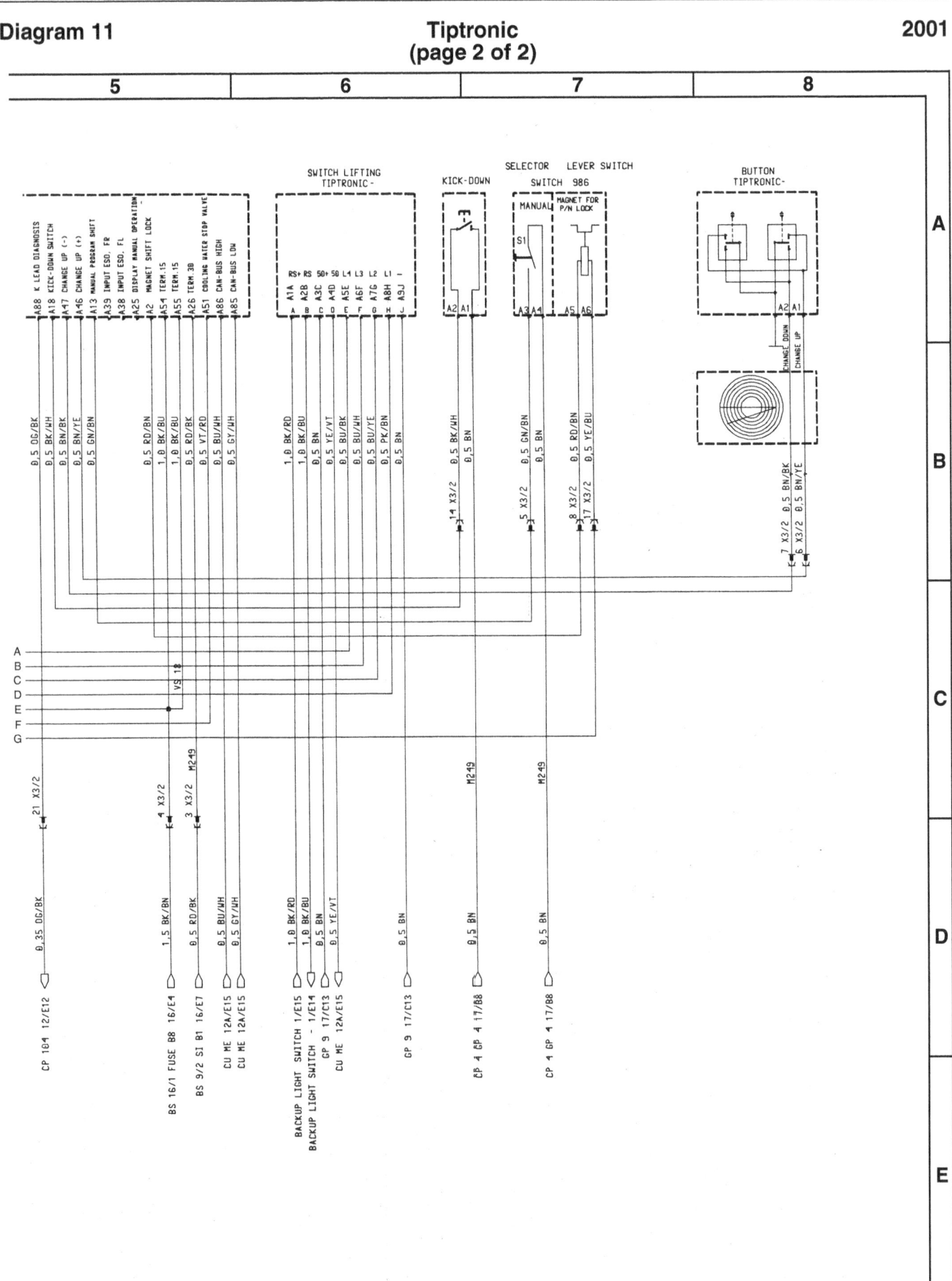

Electrical Wiring Diagrams EWD-109

2001 — Engine management (Bosch ME 7.2), vehicle (page 1 of 5) — Diagram 12

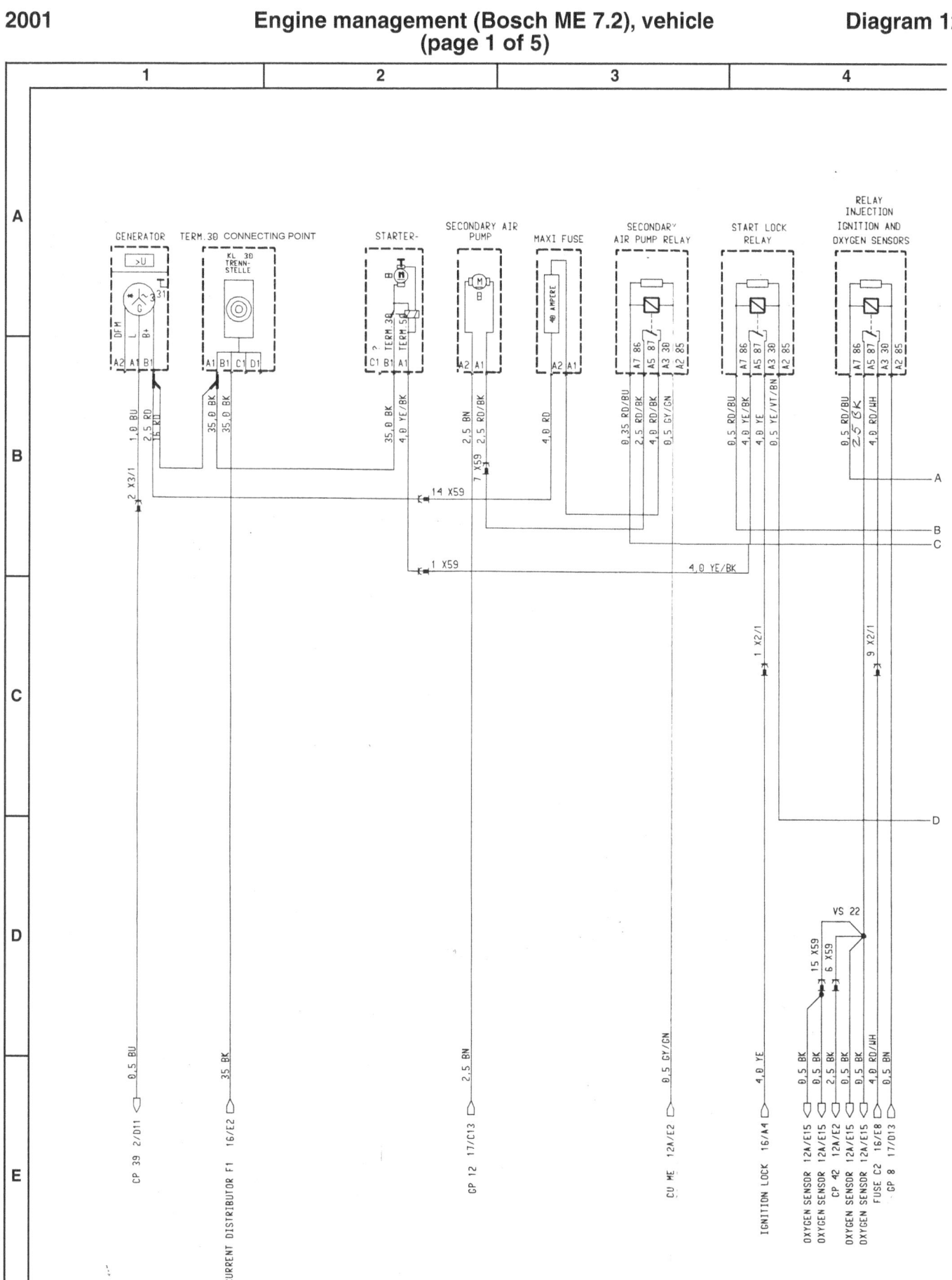

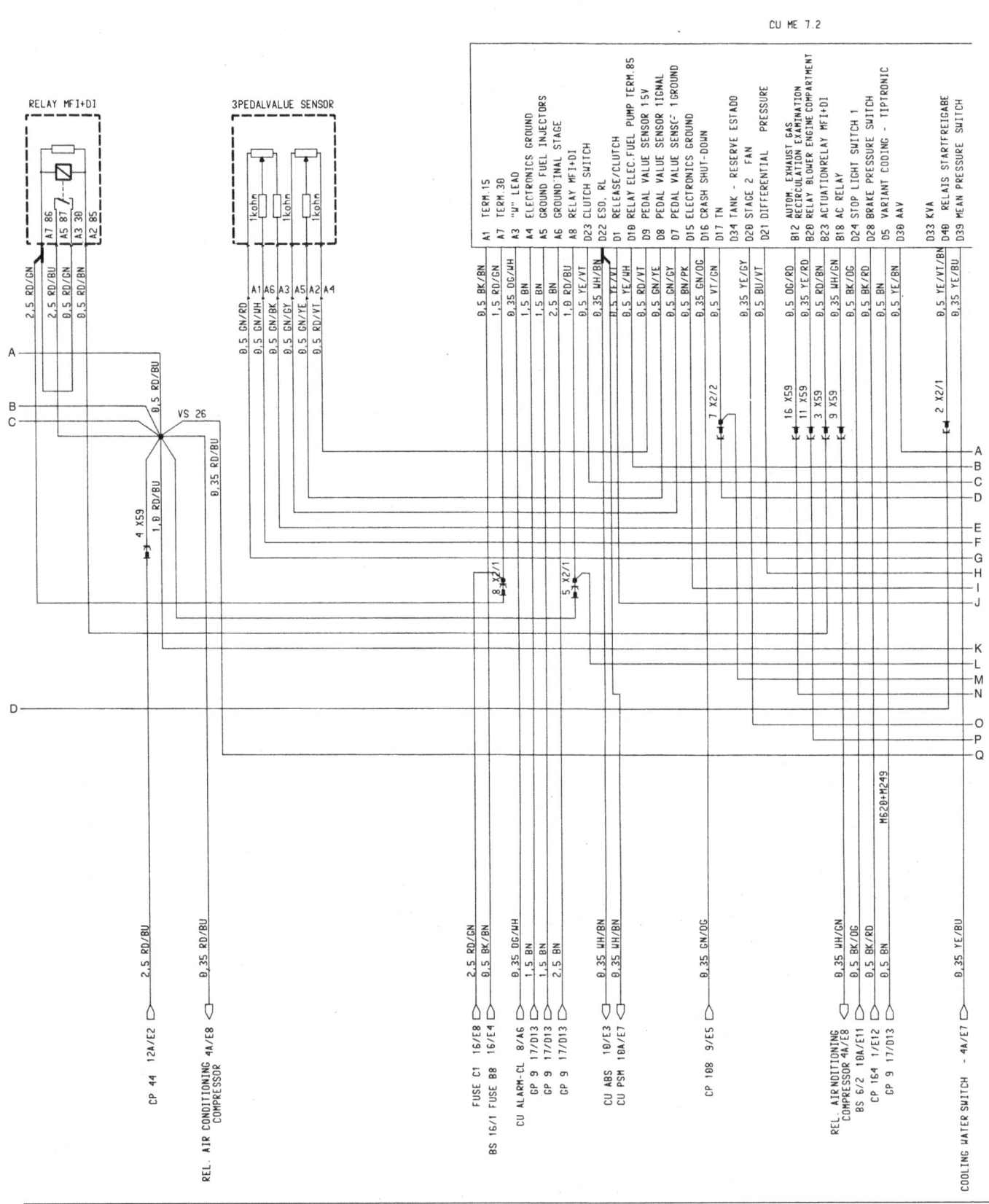

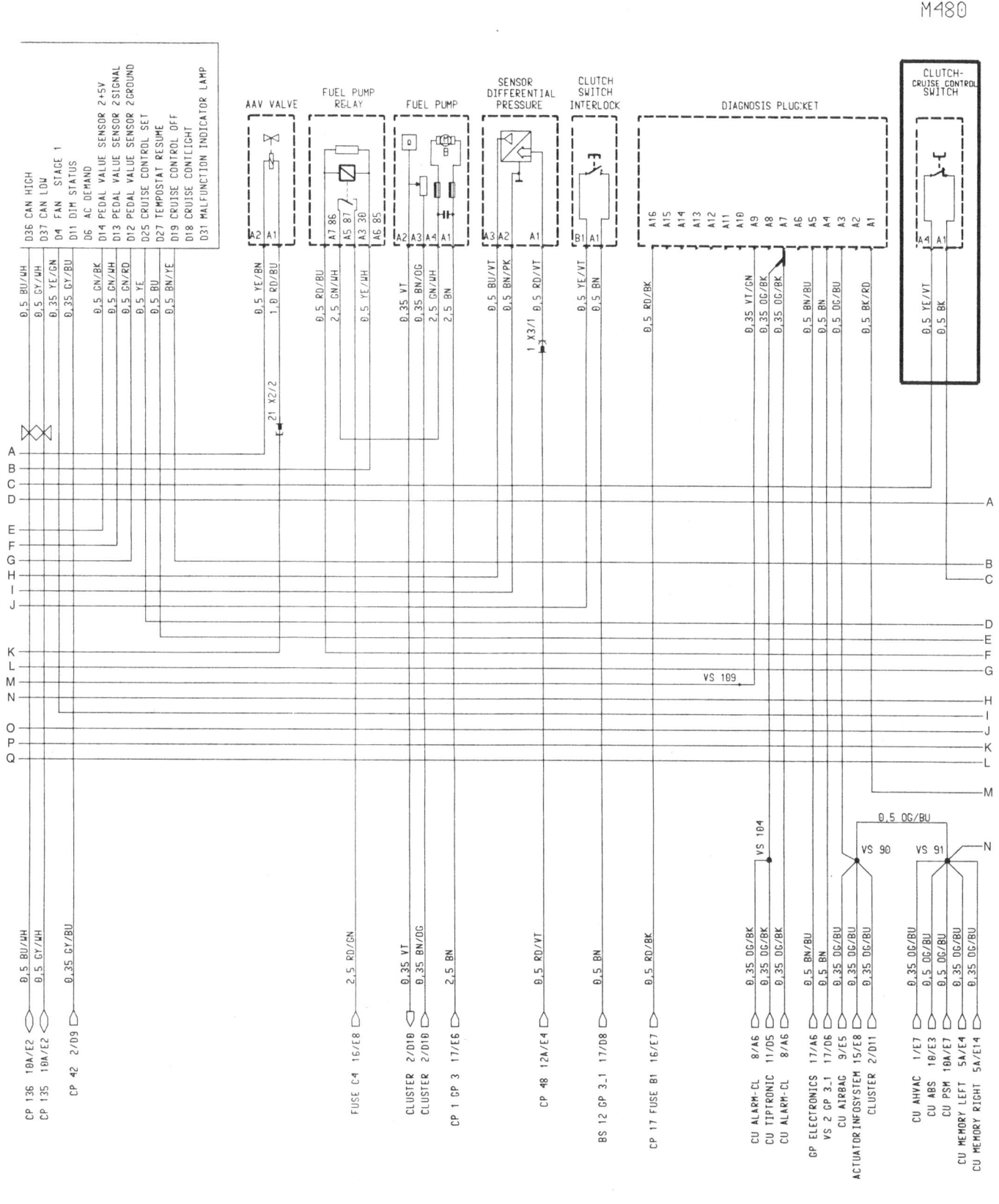

Diagram 12

Engine management (Bosch ME 7.2), vehicle (page 4 of 5) — 2001

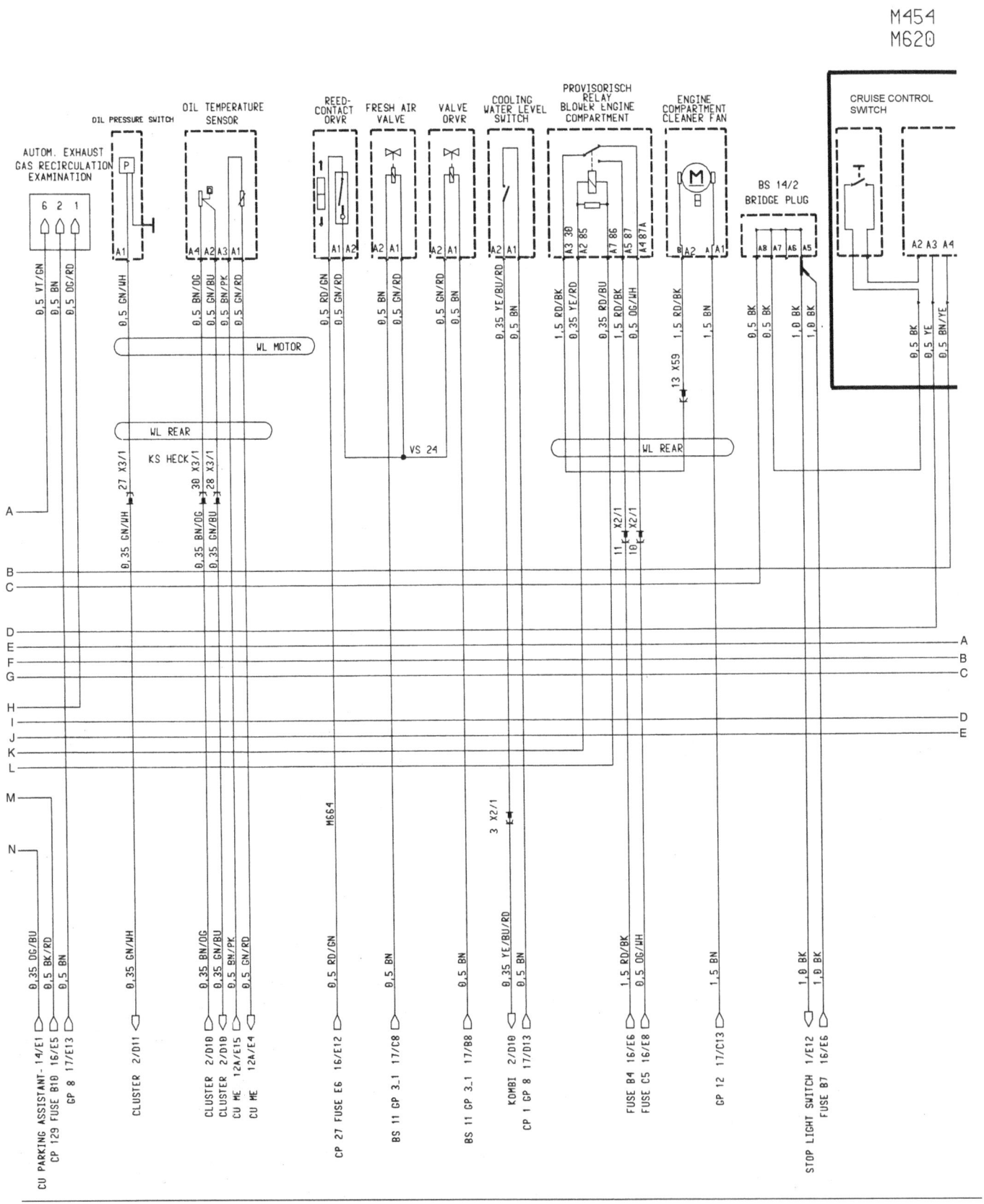

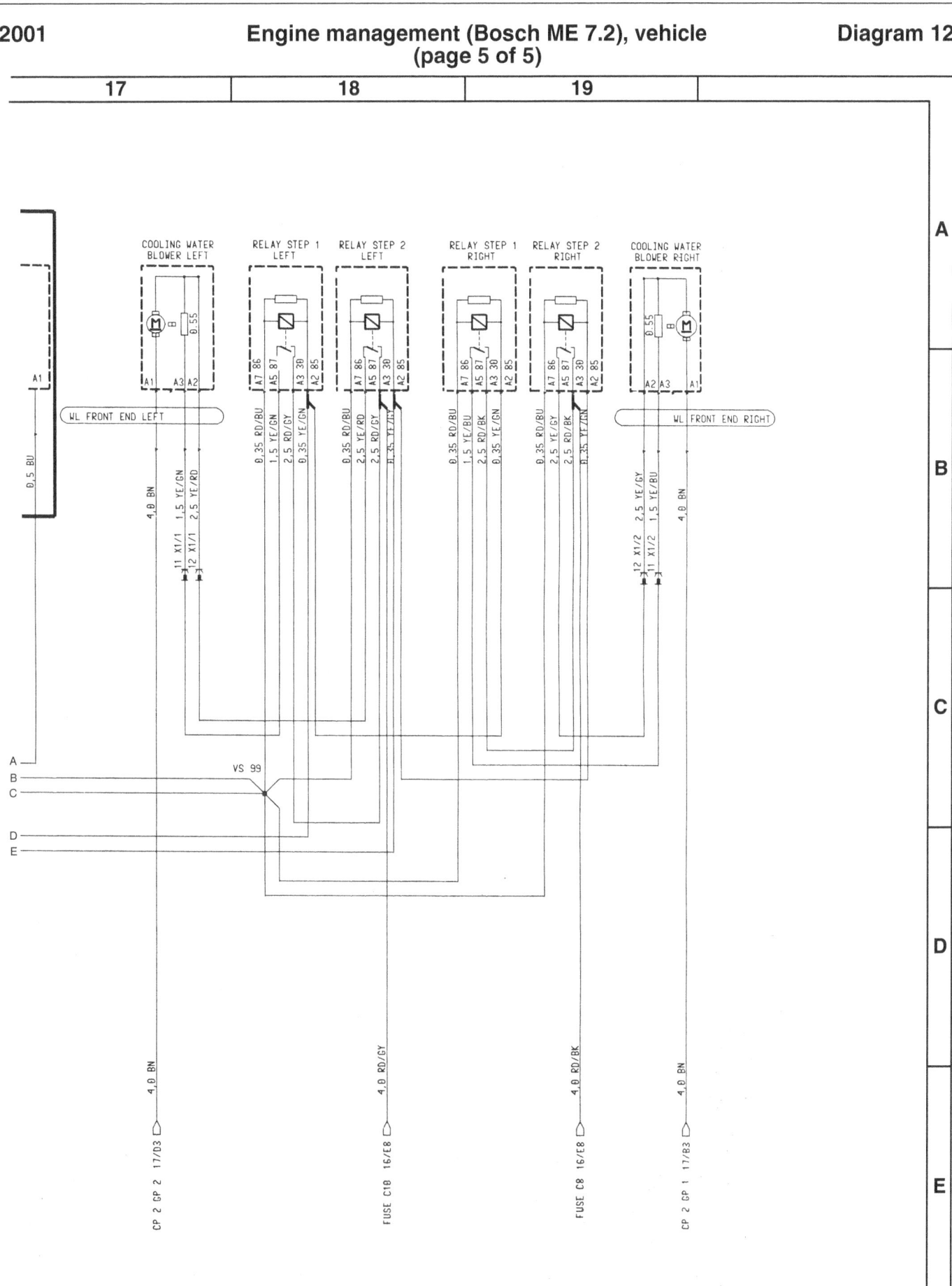

EWD-114 Electrical Wiring Diagrams

Diagram 12A — Engine management (Bosch ME 7.2), engine (page 1 of 4) — 2001

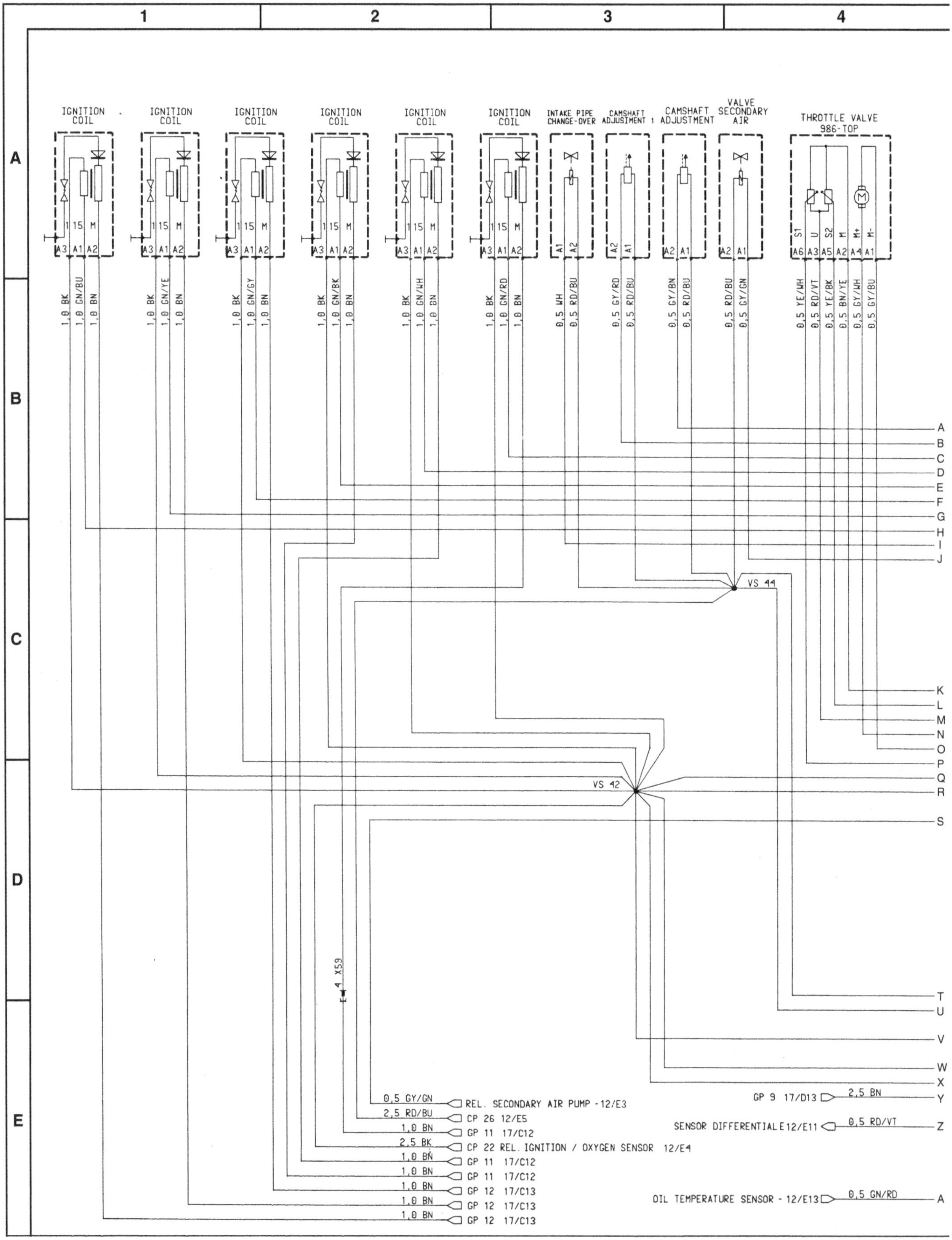

Electrical Wiring Diagrams EWD-115

2001 — Engine management (Bosch ME 7.2), engine (page 2 of 4) — **Diagram 12A**

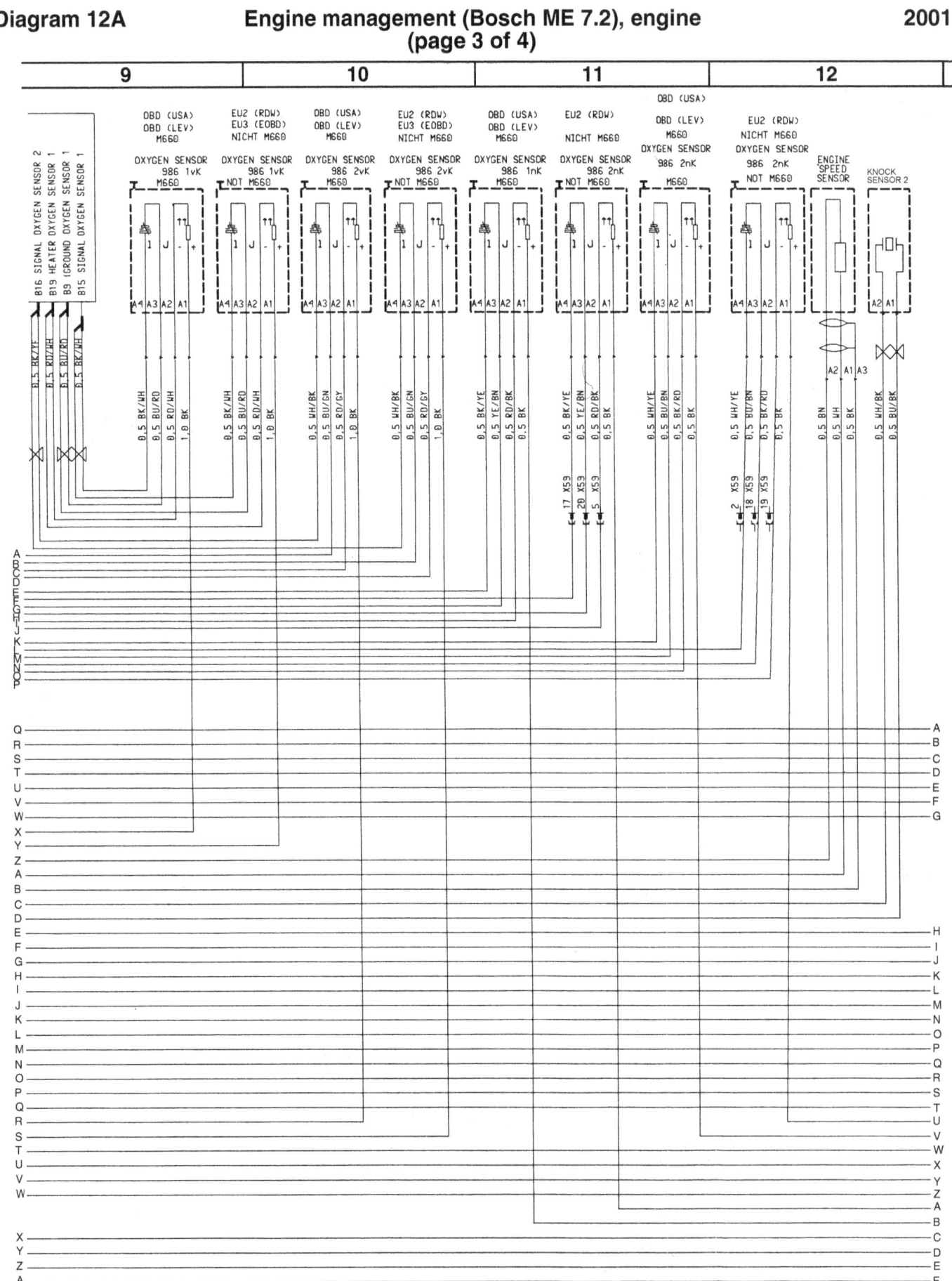

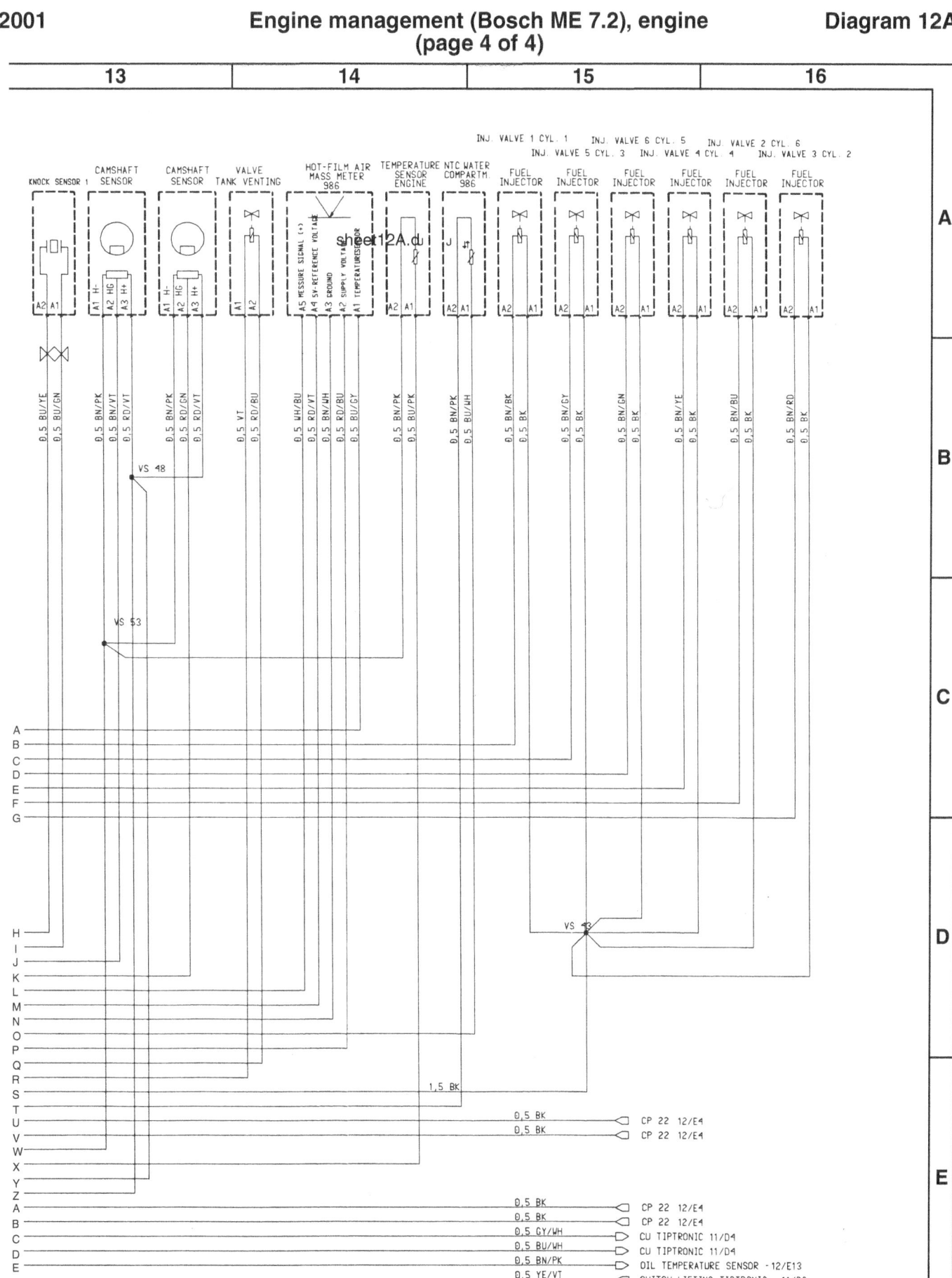

EWD-118 Electrical Wiring Diagrams

Diagram 13 — Radio — 2001

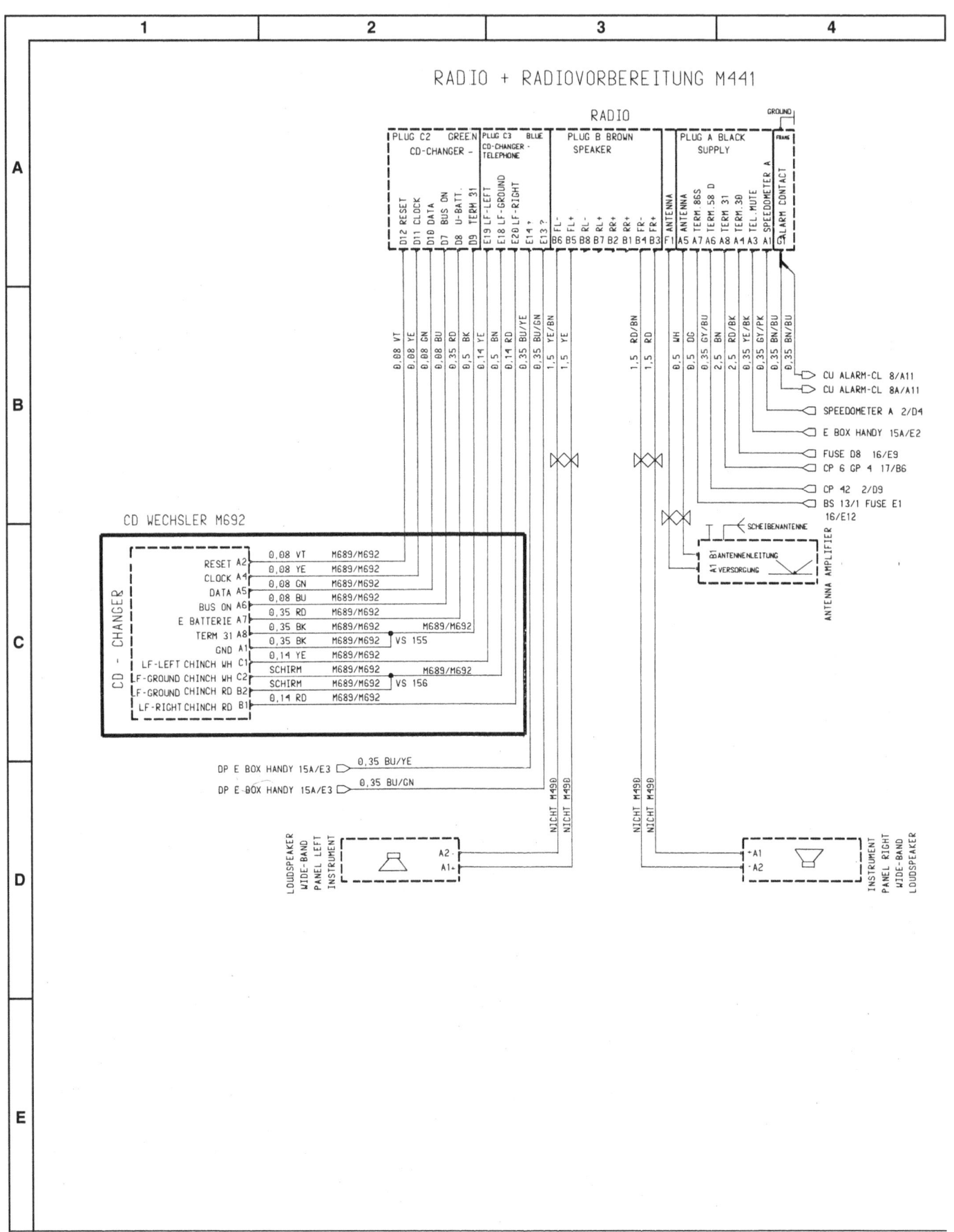

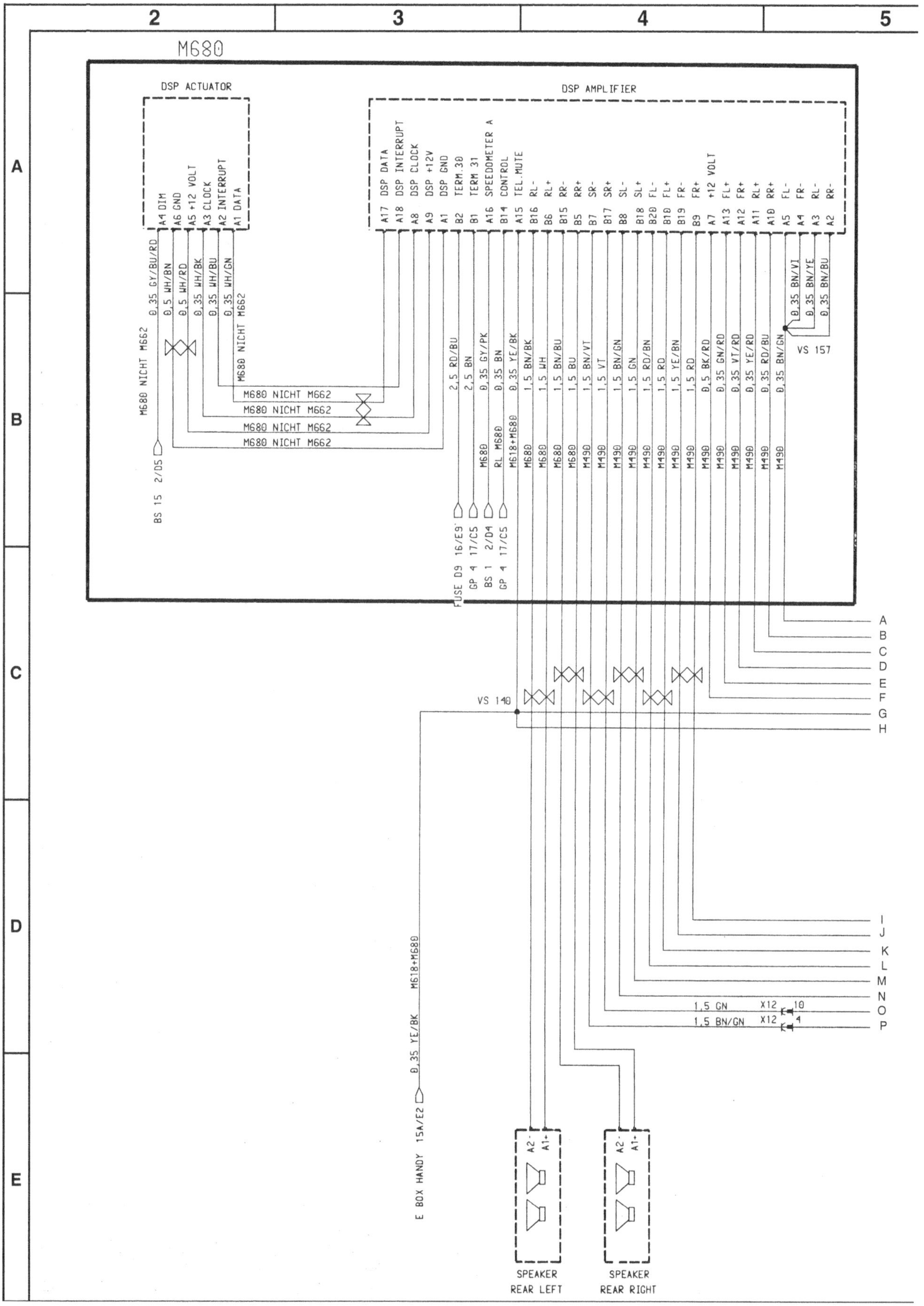

EWD-120   Electrical Wiring Diagrams

Diagram 13A   Radio, sound package, DSP   2001
                         (page 2 of 3)

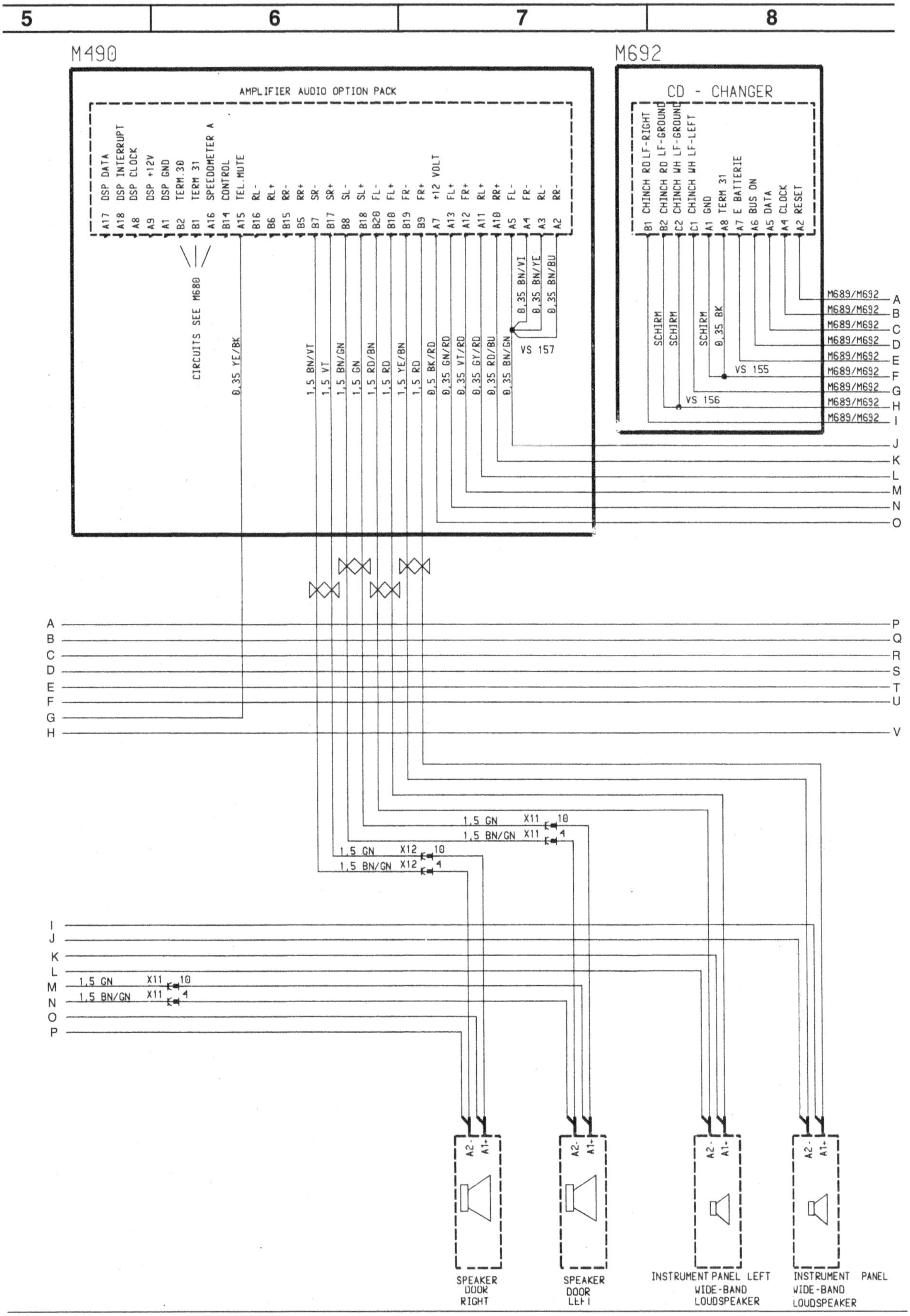

Electrical Wiring Diagrams EWD-121

2001 — Radio, sound package, DSP (page 3 of 3) — **Diagram 13A**

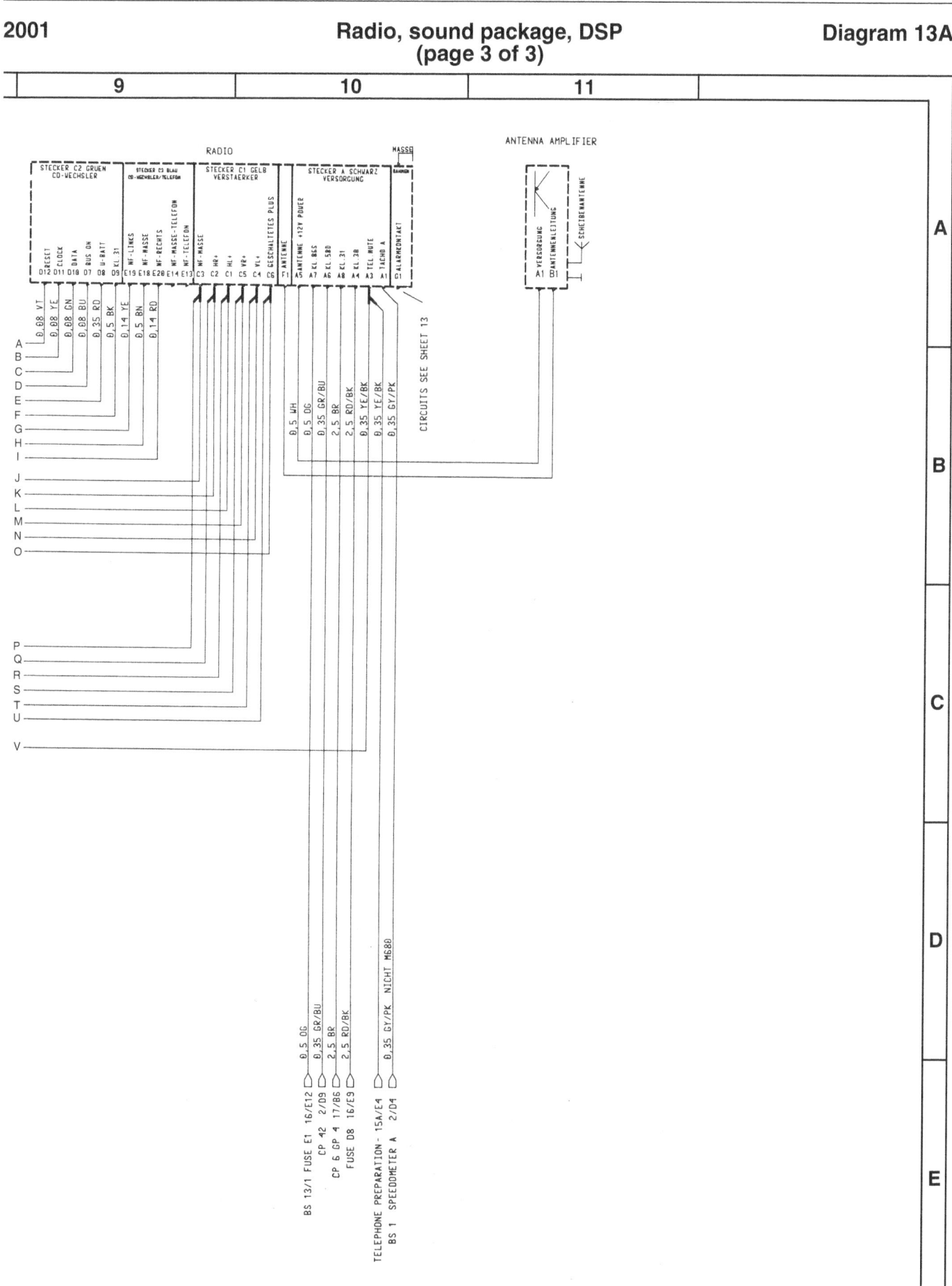

EWD-122 Electrical Wiring Diagrams

Diagram 14 — Parking assistant (page 1 of 2) — 2001

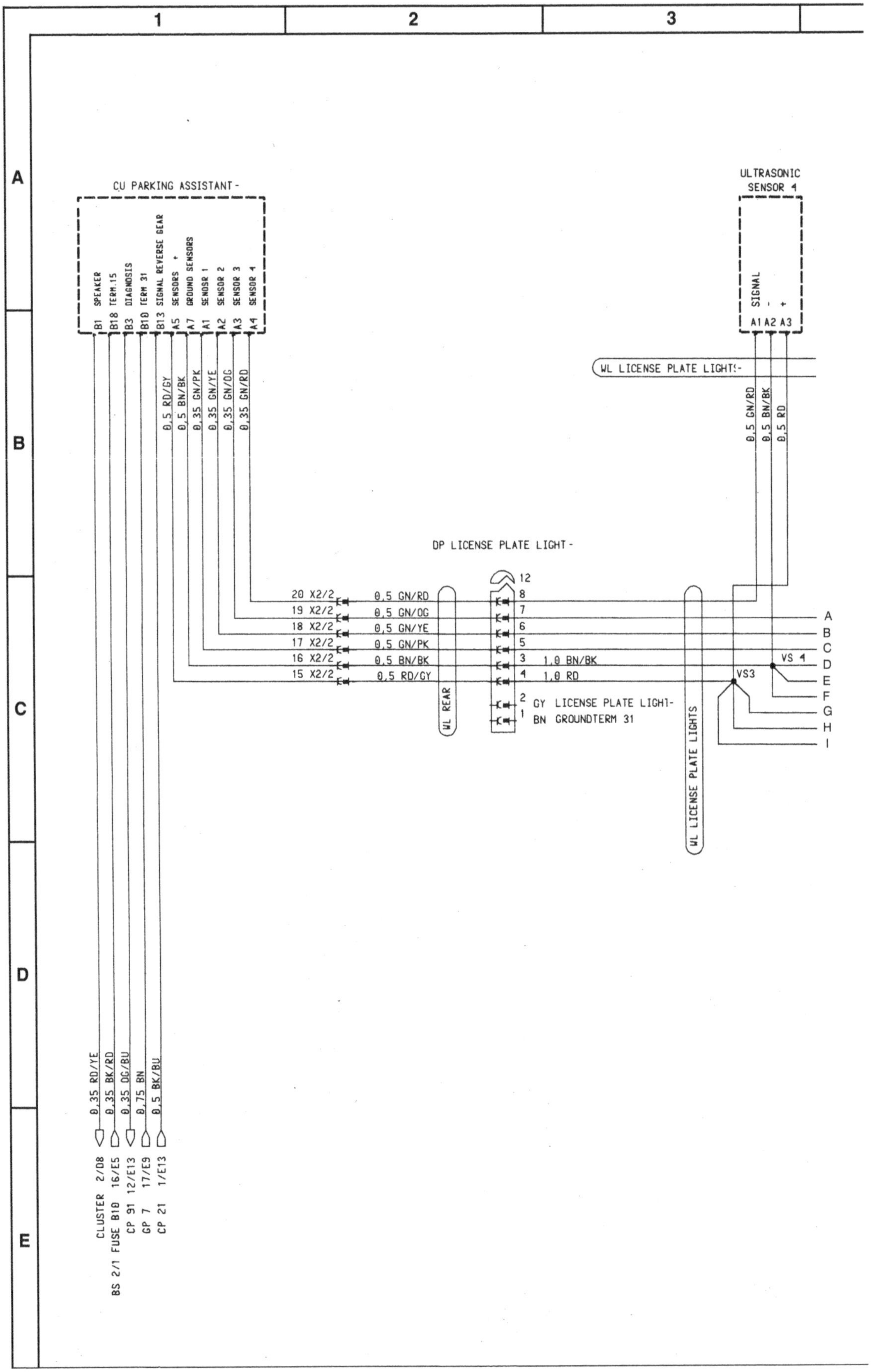

Electrical Wiring Diagrams   EWD-123

2001   **Parking assistant (page 2 of 2)**   Diagram 14

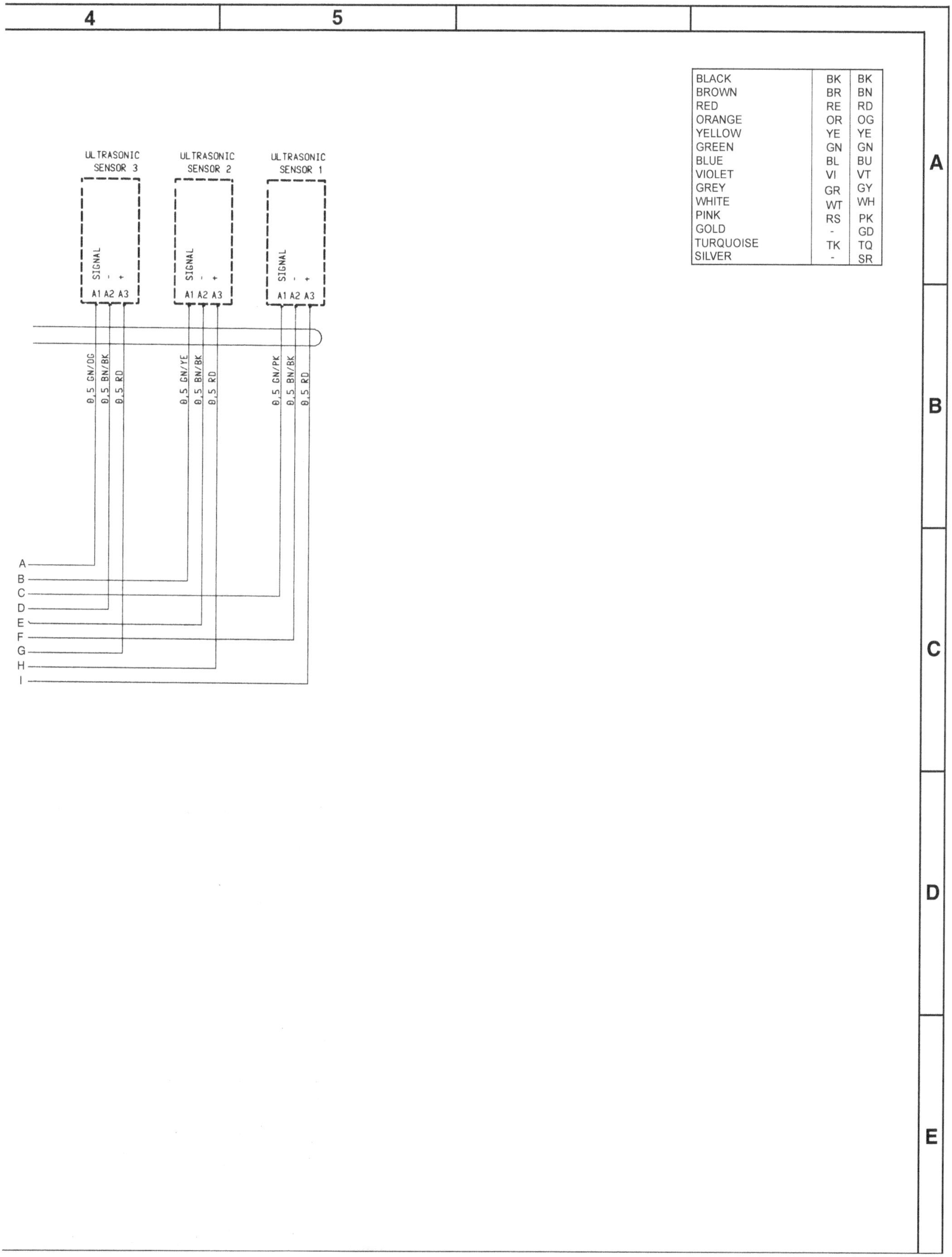

4	5	

ULTRASONIC SENSOR 3 — SIGNAL — A1 A2 A3
ULTRASONIC SENSOR 2 — SIGNAL — A1 A2 A3
ULTRASONIC SENSOR 1 — SIGNAL — A1 A2 A3

0.5 GN/OG, 0.5 BN/BK, 0.5 RD
0.5 GN/YE, 0.5 BN/BK, 0.5 RD
0.5 GN/PK, 0.5 BN/BK, 0.5 RD

A
B
C
D
E
F
G
H
I

BLACK	BK	BK
BROWN	BR	BN
RED	RE	RD
ORANGE	OR	OG
YELLOW	YE	YE
GREEN	GN	GN
BLUE	BL	BU
VIOLET	VI	VT
GREY	GR	GY
WHITE	WT	WH
PINK	RS	PK
GOLD	-	GD
TURQUOISE	TK	TQ
SILVER	-	SR

A

B

C

D

E

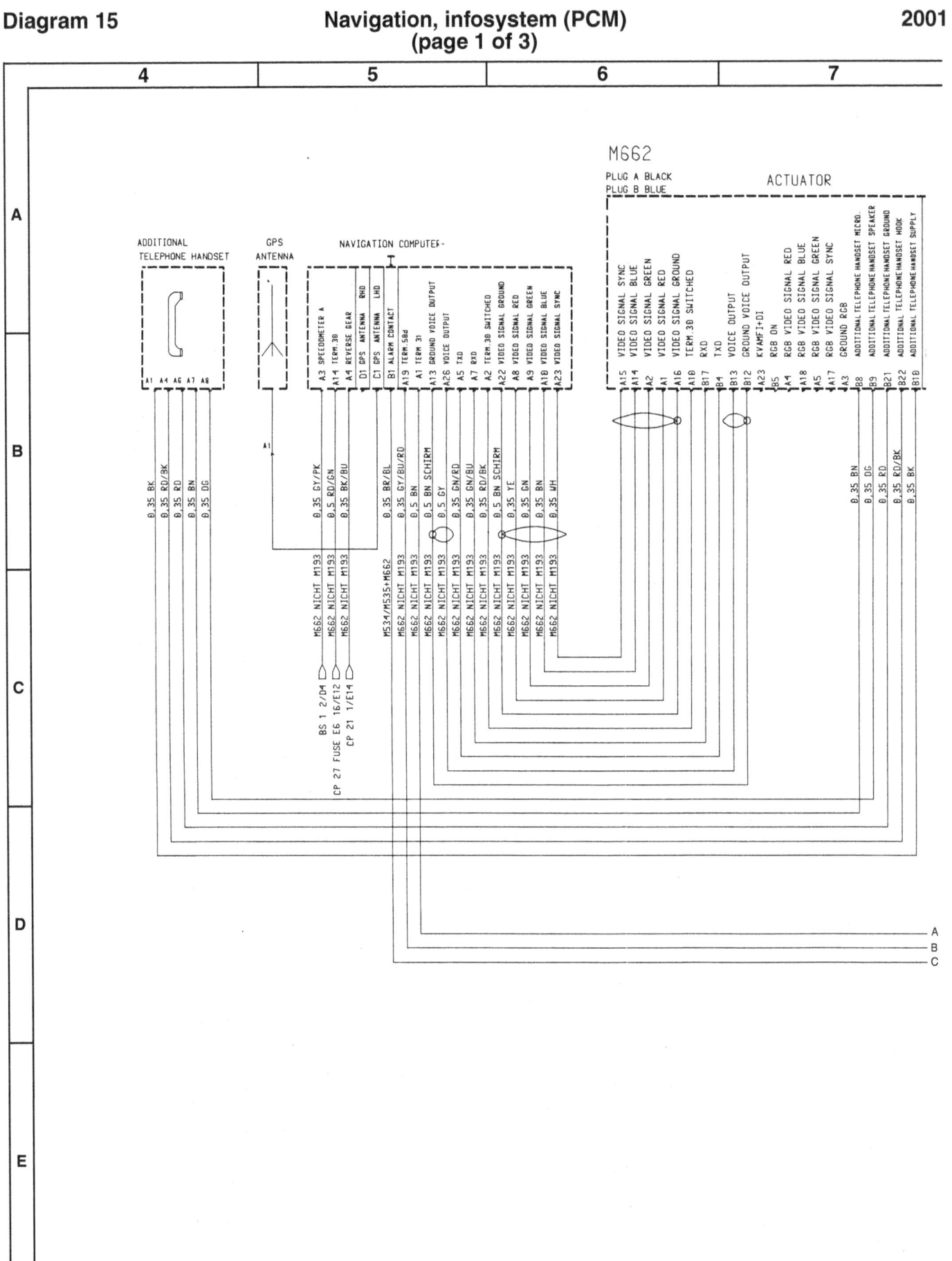

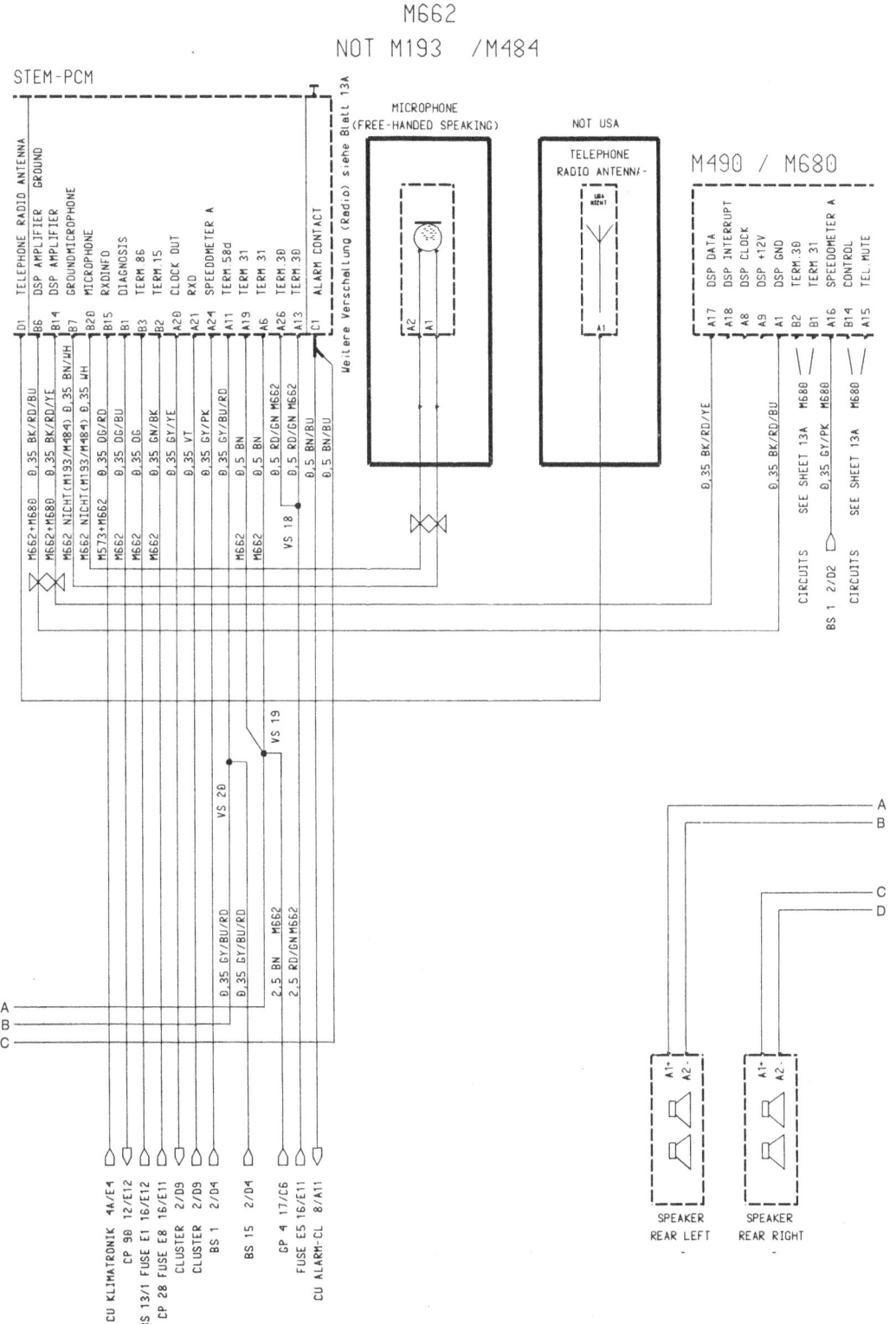

EWD-126 Electrical Wiring Diagrams

Diagram 15 — Navigation, infosystem (PCM) (page 3 of 3) — 2001

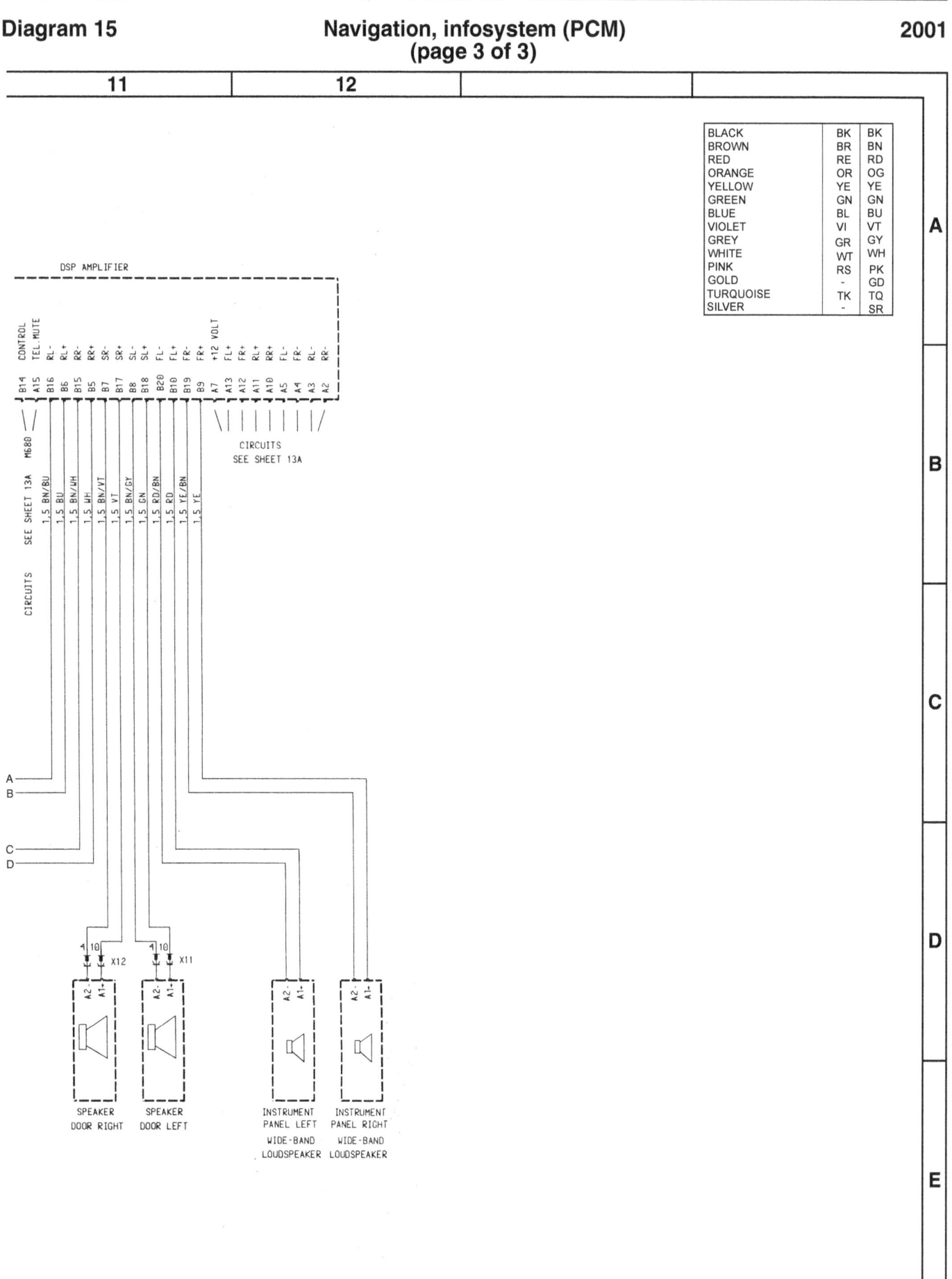

Electrical Wiring Diagrams EWD-127

2001 Telephone, mobile phone Diagram 15A

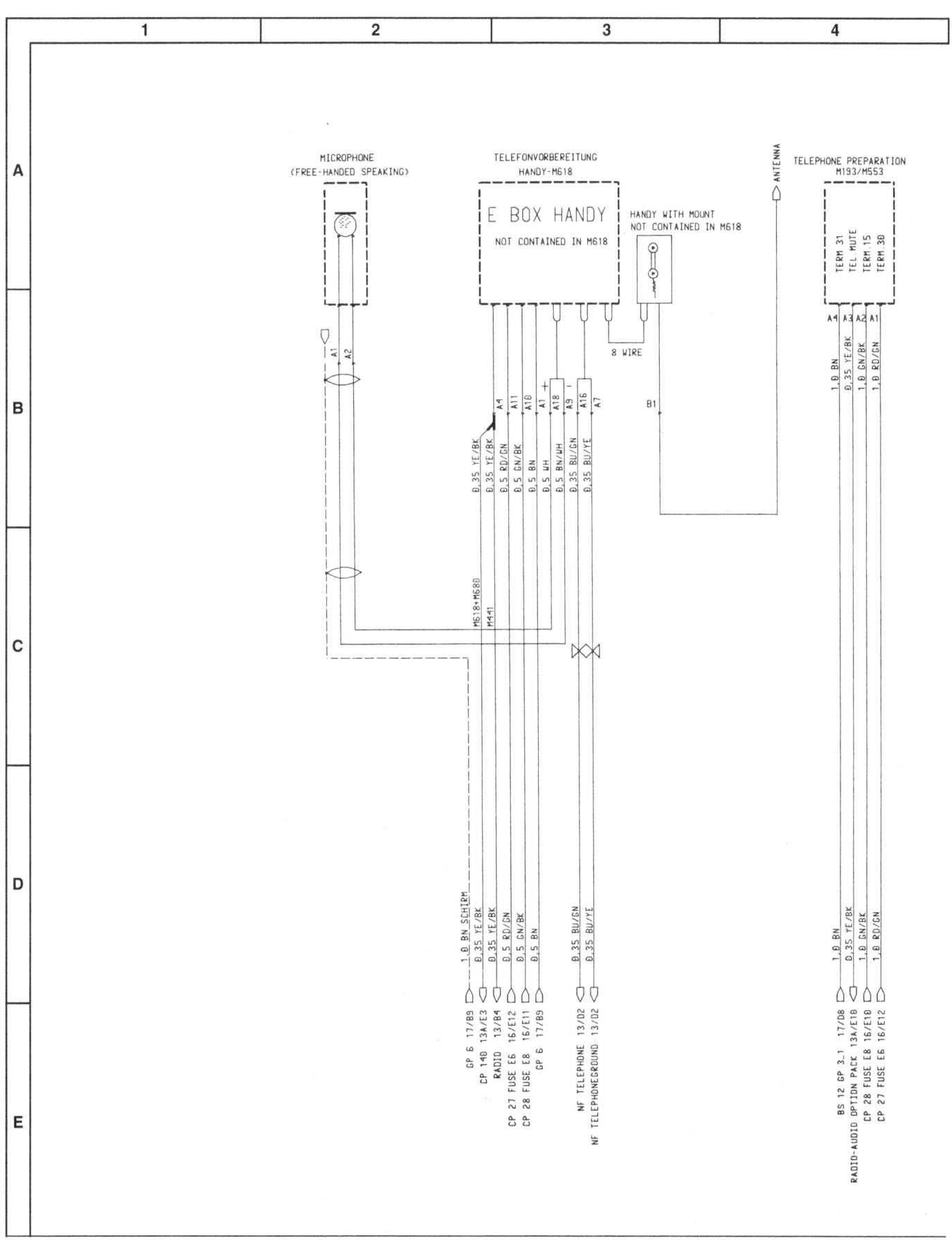

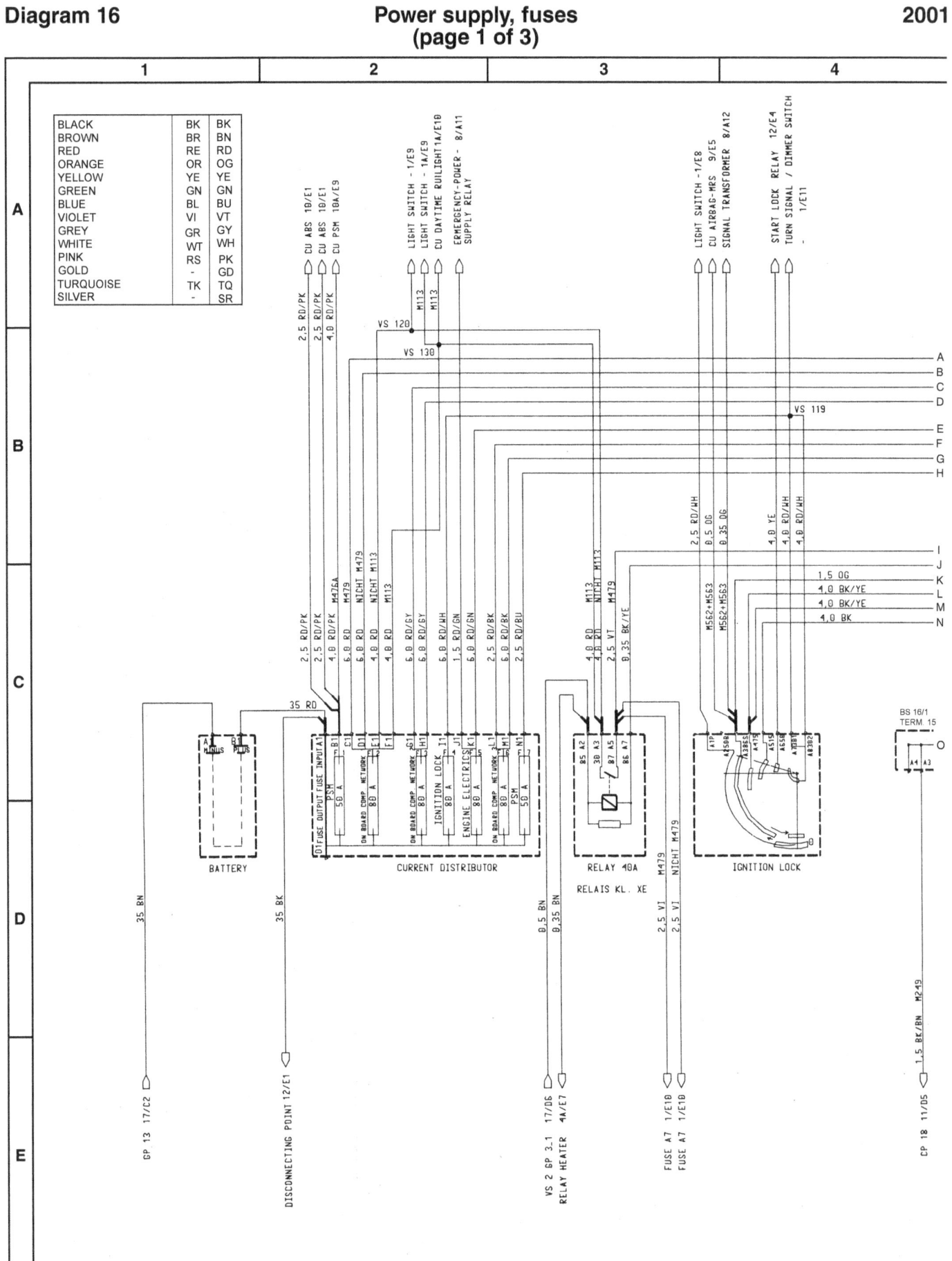

Electrical Wiring Diagrams EWD-129

2001 — Power supply, fuses (page 2 of 3) — Diagram 16

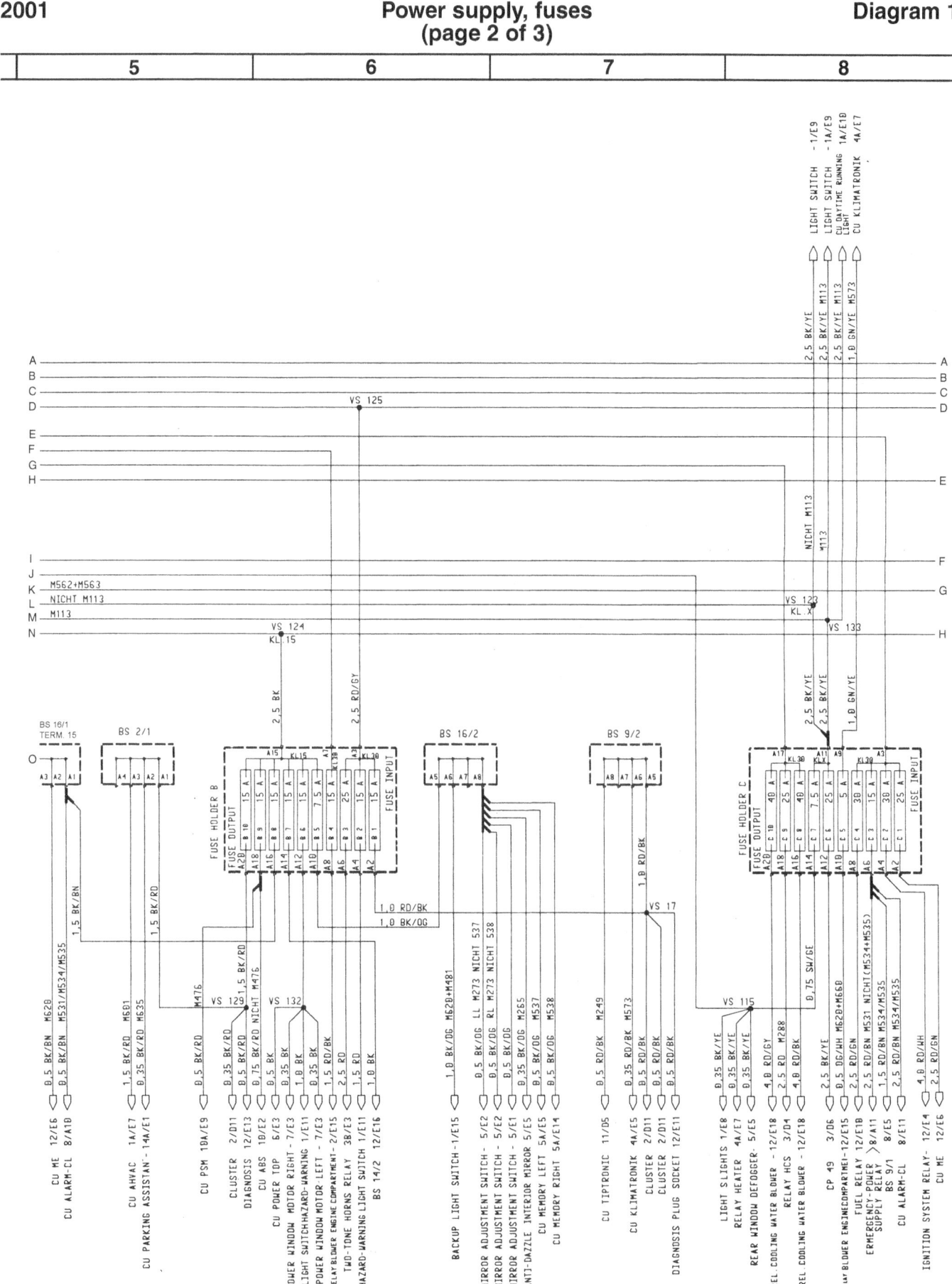

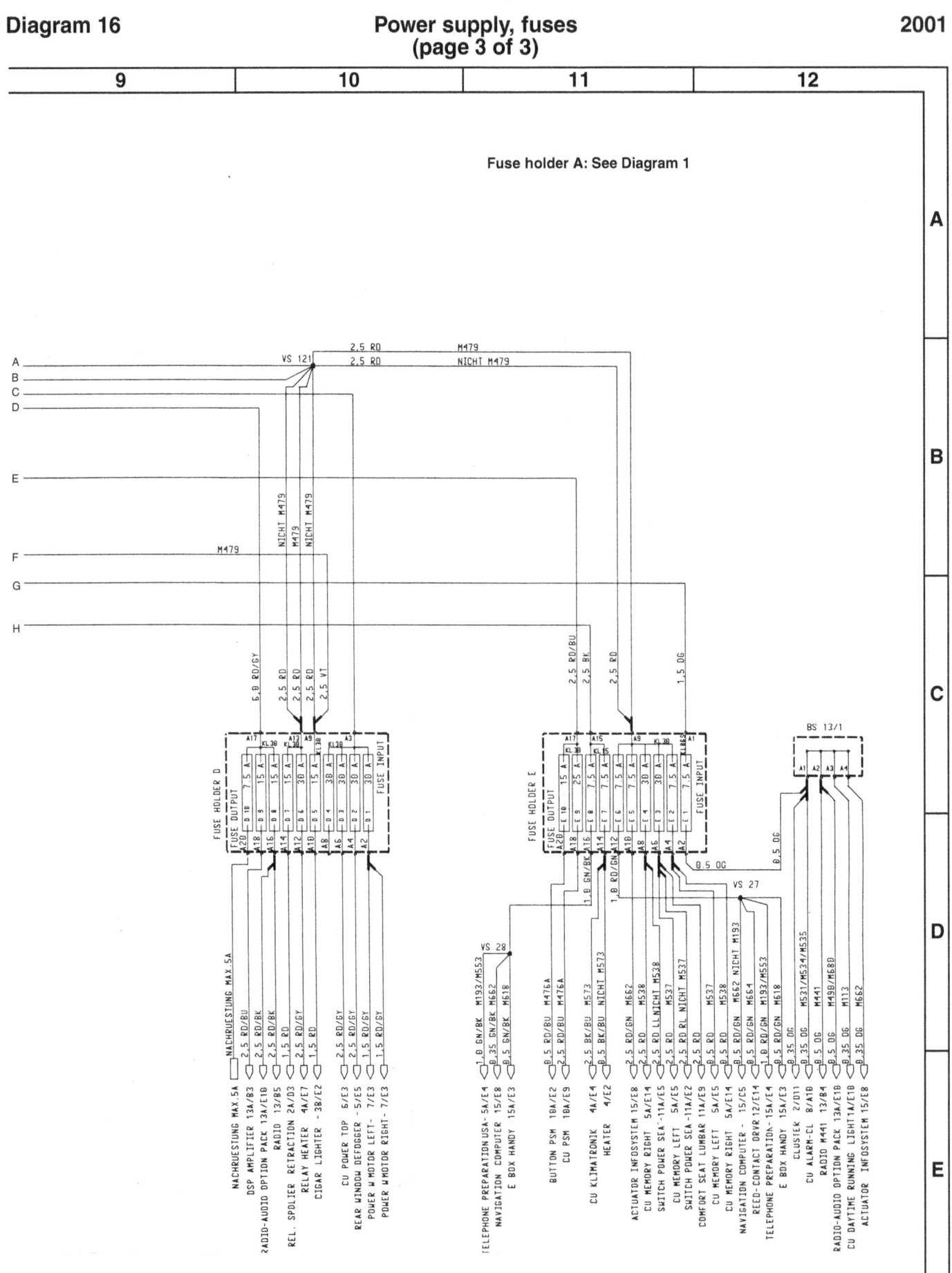

INDEX 1

WARNING

Your common sense, good judgement and general alertness are crucial to safe and successful service work. Before attempting any work on your Porsche, be sure to read **00 Warnings and Cautions** *and the copyright page at the front of the manual. Review these warnings and cautions each time you prepare to work on your car. Please also read any warnings and cautions that accompany the procedures in the manual.*

A

Abbreviations and acronyms
EWD-5

ABS (antilock brakes)
see Antilock brakes (ABS)
see also Brakes
see also Porsche Stability Management (PSM)

A/C
blower 87-8
  resistor 87-9
capacities 87-3
compressor 87-9
condenser 19-16, 87-6
control panel 87-3
evaporator 87-6
expansion valve 87-6
fault detection 87-4
housing 87-2
pressure switch 87-6
receiver-drier 87-6
refrigerant 87-3
sensor default values 87-4
troubleshooting 87-4
vent flaps and motors 87-2, 87-6
see also Heater

Accelerator linkage
10-9, 10-13

Aerodynamic lift
02-15

Air-conditioning
see A/C
see also Heater

Air filter
replacing 03-28

Airbag system (SRS)
02-18
control module 69-3, 69-8
crash sensor 69-8, 69-9
deployment 69-2
driver airbag 69-10
  contact spring 69-11
inspecting 03-33
passenger airbag 69-3, 69-13
replacement recommendations 69-4
seat occupancy sensor 69-3
side impact (door) airbag 69-14
smart 69-3
warning light 69-2

Air flow flap
87-2, 87-6

Air flow sensor
24-12

Alarm
96-7
see also Anti-theft system

Alignment
44-2

Alternator (generator)
27-10
charging system 27-10
troubleshooting 27-2, 27-10

Amplifier
91-5

Antenna
91-3, 91-4

Anti-theft system
components 96-4
with alarm 96-6
without alarm 96-5

Antifreeze (engine coolant)
03-14, 19-1

Antilock brakes (ABS)
ABS 5.3 45-3
ABS/TC 5.3 45-4
hydraulic unit / control module 45-8
identification 45-2
PSM 45-5
see also Brakes
see also Porsche Stability Management (PSM)

ATF (automatic transmission fluid)
03-19, 37-4

Automatic transmission
applications 02-13, 37-1
control module pin assignments, 37-3
electrical components 37-2
fluid 03-19, 37-4
  cooler 19-2
gear position switch 37-16
gear ratios 37-2
identification plate 37-1
removing / installing 37-7
shift knob 37-15
shift cable 37-15
torque converter 32-1

Axle joint
03-32, 42-15

B

B-pillar trim
70-6

Back-up light switch
automatic transmission 37-16
manual transmission 34-14

Battery
charging 27-5
disconnection notes 27-1
maintenance 03-34
removing / installing 27-4
service 03-34
testing 27-6
troubleshooting 27-2

Bearing, wheel
front 40-9
rear 42-11

Belt, drive
03-27, 13-3

Blade, wiper
92-2

Blower
see A/C

Board computer
90-3, 90-4, 90-7, 91-7

Body
02-15, 02-24, 02-25
dimensions 02-16

Brake fluid
03-14, 47-2

Brake light switch
94-6

2 Index

Brakes
02-23
bleeding 47-3
booster 47-6
caliper 47-6
fluid 03-15, 47-2
master cylinder 47-5
pads 03-30, 46-2
parking brake 03-38, 46-5
rotor 03-30, 46-3
troubleshooting 47-1
see also Antilock brakes (ABS)

Bridge points (BS)
EWD-6

Bulb applications
exterior 94-4
instrument 90-9
interior 96-1

Bumper
front 63-2
rear 63-5

C

Caliper, brake
47-6

Camber
44-2

Camshaft sensor
24-2, 28-3

Camshaft timing chain
02-5, 02-28

Capacities, fluid
A/C system 87-3
automatic transmission 37-4
coolant 19-1
differential 03-19
manual transmission 03-18
oil 17-1

Carbon canister, activated
24-5

Caster
44-3

Catalytic converter
02-11, 02-22, 26-3
diagnostic monitor 24-23

Cautions
00-2

CD changer
91-3

Center brake light
94-12

Center console
70-5

Central locking
96-8

Charging system
27-10

Check Engine light
02-10, 03-11, 24-3, 24-21

Climate control
see A/C
see also Heater

Clock
90-6

Clutch
bleeding 30-6
checking 03-36
fluid 03-15, 30-5
hydraulics 30-5
interlock switch 30-5
mechanical 30-3
replacing 30-2
throw-out bearing 30-2, 30-4
troubleshooting 30-2

Codes, fault
24-22

Coil
28-4

Coil spring
see Front suspension
see also Rear suspension, strut assembly

Communication
components 91-1

Component locations
97-4

Compressor
87-9

Condenser
19-16, 87-6

Constant velocity (CV) joint
03-32, 42-15

Control arm
40-3

Convertible top
02-17, 02-24
components 61-1
emergency closing 61-4
service 61-7
service position 03-24
troubleshooting 61-5

Coolant (antifreeze)
03-14, 19-1
hydrometer 19-2

Coolant pump
19-7

Cooling system
antifreeze 03-14, 19-1
automatic transmission fluid cooler 19-2
bleeding 19-5
coolant pump 19-2, 19-7
cooling fans 19-13
draining and filling 19-5
engine oil cooler 17-4
flushing 19-4
leak test 19-3
radiator 19-15
radiator air inlet duct 03-29
reservoir 19-10
thermostat 19-9
troubleshooting 19-3
see also Heater
see also Maintenance

Crankshaft seals
13-4

Crankshaft sensor
24-2, 28-4

Crash sensor
front 69-8
side impact 69-9

Cruise control
switch 48-3

CV joint
03-32, 42-15

Cylinder head cover
15-2

D

Dashboard
70-1

Deck lid
see Trunk lid

Diagnosis system
03-11, 24-21

Diagnostic monitors
24-23

Diagnostic trouble code (DTC)
03-11, 24-26
DME M 5.2.2 24-27
DME M 7.2 24-34
P-codes 24-27

Differential oil
03-19

INDEX 3

> **WARNING**
>
> *Your common sense, good judgement and general alertness are crucial to safe and successful service work. Before attempting any work on your Porsche, be sure to read* **00 Warnings and Cautions** *and the copyright page at the front of the manual. Review these warnings and cautions each time you prepare to work on your car. Please also read any warnings and cautions that accompany the procedures in the manual.*

DME (digital motor electronics)
 02-22, 02-26, 02-29
 adaptation 24-6
 applications 02-8, 24-1
 camshaft sensor 24-2
 crankshaft sensor 24-2
 engine coolant temperature (ECT) sensor 24-8
 evaporative control system 02-19, 24-5, 24-24
 FIO (fuel injector / ignition coil / oxygen sensor heater) relay 24-7
 main relay 20-1, 24-8
 mass air flow sensor 24-12
 MIL 03-11, 24-21
 overview 02-9
 oxygen sensor 26-4
 secondary air injection 02-11, 24-4, 24-24
 system overview 02-8, 24-3
 see also Engine control module (ECM)
 see also Engine management
 see also Fuel injector
 see also Fuel pressure
 see also Fuel pump
 see also Ignition system

Dome light
 96-2

Door
 57-1
 check strap, lubricating 03-36
 lock 57-6
 panel 57-3
 window
  see Door window

Door glass
 see Door window

Door lock
 57-6

Door sill trim
 70-6

Door window
 adjusting 64-5
 controls 64-3
 drop-down feature 64-2
 glass 64-4
 motor 64-9
 regulator 64-10
 standardization 64-9

Drains
 50-4

Drive axle
 42-15

Drive belt
 03-27

Drive cycle
 24-25

DTC (diagnostic trouble code)
 see Diagnostic trouble code (DTC)

E

ECM
 see Engine control module (ECM)

ECT
 24-8

Electrical components
 97-4

Electric cooling fan
 19-13

Electrical switches
 see Switches

Electrical system
 abbreviations and acronyms EWD-5
 battery disconnection notes 27-1
 component location table 97-4
 component photos 97-14
 engine 27-1
 terminal designations EWD-4
 troubleshooting 97-38
 wire color codes EWD-2

Electrical wiring diagrams
 EWD-9

Electronic immobilizer
 96-7, 96-10, 96-10

Emblems
 66-7

Emergency brake
 see Parking brake

Emergency flasher
 94-6

Engine
 02-25
 applications 02-3
 components 02-3
 crankshaft seal
  pulley seal 13-4
  rear main seal 13-5
 displacement 02-21
 drive belt 03-27, 13-3
 engine coolant temperature (ECT) sensor 24-8
 grounds 97-30
 management
  see DME (digital motor electronics)
 mount 10-16
 oil 17-1
 oil cooler 17-4
 on-board diagnostics 24-21
 pulley system 13-1
 removing / installing 10-1
 speed sensor
  see Crankshaft sensor
 see also Cooling system
 see also DME (digital motor electronics)
 see also Engine control module (ECM)
 see also Exhaust system
 see also Ignition system
 see also Lubrication system

Engine compartment blower
 19-13

Engine compartment vent
 66-6

Engine control module (ECM)
 24-12
 pin assignments
  DME M 5.2.2 24-13
  DME ME 7.2 24-15
  DME ME 7.8 24-18

Engine coolant temperature (ECT) sensor
 24-8

Engine cooling fan
 19-13

Engine covers
 front cover 03-26
 top cover 03-26

Engine management
 see DME (digital motor electronics)

Evaporative control system
 24-5
 diagnostic monitor 24-24

Evaporator
 87-6

Exhaust manifold
 26-1

4 INDEX

Exhaust system
02-11, 02-22
catalytic converter 02-11, 02-22, 26-1, 26-3, 26-4
checking 03-33
diagrams 26-7
manifold 26-1, 26-7, 26-8
oxygen sensor 26-4

Expansion valve
87-6

Exterior lights
see Lights

F

Fan
A/C blower 87-8
engine compartment blower 19-13
engine cooling fans 19-13

Fault code
03-11
see also Diagnostic trouble code (DTC)

Fault memory
03-11

Fender
front 50-2

Filters
air 03-31
fuel 03-31
oil 03-12
ventilation microfilter 03-34

Firing order
28-1

Fluids and lubricants
03-11
see also Capacities, fluid

Flywheel, dual-mass
30-1

Foglight
94-10

Front oil seal
13-4

Front suspension
02-27
coil spring applications 40-6
components 40-2
control arm 40-3
ride height 44-3
strut assembly 40-5
sway bar 40-4
wheel bearing 40-9

Fuel filter
03-31, 20-2

Fuel injection
02-9
see also DME (digital motor electronics)
see also Fuel injector

Fuel injector
02-10, 24-10
relay 24-7

Fuel level sender
20-1
see also Fuel pump

Fuel pressure
operating fuel pump for tests 20-3
pressure test 20-4
regulator
  1997- 2001 24-10
  2002 - 2004 20-2
relieving pressure 20-3
testing 20-3
volume 20-4
see also Fuel pump

Fuel pump
1997 - 2001 20-1
2002 - 2004 20-2
fuse and relay 20-1
removing / installing 20-7
troubleshooting 20-3

Fuel system
20-1
diagnostic monitor 24-24

Fuel tank
draining 20-5
filler flap 50-2

Fuel trim
see DME (digital motor electronics), adaptation

Fuses
97-10, 97-20

G

Gearshift knob
automatic transmission 37-15
manual transmission 34-14

Generator
see Alternator (generator)

Glove compartment
02-31

Grounds
97-12, 97-30

H

Hand brake
see Parking brake

Handle, door
57-6

Hard top
02-17

Hazard warning switch
94-6

Headlight
02-18, 02-20
adjusting 94-7
assembly 94-8
replacing bulb 94-9

Heater
components 80-1
core 80-3
troubleshooting 80-2
see also A/C
see also Cooling system

Hood
see Trunk lid

Horn
90-10
alarm horn 96-4

I

Ignition lock cylinder
48-5

Ignition electrical switch
48-5

Ignition system
02-12
camshaft sensor 24-2, 28-3
checking for spark 28-3
coil 28-4
  relay 24-7
crankshaft sensor 24-2, 28-4
disabling 28-2
firing order 28-1
knock sensor 24-2, 28-5
spark plugs 03-30

Immobilizer, electronic
96-7, 96-10, 96-12

Injector
see Fuel injector

Instruments
02-26
bulbs 90-9
repairs 90-8
versions 90-2

Interior
02-25
differences between Boxster and Boxster S 02-24

Interior lights
checking 03-38
see also Lights

INDEX 5

> **WARNING**
>
> *Your common sense, good judgement and general alertness are crucial to safe and successful service work. Before attempting any work on your Porsche, be sure to read* **00 Warnings and Cautions** *and the copyright page at the front of the manual. Review these warnings and cautions each time you prepare to work on your car. Please also read any warnings and cautions that accompany the procedures in the manual.*

Intermittent wiper switch
  92-4

K

Key
  96-6

Keyless entry
  02-25, 96-6

Knock sensor
  24-2, 28-5

L

Level sender, fuel
  20-1

License plate light
  94-13

Lights
  bulb applications
    exterior 94-4
    instrument 90-9
    interior 96-1
  controls 94-5
  interior 96-1
  repairs
    front 94-7
    rear 94-12

Litronic
  02-20
  see also Headlight

Locks
  door 57-6
  central locking 96-8
  trunk lid release 55-6

Lubrication system
  17-2
  air-oil separator 17-3
  oil cooler 17-2, 17-4
  oil filter 17-2
  oil level sensor 17-3
  oil pumps 15-7, 17-2

Luggage compartment lid
  see Trunk lid

M

M option codes
  see Option codes

Maintenance
  airbags, inspecting 03-33
  alignment 03-40, 44-2
  antifreeze 03-14, 19-1
  automatic transmission fluid 03-19, 37-4
  battery 03-34
  body exterior 03-35
  brake fluid 03-14, 47-2
  brake pads / rotors 03-30, 46-2
  clutch, checking 03-36
  coolant hoses 03-29
  coolant level 03-14
  CV joint boot, inspecting 03-32
  differential oil 03-19
  door check straps, hinges, lubricating 03-36
  door locks, checking 03-36
  drive belt 03-27, 13-3
  exhaust system, checking 03-33
  exterior lights, checking 03-37
  filters
    air 03-31
    fuel 03-31
    oil 03-12
  fluids and lubricants 03-11
    see also Capacities, fluid
  front lid safety hooks, checking 03-36
  headlights, adjusting 03-38
  interior lights, checking 03-38
  manual transmission fluid 03-16
  oil 03-11
  parking brake 03-38, 46-5
  power steering fluid 03-16
  radiator air inlet duct 03-29
  seat belts, checking 03-39
  spark plugs 03-30
  suspension and steering components, checking 03-33
  tables 03-7
  test drive 03-41
  tires 03-40
  trunk lid locks, checking 03-36
  ventilation microfilter 03-34
  wipers and washers 03-24

Malfunction indicator light (MIL)
  03-11, 24-3, 24-21

Manifold, exhaust
  26-1, 26-7, 26-8

Manual transmission
  applications 02-12, 34-1
  code stamp 34-2
  fluid 03-17
  identification label 34-2
  pilot bearing 30-4
  removing / installing
    5-speed 34-2
    6-speed 34-8
  shift knob 34-14

Mass air flow (MAF) sensor
  24-12

Master cylinder
  brake 47-5
  clutch 30-7

MIL (malfunction indication light)
  03-11, 24-3, 24-21

Mirror, outside rear view
  02-25, 66-1

Misfire
  diagnostic monitor 24-23

Model year news
  02-19

M-option codes
  02-33, EWD-7

MOST-bus
  02-31, 91-2, 91-9

MRS (multiple restraint system)
  69-1

Muffler
  26-2

N

Navigation
  91-8

Neutral safety switch
  see Automatic transmission, gear position switch

O

OBD (on-board diagnostics)
  03-11, 24-12

Oil cooler
  17-4

Oil level
  checking 03-11
  sensor 17-3

Oil pumps
  15-7, 17-2

6 INDEX

Oil service
03-12

On-board computer
90-3, 90-4, 90-7, 91-7

On-board diagnostics (OBD)
03-11, 24-21

Option codes
02-33, EWD-7

Outside door handle
57-6

Outside mirror switch
66-1

Oxygen sensor
26-4
diagnostic monitor 24-23
heater relay 24-7
LSF 02-29

P

P-codes
24-27

Pads, brake
03-30, 46-2

Parking assistant
02-20, 91-2

Parking brake
03-38, 46-5

Parts
03-5

Passenger compartment monitor
96-10

PCM
91-1, 91-8

Pilot bearing
30-4

Porsche Communications Management (PCM)
02-20, 02-26, 02-32, 91-1, 91-8

Porsche Stability Management (PSM)
02-25, 02-27, 45-5
precharge pump 45-5, 45-6, 45-10
rotational rate sensor 45-5, 45-7, 45-11
steering angle sensor 45-5, 45-7
see also Antilock brakes (ABS)
see also Brakes

Power mirror
66-1

Power steering
see Steering

Power window
see Door window

Precharge pump
45-5, 45-6, 45-10

PSM
see Porsche Stability Management (PSM)

Pulley system
130-1

R

Radiator
19-15

Radiator cooling fans
19-13

Radio
91-4

Rain sensor
92-1

Raising car safely
03-2

Readiness codes
24-25

Rear main seal
13-5

Rear spoiler
66-4

Rear suspension
components 42-2
control arms 42-9
reinforcement plate 02-14
ride height 44-3
strut assembly 42-3
sway bar 42-8
track arm 42-9
trailing arm 42-10
wheel bearing 42-11

Rear window
02-29

Receiver-drier
87-6

Recirculation switch
87-1

Refrigerant
87-3

Relays
97-11, 97-29

Reversing light switch
see Back-up light switch

Ride height
44-3

Roll bar
02-17

Rotational rate sensor
45-5, 45-7, 45-11

Rotor, brake
checking 03-30
replacing 46-1

S

Safety
03-2

Scan tool
24-22

Seal, crankshaft
13-4

Seat
02-25, 72-1

Seat belt
02-27, 69-1, 69-5
checking 03-39

Seat occupancy sensor
69-3

Secondary air injection
02-11, 24-4
diagnostic monitoring 24-24

Shift knob
automatic transmission 37-15
manual transmission 34-14

Side impact (door) airbag
69-14

Side turn signal light
94-11

Siren
96-4

Slave cylinder
30-7

Sound system
02-20, 02-21, 02-26, 02-28, 02-32, 91-3

Spark plug
03-30
well 15-8

Speakers
91-6

Special Edition
02-32

Speed sensor, wheel
45-7

Spoiler, rear
02-30, 66-4

Spring, coil
see Front suspension
see also Rear suspension, strut assembly

INDEX

> **WARNING**
>
> *Your common sense, good judgement and general alertness are crucial to safe and successful service work. Before attempting any work on your Porsche, be sure to read* **00 Warnings and Cautions** *and the copyright page at the front of the manual. Review these warnings and cautions each time you prepare to work on your car. Please also read any warnings and cautions that accompany the procedures in the manual.*

SRS (supplemental restraint system)
see Airbag system (SRS)

Stabilizer bar
see Sway bar

Starter
27-8
interlock switch 27-8, 30-5
removing / installing 27-9
troubleshooting 27-2, 27-8

Steering
column covers 48-4
column stalk switch 48-3
fluid 03-16, 48-9
fluid lines 48-13
ignition lock 48-5
pump 48-10
rack 48-7
wheel 48-2
tie rod 48-6

Steering angle sensor
45-5, 45-7

Stop light switch
see Brake light switch

Strut
front 40-5
rear 42-3

Suspension
see Front suspension
see also Rear suspension

Sway bar
front 40-4
rear 42-8

Switches
A/C 87-1
back-up light
  automatic transmission 37-16
  manual transmission 34-14
brake light 94-6
clutch interlock 30-5
convertible top 61-3
cruise control 48-3
hazard warning 94-6
high beam 48-3
ignition
  cylinder 48-5
  electrical switch 48-5
light 94-5
outside mirror 66-1
pedal cluster 30-5, 94-6
steering column stalk 48-3
turn signal 48-3
window 64-3
wiper / washer 48-3
see also Back-up light switch

T

Taillights
94-12

TC (traction control)
see Antilock brakes
see also Porsche Stability management (PSM)

Test drive
03-41

Thermostat
19-9

Throw-out bearing
30-2, 30-4

Tie rod
48-6

Tightening fasteners
03-4
torques 03-5

Tiptronic
control module pin assignments 37-3
see also Automatic transmission

Tires
44-1
service 03-40

Toe
44-3

Tools
03-6

Torque converter
32-1

Traction control (TC)
see Antilock brakes
see also Porsche Stability Management (PSM)

Transmission
02-12, 02-23, 02-27
mount 10-17
see also Automatic transmission
see also Manual transmission

Trim, interior
B-pillar trim 70-6
Center console 70-5
Dashboard 70-1
Doors panel 57-3
Door sill trim 70-6
Steering column covers 48-4

Trunk lid
emergency release 02-27, 02-31, 55-1
front 55-3
  components 55-2
  safety hooks, checking 03-36
  struts 55-3
rear 55-4
  components 55-4
  struts 55-5
release (latch) 55-6

Turn signal switch
48-3

V

Vacuum booster
47-6

VarioCam
02-6, 02-28, 15-3

Vehicle identification number (VIN)
01-1, 03-6

Vent flaps and motors
87-2, 87-6

Ventilation microfilter
replacing 03-34

Voltage tests
97-38

W

Warnings
00-1

Washer
92-9
see also Wiper

Washing
03-35

Water drains
50-4

8 INDEX

Water pump
19-2, 19-7

Waxing
03-35

Wheel, steering
48-2

Wheel alignment
44-2
ride height 44-3

Wheel bearing
front 40-9
rear 42-10

Wheel housing liner, front
50-1

Wheel speed sensor
45-7

Wheels
44-1

Window
see Door window

Wiper
92-2
checking 03-24
fluid 03-24
switch
  see Steering, column stalk
  switch
see also Washer

Wiring
color codes EWD-2
diagrams EWD-9
sample EWD-3

Working under car safely
03-3

X

X-connectors
EWD-8

Selected Books and Repair Information From Bentley Publishers

Motorsports

Alex Zanardi: My Sweetest Victory
Alex Zanardi and Gianluca Gasparini
ISBN 978-0-8376-1249-2

The Unfair Advantage *Mark Donohue and Paul van Valkenburgh*
ISBN 978-0-8376-0069-7

Equations of Motion - Adventure, Risk and Innovation
William F. Milliken
ISBN 978-0-8376-1570-7

Engineering

The Hack Mechanic Guide to European Automotive Electrical Systems
Rob Siegel ISBN 978-0-8376-1751-0

Physics for Gearheads *Randy Beikmann*
ISBN 978-0-8376-1615-5

Bosch Automotive Handbook
Robert Bosch GmbH
ISBN 978-0-8376-1732-9

Bosch Fuel Injection and Engine Management *Charles O. Probst, SAE*
ISBN 978-0-8376-0300-1

Maximum Boost: Designing, Testing, and Installing Turbocharger Systems
Corky Bell ISBN 978-0-8376-0160-1

Supercharged! Design, Testing and Installation of Supercharger Systems
Corky Bell ISBN 978-0-8376-0168-7

Audi

Audi A4 Service Manual: 2002-2008, 1.8L Turbo, 2.0L Turbo, 3.0L, 3.2L
Bentley Publishers
ISBN 978-0-8376-1574-5

Audi TT Service Manual: 2000-2006, 1.8L turbo, 3.2 L *Bentley Publishers*
ISBN 978-0-8376-1625-4

Audi A6 (C5 platform) Service Manual: 1998-2004 *Bentley Publishers*
ISBN 978-0-8376-1670-4

BMW

Memoirs of a Hack Mechanic
Rob Siegel ISBN 978-0-8376-1720-6

BMW X3 (E83) Service Manual: 2004-2010 *Bentley Publishers*
ISBN 978-0-8376-1731-2

BMW X5 (53) Service Manual: 2000-2006 *Bentley Publishers*
ISBN 978-0-8376-1643-8

BMW 3 Series (F30, F31, F34) Service Manual: 2012-2015
Bentley Publishers ISBN 978-0-8376-1752-7

BMW 3 Series (E90, E91, E92, E93) Service Manual: 2006-2011
Bentley Publishers ISBN 978-0-8376-1723-7

BMW 3 Series (E46) Service Manual: 1999-2005 *Bentley Publishers*
ISBN 978-0-8376-1657-5

BMW 3 Series (E36) Service Manual: 1992-1998 *Bentley Publishers*
ISBN 978-0-8376-1709-1

BMW 3 Series (E30) Service Manual: 1984-1990 *Bentley Publishers*
ISBN 978-0-8376-1647-6

BMW Z3 (E36/7) Service Manual: 1996-2002 *Bentley Publishers*
ISBN 978-0-8376-1617-9

BMW 5 Series (E60, E61) Service Manual: 2004-2010 *Bentley Publishers*
ISBN 978-0-8376-1689-6

BMW 5 Series (E39) Service Manual: 1997-2003 *Bentley Publishers*
ISBN 978-0-8376-1672-8

Corvette

Corvette: America's Star-Spangled Sports Car *Karl Ludvigsen*
ISBN 978-0-8376-1659-9

Zora Arkus-Duntov: The Legend Behind Corvette *Jerry Burton*
ISBN 978-08376-0858-7

Porsche

Porsche: Excellence Was Expected
Karl Ludvigsen ISBN 978-0-8376-0235-6

Ferdinand Porsche — Genesis of Genius
Karl Ludvigsen ISBN 978-0-8376-1557-8

Porsche — Origin of the Species
Karl Ludvigsen ISBN 978-0-8376-1331-4

Volkswagen

Volkswagen Rabbit, GTI Service Manual: 2006-2009 *Bentley Publishers*
ISBN 978-0-8376-1664-3

Volkswagen Jetta Service Manual: 2005-2010 *Bentley Publishers*
ISBN 978-0-8376-1616-2

Volkswagen Jetta, Golf, GTI Service Manual: 1999-2005 *Bentley Publishers*
ISBN 978-0-8376-1678-0

Volkswagen Jetta, Golf, GTI: 1993-1999, Cabrio: 1995-2002 Service Manual *Bentley Publishers*
ISBN 978-0-8376-1660-5

Volkswagen GTI, Golf, Jetta Service Manual: 1985-1992 *Bentley Publishers*
ISBN 978-0-8376-1637-7

Volkswagen Corrado Repair Manual: 1990-1994 *Bentley Publishers*
ISBN 978-0-8376-1699-5

Volkswagen Passat, Passat Wagon Service Manual: 1998-2005
Bentley Publishers
ISBN 978-0-8376-1669-8

MINI Repair Manuals

MINI Cooper Service Manual: 2007-2013 *Bentley Publishers*
ISBN 978-0-8376-1730-5

MINI Cooper Service Manual: 2002-2006 *Bentley Publishers*
ISBN 978-0-8376-1639-1

Mercedes-Benz

Mercedes-Benz C-Class (W202) Service Manual 1994-2000
Bentley Publishers
ISBN 978-0-8376-1692-6

Mercedes-Benz Technical Companion
Staff of The Star and members of Mercedes-Benz Club of America
ISBN 978-0-8376-1033-7

Porsche Repair Manuals

Porsche 911 Carrera (Type 993) Service Manual: 1995-1998
Bentley Publishers
ISBN 978-0-8376-1719-0

Porsche 911 Carrera Service Manual: 1984-1989
Bentley Publishers
ISBN 978-0-8376-1696-4

Porsche 911 SC Service Manual: 1978-1983
Bentley Publishers
ISBN 978-0-8376-1705-3

Porsche 911 Carrera (Type 996) Service Manual: 1999-2005
Bentley Publishers
ISBN 978-0-8376-1710-7

Bentley Publishers has published service manuals and automobile books since 1950. For more information, please contact Bentley Publishers at 1734 Massachusetts Avenue, Cambridge, MA 02138 USA, or visit our web site at
BentleyPublishers.com